Asymptotic Behaviour of Tame Harmonic Bundles and an Application to Pure Twistor *D*-Modules, Part 1

MEMOIRS
of the
American Mathematical Society

Number 869

Asymptotic Behaviour of Tame Harmonic Bundles and an Application to Pure Twistor *D*-Modules, Part 1

Takuro Mochizuki

January 2007 • Volume 185 • Number 869 (first of 2 numbers) • ISSN 0065-9266

American Mathematical Society
Providence, Rhode Island

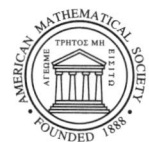

2000 *Mathematics Subject Classification.*
Primary 14C30, 32S40, 53C07, 53C43.

Library of Congress Cataloging-in-Publication Data

Mochizuki, Takuro, 1972–
Asymptotic behaviour of tame harmonic bundles and an application to pure twistor D-modules / Takuro Mochizuki.
p. cm. — (Memoirs of the American Mathematical Society, ISSN 0065-9266 ; numbers 869 and 870)
Number 869
ISBN-13: 978-0-8218-3942-3 (alk. paper)
ISBN-10: 0-8218-3942-X (alk. paper)
Number 870
ISBN-13: 978-0-8218-3943-0 (alk. paper)
ISBN-10: 0-8218-3943-8 (alk. paper)
1. Hodge theory. 2. D-modules. 3. Vector bundles. 4. Harmonic maps. I. Title.

QA564.M63 2007
514′.74—dc22 2006047813

Memoirs of the American Mathematical Society

This journal is devoted entirely to research in pure and applied mathematics.

Subscription information. The 2007 subscription begins with volume 185 and consists of six mailings, each containing one or more numbers. Subscription prices for 2007 are US$649 list, US$519 institutional member. A late charge of 10% of the subscription price will be imposed on orders received from nonmembers after January 1 of the subscription year. Subscribers outside the United States and India must pay a postage surcharge of US$38; subscribers in India must pay a postage surcharge of US$43. Expedited delivery to destinations in North America US$53; elsewhere US$130. Each number may be ordered separately; *please specify number* when ordering an individual number. For prices and titles of recently released numbers, see the New Publications sections of the *Notices of the American Mathematical Society.*

Back number information. For back issues see the *AMS Catalog of Publications.*

Subscriptions and orders should be addressed to the American Mathematical Society, P. O. Box 845904, Boston, MA 02284-5904, USA. *All orders must be accompanied by payment.* Other correspondence should be addressed to 201 Charles Street, Providence, RI 02904-2294, USA.

Memoirs of the American Mathematical Society is published bimonthly (each volume consisting usually of more than one number) by the American Mathematical Society at 201 Charles Street, Providence, RI 02904-2294, USA. Periodicals postage paid at Providence, RI. Postmaster: Send address changes to Memoirs, American Mathematical Society, 201 Charles Street, Providence, RI 02904-2294, USA.

This publication is indexed in *Science Citation Index®*, *SciSearch®*, *Research Alert®*, *CompuMath Citation Index®*, *Current Contents®/Physical, Chemical & Earth Sciences.*
Printed in the United States of America.

∞ The paper used in this book is acid-free and falls within the guidelines established to ensure permanence and durability.
Visit the AMS home page at `http://www.ams.org/`

10 9 8 7 6 5 4 3 2 1 12 11 10 09 08 07

Dedicated to Professor Kenji Fukaya,
with Deepest Veneration

Contents

Abstract

We study the asymptotic behaviour of tame harmonic bundles. First of all, we prove a local freeness of the prolongment of deformed holomorphic bundle by an increasing order. Then we obtain the polarized mixed twistor structure from the data on the divisors. As one of the applications, we obtain the norm estimate of holomorphic or flat sections by weight filtrations of the monodromies.

As another application, we establish the correspondence of semisimple regular holonomic D-modules and polarizable pure imaginary pure twistor D-modules through a tame pure imaginary harmonic bundles, which is a conjecture of C. Sabbah. Then the regular holonomic version of M. Kashiwara's conjecture follows from the results of Sabbah and us.

Received by the editor May 21, 2004, and in revised form February 18, 2005.

2000 *Mathematics Subject Classification.* 14C30, 32S40, 53C07, 53C43.

Key words and phrases. Higgs fields, harmonic bundle, variation of Hodge structure, mixed twistor structure, D-module, Hard Lefschetz Theorem.

Acknowledgement

The most part of the paper was written at Osaka City University, and it was accomplished at Kyoto University. I am heartily grateful to the colleagues for their help. All the essential ideas came to me during my stay at the Institute for Advanced Study, and the preparation for the paper was done there. I express my gratitude to their excellent hospitality. I also wish to acknowledge National Scientific Foundation for a grant DMS 9729992, although any opinions, findings and conclusions or recommendations expressed in this material do not necessarily reflect the views of the National Science Foundation. I thank the financial supports by Japan Society for the Promotion of Science and the Sumitomo Foundation.

I am indebted to many people. I express my special thanks to Mikiya Masuda for his encouragement and support. I would like to thank Yoshifumi Tsuchimoto and Akira Ishii for their constant encouragements since my undergraduate student days. I owe much thanks to Tomohide Terasoma who informed me of various things. I am also grateful to Terasoma's family for their kindness. Thanks are due to Masaki Kashiwara for some useful information. Mark Andrea de Cataldo attracted my attention to Hodge modules and a decomposition theorem of perverse sheaves. I also appreciate his great tolerance for my lack of communication ability. Much part of this paper is a result of the effort to understand the work of Carlos Simpson, especially [**81**] and [**83**]. One of his conjectures in [**83**] made me start to study tame harmonic bundles. Claude Sabbah kindly informed of the revised version of his paper [**72**]. I am glad that our works are cooperative in a solution of the regular holonomic version of Kashiwara's conjecture. I warmly appreciate Pierre Deligne for some conversation. Very special thanks go to Kari Vilonen for his interest and encouragement to this work. I am grateful to the referee for his useful advice and for his effort to read this quite long paper. I also wish to express my gratitude to the referee of my previous paper [**65**] for his kind and useful comments.

I would like to dedicate this paper to Kenji Fukaya. I hope that the paper is eligible for it.

CHAPTER 1

Introduction

1.1. Simpson's Meta-Theorem

The guiding principle of our study is the following, which we call Simpson's Meta-Theorem:

PRINCIPLE 1.1. *The theory of Hodge structure should be generalized to the theory of twistor structures.* □

In [**83**], C. Simpson stated the above principle as follows:

> **Meta-Theorem** *If the words "mixed Hodge structure" (resp. "variation of mixed Hodge structure") are replaced by the words "mixed twistor structure" (resp. "variation of mixed twistor structure") in the hypotheses and conclusions of any theorem in Hodge theory, then one obtains a true statement. The proof of the new statement will be analogous to the proof of the old statement.*

We regard it as a kind of principle. As for the study of the asymptotic behaviour of variation of pure twistor structures (harmonic bundle), it may occur that the proof of new statement is not analogous to the proof of the old statement, in our current understanding.

1.2. The purposes in this paper

We have two main purposes in this paper.

(1) In the previous paper [**65**], we discussed the behaviour of tame harmonic bundle imposed the nilpotentness and the trivial parabolic structure conditions. We would like to remove the assumption. We also improve and strengthen the arguments. In particular, we use the reduction to the Hodge theory more efficiently.
(2) We would like to apply the study on the behaviour of tame harmonic bundle to the theory of pure twistor D-module, introduced by Sabbah.

1.3. On the purpose (1)

1.3.1. A brief goal. Our principle in the study of tame harmonic bundle is as follows, which is a 'corollary' of Simpson's Meta-Theorem:

PRINCIPLE 1.2. *The asymptotic behaviour of tame harmonic bundle should be similar to the asymptotic behaviour of variation of polarized Hodge structure.* □

The variation of polarized Hodge structure were deeply studied by E. Cattani, A. Kaplan, M. Kashiwara, T. Kawai and W. Schmid and others. Many advanced results are known ([**10**], [**11**], [**12**], [**47**], [**49**], [**52**] and [**77**], for example). We can say that our goal is to show the corresponding results in the theory of harmonic bundles. In particular, we will obtain the limiting polarized mixed twistor structure and the norm estimate.

However, we do not follow closely the arguments in the above papers. Instead, we use the more differential geometric method, pioneered by Simpson. We refer the two hard points for a direct application of the classical method to the theory of harmonic bundles.

- **(a):** The nilpotent orbit theorem for harmonic bundle is not known.
- **(b):** In the case of harmonic bundles, we have non-trivial eigenvalues of the residues and non-trivial parabolic structures.

1.3.2. The difficulty (a). In the study of complex variation of polarized Hodge structure (CVHS), the nilpotent orbit theorem due to W. Schmid is quite important. It was the starting point of the later studies on CVHS. However, we do not know even the formulation of nilpotent orbit theorem for harmonic bundles, for we do not have the counterparts of the classifying space. Hence we find another starting point, and therefore the classical method cannot be applied directly.

REMARK 1.3. Now we have understood the asymptotic behaviour of tame harmonic bundle pretty well. Hence the author does not think that a generalization of nilpotent orbit theorem for harmonic bundle is necessary as a starting point, although it would be interesting. Since the harmonic bundle can be regarded as a pluri-harmonic map from a complex manifold to a symmetric space, it would be possible to obtain it after the study would be made progressed. (See Remark 12.42, for example.) □

1.3.3. The difficulty (b). We have the non-trivial eigenvalues of the residue of the Higgs field, and non-trivial parabolic structures. We have the following simple example.

EXAMPLE 1.4. Let Δ^* denote the punctured disc $\{z \in \boldsymbol{C} \,|\, 0 < |z| < 1\}$. Let us consider the holomorphic bundle $E := \mathcal{O}_{\Delta^*} \cdot e$ of rank 1 over Δ^*. We have the Higgs field $\theta := \alpha \cdot dz/z$ $(\alpha \in \boldsymbol{C})$ and the metric h determined by $h(e,e) := |z|^{-2a}$ $(a \in \boldsymbol{R})$. Then it is easy to check that the tuple (E, θ, h) is a harmonic bundle. □

In the case of CVHS, the corresponding Higgs field is always nilpotent. Hence the example cannot be the variation of Hodge structure in the case $\alpha \neq 0$. On the other hand, the example is CVHS in the case $\alpha = 0$. However if a is not rational, then the monodromy of the corresponding local system is not quasi unipotent. The local monodromy is often assumed to be quasi unipotent in the classical study of CVHS. In the case, we have only to consider unipotent CVHS essentially. Since we would like to consider the case $(a, \alpha) \not\in \boldsymbol{Q} \times \{0\}$, the classical method cannot be directly applied.

1.3.4. The starting point in the paper [65]. In our paper [65], we discussed the problem under the assumption that the difficulty (b) does not occur. Namely, we imposed the assumption of the nilpotentness of the residues and the trivial parabolic structure. We recall what was our starting point in [65], instead of the nilpotent orbit theorem (see the subsection 1.3.2).

We put $X := \Delta^n$, $D_i := \{z_i = 0\}$ and $D := \bigcup_{i=1}^n D_i$. We put $\mathcal{X} := X \times \boldsymbol{C}_\lambda$ and $\mathcal{D} := D \times \boldsymbol{C}_\lambda$. Let p denote the projection $\mathcal{X} - \mathcal{D} \longrightarrow X - D$.

Let $(E, \overline{\partial}_E, \theta, h)$ be a tame nilpotent harmonic bundle with trivial parabolic structure. We have the deformed holomorphic bundle $\mathcal{E} = \big(p^{-1}E, \overline{\partial}_E + \lambda\theta^\dagger + \overline{\partial}_\lambda\big)$ and the λ-connection $\mathbb{D}$ on $\mathcal{X} - \mathcal{D}$. For any $\boldsymbol{b} \in \boldsymbol{R}^n$, we prolong the sheaf $\mathcal{E}$ on $\mathcal{X} - \mathcal{D}$ to the sheaf ${}_{\boldsymbol{b}}\mathcal{E}$ over $\mathcal{X}$. The sections of ${}_{\boldsymbol{b}}\mathcal{E}$ over an open subset $U \subset \mathcal{X}$ is given as follows:

$${}_{\boldsymbol{b}}\mathcal{E}(U) := \Big\{ f \in \mathcal{E}(U \setminus \mathcal{D}) \,\Big|\, |f|_h = O\Big(\prod |z_i|^{-b_i - \epsilon}\Big),\ \forall \epsilon > 0 \Big\}.$$

In the case $\boldsymbol{b} = (0, \ldots, 0)$, we use the notation ${}^\diamond\mathcal{E}$. (See the section 2.2 for more detail about ${}^\diamond\mathcal{E}$ and ${}_{\boldsymbol{b}}\mathcal{E}$.) Then we proved the following.

PROPOSITION 1.5 (Theorem 4.1 and Proposition 4.9 in [**65**]). *Under the assumption of the nilpotentness and the trivial parabolic structures, the $\mathcal{O}_{\mathcal{X}}$-module ${}^\diamond\mathcal{E}$ is locally free, and $\mathbb{D}$ is a regular λ-connection in the sense $\mathbb{D}f \in {}^\diamond\mathcal{E} \otimes \Omega_X(\log D)$ for any section $f \in {}^\diamond\mathcal{E}$.* □

Then we obtain the holomorphic vector bundle $V_0 := {}^\diamond\mathcal{E}_{|\{O\}\times \boldsymbol{C}_\lambda}$ on the complex plane $\boldsymbol{C}_\lambda$. The residues $\mathrm{Res}_i(\mathbb{D})$ induce the nilpotent endomorphism $\mathcal{N}_i$.

On the other hand, we have the harmonic bundle $(E, \partial_E, \theta^\dagger, h)$ on the conjugate complex manifold $X^\dagger - D^\dagger$. We put $\mathcal{X}^\dagger := X^\dagger \times \boldsymbol{C}_\mu$ and $\mathcal{D}^\dagger := D^\dagger \times \boldsymbol{C}_\mu$. We obtain the deformed holomorphic bundle $\mathcal{E}^\dagger$ and the μ-connection $\mathbb{D}^\dagger$, and then the prolongment ${}^\diamond\mathcal{E}^\dagger$. Thus we obtain the holomorphic bundle $V_\infty := {}^\diamond\mathcal{E}^\dagger{}_{|\boldsymbol{C}_\mu}$ and the nilpotent endomorphism $\mathcal{N}_i^\dagger$ on the complex plane $\boldsymbol{C}_\mu$.

We glue $\boldsymbol{C}_\lambda$ and $\boldsymbol{C}_\mu$ by the relation $\lambda = \mu^{-1}$, and thus we obtain $\mathbb{P}^1$. Take a point $P \in X - D$, and then we obtained the isomorphisms $\Phi_{O,P} : {}^\diamond\mathcal{E}_{|\{O\}\times\boldsymbol{C}_\lambda^*} \simeq {}^\diamond\mathcal{E}_{|\{P\}\times\boldsymbol{C}_\lambda^*}$ and $\Phi_{O,P}^\dagger : {}^\diamond\mathcal{E}^\dagger{}_{|\{O\}\times\boldsymbol{C}_\mu^*} \simeq {}^\diamond\mathcal{E}^\dagger{}_{|\{P\}\times\boldsymbol{C}_\mu^*}$. They give the gluing of $(V_0, \mathcal{N}_i)$ and $(V_\infty, -\mathcal{N}_i^\dagger)$. Thus we obtained the holomorphic vector bundle $S(E, P)$ and the nilpotent maps $\mathcal{N}_i^\triangle : S(E, P) \longrightarrow S(E, P) \otimes \mathcal{O}_{\mathbb{P}^1}(2)$ over $\mathbb{P}^1$. The nilpotent map $\mathcal{N}^\triangle(\underline{n}) = \sum \mathcal{N}_i^\triangle$ induces the weight filtration W on $S(E, P)$.

PROPOSITION 1.6 (Theorem 7.2 [**65**]). *The filtered vector bundle $\big(S(E, P), W\big)$ is a mixed twistor structure.* □

Propositions 1.5 and 1.6 were the starting points of our study in the paper [**65**]. Then we obtained the constantness of the filtrations, the compatibility of the nilpotent maps, the norm estimate, the existence of the limiting CVHS and the purity theorem, by using some geometric argument.

1.3.5. The case where the difficulty (b) occurs. When the difficulty (b) occurs, we cannot use the argument in [**65**] straitforwardly. Let us see what happens in Example 1.4 in the subsection 1.3.3. In the example, ∂_E and $\theta^\dagger$ are as follows:

$$\partial_E e = e \cdot (-a)\frac{dz}{z}, \qquad \theta^\dagger = \overline{\alpha} \cdot \frac{d\overline{z}}{\overline{z}}.$$

Then we have the frame f of $\mathcal{E}$ given as follows:

$$f := \exp\big(-\overline{\alpha} \cdot \lambda \cdot \log |z|^2\big) \cdot e.$$

The λ-connection is as follows:

$$\mathbb{D}f = f \cdot \big(\alpha - a \cdot \lambda - \overline{\alpha} \cdot \lambda^2\big) \cdot \frac{dz}{z}.$$

In particular, $\mathrm{Res}(\mathbb{D}) = \alpha - a\cdot\lambda - \overline{\alpha}\cdot\lambda^2$. The norm of u with respect to h is as follows:

$$|f|_h = |z|^{-a-2\,\mathrm{Re}(\overline{\alpha}\cdot\lambda)}.$$

Hence the sheaves ${}^{\diamond}\mathcal{E}$ or ${}_{\boldsymbol{b}}\mathcal{E}$ for any $\boldsymbol{b}\in\boldsymbol{R}^n$ cannot be locally free in the case $\alpha\neq 0$. Namely Proposition 1.5 does not hold in general. (See Remark 8.20 on the explanation from the view point of the curvature.)

1.3.6. The prolongment for fixed λ and the KMS-structure. We often use the notation $\mathcal{Y}^\lambda$ to denote $Y\times\{\lambda\}$ for a complex variety Y and $\lambda\in\boldsymbol{C}$. First we discuss the prolongment for fixed λ, namely we consider the prolongment of the holomorphic bundle $\mathcal{E}^\lambda = (E, \overline{\partial}_E + \lambda\theta^\dagger)$ on $\mathcal{X}^\lambda - \mathcal{D}^\lambda$ to the sheaf ${}_{\boldsymbol{b}}\mathcal{E}^\lambda$ on $\mathcal{X}^\lambda$. We can show the local freeness by the essentially same argument as the proof of Proposition 1.6.

PROPOSITION 1.7 (Theorem 8.58). *${}_{\boldsymbol{b}}\mathcal{E}^\lambda$ is locally free.* □

Then we have the following two structures of ${}_{\boldsymbol{b}}\mathcal{E}^\lambda{}_{|\mathcal{D}_i^\lambda}$ $(i=1,\ldots,n)$.

- The parabolic filtration ${}^iF$, which is a filtration of ${}_{\boldsymbol{b}}\mathcal{E}^\lambda{}_{|\mathcal{D}_i^\lambda}$ in the category of vector bundles. It is given as follows:
$$ {}^iF_c({}_{\boldsymbol{b}}\mathcal{E}^\lambda{}_{|\mathcal{D}_i^\lambda}) := \mathrm{Im}\Big({}_{\boldsymbol{b}+(c-b_i)\boldsymbol{\delta}_i}\mathcal{E}^\lambda \longrightarrow {}_{\boldsymbol{b}}\mathcal{E}^\lambda\Big).$$
- The generalized eigen decomposition ${}^i\mathbb{E}$ of ${}_{\boldsymbol{b}}\mathcal{E}^\lambda{}_{|\mathcal{D}_i^\lambda}$ with respect to the action of the residue $\mathrm{Res}_i(\mathbb{D}^\lambda)$.

They are called the KMS-structure (Kashiwara-Malgrange-Sabbah-Simpson) in this paper. The filtration ${}^iF$ and the decomposition ${}^i\mathbb{E}$ are compatible, in the sense ${}^iF_a = \bigoplus_{\alpha\in\boldsymbol{C}} {}^i\mathbb{E}_\alpha \cap {}^iF_a$.

1.3.7. KMS-spectrum. We obtain the following data:

$$\mathcal{KMS}({}_{\boldsymbol{b}}\mathcal{E}^\lambda, i) := \big\{(a,\alpha)\in\boldsymbol{R}\times\boldsymbol{C}\,\big|\,{}^i\,\mathrm{Gr}^F_a\,{}^i\mathbb{E}({}_{\boldsymbol{b}}\mathcal{E}_{|D_i},\alpha)\neq 0\big\}\subset\boldsymbol{R}\times\boldsymbol{C}.$$

We put $\mathcal{KMS}(\mathcal{E}^\lambda,i) = \bigcup_{\boldsymbol{b}}\mathcal{KMS}({}_{\boldsymbol{b}}\mathcal{E}^\lambda,i)\subset\boldsymbol{R}\times\boldsymbol{C}$. Each element of $\mathcal{KMS}(\mathcal{E}^\lambda,i)$ is called the KMS-spectrum at λ. The number $\dim{}^i\,\mathrm{Gr}^F_a\,{}^i\mathbb{E}({}_{\boldsymbol{b}}\mathcal{E}_{|D_i},\alpha)$ is called the multiplicity of $(a,\alpha)\in\mathcal{KMS}({}_{\boldsymbol{b}}\mathcal{E},i)$.

Let (a,α) be an element of $\mathcal{KMS}(\mathcal{E}^\lambda,i)$. Then it is easy to see that $(a+n,\alpha-n\lambda)$ are also elements of $\mathcal{KMS}(\mathcal{E}^\lambda,i)$ for any $n\in\mathbb{Z}$. The multiplicities of (a,α) and $(a+n,\alpha-n\lambda)$ are same. In other words, we have the $\mathbb{Z}$-action on $\mathcal{KMS}(\mathcal{E}^\lambda,i)$.

We have the bijection $\mathfrak{k}(\lambda):\boldsymbol{R}\times\boldsymbol{C}\longrightarrow\boldsymbol{R}\times\boldsymbol{C}$. For $u=(a,\alpha)\in\boldsymbol{R}\times\boldsymbol{C}$, we put as follows:

$$\mathfrak{k}(\lambda,u) := \big(\mathfrak{p}(\lambda,u),\mathfrak{e}(\lambda,u)\big),\qquad \begin{cases}\mathfrak{p}(\lambda,u) := a + 2\,\mathrm{Re}\big(\lambda\cdot\overline{\alpha}\big),\\[1ex] \mathfrak{e}(\lambda,u) := \alpha - a\cdot\lambda - \overline{\alpha}\cdot\lambda^2.\end{cases}$$

We note that $\mathfrak{e}(\lambda,u)$ is the eigenvalue of the residue $\mathrm{Res}(\mathbb{D})$ in Example 1.4, and that $\mathfrak{p}(\lambda,u)$ is the degree of f with respect to the parabolic filtration in the example. (See the subsections 1.3.3 and 1.3.5.)

The following proposition is essentially due to Simpson ([**81**]).

PROPOSITION 1.8. *The map $\mathfrak{k}(\lambda)$ induces the one to one correspondence between $\mathcal{KMS}(\mathcal{E}^0,i)$ and $\mathcal{KMS}(\mathcal{E}^\lambda,i)$. The multiplicities are preserved.* □

Let I be a subset of $\underline{n} = \{1, \ldots, n\}$. We put $D_I := \bigcup_{i \in I} D_i$. Then we have the filtrations ${}^iF$ $(i \in I)$ and the decompositions ${}^i\mathbb{E}$ $(i \in I)$ of ${}_{\boldsymbol{b}}\mathcal{E}^\lambda{}_{|\mathcal{D}_I^\lambda}$. It can be shown that they are compatible. Then we obtain the following subset of $\boldsymbol{R}^I \times \boldsymbol{C}^I = (\boldsymbol{R} \times \boldsymbol{C})^I$:

$$\mathcal{KMS}({}_{\boldsymbol{b}}\mathcal{E}^\lambda, I) := \big\{(\boldsymbol{a}, \boldsymbol{\alpha}) \in \boldsymbol{R}^I \times \boldsymbol{C}^I \,\big|\, {}^I\operatorname{Gr}^{F}_{\boldsymbol{a}}\, {}^I\mathbb{E}\big({}_{\boldsymbol{b}}\mathcal{E}^\lambda_{|\mathcal{D}_I^\lambda}, \boldsymbol{\alpha}\big) \neq 0\big\}.$$

We put $\mathcal{KMS}(\mathcal{E}^\lambda, I) := \bigcup_{\boldsymbol{b}} \mathcal{KMS}({}_{\boldsymbol{b}}\mathcal{E}^\lambda, I)$. The element $\boldsymbol{u} = (\boldsymbol{a}, \boldsymbol{\alpha}) \in \mathcal{KMS}({}_{\boldsymbol{b}}\mathcal{E}^\lambda, I)$ is called KMS-spectrum, and the number $\dim {}^I\operatorname{Gr}^{F}_{\boldsymbol{a}}\, {}^I\mathbb{E}({}_{\boldsymbol{b}}\mathcal{E}^\lambda_{|\mathcal{D}_I^\lambda}, \boldsymbol{\alpha})$ is called the multiplicity of $\boldsymbol{u}$. Similarly to the case of $I = \{i\}$, we have the $\mathbb{Z}^I$-action on $\mathcal{KMS}(\mathcal{E}^\lambda, I)$, preserving the multiplicities.

PROPOSITION 1.9 (Proposition 8.108). *$\mathfrak{k}(\lambda)$ induces the bijective correspondence of $\mathcal{KMS}(\mathcal{E}^0, I)$ and $\mathcal{KMS}(\mathcal{E}^\lambda, I)$, preserving the multiplicities.* □

1.3.8. The graduation. For any element $\boldsymbol{u} \in \mathcal{KMS}(\mathcal{E}^0, \underline{n})$, we take an appropriate $\boldsymbol{b} \in \boldsymbol{R}^n$, and we put as follows:

$${}^{\underline{n}}\mathcal{G}^\lambda_{\boldsymbol{u}} := {}^{\underline{n}}\operatorname{Gr}^F_{\mathfrak{p}(\lambda, \boldsymbol{u})}\, {}^{\underline{n}}\mathbb{E}({}_{\boldsymbol{b}}\mathcal{E}, \mathfrak{e}(\lambda, \boldsymbol{u})).$$

The residue $\operatorname{Res}_i(\mathbb{D})$ induces the endomorphism of ${}^{\underline{n}}\mathcal{G}^\lambda_{\boldsymbol{u}}$. The unique eigenvalue of the endomorphism is $\mathfrak{e}(\lambda, u_i)$ by our construction. The nilpotent part is denoted by $\mathcal{N}_i^\lambda$.

REMARK 1.10. The tame harmonic bundle $(E, \overline{\partial}_E, \theta, h)$ is nilpotent and with the trivial parabolic structure, if and only if the set $\mathcal{KMS}(\mathcal{E}^0, \underline{n})$ is same as $(\mathbb{Z} \times \{0\})^n$. We have only to consider the spectrum $0 \in \mathcal{KMS}(\mathcal{E}^\lambda, \underline{n})$ due to the $\mathbb{Z}^n$-action, and we have ${}^{\underline{n}}\mathcal{G}^\lambda_0 = {}^\diamond\mathcal{E}_{|(O,\lambda)}$. □

Then we obtain the family $\big\{{}^{\underline{n}}\mathcal{G}^\lambda_{\boldsymbol{u}} \,\big|\, \lambda \in \boldsymbol{C}\big\}$ of vector spaces. We would like to give the structure of a holomorphic vector bundle over the complex line $\boldsymbol{C}_\lambda$. In the case where $(E, \overline{\partial}_E, \theta, h)$ is nilpotent and with trivial parabolic structure, we have the holomorphic bundle ${}^\diamond\mathcal{E}$ over $\mathcal{X}$, and thus we obtain the holomorphic bundle ${}^\diamond\mathcal{E}_{|\boldsymbol{C}_\lambda}$, which gives the structure of the holomorphic vector bundle of the family $\big\{{}^{\underline{n}}\mathcal{G}^\lambda_0 \,\big|\, \lambda \in \boldsymbol{C}\big\}$. As we have already said, the sheaves ${}^\diamond\mathcal{E}$ or ${}_{\boldsymbol{b}}\mathcal{E}$ on $\mathcal{X}$ are not locally free in general (see the subsection 1.3.3), and we need some additional argument.

1.3.9. Prolongment in the case where λ is varied. Let us pick any point $\lambda_0 \in \boldsymbol{C}_\lambda$. We put $\Delta(\lambda_0, \epsilon_0) := \big\{\lambda \in \boldsymbol{C} \,\big|\, |\lambda - \lambda_0| < \epsilon_0\big\}$ for a positive number ϵ_0. We use the notation $\mathcal{Y}(\lambda_0, \epsilon_0)$ to denote $Y \times \Delta(\lambda_0, \epsilon_0)$ for a complex variety Y.

PROPOSITION 1.11. *Pick an element $\boldsymbol{b} \in \boldsymbol{R}^n$ such that $b_i \notin \mathcal{KMS}(\mathcal{E}^{\lambda_0}, i)$ for each $i \in \underline{n}$. Take a sufficiently small positive number ϵ_0.*

- *${}_{\boldsymbol{b}}\mathcal{E}$ is a locally free sheaf on $\mathcal{X}(\lambda_0, \epsilon_0)$.*
- *We have the canonical isomorphism $\big({}_{\boldsymbol{b}}\mathcal{E}\big)_{|\mathcal{X}^\lambda} \simeq {}_{\boldsymbol{b}}\mathcal{E}^\lambda$ for any point $\lambda \in \Delta(\lambda_0, \epsilon_0)$.* □

For each $i \in \underline{n}$, we have the filtration ${}^iF^{(\lambda_0)}$ and the decomposition ${}^i\mathbb{E}$ of the vector bundle ${}_{\boldsymbol{b}}\mathcal{E}_{|\mathcal{D}_i(\lambda_0, \epsilon_0)}$. See the subsections 8.8.1 and 8.8.5 for the definitions and the properties of ${}^iF^{(\lambda_0)}$ and ${}^i\mathbb{E}$. We obtain the vector bundle ${}^{\underline{n}}\mathcal{G}^{(\lambda_0)}_{\boldsymbol{u}}$ over $\{O\} \times \Delta(\lambda_0, \epsilon_0)$ for any element $\boldsymbol{u} \in \mathcal{KMS}(\mathcal{E}^0, \underline{n})$:

$${}^{\underline{n}}\mathcal{G}^{(\lambda_0)}_{\boldsymbol{u}} = {}^{\underline{n}}\operatorname{Gr}^{F^{(\lambda_0)}}_{\mathfrak{p}(\lambda_0, \boldsymbol{u})}\, {}^{\underline{n}}\mathbb{E}^{(\lambda_0)}\Big({}_{\boldsymbol{b}}\mathcal{E}_{|\{O\} \times \Delta(\lambda_0, \epsilon_0)},\, \mathfrak{e}(\lambda_0, \boldsymbol{u})\Big).$$

Then we have ${}^{\underline{n}}\mathcal{G}^{(\lambda_0)}_{\boldsymbol{u}\,|\,\lambda}={}^{\underline{n}}\mathcal{G}^{\lambda}_{\boldsymbol{u}}$ for any point $\lambda\in\Delta(\lambda_0,\epsilon_0)$. When the intersection $S=\Delta(\lambda_0,\epsilon_0)\cap\Delta(\lambda_1,\epsilon_1)$ is not empty, we have the canonical isomorphism ${}^{\underline{n}}\mathcal{G}^{(\lambda_0)}_{\boldsymbol{u}\,|\,S}\simeq{}^{\underline{n}}\mathcal{G}^{(\lambda_1)}_{\boldsymbol{u}\,|\,S}$. Thus we obtain the global vector bundle ${}^{\underline{n}}\mathcal{G}_{\boldsymbol{u}}$ on $\boldsymbol{C}_\lambda$ such that ${}^{\underline{n}}\mathcal{G}_{\boldsymbol{u}\,|\,\Delta(\lambda_0,\epsilon_0)}\simeq{}^{\underline{n}}\mathcal{G}^{(\lambda_0)}_{\boldsymbol{u}}$ and ${}^{\underline{n}}\mathcal{G}_{\boldsymbol{u}\,|\,\lambda}\simeq{}^{\underline{n}}\mathcal{G}^{\lambda}_{\boldsymbol{u}}$. The nilpotent part of the residues induce the nilpotent endomorphisms $\mathcal{N}_i$ $(i\in\underline{n})$ of ${}^{\underline{n}}\mathcal{G}_{\boldsymbol{u}}$.

1.3.10. The induced vector bundle over $\mathbb{P}^1$. By the same construction for the tame harmonic bundle $(E,\partial_E,\theta^\dagger,h)$ on $X^\dagger-D^\dagger$, we obtain the holomorphic bundle ${}^{\underline{n}}\mathcal{G}^\dagger_{\boldsymbol{u}}$ over $\boldsymbol{C}_\mu$, for any element $\boldsymbol{u}\in\mathcal{KMS}(\mathcal{E}^{\dagger\,0},\underline{n})$. We have the induced nilpotent endomorphisms $\mathcal{N}^\dagger_i$ $(i\in\underline{n})$.

We have the morphism $\boldsymbol{R}\times\boldsymbol{C}\longrightarrow\boldsymbol{R}\times\boldsymbol{C}$ given by $(a,\alpha)\longmapsto(-a,\overline{\alpha})$. It induces the bijection $\mathcal{KMS}(\mathcal{E}^0,\underline{n})\longrightarrow\mathcal{KMS}(\mathcal{E}^{\dagger\,0},\underline{n})$. We denote the correspondence by $\boldsymbol{u}\longmapsto\boldsymbol{u}^\dagger$.

When we take a point $P\in X-D$, we have the gluing of ${}^{\underline{n}}\mathcal{G}_{\boldsymbol{u}}$ and ${}^{\underline{n}}\mathcal{G}^\dagger_{\boldsymbol{u}^\dagger}$, and thus we obtain the vector bundle $S_{\boldsymbol{u}}(E,P)$ over $\mathbb{P}^1$. We also obtain the nilpotent morphism $\mathcal{N}^\triangle_i:S_{\boldsymbol{u}}(E,P)\longrightarrow S_{\boldsymbol{u}}(E,P)\otimes\mathcal{O}(2)$. As in the previous paper, we put $\mathcal{N}^\triangle(\underline{n})=\sum_{i=1}^n\mathcal{N}^\triangle_i$, which induces the weight filtration W on $S_{\boldsymbol{u}}(E,P)$.

REMARK 1.12. We have another gluing, and the obtained vector bundle is denoted by $S^{\mathrm{can}}_{\boldsymbol{u}}(E)$. In fact, the construction of $S^{\mathrm{can}}_{\boldsymbol{u}}(E)$ is close to the traditional construction of the limiting mixed Hodge structure. On the other hand, the construction of $S_{\boldsymbol{u}}(E,P)$ is close to that in [**65**]. We will mainly use $S^{\mathrm{can}}_{\boldsymbol{u}}(E)$ in this paper. □

1.3.11. Limiting mixed twistor structure theorem. We discuss the naturally induced pairing S of $S_{\boldsymbol{u}}(E,P)$ and $S^{\mathrm{can}}_{\boldsymbol{u}}(E)$ as in the classical Hodge theory, following C. Sabbah [**72**] who considered the polarized limiting mixed twistor structures for tame harmonic bundles on a quasi projective curve in a different setting, based on the result of Biquard [**6**]. The following theorem is one of the main goals in the study of Part 2–3.

THEOREM 1.13 (Limiting Mixed Twistor theorem, Theorem 12.22). *The tuples $\big(S_{\boldsymbol{u}}(E,P),W,\boldsymbol{N}^\triangle,S\big)$ and $\big(S^{\mathrm{can}}_{\boldsymbol{u}}(E),W,\boldsymbol{N}^\triangle,S\big)$ are polarized mixed twistor structures. (See Definition* 3.50 *for polarized mixed twistor structure.)* □

1.3.12. Reduction to the classical Hodge theory. We use Theorem 1.13 by taking the associated graded objects. We have the associated graded vector bundle $V^{(0)}:=\mathrm{Gr}^W S^{\mathrm{can}}_{\boldsymbol{u}}(E)$, on which we have the naturally induced filtration $W^{(0)}$, the nilpotent maps $\boldsymbol{N}^{(0)}$ and the pairing $S^{(0)}$. Then the tuple $\big(V^{(0)},W^{(0)},\boldsymbol{N}^{(0)},S^{(0)}\big)$ is again a polarized mixed twistor structure. We can take an appropriate torus action on the tuple, and thus it is a polarized mixed Hodge structure. (See the subsection 3.4.2.) Moreover, $\big(V^{(0)},W^{(0)},\boldsymbol{N}^{(0)},S^{(0)}\big)$ is a nilpotent orbit. Since the nilpotent orbit was studied quite deeply in the classical theory of CVHS, we can say that we understand the tuple $\big(V^{(0)},W^{(0)},\boldsymbol{N}^{(0)},S^{(0)}\big)$ very well.

At this point, the study of the asymptotic behaviour of tame harmonic bundles can be reduced to the study of the asymptotic behaviour of CVHS. Much information on the tuple $\big(S^{\mathrm{can}}_{\boldsymbol{u}},W,\boldsymbol{N},S\big)$ can be obtained from $\big(V^{(0)},W^{(0)},\boldsymbol{N}^{(0)},S^{(0)}\big)$. Briefly speaking, $\big(V^{(0)},W^{(0)},\boldsymbol{N}^{(0)},S^{(0)}\big)$ is a degeneration of $\big(S^{\mathrm{can}}_{\boldsymbol{u}},W,\boldsymbol{N},S\big)$, but

$\big(V^{(0)}, W^{(0)}, \boldsymbol{N}^{(0)}, S^{(0)}\big)$ is sufficiently generic, and hence the properties are not so different.

For example, we can obtain the compatibility of the nilpotent maps and vanishing cycle theorem. We can also apply the lemmas due to Kashiwara and Saito on the nilpotent orbit to $\big(S^{\mathrm{can}}_{\boldsymbol{u}}, W, \boldsymbol{N}, S\big)$.

1.3.13. The relation with the argument in [65]. In the previous paper [**65**], we often used the argument to take a 'limit' of a sequence of tame harmonic bundles. For example, we consider the morphisms ${}^{n}\psi_m : X - D \longrightarrow X - D$ given by $(z_1, \dots, z_n) \longmapsto (z_1^m, \dots, z_n^m)$, and we consider the sequence $\big\{\psi_m^*(E, \overline{\partial}_E, \theta, h) \,\big|\, m \in \mathbb{Z}_{>0}\big\}$ of harmonic bundles. Under the assumption that $(E, \overline{\partial}_E, \theta, h)$ is nilpotent and with trivial parabolic structure, we obtained the complex variation of Hodge structure as the 'limit'.

The argument to take a limit does not work, if the residues of the Higgs field are not nilpotent. In Example 1.4 in the subsection 1.3.3, we have ${}^{1}\psi_m^*\theta = m \cdot \alpha \cdot dz/z$, and thus it is not so easy to guess what is 'limit' for $m \to \infty$.

In a sense, taking the associated graded tuple $\big(V^{(0)}, W^{(0)}, \boldsymbol{N}^{(0)}, S^{(0)}\big)$ corresponds to taking a 'limit'. Let us consider the case that $(E, \overline{\partial}_E, \theta, h)$ is nilpotent and with trivial parabolic structure. As is already mentioned, we obtain the limiting CVHS $(E^{(\infty)}, \overline{\partial}_{E^{(\infty)}}, \theta^{(\infty)}, h^{(\infty)})$ as a limit. Then it is easy to see that the limiting mixed twistor structure of $(E^{(\infty)}, \overline{\partial}_{E^{(\infty)}}, \theta^{(\infty)}, h^{(\infty)})$ is naturally isomorphic to the associated graded mixed twistor structure $\big(V^{(0)}, W^{(0)}\big)$.

Most of our argument to take a limit in the previous paper [**65**] can be replaced with the argument to consider the associated graded tuple. The only exception is the proof of the constantness of the filtration, for which we use some elementary calculus instead of taking a limit.

REMARK 1.14. As is mentioned above, we do not use the argument to take a limit in this paper. However, it seems significant to observe that CVHS appears as the 'limit'. □

1.4. On the purpose (2)

1.4.1. Pure twistor D-module and Sabbah's program. Following Simpson's Meta-Theorem, it is interesting and natural to ask whether we can construct the theory of "twistor module", which should be a generalization of the theory of Hodge module of M. Saito ([**73**] and [**75**]). In this direction, C. Sabbah has already done a remarkable work ([**72**]). He introduced regular pure twistor D-modules and proved the decomposition theorem. Since it is his motivation to attack a conjecture of Kashiwara, we recall a part of it.

CONJECTURE 1.15 (The regular holonomic version of Kashiwara's conjecture). *Let X be a smooth quasi projective variety over the complex number field $\boldsymbol{C}$, and $\mathcal{F}$ be a semisimple regular holonomic algebraic D-module on X.*

(Push-forward): *Let Y be a smooth quasi projective variety over $\boldsymbol{C}$, and let $f : X \longrightarrow Y$ be a projective morphism.*

- *Then the push-forward $Rf_+\mathcal{F}$ is isomorphic to $\bigoplus R^i f_+\mathcal{F}[-i]$ in the derived category of cohomologically holonomic complexes on Y.*
- *The hard Lefschetz theorem for $\bigoplus R^i f_+\mathcal{F}$ holds.*
- *Each $R^i f_+\mathcal{F}$ is semisimple.*

(Vanishing cycles): *Let g be a holomorphic function on X. We take the nearby cycle functor and the vanishing cycle functor along g. Then the associated graded objects with respect to the monodromy weight filtration are semisimple.* □

REMARK 1.16. The conjecture in [**50**] is stated for the projective morphisms of smooth complex algebraic varieties, not necessarily quasi projective. Although we discuss the quasi projective case, it is not an essential restriction. □

REMARK 1.17. The conjecture of Kashiwara is stronger than the statement above. In fact, he conjectured that the statement is true for semisimple holonomic D-modules which are not necessarily regular. See [**50**] for more precise. □

Sabbah's program to attack the conjecture is as follows:

Step 1.: Establish the correspondence of semisimple local systems and tame harmonic bundles on a quasi projective variety.
Step 2.: Give a definition of regular pure twistor D-module, and prove the decomposition theorem for regular pure twistor D-modules.
Step 3.: Establish the correspondence of tame harmonic bundles and regular pure twistor D-modules.

As is already remarked, Sabbah established the step 2 in [**72**]. Sabbah also proved that a harmonic bundle, without singularity, gives a regular pure twistor D-module for the step 3. Combining the results due to K. Corlette (see the subsection 1.4.2) and himself, Sabbah obtained the hard Lefschetz theorem for semisimple local system via the morphisms of projective varieties.

REMARK 1.18.

- Our pure twistor D-modules is slightly wider than Sabbah's pure twistor D-modules in [**72**]. Sabbah's pure twistor D-module is called a pure imaginary pure twistor D-module in this paper, which is most suitable for attacking Kashiwara's conjecture. But it can also be said that it is slightly narrow from the view point of Simpson's Meta Theorem.
- Although Sabbah discusses the pure twistor D-modules which are not necessarily regular, we consider only regular pure twistor D-modules in this paper. We always impose the regularity condition, even if we omit to distinguish "regular". □

We will establish Step 1 and Step 3 in Part 5 and Part 4 respectively.

1.4.2. The goal of Part 5, the characterization of semisimplicity by a pluri-harmonic metric. As for Step 1, there is known the classical result of K. Corlette who established the correspondence of semisimple local systems and harmonic bundles over a projective manifold. The result was generalized by J. Jost and K. Zuo, who proved the existence of a pluri-harmonic metric on a semisimple local system over a quasi projective manifold. In other words, there exists the structure of harmonic bundle on any semisimple local system over a quasi projective manifold. It is slightly refined in Part 5 of this paper, namely we will prove the following theorem.

THEOREM 1.19 (Theorem 25.28). *Let (E, ∇) be a flat bundle on a complex quasi projective manifold.*

- *The flat bundle (E, ∇) is semisimple if and only if there exists a tame pure imaginary pluri-harmonic metric h on (E, ∇).*
- *The tame pure imaginary pluri-harmonic metric h is uniquely determined up to the flat automorphisms of E.*

Hence we can say that the Step 1 is established.

REMARK 1.20. Theorem 1.19 seems just a combination of the results due to some people.

- The "only if" part in the first claim is hardest, which is due to J. Jost and K. Zuo [**44**], although they do not discuss the imaginary condition.
- It is technically easy to show that the obtained tame harmonic bundle is pure imaginary, once we completely understand the proof of the existence theorem of pluri-harmonic metric. However it is very important for the applications.
- The 'if' part in the first claim is proved by C. Sabbah in ([**72**]). Since it was not discussed in the older version of the paper [**72**], we proved it. But we learned the essential idea (Lemma 22.20) in the older version of [**72**].
- Since the second claim can be easily reduced to the one dimensional case, it essentially seems due to C. Simpson [**81**] or O. Biquard [**6**], although it looks that a detailed proof is omitted. We give a different proof based on Corlette's argument.

Since Theorem 1.19 is very important for our application, the author thinks it appropriate to give a detailed proof of the theorem which is available for a wide range of the readers. □

REMARK 1.21. Part 5 is rather independent of the rest parts of this paper. In fact, it was written as a separate paper [**67**], although we originally intend to include it as an appendix of this paper. That is the reason why we put it last. □

1.4.3. The goal of Part 4, the conjectures of Sabbah and Kashiwara. Let X be a complex manifold. Let Z be an irreducible closed subset of X. A tame harmonic bundle generically defined on Z is defined to be a tame harmonic bundle defined over a smooth Zariski open subset of Z. The purpose of the part 4 is to establish the step 3, namely we will prove that a tame harmonic bundle generically defined on Z gives the pure twistor D-module of weight 0. More precisely, we will prove the following correspondence:

THEOREM 1.22 (Theorem 19.6, Theorem 19.42). *We have the bijective correspondences:*

$$\mathrm{VPT_{gen}}(Z, w) \simeq \mathrm{MPT}_Z(X, w), \qquad \mathrm{VPT^{pi}_{gen}}(Z, w) \simeq \mathrm{MPT}^{pi}_Z(X, w).$$

(See the subsection 19.1.2 *for the notation.)* □

Recall that Saito established the following correspondence for Hodge modules.

PROPOSITION 1.23 (Saito, [**75**]). *Variations of pure polarized Hodge structures of weight w which are defined over Zariski open subsets of Z correspond to pure polarized Hodge modules of weight w whose strict supports are Z.* □

Theorem 1.22 is the counterpart of Proposition 1.23 in the theory of pure twistor D-module. Once we have established the resemblance of the asymptotic behaviours of tame harmonic bundles and CVHS, we can use a part of Saito's idea to prove Theorem 1.22. In fact, it can be said that the crucial ideas can be found in the sections 3.19–3.21 of [**75**]. However, it is not so clear for the author how to modify the arguments in 3.1–3.5 and the most part of 3.b of [**75**], for our purpose. Hence we will go along the other route.

Sabbah conjectured that every semisimple regular holonomic D-module on a complex projective manifold underlies a pure imaginary pure twistor D-module. Once Theorem 1.22 and Theorem 1.19 are established, it is easy to show that his conjecture is true. Namely we obtain the following theorem. (See the subsection 19.1.2 and the section 19.6 for the notation.)

THEOREM 1.24 (The conjecture of Sabbah, Theorem 19.46). *The map* Ξ_{DR} : $\mathrm{MPT}_Z^{pi}(X, 0) \longrightarrow \mathrm{RHD}_Z^{ss}(X)$ *is surjective, if* X *is a smooth projective variety.* □

As a result, we obtain the regular holonomic version of Kashiwara's conjecture, combining the results of Sabbah ([**72**]) and us.

COROLLARY 1.25 (Theorem 19.47 and Theorem 19.49). *Conjecture* 1.15 *is true.* □

REMARK 1.26. We need only algebraicity of X in Theorem 1.24, which is clear from the proof. □

REMARK 1.27.

- K. Vilonen kindly informed the author of the work of D. Gaitsgory [**30**], who proved de Jong's conjecture. G. Boeckle and C. Khare also proved de Jong's conjecture [**8**]. Since V. Drinfeld proved that de Jong's conjecture implies the regular holonomic version of Kashiwara's conjecture [**25**] (Conjecture 1.15) has been established also by their works.
- We should also mention the interesting work of M. de Cataldo and L. Migliorini [**18**], who proved a decomposition theorem of perverse sheaves of geometric origin, by using Hodge theory, but without the method of M. Saito. □

1.5. Some Remark

The first and the second versions of this paper are available on the web ([**66**]). Perhaps, this version is much longer than them. The main reasons are as follows: The preprint [**67**] is included in Part V. The formats are different. We include a summary for the theory of pure twistor D-module, by following the suggestion of the referee. Of course, we include many other additional explanations. However the arguments are same as those contained in [**66**] and [**67**].

1.6. The outline of the paper

1.6.1. Part 1, Chapter 2. The Part 1 is a preparation for the subsequent parts. The author expects that the readers can skip Part 1 until they need it.

In the chapter 2, we prepare some notation and lemmas from several areas. In the section 2.1, we prepare the notation of some sets and the functions. The maps κ_c and ν_c in the subsection 2.1.5 are used to describe the descent of sections for ramified covering. The maps $\mathfrak{k}$, $\mathfrak{e}$, and $\mathfrak{p}$ in the subsection 2.1.7 are used for the control of the KMS-structure of tame harmonic bundles.

In the section 2.2, we recall the prolongment of a holomorphic vector bundle with a hermitian metric over $X - D$ to an $\mathcal{O}_X$-sheaf on X. Here X denotes a complex manifold, and D denotes a normal crossing divisor. We give a condition for the local freeness of the prolonged sheaf (Lemma 2.8). In the section 2.3, we prepare something on the μ_c-equivariant holomorphic bundles. It will be useful when we consider the descent of holomorphic bundle.

In the section 2.4, we give a lemma for pluri-subharmonic function and convexity. It will be used in the proof of preliminary constantness of the filtrations in the subsection 12.2.4. Although the argument is elementary, it is one of the key steps. Hence we give the detail. In the section 2.5, we recall some well known lemmas related with the boundedness on a function on a disc. In the section 2.6, we consider the distributions given by a polynomials of logarithmic functions. The result will be used when we calculate the specialization of the sesqui-linear pairing of the $\mathcal{R}$-triples obtained from tame harmonic bundle (the subsection 18.4.7).

In the section 2.7, we mainly give some lemmas on a metrics on a finite dimensional vector space. They are used in the section 7.1 and the section 8.2. The notation $\mathbb{E}_\epsilon$ $(\epsilon > 0)$ is introduced.

In the section 2.8, we recall two kinds of lemmas for the acceptable bundles. One is the vanishing of the higher cohomology groups (Lemma 2.48 and Corollary 2.63). The other is Corollary 2.51 and Corollary 2.53, which control the estimate of the increasing order of holomorphic sections. We also prepare some setting, which will be used in the section 8.4.

In the section 2.9, we give a lemma for a complex of Hilbert space bundles over the disc (Lemma 2.68). It will be used to obtain a locally free prolongment of the deformed holomorphic bundle of tame harmonic bundle (the section 8.7). Although the procedure is standard, we have to care the infinite dimensionality, and we do not know an appropriate reference. Thus we give the detail. We also recall a standard lemma for embeddings of Sobolev spaces, which will be used in the section 2.10.

In the section 2.10, we give an estimate of Higgs field of harmonic bundle. In the section 2.11, we give an improvement of the convergency of a sequence of harmonic bundles, which was given in our previous paper [**65**]. Since we do not use an argument to take a limit, the reader can skip the section 2.11.

In the section 2.12, we recall the relation of Higgs field and a twisted map associated with the flat λ-connection with a hermitian metric. It will be used in the subsection 8.1.3.

1.6.2. Chapter 3. Following Simpson ([**82**]), we give the detail on the relation of Hodge structure and twistor structure. In the section 3.1, we introduce some terminology and the notation.

In the section 3.2, we recall the equivalence of the categories of equivariant holomorphic vector bundle over $\mathbb{P}^1$ and bi-filtered vector space. The equivalence is compatible with real structures.

In the section 3.3, we give the concrete description of the Tate objects, the objects $\mathcal{O}(p,q)$ and $\mathcal{O}(n)$ in the category of twistor structures. In particular, we give the isomorphism $\iota_{(p,q)} : \sigma^*\mathcal{O}(p,q) \longrightarrow \mathcal{O}(q,p)$. The isomorphism is fixed in the sequel. We also compare $\mathbb{T}(n)$ with the Tate objects in the Hodge theory.

In the section 3.4, we recall the equivalence of the category of polarized pure Hodge structures and the category of equivariant polarized pure twistor structures.

We also introduce polarized mixed twistor structure, and we see the equivalence of the category of polarized mixed Hodge structures and the category of equivariant polarized mixed twistor structures. Note that our choice of the signature of the nilpotent maps is different from that in the standard Hodge theory (see [**10**], for example).

In the section 3.5, we recall the variation of twistor structures, or more generally, the variation of $\mathbb{P}^1$-holomorphic bundles. An example given in the subsection 3.5.3 is important for our understanding of the harmonic bundles. We also see the equivalence of the category of CVHS and the category of the variation of equivariant pure twistor structures, which is compatible with some additional structures.

In the section 3.6, we introduce the twistor nilpotent orbit. We see that it is a generalization of nilpotent orbit in the Hodge theory (Proposition 3.85).

In the section 3.7, we see that a split polarized mixed twistor structure gives a nilpotent orbit in the Hodge theory (Corollary 3.110). Since we can pick an appropriate torus action on the split polarized mixed twistor structure, we can regard it as a split polarized mixed Hodge structure. Thus the result may be known in the Hodge theory, probably. However it is very important for our application, and hence we give the detail. In particular, we can always obtain the nilpotent orbit in the Hodge theory from a polarized mixed twistor structure, by taking the associated graded object. This is one of the key steps to reduce our study of tame harmonic bundle to the classical study of Hodge structures.

In the section 3.8, we see that the polarized mixed twistor structure $(V, W, \boldsymbol{N}, S)$ induces the polarized structure on the primitive part $P_h \operatorname{Gr}_h^{W(N)}(V)$ (Proposition 3.115). By using the result, we see that the tuple of nilpotent maps $\boldsymbol{N}$ is strongly sequentially compatible (Lemma 3.116).

In the section 3.9, we translate some results known for the Hodge structure, due to Kashiwara, Kawai and Saito, to the results for the twistor structure (Proposition 3.126, Corollary 3.132, Proposition 3.137 and Lemma 3.143). We use the results in 3.7. They are crucial for our study to relate tame harmonic bundles and pure twistor D-modules.

In the section 3.10, we give the concrete correspondence of the twistor structure in the sense of Simpson and those in the sense of Sabbah.

1.6.3. Chapter 4. We discuss the compatibility of decompositions, filtrations and nilpotent maps. Although we refer the definition of 'sequential compatibility' and 'strongly sequential compatibility' from our previous paper [**65**], we do not use the lemmas in [**65**] essentially.

In the section 4.1, we give the definitions of the compatibility of filtrations and decompositions on a vector space. In the section 4.2, we give those on a vector bundle. In the section 4.3, we discuss the compatibility of filtrations, decompositions and nilpotent maps.

In the section 4.4, we give some lemmas for extending a splitting given on a divisor. In the section 4.5, we give definitions of compatibility of decompositions, filtrations and nilpotent maps given on divisors. By using the result in the section 4.4, we see the existence of splitting.

1.6.4. Chapter 5. We consider a compatible tuple of filtrations on a discrete valuation ring R, such that the splitting is given on the generic point. We assume that some nice property holds on a generic point K. We also assume that the nice property holds on the associated graded vector bundle on R. Under such assumptions, we see that the nice property holds on R.

In the section 5.1, we discuss the sequential compatibility of the nilpotent maps. In the section 5.2, we discuss the strictness of the morphism with the filtrations.

The results in the chapter 5 will be useful, when we combine them with the limiting mixed twistor theorem. Briefly and imprecisely speaking, we can derive some information for the associated graded bundle of the parabolic filtrations from the limiting mixed twistor theorem. Then we can obtain the information of the original bundle by using the results in the chapter 5.

1.6.5. Chapter 6. We give easy and basic examples of harmonic bundles on a punctured disc, which we call model bundles. They are fundamental for the study of the asymptotic behaviour. In a sense, the study of the asymptotic behaviour of general tame harmonic bundles can be reduced to these basic examples. We also refer the section 3.2 in [**65**] for model bundles, although some constants in [**65**] are different from those in this paper.

1.6.6. Part 2, Chapter 7. In Part 2, we discuss the prolongment of the deformed holomorphic bundle of tame harmonic bundle.

In the chapter 7, we recall some of the results of Simpson on tame harmonic bundles over the punctured disc, with minor generalization. They play the fundamental role in the study of tame harmonic bundles on a higher dimensional complex manifold.

In the section 7.1, we give some detail on Simpson's Main estimate, that is, the estimate on the norm of Higgs field around the singularity. Since we would like to use the result in the higher dimensional case, we clarify the dependence of the constants. We also see the asymptotic orthogonality of the generalized eigen decomposition.

In the section 7.2, we recall the results on the prolongment of the deformed holomorphic bundle $\mathcal{E}^{\lambda}$ for a fixed λ. We introduce the KMS-structure of the prolongment ${}_{\boldsymbol{b}}\mathcal{E}^{\lambda}$, and we see the functoriality of the structure. In particular, we give some detail on the functoriality for pull backs via the ramified covering. It will be useful for the study in the higher dimensional case.

In the section 7.3, we recall the basic comparison due to Simpson. As a result, the KMS-structure at λ can be controlled by the KMS-structure at 0, and we see that the weight filtration at λ is equivalent to that at 0. We also obtain a rough relation of the frame at λ and the frame at 0, which will be used to show the asymptotic orthogonality.

In the section 7.4, we give some detail on the space $H(\mathcal{E}^{\lambda})$ of the multi-valued flat sections of the deformed holomorphic bundle $\mathcal{E}^{\lambda}$. We introduce the KMS-structure of $H(\mathcal{E}^{\lambda})$, and we compare it with the KMS-structure of ${}_{\boldsymbol{b}}\mathcal{E}^{\lambda}$.

In the section 7.5, we see that the situation is easy for generic λ. The observation is fundamental in the later discussions. In the subsection 7.5.1, we introduce the notion of 'generic' with respect to the KMS-structure. In the subsection 7.5.2, we see that the prolongment by an increasing order is equivalent to the quasi canonical prolongment, if λ is generic. In this sense, the prolongment for generic λ is canonically given, even if we forget the metric h.

In the section 7.6, we consider the family of the spaces $\{H(\mathcal{E}^{\lambda}) \mid \lambda \in \boldsymbol{C}_{\lambda}^{*}\}$. We introduce the decomposition $\mathbb{E}^{(\lambda_0)}$ and the filtration $\mathcal{F}^{(\lambda_0)}$ which are defined on a neighbourhood of $\lambda_0 \in \boldsymbol{C}_{\lambda}^{*}$.

In the section 7.7, we see the asymptotic orthogonality of the generalized eigen decomposition, the parabolic filtration, and the weight filtration. They are used in the proof of the limiting mixed twistor theorem in the case of curves (the section 12.1). The asymptotic orthogonality of the generalized eigen decomposition is also used for the local prolongment of $\mathcal{E}$ (the section 8.7).

In the section 7.8, we give a maximum principle for the distance of the harmonic metrics on a punctured disc. The result will be used to show a characterization of tameness in the subsection 8.1.3.

1.6.7. Chapter 8. We give the detail on the prolongment of the deformed holomorphic bundles of the tame harmonic bundle on $\Delta^* \times \Delta^{n-l}$. The chapter is one of the hearts of this monograph.

In the section 8.1, we give the remark on the constantness of the KMS-spectrum. We also see that the tame harmonic bundle of rank one is very easy to understand. We use the facts without mention. In the subsection 8.1.3, we also give a simple characterization of tameness (Corollary 8.7), which is useful when we check the tameness of a harmonic bundle. (See the subsection 19.2.1.)

In the section 8.2, we give the estimate of the Higgs bundle around the singularity in the higher dimensional case. Since we see the dependence of the constants closely in the section 7.1, the argument for the generalization to the higher dimensional case is elementary. As a consequence, we see that the tame harmonic bundle is acceptable. Thus we can apply the result in the section 2.8.

In the subsection 8.3, we give the prolongment of $\mathcal{E}^{\lambda}$ in the case that λ is generic. In this case, the situation is very easy. We see that the quasi canonical prolongment gives the prolongment by an increasing order in that case. Note that the direction of the argument is reverse to those in the one dimensional case. The results are used in the next sections.

In the section 8.4, we see the extension property of sections of $\mathcal{E}^{\lambda}$ defined over a hyperplane, by using the result in the section 2.8. The argument is essentially given in our previous paper [**65**]. However it is one of the most technical parts for the prolongment, and hence we give the detail. In the subsection 8.4.1, we give estimates of Higgs fields by using the results in the section 2.10. For a holomorphic section on a hyperplane, we construct a cocycle in the subsection 8.4.2. By using the estimate in the subsection 8.4.1, we give an estimate of the cocycle. The extension property is given in the subsection 8.4.3. We use the result in the section 2.8. In the subsection 8.4.4, we also state the extension property in the codimension one. Since this is the easier case, we give only an indication of the proof.

In the sections 8.5 and 8.6, we show that the prolongment ${}_{\boldsymbol{b}}\mathcal{E}^{\lambda}$ is locally free. As a preliminary, we show the claim under the assumption as in Lemma 8.56 in the section 8.5, by using the result in the section 8.4. Then we show the claim without the assumption in the section 8.6. We also see that the parabolic structures of the divisors give the compatible tuple of the filtrations. For that purpose, we consider the pull back of $\mathcal{E}^{\lambda}$ via the ramified covering $\psi_{\boldsymbol{c}}$ for an appropriate $\boldsymbol{c} \in \mathbb{Z}_{>0}^{l}$ in the subsection 8.6.2. Due to the result in the section 8.5, the prolongment of the pull back is locally free. Moreover we have the action of the finite abelian group, which induces the decompositions on the divisors. Due to the result in the section 7.2, we see that the decompositions give the splittings of the parabolic filtrations. In the subsection 8.6.3, we take an equivariant frame $\boldsymbol{v}$ of ${}^{\diamond}\psi_{\boldsymbol{c}}\mathcal{E}^{\lambda}$ which is compatible with the decompositions on the divisors. Then we take the descent of $\boldsymbol{v}$. We will see that the descent gives the frame of the prolongment of $\mathcal{E}^{\lambda}$ by using the result in the subsection 7.2, and we will obtain the local freeness of the prolongment.

In the section 8.7, we see that the prolongment ${}_{\boldsymbol{b}}\mathcal{E}$ is locally free on $\mathcal{X}(\lambda_0,\epsilon_0)$. First we show the extension property of holomorphic sections. For that purpose, we use the asymptotic orthogonality in the section 8.2, and the trivialization given by the argument in the section 2.9. Once the extension property is shown, it is easy to show the local freeness.

In the section 8.8, we see some structures induced on the divisors. In particular, we obtain the filtrations ${}^{i}F^{(\lambda_0)}$ (the subsection 8.8.1) and the decompositions ${}^{i}\mathbb{E}^{(\lambda_0)}$ (the subsection 8.8.5) of the vector bundle ${}_{\boldsymbol{b}}\mathcal{E}_{|\mathcal{D}_i(\lambda_0,\epsilon_0)}$. The tuples of the filtrations and the decompositions are compatible.

In particular, we obtain the induced vector bundle ${}^{\underline{l}}\mathcal{G}_{\boldsymbol{u}}^{(\lambda_0)}$ on $\mathcal{D}_{\underline{l}}(\lambda_0,\epsilon_0)$. In the section 8.9, we see that $\{{}^{\underline{l}}\mathcal{G}_{\boldsymbol{u}}^{(\lambda_0)}\,|\,\lambda_0\in \boldsymbol{C}_{\lambda}\}$ gives the holomorphic bundle over $\mathcal{D}_{\underline{l}}$, and we give some detail on the vector bundle.

In the section 8.10, we explain a method to obtain the norm estimate. It will be used in the chapters 12–13.

1.6.8. Chapter 9. We give some detail of the KMS-structure on the space of the multi-valued flat sections of $\mathcal{E}^{\lambda}$ on $\Delta^{*\,l}\times\Delta^{n-l}$.

In the section 9.1, we see some easy properties of the filtrations ${}^{i}\mathcal{F}$. In the section 9.2, we show the compatibility of the tuple of the filtrations $\big({}^{i}\mathcal{F}\,\big|\,i\in\underline{l}\big)$. The argument to deal with the filtrations is complicated a little, as usual. However it is elementary.

Then we obtain the induced object ${}^{\underline{l}}\mathcal{G}_{\boldsymbol{u}}^{(\lambda_0)}(\mathcal{H})$. In the section 9.3, we see that the family $\{{}^{\underline{l}}\mathcal{G}_{\boldsymbol{u}}^{(\lambda_0)}(\mathcal{H})\,|\,\lambda_0\in\boldsymbol{C}_{\lambda}\}$ gives a vector bundle over $\boldsymbol{C}_{\lambda}^{*}$. We also see some additional structures, the nilpotent maps and the pairing.

1.6.9. Chapter 10. In the section 10.1, we see the compatibility of the naturally defined tuple of the filtrations and the decompositions on ${}_{\boldsymbol{c}}\mathcal{E}^{\lambda}$.

In the section 10.2, we obtain the filtrations and the decompositions of the prolongment ${}_{\boldsymbol{c}}\mathcal{E}$. As a result, we obtain the induced regular λ-connection ${}^{\underline{l}}\mathcal{G}_{\boldsymbol{u}}(\mathcal{E})$ over $\mathcal{X}^{\sharp}$ in the section 10.3. By using ${}^{\underline{l}}\mathcal{G}_{\boldsymbol{u}}(\mathcal{E})$, we obtain the isomorphisms $\Phi_{\boldsymbol{u}}^{\mathrm{can}}$: ${}^{\underline{l}}\mathcal{G}_{\boldsymbol{u}}(\mathcal{H})\simeq{}^{\underline{l}}\mathcal{G}_{\boldsymbol{u}\,|\,\boldsymbol{C}_{\lambda}^{*}}$ and $\Phi_{\boldsymbol{u},P,O}:{}^{\underline{l}}\mathcal{G}_{\boldsymbol{u}}(\mathcal{E})_{|\boldsymbol{C}_{\lambda}^{*}\times\{P\}}\simeq{}^{\underline{l}}\mathcal{G}_{\boldsymbol{u}\,|\,\boldsymbol{C}_{\lambda}^{*}}$, which is described in the section 10.4.

1.6.10. Part 3, Chapter 11. In Part 3, we prove a limiting mixed twistor theorem. As an application, we obtain the norm estimate for holomorphic sections and multi-valued flat sections.

In the chapter 11, we give some detail on the construction of the vector bundle over $\mathbb{P}^1$ with the nilpotent maps and the pairing, from a tame harmonic bundle.

In the section 11.1, we recall the variation of polarized pure twistor structure induced by the harmonic bundles. In particular, we recall the conjugate of harmonic bundles. The formalism was given by Simpson in [**83**].

In the section 11.2, we see the KMS-structure and the induced objects of the conjugate. Briefly speaking, the conjugate is isomorphic to the dual. However we remark that the signature of the nilpotent maps are reversed.

In the section 11.3, we see the construction of the vector bundles $S_{\boldsymbol{u}}^{\mathrm{can}}(E)$ and $S_{\boldsymbol{u}}(E,P)$. We also see the induced nilpotent maps and the pairings on them. In particular, we obtain the weight filtration W on $S_{\boldsymbol{u}}^{\mathrm{can}}(E)$ and $S_{\boldsymbol{u}}(E,P)$ in the case $\dim(X)=1$.

In the section 11.4, we give some detail on the associated graded vector bundles $\mathrm{Gr}^W S_{\boldsymbol{u}}^{\mathrm{can}}(E)$ and $\mathrm{Gr}^W S_{\boldsymbol{u}}(E,P)$ in the case $\dim(X)=1$, which are simple.

1.6.11. Chapter 12. We prove the limiting mixed twistor theorem, which will be quite important in the study in Part 4. It is also useful to control the conjugacy classes of the nilpotent parts of the residues.

In the section 12.1, we prove the limiting mixed twistor theorem in the case $\dim(X)=1$. Although the proof is essentially same as those in [**65**], it is rather complicated to state the argument precisely.

In the section 12.2, we prove the limiting mixed twistor theorem for higher dimensional case. The different part from that in [**65**] is the proof of the constantness of the filtrations (Lemma 12.21). In our previous paper [**65**], we used the argument to take a 'limit'. Instead we use the result in the section 2.4.

In the section 12.3, we derive some consequences by using the results for polarized mixed twistor structures (the section 3.8) and the results in the section 5. We show the strongly sequential compatibility and some decomposition.

1.6.12. Chapter 13. As one of the application of the limiting mixed twistor theorem, we give a norm estimate of the holomorphic sections or flat sections of the deformed holomorphic bundles. The arguments are essentially same as those in our previous paper [**65**], i.e., we use the method of comparison explained in the section 8.10. We have only to care the parabolic structure.

In the section 13.1, we give some remark on functoriality for pull backs of deformed holomorphic bundles. In the section 13.2, we show a preliminary norm estimate for holomorphic sections. In the section 13.3, we derive a norm estimate for holomorphic sections. In the section 13.4, we reduce the norm estimate for flat sections to the norm estimate for holomorphic sections. Contrast to the holomorphic case, we do not discuss the norm estimate of the family of flat sections. It seems possible to discuss, but some additional argument is necessary, and hence we postpone it until we will need it. In the section 13.5, we give the norm estimate for primitve sections as an explanation of the results.

1.6.13. Part 4, Chapter 14. In Part 4, we apply the results in Part 2–3 to the theory of pure twistor D-modules.

For the reference of our later argument, we recall some of basic definitions in the theory of pure twistor D-modules due to Sabbah with slightly different notation in Chapter 11. We also give partial generalization in some part, although they are minor. See [**72**] for more detail and precise.

In the section 14.1, we recall the basic definitions of $\mathcal{R}$-modules. Briefly speaking, it is a generalization of filtered D-modules, and we have similar formalism for D-modules. It would be useful to see a standard text book of D-modules. In the section 14.2, we recall the nearby cycle functor and the vanishing cycle functor for $\mathcal{R}$-module with minor generalization. Since we discuss such functors for the $\mathcal{R}$-modules obtained from tame harmonic bundles in the later chapters, we give the detail.

We recall the definitions of sesqui-linear pairings and $\mathcal{R}$-triples in the section 14.3, and we recall the nearby cycle functors and the vanishing cycle functors of them in the section 14.4.

Once the nearby cycle functors and the vanishing cycle functors are prepared, the definition of polarized pure twistor D-module can be given as in the case of polarized pure Hodge modules. We recall it in the section 14.5. Our pure twistor D-module is slightly wider than Sabbah's one. His pure twistor D-module is our pure imaginary pure twistor D-module, given in the subsection 14.5.2. In the subsection 14.5.3, we recall the semisimplicity of the D-module which underlies a polarized pure imaginary pure twistor D-module, due to Sabbah. In the section 14.6, we recall the decomposition theorem of polarized pure twistor D-module due to Sabbah. Although our pure twistor D-module is slightly wider than Sabbah's one as is remarked above, the modification is minor. Hence we do not give full detail. However we will use some of the argument for the proof in the later discussion, which is explained in the subsections 14.6.2–14.6.3. We also give the detail on pure twistor D-modules on a smooth curve in Chapter 20, which is necessary for the proof of the decomposition theorem.

1.6.14. Chapter 15. We put $X = \Delta^n$, $D_i = \{z_i = 0\}$ and $D = \bigcup_{i=1}^n D_i$. Let $(E, \overline{\partial}_E, \theta, h)$ be a tame harmonic bundle on $X - D$, which naturally induce the $\mathcal{R}$-triple. We would like to prolong $\mathcal{T}(E)$ to the $\mathcal{R}$-triple on X, which is the purpose in the chapters 15–18 by using the results in Part 2–3.

In the chapter 15, we would like to prolong the deformed holomorphic bundle $\mathcal{E}$ over $\mathcal{X} - \mathcal{D}$ to the $\mathcal{R}$-modules on $\mathcal{X}$. First we give a naive prolongment $^{\square}\mathcal{E}$ in the section 15.1. It can essentially be regarded as the sheaf of meromorphic sections. Although it is not coherent, $^{\square}\mathcal{E}$ has many nice properties, and it is easy to understand $^{\square}\mathcal{E}$ algebraically. We will use $^{\square}\mathcal{E}$ as the ambient sheaf. We introduce the sheaf $^I\tilde{T}^{(\lambda_0)}(\boldsymbol{c}, \boldsymbol{d})$, and we prepare a lemma in the subsection 15.1.5, which will be used in the section 15.3.

In the section 15.2, we give the prolongment $\mathfrak{E}$, which we really need. In the subsection 15.2.1, the difference between $\mathfrak{E}$ and $^{\square}\mathcal{E}$ is explained in the simplest case. To understand $\mathfrak{E}$ more closely, we introduce the filtrations $^IV^{(\lambda_0)}$ $(I \subset \underline{l})$, and we obtain the sheaves $^IT^{(\lambda_0)}(\boldsymbol{c}, \boldsymbol{d})$. In the section 15.3, we show that $^IT^{(\lambda_0)}(\boldsymbol{c}, \boldsymbol{d})$ and $^I\tilde{T}^{(\lambda_0)}(\boldsymbol{c}, \boldsymbol{d})$ are naturally isomorphic (Lemma 15.46). Although it looks a little long and complicated, the arguments are elementary.

In the section 15.4, we show a kind of compatibility of the filtrations $^iV^{(\lambda_0)}(\mathfrak{E})$ $(i \in \underline{l})$. First goal is to show $^{\underline{l}}\mathcal{V}_S(\mathfrak{E}) = {}^{\underline{l}}\mathcal{V}_S(^{\square}\mathcal{E}) \cap \mathfrak{E}$ (Proposition 15.57). Once we prove Proposition 15.57, we can easily translate some nice properties of $^iV^{(\lambda_0)}(^{\square}\mathcal{E})$ to $^iV^{(\lambda_0)}(\mathfrak{E})$. In particular, we obtain the strictly S-decomposability of $\mathfrak{E}$ along $z_i = 0$ for $i \in \underline{l}$. We also obtain the primitive decomposition of sections of $\mathfrak{E}$.

In the section 15.5, we give a characterization of $\mathfrak{E}$ as the prolongment of the deformed holomorphic bundle $\mathcal{E}$ obtained from harmonic bundle. It will be useful when we consider the specialization of $\mathfrak{E}$ along $\prod z_i^{m_i}$.

1.6.15. Chapter 16. We give some detail on the push-forward $\mathfrak{E}[\eth_t]$ for the graph of the holomorphic functions $\prod_{i=1} z_i^{m_i}$. Following Saito [**75**], we introduce the filtration $U^{(\lambda_0)}$ in the section 16.1. Our purpose is to show that $U^{(\lambda_0)}$ gives the V-filtration along $t = 0$ at λ_0 whose associated graded module is strict, namely, $\mathfrak{E}[\eth_t]$ is strictly specializable along t_0. We will also show that the naturally induced filtrations ${}^{i}V^{(\lambda_0)}$ on $\widetilde{\psi}_{t,u}(\mathfrak{E}[\eth_t])$ gives the V-filtration along $z_i = 0$ at λ_0, whose associated graded module is strict, namely $\widetilde{\psi}_{t,u}(\mathfrak{E}[\eth_t])$ is strictly specializable along $z_i = 0$ $(i \in \underline{l})$.

In the section 16.2, we give some algorithms to describe sections of $\mathfrak{E}[\eth_t]$ in a normal (but not unique) way. In the section 16.3, we obtain the primitive decompositions of sections of $\mathrm{Gr}_b^{U^{(\lambda_0)}} \mathfrak{E}[\eth_t]$. The first goal is Proposition 16.25. Once Proposition 16.25 is established, the rest are rather formal.

We obtain the strict S-decomposability of $\mathfrak{E}[\eth_t]$ in the section 16.4. Even if it looks complicated, it is a rather formal consequence of the results in the section 16.3.

In the section 16.5, we would like to see the form of $\widetilde{\psi}_{t,u}(\mathfrak{E}[\eth_t])$ briefly. Since it is not easy to see it directly, we see ${}^{I}\,\mathrm{Gr}_{\boldsymbol{c}}^{V^{(\lambda_0)}}\,{}^{J}V_{\boldsymbol{d}}^{(\lambda_0)}(\widetilde{\psi}_{t,u}\mathfrak{E}[\eth_t])$, instead. Our goal is to relate them with the construction in the subsection 3.9.2. Compare the formulas (16.28) and (3.33).

1.6.16. Chapter 17. We consider the nilpotent map $N = t\eth_t + \mathfrak{e}(\lambda, u)$ on $\widetilde{\psi}_{t,u}\mathfrak{E}[\eth_t]$. Then we obtain the weight filtration $W(N)$. We would like to see $\mathrm{Gr}^{W(N)}\,\widetilde{\psi}_{t,u}\mathfrak{E}[\eth_t]$. It is not easy to see it directly, we introduce the filtration $\mathbb{F}^{(\lambda_0)}$ on $\widetilde{\psi}_{t,u}\mathfrak{E}[\eth_t]$, such that $\mathrm{Gr}_m^{\mathbb{F}^{(\lambda_0)}}$ is a direct sum of the locally free sheaves ${}^{I}\mathcal{L}$ on $\mathcal{D}_I(\lambda_0, \epsilon_0)$ $(|I| = n - m)$.

In the section 17.1, we see the relation of the weight filtration $W(N)$ and the induced filtrations ${}^{\underline{l}}V^{(\lambda_0)}$ on ${}^{I}\mathcal{L}$. We also see the decomposition of $P_h\,\mathrm{Gr}_h^{W(N)}$. The results easily follow from the limiting mixed twistor theorem and the results in the chapter 5.

In the section 17.2, we introduce the filtration $\mathbb{F}^{(\lambda_0)}$. We see that the exact sequences associated to $\mathbb{F}^{(\lambda_0)}$ is strict with respect to the weight filtration $W(N)$. A key observation is given in the proof of Lemma 17.26. As a result, we can understand the filtrations ${}^{I}V^{(\lambda_0)}$ on $\mathrm{Gr}^{W(N)}(\widetilde{\psi}_{t,u}\mathfrak{E}[\eth_t])$ sufficiently well, and we obtain the strictly specializability of $\mathrm{Gr}^{W(N)}(\widetilde{\psi}_{t,u}\mathfrak{E}[\eth_t])$ along $z_i = 0$ $(i \in \underline{n})$ in the section 17.3.

In the section 17.4, we obtain the strict S-decomposability of $\mathrm{Gr}^{W(N)}(\widetilde{\psi}_{t,u}\mathfrak{E}[\eth_t])$ along $z_i = 0$ $(i \in \underline{n})$, by using the results in the previous subsections and the lemmas of Kashiwara and Saito prepared in the section 3.9. As a result, we obtain the decomposition by the strict supports, as in Proposition 17.56. We see some properties of the components.

1.6.17. Chapter 18. In the section 18.1, we give the sesqui-linear pairing $\mathfrak{C}$ of $\mathfrak{E}$. As a result, we obtain the $\mathcal{R}$-triple $\mathfrak{T}(E) = (\mathfrak{E}, \mathfrak{E}, \mathfrak{C})$ from a tame harmonic bundle $(E, \overline{\partial}_E, \theta, h)$. We also see some uniqueness.

In the section 18.2, we introduce ${}^{\underline{n}}\overline{\mathcal{G}}_{\boldsymbol{u}}$ for $\boldsymbol{u} \in \overline{\mathcal{KMS}(\mathcal{E}^0, \underline{n})}$, which is a family of flat bundles on $(X - D) \times \boldsymbol{C}^*$. We see that the sesqui-linear pairing C_0 of $\mathcal{E}$ induces the pairing on ${}^{\underline{n}}\overline{\mathcal{G}}_{\boldsymbol{u}}$. By using it, we obtain the sesqui-linear pairing on ${}^{\underline{n}}\mathcal{G}_{\boldsymbol{u}'}$ for $\boldsymbol{u}' \in \mathcal{KMS}(\mathcal{E}^0, \underline{n})$ in the section 18.3. We see that it is same as the pairing for $S_{\boldsymbol{u}}^{\mathrm{can}}(E)$ considered in Part 3.

In the section 18.4, we show that the component whose support is $\{0\}$ is a polarized pure twistor structure (Proposition 18.26), by using the lemma of Kashiwara in the subsection 3.9.2. Once we know Proposition 18.26, we immediately know that the smooth part of the components of $P_h\,\mathrm{Gr}^{W(N)}\,\widetilde{\psi}_{t,u}\mathfrak{T}(E)$ together with the induced sesqui-linear pairing is a variation of pure twistor structures. By using the characterization of the prolongment, each components of $P_h \operatorname{Gr}^{W(N)} \widetilde{\psi}_{t,u}\mathfrak{T}(E)$ in the decomposition by the strict supports is isomorphic to an $\mathcal{R}$-triple obtained from a tame harmonic bundle. As a result, we arrive at the stage we can use an induction to show the existence of the prolongment as the pure twistor D-module.

1.6.18. Chapter 19. We establish the correspondence of tame harmonic bundles and pure twistor D-modules. The statement is given in the section 19.1. The argument is rather formal.

In the section 19.2, we show that every regular pure twistor D-module gives a tame harmonic bundle which is defined on a Zariski open subset of the strict support. The existence of the prolongment as pure twistor D-module is shown in the section 19.3. This is a formal consequence of the result in the section 18.4. The uniqueness is shown in the section 19.4. We need some consideration by using the uniqueness of the intermediate extension and the Riemann-Hilbert correspondence.

In the section 19.5, we see that the correspondence of Theorem 19.6 preserves the pure imaginary property. By using the results, we show that Sabbah's conjecture is true in the section 19.6, i.e., a semisimple regular holonomic D-modules underlie a pure imaginary pure twistor D-modules. As a result, we obtain the regular holonomic version of Kashiwara's conjecture by combining the results of Sabbah and us.

1.6.19. Chapter 20. This chapter is an appendix to Part 4. We see some fundamental properties of pure twistor D-modules on a smooth curves. In the section 20.1, we show the correspondence of tame harmonic bundles and regular pure twistor D-modules on a curve. It is a special case of the results in the chapter 19. However we use decomposition theorem in the chapter 19, and the correspondence on a curve is used in the proof of the decomposition theorem. Hence we include this section to remark that the argument is not circulated.

In the section 20.2, we give some arguments for the proof of the decomposition theorem for a pure twistor D-module on a projective curve. The essential part is due to Zucker [**90**]. But we need some additional argument as explained in the subsections 20.2.3–20.2.4.

1.6.20. Part 5, Chapter 21. Part 5 can be read independently, for it was written as a separate paper. It is the purpose to establish the step 1 of Sabbah's program, i.e., to establish the correspondence of tame pure imaginary harmonic bundles and semisimple local system.

In the section 21.1, we recall some standard facts just for our reference of the later discussion. In the section 21.2, we recall the elementary geometry of the symmetric space $\mathcal{PH}(r)$ of the positive definite hermitian metrics. Lemma 21.19 is one of the key lemmas, although it is elementary. We also give the comparison of the distance of the hermitian metrics in $\mathcal{PH}(r)$ and the norm of the identity with respect to the two metrics in the subsection 21.2.5.

In the section 21.3, we discuss a twisted map associated with a commuting tuple of endomorphisms. The result will be used in the section 24.2 to construct the twisted map of $(\overline{\Delta}^*)^2$ whose energy is controlled.

In the section 21.4, we recall some standard facts on twisted harmonic maps. The Bochner type formula in the subsection 21.4.5 is due to Corlette. A variant of Bochner type formula is given in the subsection 21.4.6. They will be used in the proof of the pluri-harmonicity (see the sections 25.2–25.3). It is important for our argument to consider the two kinds of Bochner type formulas.

1.6.21. Chapter 22. We give the definition of tame pure imaginary harmonic bundles, and some of useful properties. In the section 22.1, we give a definition of pure imaginary property of tame harmonic bundles. In the section 22.2, we see the estimate of the energy functions of a tame pure imaginary harmonic bundles on a punctured disc. We give a characterization of tame and pure imaginary properties by an increasing order of the energy in the subsection 22.2.3.

In the section 22.3, we show that the underlying flat bundle of a tame pure imaginary harmonic bundle is semisimple. In the section 22.4, we see the maximum principle for the distance of two tame pure imaginary harmonic metrics on a punctured disc. The result will be useful to control the energy (the chapter 24). It will be also used to show the tameness (the section 25.4).

In the section 22.5, we show the uniqueness of tame pure imaginary harmonic bundle of a flat bundle on a quasi projective variety. Note that it essentially follows from the Kobayashi-Hitchin correspondence of Simpson and Biquard ([**81**] and [**6**]). However the detailed proof for uniqueness in the case of parabolic flat bundle seems to be omitted there. Thus we give the detailed proof within our necessity. We essentially follow the argument of Corlette.

1.6.22. Chapter 23. In the section 23.1, we discuss the Dirichlet problem of a tame pure imaginary harmonic bundle on a punctured disc. The argument seems essentially due to Lohkamp [**57**] and Jost-Zuo [**44**].

In the section 23.2, we discuss the family version of the Dirichlet problem. We give the estimate of the differentials, which is given by the maximum principle (the section 22.4) and the estimate in the section 22.2. We learned the idea to use the maximum principle in [**43**]. The result will be used for the construction of the twisted map whose energy is controlled (the chapter 24).

1.6.23. Chapter 24. We construct the twisted map on the complement of a normal crossing divisor in a compact Kahler manifold. We essentially follow the method of Jost-Zuo ([**43**] and [**44**]).

In the section 24.1, we construct the twisted map around smooth points of a divisor, by solving the family of the Dirichlet problem. Then the energy of the map is controlled by the result of the section 23.2. We also give the lower bound of the energy for arbitrary twisted map around smooth points of a divisor in the subsection 24.1.5. Essentially it is a consequence of Lemma 21.19. However we need some care to control the divergent term precisely.

In the section 24.2, we construct the twisted map around the intersection of the divisors. The results in the section 21.3 and the subsection 23.2.2 are used. We also give the lower bound of the energy of arbitrary twisted maps. Again, we have to be careful to control the divergent term.

In the section 24.3, we give the decomposition of the complement of a normal crossing divisor in a compact Kahler manifold, and we obtain the twisted map whose energy is controlled. We also give the lower bound of the energy of arbitrary twisted map. They are direct consequences of the results in the sections 24.1–24.2.

1.6.24. Chapter 25. In the section 25.1, we obtain the harmonic metric of a semisimple flat bundle on a quasi projective variety. The argument is essentially same as that in the Dirichlet problem on a punctured disc (the section 23.1), except the use of the argument in [**42**] for the convergence. In the subsections 25.1.2–25.1.4, we give the detailed estimate of the energy of the resulted harmonic metric. They are necessary for the later discussion.

In the section 25.2, we show that $\overline{\partial}\theta$ and θ^2 are L^2, where θ is the associated $(1,0)$-form for the resulted harmonic metric. We use the Bochner type formula in the subsection 21.4.5. If the integral of the Bochner type formula would vanish, then we would obtain the pluri-harmonicity. However it does not seem easy to show such vanishing directly. (See the convergence (25.27), for example.)

Hence, in the section 25.3, we use another kind of Bochner type formula given in the subsection 21.4.6. It is a rather easy to show the vanishing of the integral in this time, by using the L^2-property of $\overline{\partial}\theta$ obtained in the section 25.2. As a result, we obtain the pluri-harmonicity of the harmonic metric obtained in the section 25.1.

In the section 25.4, we show the tameness and the pure imaginary property for the resulted pluri-harmonic metric. It is rather easy consequence of the estimate of the energy given in the subsection 25.1.4, the characterization of tame pure imaginary harmonic bundle on a punctured disc (the subsection 22.2.3), the maximum principle (the section 22.5) and Hartogs type theorem.

Thus we obtain the existence theorem of a tame and pure imaginary pluri-harmonic harmonic metric for any semisimple flat bundle on a quasi projective surface, or more generally, the complement of a normal crossing divisor in a compact Kahler surface.

In the section 25.5, we show the existence of tame and pure imaginary pluri-harmonic metric in the higher dimensional case, by reducing the problem to the case of quasi projective surface. Here we need the quasi projectivity. For a direct generalization of the arguments in the chapters 24–25 to those in the higher dimensional case, it seems that we need some more additional technically complicated arguments. Hence we are satisfied with the quasi projective case.

As a simple application, we show Theorem 25.30 in the section 25.6. We have only to show that the pull back of a tame pure imaginary harmonic bundle is also a tame pure imaginary harmonic bundle.

Part 1

Preliminary

CHAPTER 2

Preliminary

2.1. Notation

Some of the notation and the terminology introduced in our previous paper [**65**] will be used. We recommend the reader to see the sections 2.1.1–2.1.3, the sections 2.3.1–2.3.2, and the section 3.1.1–3.1.7, of [**65**], at least.

2.1.1. Sets. We will use the following notation:

$\mathbb{Z}$:	the set of the integers,	$\boldsymbol{Q}$:	the set of the rational numbers,
$\boldsymbol{R}$:	the set of the real numbers,	$\boldsymbol{C}$:	the set of the complex numbers,
$\underline{n}$:	the set $\{1,2,\ldots,n\}$,	$M(r)$:	the set of $r\times r$-matrices,
$\mathfrak{S}_l$:	the l-th symmetric group,	$\mathcal{H}_r$:	the set of r-hermitian matrices,

We denote the set of positive hermitian metric of a complex vector space V by $\mathcal{PH}(V)$. We often identify it with the set of the positive hermitian matrices by taking an appropriate base of V.

We put $[a,b]:=\{x\in\boldsymbol{R}\,|\,a\leq x\leq b\}$, $[a,b[:=\{x\in\boldsymbol{R}\,|\,a\leq x<b\}$, $]a,b]:=\{x\in\boldsymbol{R}\,|\,a<x\leq b\}$ and $]a,b[:=\{x\in\boldsymbol{R}\,|\,a<x<b\}$ for any $a,b\in\boldsymbol{R}$.

2.1.2. A disc, a punctured disc and some products. For any positive number $C>0$ and $z_0\in\boldsymbol{C}$, the open disc $\big\{z\in\boldsymbol{C}\,\big|\,|z-z_0|<C\big\}$ is denoted by $\Delta(z_0,C)$, and the punctured disc $\Delta(z_0,C)-\{z_0\}$ is denoted by $\Delta^*(z_0,C)$. When $z_0=0$, $\Delta(0,C)$ and $\Delta^*(0,C)$ are often denoted by $\Delta(C)$ and $\Delta^*(C)$. Moreover, if $C=1$, $\Delta(1)$ and $\Delta^*(1)$ are often denoted by Δ and Δ^*. If we emphasize the variable, we describe as Δ_z, Δ_i. For example, $\Delta_z\times\Delta_w=\{(z,w)\in\Delta\times\Delta\}$, and $\Delta_1\times\Delta_2=\{(z_1,z_2)\in\Delta\times\Delta\}$. We often use the notation $\boldsymbol{C}_\lambda$ and $\boldsymbol{C}_\mu$ to denote the complex planes $\big\{\lambda\in\boldsymbol{C}\big\}$ and $\big\{\mu\in\boldsymbol{C}\big\}$.

Unfortunately, the notation Δ is also used to denote the Laplacian. The author hopes that there will be no confusion.

We put $\overline{\Delta}(C):=\big\{z\,\big|\,|z|\leq C\big\}$ and $\overline{\Delta}^*(C):=\big\{z\,\big|\,0<|z|\leq C\big\}$. In the case $C=1$, we use $\overline{\Delta}$ and $\overline{\Delta}^*$ instead of $\overline{\Delta}(1)$ and $\overline{\Delta}^*(1)$.

For a complex manifold X, a point $\lambda_0\in\boldsymbol{C}_\lambda$ and a positive number ϵ_0, we often consider the product $X\times\Delta(\lambda_0,\epsilon_0)$ and $X\times\Delta^*(\lambda_0,\epsilon_0)$. For simplicity, we often denote them by $\mathcal{X}(\lambda_0,\epsilon_0)$ and $\mathcal{X}^*(\lambda_0,\epsilon_0)$ respectively. We also use the notation $\mathcal{X}$ and $\mathcal{X}^\sharp$ to denote the $X\times\boldsymbol{C}_\lambda$ and $X\times\boldsymbol{C}^*_\lambda$.

For a complex manifold X, we have the conjugate complex manifold, which is denoted by $X^\dagger$.

2.1.3. Projections. Let I be a finite set and J be a subset of I. In general, $q_J:X^I\longrightarrow X^J$ denotes the naturally induced projection taking the i-th components for $i\in J$. Similarly, $\pi_J:X^I\longrightarrow X^{I-J}$ denotes the projection omitting the j-th component for $j\in J$. However, we will often use π to denote some other projections. If J consists of the unique element j, we often use the notation q_j and π_j instead of $q_{\{j\}}$ and $\pi_{\{j\}}$.

2.1.4. The order on $\boldsymbol{R}^n$. We have the natural order on $\boldsymbol{R}$. Let n be a positive integer. We often use the order on $\boldsymbol{R}^n$ given as follows: For elements $\boldsymbol{a}, \boldsymbol{b} \in \boldsymbol{R}^n$, we say $\boldsymbol{a} \leq \boldsymbol{b}$ if and only if $q_i(\boldsymbol{a}) \leq q_i(\boldsymbol{b})$ for any $i \in \underline{n}$. When we consider such order, we say $\boldsymbol{a} < \boldsymbol{b}$ if and only if $q_i(\boldsymbol{a}) < q_i(\boldsymbol{b})$ for any $i \in \underline{n}$, and we say $\boldsymbol{a} \lneq \boldsymbol{b}$ if and only if $\boldsymbol{a} \leq \boldsymbol{b}$ and $\boldsymbol{a} \neq \boldsymbol{b}$.

2.1.5. κ_c and ν_c. Let c be a real number. The maps $\kappa_c : \boldsymbol{R} \longrightarrow]c-1, c]$ and $\nu_c : \boldsymbol{R} \longrightarrow \mathbb{Z}$ are defined by the following condition:

For a real number $x \in \boldsymbol{R}$, the equality $\kappa_c(x) + \nu_c(x) = x$ holds.

In the case $c = 0$, we use the notation κ and ν instead of κ_c and ν_c.

Let S be a finite subset of $]-1, 0]$ and b be a positive number. We obtain the map $\phi_b : S \longrightarrow \mathbb{Z}$ given by $\phi_b(x) = \nu(b \cdot x)$.

DEFINITION 2.1. A real number b is sufficiently large with respect to S, if the map ϕ_b is injective. □

2.1.6. ϵ-small set. Let ϵ be a real number such that $0 < \epsilon < 1$. A subset A of $]-1, 0[$ is ϵ-small, if A is contained in $]-\epsilon, 0[$.

2.1.7. $\mathfrak{p}$, $\mathfrak{e}$ and $\mathfrak{k}$. For any element $u = (a, \alpha) \in \boldsymbol{R} \times \boldsymbol{C}$, we put as follows:

$$\mathfrak{p}(\lambda, u) := a + 2 \cdot \mathrm{Re}(\lambda \cdot \bar{\alpha}),$$

$$\mathfrak{e}(\lambda, u) := \alpha - a \cdot \lambda - \bar{\alpha} \cdot \lambda^2.$$

Then we obtain the following morphism:

$$\mathfrak{k}(\lambda) = \big(\mathfrak{p}(\lambda), \mathfrak{e}(\lambda)\big) : \boldsymbol{R} \times \boldsymbol{C} \longrightarrow \boldsymbol{R} \times \boldsymbol{C}.$$

The following lemma is checked by a direct calculation.

LEMMA 2.2. *$\mathfrak{k}(\lambda)$ is bijective.*

Proof Let us consider the equation $\mathfrak{p}\big(\lambda, (a, \alpha)\big) = A$ and $\mathfrak{e}\big(\lambda, u\big) = B$ for $(B, A) \in \boldsymbol{C} \times \boldsymbol{R}$. Then we have the unique solution:

$$\alpha = \frac{\lambda \cdot A + B}{|\lambda|^2 + 1}, \qquad a = \frac{(-|\lambda|^2 + 1)A - 2\,\mathrm{Re}(\lambda \overline{B})}{|\lambda|^2 + 1}.$$

It may be useful to use the relation $\lambda \cdot \mathfrak{p}(\lambda, (a, \alpha)) + \mathfrak{e}(\lambda, (a, \alpha)) = (|\lambda|^2 + 1) \cdot \alpha$. □

REMARK 2.3. We will use Lemma 2.2 to control the KMS-structure of the prolongment of deformed holomorphic bundles. □

For any element $u = (\alpha, a) \in \boldsymbol{C} \times \boldsymbol{R}$, we put $u^\dagger = (\overline{\alpha}, -a) \in \boldsymbol{C} \times \boldsymbol{R}$.

LEMMA 2.4. *Let u be an element of $\boldsymbol{C} \times \boldsymbol{R}$. We have the following formula:*

$$-\mathfrak{p}(\lambda, u) = \mathfrak{p}\big(-\overline{\lambda}, u^\dagger\big), \qquad \overline{\mathfrak{e}(\lambda, u)} = \mathfrak{e}\big(-\overline{\lambda}, u^\dagger\big). \tag{2.1}$$

Proof It can be checked by a direct calculation. □

For any element $u = (a, \alpha) \in \boldsymbol{R} \times \boldsymbol{C}$, we put as follows:
(2.2)
$$\mathfrak{p}^f(\lambda, u) := \mathrm{Re}\big(\lambda \cdot \overline{\alpha} + \lambda^{-1} \cdot \alpha\big) = \mathfrak{p}(\lambda, u) + \mathrm{Re}\big(\lambda^{-1} \cdot \mathfrak{e}(\lambda, u)\big) \in \boldsymbol{R},$$

$$\mathfrak{e}^f(\lambda, u) := \exp\big(-2\pi\sqrt{-1}(\lambda^{-1} \cdot \alpha - a - \lambda \cdot \overline{\alpha})\big) = \exp\big(-2\pi\sqrt{-1}\lambda^{-1} \cdot \mathfrak{e}(\lambda, u)\big) \in \boldsymbol{C}^*.$$

The maps $\mathfrak{p}^f(\lambda)$ and $\mathfrak{e}^f(\lambda)$ induce the map $\mathfrak{k}^f(\lambda) : \boldsymbol{R} \times \boldsymbol{C} \longrightarrow \boldsymbol{R} \times \boldsymbol{C}^*$.

LEMMA 2.5. *Let u be an element of $\boldsymbol{R} \times \boldsymbol{C}$. We have the following formula:*
$$\mathfrak{e}^f(\lambda, u) = \mathfrak{e}^f\big(\lambda^{-1}, u^\dagger\big)^{-1}, \quad \mathfrak{p}^f(\lambda, u) = \mathfrak{p}^f\big(\lambda^{-1}, u^\dagger\big).$$

Proof We have the following equalities:
$$\mathfrak{e}^f(\lambda, u) = \exp\big(-2\pi\sqrt{-1}(\lambda^{-1}\alpha - a - \lambda\bar{\alpha})\big) = \mathfrak{e}^f\big(\lambda^{-1}, u^\dagger\big)^{-1}.$$
We also have the following equalities:
$$\mathfrak{p}^f(\lambda, u) = \mathrm{Re}\big(\lambda\bar{\alpha} + \lambda^{-1}\alpha\big) = \mathfrak{p}^f\big(\lambda^{-1}, u^\dagger\big).$$
Thus we are done. □

For any element $\boldsymbol{u} = (\boldsymbol{\alpha}, \boldsymbol{b}) \in \boldsymbol{C}^l \times \boldsymbol{R}^l$, we put $\boldsymbol{u}^\dagger := (\overline{\boldsymbol{\alpha}}, -\boldsymbol{b})$.

2.1.8. Some elements of $(\boldsymbol{R} \times \boldsymbol{C})^I$ and $\boldsymbol{R}^I$. Let I be a finite set. The element $(0, \ldots, 0) \in \boldsymbol{R}^I$ is often denoted simply by 0. For an element j of I, $\boldsymbol{\delta}_j$ often denotes the element of $\boldsymbol{R}^I$ such that $q_j(\boldsymbol{\delta}_j) = 1$ and $q_i(\boldsymbol{\delta}_j) = 0$ for $i \neq j$.

We will not distinguish $(\boldsymbol{R} \times \boldsymbol{C})^I$ and $\boldsymbol{R}^I \times \boldsymbol{C}^I$. We often use the notation $\boldsymbol{\delta}_0$ to denote the element $(1, 0) \in \boldsymbol{R} \times \boldsymbol{C}$. The element $(0, 0) \in \boldsymbol{R} \times \boldsymbol{C}$ is simply denoted by 0. Let j be any element of I. Then $\boldsymbol{\delta}_{0,j}$ denotes the element of $(\boldsymbol{R} \times \boldsymbol{C})^I$ such that $q_j(\boldsymbol{\delta}_{0,j}) = \delta_0$ and $q_i(\boldsymbol{\delta}_{0,j}) = 0$ for $(i \neq j)$.

2.2. Prolongation by an increasing order

2.2.1. Notation. Let X be a complex manifold and $D = \bigcup_{i=1}^N D_i$ be a normal crossing divisor. Let E be a holomorphic vector bundle with a hermitian metric h over $X - D$.

Let U be an open subset of X, which is admissible with respect to D, i.e., we have a coordinate $(z_1, \ldots, z_n)$ satisfying the following:
$$D \cap U = \bigcup_{k=1}^{l} (D_{i_k} \cap U), \quad D_{i_k} = \{z_k = 0\}.$$

For any section $f \in \Gamma\big(U \cap (X - D), E\big)$, let $|f|_h$ denote the norm function of f with respect to the metric h. We describe $|f|_h = O\big(\prod_{k=1}^l |z_k|^{-b_k}\big)$, if there exists a positive number C such that $|f|_h \leq C \cdot \prod_{k=1}^l |z_k|^{-b_k}$.

Recall that '$-\operatorname{ord}(f) \leq \boldsymbol{b}$' means the following:
$$|f|_h = O\Big(\prod_{k=1}^{l} |z_k|^{-b_k - \epsilon}\Big) \text{ for any positive number } \epsilon.$$

For any $\boldsymbol{b} \in \boldsymbol{R}^N$, the sheaf ${}_{\boldsymbol{b}}E$ is defined as follows:
$$\Gamma(U, {}_{\boldsymbol{b}}E) := \big\{f \in \Gamma\big(U \cap (X - D), E\big) \bigm| -\operatorname{ord}(f) \leq \boldsymbol{b}\big\}.$$

The sheaf ${}_{\boldsymbol{b}}E$ is called the prolongment of E by an increasing order $\boldsymbol{b}$. In particular, we use the notation ${}^{\diamond}E$ in the case $\boldsymbol{b} = (0, \ldots, 0)$.

2.2.2. Adaptedness and adaptedness up to log order. Let X be a C^∞-manifold, and E be a C^∞-vector bundle with a hermitian metric h. Let $\boldsymbol{v} = (v_1, \ldots, v_r)$ be a C^∞-frame of E. We obtain the $\mathcal{H}(r)$-valued function $H(h, \boldsymbol{v})$, whose (i,j)-component is given by $h(v_i, v_j)$. Recall that the frame $\boldsymbol{v}$ is called adapted, if $H(h, \boldsymbol{v})$ and $H(h, \boldsymbol{v})^{-1}$ are bounded.

Let $E = \bigoplus_i E_i$ be a C^∞-decomposition of E. The hermitian metric h of E induces the metric h_i on E_i. Then we obtain the metric $\bigoplus_i h_i$ of E.

DEFINITION 2.6. The decomposition $E = \bigoplus_i E_i$ is quasi adapted with respect to h, if h and $\bigoplus_i h_i$ are mutually bounded. □

Let us consider the case $X = \Delta^{*\,l} \times \Delta^{n-l}$. We have the coordinate $(z_1, \ldots, z_n)$. Let E, h and $\boldsymbol{v}$ be as above.

DEFINITION 2.7. A frame $\boldsymbol{v}$ is called adapted up to log order, if the following inequalities hold over X, for some positive numbers C_i $(i = 1, 2)$ and M:

$$0 < C_1 \cdot \Bigl(-\sum_{i=1}^{l} \log|z_i|\Bigr)^{-M} \leq H(h, \boldsymbol{v}) \leq C_2 \cdot \Bigl(-\sum_{i=1}^{l} \log|z_i|\Bigr)^{M}.$$

□

2.2.3. Lemmas for local freeness and the parabolic filtration. We put $X := \Delta^n$ and $D := \bigcup_{i=1}^{l} D_i$, where $D_i = \bigl\{(z_1, \ldots, z_n) \in X \bigm| z_i = 0\bigr\}$. Let E be a holomorphic bundle over $X - D$ and h be a hermitian metric of E.

LEMMA 2.8. *We assume that we have a holomorphic frame* $\boldsymbol{v} = \bigl(v_j \bigm| j = 1, \ldots, \operatorname{rank}(E)\bigr)$ *of E satisfying the following conditions:*

- *There exists* $b_i(v_j) \in]-1, 0]$ *for* $1 \leq j \leq \operatorname{rank}(E)$ *and for* $1 \leq i \leq l$.
- *The C^∞-frame* $\boldsymbol{v}' = \bigl(v'_j \bigm| j = 1, \ldots, \operatorname{rank}(E)\bigr)$, *given as follows, is adapted up to log order.*

$$v'_j := v_j \cdot \prod_{i=1}^{l} |z_i|^{b_i(v_j)}.$$

Then the following holds:

(1) *The $\mathcal{O}_X$-sheaf ${}^{\diamond}E$ is locally free.*
(2) *Each v_i is a section of ${}^{\diamond}E$, and $\boldsymbol{v}$ gives a frame of ${}^{\diamond}E$.*

Proof It is easy to see that v_j are sections of ${}^{\diamond}E$. Let f be a section of ${}^{\diamond}E$ over an open subset of $U \subset X$. We have the following development on U:

$$f = \sum f_j \cdot v_j = \sum f_j \cdot \prod_{i=1}^{l} |z_i|^{-b_i(v_j)} \cdot v'_j.$$

Here f_j are holomorphic functions on $U \cap (X - D)$. Since $\boldsymbol{v}'$ is adapted up to log order, we have the following estimate:

$$\Bigl| f_j \cdot \prod_{i=1}^{l} |z_i|^{-b_i(v_j)} \Bigr| = O\Bigl(\prod_{i=1}^{l} |z_i|^{-\epsilon}\Bigr), \quad \text{for any positive number } \epsilon.$$

In fact, the left hand side can be dominated by a polynomial of $-\log|z_i|$ ($i = 1,\dots,l$). Hence we obtain the following estimate, for any $\epsilon > 0$,

$$|f_j| = O\Big(\prod_{i=1}^{l} |z_i|^{b_i(v_j)-\epsilon}\Big).$$

Note that $b_i(v_j) - \epsilon > -1$ for any sufficiently small positive number ϵ. Hence the functions f_j are holomorphic on U. Thus $\boldsymbol{v}$ is a frame of ${}^{\diamond}E$, and ${}^{\diamond}E$ is locally free. □

Let $\boldsymbol{\delta}_i$ denote the element $(\overbrace{0,\dots,0}^{i-1},1,0,\dots,0)$. For any real number $b \leq 0$, we have the natural morphism ${}_{b\boldsymbol{\delta}_i}E \longrightarrow {}^{\diamond}E$. Thus we have the parabolic filtration ${}^{i}F$ on D_i given as follows:

$${}^{i}F_b := \mathrm{Im}\big({}_{b\boldsymbol{\delta}_i}E_{|D_i} \longrightarrow E_{|D_i}\big).$$

LEMMA 2.9. *We impose the same assumption in Lemma 2.8. Then the following holds.*

- *For each i, ${}^{i}F$ is a filtration in the category of vector bundles on D_i, namely the associated graded sheaf ${}^{i}\mathrm{Gr}^F$ is locally free on D_i.*
- *The tuple of the filtrations $\big({}^{i}F \,\big|\, i = 1,\dots,l\big)$ is compatible, in the sense of Definition 4.37.*
- *The frame $\boldsymbol{v}$ is compatible with $\big({}^{i}F \,\big|\, i = 1,\dots,l\big)$ in the sense of Definition 2.17 in* **[65]**. *We have ${}^{i}\deg^F(v_j) = b_i(v_j)$. Here ${}^{i}\deg^F(v_j)$ denotes $\deg^{{}^{i}F}(v_j)$.*

Proof Let f be a section of ${}_{b\cdot\boldsymbol{\delta}_i}E$, we have the description $f = \sum f_j \cdot v_j$ for some holomorphic functions f_j. By an argument similar to the proof of Lemma 2.8, we obtain the vanishing $f_{j\,|\,D_i} = 0$ in the case $b_i(v_j) > b$.

Thus ${}^{i}F_b$ is the vector subbundle of ${}^{\diamond}E_{|D_i}$ generated by v_i such that $b_i(v_j) \leq b$. It implies all the claims. □

2.3. Preliminary for μ_c-equivariant bundle

2.3.1. The action of the group μ_c. For any positive integer c, we put $\mu_c := \big\{z \in \boldsymbol{C} \,\big|\, z^c = 1\big\}$. We pick a generator $\omega(c)$ of μ_c. For any element $\boldsymbol{c} = (c_1,\dots,c_n) \in \mathbb{Z}_{>0}^n$, we put $\mu_{\boldsymbol{c}} := \prod_{i=1}^n \mu_{c_i}$. The element $(\overbrace{1,\dots,1}^{i-1},\omega(c_i),1,\dots,1)$ is denoted by $\omega(c_i)$ for simplicity. We have the natural inclusion $\mu_{c_i} \longrightarrow \mu_{\boldsymbol{c}}$. The image is also denoted by μ_{c_i}

We put $X = \Delta^n$, $D_i := \{z_i = 0\}$ and $D = \bigcup_{i=1}^l D_i$ for some $l \leq n$. We have the natural G_m^n-action on X, given by the componentwise multiplication. Let $\boldsymbol{c}$ be an element of $\mathbb{Z}_{>0}^n$. If we take a homomorphism $\rho : \mu_{\boldsymbol{c}} \longrightarrow G_m^n$, we obtain the $\mu_{\boldsymbol{c}}$-action ρ on X. In the following, we consider only such $\mu_{\boldsymbol{c}}$-actions on X.

2.3.2. An equivariant section and an equivariant lift. Let ρ be a $\mu_{\boldsymbol{c}}$-action on X. Let E be a ρ-equivariant holomorphic vector bundle on X.

DEFINITION 2.10. A section f of E is called ρ-equivariant, if there exists a homomorphism $\chi : \mu_{\boldsymbol{c}} \longrightarrow \boldsymbol{C}^*$ such that $g^*(f) = \chi(g) \cdot f$ for any element $g \in \mu_{\boldsymbol{c}}$. □

Let $\Gamma(X,E)$ denote the space of holomorphic sections of E over X. We have the natural μ_c-action on $\Gamma(X,E)$. Since $\mu_{\boldsymbol{c}}$ is a finite group, we have the canonical decomposition:

$$\Gamma(X,E) \simeq \bigoplus_{\eta \in Rep(\mu_c)} Hom\big(V_\eta, \Gamma(X,E)\big) \otimes V_\eta.$$

Here $\mathrm{Rep}(\mu_{\boldsymbol{c}})$ denote the set of the equivalence classes of the irreducible representations of $\mu_{\boldsymbol{c}}$, and V_η ($\eta \in \mathrm{Rep}(\mu_{\boldsymbol{c}})$) denotes an irreducible representation corresponding to η. Then a section $f \in \Gamma(X,E)$ is equivariant if and only if f is contained in one of the components in the canonical decomposition.

Let I be a subset of $\underline{n}$. We put $D_I := \bigcap_{i\in I} D_i$.

LEMMA 2.11. *Let f_0 be a holomorphic equivariant section of $E_{|D_I}$, i.e., there exists a homomorphism $\chi : \mu_{\boldsymbol{c}} \longrightarrow \boldsymbol{C}^*$ such that $g^*(f_0) = \chi(g)\cdot f_0$ for any $g \in \mu_{\boldsymbol{c}}$. Then there exists a holomorphic equivariant section f of E on X satisfying $f_{|D_I} = f_0$ and $g^*(f) = \chi(g)\cdot f$.*

Proof We have the equivariant surjection $\Gamma(X,E) \longrightarrow \Gamma(D_I, E_{|D_I})$. Since $\mu_{\boldsymbol{c}}$ is finite, we have the canonical decompositions of $\Gamma(X,E)$ and $\Gamma(D_I, E_{|D_I})$. Then we obtain the surjections of the components of the canonical decompositions. Hence we are done. □

Let $\boldsymbol{u} = (u_i)$ be an equivariant base of $E_{|O}$, i.e., there exist $\chi_i : G \longrightarrow \boldsymbol{C}^*$ such that $g^*(u_i) = \chi_i(g)\cdot u_i$ for each i.

COROLLARY 2.12. *There exists an equivariant frame $\boldsymbol{v}$ of E on a neighbourhood of O, such that $g^*(v_i) = \chi_i(g)\cdot v_i$ and $v_{i\,|\,O} = u_i$.* □

DEFINITION 2.13. A frame as in Corollary 2.12 is called an equivariant frame. □

LEMMA 2.14. *Let E_j ($j = 1,2$) be $\mu_{\boldsymbol{c}}$-vector bundles over X, and $\pi : E_1 \longrightarrow E_2$ be equivariant surjection. Let N be a $\mu_{\boldsymbol{c}}$-subbundle of E_2 and M be a $\mu_{\boldsymbol{c}}$-subbundle of $E_{1\,|\,D}$. Assume that the restriction of $\pi_{|D}$ to M gives an isomorphism of M and $N_{|D}$.*

Then there exists a $\mu_{\boldsymbol{c}}$-subbundle $\widetilde{M}$ of E_1 defined around O satisfying the following:

- *The restriction of $\pi_{|D}$ to $\widetilde{M}$ gives an isomorphism of $\widetilde{M}$ and N.*

Such $\widetilde{M}$ is called an equivariant lift of N extending M.

Proof We have the following exact sequence:

$$\Gamma(E_1) \longrightarrow \Gamma(E_2) \oplus \Gamma(E_{1\,|\,D}) \longrightarrow \Gamma(E_{2\,|\,D}) \longrightarrow 0. \tag{2.3}$$

Since $\mu_{\boldsymbol{c}}$ is finite, $\mu_{\boldsymbol{c}}$-vector spaces in the complex (2.3) have the canonical decomposition. On each component of the canonical decomposition, the complex is exact.

Let take an equivariant frame $\boldsymbol{v}_N$ of N on a neighbourhood of O. By using an isomorphism $\pi_{|D} : M \longrightarrow N_{|D}$, we take an equivariant frame $\boldsymbol{v}_M$ of M around O such that $\pi_{|D}(\boldsymbol{v}_M) = \boldsymbol{v}_{N|D}$. By using the equivariant exact sequence (2.3), we can take an equivariant sections $v_{\widetilde{M},i}$ of $\widetilde{M}$ around O such that $\pi(v_{\widetilde{M},i}) = v_{N,i}$ and $v_{\widetilde{M},i\,|\,D} = v_{M,i}$. Then $\boldsymbol{v}_{\widetilde{M}} := (v_{\widetilde{M},i})$ gives a $\mu_{\boldsymbol{c}}$-subbundle $\widetilde{M}$ of E_1, which has desired properties. □

2.3.3. Compatible frames. We put $X=\Delta^n$, $D_i=\{z_i=0\}$ and $D=\bigcup_{i=1}^{l}D_i$. Let V be a $\mu_{\boldsymbol{c}}$-vector bundle on X. Assume that we are given $\mu_{\boldsymbol{c}}$-subbundles $H_I\subset V_{|D_I}$ for any subset $I\subset \underline{l}$, satisfying $H_{I'\,|\,D_I}\supset H_I$ $(I'\subset I)$.

LEMMA 2.15. *Let v be a $\mu_{\boldsymbol{c}}$-equivariant section of $H_{\underline{l}}$. Then there exists a $\mu_{\boldsymbol{c}}$-equivariant section $\tilde{v}$ of V on a neighbourhood of $D_{\underline{l}}$, satisfying $\tilde{v}_{|D_I}\in H_I$ and $\tilde{v}_{|D_{\underline{l}}}=v$.*

Proof We construct $\tilde{v}_{|D_I}$ descending inductively on $|I|$. Assume that we have already took $\tilde{v}_{|D'_I}$ for any $I'\supsetneq I$. We put $\partial D_I:=\bigcup_{I'\supsetneq}D_{I'}$. Then $\tilde{v}_{|\partial D_I}$ is an equivariant section of $H_{I\,|\,\partial D_I}$. By an argument similar to the proof of Lemma 2.11, we can extend it to an equivariant section $\tilde{v}_{|D_I}$ of H_I. Thus the inductive construction can proceed. □

Let S be a set. For any subset $I\subset\underline{l}$, we have the set $S^I:=\{f:I\longrightarrow S\}$. For any pair $I\subset I'$, we have the naturally defined projection $q^I:S^{I'}\longrightarrow S^I$. Let us consider decompositions $V=\bigoplus_{s\in S^I}{}^IU_s$. We denote the tuple $\bigl({}^IU_s\,\big|\,s\in S^I\bigr)$ by ${}^IU$. We assume the following, for any subset $I\subset\underline{l}$ and for any element $s\in S^I$:

$$ {}^IU_{s\,|\,D_{I'}}=\bigoplus_{q_I(s')=s}{}^{I'}U_{s'}. $$

LEMMA 2.16. *Let $\boldsymbol{v}$ be an equivariant frame of $V_{|D_{\underline{l}}}$ compatible with the decomposition ${}^{\underline{l}}U$. Then we have an equivariant frame $\tilde{\boldsymbol{v}}$ of V on a neighbourhood of $D_{\underline{l}}$ with $\tilde{\boldsymbol{v}}_{|D_{\underline{l}}}=\boldsymbol{v}$, satisfying that $\tilde{\boldsymbol{v}}_{|D_I}$ is compatible with ${}^IU$.*

Proof From any element $s\in S^{\underline{l}}$, we obtain the elements $q_I(s)\in S^I$, and thus the subbundle ${}^IU_{q_I(s)}\subset V_{|D_I}$. By using Lemma 2.15, we can take a tuple of sections $\tilde{\boldsymbol{v}}$ with $\tilde{\boldsymbol{v}}_{|D_{\underline{l}}}=\boldsymbol{v}$, satisfying that $\tilde{\boldsymbol{v}}_{|D_I}$ is compatible with the decomposition ${}^IU$. On a neighbourhood of $D_{\underline{l}}$, $\tilde{\boldsymbol{v}}$ gives a desired frame. □

2.4. Some elementary preliminary for convexity

2.4.1. Preliminary. Let T^n denote an n-dimensional torus $\bigl\{(z_1,\ldots,z_n)\in\boldsymbol{C}^n\,\big|\,|z_i|=1\bigr\}$. We use the coordinate $z_i=\exp(\sqrt{-1}\theta_i)$. Let $\boldsymbol{\alpha}=(\alpha_1,\ldots,\alpha_n)$ be an element of $\boldsymbol{R}^n$ such that $\alpha_1,\ldots,\alpha_n$ are linearly independent over $\boldsymbol{Q}$. Let us consider the morphism $\psi_{\boldsymbol{\alpha}}:\boldsymbol{R}\longrightarrow T^n$, given as follows:

$$ \psi_{\boldsymbol{\alpha}}(x)=\bigl(\exp(\sqrt{-1}\alpha_i\cdot x)\,\big|\,i=1,\ldots,n\bigr). $$

Let f be an $\boldsymbol{R}$-valued C^∞-function on T^n. Then we have the Fourier decomposition of f:

$$ f=\sum_{\boldsymbol{m}\in\mathbb{Z}^n}a_{\boldsymbol{m}}\cdot\exp\Bigl(\sqrt{-1}\cdot\sum_{i=1}^{n}m_i\cdot\theta_i\Bigr). $$

Here m_i denotes the i-th component of $\boldsymbol{m}$, and $a_{\boldsymbol{m}}$ are complex numbers. Since f is $\boldsymbol{R}$-valued, we have the relation $\overline{a}_{\boldsymbol{m}}=a_{-\boldsymbol{m}}$. Thus we have the following equality:

$$ f=\sum_{\boldsymbol{m}\in\mathbb{Z}^n}\frac{1}{2}\Bigl(a_{\boldsymbol{m}}\cdot\exp\Bigl(\sqrt{-1}\sum m_i\cdot\theta_i\Bigr)+\overline{a}_{\boldsymbol{m}}\cdot\exp\Bigl(-\sqrt{-1}\sum m_i\cdot\theta_i\Bigr)\Bigr). $$

In the following, we put $\boldsymbol{m}\cdot\boldsymbol{\alpha}:=\sum_{i=1}^{n}m_i\cdot\alpha_i$. Then we have the following:

$$ \psi_{\boldsymbol{\alpha}}^{-1}(f)(x)=\sum_{\boldsymbol{m}\in\mathbb{Z}^n}\frac{1}{2}\Bigl(a_{\boldsymbol{m}}\cdot\exp(\sqrt{-1}\boldsymbol{m}\cdot\boldsymbol{\alpha}\cdot x)+\overline{a}_{\boldsymbol{m}}\exp(-\sqrt{-1}\boldsymbol{m}\cdot\boldsymbol{\alpha}\cdot x)\Bigr). $$

Let $b_{\boldsymbol{m}}$ and $c_{\boldsymbol{m}}$ denote the real part and the imaginary part of $a_{\boldsymbol{m}}$ respectively. Then we obtain the following:

$$\psi_{\boldsymbol{\alpha}}^{-1}(f)(x) = \sum_{\boldsymbol{m}\in\mathbb{Z}^n} b_{\boldsymbol{m}} \cdot \cos\bigl(\boldsymbol{m}\cdot\boldsymbol{\alpha}\cdot x\bigr) - \sum_{\boldsymbol{m}\in\mathbb{Z}_n} c_{\boldsymbol{m}} \cdot \sin\bigl(\boldsymbol{m}\cdot\boldsymbol{\alpha}\cdot x\bigr).$$

Note that $c_{\boldsymbol{m}} = 0$ in the case $\boldsymbol{m} = (0,\ldots,0)$.

LEMMA 2.17. *We have the finiteness* $\sum_{\boldsymbol{m}} |a_{\boldsymbol{m}}| < \infty$.

Proof We put $\mathbb{Z}^* := \mathbb{Z} - \{0\}$. We have the natural inclusion $\mathbb{Z}^{*\,I} \longrightarrow \mathbb{Z}^n$. Let T_I be the sub-torus of T^n determined by the condition $z_j = 1$ for any $j \in \underline{n} - I$. Then the restriction $f_{|T_I}$ is C^∞. Hence we have the following:

$$\sum_{\boldsymbol{m}\in\mathbb{Z}^{*\,I}} \prod_{i\in I} |m_i|^2 \cdot |a_{\boldsymbol{m}}|^2 < \infty.$$

Thus we obtain the following:

$$\sum_{\boldsymbol{m}\in\mathbb{Z}^{*\,I}} |a_{\boldsymbol{m}}| < \Bigl(\sum_{\boldsymbol{m}\in\mathbb{Z}^{*\,I}} \prod_{i\in I} |m_i|^{-2}\Bigr) \cdot \Bigl(\sum_{\boldsymbol{m}\in\mathbb{Z}^{*\,I}} \prod_{i\in I} |m_i|^2 \cdot |a_{\boldsymbol{m}}|^2\Bigr) < \infty.$$

On the other hand, we have the following:

$$\sum_{\boldsymbol{m}\in\mathbb{Z}^n} |a_{\boldsymbol{m}}| = \sum_{I\subset\underline{n}} \sum_{\boldsymbol{m}\in\mathbb{Z}^{*\,I}} |a_{\boldsymbol{m}}| < \infty.$$

Thus we are done. □

COROLLARY 2.18. *We have* $\sum_{\boldsymbol{m}\in\mathbb{Z}^n} |b_{\boldsymbol{m}}| < \infty$ *and* $\sum_{\boldsymbol{m}\in\mathbb{Z}^n} |c_{\boldsymbol{m}}| < \infty$. □

The next lemma can be checked by direct calculations.

LEMMA 2.19. *We have the following equalities:*

$$\int_{-N}^{N} 1\cdot\frac{dx}{2N} = 1,\quad \int_{-N}^{N} \cos(\boldsymbol{m}\cdot\boldsymbol{\alpha}\cdot x)\cdot\frac{dx}{2N} = \frac{\sin(\boldsymbol{m}\cdot\boldsymbol{\alpha}\cdot N)}{N\cdot(\boldsymbol{m}\cdot\boldsymbol{\alpha})},\quad \int_{-N}^{N} \sin(\boldsymbol{m}\cdot\boldsymbol{\alpha}\cdot x)\cdot\frac{dx}{2N} = 0.$$

Here $\boldsymbol{m} \neq (0,\ldots,0)$. □

We put as follows:

$$\Xi_N(f) := \int_{-N}^{N} \psi_{\boldsymbol{\alpha}}^{-1}(f)\cdot\frac{dx}{2N}.$$

LEMMA 2.20. *We have the following equality:*

$$\Xi_N(f) = b_0 + \sum_{\boldsymbol{m}\neq 0} \frac{\sin(\boldsymbol{m}\cdot\boldsymbol{\alpha}\cdot N)}{N\cdot(\boldsymbol{m}\cdot\boldsymbol{\alpha})}\cdot b_{\boldsymbol{m}}.$$

Proof We have the following:

$$\psi_{\boldsymbol{\alpha}}^{-1}(f) = \lim_{M\to\infty}\Bigl(\sum_{|\boldsymbol{m}|\le M} b_{\boldsymbol{m}}\cdot\cos(\boldsymbol{m}\cdot\boldsymbol{\alpha}\cdot x) - \sum_{|\boldsymbol{m}|\le M} c_{\boldsymbol{m}}\cdot\sin(\boldsymbol{m}\cdot\boldsymbol{\alpha}\cdot x)\Bigr).$$

Due to Corollary 2.18, we can change the order of the integral and the summation. Thus we obtain the following:

$$\Xi_N(f) = \sum_{\boldsymbol{m}\in\mathbb{Z}^n}\Bigl(\int_{-N}^{N} \cos(\boldsymbol{m}\cdot\boldsymbol{\alpha}\cdot x)\frac{dx}{2N}\cdot b_{\boldsymbol{m}} - \int_{-N}^{N} \sin(\boldsymbol{m}\cdot\boldsymbol{\alpha}\cdot x)\frac{dx}{2N}\cdot c_{\boldsymbol{m}}\Bigr).$$

Thus we obtain the result. □

LEMMA 2.21. *When $N \to \infty$, the sequence of the numbers $\{\Xi_N(f)\}$ is convergent. We put $\Xi(f) := \lim_{N\to\infty} \Xi_N(f)$. Then we have $\Xi(f) = b_0$.*

Proof We have only to show the following:

$$\lim_{N\to\infty}\Big(\sum_{\boldsymbol{m}\neq 0} \frac{\sin(\boldsymbol{m}\cdot\boldsymbol{\alpha}\cdot N)}{N\cdot\boldsymbol{m}\cdot\boldsymbol{\alpha}}\cdot b_{\boldsymbol{m}}\Big) = 0.$$

We have the following inequality:

$$\left|\frac{\sin(\boldsymbol{m}\cdot\boldsymbol{\alpha}\cdot N)}{N\cdot\boldsymbol{m}\cdot\boldsymbol{\alpha}}\cdot b_{\boldsymbol{m}}\right| \leq |b_{\boldsymbol{m}}|.$$

We also have $\sum_{\boldsymbol{m}\neq 0}|b_{\boldsymbol{m}}| < \infty$. Then we obtain the following:

$$\lim_{N\to\infty}\Big(\sum_{\boldsymbol{m}\neq 0} \frac{\sin(\boldsymbol{m}\cdot\boldsymbol{\alpha}\cdot N)}{N\cdot\boldsymbol{m}\cdot\boldsymbol{\alpha}}\cdot b_{\boldsymbol{m}}\Big) = \sum_{\boldsymbol{m}\neq 0}\Big(\lim_{N\to\infty} \frac{\sin(\boldsymbol{m}\cdot\boldsymbol{\alpha}\cdot N)}{N\cdot\boldsymbol{m}\cdot\boldsymbol{\alpha}}\cdot b_{\boldsymbol{m}}\Big) = 0.$$

Thus we are done. □

The morphism $\psi_{\boldsymbol{\alpha}}$ induces the morphism $\boldsymbol{R}\times\boldsymbol{R}_{>0} \longrightarrow T^n\times\boldsymbol{R}_{>0}$, which we denote also by $\psi_{\boldsymbol{\alpha}}$. Let f be an $\boldsymbol{R}$-valued C^∞-function on $T^n\times\boldsymbol{R}_{>0}$. Then we have the Fourier decomposition as before:

$$f = \sum_{\boldsymbol{m}} a_{\boldsymbol{m}}(y)\cdot\exp\Big(\sqrt{-1}\sum_{i=1}^{n} m_i\cdot\theta_i\Big).$$

Pick $y\in\boldsymbol{R}_{>0}$. Then we put as follows:

$$\Xi_N(f)(y) := \int_{-N}^{N} (\psi_{\boldsymbol{\alpha}}^{-1}f)(x,y)\frac{dx}{2N}.$$

We have the limit $\Xi(f)(y) := \lim_{N\to\infty}\Xi_N(f)(y)$, and thus we obtain the functions $\Xi(f), \Xi_N(f) : \boldsymbol{R}_{>0} \longrightarrow \boldsymbol{R}$. If we decompose $a_{\boldsymbol{m}}(y)$ into the real part $b_{\boldsymbol{m}}(y)$ and the imaginary part $c_{\boldsymbol{m}}(y)$, then we have $\Xi(f)(y) = b_0(y)$. In particular, $\Xi(f)$ is C^∞ and we have the following:

$$\Big(\frac{\partial}{\partial y}\Big)^l\Xi(f) = \frac{d^l b_0(y)}{dy^l} = \Xi\Big(\frac{\partial^l f}{\partial y^l}\Big).$$

2.4.2. Convexity. We continue to use the setting in the subsection 2.4.1. We put $F := \psi_{\boldsymbol{\alpha}}^{-1}(f)$.

LEMMA 2.22. *Assume F is subharmonic, i.e., the following inequality holds:*

$$-\frac{\partial^2 F}{\partial x^2} - \frac{\partial^2 F}{\partial y^2} \leq 0.$$

Then we have the following convexity of $\Xi(f)$:

$$-\frac{\partial^2}{\partial y^2}\Xi(f) \leq 0.$$

Proof We have the following equality:

$$-\frac{\partial^2}{\partial y^2}\Xi(f) = \Xi\Big(-\frac{\partial^2 f}{\partial y^2}\Big) = \lim_{N\to\infty}\int_{-N}^{N}\Big(-\frac{\partial^2 F}{\partial y^2}\Big)\frac{dx}{2N}.$$

We have the following inequality due to the subharmonicity of F:

$$\int_{-N}^{N}\Big(-\frac{\partial^2 F}{\partial y^2}\Big)\frac{dx}{2N}\leq\int_{-N}^{N}\frac{\partial^2 F}{\partial x^2}\frac{dx}{2N}=\frac{1}{2N}\Big(\frac{\partial F}{\partial x}(N,y)-\frac{\partial F}{\partial x}(-N,y)\Big).$$

Let V denote the vector field on T^n, given as follows:

$$V=\sum_{i=1}^{n}\alpha_i\frac{\partial}{\partial\theta_i}.$$

Then we have the following equality:

$$\frac{\partial F}{\partial x}=\psi_{\boldsymbol{\alpha}}^{-1}(Vf).$$

If we fix y, then Vf is bounded function on a compact set T^n, and thus $\partial F/\partial x$ is a bounded function on $\boldsymbol{R}$. Hence we have the following:

$$\lim_{N\to\infty}\frac{1}{2N}\Big(\frac{\partial F}{\partial x}(N,y)-\frac{\partial F}{\partial x}(-N,y)\Big)=0.$$

Thus we obtain the result. □

2.4.3. An elementary boundedness of a convex function. Lemma 2.22 will be used in the proof of preliminary constantness of the filtration in the subsection 12.2.4, together with the following lemma.

LEMMA 2.23. *Let $f:\boldsymbol{R}_{\geq 1}\longrightarrow\boldsymbol{R}$ be a C^∞-function. Assume the following:*

(1) $-\dfrac{d^2 f}{dy^2}\leq 0$.

(2) *There exist positive numbers C_1 and C_2 such that $f\leq C_1+C_2\cdot\log y$.*

Then there exists a positive number C_3 such that $f\leq C_3$.

Proof Due to the condition (1), the function f is convex below. On the other hand, the right hand side in the condition (2) is convex above. They imply that f is dominated by a constant. □

2.5. Some lemmas for functions on a disc

2.5.1. Subharmonicity and integrability. We recall Lemma 2.2 in [**81**] due to Simpson. It is one of the key points in the theory of tame harmonic bundles. Since a rather delicate estimate seems necessary, we also give the proof. Let X denote the disc $\{z\in\boldsymbol{C}\,|\,|z|<1\}$, and O denote the origin $z=0$. Let dvol denote the volume form $\sqrt{-1}dz\cdot d\bar{z}$.

Let f be a real valued C^∞-function on $X-\{O\}$, and let b be non-negative function which is locally L^1 on $X-\{O\}$. Assume the following:

- $\big(-\log|z|\big)^{-1}\cdot f(z)\longrightarrow 0$ as $|z|\longrightarrow 0$.
- Let Δ denote the Laplacian $-\partial_z\partial_{\bar{z}}$. Then the inequality $\Delta(f)\leq -b$ holds on $X-\{O\}$.

Due to the first condition, f gives a distribution on X.

PROPOSITION 2.24 (Lemma 2.2 [**81**]).

- *b is locally L^1 on X.*
- *The inequality $\Delta(f)\leq -b$ holds on X in the sense of distribution.*

Proof Let η be any non-negative test function on Δ. We have only to show the following inequality:

$$\int f \cdot \Delta\eta \cdot \mathrm{dvol} \leq -\int b \cdot \eta \cdot \mathrm{dvol}. \tag{2.4}$$

We may assume that the support of η is contained in $\{|z| < e^{-1}\}$.

Let ρ be a C^∞-function on $\boldsymbol{R}$ satisfying the following:

- $0 \leq \rho \leq 1$. $\rho(t) = 0$ $(t \geq 1)$ and $\rho(t) = 1$ $(t \leq 0)$.
- $\rho' \leq 0$, where ρ' denotes $d\rho/dt$.

We put as follows:

$$\chi_N(z) := \rho\left(\frac{1}{N} \cdot \log\bigl(-\log|z|^2\bigr)\right).$$

Since $\chi_N(z)$ is 0 around O, we have the following inequality, for any positive number a:

$$-\int (f + a \cdot \log|z|^2) \cdot \frac{\partial^2(\chi_N \cdot \eta)}{\partial z \partial \bar{z}} \cdot \mathrm{dvol} \leq -\int b \cdot (\chi_N \cdot \eta) \cdot \mathrm{dvol}.$$

Therefore we have the following for any $a > 0$:

$$\begin{aligned} \int (f + a \cdot \log|z|^2) \cdot \chi_N \cdot \Delta\eta &\leq -\int b \cdot \chi_N \cdot \eta \cdot \mathrm{dvol} \\ &+ \int (f + a \cdot \log|z|^2) \cdot \left(\frac{\partial \chi_N}{\partial z} \cdot \frac{\partial \eta}{\partial \bar{z}} + \frac{\partial \chi_N}{\partial \bar{z}} \cdot \frac{\partial \eta}{\partial z} + \frac{\partial^2 \chi_N}{\partial z \partial \bar{z}} \cdot \eta\right) \cdot \mathrm{dvol}. \end{aligned} \tag{2.5}$$

By a direct calculation, we have the following:

$$\frac{\partial \chi_N}{\partial z} = \frac{\rho'}{N \cdot z \cdot \log|z|^2}, \qquad \frac{\partial \chi_N}{\partial \bar{z}} = \frac{\rho'}{N \cdot \bar{z} \cdot \log|z|^2}, \tag{2.6}$$

$$\frac{\partial^2 \chi_N}{\partial z \partial \bar{z}} = \frac{\rho''}{N^2 \cdot \bigl(-\log|z|^2\bigr)^2 \cdot |z|^2} - \frac{\rho'}{N \cdot \bigl(-\log|z|^2\bigr)^2 \cdot |z|^2}. \tag{2.7}$$

There exists a positive constant C_1, which is independent of N, such that we have the following inequality on the region $\{0 < |z| < e^{-1}\}$:

$$\left|\frac{\partial \chi_N}{\partial z}\right| = \left|\frac{\partial \chi_N}{\partial \bar{z}}\right| \leq C_1 \cdot \frac{1}{|z| \cdot (-\log|z|^2) \cdot \log(-\log|z|^2)}. \tag{2.8}$$

Since we have $\bigl|f + a \cdot \log|z|^2\bigr| = O\bigl(-\log|z|^2\bigr)$, we obtain the following due to the Lebesgue's theorem:

$$\lim_{N \to \infty} \int (f + a \cdot \log|z|^2) \cdot \left(\frac{\partial \chi_N}{\partial z} + \frac{\partial \chi_N}{\partial \bar{z}}\right) \cdot \mathrm{dvol} = 0.$$

We remark that the second term in the right hand side of (2.7) is non-negative. We also remark that there exists a positive number $\delta(a)$ such that $f + a\log|z|^2 < 0$ for any $0 < |z| < \delta(a)$, due to our assumption on f. Hence we have the following:

(2.9)

$$\begin{aligned} \int (f + a \cdot \log|z|^2) \cdot \frac{\partial^2 \chi_N}{\partial z \partial \bar{z}} \cdot \eta \cdot \mathrm{dvol} &\leq \int (f + a \cdot \log|z|^2) \cdot \frac{\rho'' \cdot \mathrm{dvol}}{N^2 \cdot \bigl(-\log|z|^2\bigr) \cdot |z|^2} \\ &- \int_{|z| \geq \delta(a)} (f + a \cdot \log|z|^2) \frac{\rho' \cdot \mathrm{dvol}}{N \cdot \bigl(-\log|z|\bigr)^2 \cdot |z|^2}. \end{aligned}$$

There exists a positive constant C_2, which is independent of N, such that the following inequality holds on the region $\{0 < |z| < e^{-1}\}$:

$$\left|\frac{\rho''}{N^2\big(-\log|z|^2\big)\cdot|z|^2}\right| \leq C_2\cdot\frac{1}{|z|^2\cdot\big(-\log|z|^2\big)^2\cdot\big(\log(-\log|z|^2)\big)^2}.$$

Hence the right hand side of (2.9) converges to 0 due to Lebesgue's theorem.

Therefore we obtain the following inequality, by considering $N\to\infty$ in (2.5):

$$\int\big(f+a\cdot\log|z|^2\big)\cdot\Delta\eta\cdot\mathrm{dvol}\leq-\int b\cdot\eta\cdot\mathrm{dvol}.$$

Since a is arbitrary positive number, we obtain the desired inequality (2.4). □

COROLLARY 2.25. *Under the assumption of Proposition 2.24, f is subharmonic function on X. In particular, the maximum principle holds for f.*

Proof We give only a sketch of a proof. By taking a convolution product, we can take a sequence f_ϵ $(\epsilon>0)$ of subharmonic functions, which converges to f in the C^∞-sense on any compact subset of $X-\{O\}$, and which converges to f in the L^1-sense on $X-\{O\}$. Since the maximum principle holds for each f_ϵ, we obtain a constant such that $f\leq C$ near O. Then it is well known that such a subharmonic function on $X-\{O\}$ gives the subharmonic function on X (see [**69**], for example). □

2.5.2. A boundedness I.

We continue to use the notation in the previous subsection.

LEMMA 2.26. *Let f be a bounded function on X, and let g be an L^2-function on X with respect to* dvol *such that $\overline{\partial}g=f\cdot d\bar{z}$ on X. Then g is also bounded.*

Proof We put as follows:

$$h=\frac{z}{2\pi}\int_{C}\frac{w^{-1}\cdot f(w)}{w-z}\cdot\mathrm{dvol}.$$

In [**81**], it is shown that $\overline{\partial}h=f\cdot d\bar{z}$ and $|h|=O\big(|z|\cdot(-\log|z|)\big)$ around the origin. We put $\widehat{g}:=g-h$, and then $\widehat{g}$ is an L^2-holomorphic function on $X-\{O\}$. It is well known that such $\widehat{g}$ is holomorphic on X. In particular, it is bounded. Therefore we obtain the boundedness of g. □

Let E be a holomorphic bundle on a disc Δ, and h be a hermitian metric of E.

COROLLARY 2.27. *Let f be a bounded section of $E\otimes\Omega^{0,1}$, and let g be an L^2-section of E such that $\overline{\partial}_E g=f$. Then g is bounded.*

Proof Let $\boldsymbol{v}$ be a holomorphic frame of E around the origin. We may replace h with the metric h_0 for which $\boldsymbol{v}$ is orthogonal. Then we have only to apply Lemma 2.26. □

2.5.3. A boundedness II.

We put $X:=\big\{z\in\boldsymbol{C}\,\big|\,|z|\leq1\big\}$. Let O denote the origin $z=0$, and let Δ'' denote the Laplacian. The following lemma is a special case of a lemma contained in [**80**].

LEMMA 2.28. *Assume $p>1$. Let f and g be non-negative C^∞-functions on $X-\{O\}$ satisfying the following:*

- *g is L^p, and f is bounded.*

- $\Delta'' f \leq g$ *on* $X - \{O\}$.

Then the sup norm of f *is dominated by the* L^p*-norm of* g *and the maximum on the boundary, as follows*

$$\sup_{z \in X - \{O\}} f \leq C_1 \cdot ||g||_{L^p} + \max_{|z|=1} f(z).$$

Here C_1 *denotes a positive constant, which is independent of* g *and* f.

Proof Since p is assumed to be larger than 1, we have the unique solution of $\Delta''\varphi = g$ on X for the Dirichlet problem, for which we have $\sup|\varphi| \leq C_2 \cdot ||g||_{L^p}$. (See the chapter 9 of [**31**].) Then we obtain $\Delta''(f - \varphi) \leq 0$ on $X - \{O\}$. Since f is assumed to be bounded, $f - \varphi$ is bounded, and hence we can apply Corollary 2.25. Namely the maximum principle holds for $f - \varphi$. Then the claim immediately follows. □

2.6. An elementary remark on some distributions

2.6.1. Some integrals. In this subsection, ϕ denotes a test function, i.e., a C^∞-function on $\boldsymbol{R}$ whose support is compact. We denote the differential $\frac{d\phi}{dr}$ by ϕ'. We put as follows:

$$L_n := \frac{(\log r^2)^n}{n!}.$$

LEMMA 2.29. *We have the following equality:*

$$\int_0^\infty \phi \cdot L_n \cdot r^{2s-1} \cdot dr = \frac{1}{2} \sum_{i=0}^{n} (-s)^{-i-1} \int_0^\infty \phi' \cdot L_{n-i} \cdot r^{2s} dr. \tag{2.10}$$

Proof We have the following equality:

$$L_n \cdot r^{2s-1} = \frac{1}{2} \frac{d}{dr} \Big(\sum_{i=0}^{n} (-1)^i \cdot s^{-i-1} L_{n-i} \cdot r^{2s} \Big).$$

Then (2.10) immediately follows. □

Since ϕ' is C^∞ around $r = 0$, $\int_0^\infty \phi' \cdot L_k r^{2s} \cdot dr$ gives a holomorphic function for the variable s around $s = 0$.

LEMMA 2.30. *We have the following Taylor development at* $s = 0$*:*

$$\int_0^\infty \phi' \cdot L_k \cdot r^{2s} \cdot dr = \sum_{l=0}^{\infty} s^l \binom{k+l}{l} \cdot \int_0^\infty \phi' \cdot L_{k+l} \cdot dr. \tag{2.11}$$

Proof We have the following:

$$\frac{1}{l!} \left(\frac{d}{ds} \right)^l \int_0^\infty \phi' \cdot L_k \cdot r^{2s} \cdot dr = \frac{1}{l!} \int_0^\infty \phi' \cdot \frac{(\log r^2)^{k+l}}{k!} r^{2s} dr \tag{2.12}$$
$$= \binom{k+l}{l} \cdot \int_0^\infty \phi' \cdot L_{k+l} \cdot r^{2s} dr.$$

Then (2.11) immediately follows. □

LEMMA 2.31. *We have the following equality:*

$$\int_0^\infty \phi \cdot L_n \cdot r^{2s-1} \cdot dr = \frac{(-1)^n}{2} \cdot s^{-n-1} \cdot \phi(0) + \sum_{l=n+1}^{\infty} X_{n,l} \cdot s^{l-n-1} \int_0^\infty \phi' \cdot L_l \cdot dr. \tag{2.13}$$

Here we put as follows:

$$X_{n,l} := \frac{(-1)^{n-1}}{2} \sum_{h=0}^{n} (-1)^h \begin{pmatrix} l \\ h \end{pmatrix} \in \boldsymbol{R}. \tag{2.14}$$

Proof Due to the previous lemmas, we have the following equality:

$$\int \phi \cdot L_n \cdot r^{2s-1} \cdot dr = \frac{1}{2} \sum_{i=0}^{n} (-s)^{-i-1} \sum_{l=0}^{\infty} s^l \cdot \begin{pmatrix} n-i+l \\ l \end{pmatrix} \int_0^{\infty} \phi' \cdot L_{n-i+l} \cdot dr. \tag{2.15}$$

By putting $h = n - i$ and $m = n - i + l$, the right hand side can be rewritten as follows:

$$\frac{1}{2} \sum_{m=0}^{\infty} \sum_{\substack{h+l=m \\ h \le n}} (-1)^{-n+h-1} \cdot s^{-n+m-1} \cdot \begin{pmatrix} m \\ h \end{pmatrix} \cdot \int_0^{\infty} \phi' \cdot L_m \cdot dr \tag{2.16}$$

$$= \frac{1}{2} \sum_{m=0}^{\infty} \Bigg(\sum_{h \le \min(m,n)} (-1)^h \begin{pmatrix} m \\ h \end{pmatrix} \Bigg) \cdot (-1)^{n-1} s^{-n+m-1} \int_0^{\infty} \phi' \cdot L_m \cdot dr.$$

The term $m = 0$ is as follows:

$$\frac{1}{2} \cdot (-1)^{n-1} s^{-n-1} \cdot \int_0^{\infty} \phi' \cdot dr = \frac{(-1)^n}{2} \cdot s^{-n-1} \cdot \phi(0).$$

It is easy to see that the terms $1 \le m \le n$ vanish. Then we obtain (2.13). □

COROLLARY 2.32. *We have the following expansion:*

$$\int \phi \cdot L_n \cdot r^{2ms-1} dr = \frac{(-1)^n}{2} (ms)^{-n-1} \cdot \phi(0) + \frac{1}{2} \sum_{l=n+1}^{\infty} X_{n,l} \cdot (m \cdot s)^{l-n-1} \int_0^{\infty} \phi' \cdot L_k \cdot dr.$$

Here $X_{n,l}$ is given in (2.14). □

COROLLARY 2.33. *Let ϕ be a test function on the complex plane $\boldsymbol{C}$. We have the following formula:*

$$\operatorname*{Res}_{s=0} \int \phi \cdot L_n(|z|^2) \cdot |z|^{2s-2} \cdot \frac{\sqrt{-1}}{2\pi} dz \wedge d\bar{z} = \begin{cases} 1 & (n = 0), \\ 0 & (n \neq 0). \end{cases}$$

□

2.6.2. Some distributions. In this subsection ϕ denotes a test function on $\boldsymbol{C}^n$. Let us consider the function Φ on $\boldsymbol{C}^{*\,n}$ of the following form:

$$\Phi = \sum_{k=1}^{N} \sum_{\boldsymbol{n} \in S} s^k \cdot a_{\boldsymbol{n},k} \cdot \prod_{i=1}^{n} L_{n_i}(|z_i|)^{n_i}.$$

Here S denotes a finite subset of $\mathbb{Z}^n$, $a_{\boldsymbol{n},k}$ denote complex numbers, and n_i $(i = 1, \ldots, n)$ denote the i-th component of $\boldsymbol{n}$. We have the distribution $\hat{\Phi}$ defined as follows:

$$\hat{\Phi}(\phi) := \operatorname*{Res}_{s=0} \int_{\boldsymbol{C}^n} \Phi \cdot \phi \cdot \prod_{i=1}^{n} |z_i|^{2m_i \cdot s - 2} \cdot \frac{\sqrt{-1}}{2\pi} dz_i \wedge d\bar{z}_i.$$

LEMMA 2.34. *Assume that the support of the distribution* $\hat{\Phi}$ *is contained in* $\{O\}$. *Then we have the following formula:*

$$\hat{\Phi}(\phi) = \sum_{(\boldsymbol{n},k)\in S_1} a_{\boldsymbol{n},k} \cdot (-1)^{\sum n_i} \cdot \phi(0) \cdot \prod m_i^{-n_i-1}.$$

Here S_1 *denotes the set of the elements* $(\boldsymbol{n}, k) \in S \times \mathbb{Z}$ *satisfying* $\sum n_i = k - n + 1$.

Proof We have only to consider the test functions of the form $\phi(z_1, \dots, z_n) = \prod_{i=1}^n \phi_i(z_i)$. Note the following equality:

$$\int \Phi\cdot\phi\cdot\prod_{i=1}^n |z_i|^{2m_is-2}\cdot\frac{\sqrt{-1}}{2\pi}dz_i\wedge d\bar{z}_i = \sum_{i=1}^N\sum_{\boldsymbol{n}\in S} s^k\cdot a_{\boldsymbol{n},k}\cdot 2^n\cdot\prod_{i=1}^n\int_0^\infty \phi_i\cdot L_{n_i}\cdot r_i^{2m_is-1}dr_i.$$

We have the following equalities:

$$\prod_{i=1}^n \int_0^\infty \phi_i \cdot L_{n_i} \cdot r_i^{2m_is-1}dr_i \tag{2.17}$$

$$= \sum_{I\sqcup J=\underline{n}} \prod_{i\in I} m_i^{-n_i-1}s^{-|\boldsymbol{n}|-n}\frac{(-1)^{n_i}}{2}\phi_i(0) \cdot \prod_{j\in J}\sum_{l_j=n_j+1}^{\infty} X_{n_j,l_j}(m_j\cdot s)^{l_j}\int_0^\infty \phi_j'\cdot L_{l_j}\cdot dr$$

$$= \sum_{I\sqcup J=\underline{n}} \prod_{i\in I} m_i^{-n_i-1}\prod_{j\in J} m_j^{l_j}\cdot s^{-|\boldsymbol{n}|-n+|\boldsymbol{l}|}\prod_{i\in I}\frac{(-1)^{n_i}}{2}\phi_i(0)\sum_{\boldsymbol{l}\in\mathbb{Z}^J}\prod_{j\in J} X_{n_j,l_j}\int_0^\infty \phi_j'\cdot L_{l_j}\cdot dr_j.$$

Here we put $|\boldsymbol{n}| = \sum_{i=1}^n n_i$ and $|\boldsymbol{l}| = \sum_{j\in J} l_j$. We put as follows:

$$A(\boldsymbol{m}, \boldsymbol{n}, I, J, \boldsymbol{l}) := \prod_{i\in I} m_i^{-n_i-1}\cdot\prod_{j\in J} m_j^{l_j}\cdot\prod_{j\in J} X_{n_j,l_j}\cdot\prod_{i\in I}\frac{(-1)^{n_i}}{2} \in \boldsymbol{R}.$$

In the case $I = \underline{n}$, we have the following equality:

$$A(\boldsymbol{m}, \boldsymbol{n}, \underline{n}, \emptyset, 0) = \prod_{i=1}^n m_i^{-n_i-1}\cdot(-1)^{|\boldsymbol{n}|}\cdot 2^{-n}.$$

Then the right hand side can be rewritten as follows:

$$\sum_{I\sqcup J=\underline{n}}\sum_{\boldsymbol{l}\in\mathbb{Z}^J} A(\boldsymbol{m}, \boldsymbol{n}, I, J, \boldsymbol{l})\cdot s^{-|\boldsymbol{n}|-n+|\boldsymbol{l}|}\cdot\prod_{i\in I}\phi_i(0)\cdot\prod_{j\in J}\int_0^\infty \phi_j'\cdot L_{l_j}dr_j.$$

We have the following:

$$\operatorname*{Res}_{s=0}\Big(s^k\prod_{i=1}^n\int \phi_i\cdot L_{n_i}\cdot r_i^{2m_is-1}dr_i\Big) \tag{2.18}$$

$$= \sum_{I\sqcup J=\underline{n}}\sum_{\boldsymbol{l}\in S(\boldsymbol{n},n,k)} A(\boldsymbol{m}, \boldsymbol{n}, I, J, \boldsymbol{l})\cdot\prod_{i\in I}\phi_i(0)\cdot\prod_{j\in J}\int_0^\infty \phi_j'\cdot L_{l_j}\cdot dr_j.$$

Here we put $S(\boldsymbol{n},n,k) := \{\boldsymbol{l} \in \mathbb{Z}_{\geq 0}^J \mid -|\boldsymbol{n}| - n + |\boldsymbol{l}| + k = -1\}$. Then we obtain the following:

(2.19)
$$\hat{\Phi}(\phi) = \sum_{\boldsymbol{n},k} a_{\boldsymbol{n},k} \sum_{I \sqcup J = \underline{n}} \sum_{\boldsymbol{l} \in S(\boldsymbol{n},n,k)} A(\boldsymbol{m},\boldsymbol{n},I,J,\boldsymbol{l}) \cdot \prod_{i \in I} \phi_i(0) \cdot \prod_{j \in J} \int_0^\infty \phi_j' \cdot L_{l_j} \cdot dr_j$$
$$= \sum_{I \sqcup J = \underline{n}} \sum_{\boldsymbol{l} \in \mathbb{Z}^J} Y(I,J,\boldsymbol{l}) \prod_{i \in I} \phi_i(0) \cdot \prod_{j \in J} \int_0^\infty \phi_j' \cdot L_{l_j} \cdot dr_j.$$

Here we put as follows:
$$Y(I,J,\boldsymbol{l}) = \sum_{\boldsymbol{n},k} a_{\boldsymbol{n},k} \cdot A(\boldsymbol{m},\boldsymbol{n},I,J,\boldsymbol{l}) \in \boldsymbol{R}.$$

Note the tuple $\{\prod_{j \in J} L_{l_j} \mid \boldsymbol{l} \in \mathbb{Z}_{>0}^J\}$ of C^∞-functions on $\boldsymbol{C}^{*\,J}$ is linearly independent over $\boldsymbol{R}$. Since we have assumed that the support of $\hat{\Phi}$ is contained in $\{O\}$, we obtain the vanishings of the constants $Y(I,J,\boldsymbol{l})$ $(J \neq \emptyset)$. Then we obtain the following:

$$(2.20) \quad \hat{\Phi}(\phi) = \sum_{(\boldsymbol{n},k) \in S_1} a_{\boldsymbol{n},k} \cdot 2^n \cdot 2^{-n} \cdot (-1)^{|\boldsymbol{n}|} \cdot \prod_{i=1}^n m_i^{-n_i-1} \phi_i(0)$$
$$= \sum_{(\boldsymbol{n},k) \in S_1} a_{\boldsymbol{n},k} \cdot (-1)^{|\boldsymbol{n}|} \cdot \phi(0) \cdot \prod_{i=1}^n m_i^{-n_i-1}.$$

Then we obtain the result. □

2.7. Preliminary from elementary linear algebra

2.7.1. The generalized eigen decomposition. Let V be a finite dimensional vector space over $\boldsymbol{C}$, and f be an endomorphism of V. We often denote the set of eigenvalues of f by $\mathcal{S}p(f)$. For any element $\alpha \in \mathcal{S}p(f)$, we denote the generalized eigenspace corresponding to α by $\mathbb{E}(f,\alpha)$. We often denote it by $\mathbb{E}(\alpha)$ or $\mathbb{E}(V,\alpha)$, if there are no confusion. Note the compatibility of the filtration and the generalized eigen decomposition.

LEMMA 2.35. *Let V be a finite dimensional vector space, and f be an endomorphism of V. Let F be a filtration of V such that f preserves F. Then the generalized eigen decomposition of f and F is compatible.*

Proof If a subspace W of V is preserved by f, then we have $W = \bigoplus_{\alpha \in \mathcal{S}p(f)} W \cap \mathbb{E}(f,\alpha)$. Then the lemma immediately follows. □

2.7.2. The generalized eigen decomposition of the commuting tuple of the endomorphisms. Let V be a finite dimensional vector space over $\boldsymbol{C}$. Let $\boldsymbol{f} = (f_1,\ldots,f_l)$ denote a commuting tuple of endomorphisms of V. For each f_i, we have the generalized eigen decomposition $V = \bigoplus_{\omega \in \mathcal{S}p(f_i)} \mathbb{E}(f_i,\omega)$. For a tuple $\boldsymbol{\omega} = (\omega_i) \in \prod \mathcal{S}p(f_i)$, we put as follows:
$$\mathbb{E}(\boldsymbol{f},\boldsymbol{\omega}) := \bigcap_i \mathbb{E}(f_i,\omega_i).$$

Then we have the set $\mathcal{S}p(\boldsymbol{f}) \subset \prod \mathcal{S}p(f_i)$ and the decomposition of V as follows:

$$V = \bigoplus_{\boldsymbol{\omega} \in \mathcal{S}p(\boldsymbol{f})} \mathbb{E}(\boldsymbol{f}, \boldsymbol{\omega}).$$

It is called the generalized eigen decomposition for $\boldsymbol{f}$.

2.7.3. A lemma for the boundedness of the hermitian metrics. Let V be a finite dimensional vector space over $\boldsymbol{C}$ and h be a hermitian metric of V. Let $S \subset \boldsymbol{C}$ be a finite subset. Let η be a positive number such that $\overline{\Delta}(a,\eta) \cap \overline{\Delta}(b,\eta) = \emptyset$ for $a \neq b \in S$. Then we put as follows:

$$\mathcal{S}(S,\eta,C) := \bigl\{ f \in End(V) \,\big|\, |f|_h \leq C,\ \mathcal{S}p(f) \subset \bigcup_{a\in S} \overline{\Delta}(a,\eta) \bigr\}. \tag{2.21}$$

LEMMA 2.36. *The subset $\mathcal{S}(S,\eta,C) \subset End(V)$ is compact.*

Proof Due to the condition $|f|_h \leq C$, the set $\mathcal{S}(S,\eta,C)$ is bounded. We also have the closedness of the defining condition of $\mathcal{S}(S,\eta,C)$. □

For any $f \in \mathcal{S}(S,\eta,C)$, we have the decomposition of V:

$$V = \bigoplus_{a\in S} \mathbb{E}_\eta(f,a), \quad \mathbb{E}_\eta(f,a) := \bigoplus_{\substack{\alpha \in \mathcal{S}p(f),\\ |\alpha - a| < \eta}} \mathbb{E}(f,\alpha).$$

Then we obtain the hermitian metric $h(f)$ given as follows:

$$h(f) := \bigoplus_{a\in S} h_{|\mathbb{E}_\eta(f,a)}.$$

LEMMA 2.37. *The set $\bigl\{h(f) \,\big|\, f \in \mathcal{S}(S,\eta,m,C)\bigr\}$ is compact.*

Proof It immediately follows from the compactness of $\mathcal{S}(S,\eta,C)$ □

2.7.4. A lemma for ϵ-orthogonality. Let V be a finite dimensional vector space over $\boldsymbol{C}$. Let S be a finite set, and $V = \bigoplus_{a\in S} V_a$ be a decomposition. Let h be a hermitian metric of V.

DEFINITION 2.38. Let ϵ be a positive number such that $\epsilon < 1$. The decomposition $V = \bigoplus_{a\in S} V_a$ is called ϵ-orthogonal, if the inequalities $|h(u,v)| \leq \epsilon \cdot |u|_h \cdot |v|_h$ hold for any elements $u, v \in V$. □

Assume the following:

- There exist positive constants C_1 and C_2 such that the following holds for any element $v = \sum_{a\in S} v_a$ of V:
$$C_1 \cdot \sum |v_b|_h \leq |v|_h \leq C_2 \cdot \sum |v_b|_h.$$
- We assume that the decomposition $V = \bigoplus_{a\in S} V_a$ is ϵ-orthogonal with respect to h.

Let g be an element of $\bigoplus_{a\in S} End(V_a)$, and $g^\dagger$ be the adjoint of g with respect to h. We have the decomposition:

$$g^\dagger = \sum (g^\dagger)_{a\,b}, \qquad (g^\dagger)_{a\,b} \in Hom(V_a, V_b).$$

LEMMA 2.39. *There exists a positive constant C, depending only on only C_a $(a = 1,2)$, such that the following holds:*

- *The inequalities* $\left|(g^{\dagger})_{a\,b}\right|_h \leq C \cdot \epsilon \cdot |g|_h$ *hold for any* $g \in \bigoplus_{a\in S} End(V_a)$ *and for any* $a \neq b \in S$.

Proof Let v be an element of V_a. We put $w_b := g^{\dagger}_{a\,b}(v)$. Let u be an element of V_c for $c \neq a$. Then we have the following equality:

$$\big(v, g(u)\big)_h = \big(g^{\dagger}(v), u\big)_h = \sum (w_b, u)_h.$$

Hence we have the following:

$$(w_c, u)_h = \big(v, g(u)\big)_h - \sum_{b\neq c} (w_b, u)_h.$$

Then we obtain the following inequalities, due to the ϵ-orthogonality:

$$(2.22)\quad |(w_c, u)_h| \leq \epsilon \cdot |v|_h \cdot |g|_h \cdot |u|_h + \sum_{b\neq c} \epsilon \cdot |w_b|_h \cdot |u|_h$$
$$\leq \epsilon \cdot |v|_h \cdot |g|_h \cdot |u|_h + \epsilon \cdot C_1^{-1} |g^{\dagger}(v)|_h \cdot |u|_h \leq \epsilon \cdot \big(1 + C_1^{-1}\big)|v|_h \cdot |g|_h \cdot |u|_h.$$

In particular, we consider the case $u = w_c$:

$$|w_c|_h \leq \epsilon \cdot \big(1 + C_1^{-1}\big) \cdot |v|_h \cdot |g|_h.$$

Then we obtain the following inequality for any $v \in V_a$:

$$|g^{\dagger}_{a\,c}(v)|_h \leq \epsilon \cdot \big(1 + C_1^{-1}\big) \cdot |v|_h \cdot |g|_h.$$

It implies the claim. □

2.8. Preliminary from complex differential geometry

2.8.1. Some results of Andreotti-Vesentini. We recall some results of Andreotti-Vesentini in [**2**]. Let (Y, g) be a complete Kahler manifold, not necessarily compact. We denote the natural volume form of Y by dvol. Let $(E, \overline{\partial}_E, h)$ be a hermitian holomorphic bundle over Y. The hermitian metric h and the Kahler metric g induces the fiberwise hermitian metric of $E \otimes \Omega_Y^{p,q}$, which we denote by $(\cdot,\cdot)_{h,g}$. The space of C^{∞} (p,q)-forms with compact support is denoted by $A_c^{p,q}(E)$. For any $\eta_1, \eta_2 \in A_c^{p,q}(E)$, we put as follows:

$$\langle \eta_1, \eta_2 \rangle_h = \int (\eta_1, \eta_2)_{h,g} \cdot \mathrm{dvol}, \quad ||\eta||_h^2 = \langle \eta, \eta \rangle_h.$$

The completion of $A_c^{p,q}(E)$ with respect to the norm $||\cdot||_h$ is denoted by $A_h^{p,q}(E)$.

We have the operator $\overline{\partial}_E : A_c^{p,q}(E) \longrightarrow A_c^{p,q+1}(E)$, and the formal adjoint $\overline{\partial}_E^* : A_c^{p,q}(E) \longrightarrow A_c^{p,q-1}(E)$. We use the notation $\Delta'' = \overline{\partial}_E^*\overline{\partial}_E + \overline{\partial}_E\overline{\partial}_E^*$. We have the maximal closed extensions $\overline{\partial}_E : A_h^{p,q}(E) \longrightarrow A_h^{p,q+1}(E)$ and $\overline{\partial}_E^* : A_h^{p,q}(E) \longrightarrow A_h^{p,q-1}(E)$. We denote the domains of $\overline{\partial}_E$ and $\overline{\partial}_E^*$ by $Dom(\overline{\partial}_E)$ and $Dom(\overline{\partial}_E^*)$ respectively.

PROPOSITION 2.40 (Proposition 5 of [**2**]). *In* $W^{p,q} := Dom(\overline{\partial}_E) \cap Dom(\overline{\partial}_E^*)$, *the space* $A_c^{p,q}(E)$ *is dense with respect to the the graph norm:* $||\eta||_h^2 + ||\overline{\partial}_E\eta||_h^2 + ||\overline{\partial}_E^*\eta||_h^2$. *(See also* [**17**]*).* □

PROPOSITION 2.41 (Theorem 21 of [**2**]). *Assume that there exists a positive number* $c > 0$ *satisfying the following:*

For any $\eta \in W^{p,q}$, *we have* $||\overline{\partial}_E\eta||_h^2 + ||\overline{\partial}_E^*\eta||_h^2 \geq c \cdot ||\eta||_h^2$.

Then for any C^∞-element $\eta \in A_h^{p,q}(E)$ such that $\overline{\partial}_E(\eta) = 0$, we have a C^∞-solution $\rho \in A_h^{p,q-1}(E)$ satisfying the equation $\overline{\partial}_E(\rho) = \eta$. □

2.8.2. Kodaira identity. For a Kahler manifold Y, we have the operator $\Lambda : \Omega_Y^{p,q} \longrightarrow \Omega_Y^{p-1,q-1}$ (see 62 page of [**54**]). For a section f of $End(E) \otimes \Omega_Y^{p_0,q_0}$, we have the natural morphism $A_c^{p,q}(E) \longrightarrow A_c^{p+p_0,q+q_0}(E)$, defined by $\eta \longmapsto f \wedge \eta$. We denote the morphism by $e(f)$.

Let E be a holomorphic vector bundle with a hermitian metric h over Y. We have the metric connection of E induced by the holomorphic structure $\overline{\partial}_E$ and the hermitian metric h. We denote the curvature by $R(h)$.

We have the Levi-Civita connection of the tangent bundle of Y. It induces the connection of $E \otimes \Omega_Y^{0,q}$:

$$\nabla : A_c^{0,0}(E \otimes \Omega_Y^{0,q}) \longrightarrow A_c^{0,1}(E \otimes \Omega_Y^{0,q}) \oplus A_c^{1,0}(E \otimes \Omega_Y^{0,q}).$$

We denote the $(0,1)$-part of ∇ by ∇'' to distinguish with $\overline{\partial}_E : A_c^{0,q}(E) \longrightarrow A_c^{0,q+1}$. The $(1,0)$-part of ∇ is same as ∂ of $E \otimes \Omega_Y^{0,q}$. We denote the curvature of ∇ by $R(\nabla)$.

We have the Laplacian Δ'' for $A_c^{p,q}(E)$ and the equalities of the operators:

$$\Delta'' = \overline{\partial}_E\overline{\partial}_E^* + \overline{\partial}_E^*\overline{\partial}_E = \partial_E\partial_E^* + \partial_E^*\partial_E + \sqrt{-1}\big[e(R(h)), \Lambda\big].$$

In particular, we have the equality $\Delta'' = \partial_E^*\partial_E - \sqrt{-1}\Lambda \circ e(R(h))$ on $A_c^{0,q}(E)$.

On the other hand, we have the following Laplacian for $A_c^{0,0}(E \otimes \Omega_Y^{0,q})$:

$$\Delta'' = \nabla''^*\nabla'' = \partial_E^*\partial_E - \sqrt{-1}\Lambda \circ e(R(\nabla)).$$

Here we have used ∇''^* is 0 on $A_c^{0,0}(E \otimes \Omega_Y^{0,q})$. We also remark the following equality, for an element η of $A_c^{0,0}(E \otimes \Omega_Y^{0,q})$:

$$\Lambda \circ e(R(\nabla))(\eta) = \Lambda(R(\nabla)) \cdot \eta = \Lambda(R(h)) \cdot \eta + \Lambda(R(\Omega_Y^{0,q})) \cdot \eta.$$

Here we have used $R(\nabla) = R(h) + R(\Omega_Y^{0,q})$. Then we obtain the following identity:

$$\begin{aligned}(2.23)\quad \big\langle \overline{\partial}_E\eta,\, \overline{\partial}_E\eta\big\rangle_h + \big\langle \overline{\partial}_E^*\eta, \overline{\partial}_E^*\eta\big\rangle_h &= \big\langle \Delta''\eta, \eta\big\rangle_h \\ &= \big\langle \partial_E\eta,\, \partial_E\eta\big\rangle_h - \sqrt{-1}\cdot\big\langle \Lambda(R(h)\cdot\eta),\, \eta\big\rangle_h \\ = \big\langle \nabla''\eta,\, \nabla''\eta\big\rangle_h + \sqrt{-1}\cdot\big\langle \Lambda(\Omega_Y^{0,q})\cdot\eta,\, \eta\big\rangle_h &- \sqrt{-1}\cdot\big\langle \Lambda\big(R(h)\cdot\eta\big) - \Lambda R(h)\cdot\eta,\, \eta\big\rangle_h.\end{aligned}$$

Let χ be an $\boldsymbol{R}$-valued C^∞-function. If we put $\tilde{h} := h \cdot e^{-\chi}$, we obtain the following equality:

$$\begin{aligned}(2.24)\quad &\big\langle \overline{\partial}_E\eta,\, \overline{\partial}_E\eta\big\rangle_{\tilde{h}} + \big\langle \overline{\partial}_E^*\eta,\, \overline{\partial}_E^*\eta\big\rangle_{\tilde{h}} \\ &= \big\langle \nabla''\eta,\, \nabla''\eta\big\rangle_{\tilde{h}} + \sqrt{-1}\cdot\big\langle \Lambda(\Omega_Y^{0,q})\eta,\, \eta\big\rangle_{\tilde{h}} - \sqrt{-1}\cdot\big\langle \Lambda(R(h)\eta) - \Lambda(R(h))\cdot\eta,\, \eta\big\rangle_{\tilde{h}} \\ &\quad + \sqrt{-1}\big\langle \Lambda(\overline{\partial}\partial\chi\cdot\eta) - \Lambda(\overline{\partial}\partial\chi)\cdot\eta,\, \eta\big\rangle_{\tilde{h}}.\end{aligned}$$

2.8.3. Kodaira identity for sections of $A_c^{0,1}(E)$. We denote the Ricci curvature of the Kahler metric g by $\mathrm{Ric}(g)$. We can naturally regard $\mathrm{Ric}(g)$ as a section of $End(E) \otimes \Omega_Y^{1,1}$, by the natural diagonal inclusion $\boldsymbol{C} \longrightarrow End(E)$.

Let f be a section of $End(E) \otimes \Omega_Y^{1,1}$, and η be an element of $A_c^{0,1}(E)$. Then we put as follows:

(2.25)

$$\langle\langle f, \eta\rangle\rangle_h := -\sqrt{-1}\big(\xi, \eta\big)_h, \quad \xi := \Big(\Lambda \circ e(f) - e\big(\Lambda(f)\big)\Big)(\eta) = \Lambda\big(f \cdot \eta\big) - \Lambda(f) \cdot \eta.$$

We recall the following special case.

PROPOSITION 2.42 (Kodaira [55], Cornalba-Griffiths [17]). *Let η be an element of $A_c^{0,1}(E)$. We have the following equality:*

$$||\overline{\partial}_E(\eta)||_h^2 + ||\overline{\partial}_E^*(\eta)||_h^2 = ||\nabla''\eta||^2 + \int \big\langle\big\langle R(h) + Ric(g), \eta \big\rangle\big\rangle_h \mathrm{dvol}.$$

Proof See Proposition 4.5 in [65]. □

COROLLARY 2.43. *Let η be an element of $Dom(\overline{\partial}_E) \cap Dom(\overline{\partial}_E^*)$ in $A_h^{0,1}(E)$. Then we have the following inequality:*

$$||\overline{\partial}_E(\eta)||_h^2 + ||\overline{\partial}_E^*(\eta)||_h^2 \geq \int \langle\langle R(h) + \mathrm{Ric}(g), \eta\rangle\rangle_h \,\mathrm{dvol}.$$

□

2.8.4. Acceptable bundle. Let $g_{\mathbf{P}}$ be the Poincaré metric of $\Delta^n - \bigcup_{i=1}^l \{z_i = 0\}$:

$$g_{\mathbf{P}} := \sum_{j=1}^{l} q_j^* g_0 + \sum_{j=l+1}^{n} q_j^* g_1.$$

Here we put as follows:

$$g_1 = \frac{2 \cdot dz \cdot d\bar{z}}{(1 - |z|^2)^2}, \qquad g_0 = \frac{2 \cdot dz \cdot d\bar{z}}{|z|^2(-\log|z|^2)^2}.$$

We have the corresponding Kahler form:

$$\omega_{\mathbf{P}} := \sum_{j=1}^{l} q_j^* \omega_0 + \sum_{j=l+1}^{n} q_j^* \omega_1.$$

Here we put as follows:

$$\omega_1 := \frac{\sqrt{-1} dz \wedge d\bar{z}}{(1 - |z|^2)^2}, \qquad \omega_0 := \frac{\sqrt{-1} dz \wedge d\bar{z}}{|z|^2(-\log|z|^2)^2}.$$

Let X be a complex manifold, and D be a normal crossing divisor of X. Let P be a point of X and $(\mathcal{U}, \varphi)$ be an admissible coordinate around P, i.e., $\mathcal{U}$ is a neighbourhood of P, and $\varphi : \mathcal{U} \longrightarrow \boldsymbol{C}^n$ is a holomorphic open embedding, such that $\varphi(D \cap \mathcal{U}) = \varphi(\mathcal{U}) \cap \bigcup_{i=1}^l \{z_i = 0\}$ for some l. By the isomorphism $\varphi : \mathcal{U} \setminus D \simeq \Delta^{*l} \times \Delta^{n-l}$, we take the Poincaré metric $g_{\mathbf{P}}$ on $\mathcal{U} - D$. The metric h of E and the metric $g_{\mathbf{P}}$ on the tangent bundle $T(\mathcal{U} - D)$ induce the metric $(\cdot,\cdot)_{h,g_{\mathbf{P}}}$ of $End(E) \otimes \Omega_Y^{p,q}$ over $\mathcal{U} - D$. Recall the following definition.

DEFINITION 2.44. We say that $(E, \overline{\partial}_E, h)$ is acceptable at P, if the following holds:

- Let $(\mathcal{U}, \varphi)$ be any admissible coordinate around P. The norms of the curvature $R(h)$ with respect to the metric $(\cdot,\cdot)_{h,g_{\mathbf{P}}}$ is bounded over $\mathcal{U} \setminus D$.

When $(E, \overline{\partial}_E, h)$ is acceptable at any point P of X, it is called acceptable. □

2.8.5. Modification of the metric of an acceptable bundle. Cornalba and Griffiths discussed the acceptable bundles in [**17**]. We use their idea in a slightly different way. Let us consider the case $X = \Delta^n$ and $D = \bigcup_{i=1}^{l} D_i$, where we put $D_i = \{z_i = 0\}$. Let $(E, \overline{\partial}_E, h)$ be an acceptable bundle over $X - D$. We would like to apply (2.24) for $\chi = \tau(\boldsymbol{a}, N)$, where the function $\tau(\boldsymbol{a}, N)$ is as follows:

$$(2.26)\quad \tau(\boldsymbol{a},N) := -\sum_{i=1}^{l} a_i\cdot\log|z_i|^2 + N\cdot\Bigl(\sum_{i=1}^{l}\log(-\log|z_i|^2) + \sum_{i=l+1}^{n}\log(1-|z_i|^2)\Bigr).$$

We put as follows:

$$(2.27)\quad h_{\boldsymbol{a},N} := h\cdot e^{-\tau(\boldsymbol{a},N)} = h\cdot\prod_{i=1}^{l}|z_i|^{2a_i}\cdot\bigl(-\log|z_i|^2\bigr)^{-N}\cdot\prod_{i=l+1}^{n}\bigl(1-|z_i|^2\bigr)^{-N}.$$

We use the notation $|\cdot|_{\boldsymbol{a},N}$, $||\cdot||_{\boldsymbol{a},N}$, $(\cdot,\cdot)_{\boldsymbol{a},N}$ and $\langle\langle\cdot,\cdot\rangle\rangle_{\boldsymbol{a},N}$ instead of $|\cdot|_{h_{\boldsymbol{a},N}}$, $||\cdot||_{h_{\boldsymbol{a},N}}$, $(\cdot,\cdot)_{h_{\boldsymbol{a},N}}$ and $\langle\langle\cdot,\cdot\rangle\rangle_{h_{\boldsymbol{a},N}}$ for simplicity. We also use the notation $A^{p,q}_{\boldsymbol{a},N}(E)$ instead of $A^{p,q}_{h_{\boldsymbol{a},N}}(E)$.

LEMMA 2.45. *We have the following equality for any section* $\eta\in A^{0,q}_c(E)$*:*

$$\sqrt{-1}\Bigl(\bigl(\Lambda\overline{\partial}\partial\tau(\boldsymbol{a},N)\eta\bigr) - \Lambda\bigl(\overline{\partial}\partial\tau(\boldsymbol{a},N)\cdot\eta\bigr)\Bigr) = -N\cdot q\cdot\eta.$$

Proof We have $\overline{\partial}\partial\tau(\boldsymbol{a},N) = -N\cdot\sqrt{-1}\cdot\omega_{\mathbf{p}}$. Then we obtain the following:

(2.28)

$$\text{L.H.S.} = \sqrt{-1}\cdot(-N\sqrt{-1})\cdot\Bigl(\Lambda(\omega_{\mathbf{p}}\eta) - (\Lambda\omega_{\mathbf{p}})\cdot\eta\Bigr) = N\cdot\bigl((n-q)\cdot\eta - n\cdot\eta\bigr)$$
$$= -q\cdot N\cdot\eta = \text{R.H.S.}.$$

Hence we are done. □

REMARK 2.46. Let f be a holomorphic section of E on $X-D$, the boundedness $\sup|f|_{\boldsymbol{a},N} < \infty$ implies $-\operatorname{ord}(f)\le\boldsymbol{a}$ with respect to the metric h in the sense of the subsection 2.2.1. □

2.8.6. The case where N is sufficiently negative. Let $(E,\overline{\partial}_E,h)$ be an acceptable hermitian holomorphic bundle over $X-D$. We obtain the following inequality from the formula (2.24) and Lemma 2.45:

$$(2.29)\quad \bigl\langle\overline{\partial}_E\eta,\,\overline{\partial}_E\eta\bigr\rangle_{\boldsymbol{a},N} + \bigl\langle\overline{\partial}^*_E\eta,\,\overline{\partial}^*_E\eta\bigr\rangle_{\boldsymbol{a},N}$$
$$\ge\sqrt{-1}\cdot\bigl\langle\Lambda(\Omega^{0,q}_{X-D})\cdot\eta,\,\eta\bigr\rangle_{\boldsymbol{a},N} - \sqrt{-1}\cdot\Bigl\langle\Lambda\bigl(R(h)\cdot\eta\bigr) - \Lambda R(h)\cdot\eta,\,\eta\Bigr\rangle_{\boldsymbol{a},N} - N\cdot q\cdot||\eta||^2_{\boldsymbol{a},N}.$$

LEMMA 2.47. *When $(E,\overline{\partial}_E,h)$ is a general acceptable bundle, there exists a positive constant $C>0$ satisfying the following for any $q=1,\dots,n$ and for any $\eta\in A^{0,q}_c(E)$:*

$$\Bigl|\Bigl\langle\Lambda(\Omega^{0,q}_{X-D})\cdot\eta - \Lambda\bigl(R(h)\cdot\eta\bigr) + \Lambda\bigl(R(h)\bigr)\cdot\eta,\,\eta\Bigr\rangle_h\Bigr| \le C\cdot|\eta|^2_h.$$

Proof Recall that $R(h)$ is dominated by $\omega_{\mathbf{p}}$. Then the claim follows from a direct calculation of the curvature of the Poincaré metric $g_{\mathbf{p}}$ on $X-D$. □

If we take a sufficiently negative integer N such that $N < -C-1$ for the constant C as in Lemma 2.47, then we obtain the following inequalities for any $q \geq 1$ and for any $\eta \in A_c^{0,q}(E)$, due to the inequality (2.29):

$$\big\langle \overline{\partial}_E\eta,\ \overline{\partial}_E\eta\big\rangle_{\boldsymbol{a},N} + \big\langle \partial_E\eta,\ \partial_E\eta\big\rangle_{\boldsymbol{a},N} \geq ||\eta||^2_{\boldsymbol{a},N}. \tag{2.30}$$

LEMMA 2.48. *Let C be a positive constant as in Lemma 2.47. If $N < -C-1$, we have the vanishings of any higher cohomology group $H^i\big(A^{0,\cdot}_{\boldsymbol{a},N}(\mathcal{E}^\lambda), \overline{\partial}_{\mathcal{E}^\lambda}\big)$ $(i>0)$.*

Proof It follows from Proposition 2.40 and Proposition 2.41 and (2.30). □

2.8.7. The case where N is sufficiently positive. Let $\pi_i : X - D \longrightarrow D_i$ denote the natural projection for $i = 1, \ldots, n$. We put $D_i^\circ := D_i \setminus \bigcup_{j\neq i, j\leq l} D_j \cap D_i$. Let P be a point of D_i°, then we obtain the curve $\pi_i^{-1}(P)$ which is isomorphic to Δ^* $(i \leq l)$ or Δ $(l < i \leq n)$. Let $\Delta''_{\mathbf{p}}$ denote the Laplacian on $\pi_i^{-1}(P)$ with respect to the Poincaré metric $g_{\mathbf{p}\,|\,\pi_i^{-1}(P)}$. We have the restriction of the metric $h_{\boldsymbol{a},N}$ to $E_{|\pi_i^{-1}(P)}$, which is also denoted by $h_{\boldsymbol{a},N}$ for simplicity of the notation. Similarly, the restriction of $\partial_{\boldsymbol{a},N}$ and $|\cdot|_{\boldsymbol{a},N}$ to $\pi_i^{-1}(P)$ are denoted by the same notation.

LEMMA 2.49. *There exists a positive constant N_0 such that the following holds:*

- *Let P be a point of D_i°, and U be an open subset of the curve $\pi_i^{-1}(P)$. Let F be a holomorphic section of $E_{|U}$. Then the inequalities $\Delta''_{\mathbf{p}}|F|^2_{\boldsymbol{a},N} \leq 0$ and $\Delta''_{\mathbf{p}} \log|F|_{\boldsymbol{a},N} \leq 0$ hold on U for any $N \geq N_0$.*

Proof We have the following:

$$\begin{aligned}\Delta''_{\mathbf{p}}|F|^2_{\boldsymbol{a},N} &= -\big|\partial_{\boldsymbol{a},N}F\big|^2_{\boldsymbol{a},N} + \big(F, \sqrt{-1}\Lambda_{\pi_i^{-1}(P)}R(h_{\boldsymbol{a},N})F\big)_{\boldsymbol{a},N} \\ &= -\big|\partial_{\boldsymbol{a},N}F\big|^2_{\boldsymbol{a},N} + \big(F, \sqrt{-1}\Lambda_{\pi_i^{-1}(P)}R(h)F\big)_{\boldsymbol{a},N} - N\cdot\big|F\big|^2_{\boldsymbol{a},N} \\ &\leq \big(|R(h)|_{h,g_{\mathbf{p}}} - N\big)\cdot\big|F\big|^2_{\boldsymbol{a},N}.\end{aligned} \tag{2.31}$$

We also have the following:

$$\Delta''_{\mathbf{p}}\log|F|^2_{\boldsymbol{a},N} \leq \frac{\Delta''_{\mathbf{p}}|F|^2_{\boldsymbol{a},N}}{|F|^2_{\boldsymbol{a},N}} + \frac{\big|\big(\partial_{\boldsymbol{a},N}F, F\big)_{\boldsymbol{a},N}\big|^2}{|F|^4_{\boldsymbol{a},N}} \leq -N + \big|R(h)\big|_{h,g_{\mathbf{p}}}. \tag{2.32}$$

Therefore we have the constant N_0, depending only on the norm of $R(h)$ with respect to $g_{\mathbf{p}}$ and h, such that the desired inequalities hold. □

COROLLARY 2.50. *Let Δ'' denote the Laplacian of $\pi_i^{-1}(P)$ with respect to the Euclidean metric $dz\cdot d\bar{z}$. Let N_0, P, U and F be as in Lemma 2.49. Then the inequalities $\Delta''|F|^2_{\boldsymbol{a},N} \leq 0$ and $\Delta''\log|F|_{\boldsymbol{a},N} \leq 0$ hold for any $N \geq N_0$. In particular, $|F|^2_{\boldsymbol{a},N}$ and $\log|F|_{\boldsymbol{a},N} \leq 0$ are subharmonic on U.*

Proof Since Δ'' is the product of a positive function and $\Delta''_{\mathbf{p}}$. Hence it follows from Lemma 2.49. □

The following corollary makes us to derive the L^2-property on curves from the L^2-property on $X-D$.

COROLLARY 2.51. *Let F be a holomorphic function on $X-D$ such that $F \in A^{0,0}_{\boldsymbol{a},N}(E)$. Let N_0 be as in Lemma 2.49, and let M be any real number larger than*

N_0. For any $1 \leq j \leq l$ and $P \in D_j^\circ$, we have the following finiteness:

$$\int_{\pi_j^{-1}(P)} \left|F_{|\pi_j^{-1}(P)}\right|_{\boldsymbol{a},M}^2 \cdot \mathrm{dvol}_{\pi_j^{-1}(P)} < \infty.$$

Here $\mathrm{dvol}_{\pi_j^{-1}(P)}$ denotes the volume form of $\pi_j^{-1}(P)$ with respect to the restriction $g_{\mathbf{P}\,|\,\pi_j^{-1}(P)}$.

Proof We begin with a remark on the pluri-subharmonicity. Due to Corollary 2.50, the function $\left|F_{|\pi_i^{-1}(Q)}\right|_{\boldsymbol{a},M}^2$ is subharmonic for any $1 \leq i \leq n$ and $Q \in D_i^\circ$. Let $q_j : X - D \longrightarrow \Delta^*$ be the natural projection onto the j-th component, and let Q' be any point of Δ^*. It is easy to see that $\left|F_{|q_j^{-1}(Q')}\right|_{\boldsymbol{a},M}^2$ is subharmonic on the hyperplane.

By using the above remark, let us show the claim of the corollary. Let U be an appropriate small neighbourhood of P in D_j°. We have the natural isomorphism $\pi_j^{-1}(U) \simeq \pi_j^{-1}(P) \times U$. Let dvol_U denote the volume form of U with respect to the Euclidean metric. Let dvol_{X-D} denote the volume form of $X - D$ with respect to $g_{\mathbf{P}}$. There exists a positive constant C_1 such that $\mathrm{dvol}_U \cdot \mathrm{dvol}_{\pi_j^{-1}(P)} \leq C_1 \cdot \mathrm{dvol}_{X-D}$ on $\pi_j^{-1}(P) \times U = \pi_j^{-1}(U)$.

Let $Q' \in \pi_j^{-1}(P)$, and then the restriction $\left|F_{|\{Q'\}\times U}\right|_{\boldsymbol{a},M}^2$ is subharmonic, due to the above remark. Hence we have the following inequality:

$$\left|F\right|_{\boldsymbol{a},M}^2(Q,P) \leq \int_{\{Q\}\times U} \left|F_{|\{Q\}\times U}\right|_{\boldsymbol{a},M}^2 \cdot \mathrm{dvol}_U\,.$$

Therefore, we obtain the following:

$$(2.33)\quad \int_{\pi_j^{-1}(P)} \left|F_{|\pi_j^{-1}(P)}\right|_{\boldsymbol{a},M}^2 \cdot \mathrm{dvol}_{\pi_j^{-1}(P)} \leq \int_{U\times\pi_j^{-1}(P)} \left|F\right|_{\boldsymbol{a},M}^2 \cdot \mathrm{dvol}_U \cdot \mathrm{dvol}_{\pi_j^{-1}(P)}$$
$$\leq C_1 \cdot \int_{U\times\pi_j^{-1}(P)} \left|F\right|_{\boldsymbol{a},M}^2 \cdot \mathrm{dvol}_{X-D} < \infty.$$

Thus we are done. □

Next we would like to see that the growth estimates on curves implies the growth estimate on $X - D$. We are interested in the behaviour of holomorphic sections around the origin O. Hence we put $X(C) = \big\{(z_1, \ldots, z_n) \in X \,\big|\, |z_i| \leq C\big\}$, $D_i(C) = D_i \cap X(C)$ and $D(C) = D \cap X(C)$ for some $0 < C < 1$, and we consider the restrictions of holomorphic sections to $X(C) - D(C)$. As a preliminary, we prove the following lemma.

LEMMA 2.52. *Let P be a point of $D_j^\circ(C) = D_j(C) \setminus \bigcup_{i\neq j, i\leq l} D_i(C)$ for some $1 \leq j \leq l$. Let F be holomorphic section of $E_{|\pi_j^{-1}(P)\cap X(C)}$. Assume that there exist real numbers $C_1 > 0$, $C_2 > 0$, a and k such that the following holds on $\pi_j^{-1}(P) \cap X(C)$:*

$$(2.34)\qquad 0 < C_1 \leq \left|F\right|_h \cdot |z_j|^a \cdot \big(-\log|z_j|\big)^k \leq C_2.$$

Let N_0 be as in Lemma 2.49, and let M be a real number larger than N_0. Then there exists a positive number C_3, depending only on the maximum value of $|F|_h$

on the compact subset $\{|z_j| = C\}$ *of* $\pi_j^{-1}(P) \cap X(C)$*, such that the following holds:*

$$|F|_h \le C_3 \cdot \left|\frac{z_j}{C}\right|^{-a} \cdot \left(\frac{-\log|z_j|}{-\log C}\right)^M. \tag{2.35}$$

Proof We put $\mathbf{0} = (0, \ldots, 0)$, and apply Corollary 2.50. Then $\log\big|F\big|_{\mathbf{0},N}$ is subharmonic function on the punctured disc $\pi_j^{-1}(P)$. We put $G(z_j) := \log|F|_{\mathbf{0},N}(z_j) + a \cdot \log|z_j|$, which is subharmonic on $\pi_j^{-1}(P)$. Due to the assumption (2.34), we have $\big(-\log|z_j|\big)^{-1} \cdot G(z_j) \longrightarrow 0$ as $|z_j| \longrightarrow 0$. Hence we can apply Corollary 2.25, and therefore the maximum principle holds for $G(z_j)$. We obtain the following inequality:

$$\big|F(z_j)\big|_h \cdot \big(-\log|z_j|\big)^{-M} \cdot |z_j|^a \le \left(\max_{|z_j|=C} \big|F(z_j)\big|_h\right) \cdot \big(-\log C\big)^{-M} \cdot C^a.$$

Then we obtain the desired inequality (2.35). □

COROLLARY 2.53. *Let* F *be a holomorphic section of* E *over* $X - D$*. Assume the following:*

- *There exists the real numbers* a_j $(j = 1, \ldots, l)$*.*
- *For any* $1 \le j \le l$ *and* $P \in D_j(C) \setminus \bigcup_{i \ne j, i \le l} D_i(C)$*, there exist real numbers* $C_1(P) > 0$*,* $C_2(P) > 0$*,* $a(P) \le a_j$ *and* $k(P)$ *such that the following inequality holds:*

$$0 < C_1(P) \le \big|F_{|\pi_j^{-1}(P)} \cap X(C)\big|_h \cdot \big|z_j\big|^{a(P)} \cdot \big(-\log|z_k|\big)^{k(P)} \le C_2(P).$$

Let N_0 *be as in Lemma 2.49, and let* M *be any real number larger than* N_0*. Then there exists a positive number* C_3 *satisfying the following on* $X(C) - D(C)$*:*

$$\big|F\big|_h \le C_3 \cdot \prod_{j=1}^{l} \left|\frac{z_j}{C}\right|^{-a_j} \cdot \left(\frac{-\log|z_j|}{-\log C}\right)^N.$$

Here C_3 *depends only on the maximum value of* $|F|_h$ *on the following compact subset of* $X(C) - D(C)$*:*

$$\big\{(z_1, \ldots, z_n) \in X(C) - D(C) \,\big|\, |z_i| = C, \;\; (i = 1, \ldots, l)\big\}. \tag{2.36}$$

Proof We put $Y_j(C) := \big\{(z_1, \ldots, z_n) \in X \,\big|\, |z_i| = C, \; 1 \le i \le j\big\}$ for $1 \le j \le l$. Note that $Y_l(C)$ is the compact subset given in (2.36). Since $Y_l(C)$ is compact, $|F|_{\mathbf{0},M}$ is bounded on $Y_l(C)$. Let C_4 denote the maximum value.

We will prove that the following inequality holds on $Y_j(C) \setminus D(C)$, by using a descending induction on j:

$$\big|F\big|_h \le C_4 \cdot \prod_{i=j+1}^{l} \left|\frac{z_i}{C}\right|^{-a_i} \cdot \left(\frac{-\log|z_i|}{-\log C}\right)^M. \tag{2.37}$$

We have already known that (2.37) for $j = l$ holds. We have only to show (2.37) for $j - 1$ holds, assuming that (2.37) for j holds.

Let P be any point of $D_j \cap Y_{j-1}$. Due to Lemma 2.52, we have the following:

$$(2.38)\quad \left|F_{|\pi_j^{-1}(P)}(z_j)\right|_h \leq \left(\max_{|z_j|=C} \left|F_{\pi_j^{-1}(P)}(z_j)\right|_h\right) \cdot \left|\frac{z_j}{C}\right|^{-a(P)} \cdot \left(\frac{-\log|z_j|}{-\log C}\right)^M$$
$$\leq C_4 \cdot \prod_{i=j+1}^{l} \left|\frac{z_i}{C}\right|^{-a_i} \cdot \left(\frac{-\log|z_i|}{-\log C}\right)^M \cdot \left|\frac{z_j}{C}\right|^{-a_j} \cdot \left(\frac{-\log|z_j|}{-\log C}\right)^M.$$

Thus we are done. □

REMARK 2.54. Let F be a holomorphic section of E over $X - D$, which is L^2 with respect to h. Due to Corollary 2.51, we obtain the L^2-property of the restrictions of F to curves $\pi_i^{-1}(P)$. If we know that norm estimate as (2.34) holds for such L^2-holomorphic sections on any $\pi_i^{-1}(P)$, then we obtain the growth estimate of F due to Corollary 2.53. In other words, if we know the relation between L^2-property and growth estimate for the restrictions to curves, then we know that for the whole $X - D$. □

REMARK 2.55. Corollary 2.53 is the correction of Corollary 4.12 in [**65**]. The author thanks the referee for pointing out the confusion of the signature. □

2.8.8. The metrics and the curvatures of $\mathcal{O}(i)$ on $\mathbb{P}^1$. In the next few subsections, we recall the subsection 4.7.3 in [**65**], which is a preparation for the argument in the section 8.4. Let $\mathbb{P}^1$ denote the one dimensional projective space over $\boldsymbol{C}$. We use the homogeneous coordinate $[t_0 : t_1]$. The points $[0 : 1]$ and $[1 : 0]$ are denoted by 0 and ∞ respectively. We use the coordinates $t = t_0/t_1$ and $s = t_1/t_0$. We have the line bundle $\mathcal{O}(i)$ over $\mathbb{P}^1$. The coordinates of $\mathcal{O}(i)$ is given as follows: (t, ζ_1) over $\mathbb{P}^1 - \{\infty\}$, and (s, ζ_2) over $\mathbb{P}^1 - \{0\}$. The relations are given by $s = t^{-1}$ and $t^{-i} \cdot \zeta_1 = \zeta_2$.

Recall that we have the smooth metric h_i of $\mathcal{O}(i)$. Let $\xi = (t, \zeta_1) = (s, \zeta_2)$ be an element of $\mathcal{O}(i)$.

$$h_i(\xi, \xi) := |\zeta_1|^2 \cdot \left(1 + |t|^2\right)^{-i} = |\zeta_2|^2 \cdot \left(1 + |s|^2\right)^{-i}.$$

For any real numbers a and b, we have the possibly singular metrics $h_{i,(a,b)}$ of $\mathcal{O}(i)$: Let $\xi = (t, \zeta_1) = (s, \zeta_2)$ be an element of $\mathcal{O}(i)$.

$$h_{i,(a,b)}(\xi, \xi) := h_i(\xi, \xi) \cdot \left(1+|t|^{-2}\right)^{-a} \cdot \left(1+|t|^2\right)^{-b} = h_i(\xi, \xi) \cdot \left(1+|s|^2\right)^{-a} \cdot \left(1+|s|^{-2}\right)^{-b}.$$

Around $|t| = 0$, the order of $h_{i,(a,b)}$ is equivalent to $|t|^{2a}$. Around $|s| = 0$, the order of $h_{i,(a,b)}$ is equivalent to $|s|^{2b}$. The curvature $R(h_{i,(a,b)})$ is as follows:

$$(2.39)\qquad R(h_{i,(a,b)}) = (a + b + i) \cdot \frac{dt \cdot d\bar{t}}{\left(1 + |t|^2\right)^2}.$$

Take a point $P \in \mathbb{P}^1$. Then we obtain a morphism $\mathcal{O}(i) \longrightarrow \mathcal{O}(i+1)$ of coherent sheaves.

LEMMA 2.56. *The morphism is bounded with respect to the metrics $h_{i,(a,b)}$ and $h_{i+1,(a,b)}$.* □

2.8.9. Some open subset of the line bundle $\mathcal{O}(-1)$ with the complete Kahler metric. We are mainly interested in the case $i = -1$. We regard $\mathcal{O}(-1)$ as a complex manifold. The open submanifold Y is defined to be $\{\xi \in \mathcal{O}(-1) \,|\, h_{-1,(0,0)}(\xi,\xi) < 1\}$. We denote the naturally defined projection of Y onto $\mathbb{P}^1$ by π. We denote the image of the 0-section $\mathbb{P}^1 \longrightarrow Y$ by $\mathbb{P}^1$. Then we have the normal crossing divisor $D' = \mathbb{P}^1 \cup \pi^{-1}(0) \cup \pi^{-1}(\infty)$ of Y. The manifold $Y - D'$ is same as $\{(t,x) \in \boldsymbol{C}^{*\,2} \,|\, |x|^2(1+|t|^2) < 1\}$.

We have the complete Kahler metric $g := g_1 + g_2 + g_3$ of $Y - D'$ given as follows: As a contribution of the 0-section $\mathbb{P}^1$, we put $\tau_1 = -\log\Big[(1+|t|^2)\cdot|x|^2\Big]$, and as follows:

$$g_1 := \frac{1}{\tau_1^2}\left(\frac{\bar{t}\cdot dt}{1+|t|^2} + \frac{dx}{x}\right)\cdot\left(\frac{t\cdot d\bar{t}}{1+|t|^2} + \frac{d\bar{x}}{\bar{x}}\right) + \frac{1}{\tau_1}\frac{dt\cdot d\bar{t}}{\big(1+|t|^2\big)^2}.$$

LEMMA 2.57. *Note that g_1 gives the complete Kahler metric of $Y - \mathbb{P}^1$. It is equivalent to the Poincarè metric around the divisor $\mathbb{P}^1$.* □

As a contribution of $\pi^{-1}(\infty)$, we put $\tau_2 = \log(1+|t|^2)$, and as follows:

$$g_2 = \frac{1}{\tau_2}\Big(-1 + \frac{|t|^2}{\tau_2}\Big)\cdot\frac{dt\cdot d\bar{t}}{\big(1+|t|^2\big)^2}.$$

LEMMA 2.58. *We have $-1 + |t|^2\cdot\tau_2^{-1} > 0$. Around $|t| = \infty$, or equivalently, around $|s| = 0$, the behaviour of g_2 is similar to $(-|s|\log|s|)^{-2}ds\cdot d\bar{s}$. Around $|t| = 0$, we have $g_2 = \big(2^{-1} + o(|t|^2)\big)\cdot dt\cdot d\bar{t}$.*

Proof It can be checked by a direct calculation. □

As the contribution of the divisor $\pi^{-1}(0)$, we put $\tau_3 := \log(1+|t|^2) - \log|t|^2 = \log(1+|s|^2)$, where we use $s = t^{-1}$. And we put as follows:

$$g_3 = \frac{1}{\tau_3}\cdot\Big(-1 + \frac{|s|^2}{\tau_3}\Big)\frac{ds\cdot d\bar{s}}{\big(1+|s|^2\big)^2}.$$

By the symmetry, the behaviour of g_3 is similar to g_2. (See Lemma 2.58)

The following lemma can be checked directly.

LEMMA 2.59. *The metric g gives the complete Kahler metric of the complex manifold $Y - D'$. Around the divisors $\mathbb{P}^1$, $\pi^{-1}(0)$ and $\pi^{-1}(\infty)$, the behaviours of the metric g are equivalent to the Poincaré metric.* □

We note the following formulas:

$$\overline{\partial}\partial\log\tau_1 = \frac{1}{\tau_1^2}\Big(\frac{\bar{t}\cdot dt}{1+|t|^2} + \frac{dx}{x}\Big)\wedge\Big(\frac{t\cdot d\bar{t}}{1+|t|^2} + \frac{d\bar{x}}{\bar{x}}\Big) + \frac{1}{\tau_1}\frac{dt\wedge d\bar{t}}{\big(1+|t|^2\big)^2} =: \omega_1,$$

$$\overline{\partial}\partial\log\tau_2 = \frac{1}{\tau_2}\Big(-1 + \frac{|t|^2}{\tau_2}\Big)\cdot\frac{dt\wedge d\bar{t}}{\big(1+|t|^2\big)^2} =: \omega_2,$$

$$\overline{\partial}\partial\log\tau_3 = \frac{1}{\tau_3}\Big(-1 + \frac{|s|^2}{\tau_3}\Big)\cdot\frac{ds\wedge d\bar{s}}{\big(1+|s|^2\big)^2} =: \omega_3.$$

We put $\omega = \omega_1 + \omega_2 + \omega_3$. We put as follows:

$$H_0 = \frac{1}{\tau_1} + \frac{1}{\tau_2}\Big(-1 + \frac{|t|^2}{\tau_2}\Big) + \frac{1}{\tau_3}\Big(-1 + \frac{|s|^2}{\tau_3}\Big) > 0.$$

Then we have the following:

$$\omega^2 = \det(g) \cdot dt \wedge d\bar{t} \wedge dx \wedge d\bar{x} = \Big(\frac{1}{\tau_1^2 \cdot |x|^2 \cdot \big(1+|t|^2\big)^2} \times H_0 \Big) \cdot dt \wedge d\bar{t} \wedge dx \wedge d\bar{x}.$$

We put as follows:

$$H_1 := \frac{H_0}{\big(1+|t|^2\big) \cdot \big(1+|s|^2\big)}.$$

Recall that we have $\mathrm{Ric}(g) = \overline{\partial}\partial(\det(g))$.

LEMMA 2.60.

- *Let C be a number such that $0 < C < 1$. On the domain $\big\{(t,x) \in \boldsymbol{C}^{*\,2} \,\big|\, |x|^2 \cdot \big(1+|t|^2\big) \leq C\big\}$, we have the following similarity of the behaviour:*

$$H_1 \sim (\log|t|)^{-2}, \qquad (|t| \to \infty, \text{or}, |t| \to 0),$$

$$H_1 \sim (-\log|x|)^{-1}, \quad (|x| \to 0).$$

- *We have the equality: $Ric(g) - \overline{\partial}\partial \log(H_1) = -\overline{\partial}\partial \log \tau_1^2$.* □

2.8.10. The inequality and the vanishing for the acceptable bundle. We put $X = \Delta^n$, $D_i = \{z_i = 0\}$ and $D = \bigcup_{i=1}^{l} D_i$. We also put $\Delta_z^2 = \{(z_1,z_2) \,|\, |z_i| < 1\}$ and $D_i' = \{z_i = 0\} \subset \Delta_z^2$ $(i = 1,2)$. Let $\varphi : \widetilde{\Delta_z^2} \longrightarrow \Delta_z^2$ denote the blow up of Δ_z^2 at the origin $O = (0,0)$. We have the exceptional divisor $\varphi^{-1}(O)$ and the proper transforms $\widetilde{D'}_i$ of D_i' $(i = 1,2)$.

We put $\widetilde{X} = \widetilde{\Delta_z^2} \times \Delta_w^{n-2}$. Then we have the composite ψ of the natural morphisms:

$$\widetilde{X} \xrightarrow{\varphi \times id} \Delta_z^2 \times \Delta_w^{n-2} \longrightarrow \Delta_z^n.$$

Here the latter morphism is the natural isomorphism given by $w_i = z_{i+2}$. We put $\widetilde{D} := \psi^{-1}(D)$, which is same as the following:

$$\Big[\big(\varphi^{-1}(0,0) \cup \widetilde{D_1'} \cup \widetilde{D_2'}\big) \times \Delta_w^{n-2}\Big] \cup \Bigg[\widetilde{\Delta_z^2} \times \Big(\bigcup_{i=1}^{l-2} \{w_i = 0\}\Big)\Bigg].$$

The restriction of ψ to $\widetilde{X} - \widetilde{D}$ gives an isomorphism $\widetilde{X} - \widetilde{D} \simeq X - D$.

We can take a holomorphic embedding ι of Y, given in the subsection 2.8.9, to $\widetilde{\Delta}^2$ satisfying the following:

- The image of the 0-section $\mathbb{P}^1$ is the exceptional divisor $\varphi^{-1}(O)$.
- We have $\iota^{-1}(D_1') = \pi^{-1}(\infty)$ and $\iota^{-1}(D_2') = \pi^{-1}(0)$.

We put $\overline{X} := Y \times \Delta_w^{n-2}$. Then we have the naturally induced morphism $\overline{X} \longrightarrow \widetilde{\Delta_z^2} \times \Delta_w^{n-2}$, which we also denote by ι. We have the point $[1:1] \in \mathbb{P}^1$, and we put as follows:

$$\overline{D} := \iota^{-1}(\widetilde{D}), \quad \overline{X}^{(1)} := \pi^{-1}([1:1]) \times \Delta_w^{n-2}, \quad \overline{D}^{(1)} := \overline{X}^{(1)} \cap \overline{D}.$$

The composite $\psi \circ \iota$ is denoted by ψ_1. The metric $g_{\overline{X}-\overline{D}}$ is induced from the metric g of $Y - D'$ (see the subsection 2.8.9) and the Poincaré metric of $\Delta_w^{n-2} \setminus \bigcup_{i=1}^{l-2} \{w_i = 0\}$.

Let $(E, \overline{\partial}_E)$ be a holomorphic bundle with a hermitian metric h over $X - D$. We assume that $(E, \overline{\partial}_E, h)$ is acceptable in the sense of Definition 2.44. We denote the curvature of $\psi_1^{-1}(E, \overline{\partial}_E, h)$ by $\psi_1^{-1}R(h)$.

Let ϵ_i $(i = 1, 2)$ be positive numbers such that $\epsilon_1 + \epsilon_2 < 1$.

LEMMA 2.61. *We can pick positive numbers ϵ, a and b satisfying the following:*

$$a + b = 1, \quad 0 < a + 2\epsilon < 1 - \epsilon_1, \quad 0 < b + 2\epsilon < 1 - \epsilon_2.$$

Proof It can be checked elementarily. □

Let ϵ, a and b be as in Lemma 2.61. We put $\boldsymbol{\delta} = (1, \ldots, 1) \in \boldsymbol{R}^l$, and we put $h_{\epsilon\cdot\boldsymbol{\delta},N} := h \cdot e^{-\tau(\epsilon\cdot\boldsymbol{\delta},N)}$ (see the formula (2.27)). The metric $\tilde{h}$ of $\psi_1^{-1}(E)(-\overline{X}^{(1)}) := \psi_1^{-1}(E) \otimes \pi^{-1}\mathcal{O}_{\mathbb{P}}(-1)$ over the complex manifold $\overline{X} - \overline{D}$ is given as follows:

$$\tilde{h}_{N,\epsilon,a,b} := \psi_1^{-1}h_{\epsilon\cdot\boldsymbol{\delta},N} \cdot h_{-1,a,b} \cdot H_1^{-1} \cdot \tau_1^{2+\epsilon}\bigl(\tau_2 \cdot \tau_3\bigr)^{\epsilon}. \tag{2.40}$$

For simplicity, we use the notation $\tilde{h}$ instead of $\tilde{h}_{N,\epsilon,a,b}$.

LEMMA 2.62. *Let ϵ, a and b be as in Lemma 2.61. When N is sufficiently negative such that* (2.30) *holds, then the following inequality holds for any $\eta \in A_c^{0,1}(\psi_1^{-1}E(-\overline{X}^{(1)}))$:*

$$\langle\langle R(\tilde{h}) + \mathrm{Ric}(g), \eta\rangle\rangle_{\tilde{h}} \geq \epsilon||\eta||_{\tilde{h}}^2.$$

(See the formula (2.25) *for the definition of $\langle\langle\cdot,\cdot\rangle\rangle_{\tilde{h}}$.)*

Proof Recall that $\overline{X} - \overline{D}$ is isomorphic to the product of $Y - D'$ and $\Delta_w^{n-2} - \bigcup_{i=1}^{l-2}\{w_i = 0\}$, which is compatible with the metrics. We can apply the inequality (2.30) to the $\Bigl(\Delta_w^{n-2} - \bigcup_{i=1}^{l-2}\{w_i = 0\}\Bigr)$-direction. Hence we have only to consider $(Y - D')$-direction, i.e., we may and will assume $n = 2$. We have the following equality:

$$\begin{aligned} R(\tilde{h}) + \mathrm{Ric}(g) &= R(\psi_1^{-1}h_{\epsilon\cdot\boldsymbol{\delta},N}) + R(h_{-1,a,b}) - \overline{\partial}\partial\log H_1 \\ &\quad + (2+\epsilon)\cdot\overline{\partial}\partial\log\tau_1 + \epsilon\cdot\overline{\partial}\partial(\log\tau_2 + \log\tau_3) + \mathrm{Ric}(g) \\ &= R(\psi_1^{-1}h_{\epsilon\cdot\boldsymbol{\delta},N}) + \epsilon\cdot(\omega_1 + \omega_2 + \omega_3). \end{aligned} \tag{2.41}$$

Here we have used $R(h_{-1,a,b}) = 0$ due to (2.39) and our choice of a and b. By taking sufficiently negative N, we can assume the following inequality for any $\eta \in A_c^{0,1}(E)$ on $X - D$ (the subsection 2.8.6):

$$\langle\langle R(h_{\epsilon\cdot\boldsymbol{\delta},N}), \eta\rangle\rangle_{\epsilon\cdot\boldsymbol{\delta},N} \geq 0. \tag{2.42}$$

Then, by a fiberwise linear algebraic argument, it is easy to see that the following inequality holds for any $\eta \in A_c^{0,1}(\psi_1^{-1}(E))$:

$$\langle\langle\psi^{-1}R(h_{\epsilon\cdot\boldsymbol{\delta},N}), \eta\rangle\rangle_{\tilde{h}} \geq 0.$$

On the other hand, we obtain $\bigl\langle\bigl\langle\omega_1 + \omega_2 + \omega_3, \eta\bigr\rangle\bigr\rangle_{\tilde{h}} \geq \epsilon\cdot||\eta||_{\tilde{h}}$, which can be checked directly from definition. Thus we are done. □

COROLLARY 2.63. *Let ϵ, a and b be as in Lemma 2.61. If N is sufficiently negative such that* (2.30) *holds, then the first cohomology $H^1\bigl(A_{\tilde{h}}^{0,\cdot}(\psi_1^{-1}E(-\overline{X}^{(1)}))\bigr)$ vanishes.*

Proof It immediately follows from Lemma 2.62, Proposition 2.40 and Proposition 2.41. □

We remark the following.

LEMMA 2.64. *The contribution of* $h_{-1,a,b} \cdot H_1^{-1} \cdot \tau_1^{2+\epsilon} \cdot \big(\tau_2 \cdot \tau_3\big)^{\epsilon}$ *to the metric* $\tilde{h}$ *is equivalent to the following, around* $|t| = 0$*:*

$$|t|^{2a} \cdot \big(-\log|t|\big)^2 \cdot |t|^{2\epsilon}\big(-\log|t|\big)^{2\epsilon} = |t|^{2(a+\epsilon)} \cdot \big(-\log|t|\big)^{2+2\epsilon}.$$

We have a similar estimate around $|s| = 0$*. The contribution is equivalent to* $(-\log|x|)^{3+\epsilon}$ *around* $x = 0$*.*

Proof We have the following equivalences:

$$h_{-1,a,b} \sim |t|^{2a}, \quad H_1^{-1} \sim (-\log|t|)^2,$$

$$\tau_1^{2+\epsilon} \sim \big(-\log|x|^2\big)^{2+\epsilon}, \quad \tau_2^{\epsilon} \sim |t|^{2\epsilon}, \quad \tau_3^{\epsilon} \sim (-\log|t|)^{2\epsilon}.$$

Thus we are done. □

2.9. Preliminary from functional analysis

2.9.1. On a family of complexes of Hilbert spaces.

LEMMA 2.65. *Let* H_i $(i = 1, 2)$ *be Hilbert spaces. Let* $F(\lambda) : H_1 \longrightarrow H_2$ *be bounded morphisms depending on* $\lambda \in \Delta(\lambda_0, \epsilon)$*. Assume the following:*

1. $F(\lambda)$ *is bounded for any* $\lambda \in \Delta(\lambda_0, \epsilon)$*.*
2. *There exists a positive constant* C *such that* $||F(\lambda) - F(\lambda')|| \leq C \cdot |\lambda - \lambda'|$ *for any* $\lambda, \lambda' \in \Delta(\lambda_0, \epsilon)$*.*
3. F *is holomorphic with respect to* λ *in the following sense: For any* $v \in H_1$*,* $F(\lambda)(v)$ *gives a holomorphic function from* $\Delta(\lambda_0, \epsilon) \longrightarrow H_2$*. (See* **[59]** *for Hilbert space valued holomorphic functions, for example.)*
4. $F(\lambda_0)$ *is surjective.*

Then there exists $\eta > 0$ *such that the following holds:*

- $F(\lambda)$ *is surjective for any* $\lambda \in \Delta(\lambda_0, \eta)$*.*
- *There exists a family of bounded morphisms* $\Psi(\lambda) : H_1 \longrightarrow H_2$ *for* $\lambda \in \Delta(\lambda_0, \eta)$*, and the following holds:*
  1. $\Psi(\lambda)$ *is homeomorphic for any* $\lambda \in \Delta(\lambda_0, \eta)$*.*
  2. *There exists a positive constant* C' *such that the following holds:*

$$\max\Big\{ ||\Psi(\lambda) - \Psi(\lambda')||, \quad ||\Psi(\lambda)^{-1} - \Psi(\lambda')^{-1}|| \Big\} \leq |\lambda - \lambda'| \cdot C'.$$

  3. $\Psi(\lambda)$ *is holomorphic with respect to* λ*.*
  4. *The following diagramm is commutative:*

$$\begin{array}{ccc} H_1 & \xrightarrow{F(\lambda_0)} & H_2 \\ {\scriptstyle\Psi}\downarrow & & \downarrow{\scriptstyle id} \\ H_1 & \xrightarrow{F(\lambda)} & H_2. \end{array}$$

Proof We put $C_1 := \operatorname{Ker} F(\lambda_0)$ and $C_2 = \operatorname{Ker} F(\lambda_0)^{\perp}$. We put $\varphi := F(\lambda_0)_{|C_2} : C_2 \longrightarrow H_2$ is bijective and bounded. Hence it is homeomorphic, i.e., φ^{-1} is also bounded.

We have the bounded morphisms $a(\lambda) : C_1 \longrightarrow H_2$ and $b(\lambda) : C_2 \longrightarrow H_2$ defined by $a(\lambda) = F(\lambda)_{|C_1}$ and $b(\lambda) = F(\lambda)_{|C_2} - \varphi$. It is easy to check the following:

- a and b are holomorphic with respect to λ.
- There exists positive constants C_2 such that $|a(\lambda) - a(\lambda')| \leq C_2 \cdot |\lambda - \lambda'|$ and $|b(\lambda) - b(\lambda')| \leq C_2 \cdot |\lambda - \lambda'|$.

- $a(\lambda_0) = b(\lambda_0) = 0$.

Since φ is homeomorphic, there exists $\eta > 0$ such that $\varphi + b(\lambda)$ is homeomorphic for any $\lambda \in \Delta(\lambda_0, \eta)$. In particular, $F(\lambda)$ is surjective.

The morphism $\Psi : C_1 \oplus C_2 \longrightarrow C_1 \oplus C_2$ can be given as follows:

$$\begin{pmatrix} 1 & (\varphi + b(\lambda))^{-1} \circ a(\lambda) \\ 0 & (\varphi + b(\lambda))^{-1} \circ \varphi(\lambda) \end{pmatrix}.$$

Then it is easy to check that Ψ has the desired properties. □

COROLLARY 2.66. *Let F and H_i $(i = 1, 2)$ be as in Lemma 2.65. There exists a positive constant $\eta > 0$ and the linear morphism $G(\lambda) : \operatorname{Ker} F(\lambda_0) \longrightarrow H_1$ depending on $\lambda \in \Delta(\lambda_0, \eta)$ with the following properties:*

- *It satisfies the conditions (1), (2) and (3) in Lemma 2.65.*
- *$G(\lambda)$ gives a homeomorphism $\operatorname{Ker} F(\lambda_0) \simeq \operatorname{Ker} F(\lambda)$ for any $\lambda \in \Delta(\lambda_0, \eta)$.*

Namely it gives the trivialization of the family of Hilbert spaces $\{\operatorname{Ker} F(\lambda) \mid \lambda \in \Delta(\lambda_0, \eta)\}$.

Proof We have only to restrict Ψ to $\operatorname{Ker}(F(\lambda_0))$. □

Reduction procedure We have the standard reduction procedure of the family of the complexes of Hilbert spaces, as is explained in the following. Let us consider the following family of complexes of Hilbert spaces H_i $(i = 0, 1, 2)$ depending on $\lambda \in \Delta(\lambda_0, \epsilon)$:

$$H_0 \xrightarrow{F_0(\lambda)} H_1 \xrightarrow{F_1(\lambda)} H_2. \tag{2.43}$$

Assume the family satisfies the following conditions:

- F_0 and F_1 satisfy the conditions (1), (2) and (3) in Lemma 2.65.
- F_1 satisfies the condition (4) in Lemma 2.65.
- The complex at λ_0 is exact, i.e., $\operatorname{Ker}(F_1(\lambda_0)) = \operatorname{Im} F_0(\lambda_0)$.

Then we can take the family of the morphisms $\Psi(\lambda) : H_1 \longrightarrow H_1$ depending on $\lambda \in \Delta(\lambda_0, \eta)$ as in Lemma 2.65. We put $F_0'(\lambda) := \Psi^{-1} \circ F_0(\lambda)$.

LEMMA 2.67. *The image of $F_0'(\lambda)$ is contained in $\operatorname{Ker} F_1(\lambda_0)$.*

Proof We have $F_1(\lambda_0) \circ F_0'(\lambda) = F_1(\lambda) \circ \Psi \circ \Psi^{-1} \circ F_0(\lambda) = F_1(\lambda) \circ F_0(\lambda) = 0$. □

Hence we obtain the family of morphisms $F_0''(\lambda) : H_0 \longrightarrow \operatorname{Ker}(F(\lambda_0))$ depending on $\lambda \in \Delta(\lambda_0, \eta)$. By our construction, F_0'' satisfies the conditions (1), (2) and (3) in Lemma 2.65. Since the complex is exact at λ_0, the morphism F_0'' satisfies the condition (4) in Lemma 2.65, too.

LEMMA 2.68. *Let H_i $(i = 0, \dots, n)$ be Hilbert spaces. For simplicity, we put $H_{n+1} = 0$. Let $d_i(\lambda) : H_i \longrightarrow H_{i+1}$ be family of linear morphisms satisfying the conditions (1), (2) and (3) in Lemma 2.65. Assume the following:*

- *The higher cohomology groups vanishes at λ_0. Namely, $\operatorname{Ker}(d_{i+1}(\lambda_0)) = \operatorname{Im}(d_i(\lambda_0))$ hold for $i = 0, \dots, n-1$.*

Then there exists a positive constant η and the family of linear morphisms $G(\lambda) : \operatorname{Ker} d_0(\lambda_0) \longrightarrow H^0$ $(\lambda \in \Delta(\lambda_0, \eta))$ satisfying the following:

- *The higher cohomology groups vanishes at any $\lambda \in \Delta(\lambda_0, \eta)$.*

- *$G(\lambda)$ satisfies the conditions* (1), (2) *and* (3) *in Lemma* 2.65, *and it gives the trivialization of the family* $\{\mathrm{Ker}\, d_0(\lambda) \,|\, \lambda \in \Delta(\lambda_0, \eta)\}$, *namely,* $G(\lambda)$ *gives the homeomorphism of* $\mathrm{Ker}\, d_0(\lambda_0)$ *and* $\mathrm{Ker}\, d_0(\lambda)$ *for* $\lambda \in \Delta(\lambda_0, \eta)$.

Proof We have only to use the above reduction procedure successively and Corollary 2.66. □

2.9.2. Sobolev spaces. We recall the following theorem. (See Theorem 9.1 in [**70**], for example.)

PROPOSITION 2.69. *Let p and q be real numbers such that $p, q \geq 1$. Let k and l be real numbers satisfying the following inequality: $k - \frac{m}{p} \geq l - \frac{m}{q}$. Then we have the natural inclusion $L^p_k(\boldsymbol{R}^m) \subset L^q_l(\boldsymbol{R}^m)$, and it is continuous. If the inequality $k > l$ and $k - \frac{m}{p} > l - \frac{m}{q}$ hold, then the inclusion is compact.*

LEMMA 2.70. *Let p be a real number such that $p > 2d$. We put as follows for $i = 1, \ldots, d$:*

$$q_i := \begin{cases} \dfrac{2d}{d-i}, & (i < d), \\ p, & (i = d). \end{cases}$$

Then we have the continuous inclusion $L^{q_{i-1}}_2(\boldsymbol{R}^{2d}) \subset L^{q_i}_1(\boldsymbol{R}^{2d})$.

Proof We have the following relation from our choice of q_i:

$$2 - \frac{2n}{q_{i-1}} \geq 1 - \frac{2n}{q_i}.$$

Then the lemma immediately follows from Proposition 2.69. □

2.10. An estimate of the norms of Higgs field and the conjugate

2.10.1. Preliminary. Let X_0 be an open subset of $\boldsymbol{C}^m_z$. Assume that we are given the holomorphic one forms $\eta_i \in \Omega^{1,0}_{X_0}$ $(i = 1, \ldots, m)$, which give a frame. For example, we consider $dz_1, \ldots, dz_m$ or $z_1^{-1} \cdot dz_1, \ldots, z_m^{-1} \cdot dz_m$. We put $X = \Delta^d_\zeta \times X_0$.

Let $(E, \overline{\partial}_E, \theta, h)$ be a harmonic bundle over X of rank r. Then we have the deformed holomorphic bundle $\mathcal{E}^\lambda$ with the holomorphic structure $\overline{\partial}_{\mathcal{E}^\lambda}$ and the λ-connection $\mathbb{D}^\lambda$. (See the section 3.1 in our previous paper [**65**]).

Let $\boldsymbol{v}$ be a C^∞-frame of $E = \mathcal{E}^\lambda$. The $(0,1)$-form $K \in C^\infty(X, \Omega^{0,1}_X \otimes M(r))$ is determined by the following relation:

$$\overline{\partial}_{\mathcal{E}^\lambda} \boldsymbol{v} = \boldsymbol{v} \cdot K.$$

The $(1,0)$-form $A \in C^\infty(X, \Omega^{1,0} \otimes M(r))$ is determined by the following relation:

$$\mathbb{D}^\lambda \boldsymbol{v} = \boldsymbol{v} \cdot (A + K).$$

We also have $\Theta \in C^\infty(X, M(r) \otimes \Omega^{1,0}_X)$ and $\Theta^\dagger \in C^\infty(X, M(r) \otimes \Omega^{0,1}_X)$ given by the following relation:

$$\theta \boldsymbol{v} = \boldsymbol{v} \cdot \Theta, \qquad \theta^\dagger \boldsymbol{v} = \boldsymbol{v} \cdot \Theta^\dagger.$$

We decompose as follows:

$$\Theta = \sum \Theta_{\zeta_i} \cdot d\zeta_i + \sum \Theta_{\eta_j} \cdot \eta_j, \qquad \Theta^\dagger = \sum \Theta^\dagger_{\zeta_i} \cdot d\overline{\zeta}_i + \sum \Theta^\dagger_{\eta_j} \cdot \overline{\eta}_j.$$

LEMMA 2.71. *We have the following relation:*

$$\overline{\partial}\Theta + \big[K - \lambda\Theta^{\dagger}, \Theta\big] = 0.$$

Proof It follows from the relations $\overline{\partial}_E = \overline{\partial}_{\mathcal{E}^\lambda} - \lambda \cdot \theta^\dagger$ and $\overline{\partial}_E \theta = 0$. □

LEMMA 2.72. *We have the following relation:*

$$\partial\Theta^{\dagger} + \lambda^{-1}\big[A - \Theta, \Theta^{\dagger}\big] = 0.$$

Proof It follows from the relation $\partial_E = \lambda^{-1} \cdot \big(\mathbb{D}^\lambda - \overline{\partial}_{\mathcal{E}^\lambda} - \theta\big)$ and $\partial_E \theta^\dagger = 0$. □

Let φ be a C^∞-function.

LEMMA 2.73. *We have the following formula:*

$$(2.44)\qquad \begin{aligned} \overline{\partial}(\varphi^{2b} \cdot \Theta) &= -\big[\varphi^b K - \lambda\cdot\varphi^b\Theta^\dagger,\ \varphi^b\Theta\big] - 2 \cdot \overline{\partial}\varphi^b \cdot \varphi^b\Theta, \\ \partial(\varphi^{2b}\Theta^\dagger) &= -\lambda^{-1} \cdot \big[\varphi^b A - \varphi^b\Theta,\ \varphi^b\Theta^\dagger\big] - 2 \cdot \partial\varphi^b \cdot \varphi^b\Theta^\dagger. \end{aligned}$$

Proof It immediately follows from Lemma 2.71 and Lemma 2.72. □

LEMMA 2.74. *We have the following relation for $\zeta = \zeta_j$ and $a = \zeta_j, z_k$:*

$$(2.45)\qquad \begin{aligned} \overline{\partial}_\zeta(\varphi^{2b}\Theta_a) &= -\big[\varphi^b K_\zeta - \lambda \cdot \varphi^b\Theta_\zeta^\dagger,\ \varphi^b\Theta_a\big] - 2 \cdot \overline{\partial}_\zeta\varphi^b \cdot \varphi^b\Theta_a, \\ \partial_\zeta(\varphi^{2b}\Theta_a^\dagger) &= -\lambda^{-1} \cdot \big[\varphi^b A_\zeta - \varphi^b\Theta_\zeta,\ \varphi^b\Theta_a^\dagger\big] - 2 \cdot \overline{\partial}_\zeta\varphi^b \cdot \varphi^b\Theta_a^\dagger. \end{aligned}$$

Proof It immediately follows from (2.44) □

LEMMA 2.75. *We have the following formulas:*

$$(2.46)$$
$$\begin{aligned} \partial_\zeta\overline{\partial}_\zeta(\varphi^{2b}\Theta) =& -\big[\partial_\zeta(\varphi^b K_\zeta) - \lambda \cdot \partial_\zeta\varphi^b\Theta_\zeta^\dagger,\ \varphi^b\Theta_a\big] - \big[\varphi^b K_\zeta - \lambda \cdot \varphi^b\Theta_\zeta^\dagger,\ \partial_\zeta\varphi^b\Theta_a\big] \\ & - 2 \cdot \partial_\zeta\overline{\partial}_\zeta\varphi^b \cdot \varphi^b\Theta_a - 2 \cdot \overline{\partial}_\zeta\varphi^b \cdot \partial_\zeta(\varphi^b\Theta_a), \\ \overline{\partial}_\zeta\partial_\zeta(\varphi^{2b}\Theta^\dagger) =& -\lambda^{-1} \cdot \big[\overline{\partial}_\zeta(\varphi^b A_\zeta) - \overline{\partial}_\zeta(\varphi^b\Theta_\zeta),\ \varphi^b\Theta^\dagger\big] \\ & - \lambda^{-1} \cdot \big[\varphi^b A_\zeta - \varphi^b\Theta_\zeta,\ \overline{\partial}_\zeta(\varphi^b\Theta^\dagger)\big] - 2 \cdot \overline{\partial}_\zeta\partial_\zeta\varphi^b \cdot \varphi^b\Theta^\dagger \\ & - 2 \cdot \partial_\zeta\varphi^b \cdot \overline{\partial}_\zeta(\varphi^b\Theta^\dagger). \end{aligned}$$

Proof It immediately follows from (2.45). □

LEMMA 2.76. *Let f be a compact support C^∞-function on X. Then we have the following equality:*

$$\int (\overline{\partial} f, \overline{\partial} f) = \int (\partial f, \partial f).$$

Proof It follows from the following:

$$\int (\overline{\partial} f, \overline{\partial} f) = -\int (f, \Delta(f)) = \int (\partial f, \partial f).$$

Here we have used the Kahler identity $\overline{\partial}^*\overline{\partial} = \Delta = \partial^*\partial$. □

2.10.2. An estimate. Let P be a point of X_0. An L^p_k-function spaces on $\Delta^d_\zeta \times \{P\}$ is denoted by $L^p_{k,P}$. If $k=0$, we use the notation L^p_P. For a C^∞-function F on X, we obtain the restriction $F_{|P} := F_{|\Delta^d_\zeta \times \{P\}}$, and the norm $\big|\big|F_{|P}\big|\big|_{L^p_k}$. For simplicity, we denote it by $\big|\big|F\big|\big|_{L^p_{k\,P}}$. We put as follows:

$$Q(P) := ||\Theta||_{L^\infty_P} + ||\Theta^\dagger||_{L^\infty_P} + ||K||_{L^\infty_P} + ||A||_{L^\infty_P} + 1. \tag{2.47}$$

Let us pick a C^∞-function φ on Δ^d_ζ satisfying the following:

$$0 \leq \varphi(P_1) \leq 1, \qquad \varphi(\zeta) = \begin{cases} 1, & (P_1 \in \Delta^d_\zeta(1/3)), \\ 0, & (P_1 \not\in \Delta^d_\zeta(2/3)). \end{cases}$$

LEMMA 2.77. *For any $b \in \mathbb{Z}_{>0}$, we have the following:*

$$\max\Big\{||\varphi^b\Theta||_{L^\infty_P},\ ||\varphi^b\Theta^\dagger||_{L^\infty_P}\Big\} \leq Q(P).$$

Proof It immediately follows from our choice of $Q(P)$. □

LEMMA 2.78. *For any $b \in 2\cdot\mathbb{Z}_{>0}$, there exists a positive constant $C_1(b)$ such that the following holds:*

$$\max\Big\{||\overline{\partial}_\zeta(\varphi^b\Theta)||_{L^2_P},\ ||\partial_\zeta(\varphi^b\Theta^\dagger)||_{L^2_P}\Big\} \leq C_1(b)\cdot Q(P)^2.$$

Proof It follows from (2.45) and Lemma 2.77. □

LEMMA 2.79. *For any $b \in 2\cdot\mathbb{Z}_{>0}$, there exists a positive constant $C_1(b)$ such that the following holds:*

$$\max\Big\{||\partial_\zeta(\varphi^b\Theta)||_{L^2_P},\ ||\overline{\partial}_\zeta(\varphi^b\Theta^\dagger)||_{L^2_P}\Big\} \leq C_1(b)\cdot Q(P)^2.$$

Proof It follows from Lemma 2.78 and Lemma 2.76. □

LEMMA 2.80. *For any $b \in 2^2\cdot\mathbb{Z}_{>0}$, there exists a positive constant $C_2(b) > 0$ such that the following holds:*

$$\max\Big\{||\varphi^b\Theta||_{L^2_{2,P}},\ ||\varphi^b\Theta^\dagger||_{L^2_{2,P}}\Big\} \leq C_2(b)\cdot Q(P)^3.$$

Proof From (2.46), Lemma 2.79 and Lemma 2.77, we obtain the following for any $b \in 2^2\cdot\mathbb{Z}_{>0}$

$$\max\Big\{||\partial_\zeta\overline{\partial}_\zeta(\varphi^b\Theta)_a||_{L^2},\ ||\partial_\zeta\overline{\partial}_\zeta(\varphi^b\Theta^\dagger_a)||_{L^2}\Big\} \leq C'(b)\cdot Q(P)^3.$$

Thus we are done. □

LEMMA 2.81. *Let b be an integer contained in $2^{2+i}\cdot\mathbb{Z}_{>0}$. Let q_i $(i=1,\ldots,d)$ be the numbers given in Lemma 2.70. Then there exist positive constants $C_{2+i}(b)$ satisfying the following:*

$$\max\Big\{||\varphi^b\cdot\Theta_a||_{L^{q_i}_{2,P}},\ ||\varphi^b\cdot\Theta^\dagger_a||_{L^{q_i}_{2,P}}\Big\} \leq C_{2+i}(b)\cdot Q(P)^{3+i}. \tag{2.48}$$

Proof We use an induction on i. We have already known that the claim for $i=0$ holds. We assume that the claim for i holds, and we will show that the claim for $i+1$ holds.

Due to the hypothesis of our induction, we have the following inequality:

$$\max\Big\{||\varphi^b\cdot\Theta_a||_{L^{q_i}_{2,P}},\ ||\varphi^b\cdot\Theta^\dagger_a||_{L^{q_i}_{2,P}}\Big\} \leq C_{2+i}(b)\cdot Q(P)^{3+i}.$$

Recall that we have the continuous inclusion $L_2^{q_i} \subset L_1^{q_{i+1}}$ due to Lemma 2.70. Thus we obtain the following:

$$\max\Big\{||\varphi^b \cdot \Theta_a||_{L_{1,P}^{q_{i+1}}},\ ||\varphi^b \cdot \Theta_a^\dagger||_{L_{1,P}^{q_{i+1}}}\Big\} \leq C'(b) \cdot Q(P)^{3+i}.$$

In particular, we obtain the following:

$$\max\Big\{||\partial_\zeta \varphi^b \cdot \Theta_a||_{L_P^{q_{i+1}}},\ ||\overline{\partial}_\zeta \varphi^b \cdot \Theta_a^\dagger||_{L_P^{q_{i+1}}}\Big\} \leq C''(b) \cdot Q(P)^{3+i}.$$

Then we obtain the following due to Lemma 2.77 and (2.46):

$$\max\Big\{||\overline{\partial}\partial_\zeta \varphi^{2b} \cdot \Theta_a||_{L_P^{q_{i+1}}},\ ||\partial\overline{\partial}_\zeta \varphi^{2b} \cdot \Theta_a^\dagger||_{L_P^{q_{i+1}}}\Big\} \leq C^{(3)}(b) \cdot Q(P)^{3+i+1}.$$

Thus we obtain the inequality (2.48) for $i+1$, and the induction can proceed. □

COROLLARY 2.82. *Let p be a real number such that $p > 2d = 2\dim_{\boldsymbol{C}} X$. Let b be an integer contained in $2^{2+d} \cdot \mathbb{Z}_{>0}$. Then there exists a positive constant C_{2+d} satisfying the following:*

$$\max\Big\{||\varphi^b \cdot \Theta_a||_{L_{2,P}^p},\ ||\varphi^b \cdot \Theta_a^\dagger||_{L_{2,P}^p}\Big\} \leq C_{2+d}(b) \cdot Q(P)^{3+d}.$$

Proof It immediately follows from Lemma 2.81. □

LEMMA 2.83. *For any $b \in 2^{d+k} \cdot \mathbb{Z}_{>0}$, there exists a positive constant $C_{d+k}(b)$ satisfying the following:*

$$\max\Big\{||\varphi^b \Theta||_{L_{k,P}^p}, ||\varphi^b \Theta^\dagger||_{L_{k,P}^p}\Big\} \leq C_{k+d}(b) \cdot Q(P)^{d+k+1}.$$

Proof We can show it by a standard boot strapping argument. □

We put $K_0 := \overline{\Delta}_\zeta(1/3)^d$.

COROLLARY 2.84. *Let k be an integer, and p be any sufficiently large number. There exist positive constants M and C such that the L_k^p-norms and C^k-norms of $\Theta_{|K_0 \times \{P\}}$ and $\Theta^\dagger_{|K_0 \times \{P\}}$ are dominated by $C \cdot Q(P)^M$.*

Proof Let us pick a real number k_1 such that we have the continuous inclusion $L_{k_1}^p(\boldsymbol{R}^d) \subset C^k(\boldsymbol{R}^d)$. Then we can pick b_1, C_1 and M_1 such that the following holds, due to Lemma 2.83:

$$\max\Big\{||\varphi^{b_1} \cdot \Theta||_{L_{k_1,P}^p}, ||\varphi^{b_1} \cdot \Theta^\dagger||_{L_{k_1,P}^p}\Big\} \leq C_1 \cdot Q(P)^{M_1}.$$

Then we obtain the estimate for C^k-norms of $\varphi^{b_1}\Theta$ and $\varphi^{b_1}\Theta^\dagger$. Since φ is identically 1 on a neighbourhood of K_0 by our choice of φ. Thus we obtain the result. □

2.10.3. Estimate for the differential of $\Theta^\dagger_\zeta$. Let us consider the case $\eta_i = dz_i/z_i$ $(i = 1, \ldots, m)$. We have the vector fields $V_i := z_i\partial/\partial z_i$, and $\partial f = \sum V_i(f) \wedge \eta_i$ and $\overline{\partial} f = \sum \overline{V}_i(f) \wedge \overline{\eta}_i$.

LEMMA 2.85. *We have the following relation:*

$$\overline{\partial}\Theta^\dagger + \big[K - \lambda \cdot \Theta^\dagger,\ \Theta^\dagger\big] = 0.$$

Proof It follows from the relation $\overline{\partial}_E \theta^\dagger = 0$ and $\overline{\partial}_E = \overline{\partial}_{\mathcal{E}^\lambda} - \lambda\theta^\dagger$. □

We obtain the following relation for any $\zeta = \zeta_j$ and $i = 1, \ldots, m$:

$$\overline{V}_i \Theta_\zeta^\dagger - \overline{\partial}_\zeta \Theta_{\eta_i}^\dagger + \big[K_{\eta_i} - \lambda\Theta_{\eta_i}^\dagger,\ \Theta_\zeta^\dagger\big] - \big[K_\zeta - \lambda\Theta_\zeta^\dagger,\ \Theta_{\eta_i}^\dagger\big] = 0.$$

Hence we obtain the following:

LEMMA 2.86. *There exist positive constants C and M such that the functions $\overline{V}_i\Theta^{\dagger}_{\zeta_j}$ $(i=1,\ldots,m,\ j=1,\ldots,d)$ are dominated by $C\cdot Q(P)^M$ on $K_0\times X_0$.* □

2.11. Convergency of the sequence of harmonic bundles

Although we do not use the argument to take a 'limit' of a sequence of harmonic bundles, contrary to the previous paper [**65**], the author thinks that such convergency seems significant for the study of harmonic bundles. We can improve the argument for the convergency in [**65**] by using the estimate in the subsection 2.10.2. We explain it in this section. In this section, we assume that $\lambda\neq 0$.

2.11.1. Convergency of the sequences of the Higgs fields $\Theta^{(n)}$ and the adjoint maps $\Theta^{\dagger\,(n)}$. Let X be $\Delta^{*\,l}\times\Delta^{d-l}$, and $(E^{(n)},\theta^{(n)},h^{(n)})$ be a harmonic bundles on X such that $\mathrm{rank}(E^{(n)})=r$. Recall that we have the deformed holomorphic bundles $(\mathcal{E}^{\lambda\,(n)},d''^{\,\lambda\,(n)},\mathbb{D}^{\lambda\,(n)},h^{(n)})$ on $\mathcal{X}^{\lambda}=\{\lambda\}\times X\subset\mathcal{X}$. In this section, the metric and the measure of X are $\sum_{i=1}^{n}dz_i\cdot d\bar{z}_i$ and $\prod_{i=1}^{n}|dz_i\cdot d\bar{z}_i|$.

Assume that we are given holomorphic frames $\boldsymbol{w}^{(n)}=(w_1^{(n)},\ldots,w_r^{(n)})$ of $\mathcal{E}^{\lambda\,(n)}$. Then we have the elements $\Theta^{(n)}\in C^{\infty}\bigl(X,M(r)\otimes\Omega^{1,0}\bigr)$ determined by $\theta^{(n)}\boldsymbol{w}^{(n)}=\boldsymbol{w}^{(n)}\cdot\Theta^{(n)}$. We also have the elements $\Theta^{\dagger\,(n)}\in C^{\infty}(X,M(r)\otimes\Omega^{0,1}_X)$ determined by $\theta^{(n)\dagger}\boldsymbol{w}^{(n)}=\boldsymbol{w}^{(n)}\Theta^{\dagger\,(n)}$.

We assume the following condition.

CONDITION 2.87.

(1) We have the holomorphic λ-connection forms $A^{(n)}\in\Gamma(X,M(r)\otimes\Omega^{1,0}_X)$ determined by $\mathbb{D}^{\lambda\,(n)}\boldsymbol{w}^{(n)}=\boldsymbol{w}^{(n)}\cdot A^{(n)}$. Then the sequence $\{A^{(n)}\}$ converges to $A^{(\infty)}\in\Gamma(X,M(r)\otimes\Omega^{1,0}_X)$ on any compact subset $Y\subset X$.
(2) The sequences $\{\Theta^{(n)}\}$ and $\{\Theta^{\dagger\,(n)}\}$ are bounded independently of n on any compact subset $Y\subset X$. □

Let Y be a compact subset of X. We put as follows:

$$Q_Y^{(n)}:=||\Theta^{(n)}||_{L^{\infty}(Y)}+||\Theta^{\dagger\,(n)}||_{L^{\infty}(Y)}+||A^{(n)}||_{L^{\infty}(Y)}.$$

Then we have a positive constant C_Y independent of n such that $Q_Y^{(n)}\leq C_Y$ for any integer n due to Condition 2.87.

LEMMA 2.88. *Let k be a positive number. There exist positive constants C and M such that L^p_k-norms and C^k-norms of $\Theta^{(n)}_{|Y}$ and $\Theta^{(n)\dagger}_{|Y}$ are dominated by $C\cdot C_Y^M$.*

Proof Since we have the boundedness of $Q_Y^{(n)}$, we can use Corollary 2.84. Note that we use the holomorphic frames $\boldsymbol{w}^{(n)}$, the term $||K||_{L^{\infty}}$ is trivial. □

COROLLARY 2.89. *There exists a subsequence $\{n_i\}$ such that $\{\Theta^{(n_i)}_{|Y}\}$ and $\{\Theta^{(n_i)\dagger}_{|Y}\}$ are convergent in the C^{∞}-sense.*

Proof We have only to use Lemma 2.88 and an easy diagonal argument. □

PROPOSITION 2.90. *There exists a subsequence $\{n_i\}$ such that $\{\Theta^{(n_i)}\}$ and $\{\Theta^{(n_i)\dagger}\}$ are convergent on any compact subset $Y\subset X$ in the C^{∞}-sense.*

Proof We have only to use Corollary 2.89 and an easy diagonal argument. □

2.11.2. The convergency of $\{H^{(n)}\}$. Besides Condition 2.87, we consider the following additional condition:

CONDITION 2.91.

- We put $H^{(n)} := H(h^{(n)}, \boldsymbol{w}^{(n)}) \in C^{\infty}(X, \mathcal{H}(r))$. On any compact subset $Y \subset X$, $\{H^{(n)}\}$ and $\{H^{(n)\,-1}\}$ are bounded independently of n. Namely we have a constant C_Y depending on Y such that $|H^{(n)}| < C_Y$ and $|H^{(n)\,-1}| < C_Y$. □

Due to Proposition 2.90, we may assume that $\{\Theta^{(n)}\}$ and $\{\Theta^{\dagger\,(n)}\}$ are convergent on any compact subset Y in the C^{∞}-sense, by picking a subsequence.

We have the unitary connection $\nabla^{(n)} := \overline{\partial}_{\mathcal{E}^{\lambda\,(n)}} + \partial_{\mathcal{E}^{\lambda\,(n)}}$ of $\mathcal{E}^{\lambda\,(n)}$. Let $B^{(n)}$ denote the connection form of $\nabla^{(n)}$, i.e., $\nabla^{(n)}\boldsymbol{w}^{(n)} = \boldsymbol{w}^{(n)} \cdot B^{(n)}$. Then we have the following by a standard theory:

$$B^{(n)} = H^{(n)\,-1}\partial H^{(n)}. \tag{2.49}$$

On the other hand, we have $\nabla^{(n)} = \mathbb{D}^{\lambda\,(n)} - 2\theta^{(n)}$. Hence we obtain the following relation:

$$B^{(n)} = A^{(n)} - \lambda^{-1}\big(1 + |\lambda|^2\big) \cdot \Theta^{(n)}. \tag{2.50}$$

Let Y be a compact subset of X, and we pick the function φ as in the previous subsection. Let us pick a compact subset $Y \subset X$, and pick compact subsets Y_1 and Y_2 of X such that Y is contained in the interior of Y_1, and that Y_1 is contained in the interior of Y_2. We can pick an element $\varphi \in C^{\infty}(X, \boldsymbol{R})$, satisfying the following:

$$0 \le \varphi(x) \le 1, \quad \varphi(x) = \begin{cases} 1, & (x \in Y), \\ 0, & (x \notin Y_2). \end{cases}$$

LEMMA 2.92. *We have the following formula:*

$$\partial\big(\varphi^{2b} H^{(n)}\big) = \varphi^b \cdot H^{(n)} \cdot \big(\varphi^b A^{(n)} - \varphi^b \lambda^{-1}(1 + |\lambda|^2)\Theta^{(n)}\big) + 2\partial\varphi^b \cdot \varphi^b H^{(n)}. \tag{2.51}$$

Proof We obtain the following relation due to (2.49) and (2.50):

$$\partial H^{(n)} = H^{(n)} \cdot \big(A^{(n)} - 2 \cdot \lambda^{-1}(1 + |\lambda|^2)\Theta^{(n)}\big).$$

Then (2.51) follows immediately. □

LEMMA 2.93. *Let $\{n_i\}$ be a subsequence of $\{n\}$.*

- *Assume $\{\varphi^b \cdot H^{(n_i)}\}$ is bounded in L^p_k, and $\{\partial(\varphi^b \cdot H^{(n_i)})\}$ is bounded in L^p_k. Then $\{\varphi^b \cdot H^{(n_i)}\}$ is bounded in L^p_{k+1}.*
- *Assume $\{\varphi^b \cdot H^{(n_i)}\}$ is bounded in L^p_k, and $\{\partial(\varphi^b \cdot H^{(n_i)})\}$ is convergent in L^p_k. Then $\{\varphi^b \cdot H^{(n_i)}\}$ is convergent in L^p_{k+1}.*

Proof Note that we have $\overline{\varphi \cdot H_{i\,j}} = \varphi \cdot H_{j\,i}$. Hence we have the following:

$$\overline{\overline{\partial}(\varphi \cdot H_{i\,j})} = \partial(\varphi \cdot H_{j\,i}).$$

Thus the estimate for ∂H implies the estimate for $\overline{\partial} H$. Thus we obtain the result. □

LEMMA 2.94. *There exists a subsequence $\{n_i\}$ such that $\{\varphi^{2b} \cdot H^{(n_i)}\}$ are convergent in L^p_0 for any $b \in 2 \cdot \mathbb{Z}$.*

Proof Due to (2.51) and Lemma 2.93, we have the boundedness of the family $\{\varphi^{2b}\cdot H^{(n_i)}\}$ are bounded in L_1^p. Then we obtain the result due to the compactness of the inclusion $L_1^p \subset L_0^p$. □

LEMMA 2.95. *Let $\{n_i\}$ be the subsequence as in Lemma 2.94. Then $\{H^{(n_i)}\}$ are convergent on the compact subset Y in the L_l^p-sense for any l.*

Proof Let k be a positive integer. For any sufficiently large integer b, the family $\{\varphi^b\cdot H^{(n_i)}\}$ are convergent in L_k^p-sense, by using (2.51) and the bootstrapping argument. Since $\varphi = 1$ on a neighbourhood of K, we obtain the result. □

PROPOSITION 2.96. *Assume the conditions in Condition 2.87 and Condition 2.91 are satisfied. There exists a subsequence $\{n_i\}$ such that $\{H^{(n_i)}\}$, $\{\Theta^{(n_i)}\}$ and $\{\Theta^{\dagger\,(n_i)}\}$ are convergent on any compact subset Y in the C^∞-sense.*

Proof We have already seen the convergency of $\{\Theta^{(n)}\}$ and $\{\Theta^{\dagger\,(n)}\}$ for some subsequence in Proposition 2.90. We have seen the convergency of $\{H^{(n)}\}$ on any compact subset for some subsequence in Lemma 2.95. Then we have only to use the diagonal argument. □

2.12. Higgs field and twisted map

2.12.1. A description of Higgs field. Let X be a complex manifold, and let $(E,\overline{\partial}_E,\theta,h)$ be a harmonic bundle over X. Let $(\mathcal{E}^\lambda,\mathbb{D}^\lambda)$ be the corresponding λ-connection, and $\mathbb{D}^{\lambda\,f}$ denote the associated flat connection. We have the decomposition $\mathbb{D}^{\lambda\,f} = d' + d''$ into the $(1,0)$-part and the $(0,1)$-part. Let δ' denote the $(1,0)$-operator such that $d''+\delta'$ is a unitary connection. Let δ'' denote the $(0,1)$-operator such that $d'+\delta''$ is a unitary connection. Then we have the following formula:

$$d' = \partial + \lambda^{-1}\cdot\theta,\quad d'' = \overline{\partial}+\lambda\cdot\theta^\dagger,\quad \delta' = \partial - \overline{\lambda}\cdot\theta,\quad \delta'' = \overline{\partial}-\overline{\lambda}^{-1}\cdot\theta^\dagger.$$

It can be rewritten as follows:

$$\partial = \lambda\cdot\bigl(1+|\lambda|^2\bigr)^{-1}\cdot\bigl(\overline{\lambda}\cdot d' + \lambda^{-1}\cdot\delta'\bigr),\quad \overline{\partial} = \overline{\lambda}\cdot\bigl(1+|\lambda|^2\bigr)^{-1}\cdot\bigl(\overline{\lambda}^{-1}\cdot d'' + \lambda\cdot\delta''\bigr),$$
$$\theta = \lambda\cdot\bigl(1+|\lambda|^2\bigr)^{-1}\cdot\bigl(d'-\delta'\bigr),\qquad \theta^\dagger = \overline{\lambda}\cdot\bigl(1+|\lambda|^2\bigr)^{-1}\cdot\bigl(d''-\delta''\bigr).$$

Let $\boldsymbol{v}$ be a flat frame with respect to the flat connection $\mathbb{D}^{\lambda\,f}$. We put $H = H(h,\boldsymbol{v})$. We have the C^∞-section Θ and $\Theta^\dagger$ of $M(r)\otimes\Omega^{1,0}$ and $M(r)\otimes\Omega^{0,1}$ respectively, determined by the relations $\theta\cdot\boldsymbol{v} = \boldsymbol{v}\cdot\Theta$ and $\theta^\dagger\cdot\boldsymbol{v} = \boldsymbol{v}\cdot\Theta^\dagger$ respectively. Then we have the following relation:

$$\begin{aligned}(2.52)\quad \overline{\partial}H_{i\,j} &= \overline{\partial}h(v_i,v_j) = h(v_i,\delta' v_j) = -h\bigl(v_i,(d'-\delta')v_j\bigr)\\ &= -h\bigl(v_i,\lambda^{-1}\cdot(1+|\lambda|^2)\theta\cdot v_j\bigr) = -\overline{\lambda}^{-1}\cdot\bigl(1+|\lambda|^2\bigr)\cdot H_{i\,j}\cdot\overline{\Theta_{k\,j}}.\end{aligned}$$

Hence we obtain the relation $\Theta = -\lambda\cdot\bigl(1+|\lambda|^2\bigr)^{-1}\cdot\overline{H}\cdot\partial\overline{H}$. Similarly, we obtain the relation $\Theta^\dagger = -\overline{\lambda}\cdot\bigl(1+|\lambda|^1\bigr)^{-1}\cdot\overline{H}^{-1}\cdot\overline{\partial}\overline{H}$. Therefore, we have the following relation:

$$\overline{\lambda}\cdot\Theta + \lambda\cdot\Theta^\dagger = |\lambda|^2\cdot\bigl(1+|\lambda|^2\bigr)^{-1}\cdot\overline{H}^{-1}\cdot d\overline{H}.$$

Let v be a real vector field on X. Hence we obtain the following equality, by using Lemma 21.4 (it can be easily shown directly):

(2.53)

$$\big|\overline{\lambda}\cdot\theta(v)+\lambda\cdot\theta^{\dagger}(v)\big|_h^2=|\lambda|^2\cdot\big(1+|\lambda|^2\big)^{-1}\cdot\operatorname{tr}\Big(\overline{H}^{-1}\cdot d\overline{H}(v)\cdot\overline{H}^{-1}\cdot d\overline{H}(v)\cdot\overline{H}^{-1}\cdot\overline{H}\Big)$$
$$=|\lambda|^2\cdot\big(1+|\lambda|^2\big)^{-1}\cdot\operatorname{tr}\Big(H^{-1}\cdot dH(v)\cdot H^{-1}\cdot dH(v)\Big).$$

2.12.2. Higgs field as the differential of the twisted map. Let $\pi:\tilde{X}\longrightarrow X$ be a universal covering. Recall that $\mathcal{PH}(r)$ denote the sets of the $(r\times r)$-positive definite hermitian matrices. By taking a flat trivialization of the flat bundle $\pi^{-1}\big(\mathcal{E}^{\lambda},\mathbb{D}^{\lambda f}\big)$, we obtain the equivariant map $\Psi_h:\tilde{X}\longrightarrow\mathcal{PH}(r)$, which is essentially independent of a choice of a flat trivialization.

We have the $\pi_1(X)$-action on $\tilde{X}$. The monodromy induces endomorphism $\rho:\pi_1(X)\longrightarrow GL(r)$, which induces the $\pi_1(X)$-action on $\mathcal{PH}(r)$. The map Ψ_h is equivariant with respect to the actions of the fundamental group $\pi_1(X)$.

The map Ψ_h is called the twisted map associated with $(\mathcal{E}^{\lambda},\mathbb{D}^{\lambda f},h)$. We often regard it as a map $\Psi_h:X\longrightarrow\mathcal{PH}(r)/\pi_1(X)$, although $\mathcal{PH}(r)/\pi_1(X)$ is not a good topological space in general.

Then the equality (2.53) can be reformulated as follows, by using Lemma 21.11 (see the subsection 21.2.1):

$$\big|\overline{\lambda}\cdot\theta(v)+\lambda\cdot\theta^{\dagger}(v)\big|^2=|\lambda|^4\cdot\big(1+|\lambda|^2\big)^{-1}\cdot\big|d\Psi_h(v)\big|^2. \tag{2.54}$$

CHAPTER 3

Preliminary for Mixed Twistor Structure

3.1. $\mathbb{P}^1$-holomorphic vector bundle over $X \times \mathbb{P}^1$

3.1.1. Y-holomorphic structure. Let X be a C^∞-manifold, Y be a complex manifold, and p_Y denote the projection of $X \times Y$ onto Y. Let V be a C^∞-vector bundle over $X \times Y$.

DEFINITION 3.1. A Y-holomorphic structure of V is defined to be a differential operator $d''_{V,Y} : C^\infty(V) \longrightarrow C^\infty(V \otimes p_Y^* \Omega_Y^{0,1})$ satisfying the following:

- $(d''_{V,Y})^2 = 0$.
- For any function $f \in C^\infty\big(X \times Y\big)$ and any section $s \in C^\infty\big(X \times Y, V\big)$, the following holds:

$$d''_{V,Y}(f \cdot s) = \overline{\partial}_Y(f) \cdot s + f \cdot d''_{V,Y} s.$$

The pair $(V, d''_{V,Y})$ or V is called a Y-holomorphic vector bundle. We often denote d'' instead of $d''_{V,Y}$. □

DEFINITION 3.2. Let (V, d'') be a Y-holomorphic vector bundle. A C^∞-section s of V is caller Y-holomorphic, if $d''(s) = 0$ holds. □

Let $(V, d''_{V,Y})$ be a Y-holomorphic vector bundle. Then the Y-holomorphic structure on the dual $V^\vee$ is naturally defined. Let $(V^{(i)}, d''_{V^{(i)},Y})$ $(i = 1, 2)$ be Y-holomorphic vector bundles over $X \times Y$. Then the tensor product and the direct sum of $(V^{(i)}, d''_{V^{(i)},Y})$ $(i = 1, 2)$ are naturally defined.

DEFINITION 3.3. Let $(V^{(i)}, d''_{V^{(i)},Y})$ $(i = 1, 2)$ be Y-holomorphic vector bundles over $X \times Y$. A morphism of $(V^{(1)}, d''_{V^{(1)},Y})$ to $(V^{(2)}, d''_{V^{(2)},Y})$ is defined to be a Y-holomorphic section of $Hom(V^{(1)}, V^{(2)})$. □

3.1.2. Some description of $\mathbb{P}^1$-holomorphic vector bundle over $X \times \mathbb{P}^1$. We are particularly interested in the $\mathbb{P}^1$-holomorphic vector bundle over $X \times \mathbb{P}^1$. It is often useful to consider the following object. In the following, we regard $\mathbb{P}^1$ as the gluing of $\boldsymbol{C}_\lambda$ and $\boldsymbol{C}_\mu$ by the relation $\lambda = \mu^{-1}$.

DEFINITION 3.4. A patched object on X is defined to be a tuple $(V_\lambda, V_\mu, \varphi)$ as follows:

- V_a is a $\boldsymbol{C}_a$-holomorphic vector bundle over $X \times \boldsymbol{C}_a$ for $a = \lambda, \mu$.
- We have the $\boldsymbol{C}_\lambda^*$-holomorphic vector bundles $V_{\lambda\,|\,X \times \boldsymbol{C}_\lambda^*}$ and $V_{\mu\,|\,X \times \boldsymbol{C}_\mu^*}$ over $X \times \boldsymbol{C}_\lambda^*$. Then φ is a $\boldsymbol{C}_\lambda^*$-holomorphic isomorphism of $V_{\lambda\,|\,X \times \boldsymbol{C}_\lambda^*}$ to $V_{\mu\,|\,X \times \boldsymbol{C}_\mu^*}$. □

DEFINITION 3.5. Let $(V_\lambda^{(i)}, V_\mu^{(i)}, \varphi^{(i)})$ $(i = 1, 2)$ be patched objects over X. A morphism f of $(V_\lambda^{(1)}, V_\mu^{(1)}, \varphi^{(1)})$ to $(V_\lambda^{(2)}, V_\mu^{(2)}, \varphi^{(2)})$ is defined to be a tuple (f_λ, f_μ), where f_a $(a = \lambda, \mu)$ is a $\boldsymbol{C}_a$-holomorphic morphism $V_a^{(1)} \longrightarrow V_a^{(2)}$ which is compatible with $\varphi^{(1)}$ and $\varphi^{(2)}$. □

A direct sum, a tensor product and a dual are naturally defined.

Equivalence

Let $(V_\lambda, V_\mu, \varphi)$ be a patched object over X. By gluing them, we obtain the $\mathbb{P}^1$-holomorphic vector bundle over $X \times \mathbb{P}^1$. On the other hand, let V be a $\mathbb{P}^1$-holomorphic vector bundle over $X \times \mathbb{P}^1$, we put $V_a := V_{|\boldsymbol{C}_a}$ for $a = \lambda, \mu$. We have the naturally defined isomorphism id : $V_{\lambda\,|\,\boldsymbol{C}_\lambda^*} \longrightarrow V_{\mu\,|\,\boldsymbol{C}_\mu^*}$. Thus we obtain the patched object. It is easy to check the equivalence of the category of the $\mathbb{P}^1$-holomorphic vector bundle over $X \times \mathbb{P}^1$ and the category of the patched object over X. The equivalence preserves a tensor product, a direct sum and a dual.

Another description

We can also consider another kind of patched objects $(V_0, V_1, V_\infty; \alpha_0, \alpha_\infty)$ as follows: For simplicity, we put $Y_0 := \boldsymbol{C}_\lambda$, $Y_\infty := \boldsymbol{C}_\mu$ and $Y_1 := \boldsymbol{C}_\lambda^*$.

- V_a be a Y_a-holomorphic vector bundle over $X \times Y_a$.
- We have the Y_1-holomorphic vector bundles $V_{a\,|\,Y_1}$ $(a = 0, 1, \infty)$. Then α is a Y_1-holomorphic morphism $V_{a\,|\,Y_1} \longrightarrow V_1$ $(a = 0, \infty)$.

We can naturally define the morphism of such patched objects. We also have a tensor product, a direct sum and a dual of such objects. As before we have the naturally defined equivalence of the category of such patched objects and the category of the $\mathbb{P}^1$-holomorphic vector bundle over $X \times \mathbb{P}^1$.

Due to the equivalences explained above, we will often use the descriptions $(V_\lambda, V_\mu, \varphi)$ or $(V_0, V_\infty, V_1; \alpha, \beta)$ to denote a $\mathbb{P}^1$-holomorphic vector bundle over $X \times \mathbb{P}^1$, in the following.

3.1.3. The involution σ and the induced bundles. Let σ denote the morphism $\mathbb{P}^1 \longrightarrow \mathbb{P}^1$ given by $[z_0 : z_\infty] \longmapsto [\bar{z}_\infty : -\bar{z}_0]$. We also use the notation σ to denote the following induced morphisms:

- $\boldsymbol{C}_\lambda \longrightarrow \boldsymbol{C}_\mu$ given by $\sigma^*\mu = -\overline{\lambda}$.
- $\boldsymbol{C}_\mu \longrightarrow \boldsymbol{C}_\lambda$ given by $\sigma^*\lambda = -\overline{\mu}$.
- $\boldsymbol{C}_\lambda^* \longrightarrow \boldsymbol{C}_\lambda^*$ given by $\lambda \longmapsto -\overline{\lambda}^{-1}$.

We take the anti-linear isomorphism $\varphi : C^\infty(X \times \boldsymbol{C}_\mu) \longrightarrow C^\infty(X \times \boldsymbol{C}_\lambda)$ by $\varphi(f) = \sigma^*(\overline{f})$. It is induced by the conjugate map $\varphi_0 : \sigma^*\boldsymbol{C}_{X \times \boldsymbol{C}_\mu} \longrightarrow \boldsymbol{C}_{X \times \boldsymbol{C}_\lambda}$ given by $a \longmapsto \bar{a}$, where $\boldsymbol{C}_Y$ denotes the trivial line bundle over a C^∞-manifold Y.

Since the morphism σ is anti-holomorphic, we have the naturally induced isomorphism $\sigma^*\Omega^{0,1}_{\boldsymbol{C}_\mu} \longrightarrow \Omega^{1,0}_{\boldsymbol{C}_\lambda}$ given by $\sigma^*(a \cdot d\overline{\mu}) \longmapsto -\sigma^*(a) \cdot d\lambda$. On the other hand, we have the conjugate $\Omega^{1,0}_{\boldsymbol{C}_\lambda} \longrightarrow \Omega^{0,1}_{\boldsymbol{C}_\lambda}$ given by $a \cdot d\lambda \longmapsto \bar{a} \cdot d\overline{\lambda}$, which is anti-linear. As the composite, we obtain the linear morphism $\varphi_0 : \sigma(\Omega^{0,1}_{\boldsymbol{C}_\mu}) \longrightarrow \Omega^{0,1}_{\boldsymbol{C}_\lambda}$. They induce the morphism $\varphi : C^\infty\big(X \times \boldsymbol{C}_\mu, \Omega^{0,1}_{\boldsymbol{C}_\mu}\big) \longrightarrow C^\infty\big(X \times \boldsymbol{C}_\lambda, \Omega^{0,1}_{\boldsymbol{C}_\lambda}\big)$ given by $f \cdot d\overline{\mu} \longmapsto -\sigma^*(\overline{f}) \cdot d\overline{\lambda}$.

Let f be an element of $C^\infty(X \times \boldsymbol{C}_\mu)$. Then we obtain the elements $\varphi(f) \in C^\infty(X \times \boldsymbol{C}_\lambda)$ and $\overline{\partial}_\lambda \varphi(f) \in C^\infty\big(X \times \boldsymbol{C}_\lambda, \Omega^{0,1}_{\boldsymbol{C}_\lambda}\big)$. On the other hand, we have the elements $\overline{\partial}_\mu f \in C^\infty\big(X \times \boldsymbol{C}_\mu, \Omega^{0,1}_{\boldsymbol{C}_\mu}\big)$. We also have the element $\varphi\big(\overline{\partial}_\mu f\big) \in C^\infty\big(X \times \boldsymbol{C}_\lambda, \Omega^{0,1}_{\boldsymbol{C}_\lambda}\big)$.

LEMMA 3.6. *We have the equality* $\overline{\partial}_\lambda \varphi(f) = \varphi\big(\overline{\partial}_\mu f\big)$. *Namely, we have* $\overline{\partial}_\lambda \circ \varphi = \varphi \circ \overline{\partial}_\mu$.

Proof We have the following:

$$\varphi\Big(\frac{\partial f}{\partial \overline{\mu}} d\overline{\mu}\Big) = -\sigma^* \overline{\Big(\frac{\partial f}{\partial \overline{\mu}}\Big)} \cdot d\overline{\lambda} = -\sigma^* \left(\frac{\partial \overline{f}}{\partial \mu}\right) d\overline{\lambda}. \tag{3.1}$$

We have the following equalities:

$$\sigma^* \left(\frac{\partial \overline{f}}{\partial \mu}\right)(z, \lambda) = \frac{\partial \overline{f}}{\partial \mu}(z, -\overline{\lambda}) = -\frac{\partial \overline{f}(z, -\overline{\lambda})}{\partial \overline{\lambda}} = -\frac{\partial}{\partial \overline{\lambda}}\big(\varphi(f)(z, \lambda)\big). \tag{3.2}$$

From (3.1) and (3.2), we obtain the result. □

Let V be a C^∞-vector bundle over $\boldsymbol{C}_\mu \times X$. Then we obtain the pull back $\sigma^* V$ over $\boldsymbol{C}_\lambda \times X$. We change the $\boldsymbol{C}$-vector bundle structure of $\sigma^* V$ as follows: For a complex number $a \in \boldsymbol{C}$ and an element $\sigma^*(v) \in \sigma^* V_{|(\lambda,P)}$, $a \cdot \sigma^*(v)$ is defined to be $\sigma^*(\overline{a} \cdot v)$. We denote the resulted C^∞-vector bundle by $\sigma(V)$. The anti-linear morphisms φ_0 given above can be regarded as the linear morphisms $\sigma \boldsymbol{C}_{X \times \boldsymbol{C}_\mu} \longrightarrow \boldsymbol{C}_{X \times \boldsymbol{C}_\lambda}$ or $\sigma \Omega^{0,1}_{\boldsymbol{C}_\mu} \longrightarrow \Omega^{0,1}_{\boldsymbol{C}_\lambda}$.

Then the morphism φ_0 induces the following isomorphism

$$\varphi_0 : \sigma\big(V \otimes \Omega^{0,1}_{\boldsymbol{C}_\mu}\big) \longrightarrow \sigma\big(V\big) \otimes \Omega^{0,1}_{\boldsymbol{C}_\lambda}, \quad \varphi_0\big(\sigma(v \otimes d\overline{\mu})\big) = -\sigma(v) \otimes d\overline{\lambda}.$$

It induces the morphism $\varphi_0 : C^\infty\Big(X \times \boldsymbol{C}_\lambda, \sigma\big(V \otimes \Omega^{0,1}_{\boldsymbol{C}_\mu}\big)\Big) \longrightarrow C^\infty\Big(X \times \boldsymbol{C}_\lambda, \sigma\big(V\big) \otimes \Omega^{0,1}_{\boldsymbol{C}_\lambda}\Big)$.

Let $(V, \overline{\partial}_\mu)$ be a $\boldsymbol{C}_\mu$-holomorphic vector bundle over $X \times \boldsymbol{C}_\mu$. Then the differential operator $\overline{\partial}_\lambda$ on $\sigma(V)$ is defined as follows:

$$\overline{\partial}_\lambda(\sigma(v)) = \varphi_0\big(\sigma(\overline{\partial}_\mu v)\big).$$

LEMMA 3.7. *The operator* $\overline{\partial}_\lambda$ *gives the* $\boldsymbol{C}_\lambda$*-holomorphic structure of* $\sigma(V)$.

Proof We have the following:

(3.3)

$$\overline{\partial}_\lambda\big(f \cdot \sigma(v)\big) = \overline{\partial}_\lambda\big(\sigma(\varphi(f) \cdot v)\big) = \varphi_0\big(\sigma\big(\overline{\partial}_\mu(\varphi(f) \cdot v)\big)\big) = \varphi_0\big(\sigma\big(\overline{\partial}_\mu \varphi(f) \cdot v + \varphi(f) \cdot \overline{\partial}_\mu v\big)\big)$$

We have the following:

$$\varphi_0 \sigma\big(\overline{\partial}_\mu \varphi(f) \cdot v\big) = \varphi_0 \sigma\big(\varphi(\overline{\partial}_\lambda f)\big) \cdot \sigma(v) = \overline{\partial}_\lambda f \cdot \sigma(v). \tag{3.4}$$

We also have the following:

$$\varphi_0\big(\sigma\big(\varphi(f) \cdot \overline{\partial}_\mu v\big)\big) = f \cdot \varphi_0\big(\sigma\big(\overline{\partial}_\mu v\big)\big).$$

It is easy to check that $\overline{\partial}_\lambda^2 = 0$. □

Namely we obtain the induced $\boldsymbol{C}_\lambda$-holomorphic vector bundle $\sigma(V)$ from a $\boldsymbol{C}_\mu$-holomorphic vector bundle V. Similarly, a Y-holomorphic vector bundle V induces the $\sigma(Y)$-holomorphic vector bundle $\sigma(V)$ for $Y = \boldsymbol{C}_\lambda$, $\boldsymbol{C}_\mu$, $\boldsymbol{C}_\lambda^*$ and $\mathbb{P}^1$.

3.2. Equivariant $\mathbb{P}^1$-holomorphic bundle over $X \times \mathbb{P}^1$

3.2.1. Torus action. Let ρ_0 denote the G_m-action on $\mathbb{P}^1$ given as follows:

$$\rho_0\big(t, [z_0 : z_\infty]\big) = [t \cdot z_0 : z_\infty].$$

It induces the G_m-action on $X \times \mathbb{P}^1$, which we also denote by ρ_0. We have the open subsets $X \times \boldsymbol{C}_\lambda$, $X \times \boldsymbol{C}_\mu$ and $X \times \boldsymbol{C}_\lambda^*$ of $X \times \mathbb{P}^1$. They are stable with respect to ρ_0.

Let V be a C^∞-vector bundle over $X \times Y$ for $Y = \boldsymbol{C}_\lambda$, $\boldsymbol{C}_\mu$, $\boldsymbol{C}_\lambda^*$ or $\mathbb{P}^1$

When we consider the G_m-action on V, we consider only the action which is a lift of ρ_0. Assume that V is holomorphic along the $\mathbb{P}^1$-direction. Let ρ be a G_m-action on V. It is called holomorphic if $t^*\overline{\partial}_\lambda = \overline{\partial}_\lambda$ for any $t \in G_m$.

3.2.2. Rees bundle (one filtration). Let p_0, p_∞ and p_1 denote the projection of $X \times \boldsymbol{C}_\lambda$, $X \times \boldsymbol{C}_\mu$ and $X \times \boldsymbol{C}_\lambda^*$ onto X. Let H be a C^∞-bundle over X, and F be a decreasing C^∞-filtration of H. Let us pick a point of P and a small neighbourhood U of P. We may assume that we have a frame $\boldsymbol{v} = (v_i)$ of $H_{|U}$ which is compatible with F. We put $b_i = \deg_F(v_i) = \max\{h \,|\, v_i \in F_h\}$. We put $\tilde{v}_i := \lambda^{-b_i} \cdot p_1^*(v_i)$. Then we obtain the frame $\tilde{\boldsymbol{v}} := (\tilde{v}_i)$ of $p_1^* H_{|U}$ over $\boldsymbol{C}_\lambda^* \times U$. The frame $\tilde{\boldsymbol{v}}$ gives a prolongation of $p_1^* H_{|U}$ to a vector bundle $\xi(H_{|U}; F)$ over $U \times \boldsymbol{C}_\lambda$.

LEMMA 3.8. *The C^∞-bundle $\xi(H_{|U}, F)$ is independent of a choice of compatible frame $\boldsymbol{v}$.*

Proof Assume that $\boldsymbol{u}$ is other compatible frame. We have the following relation:

$$u_i = \sum b_{j\,i} \cdot v_j.$$

Here we have $b_{j\,i} = 0$ in the case $\deg_F(u_i) > \deg_F(v_j)$. Then we obtain the following relation.

$$\tilde{u}_i = \lambda^{-b(u_i)} \cdot u_i = \sum b_{j\,i} \cdot \lambda^{-b(u_i)+b(v_j)} \cdot \lambda^{-b(v_j)} \cdot v_j = \sum b_{j\,i} \cdot \lambda^{-b(u_i)+b(v_j)} \cdot \tilde{v}_j.$$

It implies the well definedness of $\xi(H_{|U}; F)$. □

COROLLARY 3.9. *We obtain the global C^∞-vector bundle $\xi(H; F)$, which is $\boldsymbol{C}_\lambda$-holomorphic.* □

It is characterized locally as follows: We may assume that $U = X$. We have the ring $C^\infty(X)[\lambda]$. We have the $C^\infty(X)[\lambda]$-module $C^\infty(X, H)[\lambda, \lambda^{-1}]$. We have the submodule of $C^\infty(X, H)[\lambda, \lambda^{-1}]$ given as follows:

$$\sum_{p \in \mathbb{Z}} \lambda^{-p} \cdot C^\infty(X, F^p)[\lambda].$$

By taking $\tilde{\boldsymbol{v}}$ as above, we can show that it is locally free, and $\xi(H_{|U}, F)$ is the corresponding vector bundle. The restriction $\xi(H; F)_{|\mathcal{X}^0}$ is naturally isomorphic to the associated graded vector bundle $\mathrm{Gr}_F(H) = \bigoplus_p \mathrm{Gr}_F^p(H)$.

The ρ_0 can be naturally lifted to the action on $p_1^*(H)$. It is easy to check that the action can be prolonged to the action on $\xi(H; F)$. Since $\mathcal{X}^0 := X \times \{0\}$ is fixed by the torus action, we obtain the weight decomposition of $\xi(H; F)_{|\mathcal{X}^0}$. It is given by the decomposition $\mathrm{Gr}_F(H) = \bigoplus_p \mathrm{Gr}_F^p(H)$, i.e., the weight on $\mathrm{Gr}_F^p(H)$ is p.

Let $\mathrm{Filt}(X)$ be the category of filtered C^∞-vector bundle over X. For two filtered bundles (H_i, F_i) $(i = 1, 2)$, a morphism $F : (H_1, F_1) \longrightarrow (H_2, F_2)$ is defined to be a morphism $H_1 \longrightarrow H_2$ preserving filtrations. Let $\mathrm{Equi}(X \times \boldsymbol{C}_\lambda)$ be the C^∞-vector bundle with G_m-action over $X \times \boldsymbol{C}_\lambda$. For equivariant bundles V_i $(i = 1, 2)$, a morphism $f : V_1 \longrightarrow V_2$ is defined to be an equivariant $\boldsymbol{C}_\lambda$-holomorphic section of the equivariant bundle $Hom(V_1, V_2)$.

Let $f : (H_1, F) \longrightarrow (H_2, F)$ be a morphism. Let v be a section of $F^p(H_1)$. Then $f(v)$ is contained in $F^p(H_2)$. Hence $f(\lambda^{-p} \cdot v)$ gives a section of $\xi(H_2, F)$. Thus we obtain the section $\xi(f)$ of $Hom(V_1, V_2)$. It is easy to check that $\xi(f)$ is equivariant with respect to the torus action. Therefore we obtain the equivariant morphism $\xi(f) : V_1 \longrightarrow V_2$.

Let (H_i, F) $(i = 1, 2)$ be a filtered vector bundle. Recall that the filtration of the tensor product $H_1 \otimes H_2$ is defined as follows:

$$F^p(H_1 \otimes H_2) := \sum_{r+s \geq p} F^r(H_1) \otimes F^s(H_2).$$

Let (H, F) be a filtered vector bundle. Recall that the dual of (H, F) is defined as follows:

$$F^p(H^\vee) = \mathrm{Ker}\bigl(H^\vee \longrightarrow (F^{-p+1})^\vee\bigr).$$

LEMMA 3.10. *ξ gives the fully faithful functor from* $\mathrm{Filt}(X)$ *to* $\mathrm{Equi}(X)$. *It preserves direct sums, tensor products and duals.*

Proof It is clear that ξ is a functor and that ξ preserves direct sums.

We pick a splittings $H_1 = \bigoplus_p U_1^p$ and $H_2 = \bigoplus_p U_2^p$ of the filtrations. Then the decomposition $H_1 \otimes H_2 = \bigoplus_p \bigoplus_{r+s=p} U_1^r \otimes U_2^s$ gives the splitting of the filtration of $H_1 \otimes H_2$. Then it is easy to see that the tensor product $\xi(H_1, F) \otimes \xi(H_2, F)$ is isomorphic to $\xi\bigl((H_1, F) \otimes (H_2, F)\bigr)$.

We have the perfect pairing $H \otimes H^\vee \longrightarrow \boldsymbol{C}$. Due to the definition of $F(H^\vee)$, the composite of the following morphisms is trivial for $i \geq -p + 1$:

$$F^i(H) \otimes F^p(H^\vee) \longrightarrow F^{-p+1}(H) \otimes F^p(H^\vee) \longrightarrow \boldsymbol{C}.$$

For any $v \in F^i(H)$ and $v^\vee \in F^p(H^\vee)$, the pairing $\langle \lambda^{-p} \cdot v^\vee, \lambda^{-i} \cdot v \rangle$ is contained in $\boldsymbol{C}[\lambda]$. It implies we have the equivariant morphism $\xi(H^\vee, f) \longrightarrow \xi(H, f)^\vee$. By using the perfectness of $\mathrm{Gr}_F(H) \otimes \mathrm{Gr}_F(H^\vee) \longrightarrow \boldsymbol{C}$, we obtain that $\xi(H^\vee, F)$ is isomorphic to $\xi(H, F)^\vee$.

Let $\phi : \xi(H_1, F) \longrightarrow \xi(H_2, F)$ be an equivariant morphism. Then we obtain the morphism $f = \phi_{|\{1\} \times X} : H_1 \longrightarrow H_2$. Since ϕ is equivariant, we have $\phi(\lambda^{-p} \cdot v) = \lambda^{-p} f(v)$. Let v be a section of $F^p(H_1)$. Since $\phi(\lambda^{-p} \cdot v) = \lambda^{-p} \cdot f(v)$ is a section of $\xi(H_2, F)$, $f(v)$ is contained in $F^p(H_2)$. Thus we are done. □

REMARK 3.11. In fact, the functor ξ gives the equivalence of two categories. □

3.2.3. Rees bundle (bi-filtration). Let H be a C^∞-vector bundle over X. Let F and G be filtrations of H in the C^∞-category. As we have already seen, we obtain the equivariant bundle $\xi(H; F)$ over $X \times \boldsymbol{C}_\lambda$. By a similar way, we obtain the equivariant bundle $\xi(H; G)$ over $X \times \boldsymbol{C}_\mu$. Note that they are isomorphic to $p_1^* H$ on $X \times \boldsymbol{C}_\lambda^*$. Hence we can glue them, and we obtain the equivariant vector bundle $\xi(H; F, G)$ over $X \times \mathbb{P}^1$. Or, we can say that we obtain the patched object $\bigl(\xi(H; F), \xi(H; G), p_1^* H; \mathrm{id}, \mathrm{id}\bigr)$.

Let $\mathrm{Bifilt}(X)$ be the category of bi-filtered C^∞-vector bundle over X. Let (H_i, F, G) $(i = 1, 2)$ be bi-filtered vector bundles over X. Naturally, a morphism

$f : (H_1, F, G) \longrightarrow (H_2, F, G)$ is defined to be a morphism $f : H_1 \longrightarrow H_2$ preserving the filtrations F and G.

Let $\mathrm{Equi}(X \times \mathbb{P}^1)$ be the category of equivariant $\mathbb{P}^1$-holomorphic vector bundle over $X \times \mathbb{P}^1$. Let V_i $(i = 1, 2)$ be equivariant vector bundles over $X \times \mathbb{P}^1$. A morphism $f : V_1 \longrightarrow V_2$ is defined to be an equivariant section of the equivariant bundle $Hom(V_1, V_2)$.

Let $f : (H_1, F, G) \longrightarrow (H_2, F, G)$ be a morphism of bi-filtered vector bundles. Then we obtain the morphism $\xi(f) : \xi(H_1, F, G) \longrightarrow \xi(H_2, F, G)$.

PROPOSITION 3.12. *The functor ξ gives an equivalence of two categories. It preserves direct sums, tensor products and duals. It is functorial for a C^∞-morphism $Y \longrightarrow X$.*

Proof We only show the equivalence of the categories, because the rests are easy. It follows easily from Lemma 3.10 that ξ is fully faithful. Thus we have only to show that ξ is essentially surjective.

Let V be an equivariant vector bundle over $X \times \mathbb{P}^1$. We will construct the bi-filtered vector bundle (H, F, G). We put $H = V_{|\mathcal{X}^1}$, where we put $\mathcal{X}^1 := X \times \{1\}$. We will construct the filtrations F and G. We have only to construct them locally on X.

Pick a point P of X. The following lemmas are standard.

LEMMA 3.13. *There exists a number n_0, such that the following holds for any $n \geq n_0$:*

(1) *The following morphism is surjective:*

$$H^0\big(V \otimes \mathcal{O}(0, n)_{|\{P\} \times \mathbb{P}^1}\big) \longrightarrow V \otimes \mathcal{O}(0, n)_{|\{P\} \times \{0\}} = V_{|\{P\} \times \{0\}}.$$

(2) $H^1\big(V \otimes \mathcal{O}(0, n)_{|\{P\} \times \mathbb{P}^1}\big) = 0.$ □

LEMMA 3.14. *We have a neighbourhood U of P in X satisfying the following:*

- *The properties* (1) *and* (2) *in Lemma* 3.13 *for any point $Q \in U$.*
- $\big\{H^0\big(V \otimes \mathcal{O}_{\mathbb{P}^1}(0, n)_{|Q \times \mathbb{P}^1}\big) \,\big|\, Q \in U\big\}$ *forms a C^∞-vector bundle $\mathcal{E}_n$ on U.*

Proof It can be shown by arguments similar to those in the subsection 2.9.1. □

Let us return to the proof of Proposition 3.12. Then we have the equivariant surjective morphism $\pi : \mathcal{E}_n \longrightarrow V_{|\mathcal{X}^0}$ over X. If $n < n'$, then we have the commutativity $\mathcal{E}_n \subset \mathcal{E}_{n'} \longrightarrow V_{|\mathcal{X}^0}$.

Let $\boldsymbol{v} = (v_i)$ be a C^∞-frame of $V_{|\mathcal{X}^0}$, which is compatible with the weight decomposition $V_{|\mathcal{X}^0} = \bigoplus U_h$. We denote the weight of v_i by w_i. Let $\tilde{v}_i$ be a C^∞-section of the weight w_i-part of $\mathcal{E}_n$ such that $\pi(\tilde{v}_i) = v_i$. Note that $\tilde{v}_i$ naturally give sections of $V \otimes \mathcal{O}(0, n)$, and they are $\boldsymbol{C}_\lambda$-holomorphic. Then there exists a neighbourhood U_1 of O in $\boldsymbol{C}_\lambda$, and a neighbourhood U_2 of P in X, such that $\tilde{\boldsymbol{v}} = (\tilde{v}_i)$ gives a frame of $V \otimes \mathcal{O}(0, n)_{|U_1 \times U_2}$.

Let $\mathcal{E}_{n,h}$ denote the weight h-part of $\mathcal{E}_n$. The decreasing filtration F is defined as follows:

$$F^h := \mathrm{Im}\Big(\bigoplus_{k \geq h} \mathcal{E}_{n,k} \longrightarrow H = V_{|\{1\} \times X}\Big).$$

LEMMA 3.15. *We have $\mathrm{Gr}_F^h \simeq U_h$.*

Proof Let f be a section of the weight h-part of $\mathcal{E}_n$. Then there exist C^∞-functions $a_i(x,\lambda)$ which are $\boldsymbol{C}_\lambda$-holomorphic, such that the following holds:

$$f = \sum a_i(x,\lambda) \cdot \tilde{v}_i.$$

Due to the condition on the weight, the functions $a_i(x,\lambda)$ are described as $\tilde{a}_i(x) \cdot \lambda^{w_i - h}$ for C^∞-functions $\tilde{a}_i(x)$. Note that $\tilde{a}_i(x) = 0$ if $w_i - h < 0$. Then Lemma 3.15 follows. □

LEMMA 3.16. *The construction of the subbundle F^h of H is independent of a choice of n.*

Proof Recall that we have $\mathcal{E}_n \longrightarrow \mathcal{E}_{n'} \longrightarrow V_{\mathcal{X}^\lambda}$ for any $n \leq n'$ and for any λ. The lemma follows easily. □

Thus we obtain the well defined filtration F of H. We denote the image of $\tilde{v}_i$ by $\bar{v}_i$. We have the equivariant morphism $p_1^* H \longrightarrow V_{|X \times \boldsymbol{C}_\lambda^*}$, given by $\lambda^{-w_i} \cdot \bar{v}_i \longmapsto \tilde{v}_{i\,|\,\boldsymbol{C}_\lambda^* \times X}$. It is prolonged to the equivariant isomorphism $\xi(H,F) \longrightarrow V_{|\boldsymbol{C}_\lambda \times X}$ over $\boldsymbol{C}_\lambda \times X$.

Similarly, we obtain the filtration G by considering $V \otimes \mathcal{O}(n,0)$ and $V_{|\lambda=\infty}$, and we have the natural equivariant isomorphism $\xi(H,G) \longrightarrow V_{|X \times \boldsymbol{C}_\mu}$. Then we obtain the isomorphism $\xi(H,F,G) \longrightarrow V$. Thus the proof of Proposition 3.12 is accomplished. □

LEMMA 3.17. *An equivariant subbundle W of $V = \xi(H,F,G)$ corresponds to a subbundle H' of H with strict filtrations $F' = F \cap H'$ and $G' = G \cap H'$.*

Proof Let H' be a subbundle of H. We put $F' = F \cap H'$ and $G' = G \cap H'$. Then it is easy to check that $\xi(H',F',G')$ gives a subbundle of V.

Let W be a subbundle of V. We put $H' = W_{|\lambda=1}$, and then we obtain $\xi(H',F',G')$ as above. The restrictions of W and $\xi(H',F',G')$ to $X \times \boldsymbol{C}_\lambda^*$ are same, due to the equivariance. It implies W and $\xi(H',F',G')$ are same. Thus we are done. □

3.2.4. Real structure of equivariant bundles. Let (V,ρ) be an equivariant vector bundle over $X \times \mathbb{P}^1$. We have the G_m-action $\sigma(\rho)$ on $\sigma(V)$ defined as follows:

$$\sigma(\rho)(t) \cdot \big(\sigma(v)\big) := \sigma\big(\rho(\bar{t})^{-1} \cdot v\big).$$

On the other hand, we have the bi-filtered vector space (H,F,G) and the conjugate $(H^\dagger, G^\dagger, F^\dagger)$.

LEMMA 3.18. *If $(V,\rho) = \xi(H,F,G)$, we have $\big(\sigma(V),\sigma(\rho)\big) = \xi\big(H^\dagger,G^\dagger,F^\dagger\big)$.*

Proof It directly follows from the definition. □

DEFINITION 3.19. A real structure on an equivariant bundle (V,ρ) is defined to be an equivariant isomorphism $\iota_V : \big(\sigma(V),\sigma(\rho)\big) \simeq (V,\rho)$, such that $\iota_V \circ \sigma(\iota_V) = \mathrm{id}_V$. □

REMARK 3.20. In this paper, we do not consider the real structure of non-equivariant $\mathbb{P}^1$-holomorphic vector bundle. See the section 2 in [**83**]. □

A real structure of (H,F,G) is defined to be an isomorphism $\iota_H : (H,F,G) \longrightarrow (H^\dagger,G^\dagger,F^\dagger)$. In other words, ι_H is anti-isomorphism $H \longrightarrow H$ and we have $G = F^\dagger$.

LEMMA 3.21. *In the case $(V,\rho) = \xi(H,F,G)$, a real structure ι_V of (V,ρ) and a real structure ι_H of $\xi(H,F,G)$ corresponds by the relation $\iota_V = \xi(\iota_H)$.* □

DEFINITION 3.22. Let (W,ρ_W) be a vector subbundle of (V,ρ). If we have $\iota_V\big(\sigma(W)\big) \subset W$, the subbundle W is called defined over $\boldsymbol{R}$. □

LEMMA 3.23. *An equivariant subbundle (W,ρ_W) is defined over $\boldsymbol{R}$ if and only if the subbundle $H' = W_{|\lambda=1}$ of H is defined over $\boldsymbol{R}$.*

Proof We have the action of ι_V on $C^\infty(X,H)[\lambda,\lambda^{-1}]$, it is equivalent with $\iota_H\otimes\iota_0$. Here ι_0 is given by $\iota_0(\lambda) = -\lambda^{-1}$. Let us consider the following property:

$C^\infty(X,H')[\lambda,\lambda^{-1}]$ is preserved by $\iota_V = \iota_H \otimes \iota_0$.

Then it is easy to see that three properties are equivalent. □

DEFINITION 3.24. Let (V_i,ρ_i,ι_i) $(i=1,2)$ be equivariant vector bundles defined over $\boldsymbol{R}$. Let $f : (V_1,\rho_1) \longrightarrow (V_2,\rho_2)$ be an equivariant morphism. It is called defined over $\boldsymbol{R}$ if $\iota_2\circ\sigma(f) = f\circ\iota_1$. □

LEMMA 3.25. *Let (H_i,F_i,G_i) $(i=1,2)$ be bi-filtered vector bundles with real structures ι_{H_i}. We put $(V_i,\rho_i) = \xi(H_i,F_i,G_i)$. A morphism $f : (V_1,\rho_1) \longrightarrow (V_2,\rho_2)$ is defined over $\boldsymbol{R}$ if and only if $f = \xi(N)$ and $N : (H_1,F_1,G_1) \longrightarrow (H_2,F_2,G_2)$ are defined over $\boldsymbol{R}$.*

Proof We obtain the morphism $f' : C^\infty(X,H_1)[\lambda,\lambda^{-1}] \longrightarrow C^\infty(X,H_2)[\lambda,\lambda^{-1}]$. from f. Then the both properties are equivalent to the property that f' is compatible with $\boldsymbol{R}$-structure. □

3.3. Tate objects and $\mathcal{O}(p,q)$

3.3.1. Tate object. The following patched object is called the Tate object:

$$\mathbb{T}(n) = \Big(\mathcal{O}_{\boldsymbol{C}_\lambda}\cdot t_0^{(n)},\ \mathcal{O}_{\boldsymbol{C}_\mu}\cdot t_\infty^{(n)},\ \mathcal{O}_{\boldsymbol{C}_\lambda^*}\cdot t_1^{(n)},\ \alpha_0^{(n)},\ \alpha_\infty^{(n)}\Big).$$

Here the morphisms $\alpha_0^{(n)}$ and $\alpha_\infty^{(n)}$ are given as follows:

$$\alpha_0^{(n)}(t_0^{(n)}) = (\sqrt{-1}\lambda)^n\cdot t_1^{(n)},\quad \alpha_\infty^{(n)}(t_\infty^{(n)}) = \big(-\sqrt{-1}\mu\big)^n\cdot t_1^{(n)}.$$

The vector bundle corresponding to the patched object above can be regarded as the gluing of $\mathcal{O}_{\boldsymbol{C}_\lambda}\cdot t_0^{(n)}$ and $\mathcal{O}_{\boldsymbol{C}_\mu}\cdot t_\infty^{(n)}$ by the relation $(\sqrt{-1}\lambda)^{-n}\cdot t_0^{(n)} = (-\sqrt{-1}\mu)^{-n}\cdot t_\infty^{(n)}$, i.e., $(\sqrt{-1}\lambda)^{-2n}\cdot t_0^{(n)} = t_\infty^{(n)}$. Hence it is isomorphic to $\mathcal{O}_{\mathbb{P}^1}(-2n)$. We denote the corresponding vector bundle also by $\mathbb{T}(n)$ or $\mathcal{O}(-2n)$ for simplicity.

By the projection $\pi : X\times\mathbb{P}^1 \longrightarrow \mathbb{P}^1$, we obtain the patched object $\pi^*\mathbb{T}(n)$ over X, which we denote by $\mathbb{T}(n)_X$ or $\mathbb{T}(n)$ for simplicity.

We have the torus action $\rho_{\mathbb{T}(n)}$ on $\mathbb{T}(n)$ given as follows:

$$(t,t_0^{(n)}) \longmapsto t^{-n}\cdot t_0^{(n)},\quad (t,t_\infty^{(n)}) \longmapsto t^n\cdot t_\infty^{(n)},\quad (t,t_1^{(n)}) \longmapsto t_1^{(n)}.$$

We have the isomorphism $\iota_{\mathbb{T}(n)} : \sigma^*\mathbb{T}(n) \longrightarrow \mathbb{T}(n)$ given as follows:

$$\sigma^*(t_0^{(n)}) \longmapsto (-1)^n\cdot t_\infty^{(n)},\quad \sigma^*(t_\infty^{(n)}) \longmapsto (-1)^n\cdot t_0^{(n)},\quad \sigma^*(t_1^{(n)}) \longmapsto t_1^{(n)}.$$

It means $\iota_{\mathbb{T}(n)}\big(\sigma^*(f\cdot t_0^{(n)})\big) = (-1)^n\cdot\overline{\sigma^*(f)}\cdot t_\infty^{(n)}$. Note that it is well defined as it can be checked as follows:

$$\sigma^* t_\infty^{(n)} = (\sqrt{-1}\mu)^{-2n}\cdot\sigma^* t_0^{(n)} \longmapsto (\sqrt{-1}\mu)^{-2n}\cdot(-1)^n\cdot t_\infty^{(n)} = (-1)^n\cdot t_0^{(n)},$$

$$\sigma^* t_1^{(n)} = (\sqrt{-1}\mu)^{-n}\cdot\sigma^* t_0^{(n)} \longmapsto (\sqrt{-1}\mu)^{-n}\cdot(-1)^n\cdot t_\infty^{(n)} = t_1^{(n)}.$$

LEMMA 3.26. *We have $\iota_{\mathbb{T}(n)} \circ \sigma^*\bigl(\iota_{\mathbb{T}(n)}\bigr) = \mathrm{id}_{\mathbb{T}(n)}$. Namely, the morphism $\iota_{\mathbb{T}(n)}$ gives the real structure of $\mathbb{T}(n)$.*

Proof It can be checked by a direct calculation. □

We have the corresponding real structure on $\mathbb{T}(n)_{|\lambda=1}$. In the following, the real base $t^{(n)}_{1\,|\,\lambda=1}$ is fixed.

3.3.2. $\mathcal{O}(p,q)$ and $\mathcal{O}(n)$. We have the following patched object:

$$\mathcal{O}_{\mathbb{P}^1}(p,q) = \Bigl(\mathcal{O}_{\boldsymbol{C}_\lambda}\cdot f_0^{(p,q)},\ \mathcal{O}_{\boldsymbol{C}_\mu}\cdot f_\infty^{(p,q)},\ \mathcal{O}_{\boldsymbol{C}_\lambda^*}\cdot f_1^{(p,q)},\ \alpha_0^{(p,q)},\ \alpha_\infty^{(p,q)}\Bigr).$$

Here the morphism $\alpha_0^{(p,q)}$ and $\alpha_\infty^{(p,q)}$ are given as follows:

$$\alpha_0^{(p,q)}\bigl(f_0^{(p,q)}\bigr) = \bigl(\sqrt{-1}\lambda\bigr)^{-p}\cdot f_1^{(p,q)}, \qquad \alpha_\infty^{(p,q)}\bigl(f_\infty^{(p,q)}\bigr) = \bigl(-\sqrt{-1}\mu\bigr)^{-q}\cdot f_1^{(p,q)}.$$

Since $\mathcal{O}(p,q)$ is the vector bundle over $\mathbb{P}^1$ obtained as the gluing of $\mathcal{O}_{\boldsymbol{C}_\lambda}\cdot f_0^{(p,q)}$ and $\mathcal{O}_{\boldsymbol{C}_\mu}\cdot f_\infty^{(p,q)}$ by the relation $\bigl(\sqrt{-1}\lambda\bigr)^{p+q}\cdot f_0^{(p,q)} = f_\infty^{(p,q)}$, it is isomorphic to $\mathcal{O}_{\mathbb{P}^1}(p+q)$. We also note that we have the canonical isomorphism $\phi_{(p,q),(p',q')}$: $\mathcal{O}(p,q)\simeq\mathcal{O}(p',q')$ in the case $p+q=p'+q'$, which is given as follows:

$$f_0^{(p,q)} \longmapsto f_0^{(p',q')}, \qquad f_\infty^{(p,q)} \longmapsto f_\infty^{(p',q')}. \tag{3.5}$$

In this sense, we may also use $f_0^{(n)}$ and $f_\infty^{(n)}$ instead of $f_0^{(p,q)}$ and $f_\infty^{(p,q)}$, when we forget the torus action and we have $p+q=n$.

Let $\pi_X : X\times\mathbb{P}^1 \longrightarrow \mathbb{P}^1$ denote the projection. The induced patched objects $\pi_X^*\mathcal{O}(p,q)$ are denoted by $\mathcal{O}(p,q)_X$ or simply by $\mathcal{O}(p,q)$.

We have the torus action $\rho_{(p,q)}$ on $\mathcal{O}(p,q)$ given as follows:

$$\bigl(t, f_0^{(p,q)}\bigr) \longmapsto t^p\cdot f_0^{(p,q)}, \quad \bigl(t, f_\infty^{(p,q)}\bigr) \longmapsto t^{-q}\cdot f_\infty^{(p,q)}, \quad \bigl(t, f_1^{(p,q)}\bigr) \longmapsto f_1^{(p,q)}.$$

The isomorphism $\phi_{(p,q),(p',q')}$ is not compatible with the torus actions, if $(p,q)\neq(p',q')$.

We have the isomorphism $\iota_{(p,q)} : \sigma^*\mathcal{O}(p,q)\simeq\mathcal{O}(q,p)$ given as follows:

$$\sigma^*\bigl(f_0^{(p,q)}\bigr) \longmapsto \sqrt{-1}^{\,p+q} f_\infty^{(q,p)},$$

$$\sigma^*\bigl(f_\infty^{(p,q)}\bigr) \longmapsto (-\sqrt{-1})^{p+q} f_0^{(q,p)},$$

$$\sigma^*\bigl(f_1^{(p,q)}\bigr) \longmapsto \sqrt{-1}^{\,q-p} f_1^{(q,p)}.$$

LEMMA 3.27. *The morphism $\iota_{(p,q)}$ is well defined.*

Proof We have only to check that the second and the third correspondences are induced by the first correspondence. We have the following correspondences:

$$\begin{aligned}\sigma^*\bigl(f_\infty^{(p,q)}\bigr) &= \sigma^*\bigl((\sqrt{-1}\lambda)^{p+q}\cdot f_0^{(p,q)}\bigr) = (\sqrt{-1}\mu)^{p+q}\cdot\sigma^*\bigl(f_0^{(p,q)}\bigr)\\ &\longmapsto (-1)^{p+q}(\sqrt{-1}\lambda)^{-p-q}\cdot(\sqrt{-1})^{p+q}\cdot f_\infty^{(q,p)} = \bigl(-\sqrt{-1}\bigr)^{p+q}\cdot f_0^{(q,p)}.\end{aligned} \tag{3.6}$$

We also have the following:

$$\begin{aligned}\sigma^*\bigl(f_1^{(p,q)}\bigr) &= \sigma^*\bigl((\sqrt{-1}\lambda)^p\cdot f_0^{(p,q)}\bigr) = (\sqrt{-1}\mu)^p\cdot\sigma^* f_0^{(p,q)}\\ &\longmapsto (\sqrt{-1}\mu)^p\sqrt{-1}^{\,p+q}\cdot f_\infty^{(q,p)} = \sqrt{-1}^{\,q-p}\cdot f_1^{(q,p)}.\end{aligned} \tag{3.7}$$

Thus the morphism is well defined. □

LEMMA 3.28. *The morphism $\iota_{(p,q)}$ is compatible with $\phi_{(p,q),(p',q')}$, i.e., we have the following:*

$$\iota_{(p',q')} \circ \sigma^*\big(\phi_{(p,q),(p',q')}\big) = \phi_{(p,q),(p',q')} \circ \iota_{(p',q')}.$$

Proof It can be checked by a direct calculation. □

LEMMA 3.29. *We have the natural isomorphism:*

$$\mathcal{O}(p_1,q_1) \otimes \mathcal{O}(p_2,q_2) \longmapsto \mathcal{O}(p_1+p_2,q_1+q_2).$$

It is given by $f_a^{(p_1,q_1)} \otimes f_a^{(p_2,q_2)} \longmapsto f_a^{(p_1+p_2,q_1+q_2)}$ for $a = 0, 1, \infty$.

It is compatible with the morphisms $\phi_{(p,q),(p',q')}$, $\rho_{(p,q)}$ and $\iota_{(p,q)}$. □

LEMMA 3.30. *The isomorphism $\mathcal{O}(-n,-n) \simeq \mathbb{T}(n)$ is given by $f_a^{(-n,-n)} \longmapsto t_a^{(n)}$ for $a = 0, 1, \infty$.* □

REMARK 3.31. Let us fix a complex number a such that $a^2 = \sqrt{-1}$. If we put $\mathfrak{f}_0^{(p,q)} := a^{(p+q)} \cdot f_0^{(p,q)}$, $\mathfrak{f}_\infty^{(p,q)} := a^{-(p+q)} \cdot f_\infty^{(p,q)}$ and $\mathfrak{f}_1^{(p,q)} := a^{-p+q} \cdot \mathfrak{f}_1^{(p,q)}$. We also put $\mathfrak{t}_0^{(n)} := \sqrt{-1}^{-n} \cdot t_0^{(n)}$, $\mathfrak{t}_\infty^{(n)} := t_1^{(n)}$ and $\mathfrak{t}_1^{(n)} := t_\infty^{(n)}$. Then the number of "$\sqrt{-1}$" in the formalism would be reduced, and it would be easier to see the connection with the Sabbah's formalism in the section 3.10. □

3.3.3. The description as the Rees bundle. We have the following description as the Rees bundle. We put $\boldsymbol{C}(p,q) = \boldsymbol{C} \cdot e^{(p,q)}$. The decreasing filtrations ${}_aF_{(p,q)}$ $(a = 1,2)$ are defined as follows:

$${}_1F^i_{(p,q)} = \begin{cases} 0, & (i > p), \\ \boldsymbol{C}(p,q), & (i \le p), \end{cases} \qquad {}_2F^i_{(p,q)} = \begin{cases} 0, & (i > q), \\ \boldsymbol{C}(p,q), & (i \le q). \end{cases}$$

We have the Rees bundle $\xi\big(\boldsymbol{C}(p,q), {}_1F_{(p,q)}, {}_2F_{(p,q)}\big)$. The correspondence $e^{(p,q)} \longmapsto f_1^{(p,q)}$ induces the equivariant isomorphism $\xi\big(\boldsymbol{C}(p,q), {}_1F_{(p,q)}, {}_2F_{(p,q)}\big) \longrightarrow \mathcal{O}_{\mathbb{P}^1}(p,q)$

We have the isomorphism $\boldsymbol{C}(p,q)^\dagger \longrightarrow \boldsymbol{C}(q,p)$ given by $a \cdot \overline{e^{p,q}} \longmapsto \sqrt{-1}^{q-p} \cdot a \cdot e^{q,p}$. It induces the isomorphism $\iota_{(p,q)} : \sigma^*\mathcal{O}(p,q) \longrightarrow \mathcal{O}(q,p)$.

We have the pairing $\langle \cdot,\cdot \rangle_{(p,q)} : \boldsymbol{C}(p,q) \otimes \boldsymbol{C}(p,q)^\dagger \longrightarrow \boldsymbol{C}(n,n)$, which is given as follows:

$$e^{(p,q)} \otimes \overline{e^{(p,q)}} \longmapsto \sqrt{-1}^{q-p} \cdot e^{(n,n)}.$$

Since the real base of $\boldsymbol{C}(n,n)$ is given by $e^{(n,n)}$, the pairing $\sqrt{-1}^{p-q} \cdot \langle \cdot,\cdot \rangle_{(p,q)}$ is positive definite. Note that the pairing $\langle \cdot,\cdot \rangle$ corresponds to $S_{(p,q)}$ below.

3.3.4. Polarization of twistor structure. In the case $p+q=n$, the canonical pairing $S_{(p,q)} : \mathcal{O}(p,q) \otimes \sigma^*\mathcal{O}(p,q) \longrightarrow \mathbb{T}(-n)$ is defined to be the composite of the morphisms $\mathcal{O}(p,q) \otimes \sigma^*\mathcal{O}(p,q) \longrightarrow \mathcal{O}(p,q) \otimes \mathcal{O}(q,p) \longrightarrow \mathcal{O}(p+q,p+q) = \mathbb{T}(-n)$. More precise correspondence is as follows:

$$(3.8)\qquad \begin{aligned} f_1^{(p,q)} \otimes \sigma^*(f_1^{(p,q)}) &\longmapsto \sqrt{-1}^{q-p} \cdot f_1^{(p,q)} \otimes f_1^{(q,p)} \longmapsto \sqrt{-1}^{q-p} \cdot t_1^{(-n)}, \\ f_0^{(p,q)} \otimes \sigma^*(f_\infty^{(p,q)}) &\longmapsto (-\sqrt{-1})^n \cdot f_0^{(p,q)} \otimes f_0^{q,p} \longmapsto (-\sqrt{-1})^n \cdot t_0^{(-n)}, \\ f_\infty^{(p,q)} \otimes \sigma^*(f_0^{(p,q)}) &\longmapsto \sqrt{-1}^n \cdot f_\infty^{(p,q)} \otimes f_\infty^{(q,p)} \longmapsto \sqrt{-1}^n \cdot t_\infty^{(-n)}. \end{aligned}$$

LEMMA 3.32. *Let us consider the case $p+q = p'+q' = n$. Under the isomorphism $\mathcal{O}(p,q) \simeq \mathcal{O}(p',q')$ given by (3.5), we have $S_{(p,q)} = S_{(p',q')}$.*

Proof It immediately follows from (3.8). □

When we forget the torus action, we use the notation $S_{(n)}$ instead of $S_{(p,q)}$.

DEFINITION 3.33. Let V be a vector bundle on $\mathbb{P}^1$. A pairing $S: V\otimes\sigma^*V \longrightarrow \mathbb{T}(-n)$ and the isomorphism $\sigma^*\mathbb{T}(-n)\simeq\mathbb{T}(-n)$ induce the pairing $\sigma(S):\sigma^*V\otimes V\longrightarrow\mathbb{T}(-n)$. If $S\big(a\otimes\sigma(b)\big)=(-1)^n\sigma(S)\big(\sigma(b)\otimes a\big)$ holds, the paring S is called $(-1)^n$-symmetric. □

LEMMA 3.34. *The pairing $S_{(n)}$ is $(-1)^n$-symmetric,*

Proof We have only to compare $S_{(n)}(f_0^{(n)}\otimes\sigma(f_\infty^{(n)}))$ and $\sigma S_{(n)}\big(\sigma(f_\infty)\otimes f_0^{(n)}\big)$. We have $S_{(n)}\big(f_0^{(n)}\otimes\sigma(f_\infty^{(n)})\big)=(\sqrt{-1})^{-n}\cdot t_0^{(-n)}$ and $S_{(n)}\big(f_\infty^{(n)}\otimes\sigma(f_0^{(n)})\big)=(\sqrt{-1})^n t_\infty^{(-n)}$. We have the following correspondence:

$$\begin{aligned}(3.9)\quad \sigma\Big(S_{(n)}\big(f_\infty^{(n)}\otimes\sigma(f_0^{(n)})\big)\Big)&=\sigma\big(\sqrt{-1}\cdot t_\infty^{(-n)}\big)\\ &\longrightarrow(\sqrt{-1})^{-n}\cdot(-1)^n\cdot t_0^{(-n)}=\sigma(S_{(n)})\big(\sigma(f_\infty^{(n)})\otimes f_0^{(n)}\big).\end{aligned}$$

Hence we obtain $\sigma(S_{(n)})\big(\sigma(f_\infty^{(n)})\otimes f_0^{(n)}=(-1)^n\cdot S_{(n)}\big(f_0^{(n)}\otimes\sigma(f_\infty^{(n)})\big)$. Thus we are done. □

Let V be a $\mathbb{P}^1$-holomorphic bundle, and let $S: V\otimes\sigma(V)\longrightarrow\mathbb{T}(-n)$ be a pairing. Then the pairing $S_{(i)}$ induces the following pairing:

$$S(i):\big(V\otimes\mathcal{O}(i)\big)\otimes\sigma\big(V\otimes\mathcal{O}(i)\big)\longrightarrow\mathbb{T}(-n-i).$$

If the pairing S is $(-1)^n$-symmetric, then the induced pairing $S(i)$ is $(-1)^{n+i}$-symmetric.

DEFINITION 3.35 (Simpson). Let V be a pure twistor of weight n, and let $S: V\otimes\sigma(V)\longrightarrow\mathbb{T}(-n)$ is a $(-1)^n$-symmetric pairing.

- In the case $n=0$, the pairing S is called a polarization if the indueed hermitian pairing on $H^0(\mathbb{P}^1,V)$ is positive definite.
- For any n, the pairing S is called a polarization if the induced pairing $S(-n):\big(V\otimes\mathcal{O}(-n)\big)\otimes\sigma\big(V\otimes\mathcal{O}(-n)\big)\longrightarrow\mathbb{T}(0)$ is a polarization of the pure twistor structure $V\otimes\mathcal{O}(-n)$ of weight 0. □

3.3.5. The polarizations of the conjugate and the dual. Let V be a pure twistor structure of weight n. Recall that the polarization $S: V\otimes\sigma(V)\longrightarrow\mathbb{T}(-n)$ and the isomorphism $\sigma(\mathbb{T}(-n))\simeq\mathbb{T}(-n)$ induces the pairing $\sigma(S):\sigma(V)\otimes V\longrightarrow\mathbb{T}(-n)$ of the conjugates. We also have the induced pairing $V^\vee\otimes\sigma(V^\vee)\longrightarrow\mathbb{T}(n)$, which is obtained as follows: The perfect pairing S induces the isomorphisms $V^\vee\simeq\sigma(V)\otimes\mathbb{T}(n)$ and thus $\sigma(V^\vee)\simeq V\otimes\mathbb{T}(n)$. Hence we obtain the pairing $S_1:\sigma(V^\vee)\otimes V^\vee\longrightarrow\mathbb{T}(n)$ and thus $S^\vee:=\sigma(S_1):V^\vee\otimes\sigma(V^\vee)\longrightarrow\mathbb{T}(n)$. We show that $\sigma(S)$ and $S^\vee$ are polarizations.

First let us show the claim for the polarization $S_{(n)}:\mathcal{O}(n)\otimes\sigma\mathcal{O}(n)\longrightarrow\mathbb{T}(-n)$.

LEMMA 3.36. *The pairing $\sigma(S_{(n)})$ is a polarization of $\sigma\mathcal{O}(n)$.*

Proof We put $S := S_{(n)}$. Since S is $(-1)^n$-symmetric, $\sigma(S)$ is also $(-1)^n$-symmetric. The pairing $\sigma(S)$ is the composite of the following correspondences:

$$\sigma(f_0^{(n)})\otimes f_\infty^{(n)} \longmapsto \sigma\big((-\sqrt{-1})^n\cdot t_0^{(-n)}\big)\longmapsto (-\sqrt{-1})^n\cdot t_\infty^{(-n)},$$

$$\sigma(f_\infty^{(n)})\otimes f_0^{(n)} \longmapsto \sigma\big((\sqrt{-1})^n\cdot t_\infty^{(-n)}\big)\longmapsto (-\sqrt{-1})^{-n}\cdot t_0^{(-n)}.$$

Hence the pairing $\sigma(S)(-n)$ is given by the composite of the following correspondences:

$$\big(\sigma(f_0^{(n)})\otimes f_\infty^{(-n)}\big)\otimes\big(f_\infty^{(n)}\otimes\sigma(f_0^{(-n)})\big)\longmapsto(-1)^n\cdot t_\infty^{(-n)}\cdot t_\infty^{(n)}=(-1)^n\cdot t_\infty^{(0)},$$

$$\big(\sigma(f_\infty^{(n)})\otimes f_0^{(-n)}\big)\otimes\big(f_0^{(n)}\otimes\sigma(f_\infty^{(-n)})\big)\longmapsto(-1)^n\cdot t_0^{(-n)}\cdot t_0^{(n)}=(-1)^n\cdot t_0^{(0)}.$$

A global section s of $\sigma\big(\mathcal{O}(n)\big)\otimes\mathcal{O}(-n)$ is given as follows:

$$s=\sigma\big(f_0^{(n)}\big)\otimes f_\infty^{(-n)}=(-1)^n\cdot\sigma(f_\infty^{(n)})\otimes f_0^{(-n)}.$$

Hence we obtain $\sigma(S)(-n)\big(s\otimes\sigma(s)\big)=1$, which means $\sigma(S)$ is a polarization. □

LEMMA 3.37. *The pairing $S_{(n)}^\vee$ is a polarization of $\mathcal{O}(-n)=\mathcal{O}(n)^\vee$.*

Proof We put $S_{(n)}=S$. We recall $S\big(f_0^{(n)}\otimes\sigma(f_\infty^{(n)})\big)=(-\sqrt{-1})^n\cdot t_0^{(-n)}$. Hence the isomorphisms $\mathcal{O}(-n)\simeq\mathbb{T}(n)\otimes\sigma\mathcal{O}(n)$ and $\sigma\mathcal{O}(-n)\simeq\mathbb{T}(n)\otimes\mathcal{O}(n)$ induced by S gives the following correspondences:

$$f_0^{(-n)}\longmapsto\sqrt{-1}^{\,n}\cdot t_0^{(n)}\cdot\sigma(f_\infty^{(n)}),$$
$$\sigma(f_\infty^{(-n)})\longmapsto\big(-\sqrt{-1}\big)^n\cdot t_0^{(n)}\cdot f_0^{(n)}.$$

Hence the induced pairing $S_1:\sigma\mathcal{O}(-n)\otimes\mathcal{O}(-n)\longrightarrow\mathbb{T}(n)$ gives the following correspondence:

$$\sigma(f_\infty^{(-n)})\otimes f_0^{(-n)}\longmapsto\big(-\sqrt{-1}\big)^n\cdot t_0^{(n)}.$$

It is same as $\sigma\big(S(-n)\big)$. Hence we are done. □

LEMMA 3.38. *Let V be a pure twistor of weight n, and $S:V\otimes\sigma\big(V\big)\longrightarrow\mathbb{T}(-n)$ is a polarization. Then the following induced pairings are also the polarization:*

$$\sigma(S):\sigma(V)\otimes V\longrightarrow\mathbb{T}(-n),\qquad S^\vee:V^\vee\otimes\sigma(V^\vee)\longrightarrow\mathbb{T}(n).$$

Proof Since any polarized pure twistor of weight n is isomorphic to a direct sum of $\big(\mathcal{O}(n),S\big)$, the lemma can be reduced to the case $\big(\mathcal{O}(n),S\big)$. It has been already checked above. □

3.3.6. The equivariant case. Let (H,F,G) be a bi-filtered vector bundle, and we put $(V,\rho)=\xi(H,F,G)$. A pairing $S:(V,\rho)\otimes\sigma(V,\rho)\longrightarrow\mathbb{T}(-n)$ corresponds to the pairing $\langle\cdot,\cdot\rangle:(H,F,G)\otimes(H^\dagger,G^\dagger,F^\dagger)\longrightarrow\boldsymbol{C}(n,n)$. The correspondence is given as follows: S induces $C^\infty(X,H)[\lambda,\lambda^{-1}]\otimes C^\infty(X,H^\dagger)[\lambda,\lambda^{-1}]\longrightarrow C^\infty\big(X,\boldsymbol{C}(n,n)\big)[\lambda,\lambda^{-1}]$, and thus $H\otimes H^\dagger\longrightarrow\boldsymbol{C}(n,n)$.

LEMMA 3.39. *Assume that (H,F,G) is defined over $\boldsymbol{R}$. Then S is defined over $\boldsymbol{R}$ if and only if $\langle\cdot,\cdot\rangle$ is defined over $\boldsymbol{R}$.* □

LEMMA 3.40. *A morphism $V\longrightarrow V\otimes\mathbb{T}(-1)$ corresponds to a morphism $(H,F,G)\longrightarrow(H,F,G)\otimes\boldsymbol{C}(1,1)$.*

Proof It follows from $\mathbb{T}(-1) = \xi(\boldsymbol{C}(1,1))$ and the equivalence of the categories (Proposition 3.12). □

Let f be a morphism $(V,\rho) \longrightarrow (V,\rho) \otimes \mathbb{T}(-1)$, which corresponds to $f_{|1} := f_{|\lambda=1} : (H,F,G) \longrightarrow (H,F,G) \otimes \boldsymbol{C}(1,1)$.

LEMMA 3.41. *$S(f\otimes\mathrm{id})+S\big(\mathrm{id}\otimes\sigma(f)\big) = 0$ if and only if $\langle f_{|1}\otimes\mathrm{id}\rangle+\langle\mathrm{id}\otimes\sigma(f)_{|1}\rangle = 0$.* □

We remark that the torus action identifies $V_{|\lambda=1} \simeq V_{|\lambda=-1}$, and we have $f_{|1} = \sigma(f)_{|1}$ under the identification. Recall that we have the isomorphism $\boldsymbol{C}(1,1) \otimes \boldsymbol{C}(-1,-1) \longrightarrow \boldsymbol{C}(0,0)$, given by the following:

$$\big(t^{(-1)}_{1\,|\,\lambda=1}\big) \otimes \big(t^{(1)}_{1\,|\,\lambda=1}\big) \longmapsto t^{(0)}_{1\,|\,\lambda=1}.$$

The multiplication $t^{(1)}_{1\,|\,\lambda=1}$ gives the isomorphism of the vector spaces $\boldsymbol{C}(1,1)$ and $\boldsymbol{C}(0,0)$, which preserves the real bases. Under the isomorphism, we can identify the morphisms $\big(f \cdot t_1^{(1)}\big)_{|\,\lambda=1}$ and $f_{|\lambda=1}$. We put $N := -(f \cdot t_1^{(1)})_{|\lambda=1}$.

Let $\sigma(N)$ denote the naturally induced endomorphism of $H^{\dagger}$ by N, which is same as N as a map of sets. We also have $-\sigma(f \cdot t_1^{(1)})_{|1} = N^{\dagger}$.

LEMMA 3.42. *Under the identification, we have the following identities:*

$$\big\langle f_{|1} \otimes \mathrm{id}\big\rangle + \big\langle \mathrm{id} \otimes\sigma(f)_{|1}\big\rangle = -\Big(\big\langle N \otimes \mathrm{id}\big\rangle + \big\langle \mathrm{id}\otimes\sigma(N)\big\rangle\Big).$$

$$(-1)^h \cdot \big\langle \mathrm{id}\otimes\sigma\big(f^h\big)_{|\lambda=1}\big\rangle = \big\langle \mathrm{id}\otimes\sigma(N^h)\big\rangle.$$

Proof We have the following:

$$\begin{aligned}(3.10)\quad \big\langle f_{|1} \otimes \mathrm{id}\big\rangle + \big\langle \mathrm{id}\otimes\sigma(f)_{|1}\big\rangle &= -\big\langle\big(f \otimes t_1^{(1)}\big)_{|1} \otimes \mathrm{id}\big\rangle - \big\langle \mathrm{id}\otimes\sigma(f \otimes t_1^{(1)})_{|1}\big\rangle \\ &= -\big(\big\langle N \otimes \mathrm{id}\big\rangle + \big\langle \mathrm{id}\otimes\sigma(N)\big\rangle\big).\end{aligned}$$

We also have the following:

$$(-1)^h \cdot \big\langle \mathrm{id}\otimes\sigma(f)_{|\lambda=1}\big\rangle = \big\langle \mathrm{id}\otimes\sigma\big(\big(-f \cdot t_1^{(1)}\big)^h\big)_{|\lambda=1}\big\rangle = \big\langle \mathrm{id}\otimes\sigma(N^h)\big\rangle.$$

Thus we are done. □

3.4. Equivalence of some categories

3.4.1. Complex Hodge structure and the equivariant twistor structure. In this subsection, "bundle" means a family in the C^{∞}-category. We put $(V,\rho) := \xi(H,F,G)$ for a bi-filtered vector bundle (H,F,G) bundle over a C^{∞}-manifold X.

LEMMA 3.43. *V is a pure twistor bundle of weight n if and only if (H,F,G) is complex Hodge structure bundle, i.e., F and G are n-opposed.*

Proof If F and G are n-opposed, the decomposition $H = \bigoplus_{p+q=n} H^{p,q}$ is given, where we put $H^{p,q} = F^p \cap G^q$. Then $\xi(H,F,G) \simeq \bigoplus_{p+q=n} H^{p,q} \otimes \mathcal{O}(p,q)$. Hence it is pure twistor of weight n.

We can assume X is a point. By considering $V \otimes \mathcal{O}(-n,0)$, we can reduce the problem to the case $n = 0$. We have the weight decomposition

$$H^0(\mathbb{P}^1, V) = \bigoplus_h U_h.$$

Here U_h denotes the weight h-space. Then U_h gives the subbundle $U_h \otimes \mathcal{O}_{\mathbb{P}^1} \subset V$, which is isomorphic to a direct sum of $\mathcal{O}(h,-h)$, and $(V,\rho) = \bigoplus_h U_h \otimes \mathcal{O}(h,-h)$. Then it can be checked that the corresponding filtrations F and G are n-opposed. □

COROLLARY 3.44. *The functor ξ gives the equivalence of the following two categories:*

- *The category of equivariant pure twistor bundle of weight n.*
- *The category of complex pure Hodge structure bundle of weight n.*

The functor ξ gives the equivalence of the following two categories:

- *The category of equivariant pure twistor bundle of weight n defined over $\boldsymbol{R}$.*
- *The category of real pure Hodge structure bundle of weight n.* □

We recall that a filtered holomorphic bundle (V,W) is called a mixed twistor structure, if $\mathrm{Gr}_n^W(V)$ is a pure twistor of weight n (see [**83**] or [**65**]).

COROLLARY 3.45. *The functor ξ gives the equivalence of the following two categories:*

- *The category of equivariant mixed twistor bundle.*
- *The category of complex mixed Hodge structure bundle.*

The functor ξ gives the equivalence of the following two categories:

- *The category of equivariant mixed twistor bundle defined over $\boldsymbol{R}$.*
- *The category of real mixed Hodge structure bundle.* □

We would like to compare the polarizations of equivariant pure twistor structures and pure Hodge structures. Let $S : (V,\rho)\otimes\sigma(V,\rho) \longrightarrow \mathbb{T}(-n)$ be an equivariant pairing. Let $\langle\cdot,\cdot\rangle : (H,F,G)\otimes(H^\dagger,G^\dagger,F^\dagger) \longrightarrow \boldsymbol{C}(n,n)$ be the corresponding pairing.

LEMMA 3.46. *The pairing S is a polarization of twistor structure if and only if the induced pairing $\langle\cdot,\cdot\rangle$ is a polarization of Hodge structure.*

Proof We may assume that X is a point. We have the following:

$$\sigma^* S\big(a\cdot u\otimes\sigma^*(b\cdot v)\big) = \sigma^*\Big(a\cdot\overline{\sigma^*(b)}\cdot\langle u,v\rangle\Big) = \overline{\sigma^*(a)}\cdot b\cdot\overline{\langle u,v\rangle}.$$

We also have the following equality:

$$S\big(b\cdot v\otimes\sigma^*(a\cdot u)\big) = \overline{\sigma^*(a)}\cdot b\cdot\langle v,u\rangle.$$

Hence S is $(-1)^n$-symmetric if and only if $\langle\cdot,\cdot\rangle$ is $(-1)^n$-hermitian symmetric.

To compare the positive definiteness conditions, note the following:

$$H^0\big(\mathbb{P}^1, V\otimes\mathcal{O}(-n)\big) = \bigoplus_{p+q=n}\big\{\lambda^{-p+n}\cdot u \,\big|\, u\in H^{p,q}\big\}.$$

Then we have the following:

$$\text{(3.11)}\quad S(-n)\Big(\lambda^{-p+n}\cdot u\otimes f_1^{(-n,0)}\otimes\sigma\big(\lambda^{-p+n}\cdot u\otimes f_1^{(-n,0)}\big)\Big)$$
$$= \langle u,u\rangle\cdot\lambda^{-p+n}\cdot(-\lambda)^{p-n}\cdot f_1^{(-n,0)}\otimes\big(\sqrt{-1}^{\,n}\cdot f_1^{(0,-n)}\big) = \langle u,u\rangle\cdot t_1^{(n)}\cdot\sqrt{-1}^{\,p-q}.$$

Here we have used the equality $(-1)^{p-n}\cdot\sqrt{-1}^{n}=\sqrt{-1}^{p-q}$. Thus $H^0(S(-n))$ is positive definite if and only if $\sqrt{-1}^{p-q}\cdot\langle\cdot,\cdot\rangle$ is positive definite. Note that the real base of $\boldsymbol{C}(n,n)$ is fixed as $t_1^{(-n)}$. □

COROLLARY 3.47. *The functor ξ gives an equivalence of the following categories:*

- *The category of the equivariant polarized pure twistor bundle of weight n.*
- *The category of polarized complex pure Hodge structure bundle of weight n.*

The functor ξ gives an equivalence of the following categories:

- *The category of the equivariant polarized pure twistor bundle of weight n defined over $\boldsymbol{R}$.*
- *The category of polarized real pure Hodge structure bundle of weight n.*

□

3.4.2. Polarized mixed twistor structure.

DEFINITION 3.48. A polarized mixed twistor structure of weight n in one variable is a tuple (V,W,f,S) as follows:

(1) (V,W) is a mixed twistor structure.
(2) $f:V\longrightarrow V\otimes\mathbb{T}(-1)$ is a morphism of mixed twistors, and it is nilpotent. The weight filtration of f is denoted by $W(f)$.
(3) $S:V\otimes\sigma(V)\longrightarrow\mathbb{T}(-n)$ is a $(-1)^n$-symmetric pairing of mixed twistor satisfying the following:
$$S(f\otimes\mathrm{id})+S(\mathrm{id}\otimes\sigma(f))=0.$$
Note that we obtain the induced morphism:
$$S:\mathrm{Gr}^W_{h+n}\otimes\mathrm{Gr}^W_{-h+n}\longrightarrow\mathbb{T}(-n).$$
(4) We have $W_h=W(f)_{h-n}$ for any h.
(5) The induced pairing $S(f^h\otimes\mathrm{id})=(-1)^h\cdot S\bigl(\mathrm{id}\otimes\sigma(f^h)\bigr)$ gives the polarization of the primitive part $P\,\mathrm{Gr}^W_{h+n}=\mathrm{Ker}\bigl(N^{h+1}:\mathrm{Gr}^W_{h+n}\longrightarrow\mathrm{Gr}^W_{n-h-2}\bigr)$.

If a tuple (V,W,f,S) satisfies the conditions (1)–(4), then it is called a pseudo-polarized mixed twistor structure of weight n in one variable. □

Assume (V,W,f,S) is a pseudo-polarized mixed twistor structure of weight n in one variable. We put as follows:
$$V^{(0)}:=\mathrm{Gr}^W(V),\quad W^{(0)}_h=\bigoplus_{i\leq h}\mathrm{Gr}^W_i(V).$$
We obtain the induced morphism $f^{(0)}:\mathrm{Gr}^W_i\longrightarrow\mathrm{Gr}^W_{i-2}\otimes\mathbb{T}(-1)$, and $f^{(0)}:V^{(0)}\longrightarrow V^{(0)}$. We also obtain the induced morphism $S^{(0)}:V^{(0)}\otimes\sigma(V^{(0)})\longrightarrow\mathbb{T}(-n)$. Then it is easy to check that $\bigl(V^{(0)},W^{(0)},f^{(0)},S^{(0)}\bigr)$ is also a pseudo-polarized mixed twistor structure of weight n in one variable.

The following lemma is clear.

LEMMA 3.49. *Let (V,W,f,S) be a pseudo polarized mixed twistor structure of weight n in one variable. Then it is a polarized mixed twistor structure of weight n in one variable, if and only if the induced tuple $(V^{(0)},W^{(0)},f^{(0)},S^{(0)})$ is a polarized mixed twistor structure of weight n in one variable.* □

DEFINITION 3.50. A polarized mixed twistor structure of weight n in l variables is a tuple $(V, W, \boldsymbol{f}, S)$ as follows:

- (V, W) is a mixed twistor structure.
- $\boldsymbol{f}$ is a commuting tuple of nilpotent morphisms $f_i : V \longrightarrow V \otimes \mathbb{T}(-1)$ $(i = 1, \ldots, l)$ of mixed twistor structures.
- $S : V \otimes \sigma(V) \longrightarrow \mathbb{T}(-n)$ is a $(-1)^n$-symmetric pairing of mixed twistor structures satisfying the following:

$$S(f_i \otimes \mathrm{id}) + S\bigl(\mathrm{id} \otimes \sigma(f_i)\bigr) = 0.$$

- For any element $\boldsymbol{a} \in \boldsymbol{C}^l$, we put $f(\boldsymbol{a}) := \sum a_i \cdot f_i$. Then $\bigl(V, W, f(\boldsymbol{a}), S\bigr)$ is a polarized mixed twistor structure in one variable of weight n for any element $\boldsymbol{a} \in \boldsymbol{R}^l_{>0}$.

For simplicity, polarized mixed twistor structure of weight n in l-variables is abbreviated to Pol-MTS of (n, l)-type.

If a tuple $(V, W, \boldsymbol{f}, S)$ satisfies the first three conditions, then it is called pseudo polarized mixed twistor structure of weight n in l-variable. Similarly, pseudo polarized mixed twistor structure of weight n in l-variable is abbreviated to Ψ-Pol-MTS of (n, l)-type. □

Let $(V, W, \boldsymbol{f}, S)$ is a Ψ-Pol-MTS of (n, l)-type. Then we put as follows:

$$V^{(0)} := \mathrm{Gr}^W(V), \quad W_h^{(0)} := \bigoplus_{i \leq h} \mathrm{Gr}_i^W(V).$$

We have the induced morphism $f_j^{(0)} : \mathrm{Gr}_i^W(V) \longrightarrow \mathrm{Gr}_{i-2}^W(V) \otimes \mathbb{T}(-1)$ and $f_j^{(0)} : V^{(0)} \longrightarrow V^{(0)} \otimes \mathbb{T}(-1)$. We also have the induced morphism:

$$S^{(0)} : V^{(0)} \otimes \sigma(V^{(0)}) \longrightarrow \mathbb{T}(-n),$$

$$S^{(0)} : \mathrm{Gr}_{i+n}^W(V) \otimes \sigma\bigl(\mathrm{Gr}_{-i+n}^W(V)\bigr) \longrightarrow \mathbb{T}(-n).$$

Then it is easy to check that $\bigl(V^{(0)}, W^{(0)}, \boldsymbol{f}^{(0)}, S^{(0)}\bigr)$ is also a Ψ-Pol-MTS of (n, l)-type.

LEMMA 3.51. *Let $(V, W, \boldsymbol{f}, S)$ is a Ψ-Pol-MTS of (n, l)-type. It is a Pol-MTS of (n, l)-type, if and only if the induced tuple $(V^{(0)}, W^{(0)}, \boldsymbol{f}^{(0)}, S^{(0)})$ is a Pol-MTS of (n, l).*

Proof It follows from Lemma 3.49. □

LEMMA 3.52. *The functor ξ gives the following categories:*

- *The category of polarized equivariant mixed twistor structure.*
- *The category of polarized mixed Hodge structures defined over $\boldsymbol{C}$.*

The functor ξ gives the following categories:

- *The category of equivariant polarized mixed twistor structure defined over $\boldsymbol{R}$.*
- *The category of polarized mixed Hodge structure defined over $\boldsymbol{R}$.*

Proof It follows from Lemma 3.41, Lemma 3.42 and Corollary 3.47. □

3.5. Variation of $\mathbb{P}^1$-holomorphic bundles

3.5.1. Definition and some functorial properties. We put as follows:

$$\xi\Omega^1_X := \Omega^{1,0}_X \otimes \mathcal{O}_{\mathbb{P}^1}(1,0) \oplus \Omega^{0,1}_X \otimes \mathcal{O}_{\mathbb{P}^1}(0,1), \qquad \xi\Omega^h_X := \bigwedge^h \bigl(\xi\Omega^1_X\bigr).$$

We have the differential operator $\mathbb{D}^{\triangle}_X$ defined as follows:

$$\mathbb{D}^{\triangle}_X : C^\infty(X) \longrightarrow C^\infty(X, \xi\Omega^1_X), \quad g \longmapsto \partial_X(g) \otimes f^{(1,0)}_\infty + \overline{\partial}_X(g) \otimes \bigl(\sqrt{-1} \cdot f^{(0,1)}_0\bigr).$$

When we forget the torus action, we can use the notation $f^{(1)}_x$ $(x = 0, 1, \infty)$ instead of $f^{(1,0)}_x$ or $f^{(0,1)}_x$. On the open subsets $X \times \boldsymbol{C}_\lambda$, $X \times \boldsymbol{C}_\mu$ and $X \times \boldsymbol{C}^*_\lambda$, it can be regarded as follows: On $X \times \boldsymbol{C}_\lambda$, we take the base $f^{(1)}_0$ of $\mathcal{O}(1)_{|\boldsymbol{C}_\lambda}$. Then $\mathbb{D}^{\triangle}_X$ induces the operator $(\overline{\partial}_X + \lambda \cdot \partial_X) \otimes (\sqrt{-1} \cdot f^{(1)}_0)$. On $X \times \boldsymbol{C}_\mu$, we take the base $f^{(1)}_\infty$ of $\mathcal{O}(1)_{|\boldsymbol{C}_\mu}$. Then $\mathbb{D}^{\triangle}_X$ induces the operator $(\partial_X + \mu\overline{\partial}_X) \otimes f^{(1)}_\infty$. On $X \times \boldsymbol{C}^*_\lambda$, we take the base of $f^{(1,0)}_\infty$ of $\mathcal{O}(1,0)$ and $\sqrt{-1} \cdot f^{(0,1)}_0$ of $\mathcal{O}(0,1)$, and then $\mathbb{D}^{\triangle}$ induces the operator $d_X = \overline{\partial}_X + \partial_X$. The following lemma can be checked easily.

LEMMA 3.53.

(1) *The Leibniz rule holds in the following sense:*

$$\mathbb{D}^{\triangle}_X(f \cdot g) = f \cdot \mathbb{D}^{\triangle}_X(g) + \mathbb{D}^{\triangle}_X(f) \cdot g.$$

(2) *We have the induced operator $\mathbb{D}^{\triangle}_X$ on $C^\infty(X \times \mathbb{P}^1, \xi\Omega^{\cdot}_X)$, and we have the flatness $(\mathbb{D}^{\triangle}_X + \overline{\partial}_{\mathbb{P}^1})^2 = 0$.* □

Let V be a $\mathbb{P}^1$-holomorphic vector bundle over $X \times \mathbb{P}^1$.

DEFINITION 3.54. A variation of $\mathbb{P}^1$-holomorphic vector bundle over $X \times \mathbb{P}^1$ is defined to be a differential operator $\mathbb{D}^{\triangle}_V : C^\infty(X \times \mathbb{P}^1, V) \longrightarrow C^\infty\bigl(X \times \mathbb{P}^1, V \otimes \xi\Omega^1_X\bigr)$ satisfying the following conditions:

$$\bigl(\mathbb{D}^{\triangle}_V + d''_V\bigr)^2 = 0, \qquad \mathbb{D}^{\triangle}_V(f \cdot v) = f \cdot \mathbb{D}^{\triangle}_V(v) + (\mathbb{D}^{\triangle}_X f) \cdot v.$$

□

A tensor product, a direct sum and a dual for variations of $\mathbb{P}^1$-holomorphic vector bundles are naturally defined. Let $(V^{(i)}, \mathbb{D}^{\triangle}_{V^{(i)}})$ $(i = 1, 2)$ be variation of $\mathbb{P}^1$-holomorphic vector bundles. A morphism of $(V^{(1)}, \mathbb{D}^{\triangle}_{V^{(1)}})$ to $(V^{(2)}, \mathbb{D}^{\triangle}_{V^{(2)}})$ is defined to be a $\mathbb{P}^1$-holomorphic and $\mathbb{D}^{\triangle}_V$-flat section of $Hom(V^{(1)}, V^{(2)})$.

REMARK 3.55. If we use the basis $\mathfrak{f}^{(p,q)}_x$ $(x = 0, \infty)$ as in Remark 3.31, it is appropriate to use $a^{-1} \cdot \mathbb{D}^{\triangle} = \partial_X \otimes \mathfrak{f}^{(1)}_\infty + \overline{\partial}_X \otimes \mathfrak{f}^{(1)}_0$. □

3.5.2. Some description of a variation of $\mathbb{P}^1$-holomorphic bundles. Let V be a $\mathbb{P}^1$-holomorphic vector bundle with variation $\mathbb{D}^{\triangle}_V$. On $X \times \boldsymbol{C}_\lambda$, we take the base $f^{(1)}_0$ of $\mathcal{O}(1)$, and then $\mathbb{D}^{\triangle}_V$ induces the λ-connection $\mathbb{D}_V : C^\infty(X \times \boldsymbol{C}_\lambda, V_0) \longrightarrow C^\infty(X \times \boldsymbol{C}_\lambda, V_0 \otimes \Omega^1_X)$, and we have the flatness $(\mathbb{D}_{V_0} + d'')^2 = 0$. On $X \times \boldsymbol{C}_\mu$, we take the base $f^{(1)}_\infty$ of $\mathcal{O}(1)$, and then $\mathbb{D}^{\triangle}_{V_\infty}$ induces the μ-connection $\mathbb{D}^{\dagger}_{V_\infty} : C^\infty\bigl(X^\dagger \times \boldsymbol{C}_\mu, V_\infty\bigr) \longrightarrow C^\infty\bigl(X^\dagger \times \boldsymbol{C}_\mu, V_\infty \otimes \Omega^1_{X^\dagger}\bigr)$, and we have the flatness $(\mathbb{D}^{\dagger}_{V_\infty} + d'')^2 = 0$. On $X \times \boldsymbol{C}^*_\lambda$, we take the bases $f^{(1,0)}_\infty$ of $\mathcal{O}(1,0)$ and $f^{(0,1)}_0$ of

$\mathcal{O}(0,1)$, and then $\mathbb{D}^{\triangle}$ induces the family of holomorphic connections $\mathbb{D}^f_{V_1} : C^\infty\bigl(X \times \boldsymbol{C}^*_\lambda, V_1\bigr) \longrightarrow C^\infty\bigl(X \times \boldsymbol{C}^*_\lambda, V_1 \otimes \Omega^1_X\bigr)$, and we have the flatness $(\mathbb{D}^f_{V_1} + d'')^2 = 0$.

On the other hand, we can consider a patched object $(V_0, V_\infty, V_1; \alpha_0, \alpha_\infty)$:

- V_0 is a $\boldsymbol{C}_\lambda$-holomorphic vector bundle over $X \times \boldsymbol{C}_\lambda$, which is equipped with the λ-connection $\mathbb{D}_{V_0}$ such that $(\mathbb{D}_{V_0} + d'')^2 = 0$.
- V_∞ is a $\boldsymbol{C}_\mu$-holomorphic vector bundle over $X \times \boldsymbol{C}_\mu$, which is equipped with the μ-connection $\mathbb{D}_{V_\infty}$ such that $(\mathbb{D}_{V_\infty} + d'')^2 = 0$.
- V_1 is a $\boldsymbol{C}^*_\lambda$-holomorphic vector bundle over $X \times \boldsymbol{C}^*_\lambda$, which is equipped with the holomorphic family of the flat connections $\mathbb{D}^f_{V_1}$.
- We have the induced families of flat connections $(V_{0\,|\,X\times\boldsymbol{C}^*_\lambda}, \mathbb{D}^f_{V_0})$ and $(V_{\infty\,|\,X\times\boldsymbol{C}^*_\lambda}, \mathbb{D}^{\dagger f}_{V_\infty})$. Then α_a $(a = 0, \infty)$ are isomorphisms $V_{a\,|\,X\times\boldsymbol{C}^*_\lambda} \longrightarrow V_{1\,|\,X\times\boldsymbol{C}^*_\lambda}$, which are compatible with the $\boldsymbol{C}^*_\lambda$-holomorphic structure and the family of flat connections.

Once we are given such a patched object $(V_0, V_\infty, V_1; \alpha_0, \alpha_\infty)$, then we have the $\mathbb{P}^1$-holomorphic vector bundle V by gluing.

LEMMA 3.56. *We have the well defined differential operator $\mathbb{D}^{\triangle}_V$ given as follows:*

$$\mathbb{D}^{\triangle}_V(f) = \begin{cases} \mathbb{D}_{V_0}(f) \otimes \bigl(\sqrt{-1}\cdot f_0^{(1)}\bigr), & (\text{on } X \times \boldsymbol{C}_\lambda), \\ \mathbb{D}^\dagger_{V_\infty}(f) \otimes f_\infty^{(1)}, & (\text{on } X \times \boldsymbol{C}_\mu). \end{cases} \tag{3.12}$$

The operator $\mathbb{D}^{\triangle}_V$ gives a variation of $\mathbb{P}^1$-holomorphic vector bundles.

Proof The well definedness follows from the compatibility of α_a $(a = 0, \infty)$ with the flat connections. It is easy to see that $\mathbb{D}^{\triangle}_V$ gives a variation. □

COROLLARY 3.57. *The category of the patched objects is equivalent to the category of the variations of $\mathbb{P}^1$-holomorphic vector bundles, by the correspondence given above.* □

We can also consider patched objects $(V_0, V_\infty; \psi)$:

- V_0 is a $\boldsymbol{C}_\lambda$-holomorphic vector bundle equipped with λ-connection $\mathbb{D}_{V_0}$.
- V_∞ is a $\boldsymbol{C}_\mu$-holomorphic vector bundle equipped with μ-connection $\mathbb{D}^\dagger_{V_\infty}$.
- ψ is an isomorphism of the induced family of the flat bundles $V_{0\,|\,\boldsymbol{C}^*_\lambda\times X}$ and $V_{1\,|\,\boldsymbol{C}^*_\lambda\times X}$.

As before, the category of such patched objects and the category of variations of $\mathbb{P}^1$-holomorphic vector bundles are naturally equivalent.

We often use the descriptions $\bigl(V_0, V_\infty, V_1; \alpha_0, \alpha_\infty\bigr)$ or (V_0, V_∞, ψ) to denote the variation of $\mathbb{P}^1$-holomorphic vector bundles in the following.

3.5.3. An example of variation of the $\mathbb{P}^1$-holomorphic vector bundle. Let V be a vector bundle over $\mathbb{P}^1$, and $f_i : V \longrightarrow V \otimes \mathbb{T}(-1)$ $(i = 1, \ldots, n)$ be nilpotent morphisms such that f_i and f_j are commutative. We put $X := \boldsymbol{C}^n$, $D_i := \{z_i = 0\}$, and $D = \bigcup_{i=1}^l D_i$. We will construct the $\mathbb{P}^1$-holomorphic vector bundle $\mathcal{V}$ and the variation $\mathbb{D}^{\triangle}_{\mathcal{V}}$ over $X - D$.

We put $V_0 := V_{|\boldsymbol{C}_\lambda}$ and $V_\infty := V_{|\boldsymbol{C}_\mu}$. We put as follows:

$$\mathcal{V}_0 := V_0 \otimes \mathcal{O}_{\mathcal{X}-\mathcal{D}}, \qquad \mathcal{V}_\infty := V_\infty \otimes \mathcal{O}_{\mathcal{X}^\dagger-\mathcal{D}^\dagger}.$$

From the morphism $f_i \in Hom(V, V \otimes \mathbb{T}(-1))$, we obtain the morphism $f_i \in Hom(V_0, V_0 \otimes \mathbb{T}(-1))$ on $\boldsymbol{C}_\lambda$, and then we have the endomorphism $f_i \otimes t_0^{(1)} \in \mathrm{End}(V_0)$. Then the λ-connection $\mathbb{D}_{\mathcal{V}_0}$ is given. Namely, for any $v \in \Gamma(\boldsymbol{C}_\lambda, V_0)$ and $g \in C^\infty(\mathcal{X} - \mathcal{D})$, we put as follows:

$$\mathbb{D}_{\mathcal{V}_0}(g \cdot v) := \sum_{i=1}^{l} g \cdot f_i(v) \cdot t_0^{(1)} \cdot \frac{dz_i}{z_i} + \big(\lambda \cdot \partial_X(g) + \overline{\partial}_X(g)\big) \cdot v. \tag{3.13}$$

We also have the endomorphism $f_i \otimes t_\infty^{(1)} \in \mathrm{End}(V_\infty)$ on $\boldsymbol{C}_\mu$. The μ-connection $\mathbb{D}^\dagger_{\mathcal{V}_\infty}$ is given. Namely, for any $v^\dagger \in \Gamma(\boldsymbol{C}_\mu, V_\infty)$ and $g \in C^\infty(\mathcal{X}^\dagger - \mathcal{D}^\dagger)$, we put as follows:

$$\mathbb{D}^\dagger_{\mathcal{V}_\infty}(g \cdot v^\dagger) := \sum_{i=1}^{l} g \cdot f_i(v^\dagger) \cdot t_\infty^{(1)} \cdot \frac{d\bar{z}_i}{\bar{z}_i} + \big(\mu \cdot \overline{\partial}_X(g) + \partial_X(g)\big) \cdot v^\dagger. \tag{3.14}$$

We will give the isomorphism $\Psi : \mathcal{V}_{0|\mathcal{X}^\sharp - \mathcal{D}^\sharp} \longrightarrow \mathcal{V}_{\infty|\mathcal{X}^{\dagger\sharp} - \mathcal{D}^{\dagger\sharp}}$. Let $\lambda = \mu^{-1}$ be an element of $\boldsymbol{C}^*_\lambda$ and P be a point of $X - D$. Let v be an element of $V_{|\lambda}$. It gives the elements $v_0 \in V_{0\,|\,\lambda}$ and $v_\infty \in V_{\infty\,|\,\mu}$. Then we naturally obtain the following elements:

$$v_{0\,|\,P} \in \mathcal{V}_{0|(\lambda,P)}, \qquad v^\dagger_{\infty\,|\,P} \in \mathcal{V}_{\infty|(\mu,P)}.$$

Then Ψ is defined as follows:

$$\Psi\Big(\exp\Big(-\sum_{i=1}^{n} \lambda^{-1} \cdot \log|z_i(P)|^2 \cdot f_i \otimes t_0^{(1)}\Big) \cdot v_{0|(\lambda,P)}\Big) = v_{\infty|(\mu,P)}, \tag{3.15}$$

or equivalently,

$$\Psi(v_{0\,|\,P}) = \exp\Big(-\sum_{i=1}^{n} \mu^{-1} \cdot \log|z_i(P)|^2 \cdot f_i \otimes t_\infty^{(1)}\Big) \cdot v_{\infty\,|\,P}.$$

Here we have used the relation $\lambda^{-1} \cdot t_0^{(1)} = -\mu^{-1} \cdot t_\infty^{(1)}$.

Corresponding to $v \in V_{|\lambda}$, we have the multi-valued flat sections of $\mathcal{V}_{0\,|\,\mathcal{X}^\lambda - \mathcal{D}^\lambda}$ and $\mathcal{V}_{\infty\,|\,\mathcal{X}^{\dagger\mu} - \mathcal{D}^{\dagger\mu}}$:

$$\tilde{s}_0 := \exp\Big(-\sqrt{-1}\sum_{i=1}^{n} \log z_i \cdot f_i \otimes t_1^{(1)}\Big) \cdot v_0,$$

$$\tilde{s}_\infty := \exp\Big(\sqrt{-1}\sum_{i=1}^{n} \log \overline{z}_i \cdot f_i \otimes t_1^{(1)}\Big) \cdot v_\infty.$$

LEMMA 3.58. *We have the relation $\tilde{s}_0 = \tilde{s}_\infty$. The gluing Ψ is characterized by this property.*

Proof It can be easily checked by a direct calculation. □

LEMMA 3.59. *Ψ is compatible with $\mathbb{D}_{\mathcal{V}_0}$ and $\mathbb{D}^\dagger_{\mathcal{V}_\infty}$.*

Proof We put as follows:

$$\tilde{v} := \exp\Big(-\sum_{i=1}^{n} \lambda^{-1} \cdot \log|z_i|^2 \cdot f_i \otimes t_0^{(1)}\Big) v_0.$$

Then we have the following:

$$(3.16)\quad \mathbb{D}^f_{\mathcal{V}_0}(\tilde{v}) = \sum \lambda^{-1}\cdot f_i(\tilde{v})\otimes t_0^{(1)}\frac{dz_i}{z_i} - \sum\lambda^{-1}\Big(\frac{dz_i}{z_i}+\frac{d\bar{z}_i}{\bar{z}_i}\Big)\cdot f_i(\tilde{v})\otimes t_0^{(1)}$$
$$= -\sum\lambda^{-1} f_i(\tilde{v})\otimes t_0^{(1)}\frac{d\bar{z}_i}{\bar{z}_i} = \mu^{-1}\sum f_i(\tilde{v})\otimes t_\infty^{(1)}\frac{d\bar{z}_i}{\bar{z}_i}.$$

Hence it is compatible with the definition of $\mathbb{D}^{\dagger f}_{\mathcal{V}_\infty}$ given in (3.14). □

We put $N_i := -\big(f_i\otimes t_1^{(1)}\big)_{|\lambda=1}$.

LEMMA 3.60. *The endomorphism* $\exp\big(2\pi\cdot N_i\big)$ *is the monodromy of the loop* $\{0\le t\le 1\}\longrightarrow \boldsymbol{C}^{*\,n}$ *given by the following:*

$$t\longmapsto\big(z_1,\ldots,z_{i-1},\exp(-2\pi\sqrt{-1}t)\cdot z_i, z_{i+1},\ldots,z_n\big).$$

Proof It can be checked by a direct calculation. □

REMARK 3.61. When we consider the monodromy in this paper, we usually use the loop with the inverse direction:

$$t\longmapsto\big(z_1,\ldots,\cdot z_{i-1},\exp(2\pi\sqrt{-1}t)\cdot z_i, z_{i+1},\ldots,z_n\big).$$

On the other hand, it seems standard to use the loop given in Lemma 3.60 in the classical Hodge theory. It raises the difference of the signatures of the nilpotent maps. □

3.5.4. The involution and the induced variation. We have the isomorphisms $\iota_{1,0}:\sigma\mathcal{O}(1,0)\longrightarrow\mathcal{O}(0,1)$ and $\iota_{0,1}:\sigma\mathcal{O}(0,1)\longrightarrow\mathcal{O}(1,0)$. We take the isomorphisms $\sigma\Omega_X^{1,0}\longrightarrow\Omega_X^{0,1}$ and $\sigma\Omega_X^{0,1}\longrightarrow\Omega_X^{1,0}$ given by the following:

$$dz_i\longmapsto -d\bar{z}_i,\qquad d\bar{z}_i\longmapsto dz_i.$$

Then we obtain the morphisms $\sigma^*\big(\Omega_X^{1,0}\otimes\mathcal{O}(1,0)\big)\simeq\Omega_X^{0,1}\otimes\mathcal{O}(0,1)$ and $\sigma^*\big(\Omega_X^{0,1}\otimes\mathcal{O}(0,1)\big)\simeq\Omega_X^{1,0}\otimes\mathcal{O}(1,0)$. We denote them by φ_0.

LEMMA 3.62. *We have the following equalities:*

$$\varphi_0\big(\sigma^*\big(dz_i\cdot f_\infty^{(1,0)}\big)\big) = d\bar{z}_i\cdot\sqrt{-1}f_0^{(0,1)},\quad \varphi_0\big(\sigma^*\big(d\bar{z}_i\cdot f_\infty^{(1,0)}\big)\big) = -dz_i\cdot\sqrt{-1}f_0^{(0,1)},$$

$$\varphi_0\big(\sigma^*\big(d\bar{z}_i\cdot\sqrt{-1}\cdot f_0^{(0,1)}\big)\big) = dz_i\cdot f_\infty^{(1,0)},\quad \varphi_0\big(\sigma^*\big(dz_i\cdot\sqrt{-1}\cdot f_0^{(0,1)}\big)\big) = -d\bar{z}_i\cdot f_\infty^{(1,0)}.$$

Proof It can be checked by a direct calculation. □

The morphism φ_0 induces the following morphisms:

$$C^\infty\big(X\times\mathbb{P}^1,\Omega_X^{0,1}\otimes\mathcal{O}(1,0)\big)\simeq C^\infty\big(X\times\mathbb{P}^1,\Omega_X^{0,1}\otimes\mathcal{O}(0,1)\big),$$

$$C^\infty\big(X\times\mathbb{P}^1,\Omega_X^{0,1}\otimes\mathcal{O}(0,1)\big)\simeq C^\infty\big(X\times\mathbb{P}^1,\Omega_X^{1,0}\otimes\mathcal{O}(1,0)\big).$$

We denote them by φ.

LEMMA 3.63. *We have the following:*

$$\varphi\big(g\cdot dz\otimes f_\infty^{(1,0)}\big) = \sigma^*(\bar{g})\cdot d\bar{z}\otimes\sqrt{-1}\cdot f_0^{(0,1)},$$
$$\varphi\big(g\cdot d\bar{z}\otimes\sqrt{-1}\cdot f_0^{(0,1)}\big) = \sigma^*(\bar{g})\cdot dz\otimes f_\infty^{(1,0)}.$$

Proof It can be checked by a direct calculation. □

Recall that we put $\varphi(f) := \sigma^*(\overline{f})$ for a function f.

LEMMA 3.64. *We have* $\varphi \circ \mathbb{D}_X^{\triangle} = \mathbb{D}_X^{\triangle} \circ \varphi$.

Proof We have the following:

$$\varphi(\mathbb{D}_X^{\triangle} f) = \varphi(\partial_X f \otimes f_\infty^{(1)} + \overline{\partial}_X f \otimes \sqrt{-1} f_0^{(1)}).$$

We have the following, by using Lemma 3.63:

$$\varphi\Bigl(\frac{\partial f}{\partial z_i} dz_i \otimes f_\infty^{(1)}\Bigr) = \varphi\Bigl(\frac{\partial f}{\partial z_i}\Bigr) \cdot d\bar{z}_i \otimes \sqrt{-1} f_0^{(1)} = \frac{\partial \varphi(f)}{\partial \bar{z}_i} \cdot d\bar{z}_i \otimes \sqrt{-1} f_0^{(1)}.$$

Similarly we have the following:

$$\varphi\Bigl(\frac{\partial f}{\partial \bar{z}_i} \cdot d\bar{z}_i \otimes \sqrt{-1} \cdot f_0^{(1)}\Bigr) = \frac{\partial \varphi(f)}{\partial z_i} \cdot dz_i \otimes f_\infty^{(1)}.$$

It implies the commutativity $\varphi \circ \mathbb{D}_X^{\triangle} = \mathbb{D}_X^{\triangle} \circ \varphi$. □

Let V be a $\mathbb{P}^1$-holomorphic vector bundle over $X \times \mathbb{P}^1$. Let $\mathbb{D}_V^{\triangle}$ be a variation of $\mathbb{P}^1$-holomorphic vector bundles on V. We have the $\mathbb{P}^1$-holomorphic bundle $\sigma(V)$. Then we have the operator $\mathbb{D}_{\sigma(V)}^{\triangle}$ on $\sigma(V)$ defined as follows:

$$\mathbb{D}_{\sigma(V)}^{\triangle}\bigl(\sigma(v)\bigr) = \varphi_0 \circ \sigma\bigl(\mathbb{D}_V^{\triangle} v\bigr).$$

LEMMA 3.65. $\mathbb{D}_{\sigma(V)}^{\triangle}$ *is a variation of* $\mathbb{P}^1$*-holomorphic vector bundles.*

Proof We have the following:

$$(3.17)\quad \mathbb{D}_{\sigma(V)}^{\triangle}\bigl(f \cdot \sigma(v)\bigr) = \mathbb{D}_{\sigma(V)}^{\triangle}\bigl(\sigma(\varphi(f) \cdot v)\bigr) = \varphi_0 \circ \sigma\bigl(\mathbb{D}_V^{\triangle}(\varphi(f) \cdot v)\bigr)$$
$$= \varphi_0 \circ \sigma\Bigl(\mathbb{D}_X^{\triangle}(\varphi(f)) \cdot v + \varphi(f) \cdot \mathbb{D}_V^{\triangle}(v)\Bigr) = \mathbb{D}_X^{\triangle}(f) \cdot \sigma(v) + f \cdot \mathbb{D}_{\sigma(V)}^{\triangle}(\sigma(v)).$$

Thus we are done. □

3.5.5. Equivariant variation of $\mathbb{P}^1$-holomorphic vector bundles.

The following lemma is easy to see.

LEMMA 3.66. $\mathbb{D}_X^{\triangle}$ *is equivariant with respect to the torus action. We have* $\mathbb{D}_X^{\triangle}(\sigma^* g) = \sigma^*\bigl(\mathbb{D}_X^{\triangle} g\bigr)$. □

Let (V,ρ) be an equivariant $\mathbb{P}^1$-holomorphic vector bundle over $X \times \mathbb{P}^1$. Let $\mathbb{D}_V^{\triangle}$ be the variation of $\mathbb{P}^1$-holomorphic vector bundles. We put $H := V_{|\{1\} \times X}$. Then we have the flat connection $\mathbb{D}_0 := \mathbb{D}_{V\,|\,\{1\}\times X}^{\triangle}$. Since V is equivariant, we have the two filtrations F and G on H, such that $\xi(H,F,G) \simeq (V,\rho)$.

LEMMA 3.67. $\mathbb{D}_V^{\triangle}$ *is equivariant, if and only if* $\mathbb{D}_0$ *satisfies the Griffiths transversality in the following sense:*

$$\mathbb{D}_0^{(0,1)}\Bigl(C^\infty(F^p)\Bigr) \subset C^\infty(F^p \otimes \Omega_X^{0,1}), \qquad \mathbb{D}_0^{(1,0)}\Bigl(C^\infty(F^p)\Bigr) \subset C^\infty(F^{p-1} \otimes \Omega_X^{1,0}),$$

$$\mathbb{D}_0^{(0,1)}\Bigl(C^\infty(G^p)\Bigr) \subset C^\infty(G^{p-1} \otimes \Omega_X^{0,1}), \quad \mathbb{D}_0^{(1,0)}\Bigl(C^\infty(G^p)\Bigr) \subset C^\infty(G^p \otimes \Omega_X^{1,0}).$$

Proof Let s be a section of F^p. Then $\lambda^{-p}\cdot s$ is a section of V on $X\times \boldsymbol{C}_\lambda$. We have the following:

$$(3.18)\quad \mathbb{D}_V^{\triangle}(\lambda^{-p}\cdot s)=\lambda^{-p}\cdot\left(\mathbb{D}^{(1,0)}s\otimes f_\infty^{(1,0)}+\mathbb{D}^{(0,1)}s\otimes\sqrt{-1}\cdot f_0^{(0,1)}\right)$$
$$=\lambda^{-p+1}\cdot\mathbb{D}_0^{(1,0)}(s)\otimes\sqrt{-1}\cdot f_0^{(1,0)}+\lambda^{-p}\cdot\mathbb{D}_0^{(0,1)}(s)\otimes\sqrt{-1}\cdot f_0^{(0,1)}.$$

Thus $\mathbb{D}_V^{\triangle}(\lambda^{-p}\cdot s)$ is a section of $V\otimes\xi\Omega_X^1$ on $X\times\boldsymbol{C}_\lambda$ if and only if the following is satisfied:

$$\mathbb{D}_0^{(1,0)}(s)\in F^{p-1}\otimes\Omega_X^{1,0},\qquad \mathbb{D}_0^{(0,1)}(s)\in F^p\otimes\Omega_X^{0,1}.$$

Similar things for G hold. Thus we are done. □

Let (W,ρ) be an equivariant subbundle of (V,ρ). We have the corresponding vector subbundle H_W of H.

LEMMA 3.68. *We have $\mathbb{D}_V^{\triangle}\big(C^\infty(X,W)\big)\subset C^\infty(X,W\otimes\xi\Omega_X^1)$ if and only if H_W is a flat vector subbundle of H with respect to the flat connection $\mathbb{D}_0$.*

Proof $\mathbb{D}_V$ induces the morphism $C^\infty(X,H)[\lambda,\lambda^{-1}]\longrightarrow C^\infty(X,H\otimes\Omega_X^1)[\lambda,\lambda^{-1}]$, which is same as $\mathbb{D}_0\otimes\mathrm{id}$. The both claims are equivalent to the following:

$$\mathbb{D}_V\Big(C^\infty(X,H_W)[\lambda,\lambda^{-1}]\Big)\subset C^\infty(X,H_W\otimes\Omega_X^1)[\lambda,\lambda^{-1}].$$

Then the lemma follows. □

Let f be an equivariant morphism $(V_1,\rho_1)\longrightarrow(V_2,\rho_2)$ corresponding to $f_{|1}$: $(H_1,F_1,G_1)\longrightarrow(H_2,F_2,G_2)$.

LEMMA 3.69. $\mathbb{D}_{V_2}^{\triangle}\circ f=f\circ\mathbb{D}_{V_1}^{\triangle}$ *if and only if* $\mathbb{D}_{0\,H_2}\circ f_{|1}=f_{|1}\circ\mathbb{D}_{0\,H_1}$.

Proof f induces the morphism $\tilde{f}:C^\infty(X,H_1)[\lambda,\lambda^{-1}]\longrightarrow C^\infty(X,H_2)[\lambda,\lambda^{-1}]$. Then the both claims are equivalent to the compatibility of $\tilde{f}$, $\mathbb{D}_{V_2}^{\triangle}$ and $\mathbb{D}_{V_1}^{\triangle}$. □

COROLLARY 3.70.

- *ι_V is flat with respect to $\mathbb{D}_V^{\triangle}$ if and only if ι_H is flat with respect to $\mathbb{D}_0$.*
- *S is flat if and only if $\langle\cdot,\cdot\rangle$ is flat.*

Proof It follows from the previous lemma. □

COROLLARY 3.71. *ξ gives the equivalence of the following categories:*

- *The category of variation of equivariant pure twistor structures.*
- *The category of variation of complex pure Hodge structure.*

It is compatible with real structures and polarizations. □

COROLLARY 3.72. *ξ gives the equivalence of the following categories:*

- *The category of variation of equivariant mixed twistor structures.*
- *The category of variation of complex mixed Hodge structure.*

It is compatible with the real structures and the polarizations. □

REMARK 3.73. When we care the torus action, we change the formula (3.12) of Lemma 3.56 with the following:

$$\mathbb{D}_V^{\triangle}(f)=\begin{cases}\mathbb{D}_{V_0}^{(1,0)}(f)\otimes\big(\sqrt{-1}f_0^{(1,0)}\big)+\mathbb{D}_{V_0}^{(0,1)}(f)\otimes\big(\sqrt{-1}f_0^{(0,1)}\big), & (\text{on } X\times\boldsymbol{C}_\lambda),\\[2ex] \mathbb{D}_{V_\infty}^{\dagger\,(1,0)}(f)\otimes f_\infty^{(1,0)}+\mathbb{D}_{V_\infty}^{\dagger\,(0,1)}(f)\otimes f_\infty^{(0,1)}, & (\text{on } X\times\boldsymbol{C}_\mu).\end{cases}$$

Here $\mathbb{D}_{V_0}^{(p,q)}$ and $\mathbb{D}_{V_\infty}^{\dagger\,(p,q)}$ denote the (p,q)-parts of $\mathbb{D}_{V_0}$ and $\mathbb{D}_{V_\infty}^{\dagger}$. See (3.18), for example. □

3.6. The twistor nilpotent orbit

3.6.1. Pairing. We put $X = \boldsymbol{C}^n$, $D_i := \{z_i = 0\}$ and $D = \bigcup_{i=1}^n D_i$. Let V be a holomorphic vector bundle over $\mathbb{P}^1$ and $\boldsymbol{f}$ be a commuting tuple of nilpotent maps $f_i : V \longrightarrow V \otimes \mathbb{T}(-1)$ $(i = 1, \ldots, n)$. From $(V, \boldsymbol{f})$, we obtain the $\mathbb{P}^1$-holomorphic vector bundle $\mathcal{V}$ and the variation $\mathbb{D}_{\mathcal{V}}^{\triangle}$ over $\mathbb{P}^1 \times (X - D)$ (the subsection 3.5.3).

Let $S : V \otimes \sigma(V) \longrightarrow \mathbb{T}(0)$ be a pairing such that $S(f_i \otimes \mathrm{id}) + S(\mathrm{id} \otimes \sigma(f_i)) = 0$. Then the pairing $\tilde{S} : \mathcal{V}_0 \otimes \sigma\mathcal{V}_\infty \longrightarrow \mathcal{O}_{\mathcal{X}-\mathcal{D}}$ is given. Namely, for any sections $u \in V_0$, $v \in V_\infty$, $a \in C^\infty(\mathcal{X} - \mathcal{D})$ and $b \in C^\infty(\mathcal{X}^\dagger - \mathcal{D}^\dagger)$, we put as follows:

$$\tilde{S}\big(a \cdot u \otimes \sigma(b \cdot v)\big) = a \cdot \overline{\sigma^*(b)} \cdot S\big(u \otimes \sigma(v)\big).$$

Then we obtain the morphism $\tilde{S} : \mathcal{V}(V, \boldsymbol{f})_0 \otimes \sigma\mathcal{V}(V, \boldsymbol{f})_\infty \longrightarrow \mathcal{O}_{\mathcal{X}}$.

LEMMA 3.74. *$\tilde{S}$ is a morphism of λ-connections.*

Proof Let u and v be sections of V_0 and $\sigma(V_\infty)$ respectively. They naturally induce the holomorphic sections of $\mathcal{V}_0$ and $\mathcal{V}_\infty$ respectively, which are also denoted by the same notation. We have $\mathbb{D}_X \tilde{S}\big(u, \sigma^*(v)\big) = 0$. On the other hand, we have the following equality on $\mathcal{X} - \mathcal{D}$:

$$\tilde{S}(\mathbb{D}^{\triangle} u \otimes \sigma(v)) = \sum_i S\big(f_i(u) \cdot t_0^{(1)} \otimes v\big) \cdot \frac{dz_i}{z_i} \cdot \sqrt{-1} \cdot f_0^{(1)}.$$

We also have the following:

$$\begin{aligned}
(3.19)\quad \tilde{S}\big(u \otimes \mathbb{D}^{\triangle}\sigma(v)\big) &= \sum_i \tilde{S}\Big(u \otimes \varphi_0\sigma\Big(f_i(v) \cdot t_\infty^{(1)} \cdot f_\infty^{(1)} \cdot \frac{d\bar{z}_i}{\bar{z}_i}\Big)\Big) \\
&= \sum_i \tilde{S}\Big(u \otimes \varphi_0\sigma\big(f_i(v)\big)\Big) \cdot (-t_0^{(1)}) \cdot (-\sqrt{-1}) \cdot f_0^{(1)} \cdot \frac{dz_i}{z_i} \\
&= \sum_i S\Big(u \otimes \sigma\big(f_i(v)\big)\Big) \cdot t_0^{(1)} \cdot \frac{dz_i}{z_i} \cdot \sqrt{-1} \cdot f_0^{(1)}.
\end{aligned}$$

Thus we obtain the equality $\tilde{S}(\mathbb{D}_{\mathcal{V}_0} \otimes \mathrm{id}) + \tilde{S}(\mathrm{id} \otimes \mathbb{D}_{\sigma(\mathcal{V}_\infty)}) = \mathbb{D}_X \circ \tilde{S}$. □

On the plane $\mathcal{X}^\dagger - \mathcal{D}^\dagger$, we have the pairing $\tilde{S} : \mathcal{V}_\infty \otimes \sigma\mathcal{V}_0 \longrightarrow \mathcal{O}_{\mathcal{X}^\dagger - \mathcal{D}^\dagger}$.

LEMMA 3.75. *$\tilde{S}$ is a morphism of μ-connections.*

Proof Note the following equality on $\mathcal{X}^\dagger - \mathcal{D}^\dagger$:

$$\tilde{S}\big(\mathbb{D}_{\mathcal{V}_\infty}^{\triangle} u \otimes \sigma(v)\big) = \sum_i S\big(f_i(u) \cdot t_\infty^{(1)} \otimes \sigma(v)\big) \otimes f_\infty^{(1)} \cdot \frac{d\bar{z}_i}{\bar{z}_i}.$$

We also have the following:

$$(3.20)\quad \tilde{S}\big(u\otimes \mathbb{D}^{\dagger}_{\sigma(\mathcal{V}_0)}\sigma(v)\big)=\sum_i S\Big(u\otimes \varphi_0\sigma\Big(f_i(v)\cdot t_0^{(1)}\cdot\sqrt{-1}f_0^{(1,0)}\cdot\frac{dz_i}{z_i}\Big)\Big)$$
$$=\sum_i S\big(u\otimes\sigma\big(f_i(v)\big)\big)\cdot\big(-t_\infty^{(1)}\big)\cdot f_\infty^{(1)}\Big(-\frac{d\bar{z}_i}{\bar{z}_i}\Big)$$
$$=\sum_i S\big(u\otimes\sigma\big(f_i(v)\big)\big)\cdot t_\infty^{(1)}\cdot f_\infty^{(1)}\cdot\frac{d\bar{z}_i}{\bar{z}_i}.$$

Thus we obtain the equality $\tilde{S}\big(\mathbb{D}^{\dagger}_{\mathcal{V}_\infty}(u)\otimes v\big)+\tilde{S}\big(u\otimes\mathbb{D}_{\sigma(\mathcal{V}_0)}\sigma(v)\big)=0$ for any $u,v\in\Gamma(\boldsymbol{C}_\mu,V_\infty)$. □

LEMMA 3.76. *We obtain the pairing $S_{\mathcal{V}}:\mathcal{V}\otimes\sigma\mathcal{V}\longrightarrow\mathbb{T}(0)$.*

Proof We have only to check the pairings on the planes $\mathcal{X}-\mathcal{D}$ and $\mathcal{X}^\dagger-\mathcal{D}^\dagger$ are preserved by the gluing morphism. Note we have the following:

$$S\Big(\big(-\lambda^{-1}\cdot f_i\cdot t_0^{(1)}\big)\otimes\mathrm{id}\Big)+S\Big(\mathrm{id}\otimes\sigma\big(-\mu^{-1}\cdot f_i\cdot t_\infty^{(1)}\big)\Big)=0.$$

Then we obtain the following compatibility.

$$(3.21)\quad S(u\otimes\sigma^*v)=$$
$$S\Big(\exp\Big(-\sum\lambda^{-1}\cdot\log|z_i|^2f_i\cdot t_0^{(1)}\Big)u\otimes\sigma^*\exp\Big(-\sum\mu^{-1}\cdot\log|z_i|^2f_i\cdot t_\infty^{(1)}\Big)v\Big).$$

Thus we are done. □

3.6.2. Definition of twistor nilpotent orbit.

DEFINITION 3.77. $(V,\boldsymbol{f},S)$ is called a twistor nilpotent orbit, if there exists a positive constant $C>0$ such that $\big(\mathcal{V},\mathbb{D}^{\triangle}_{\mathcal{V}},S_{\mathcal{V}}\big)$ is a variation of polarized pure twistor structure over $\Delta(C)^{*\,n}$. □

LEMMA 3.78. *The resulted harmonic bundle over $\Delta(C)^{*\,n}$ is tame. The eigenvalues of the residues of Higgs field are trivial. The parabolic structure is trivial.*

Proof The first two claims are clear from our construction of the variation. By our construction, it is clear that the eigenvalue is trivial. By seeing the eigenvalues of λ-connections for any λ, we obtain the triviality of the parabolic structures. □

LEMMA 3.79. *The tuple $\big(S^{\mathrm{can}}(Pat(V,\boldsymbol{f},S)),\mathrm{Res}_i,S\big)$ is isomorphic to the original $(V,\boldsymbol{f},S)$. (See the subsection 11.3.4.)*

Proof Note that the prolongment of $\mathcal{V}_0$ is $V_0\otimes\mathcal{O}_{\mathcal{X}}$. On $X\times\boldsymbol{C}^*_\lambda$, it is clear. Then by using the Hartogs Theorem, we obtain the coincidence ${}^\diamond\mathcal{V}_0=V_0\otimes\mathcal{O}_{\mathcal{X}}$. Similarly, we obtain ${}^\diamond\mathcal{V}_\infty=V_\infty\otimes\mathcal{O}_{\mathcal{X}^\dagger}$. Then we obtain the isomorphisms:

$$S^{\mathrm{can}}_{|\boldsymbol{C}_\lambda}\simeq V_0,\qquad S^{\mathrm{can}}_{|\boldsymbol{C}_\mu}\simeq V_\infty.$$

Let us compare the gluing. Let λ be a point of $\boldsymbol{C}^*_\lambda$. Let v be an element of $V_{|\lambda}$. Let us consider the multi-valued flat section $\tilde{v}_1$ of $\mathcal{V}_{0|\lambda}$ and the multi-valued flat section $\tilde{v}_2$ of $\mathcal{V}_{\infty|\lambda^{-1}}$ are given as follows:

$$\tilde{v}_1:=\exp\Big(-\sqrt{-1}\sum_{i=1}^n\log z_i\cdot f_i\otimes t_1^{(1)}\Big)v,\quad \tilde{v}_2:=\exp\Big(\sqrt{-1}\sum_{i=1}^n\log\bar{z}_i\cdot f_i\otimes t_1^{(1)}\Big)v.$$

Then $\tilde{v}_1$ gives a flat section of $V_{0\,|\,\lambda}$, and $\tilde{v}_2$ gives a flat section of $V_{\infty\,|\,\lambda^{-1}}$. Then the gluing of S^{can} is obtained by the following relations:

$$\tilde{v}_1 = \tilde{v}_2, \quad \tilde{v}_1 \longmapsto v \in V_{0\,|\,\lambda}, \quad \tilde{v}_2 \longmapsto v \in V_{\infty\,|\,\lambda^{-1}}.$$

Here recall Φ^{can} is obtained by taking the degree 0-part of the polynomials $\sum v_J \cdot (\log z)^J$, which gives the second correspondence. The third correspondence can be obtained similarly. Hence the gluing of S^{can} is same as the gluing of V.

The comparison of f_i and Res_i and the pairings can easily be checked. □

3.6.3. A lemma for the restriction of twistor nilpotent orbit. We put $X = \boldsymbol{C}^l$, $D_i = \{z_i = 0\}$, $D = \bigcup_{i=1}^{l} D_i$, $D_{\underline{m}} := \bigcup_{i \in \underline{m}} D_i$ and $D_{\underline{m}}^{\circ} := D_{\underline{m}} \setminus \bigcup_{j>m} D_j \cap D_{\underline{m}}$. Let $\pi : X \longrightarrow D_{\underline{m}}$ denote the projection $(z_1, \ldots, z_l) \longmapsto (z_{m+1}, \ldots, z_l)$. For any point $Q \in D_{\underline{m}}^{\circ}$, $\pi^{-1}(Q)$ is naturally isomorphic to Δ^m.

From $(V, \boldsymbol{f})$, we obtain the variation $(\mathcal{V}, \mathbb{D}^{\triangle})$ of $\mathbb{P}^1$-holomorphic bundles over $X - D$. Let us consider the restriction $\mathcal{V}_{|\pi^{-1}(Q)}$.

We put $\tilde{Q} = (\overbrace{1, \ldots, 1}^{m}, Q) \in X - D$. We put $V' := \mathcal{V}_{|\{\tilde{Q}\} \times \mathbb{P}^1}$. The vector bundle V' is a twist of V by the following endomorphism of $V_{|\boldsymbol{C}_\lambda^*}$:

$$\exp\Bigl(-\sum_{i=m+1}^{l} f_i \cdot \log|z_i(Q)|^2 \cdot t_1^{(1)}\Bigr).$$

The tuple $\boldsymbol{f}'$ of the morphisms $f_i' : V' \longrightarrow V' \otimes \mathbb{T}(-1)$ $(i = 1, \ldots, m)$ are naturally defined. Then it is easy to see that we have the isomorphism:

$$\mathcal{V}(V', \boldsymbol{f}') \simeq \mathcal{V}(V, \boldsymbol{f})_{|\pi^{-1}(Q)}.$$

Thus we obtain the following lemma.

LEMMA 3.80. *Let $(E, \overline{\partial}_E, \theta, h)$ be a harmonic bundle over $\Delta^{*\,l}$ corresponding to a twistor nilpotent orbit. Let Q be as above. Then the restriction $(E, \overline{\partial}_E, \theta, h)_{|\pi^{-1}(Q)}$ is also a harmonic bundle corresponding to a twistor nilpotent orbit.* □

3.6.4. Twist of Rees bundles. We would like to see the relation of twistor nilpotent orbit and the nilpotent orbit in the Hodge theory. We give a preparation in this subsection.

Let H be a vector space with a decreasing filtration $F = (F^p)$. For an endomorphism g of H, we put $g \cdot F := (g \cdot F^p)$, which is called the twist of F by g. Then we obtain the left $Aut(H)$-action on the set of filtrations of H.

Let p_1 be the projection of $\boldsymbol{C}_\lambda^*$ to a point. We have the natural isomorphism $i_F : \xi(H, F)_{|\boldsymbol{C}_\lambda^*} \simeq p_1^* H$. For an element $g \in Aut(H)$, we have the natural isomorphism $g : (H, F) \longrightarrow (H, g \cdot F)$. Then it induces the isomorphism $\phi_g : \xi(H, F) \longrightarrow \xi(H, g \cdot F)$.

Let $Aut_{eq}(p_1^* H)$ be equivariant automorphisms of $p_1^* H$. Clearly we have the natural isomorphism $Aut_{eq}(p_1^* H) \simeq Aut(H)$. We do not distinguish them.

Let g be an element of $Aut(H)$. Then we have the twist of i_F, i.e., $g \circ i_F : \xi(H,F)_{|\boldsymbol{C}^*_\lambda} \simeq p_1^*(H)$. Then we have the following commutative diagramm:

$$\begin{array}{ccc} \xi(H,F)_{|\boldsymbol{C}^*_\lambda} & \xrightarrow{g \circ i_F} & p_1^* H \\ \phi_g \big\downarrow & & \mathrm{id} \big\downarrow \\ \xi(H, g \cdot F)_{|\boldsymbol{C}^*_\lambda} & \xrightarrow{i_F} & p_1^* H. \end{array}$$

Let (H,F,G) be bi-filtered vector bundle. The Rees bundle $\xi(H,F,G)$ is obtained by the following gluing:

$$\xi(H,F)_{|\boldsymbol{C}^*_\lambda} \xrightarrow{i_F} p_1^* H \xleftarrow{i_G} \xi(H,G)_{|\boldsymbol{C}^*_\mu}.$$

Let g_i $(i = 1,2)$ be elements of $Aut(H)$. Then the vector bundle $\xi(H,F,G,g_1,g_2)$, is obtained as the twisting $\xi(H,F,G)$ by g_i:

$$\xi(H,F)_{|\boldsymbol{C}^*_\lambda} \xrightarrow{g_1 \circ i_F} p_1^* H \xleftarrow{g_2 \circ i_G} \xi(H,G)_{|\boldsymbol{C}^*_\mu}.$$

LEMMA 3.81. *$\xi(H,F,G,g_1,g_2)$ is naturally isomorphic to $\xi(H, g_1 \cdot F, g_2 \cdot G)$.*

Proof We have the following commutative diagram:

$$\begin{array}{ccccc} \xi(H,F)_{|\boldsymbol{C}^*_\lambda} & \xrightarrow{g_1 \circ i_F} & p_1^* H & \xleftarrow{g_2 \circ i_G} & \xi(H,G)_{|\boldsymbol{C}^*_\mu} \\ \phi_{g_1} \big\downarrow & & \mathrm{id} \big\downarrow & & \phi_{g_2} \big\downarrow \\ \xi(H, g_1 \cdot F)_{|\boldsymbol{C}^*_\lambda} & \xrightarrow{i_F} & p_1^* H & \xleftarrow{i_G} & \xi(H, g_2 \cdot G)_{|\boldsymbol{C}^*_\mu}. \end{array}$$

It gives the isomorphism desired. □

3.6.5. The induced variation from an equivariant nilpotent tuple. We put $X = \boldsymbol{C}^n$, $D_i = \{z_i = 0\}$ and $D = \bigcup_{i=1}^n D_i$. We put $\tilde{X} = \boldsymbol{C}^n$, and then we have the universal covering $\pi : \tilde{X} \longrightarrow X - D$, given by the correspondence $\zeta_i \longmapsto \exp(\sqrt{-1}\zeta_i)$.

From $(V, \boldsymbol{f})$, we obtain the $\mathbb{P}^1$-holomorphic vector bundle $\mathcal{V}$ and the variation $\mathbb{D}^\triangle_\mathcal{V}$ over $\mathbb{P}^1 \times (X - D)$ (the subsection 3.5.3).

LEMMA 3.82. *If $(V, \boldsymbol{f})$ is equivariant, then the vector bundle $\mathcal{V}$ and the variation $\mathbb{D}^\triangle_\mathcal{V}$ are naturally equivariant.*

Proof Since $f_i \otimes t_1^{(1)}$ is equivariant, the gluing (3.15) is equivariant. Thus we have the natural torus action on $\mathcal{V}$. The following sections, appearing in (3.13) and (3.14) (see also Remark 3.73) are invariant with respect to the torus action:

$$f_i \otimes t_0^{(1)} \otimes \frac{dz_i}{z_i} \cdot f_0^{(1,0)}, \quad f_i \otimes t_\infty^{(1)} \otimes \frac{d\bar{z}_i}{\bar{z}_i} \cdot f_\infty^{(0,1)}.$$

Hence $\mathbb{D}^\triangle_\mathcal{V}$ is also equivariant. □

We put $\mathcal{H} := \mathcal{V}_{|\{1\} \times (X-D)}$. Then we obtain the two filtrations $\mathcal{F}$ and $\mathcal{G}$ such that $\xi(\mathcal{H}, \mathcal{F}, \mathcal{G}) \simeq \mathcal{V}$. We also have the flat connection $\mathbb{D}_0$ on $\mathcal{H}$.

On $P_0 := \overbrace{(1, \ldots, 1)}^{n} \in X - D$, the fiber $\mathcal{V}_{|\mathbb{P}^1 \times \{P_0\}}$ is naturally identified with V, and $\mathcal{H}_{|(1,P_0)} \simeq H$.

Let us consider $\pi^* \mathcal{H}$. We have the flat connection $\pi^* \mathbb{D}_0$ and the isomorphism $\pi^* \mathcal{H}_{|O} \simeq H$. They induce the isomorphism $\pi^* \mathcal{H} \simeq p^* H$, where p denotes the natural morphism $\boldsymbol{C}^n \longrightarrow \{(1, P_0)\} \subset \mathbb{P}^1 \times (X - D)$. We have the two filtrations $\pi^*\mathcal{F}$ and p^*F. We also have $\pi^*\mathcal{G}$ and p^*G.

We put $N_i := -(f_i \otimes t_1^{(1)})_{|\lambda=1}$. Recall Lemma 3.60 and Remark 3.61.

LEMMA 3.83. *The following equalities of filtrations hold:*

$$\pi^*\mathcal{F} = \exp\Big(\sum \zeta_i N_i\Big) \cdot p^*F, \qquad \pi^*\mathcal{G} = \exp\Big(\sum \overline{\zeta}_i N_i\Big) \cdot p^*G.$$

Proof We put as follows:

$$g_0 := \exp\big(-\sum_i \zeta_i \cdot N_i\big), \quad g_\infty := \exp\big(-\sum_i \overline{\zeta}_i \cdot N_i\big).$$

Let v be an element of $H \simeq \mathcal{V}_{|(O,\lambda)}$. We put $\tilde{v}_0 := g_0 \cdot v$, and then it is the flat section of $\pi^*\mathcal{V}_{0\,|\,\mathcal{X}^\lambda}$ such that $\tilde{v}_{0|(O,\lambda)} = v$. Similarly, we put $\tilde{v}_\infty := g_\infty \cdot v$, and then it is the flat section of $\pi^*\mathcal{V}_{\infty\,|\,\mathcal{X}^{\dagger\lambda}}$ such that $\tilde{v}_{\infty\,|\,(O,\lambda)} = v$.

The construction of the vector bundle $\mathcal{V}$ is given by the relation $\tilde{v}_0 = \tilde{v}_\infty$, due to Lemma 3.58. Hence the vector bundle $\mathcal{V}$ is given by the following gluing:

$$\xi\big(p^*H, p^*F\big)_{|\tilde{X}\times\boldsymbol{C}^*_\lambda} \xrightarrow{g_0^{-1}\circ i_F} p_1^*H \xleftarrow{g_\infty^{-1}\circ i_G} \xi\big(p^*H, p^*G\big)_{|\tilde{X}\times\boldsymbol{C}^*_\mu}.$$

Here p_1 denote the canonical morphism of $\tilde{X} \times \boldsymbol{C}^*_\lambda$ to a point. Then we obtain the following equality, due to Lemma 3.81

$$\xi\big(\pi^*\mathcal{H}, \pi^*\mathcal{F}, \pi^*\mathcal{G}\big) \simeq \xi\Big(p^*H,\, g_0^{-1}\cdot p^*F,\, g_\infty^{-1}\cdot p^*G\Big).$$

Thus we are done. □

3.6.6. Reword. Let (H, F, G) be a bi-filtered vector space, and N_i be a morphism $(H, F, G) \longrightarrow (H, F, G) \otimes \boldsymbol{C}(1, 1)$. We put $V = \xi(H, F, G)$, and then we have $f_i : V \longrightarrow V \otimes \mathbb{T}(-1)$.

We have the trivial local system p^*H on $\boldsymbol{C}^n$. Then we obtain the C^∞-bundle $\mathcal{H}^{(1)}$ with the natural flat connection $\mathbb{D}^{(1)}$. We have the $\mathbb{Z}^n$-action on $\boldsymbol{C}^n$ by $\boldsymbol{\zeta} \longmapsto \boldsymbol{\zeta} + 2\pi\boldsymbol{n}$. We lift it to the action on $\mathcal{H}^{(1)}$ as follows:

$$(\boldsymbol{\zeta}, v) \longmapsto \Big(\boldsymbol{\zeta} + 2\pi \cdot \boldsymbol{n},\ \prod \exp\big(2\pi \cdot n_i \cdot N_i\big) \cdot v\Big).$$

Here we regard N_i as the endomorphism of H by using the isomorphism of $\boldsymbol{C}(1, 1) \simeq \boldsymbol{C}(0, 0)$ given by $t^{(-1)}_{1\,|\,\lambda=1} \longmapsto t^{(0)}_{1\,|\,\lambda=1}$. We have two filtrations on $\mathcal{H}^{(1)}$:

$$\exp\Big(\sum_i \zeta_i \cdot N_i\Big) \cdot p^*F, \qquad \exp\Big(\sum \overline{\zeta}_i \cdot N_i\Big) \cdot p^*G.$$

We obtain the C^∞-bundle $\mathcal{H}$ with the flat connection on $\boldsymbol{C}^{*\,n}$, and the filtrations $\mathcal{F}$ and $\mathcal{G}$. They satisfy the Griffiths transversality, and $\xi(\mathcal{H}, \mathcal{F}, \mathcal{G}) \simeq Pat(V, \boldsymbol{f})$.

Let $(H, F, \overline{F})$ be a bi-filtered vector space defined over $\boldsymbol{R}$. We put $V := \xi(H, F, \overline{F})$. Let $N_i : (H, F, \overline{F}) \longrightarrow (H, F, \overline{F}) \otimes \boldsymbol{C}(1, 1)$ be morphisms. The morphisms N_i induce the morphisms from $V \longrightarrow V \otimes \mathbb{T}(-1)$, which we also denote by N_i. We put $f_i := -N_i$. Then we obtain the tuple of endomorphisms $\boldsymbol{f} = (f_i)$.

COROLLARY 3.84. *We have the isomorphism* $\xi(\mathcal{H}, \mathcal{F}, \overline{\mathcal{F}}) \simeq \mathcal{V}(V, \boldsymbol{f})$.

Proof This is a reformulation of Lemma 3.83. □

PROPOSITION 3.85. *The functor ξ gives an equivalence of the following categories:*

- *The equivariant twistor nilpotent orbit defined over* $\boldsymbol{R}$.
- *The nilpotent orbit in the category of Hodge theory in the sense of Schmid (Definition* 1.14 *in* [**11**], *for example).*

Proof It follows from the various equivalences (Corollary 3.71 and Corollary 3.84). □

3.7. Split polarized mixed twistor structure and the nilpotent orbit

3.7.1. Definition.

DEFINITION 3.86. Let $(V, W, \boldsymbol{f}, S)$ be a Pol-MTS of type (n, l). Assume that the grading $V = \bigoplus V_h$ is given, such that the following holds:

- V_h is a pure twistor structure of weight h.
- $W_h = \bigoplus_{i \leq h} V_i$.
- f_j preserves the grading.
- The restriction of S to $V_i \otimes V_j$ is 0 unless $i + j = n$.

In that case, $(V, W, \boldsymbol{f}, S)$ is called a split Pol-MTS of type (n, l).

If $(V, W, \boldsymbol{f}, S)$ is a Ψ-Pol-MTS of type (n, l), and if the grading satisfying the above conditions is given, then $(V, W, \boldsymbol{f}, S)$ is called a split Ψ-Pol-MTS of type (n, l) □

3.7.2. Preliminary on the split Pol-MTS in one variable of rank two. We put $V^{[2]} := \mathcal{O}(1,0) \oplus \mathcal{O}(0,-1)$. We have the filtration given by $W_{-1} = \mathcal{O}(0,-1) \subset W_1 = V^{[2]}$. Then $(V^{[2]}, W)$ is a mixed twistor structure.

Let $F^{[2]} : V^{[2]} \longrightarrow V^{[2]} \otimes \mathbb{T}(-1)$ be the morphism given by $f_x^{(1,0)} \longmapsto f_x^{(0,-1)} \otimes t_x^{(-1)}$ and $f_x^{(0,-1)} \longmapsto 0$ for $x = 0, 1, \infty$. We put $N = F^{[2]} \otimes t_1^{(1)}$.

We put $V_0^{[2]} := V^{[2]}_{|\boldsymbol{C}_\lambda}$ and $V_\infty^{[2]} := V^{[2]}_{|\boldsymbol{C}_\mu}$. We have the frames $\big(f_0^{(1,0)}, f_0^{(0,-1)}\big)$ and $\big(f_\infty^{(1,0)}, f_\infty^{(0,-1)}\big)$ of $V_0^{[2]}$ and $V_\infty^{[2]}$ respectively. We have the frame $\big(f_1^{(1,0)}, f_1^{(0,-1)}\big)$ of $V^{[2]}_{0\,|\,\boldsymbol{C}_\lambda^*} = V^{[2]}_{\infty\,|\,\boldsymbol{C}_\mu^*}$. We use the notation $f_1^{(1,0)\,\dagger}$ and $f_1^{(0,-1)\,\dagger}$, when we consider them as the frame of $V^{[2]}_{\infty\,|\,\boldsymbol{C}_{\mu*}}$.

We twist the gluing as follows: Let λ be a point of $\boldsymbol{C}_\lambda^*$. Let v be an element of $V^{[2]}_{|\lambda}$. It induces the element of $v \in V^{[2]}_{0\,|\,\lambda}$ and $v^\dagger \in V^{[2]}_{\infty\,|\,\mu}$, where we put $\mu = \lambda^{-1}$. The original gluing is given by $v = v^\dagger$. The twisted gluing is given by the following relation:

$$\exp\big(\sqrt{-1}y \cdot N\big)v = v^\dagger.$$

Note that $y \cdot N$ gives the following correspondence:

$$f_1^{(1,0)} \longmapsto y \cdot f_1^{(0,-1)}, \qquad f_1^{(0,-1)} \longmapsto 0.$$

Thus the gluing is given by the following:

$$\big(f_1^{(1,0)}, f_1^{(0,-1)}\big) \cdot \begin{pmatrix} 1 & 0 \\ \sqrt{-1}y & 1 \end{pmatrix} = \big(f_1^{(1,0)\,\dagger}, f_1^{(0,-1)\,\dagger}\big).$$

Then we obtain the following relation:

$$\big(f_0^{(1,0)}, f_0^{(0,-1)}\big) \cdot \begin{pmatrix} \sqrt{-1}\lambda & 0 \\ \sqrt{-1}y & -\sqrt{-1}\mu \end{pmatrix} = \big(f_\infty^{(1,0)}, f_\infty^{(0,-1)}\big).$$

Let $\widetilde{V}_y^{[2]}$ denote the resulted vector bundle.

LEMMA 3.87. *Assume that $y > 0$. The vector bundle $\widetilde{V}_y^{[2]}$ is a pure twistor of weight 0 of rank 2. The tuple of global sections $(\widetilde{s}_1, \widetilde{s}_2)$, which are given as follows, is a base of the space of the global sections:*

$$\widetilde{s}_1 := \sqrt{-1}\lambda \cdot f_0^{(1,0)} + \sqrt{-1}y \cdot f_0^{(0,-1)} = f_\infty^{(1,0)}, \tag{3.22}$$
$$\widetilde{s}_2 := f_0^{(1,0)} = -\sqrt{-1}\mu \cdot f_\infty^{(1,0)} - \sqrt{-1}y \cdot f_\infty^{(0,-1)}.$$

Proof It can be checked by direct calculations. □

The pairing $\eta_1 : \mathcal{O}(1,0) \otimes \sigma(\mathcal{O}(0,-1)) \longrightarrow \mathbb{T}(0)$ is given as the composite of the following naturally defined morphisms:

$$\mathcal{O}(1,0) \otimes \sigma(\mathcal{O}(0,-1)) \longrightarrow \mathcal{O}(1,0) \otimes \mathcal{O}(-1,0) \longrightarrow \mathbb{T}(0).$$

The pairing $\eta_2 : \mathcal{O}(0,-1) \otimes \sigma(\mathcal{O}(1,0)) \longrightarrow \mathbb{T}(0)$ is given similarly.

Let us consider the pairing $S^{[2]} : V^{[2]} \otimes V^{[2]} \longrightarrow \mathbb{T}(0)$ given as follows: The restriction of $S^{[2]}$ to $\mathcal{O}(1,0) \otimes \sigma(\mathcal{O}(1,0)) \oplus \mathcal{O}(0,-1) \otimes \sigma(\mathcal{O}(0,-1))$ is defined to be trivial. The restriction of $S^{[2]}$ to $\mathcal{O}(1,0) \otimes \sigma(\mathcal{O}(0,-1))$ is defined to be $-\eta_1$. The restriction of $S^{[2]}$ to $\mathcal{O}(0,-1) \otimes \sigma(\mathcal{O}(1,0))$ is defined to be η_2.

LEMMA 3.88. *We have $S^{[2]}\big(f_0^{(1,0)} \otimes \sigma(f_\infty^{(0,-1)})\big) = -\sqrt{-1}t_0^{(0)}$ and $S^{[2]}\big(f_0^{(0,-1)} \otimes \sigma(f_\infty^{(1,0)})\big) = -\sqrt{-1}t_0^{(0)}$.*

Proof It can be checked by a direct calculation. □

LEMMA 3.89.

- *The pairing $S^{[2]}$ is symmetric.*
- *We have the relation $S^{[2]}\big(\mathrm{id} \otimes \sigma(F^{[2]})\big) + S^{[2]}\big(F^{[2]} \otimes \mathrm{id}\big) = 0$.*
- *The induced pairing $-S^{[2]}(\mathrm{id} \otimes \sigma(F^{[2]}))$ on $\mathcal{O}(1,0)$ gives a polarization of weight 1.*

Proof By a direct calculation, we have $S^{[2]}\big(f_0^{(1,0)} \otimes \sigma(f_\infty^{(0,-1)})\big) = -\big(\sqrt{-1} \cdot t_0^{(0)}\big)$ and $S^{[2]}\big(f_\infty^{(0,-1)} \otimes \sigma(f_0^{(1,0)})\big) = \sqrt{-1}t_\infty^{(0)}$. Thus we obtain $\sigma\big(S^{[2]}\big(f_0^{(1,0)} \otimes \sigma(f_\infty^{(0,-1)})\big)\big) = S^{[2]}\big(f_\infty^{(0,-1)} \otimes \sigma(f_0^{(1,0)})\big)$. Hence the pairing $S^{[2]}$ is symmetric.

Let us show the second claim. The morphism $S^{[2]}\big(\mathrm{id} \otimes \sigma(F^{[2]})\big)$ gives the composite of the following correspondence:

$$f_0^{(1,0)} \otimes \sigma\big(f_\infty^{(1,0)}\big) \longmapsto f_0^{(1,0)} \otimes \sigma\big(f_\infty^{(0,-1)} \cdot t_\infty^{(-1)}\big) \longmapsto -\sqrt{-1} \cdot \big(-t_0^{(-1)}\big) = \sqrt{-1} \cdot t_0^{(-1)}.$$

On the other hand, $S^{[2]}\big(F^{[2]} \otimes \mathrm{id}\big)$ gives the composite of the following correspondence:

$$f_0^{(1,0)} \otimes \sigma\big(f_\infty^{(1,0)}\big) \longmapsto f_0^{(0,-1)} \cdot t_0^{(-1)} \otimes \sigma\big(f_\infty^{(1,0)}\big) \longmapsto -\sqrt{-1} \cdot t_0^{(-1)}.$$

Thus we obtain the second claim.

Let us show the third claim. A base of the space $H^0\big(\mathcal{O}(1,0) \otimes \mathcal{O}(-1,0)\big)$ is given by $s = f_0^{(1,0)} \otimes f_0^{(-1,0)} = f_\infty^{(1,0)} \otimes f_\infty^{(-1,0)}$. We have the following:

$$S^{[2]}\big(F^{[2]}(s), \sigma(s)\big) = S^{[2]}\big(f_\infty^{(0,-1)} \cdot t_\infty^{(-1)} \otimes f_\infty^{(-1,0)}, \sigma\big(f_0^{(1,0)} \otimes f_0^{(-1,0)}\big)\big) = 1.$$

Thus we are done. □

LEMMA 3.90. *The tuple $(V^{[2]}, W, F^{[2]}, S^{[2]})$ is a split polarized mixed twistor structure of weight 0.*

Proof It immediately follows from Lemma 3.89. □

On the other hand, the pairing $S^{[2]}$ induces the pairing $\widetilde{S}^{[2]}$ on $\widetilde{V}_y^{[2]}$, for we have the relation:

$$S^{[2]}\big(\mathrm{id}\otimes\sigma(F^{[2]}\otimes t_1^{(1)})\big)+S^{[2]}\big(F^{[2]}\otimes t_1^{(1)}\otimes\mathrm{id}\big)=0.$$

LEMMA 3.91. *The pairing $\widetilde{S}^{[2]}$ is a polarization of $\widetilde{V}_y^{[2]}$ of weight* 0.

Proof We have only to show the positivity $\widetilde{S}^{[2]}(\widetilde{s}_i,\widetilde{s}_i)>0$ for the sections $\widetilde{s}_i$ $(i=1,2)$ given in (3.22). As for $\widetilde{s}_1$, we have the following:

$$\begin{aligned}(3.23)\quad \widetilde{S}^{[2]}(\widetilde{s}_1,\widetilde{s}_1)&=S^{[2]}\big(\sqrt{-1}\lambda\cdot f_0^{(1,0)}+\sqrt{-1}y\cdot f_0^{(0,-1)},\sigma(f_\infty^{(1,0)})\big)\\&=\sqrt{-1}y\cdot S^{(0)}\big(f_0^{(0,-1)},\sigma(f_\infty^{(1,0)})\big)=y\cdot\sqrt{-1}\cdot\sqrt{-1}^{-1}=y.\end{aligned}$$

As for $\widetilde{s}_2$, we have the following:

$$\begin{aligned}(3.24)\quad \widetilde{S}^{[2]}(\widetilde{s}_2,\widetilde{s}_2)&=S^{[2]}\big(-\sqrt{-1}\mu\cdot f_\infty^{(1,0)}-\sqrt{-1}y\cdot f_\infty^{(0,-1)},\sigma(f_0^{(1,0)})\big)\\&=S^{[2]}\big(-\sqrt{-1}y\cdot f_\infty^{(0,-1)},\sigma(f_0^{(1,0)})\big)=-\sqrt{-1}y\cdot S^{[2]}\big(f_\infty^{(0,-1)},\sigma(f_0^{(1,0)})\big)\\&=-\sqrt{-1}y\cdot\sqrt{-1}=y.\end{aligned}$$

Hence we have $\widetilde{S}^{[2]}(\widetilde{s}_i,\widetilde{s}_i)>0$ for $i=1,2$. Thus we are done. □

3.7.3. Preliminary on the split Pol-MTS of rank h. Let $h>1$ be an integer. We put $V^{(1)}:=V^{[2]\otimes h-1}$. We have the naturally defined pairing $S^{(1)}:V^{(1)}\otimes\sigma(V^{(1)})\longrightarrow\mathbb{T}(0)$, given as follows:

$$S^{(1)}\Big(\bigotimes_{i=1}^{h-1}f_x^{(p_i,q_i)},\bigotimes_{i=1}^{h-1}\sigma(f_x^{(\widetilde{p}_i,\widetilde{q}_i)})\Big)=\prod_{i=1}^{h-1}S^{[2]}\big(f_x^{(p_i,q_i)},\sigma(f_x^{(\widetilde{p}_i,\widetilde{q}_i)})\big).$$

Here (p_i,q_i) and $(\widetilde{p}_i,\widetilde{q}_i)$ denote $(1,0)$ or $(0,-1)$. We also have the morphism $F^{(1)}:V^{(1)}\longrightarrow V^{(1)}\otimes\mathbb{T}(-1)$, which is induced from $F^{[2]}$ by the Leibniz rule.

We have the natural $\mathfrak{S}_{h-1}$-action on $V^{(1)}$. It preserves $S^{(1)}$ and $F^{(1)}$. Then we obtain the invariant part $V^{[h]}=\mathrm{Sym}^{h-1}(V^{(0)})$ and the induced pairing $S^{[h]}$ and the induced morphism $F^{[h]}$. We also have the induced filtration W of $V^{[h]}$. The following lemma is clear.

LEMMA 3.92. *We have the natural grading:*

$$(3.25)\qquad V^{[h]}\simeq\bigoplus_{\substack{p+q=h-1,\\p,q\geq 0}}\mathcal{O}(1,0)^{\otimes p}\otimes\mathcal{O}(0,-1)^{\otimes q}.$$

The filtration of the left hand side in (3.25) *is isomorphic to the following filtration of the right hand side:*

$$W_a=\bigoplus_{p-q\leq a}\mathcal{O}(1,0)^{\otimes p}\otimes\mathcal{O}(0,-1)^{\otimes q}.$$

In particular, the vector bundle $V^{[h]}$ with the filtration W is a mixed twistor structure. □

LEMMA 3.93. *The tuple $\big(V^{[h]},W,F^{[h]},S^{[h]}\big)$ is a split polarized mixed twistor structure of weight* 0.

Proof The condition $S^{(1)}\big(\mathrm{id}\otimes\sigma(F^{(1)})\big)+S^{(1)}\big(F^{(1)}\otimes\mathrm{id}\big)=0$ can be checked easily. Then $S^{[h]}\big(\mathrm{id}\otimes\sigma(F^{[h]})\big)+S^{[h]}\big(F^{[h]}\otimes\mathrm{id}\big)=0$ immediately follows.

It is easy to see that there exists a positive number B such that the following holds:

$$\big(F^{[h]}\big)^{h-1}\big(f_x^{(1,0)\otimes h-1}\big)=B\cdot f_x^{(0,-1)\otimes h-1}.$$

A base of the space of the global sections of $\mathrm{Gr}^W_{h-1}\otimes\mathcal{O}(-h+1,0)$ is given by the following:

$$s=f_0^{(1,0)\otimes h-1}\otimes f_0^{(-h+1,0)}=f_\infty^{(1,0)\otimes h-1}\otimes f_\infty^{(-h+1,0)}.$$

We have the following:

$$S^{[h]}\big(F^{[h]\,h-1}(s),\sigma(s)\big)=\tag{3.26}$$

$$B\cdot S^{[h]}\Big(f_\infty^{(0,-1)\otimes h-1}\otimes f_\infty^{(-h+1,0)},\ \sigma\big(f_0^{(1,0)\otimes h-1}\otimes f_0^{(-h+1,0)}\big)\Big)=B.$$

Thus $S^{[h]}\big(F^{[h]\,h-1}\otimes\mathrm{id}\big)$ is a polarization on $\mathrm{Gr}^W_{h-1}=P\,\mathrm{Gr}^W_{h-1}$. □

From the vector bundles with the nilpotent maps $(V^{(1)},F^{(1)})$ and $(V^{[h]},F^{[h]})$, we obtain the vector bundles $\widetilde{V}_y^{(1)}$ and $\widetilde{V}_y^{[h]}$ for $y>0$, as in the subsection 3.7.2. We have the induced pairings $\widetilde{S}^{(h)}$ and $\widetilde{S}^{[h]}$ on $\widetilde{V}_y^{(1)}$ and $\widetilde{V}_y^{[h]}$ respectively.

LEMMA 3.94. *$\widetilde{S}^{[h]}$ is a polarization of $\widetilde{V}_y^{[h]}$.*

Proof It is easy to see that $\widetilde{S}^{(1)}$ is a polarization of $\widetilde{V}_y^{(1)}$. Then the lemma immediately follows. □

3.7.4. Classification of split Pol-MTS in one variable. Let us consider the vector bundle $V=\bigoplus_{p+q=h,0\leq p,q\leq h}\mathcal{O}(p,-q)$ over $\mathbb{P}^1$. The filtration W is given as follows:

$$W_a:=\bigoplus_{p-q\leq a}\mathcal{O}(p,-q).$$

Then (V,W) is a mixed twistor structure.

LEMMA 3.95. *Let (V,W) be as above. Let $F:V\longrightarrow V\otimes\mathbb{T}(-1)$ be a morphism of mixed twistor structure preserving the grading. Let $S:V\otimes\sigma(V)\longrightarrow\mathbb{T}(0)$ be a morphism of mixed twistor structures. Assume that (V,W,F,S) be a split polarized mixed twistor structure of weight 0. Then F and S are unique up to isomorphisms.*

Proof Up to isomorphisms, we may assume that $F:\mathcal{O}(p,-q)\longrightarrow\mathcal{O}(p-1,-q-1)\otimes\mathbb{T}(-1)$ is given by $f_x^{(p,-q)}\longmapsto f_x^{(p-1,-q-1)}\otimes t_x^{(-1)}$ for $x=0,1,\infty$. Since we have the relation $S\big(\mathrm{id}\otimes\sigma(F)\big)+S\big(F\otimes\mathrm{id}\big)=0$, we obtain the following:

$$S\big(f_\infty^{(p-1,-q-1)},\ \sigma(f_0^{(q+1,-p+1)})\big)-S\big(f_\infty^{(p,-q)},\ \sigma(f_0^{(q,-p)})\big)=0.$$

Hence S is determined by the number $C=S\big(f_\infty^{(0,-h)}\otimes\sigma(f_0^{(h,0)})\big)$.

We have the following:

$$S\Big(F^h\big(f_\infty^{(h,0)}\otimes f_\infty^{(-h,0)}\big),\ \sigma\big(f_0^{(h,0)}\otimes f_0^{(-h,0)}\big)\Big)=C\cdot(-\sqrt{-1})^h.$$

Hence we obtain $C\cdot(-\sqrt{-1})^h>0$. We may assume that $C\cdot(-\sqrt{-1})^h=1$ up to isomorphisms again. Thus we are done. □

COROLLARY 3.96. *Let V, W, F and S be as above. If (V,W,F,S) is a split polarized mixed twistor structure, then it is isomorphic to a polarized mixed twistor structure of the form $(V^{[h]},W,F^{[h]},S^{[h]})$, given in the subsection* 3.7.3.

Proof It immediately follows from Lemma 3.94 and Lemma 3.95. □

COROLLARY 3.97. *Let (V,W,F,S) be a split polarized mixed twistor structure of weight n. Then it is isomorphic to a split Pol-MTS of weight n of the following form:*

$$\bigoplus_i (V^{[h_i]},W,F^{[h_i]},S^{[h_i]})\otimes\mathcal{O}(n).$$

Proof We have only to consider the case $n=0$. By taking the primitive decomposition, we can reduce the problem to the case $V=\bigoplus_{p+q=h,0\leq p,q\leq h}\mathcal{O}(p,-q)$. Then the lemma immediately follows from Corollary 3.96.

□

LEMMA 3.98. *Let (V,W,F,S) be a split polarized mixed twistor structure of weight n. Then the twisted vector bundle $\widetilde{V}$ is a pure twistor of weight n.*

Proof It immediately follows from Corollary 3.97 and Lemma 3.94. □

3.7.5. Polarized mixed twistor structure and nilpotent orbit in one variable. We will show the equivalence of polarized mixed twistor structure and twistor nilpotent orbit, in the one variable case.

LEMMA 3.99. *Let (V,W,f,S) be a Pol-MTS of type $(n,1)$. Then it is a nilpotent orbit of type $(n,1)$.*

Proof We put $V^{(0)}:=\mathrm{Gr}^W(V)$. Then tuple $(V^{(0)},W^{(0)},f^{(0)},S^{(0)})$ is induced. We put as follows:

$$V_0:=V_{|\boldsymbol{C}_\lambda},\quad V_\infty:=V_{|\boldsymbol{C}_\mu},\quad V_0^{(0)}:=V^{(0)}_{|\boldsymbol{C}_\lambda},\quad V_\infty^{(0)}:=V^{(0)}_{|\boldsymbol{C}_\mu}.$$

Let us take frames $\boldsymbol{u}_0^{(0)}=(u_{0\,i}^{(0)})$ and $\boldsymbol{u}_\infty^{(0)}=(u_{\infty\,i}^{(0)})$ of $V_0^{(0)}$ and $V_\infty^{(0)}$ respectively, which are compatible with the natural grading. We take frames $\boldsymbol{u}_0=(u_{0\,i})$ and $\boldsymbol{u}_\infty=(u_{\infty\,j})$ of V_0 and V_∞ respectively, such that they induce $\boldsymbol{u}_0^{(0)}$ and $\boldsymbol{u}_\infty^{(0)}$ respectively. We put as follows:

$$K(i):=\deg^W(u_{0\,i})=\deg^W(u_{0\,i}^{(0)}),\qquad L(i):=\deg^W(u_{\infty\,i})=\deg^W(u_{\infty\,i}^{(0)}).$$

We have the relations:

$$u_{\infty\,i}^{(0)}=\sum B_{j\,i}^{(0)}\cdot u_{0\,j}^{(0)},\quad u_{\infty\,i}=\sum B_{j\,i}\cdot u_{0\,j}.$$

LEMMA 3.100.

- *We have $B_{j\,i}^{(0)}=0$ unless $L(i)=K(j)$.*
- *We have $B_{j\,i}=0$ unless $L(i)\geq K(j)$.*
- *We have $B_{j\,i}=B_{j\,i}^{(0)}$ in the case $L(i)=K(j)$.*

Proof It immediately follows from our choice of the frames. □

We have the following relations:

$$f^{(0)}\otimes t_1^{(-1)}(u_{0\,i}^{(0)})=\sum A_{j\,i}^{(0)}\cdot u_{0\,j}^{(0)},\quad f\otimes t_1^{(-1)}(u_{0\,i})=\sum A_{j\,i}\cdot u_{0\,j}.$$

LEMMA 3.101.

- *We have* $A^{(0)}_{j\,i}=0$ *unless* $K(j)=K(i)-2$.
- *We have* $A_{j\,i}=0$ *unless* $K(j)\leq K(i)-2$.
- *In the case* $K(j)=K(i)-2$, *we have* $A^{(0)}_{j\,i}=A_{j\,i}$.

Proof It immediately follows from our choice of the frames. □

Let us pick a point $P\in\Delta^*$, and we put $y:=-\log|z(P)|^2>0$. The restrictions of $\mathrm{Pat}(V^{(0)},f^{(0)})$ and $\mathrm{Pat}(V,f)$ to $\mathbb{P}^1\times\{P\}$ are given by the following gluings:

$$\boldsymbol{u}^{(0)}_{\infty}=\boldsymbol{u}^{(0)}_{0}\cdot B^{(0)}\cdot\exp\bigl(\sqrt{-1}y\cdot A^{(0)}\bigr),\quad \boldsymbol{u}_{\infty}=\boldsymbol{u}_0\cdot B\cdot\exp\bigl(\sqrt{-1}y\cdot A\bigr).$$

Let us consider the frames $\boldsymbol{u}_{\infty}(y)$, $\boldsymbol{u}_0(y)$, $\boldsymbol{u}^{(0)}_{\infty}(y)$ and $\boldsymbol{u}^{(0)}_{0}(y)$ given as follows:

$$u_{\infty\,i}(y):=y^{-L(i)/2}\cdot u_{\infty\,i},\quad u_{0\,i}(y):=y^{-K(i)/2}\cdot u_{0\,i},$$

$$u^{(0)}_{\infty\,i}(y):=y^{-L(i)/2}\cdot u^{(0)}_{\infty\,i},\quad u^{(0)}_{0\,i}(y):=y^{-K(i)/2}\cdot u^{(0)}_{0\,i}.$$

Then it is easy to see that we have the following relation:

$$\boldsymbol{u}^{(0)}_{\infty}(y)=\boldsymbol{u}^{(0)}_{0}(y)\cdot B^{(0)}\cdot\exp\bigl(\sqrt{-1}A^{(0)}\bigr).\tag{3.27}$$

The vector bundle whose gluing is given by (3.27) is pure twistor of weight 0, due to Lemma 3.94 and Corollary 3.96. On the other hand, we have the following relation:

$$\boldsymbol{u}_{\infty}(y)=\boldsymbol{u}_0(y)\cdot B(y)\cdot\exp\bigl(\sqrt{-1}A(y)\bigr).$$

Here $B(y)_{i\,j}$ and $A(y)_{i\,j}$ are given as follows:

$$B(y)_{i\,j}:=y^{(K(j)-L(i))/2}\cdot B_{i\,j},\quad A(y)_{i\,j}:=y^{(K(j)+2-L(i))/2}\cdot A_{i\,j}.$$

LEMMA 3.102. *We have the following:*

$$\lim_{y\to\infty}B(y)\cdot\exp\bigl(\sqrt{-1}A(y)\bigr)=B^{(0)}\cdot\exp\bigl(\sqrt{-1}A^{(0)}\bigr).$$

Proof It immediately follows from Lemma 3.100 and Lemma 3.101. □

In particular, there exists a positive constant ϵ such that the restriction of $\mathrm{Pat}(V,f)$ to $\Delta^*(\epsilon)$ is a variation of pure twistor structures.

Let us consider the pairing $\tilde{S}$ on $\mathrm{Pat}(V,f)$. Note that it gives the non-degenerate hermitian pairing of $H^0\bigl(\mathrm{Pat}(V,f)_{|\mathbb{P}^1\times\{P\}}\bigr)$ for any point $P\in\Delta^*(\epsilon)$. We have the following:

$$S\bigl(u_{0\,i}(y),\sigma\bigl(u_{\infty,j}(y)\bigr)\bigr)=y^{-(L(i)+K(j))/2}\cdot S\bigl(u_{0\,i},\sigma(u_{\infty,j})\bigr).$$

Hence the limit of S is same as $S^{(0)}$ when we take the limit $y\to\infty$. Hence we obtain the positive definiteness of $\tilde{S}$. Thus the proof of Lemma 3.99 is accomplished. □

LEMMA 3.103. *A split nilpotent orbit* (V,W,f,S) *of weight* n *is a split Pol-MTS of type* $(n,1)$.

Proof By taking the primitive decomposition, we may assume the following:

$$V=\bigoplus_{p+q=n,0\leq p,q}\mathcal{O}(p,-q),\quad W_a=\bigoplus_{p-q\leq a}\mathcal{O}(p,-q).$$

By taking an appropriate isomorphisms, we may assume that f is naturally defined morphism. Recall that S is determined up to constant multiplication. (See the proof of Lemma 3.95.) Then we may derive that (V,W,f,S) is isomorphic to a mixed twistor of the form $(V^{[h]},W^{[h]},f^{[h]},S^{[h]})$, given in the subsection 3.7.3. Thus (V,W,f,S) is a Pol-MTS. □

LEMMA 3.104. *A nilpotent orbit (V, W, f, S) of type $(n, 1)$ is a Pol-MTS of type $(n, 1)$.*

Proof We put $V^{(0)} := \mathrm{Gr}^W(V)$, and then the tuple $(V^{(0)}, W^{(0)}, f^{(0)}, S^{(0)})$ is induced. We use the notation in the proof of Lemma 3.99. We may assume that the restriction of $\mathrm{Pat}(V, f)$ to $\Delta^*(C)$ is a variation of pure twistor for some $0 < C < 1$.

Let us consider the frames $\boldsymbol{u}_\infty(m)$, $\boldsymbol{u}_0(m)$, $\boldsymbol{u}_\infty^{(0)}(m)$ and $\boldsymbol{u}_0^{(0)}(m)$, given as follows:

$$u_{\infty\, i}(m) := m^{-L(i)/2} \cdot u_{\infty\, i}, \quad u_{0\, i}(m) := m^{-K(i)/2} \cdot u_{0\, i},$$

$$u_{\infty\, i}^{(0)}(m) := m^{-L(i)/2} \cdot u_{\infty\, i}^{(0)}, \quad u_{0\, i}^{(0)}(m) := m^{-K(i)/2} \cdot u_{0\, i}^{(0)}.$$

By an argument similar to the proof of Lemma 3.99, the limit vector bundle is naturally isomorphic to $\mathrm{Pat}(V^{(0)}, f^{(0)})$, when we take the limit $m \to \infty$. Note that $\mathrm{Pat}(V^{(0)}, f^{(0)})$ on Δ^* is always a variation of pure twistors.

Pick a point $P \in \Delta^*(C)$. The pairings $\tilde{S}$ and $\tilde{S}^{(0)}$ induce the perfect hermitian product of $H^0(\mathrm{Pat}(V, f)_{|P})$ and $H^0(\mathrm{Pat}(V^{(0)}, f^{(0)})_{|P})$. Since $\tilde{S}$ degenerates to $\tilde{S}^{(0)}$, and since $\tilde{S}$ is a positive definite, we obtain the positive definiteness of $\tilde{S}^{(0)}$. It means that $(V^{(0)}, W^{(0)}, f^{(0)}, S^{(0)})$ is a nilpotent orbit. Due to Lemma 3.103, $(V^{(0)}, W^{(0)}, f^{(0)}, S^{(0)})$ is a Pol-MTS. It implies that (V, W, f, S) is a Pol-MTS. □

In all, we obtain the following:

PROPOSITION 3.105. *Let (V, W, f, S) be a Ψ-Pol-MTS of type $(n, 1)$. It is a nilpotent orbit of type $(n, 1)$ if and only if it is a Pol-MTS of type $(n, 1)$.* □

3.7.6. The twistor nilpotent orbit of split type.

PROPOSITION 3.106. *Let $(V, W, \boldsymbol{f}, S)$ be a split Ψ-Pol-MTS of type (n, l). In this case, the induced variation of $\mathbb{P}^1$-holomorphic bundles $(\mathcal{V}, \mathbb{D}_{\mathcal{V}}^{\triangle})$ is a variation of pure twistors of weight 0 over $\Delta^{*\, l}$.*

Moreover, let S be a pairing of V above such that $(V, S, \sum a_i f_i)$ is a split Pol-MTS. Then the induced pairing $S_{\mathcal{V}}$ of $\mathcal{V}$ gives a polarization of $(\mathcal{V}, \mathbb{D}_{\mathcal{V}}^{\triangle})$.

Proof We have only to show the following:

- Let P be a point of $\Delta^{*\, l}$. Then $\mathcal{V}_{|\{P\}\times\mathbb{P}^1}$ is isomorphic to a trivial bundle.
- Let S be as above. Then $S_{\mathcal{V}\,|\,\{P\}\times\mathbb{P}^1}$ gives a polarization of the pure twistor $\mathcal{V}_{|\{P\}\times\mathbb{P}^1}$.

Recall that $\mathcal{V}_{|\{P\}\times\mathbb{P}^1}$ is obtained as the twisting of the gluing of V by the following:

$$\exp\Big(\sqrt{-1}\sum_{i=1}^{l} -\log|z_i(P)|^2 \cdot f_i \otimes t_1^{(1)}\Big).$$

Note that we have $-\log|z_i(P)|^2 > 0$. Then the proposition immediately follows from Lemma 3.98. □

COROLLARY 3.107. *Let $(V, W, \boldsymbol{f}, S)$ be a split Pol-MTS of type (n, l). Then it is a twistor nilpotent orbit.* □

3.7.7. A split Pol-MTS and a nilpotent orbit in the Hodge theory.

Let $(V, W, \boldsymbol{f}, S)$ be a split Pol-MTS of type (n, l).

LEMMA 3.108. *We can pick a G_m-action ρ_1 on $(V, \boldsymbol{f}, S)$. If we have a real structure ι of V preserving the grading, then ρ_1 can be compatible with ι.*

Proof For an integer h, the integers $p(h)$ and $q(h)$ are given as follows:

- If we have $h = 2m$ for some integer m, then we put $p(h) = q(h) := m$.
- If we have $h = 2m+1$ for some integer m, then we put $p(h) := m+1$ and $q(h) := m$.

Then the G_m-action ρ_1 on V can be given by the isomorphism $V \simeq \bigoplus_h V_h \otimes \mathcal{O}\big(p(h), q(h)\big)$.

The equivariance of $\boldsymbol{f}$ and ι with respect to ρ_1 is clear. To see that S is equivariant, we have only to check the equivariance of the following morphism:

$$\mathcal{O}\big(p(h_1), q(h_1)\big) \otimes \sigma^* \mathcal{O}\big(p(h_2), q(h_2)\big) \longrightarrow \mathbb{T}(-n).$$

Here we put $2n = h_1 + h_2$. But it is clear, for we have $p(h_1)+q(h_2) = p(h_2)+q(h_1) = n$. □

COROLLARY 3.109. *Let $(V, \boldsymbol{f}, S)$ be a split Pol-MTS. Let ι be a real structure of $(V, \boldsymbol{f}, S)$. It gives a nilpotent orbit in the Hodge theory, when we take an appropriate G_m-action of $(V, \boldsymbol{f}, S)$.* □

COROLLARY 3.110. *Let $(V, \boldsymbol{f}, S)$ be a split Pol-MTS. Then the tuple $(V, \boldsymbol{f}, S) \oplus \sigma(V, \boldsymbol{f}, S)$ gives a nilpotent orbit in the Hodge theory, when we take an appropriate G_m-action.*

Proof Since we have the canonical real structure on $(V, \boldsymbol{f}, S) \oplus \sigma(V, \boldsymbol{f}, S)$, it immediately follows from Corollary 3.109. □

3.8. The induced tuple on the divisor

3.8.1. The nilpotent orbit on the divisors. Let $(V, W, \boldsymbol{f}, S)$ be a nilpotent orbit of type (n, l). We put $V_h^{(1)} := P\,\mathrm{Gr}_h^{W(f_1)}(V)$. Then we have the induced filtration $W^{(1)}$, the induced morphisms $f_i^{(1)} : V_h^{(1)} \longrightarrow V_h^{(1)} \otimes \mathbb{T}(-1)$ $(i = 2, \ldots, l)$. The pairing $S\big(f_1^h \otimes \mathrm{id}\big)$ induces the pairing $S_h^{(1)}$. Thus we obtain the induced tuple $\big(V_h^{(1)}, W^{(1)}, \boldsymbol{f}^{(1)}, S_h^{(1)}\big)$, which is a Ψ-Pol-MTS of type $(n+h, l-1)$.

LEMMA 3.111. *The tuple $\big(V_h^{(1)}, W^{(1)}, \boldsymbol{f}^{(1)}, S_h^{(1)}\big)$ is a nilpotent orbit of type $(n+h, l-1)$.*

Proof We may assume $n = 0$. It suffices to show that $\mathrm{Pat}(V_h^{(1)}, W^{(1)}, \boldsymbol{f}^{(1)}, S_h^{(1)})$ is a variation of polarized pure twistor of weight h, on $\Delta^{*\,n-1}$. We identify $\Delta^{*\,n-1}$ and D_1°, naturally.

Let us take a point $P \in D_1^\circ$. Let $\tilde{P}$ denote the point of $\boldsymbol{C}^l$ such that $\{\tilde{P}\} = q_1^{-1}(1) \cap \pi_1^{-1}(P)$. We put $V(P) := \mathrm{Pat}(V, \boldsymbol{f})_{|\tilde{P}}$. We have the induced filtration $W(P)$, the induced morphism f_1, and the induced pairing S on $V(P)$. Due to Lemma 3.80, the tuple $(V(P), W, f_1, S)$ is a nilpotent orbit. In particular, $(V(P), W, f_1, S)$ is a polarized MTS. It implies that $\mathrm{Pat}(V_h^{(1)}, \boldsymbol{f}^{(1)})_{|P}$ is pure twistor of weight h, and $S_h^{(1)}$ is a polarization of $\mathrm{Pat}(V_h^{(1)}, \boldsymbol{f}^{(1)})_{|P}$. Thus we are done. □

3.8.2. The Pol-MTS on the divisor. Let $(V, W, \boldsymbol{N}, S)$ be a Pol-MTS of type (n, l). We put $V_h^{(1)} := P\,\mathrm{Gr}_h^{W(N_1)}(V)$, and we obtain the mixed twistor structure $W^{(1)}$ and the tuple of the induced nilpotent morphisms $N_2^{(1)}, \ldots, N_l^{(1)}$ on $V^{(1)}$. We also obtain the pairing $S_h^{(1)} : V_h^{(1)} \otimes \sigma(V_h^{(1)}) \longrightarrow \mathbb{T}(-n-h)$ by putting $S_h^{(1)} := S(N_1^h \otimes \mathrm{id})$. Then we obtain the tuple $\big(V_h^{(1)}, W^{(1)}, \boldsymbol{N}^{(1)}, S_h^{(1)}\big)$.

LEMMA 3.112. *If the tuple $(V, W, \boldsymbol{N}, S)$ is a split Pol-MTS of type (n, l), the induced tuple $(V_h^{(1)}, W^{(1)}, \boldsymbol{N}^{(1)}, S^{(1)})$ above is a split Pol-MTS of type $(n+h, l-1)$.*

Proof It is easy to check that $\big(V_h^{(1)}, W^{(1)}, \boldsymbol{N}^{(1)}, S^{(1)}\big)$ is a split Ψ-Pol-MTS of type $(n+h, l-1)$. Since $(V, W, \boldsymbol{N}, S)$ is a nilpotent orbit, the tuple $(V_h^{(1)}, W^{(1)}, \boldsymbol{N}^{(1)}, S^{(1)})$ is also nilpotent orbit, due to Lemma 3.111. In particular, it is a Pol-MTS of type $(n+h, l-1)$. Thus we are done. □

Let $(V, W, \boldsymbol{N}, S)$ be a Ψ-Pol-MTS of type (n, l). Let $(V_b^{(1)}, W^{(1)}, \boldsymbol{N}^{(1)}, S^{(1)})$ be the induced Ψ-Pol-MTS of type $(n+b, l-1)$. Then we obtain the following induced Ψ-Pol-MTS of type $(n+b, l-1)$:

$$\Big((V_b^{(1)})^{(0)}, (W^{(1)})^{(0)}, (\boldsymbol{N}^{(1)})^{(0)}, (S^{(1)})^{(0)}\Big).$$

On the other hand, we have the induced Ψ-Pol-MTS $(V^{(0)}, W^{(0)}, \boldsymbol{N}^{(0)}, S^{(0)})$ of type (n, l) obtained from $(V, W, \boldsymbol{N}, S)$. Then we obtain the following induced Ψ-Pol-MTS of type $(n+b, l-1)$ as is given above:

$$\Big((V^{(0)})_b^{(1)}, (W^{(0)})^{(1)}, (\boldsymbol{N}^{(0)})^{(1)}, (S^{(0)})^{(1)}\Big).$$

LEMMA 3.113. *We have the natural isomorphism:*

$$\Big((V_b^{(1)})^{(0)}, (W^{(1)})^{(0)}, (\boldsymbol{N}^{(1)})^{(0)}, (S^{(1)})^{(0)}\Big) \simeq \Big((V^{(0)})_b^{(1)}, (W^{(0)})^{(1)}, (\boldsymbol{N}^{(0)})^{(1)}, (S^{(0)})^{(1)}\Big).$$

Proof We have the two filtrations on Gr_h^W, i.e., $W(N_1^{(0)})$ and $W(N_1)^{(0)}$. We use the following lemma.

LEMMA 3.114. *We have $W(N_1^{(0)}) = W(N_1)^{(0)}$.*

Proof We have only to check that the filtration $W(N_1)^{(0)}$ satisfies the axioms of weight filtrations for $N_1^{(0)}$. It is easy to see $N_1^{(0)} W(N_1)_h^{(0)} \subset W(N_1)_{h-2}^{(0)} \otimes \mathbb{T}(-1)$. We have $\mathrm{Gr}_b^{W(N_1)^{(0)}}\big(\mathrm{Gr}_h^W\big) \simeq \mathrm{Gr}_h^W\big(\mathrm{Gr}_b^{W(N_1)}\big)$. Since N_1 is a morphism of mixed twistor structures, the following morphism is isomorphic:

$$N_1^b : \mathrm{Gr}_h^W\big(\mathrm{Gr}_b^{W(N_1)}\big) \longrightarrow \mathrm{Gr}_{h-2b}^W\big(\mathrm{Gr}_{-b}^{W(N_1)}\big) \otimes \mathbb{T}(-b).$$

Hence the following morphism is isomorphic:

$$(N_1^{(0)})^b : \bigoplus_h \mathrm{Gr}_b^{W(N_1)^{(0)}}\big(\mathrm{Gr}_h^W\big) \simeq \bigoplus_h \mathrm{Gr}_{-b}^{W(N_1)^{(0)}}\big(\mathrm{Gr}_h^W \otimes \mathbb{T}(-b)\big).$$

Thus we obtain Lemma 3.114. □

Therefore, we have the canonical isomorphism $\mathrm{Gr}_b^{W(N_1)}\big(\mathrm{Gr}^W\big) \simeq \mathrm{Gr}^W \mathrm{Gr}_b^{W(N_1)}$. We have the following commutative diagramm:

$$\begin{array}{ccc}
\mathrm{Gr}_b^{W(N_1)}\big(\mathrm{Gr}^W\big) & \xrightarrow{\simeq} & \mathrm{Gr}^W \mathrm{Gr}_b^{W(N_1)} \\
{\scriptstyle N_1^{b+1}}\big\downarrow & & {\scriptstyle N_1^{b+1}}\big\downarrow \\
\mathrm{Gr}_{-b-2}^{W(N_1^{(0)})} \mathrm{Gr}^W \otimes \mathbb{T}(-b-1) & \xrightarrow{\simeq} & \mathrm{Gr}^W \mathrm{Gr}_{-b-2}^{W(N_1)} \otimes \mathbb{T}(-b-1).
\end{array}$$

The kernel of the left vertical arrow is the primitive part $P\,\mathrm{Gr}_b^{W(N_1)} \mathrm{Gr}^W$ by definition. On the other hand, it is easy to see that the kernel of the right vertical arrow is naturally isomorphic to $\mathrm{Gr}^W P\,\mathrm{Gr}_b^{W(N_1)}$, by using the primitive decomposition of

$\mathrm{Gr}_b^{W(N_1)}$. Thus we obtain the canonical isomorphism $\big(V^{(0)}\big)_b^{(1)} \simeq \big(V_b^{(1)}\big)^{(0)}$. Once we obtain the isomorphism of vector bundles, then it is easy to see the coincidence of the filtration, the nilpotent maps and the pairings. Thus we obtain Lemma 3.113. □

PROPOSITION 3.115. *Let $(V, W, \boldsymbol{N}, S)$ be a Pol-MTS of type (n, l). Then the induced tuple $(V_h^{(1)}, W^{(1)}, \boldsymbol{N}^{(1)}, S^{(1)})$ is a Pol-MTS of type $(n+h, l-1)$.*

Proof It suffices to show that the tuple $((V_h^{(1)})^{(0)}, (W^{(1)})^{(0)}, (\boldsymbol{N}^{(1)})^{(0)}, (S^{(1)})^{(0)})$ is a Pol-MTS of type $(n+h, l-1)$. Since the tuple $(V^{(0)}, W^{(0)}, \boldsymbol{N}^{(0)}, S^{(0)})$ is a split Pol-MTS of type (n, l), the induced tuple $((V_h^{(0)})^{(1)}, (W^{(0)})^{(1)}, (\boldsymbol{N}^{(0)})^{(1)}, (S^{(0)})^{(1)})$ is a Pol-MTS of type $(n+h, l-1)$ due to Corollary Lemma 3.112. Then we obtain the result from Lemma 3.113. □

3.8.3. Strongly sequential compatibility.

LEMMA 3.116. *Let $(V, W, \boldsymbol{N}, S)$ be a Pol-MTS of type (n, l). Then $\boldsymbol{N}$ is strongly sequentially compatible.*

Proof We use an induction on l. Due to the hypothesis of the induction and Proposition 3.115, the tuple $\boldsymbol{N}^{(1)} = (N_2^{(1)}, \ldots, N_l^{(1)})$ on $V^{(1)} = \bigoplus V_h^{(1)}$ is sequentially compatible.

Since $(V^{(0)}, W^{(0)}, \boldsymbol{N}^{(0)}, S^{(0)})$ is a nilpotent orbit, $\boldsymbol{N}^{(0)}$ is strongly sequentially compatible, which was shown in our previous paper [**65**]. Hence the conjugacy classes of $N^{(0)}(\boldsymbol{a})$ $(\boldsymbol{a} \in \boldsymbol{R}^I_{>0})$ are constant if we fix a subset $I \subset \underline{l}$. Since the conjugacy classes of $N(\boldsymbol{a})$ and $N^{(0)}(\boldsymbol{a})$ are same, we obtain the constantness of the conjugacy classes of $N(\boldsymbol{a})$ for $\boldsymbol{a} \in \boldsymbol{R}^I_{>0}$.

LEMMA 3.117. *We have $W_{h+b}\big(N(\underline{i})\big)^{(1)} \cap \mathrm{Gr}_h^{W(N_1)} = W_b\big(N(\underline{i})^{(1)}\big) \cap \mathrm{Gr}_h^{W(N_1)}$.*

Proof Let $W^{(1)}$ denote the induced mixed twistor structure on $\mathrm{Gr}_h^{W(N_1)}$. Since $W^{(1)}$ gives the mixed twistor structure, we have the following:
(3.28)
$$\mathrm{Gr}^{W^{(1)}}\big(W_{h+b}\big(N(\underline{i})\big)^{(1)} \cap \mathrm{Gr}_h^{W(N_1)}\big) = W_{h+b}\big(N(\underline{i})\big)^{(1)\,(0)} \cap \mathrm{Gr}^{W^{(1)}} \mathrm{Gr}_h^{W(N_1)},$$
$$\mathrm{Gr}^{W^{(1)}}\big(W_b\big(N(\underline{i})^{(1)}\big) \cap \mathrm{Gr}_h^{W(N_1)}\big) = W_b\big(N(\underline{i})^{(1)}\big)^{(0)} \cap \mathrm{Gr}^{W^{(1)}} \mathrm{Gr}_h^{W(N_1)}.$$
Since $\boldsymbol{N}^{(0)}$ is strongly sequentially compatible, we have the following:
$$W_{h+b}\big(N(\underline{i})^{(0)}\big)^{(1)} \cap \mathrm{Gr}_h^{W(N_1)^{(0)}} \mathrm{Gr}^W = W_b\big(N(\underline{i})^{(0)\,(1)}\big) \cap \mathrm{Gr}_h^{W(N_1)^{(0)}} \mathrm{Gr}^W. \tag{3.29}$$

LEMMA 3.118. *Under the isomorphism $\mathrm{Gr}^W \mathrm{Gr}_h^{W(N_1)} \simeq \mathrm{Gr}_h^{W(N_1)^{(0)}} \mathrm{Gr}^W$, we have the isomorphisms*
(3.30)
$$W_{h+b}(N(\underline{i})^{(0)})^{(1)} \simeq W_{h+b}(N(\underline{i}))^{(1)\,(0)}, \qquad W_b\big(N(\underline{i})^{(0)\,(1)}\big) \simeq W_b\big(N(\underline{i})^{(1)}\big)^{(0)}.$$

Proof We have $W_{h+b}\big(N(\underline{i})^{(0)}\big)^{(1)} = W_{h+b}\big(N(\underline{i})\big)^{(0)\,(1)}$, due to the mixed twistor property of W and Lemma 3.120 (below). Then the first isomorphism is equivalent to the isomorphism $\mathrm{Gr}^W \mathrm{Gr}_h^{W(N_1)}(W_{h+b}(N(\underline{i}))) \simeq \mathrm{Gr}_h^{W(N_1)} \mathrm{Gr}^W\big(W_{h+b}(N(\underline{i}))\big)$, which always holds.

Since the filtration $W^{(1)}$ gives mixed twistor structure, we have the equality $W_b\big(N(\underline{i})^{(1)}\big)^{(0)} = W_b\big(N(\underline{i})^{(1)\,(0)}\big)$, due to Lemma 3.120 (below). Thus the second isomorphism is equivalent to $W_b\big(N(\underline{i})^{(0)\,(1)}\big) \simeq W_b\big(N(\underline{i})^{(1)\,(0)}\big)$. It follows from $N(\underline{i})^{(0)\,(1)} = N(\underline{i})^{(1)(0)}$ under the isomorphism $\mathrm{Gr}^{W^{(1)}} \mathrm{Gr}_h^{W(N_1)} \simeq \mathrm{Gr}_h^{W(N_1)^{(0)}} \mathrm{Gr}^{W}$. Thus we obtain Lemma 3.118. □

From (3.28), (3.29) and (3.30), we obtain the following:

$$\mathrm{Gr}^{W^{(1)}}\big(W_{h+b}\big(N(\underline{i})^{(1)}\big) \cap \mathrm{Gr}_h^{W(N_1)}\big) = \mathrm{Gr}^{W^{(1)}}\big(W_b\big(N(\underline{i})^{(1)}\big) \cap \mathrm{Gr}_h^{W(N_1)}\big). \tag{3.31}$$

Then Lemma 3.117 follows from (3.31) and Lemma 3.121 (below). □

LEMMA 3.119. *We have $W_{h+b}(N(\underline{i}))^{(1)} \cap V_h^{(1)} = W_b(N(\underline{i})^{(1)}) \cap V_h^{(1)}$.*

Proof Since $N(\underline{i})^{(1)}$ is compatible with the primitive decomposition of $\mathrm{Gr}_b^{W(N_1)}$, the weight filtration $W(N(\underline{i})^{(1)})$ is also compatible with it. Then Lemma 3.119 follows from Lemma 3.117. □

We would like to show the surjectivity of the morphism

$$\bigcap_{i=1}^{l} W_{h_i}(\underline{i}) \longrightarrow \bigcap_{i=2}^{l} W_{h_i}^{(1)}(\underline{i}) \cap \mathrm{Gr}_{h_1}^{W(N_1)}. \tag{3.32}$$

The morphism (3.32) induces the morphism on the associated graded vector bundles for $W = W(\underline{l})$, and the induced morphism is surjective. Thus the morphism (3.32) itself is surjective. Similarly, we can show the surjectivity of the following morphism:

$$\operatorname{Ker} N_1^{h_1} \cap \bigcap_{i=1}^{l} W_{h_i}(\underline{i}) \longrightarrow \bigcap_{i=2}^{l} W_{h_i}^{(1)}(\underline{i}) \cap P_{h_1} \mathrm{Gr}_{h_1}^{W(\underline{1})}.$$

Thus we obtain the strongly sequential compatibility, i.e., the proof of Lemma 3.116 is accomplished. □

3.8.4. Lemmas for mixed twistor structure.

LEMMA 3.120. *Let (V, W) be a mixed twistor structure. Let $f : V \longrightarrow V \otimes \mathbb{T}(-1)$ be a nilpotent morphism of mixed twistor structure. Let $W(f)$ be the weight filtration of f. We have the induced mixed twistor structure $V^{(0)} := \mathrm{Gr}^W(V)$ and the induced morphism $f^{(0)} : V^{(0)} \longrightarrow V^{(0)} \otimes \mathbb{T}(-1)$. We also have the induced filtration $W(f)^{(0)}$. Then we have $W(f^{(0)}) = W(f)^{(0)}$.*

Proof In the case $f^{h+1} = 0$, we have $W_{-h}(f) = \mathrm{Im}(f^h)$ and $W_h(f) = \mathrm{Ker}(f^h)$. Since f is strict with respect to the filtration W, we obtain $W_{-h}(f)^0 = W_{-h}(f^{(0)})$ and $W_h(f)^{(0)} = W_h(f^{(0)})$. Due to the recursive construction of the weight filtration (see [**20**]), we obtain $W_k(f)^{(0)} = W_k(f^{(0)})$ for any k. □

LEMMA 3.121. *Let (V, W) be a MTS. Let $V_i \subset V$ $(i = 1, 2)$ be sub MTS. Assume that $\mathrm{Gr}^W(V_1) = \mathrm{Gr}^W(V_2)$ in $\mathrm{Gr}^W(V)$. Then we have $V_1 = V_2$.*

Proof We have only to show $W_h \cap V_1 = W_h \cap V_2$ for any h.

1. We put $h_0 := \min\{h \mid \dim W_h \cap V_1 \neq 0\}$. Then $W_{h_0} \cap V_i \simeq \mathrm{Gr}_{h_0}^W(V_i) \subset \mathrm{Gr}_{h_0}^W(V)$. We put $U := \mathrm{Gr}_{h_0}^W(V_1) = \mathrm{Gr}_{h_0}^W(V_2)$. We have the natural isomorphisms $\phi_i : U \longrightarrow W_{h_0} \cap V_i \subset V$. Then $(\phi_1 - \phi_2)(U) \subset W_{h_0-1}(V)$. Since U is pure twistor of weight h_0, we obtain $\phi_1 - \phi_2 = 0$, i.e., $W_{h_0} \cap V_1 = W_{h_0} \cap V_2$.

2. We assume that the claim holds for $h-1$, and then we will show the claim for h. We put $K := W_{h-1} \cap V_1 = W_{h-1} \cap V_2$. It is sub MTS of V. Thus V/K is also a MTS, and $V_i/K \subset V/K$ $(i=1,2)$ satisfies the condition. Thus the problem is reduced to the previous case. □

REMARK 3.122. See also [**82**] and [**65**] for some more properties of mixed twistor structure. □

3.9. Translation of some results due to Kashiwara, Kawai and Saito

3.9.1. Vanishing cycle theorem. Let $(V, W, \boldsymbol{N}, S)$ be a Pol-MTS of type (n, l). We put $V^{(1)} := \mathrm{Im}(N_1)$. Since $N_1 : V \longrightarrow V \otimes \mathbb{T}(-1)$ is a morphism of mixed twistors, $V^{(1)}$ is a sub mixed twistor of $V \otimes \mathbb{T}(-1)$. The filtration $W^{(1)}$ and the tuple of the nilpotent maps $\boldsymbol{N}^{(1)}$ on $V^{(1)}$ are naturally induced. Since $V^{(1)}$ is a subbundle of $V \otimes \mathbb{T}(-1)$, we have the naturally induced pairing $S' : V^{(1)} \otimes \sigma(V) \longrightarrow \mathbb{T}(-n-1)$. It is easy to see that S' vanishes on $V^{(1)} \otimes \sigma(\mathrm{Ker}\, N_1)$. Hence we obtain the induced pairing $S^{(1)} : V^{(1)} \otimes \sigma(V^{(1)}) \longrightarrow \mathbb{T}(-n-1)$.

REMARK 3.123. Although the way of the definition of the induced pairing $S^{(1)}$ looks slightly different from the way in [**52**], the pairings are same. We remark that our choice of the signature of the nilpotent maps are opposite from the standard one. □

LEMMA 3.124. *Let $(V, W, \boldsymbol{N}, S)$ be a split Pol-MTS of type (n, l). Then the induced tuple $(V^{(1)}, W^{(1)}, \boldsymbol{N}^{(1)}, S^{(1)})$ is a split Pol-MTS of type $(n+1, l)$.*

Proof Recall that we can pick a torus action on $(V, W, \boldsymbol{N}, S)$, and then it is regarded as a nilpotent orbit in the Hodge theory (See Corollary 3.110). Thus the lemma is a consequence of the vanishing cycle theorem due to Kashiwara-Kawai (Theorem 2.15 in [**52**]). □

For a polarized mixed twistor structure $(V, W, \boldsymbol{N}, S)$, we put $V^{(0)} := \mathrm{Gr}^W(V)$. Then we have the induced split Pol-MTS $(V^{(0)}, W^{(0)}, \boldsymbol{N}^{(0)}, S^{(0)})$ of type (n, l). By applying the construction above, we obtain $(V^{(0)\,(1)}, W^{(0)\,(1)}, \boldsymbol{N}^{(0)\,(1)}, S^{(0)\,(1)})$.

On the other hand, we obtain the tuple $(V^{(1)}, W^{(1)}, \boldsymbol{N}^{(1)}, S^{(1)})$. It is easy to check that the tuple is a Ψ-Pol MTS of type (n, l). Hence we obtain the induced tuple $(V^{(1)\,(0)}, W^{(1)\,(0)}, \boldsymbol{N}^{(1)\,(0)}, S^{(1)\,(0)})$.

LEMMA 3.125. *The following tuples are naturally isomorphic:*

$$\big(V^{(0)\,(1)}, W^{(0)\,(1)}, \boldsymbol{N}^{(0)\,(1)}, S^{(0)\,(1)}\big), \quad \big(V^{(1)\,(0)}, W^{(1)\,(0)}, \boldsymbol{N}^{(1)\,(0)}, S^{(1)\,(0)}\big).$$

Proof It is clear from our constructions. □

PROPOSITION 3.126. *Let $(V, W, \boldsymbol{N}, S)$ be a Pol-MTS of type (n, l). We put $V^{(1)} := \mathrm{Im}(N_1)$. Then the naturally induced tuple $(V^{(1)}, W^{(1)}, \boldsymbol{N}^{(1)}, S^{(1)})$ is a Pol-MTS of type (n, l).*

Proof It follows from Lemma 3.124 and Lemma 3.125. □

3.9.2. Kashiwara's Lemma. Let V be a vector bundle over $\mathbb{P}^1$ and $\boldsymbol{N}$ be a tuple of nilpotent maps $N_i : V \longrightarrow V \otimes \mathbb{T}(-1)$ $(i = 1, \ldots, l)$. Let N be a variable, and we put as follows:

$$V[N] := \bigoplus_{j=0}^{\infty} V \otimes \mathbb{T}(j) \cdot N^j.$$

We have the natural mixed twistor structure W on the infinite dimensional vector bundle $V[N]$. We have the natural morphism of mixed twistor structures:

$$N : V[N] \longrightarrow V[N] \otimes \mathbb{T}(-1), \quad u \cdot N^j \longmapsto u \cdot N^{j+1}.$$

For a subset $I \subset \underline{l}$, we have the morphism:

$$\prod_{i \in I}(N - N_i) : V[N] \otimes \mathbb{T}(|I|) \longrightarrow V[N].$$

Then we put as follows:

$$\mathcal{V}_I(V, \boldsymbol{N}) := V[N] \Big/ \prod_{i \in I}(N - N_i) \simeq \bigoplus_{j=0}^{|I|-1} V \otimes \mathbb{T}(j) \cdot N^j. \tag{3.33}$$

We often omit to denote $\boldsymbol{N}$.

Let (V, W) be a mixed twistor structure and $\boldsymbol{N}$ be a tuple of morphisms $N_i : V \longrightarrow V \otimes \mathbb{T}(-1)$ of mixed twistors. We have the induced filtration W and the induced morphisms N, N_i on $(\mathcal{V}_I, W)$. The following lemma is clear from the construction.

LEMMA 3.127. *The morphisms N and N_i are morphisms of mixed twistor structures.* □

COROLLARY 3.128. *The conjugacy classes of $aN + \sum b_i N_i$ are independent of $\lambda \in \mathbb{P}^1$.* □

We put $\tilde{\boldsymbol{N}} := (N, N_1, \ldots, N_l)$. Thus we obtain the mixed twistor structure $(\mathcal{V}_I, W)$ and the tuple of nilpotent maps $\tilde{\boldsymbol{N}}$.

Let $(V, W, \boldsymbol{N}, S)$ be a Pol-MTS(n, l). We have the induced object $(\mathcal{V}_I, \mathcal{W}, \tilde{\boldsymbol{N}})$. We have the induced pairing $V(i) \otimes \sigma\big(V(j)\big) \longrightarrow \mathbb{T}(-n + i + j)$, which we also denote by S. We have the induced morphism $S' : V[N, N^{-1}] \otimes \sigma(V[N, N^{-1}]) \longrightarrow \mathbb{T}(-n)[N, N^{-1}]$, defined as follows:

$$S\big(u \cdot N^i, \sigma(v \cdot N^j)\big) := (-1)^j \cdot S\big(u, \sigma(v)\big) \cdot N^{i+j}.$$

Since N_i are nilpotent, we have the morphism:

$$\prod_{i \in I}(N - N_i)^{-1} : V[N, N^{-1}] \longrightarrow V[N, N^{-1}] \otimes \mathbb{T}(|I|).$$

Then we have the following element of $\mathbb{T}(-n+|I|-k)[N, N^{-1}]$ for any $f_i \in V \otimes \mathbb{T}(i)$ and $g_j \in V \otimes \mathbb{T}(j)$, and for any $k \in \mathbb{Z}$:

$$\begin{aligned} &S'\Big(\prod_{i \in I}(N - N_i)^{-1} f_i \otimes N^{i+k},\ \sigma(g_j \cdot N^j)\Big) = \\ &\qquad \sum_{\boldsymbol{n} \in \mathbb{Z}^I} (-1)^j \cdot N^{-|I|+i+j+k-|\boldsymbol{n}|} S\Big(\prod_{i \in I} N_a^{n_a} f_i, \sigma(g_j)\Big) \in \mathbb{T}(-n + |I| - k)[N]. \end{aligned} \tag{3.34}$$

By taking the residue, i.e., the $N^{-1}\cdot\mathbb{T}(-n+|I|-k-1)$-component, we obtain the morphism $\mathcal{V}_I(S): V[N,N^{-1}]\otimes\sigma(V[N,N^{-1}])\longrightarrow\mathbb{T}(-n+|I|-k-1)$:

$$\mathcal{V}_I(S)\big(f(N)\cdot N^k,\sigma\big(g(N)\big)\big)=\operatorname*{Res}_{N=0}S'\Big(\tilde{N}_I^{-1}\cdot f(N)\cdot N^k,\sigma\big(g(N)\big)\Big). \tag{3.35}$$

Here we put $\tilde{N}_I:=\prod_{i\in I}(N-N_i)$.

LEMMA 3.129. *We have the following formula for $f=\sum f_i\cdot N^i$ and $\sum g_j\cdot N^j$ and for any $k\in\mathbb{Z}$:*

$$\begin{aligned}(3.36)\quad &\mathcal{V}_I(S)\big(f(N)\cdot N^k,\ \sigma\big(g(N)\big)\big)=\\ &\sum_{i,j}\ \sum_{-|I|-|\boldsymbol{n}|+i+j+k=-1}(-1)^j\cdot S\Big(\prod_{a\in I}N_a^{n_a}\cdot f_i,\ \sigma(g_j)\Big)\in\mathbb{T}(-n+|I|-1-k).\end{aligned}$$

Proof It can be checked by a direct calculation. □

In the case $k=0$, we obtain $\mathcal{V}_I(S):\mathcal{V}_I(V)\otimes\sigma(\mathcal{V}_I(V))\longrightarrow\mathbb{T}(-n+|I|-1)$. Thus we obtain the tuple $\big(\mathcal{V}_I(V),\mathcal{W},\boldsymbol{N},\mathcal{V}_I(S)\big)$.

The signature of the construction in [**75**] looks slightly different from that in the above construction at a sight. Let us see that they are same, in fact. Recall the construction in [**75**]. For distinction, we use the variable s. We also denote the given nilpotent maps of V by s_i. Note the relation $s_i=-N_i$. (See Remark 3.61, for example.)

We obtain the vector bundle $\mathcal{V}_I(V)^{(1)}:=V[s]/\prod_{i\in I}(s-s_i)$. The extended pairing $S^{(1)}: V[s,s^{-1}]\otimes\sigma(V)[s,s^{-1}]\longrightarrow\mathbb{T}(-n)[s,s^{-1}]$ given in [**75**] is as follows:

$$S^{(1)}(u\cdot s^i,v\cdot s^j)=(-1)^i\cdot S(u,v)\cdot s^{i+j}.$$

Then we obtain the pairing $\tilde{S}^{(1)}$, given as follows:

$$\tilde{S}^{(1)}(s^i\cdot u,s^j\cdot v):=\operatorname{Res}_{s=0}S^{(1)}\big(u\cdot s^i,\ \tilde{s}_I^{-1}\cdot v\cdot s^j\big).$$

Here $\tilde{s}_I$ denote $\prod_{i\in I}(s-s_i)$.

Since we have the relation $s_i=-N_i$, the correspondence $N\longmapsto -s$ induces the isomorphism $V[N]\simeq V[s]$, and $\mathcal{V}_I(V)\simeq\mathcal{V}_I(V)^{(1)}$.

LEMMA 3.130. *Under the isomorphism, we have $\tilde{S}(N^iu,N^jv)=\tilde{S}^{(1)}(s^iu,s^jv)$.*

Proof We put $\mathcal{S}:=\big\{\boldsymbol{l}\in\mathbb{Z}^I\bigm| -|I|-|\boldsymbol{l}|+i+j=-1\big\}$. By a direct calculation, we have the following:

$$\tilde{S}(N^iu,N^jv)=\sum_{\boldsymbol{l}\in\mathcal{S}}(-1)^j\cdot S\Big(\prod N_k^{l_k}u,v\Big). \tag{3.37}$$

We also have the following:

$$\tilde{S}^{(1)}(s^iu,s^jv)=\sum_{\boldsymbol{l}\in\mathcal{S}}(-1)^i\cdot S\Big(u,\prod s_k^{l_k}v\Big). \tag{3.38}$$

Let us substitute $s=-N$ and $s_i=-N_i$ in (3.38), then we obtain the following:

$$(-1)^{i+j}\tilde{S}^{(1)}(N^iu,N^jv)=\sum_{\boldsymbol{l}\in\mathcal{S}}(-1)^i\cdot S\Big(u,\prod(-N_k)^{l_k}v\Big)=\sum_{\boldsymbol{l}\in\mathcal{S}}(-1)^i\cdot S\Big(\prod N_k^{l_k}u,v\Big).$$

Since it is same as (3.37), Lemma 3.130 is shown. □

LEMMA 3.131. *Let $(V,W,\boldsymbol{N},S)$ be a split Pol-MTS of type (n,l). Then the induced tuple $\big(\mathcal{V}_I(V),\mathcal{W},\boldsymbol{N},\mathcal{V}_I(S)\big)$ is a split Pol-MTS of type $(n-|I|+1,l+1)$.*

Proof Clearly we have a splitting. We put $(V', W', \boldsymbol{N}', S') = (V, W, \boldsymbol{N}, S) \oplus \sigma(V, W, \boldsymbol{N}, S)$, which gives an $\boldsymbol{R}$-nilpotent orbit. Then $\big(\mathcal{V}_I(V'), \mathcal{W}, \boldsymbol{N}, \mathcal{V}_I(S')\big)$ is an $\boldsymbol{R}$-nilpotent orbit. Since we have the following:

$$\big(\mathcal{V}_I(V'), \mathcal{W}, \boldsymbol{N}, \mathcal{V}_I(S')\big) = \big(\mathcal{V}_I(V), \mathcal{W}, \boldsymbol{N}, \mathcal{V}_I(S)\big) \oplus \sigma\big(\mathcal{V}_I(V), \mathcal{W}, \boldsymbol{N}, \mathcal{V}_I(S)\big).$$

Thus it follows from the original Kashiwara's lemma (Proposition 3.19 [**75**] and Appendix [**75**]). Or, it is not difficult to apply Kashiwara's argument in Appendix in [**75**] to the nilpotent twistor orbits. □

COROLLARY 3.132. *Let $(V, W, \boldsymbol{N}, S)$ be a Pol-MTS of type (n, l). Then the tuple $\big(\mathcal{V}_I(V), \mathcal{W}, \boldsymbol{N}, \mathcal{V}_I(S)\big)$ is a Pol-MTS of type $(n - |I| + 1, l + 1)$.*

Proof It is easy to check that the tuple $\big(\mathcal{V}_I(V), \mathcal{W}, \boldsymbol{N}, \mathcal{V}_I(S)\big)$ satisfies the first three conditions. Then the tuple $\big(\mathcal{V}_I(V)^{(0)}, \mathcal{W}^{(0)}, \boldsymbol{N}^{(0)}, \mathcal{V}_I(S)^{(0)}\big)$ is naturally isomorphic to the tuple obtained from $\big(\mathcal{V}_I(V^{(0)}), \mathcal{W}, \boldsymbol{N}^{(0)}, \mathcal{V}_I(S^{(0)})\big)$. Thus we are done. □

3.9.3. Preliminary lemma. In the subsections 3.9.3–3.9.5, we recall the argument given by Saito in the page 302 in [**75**]. See also the original argument.

Let $(V, W, \boldsymbol{N}, S)$ be an equivariant $\boldsymbol{R}$-nilpotent orbit of (n, l)-type, and let $(\hat{V}, \hat{W}, \hat{\boldsymbol{N}}, \hat{S})$ be an equivariant $\boldsymbol{R}$-nilpotent orbit of $(n-1, l)$-type. Let $f : V \longrightarrow \hat{V} \otimes \mathbb{T}(-1)$ and $g : \hat{V} \longrightarrow V$ be morphisms of MTS with the following conditions:

- $g \circ f = N_2 : V \longrightarrow V \otimes \mathbb{T}(-1)$.
- $f \circ g = \hat{N}_2 : \hat{V} \longrightarrow \hat{V} \otimes \mathbb{T}(-1)$.
- $\hat{S}(f \otimes \mathrm{id}) = S(\mathrm{id} \otimes g)$.

Then we have the induced morphisms

$$f^{(1)} : P_h \operatorname{Gr}_h^{W(N_1)}(V) \longrightarrow P_h \operatorname{Gr}_h^{W(\hat{N}_1)}(\hat{V}) \otimes \mathbb{T}(-1),$$

$$g^{(1)} : P_h \operatorname{Gr}_h^{W(\hat{N}_1)}(\hat{V}) \longrightarrow P_h \operatorname{Gr}_h^{W(N_1)}(V).$$

LEMMA 3.133. *In the case $l = 2$, we have the following decomposition:*

$$P_h \operatorname{Gr}^{W(\hat{N}_1)}(\hat{V}) = \operatorname{Ker} g^{(1)} \oplus \Big(\operatorname{Im}(f^{(1)}) \otimes \mathbb{T}(1)\Big).$$

Proof It is easy to see that $f^{(1)}$ and $g^{(1)}$ are morphisms of mixed twistor structures, and that we have $f^{(1)} \circ g^{(1)} = \hat{N}_2^{(1)}$ and $g^{(1)} \circ f^{(1)} = N_2^{(1)}$. We also have $\hat{S}^{(1)}(f^{(1)} \otimes 1) = S^{(1)}(1 \otimes g^{(1)})$.

We have the induced morphisms

$$f^{(2)} : \operatorname{Gr}_a^W P_h \operatorname{Gr}_h^{W(N_1)}(V) \longrightarrow \operatorname{Gr}_{a-2}^W\big(P_h \operatorname{Gr}_h^{W(\hat{N}_1)}(\hat{V})\big) \otimes \mathbb{T}(-1),$$

$$g^{(2)} : \operatorname{Gr}_a^W P_h \operatorname{Gr}_h^{W(\hat{N}_1)}(\hat{V}) \longrightarrow \operatorname{Gr}_a^W P_h \operatorname{Gr}_h^{W(N_1)}(V).$$

In all, we have the following:

$$f^{(2)} : \operatorname{Gr}^W P_h \operatorname{Gr}_h^{W(N_1)}(V) \longrightarrow \operatorname{Gr}^W\big(P_h \operatorname{Gr}_h^{W(\hat{N}_1)}(\hat{V})\big) \otimes \mathbb{T}(-1),$$

$$g^{(2)} : \operatorname{Gr}^W P_h \operatorname{Gr}_h^{W(\hat{N}_1)}(\hat{V}) \longrightarrow \operatorname{Gr}^W P_h \operatorname{Gr}_h^{W(N_1)}(V).$$

We have only to show $\operatorname{Im}(f^{(2)}) \otimes \mathbb{T}(1) \oplus \operatorname{Ker} g^{(2)} = \operatorname{Gr}^W P_h \operatorname{Gr}_h^{W(\hat{N}_1)}(\hat{V})$.

Note the following equalities:

$$\operatorname{Gr}_a^W P_h \operatorname{Gr}_h^{W(N_1)}(V) = \operatorname{Gr}_{a-n-h}^{W(N_2^{(1)})} P_h \operatorname{Gr}_h^{W(N_1)}(V),$$

$$\operatorname{Gr}_a^W P_h \operatorname{Gr}_h^{W(\hat{N}_1)}(\hat{V}) = \operatorname{Gr}_{a-n+1-h}^{W(\hat{N}_2^{(1)})} P_h \operatorname{Gr}_h^{W(\hat{N}_1)}(V).$$

Thus $f^{(2)}$ and $g^{(2)}$ can be regarded as the morphisms:

$$f^{(2)} : \operatorname{Gr}_a^{W(N_2^{(1)})} P_h \operatorname{Gr}_h^{W(N_1)}(V) \longrightarrow \operatorname{Gr}_{a-1}^{W(\hat{N}_2^{(1)})}(P_h \operatorname{Gr}_h^{W(\hat{N}_1)}(\hat{V})) \otimes \mathbb{T}(-1),$$

$$g^{(2)} : \operatorname{Gr}_a^{W(\hat{N}_2^{(1)})} P_h \operatorname{Gr}_h^{W(\hat{N}_1)}(\hat{V}) \longrightarrow \operatorname{Gr}_{a-1}^{W(N_2^{(1)})}(P_h \operatorname{Gr}_h^{W(N_1)}(V)).$$

We put $n + h + 1 = n'$ and as follows:

$$H^a := \operatorname{Gr}_a^{W(N_2^{(1)})} P_h \operatorname{Gr}_h^{W(N_1)}(V), \qquad (\text{ weight } a + n' - 1),$$

$$\hat{H}^a := \operatorname{Gr}^{W(\hat{N}_2^{(1)})} P_h \operatorname{Gr}_h^{W(\hat{N}_1)}(\hat{V}) \otimes \mathbb{T}(-1), \quad (\text{ weight } a + n').$$

Then we obtain the following morphisms:

$$f^{(2)} : H^a \longrightarrow \hat{H}^{a-1},$$

$$g^{(2)} : \hat{H}^a \longrightarrow H^{a-1} \otimes \mathbb{T}(-1).$$

We also have the pairings:

$$S^{(2)} : H^a \otimes \sigma H^{-a} \longrightarrow \mathbb{T}(-n'+1),$$

$$\hat{S}^{(2)} : \hat{H}^a \otimes \sigma \hat{H}^{-a} \longrightarrow \mathbb{T}(-n').$$

We have $\hat{S}^{(2)}(f^{(2)} \otimes \mathrm{id}) = S^{(2)}(1 \otimes g^{(2)})$, and we have the polarizations $S^{(2)}(N_2^{(2)\,a} \otimes \mathrm{id})$ and $\hat{S}^{(2)}(\hat{N}_2^{(2)\,a} \otimes \mathrm{id})$. Then the problem is reduced to Lemma 5.2.15 of [**73**]. □

PROPOSITION 3.134. *We have the following decomposition for any l:*

$$P_h \operatorname{Gr}^{W(\hat{N}_1)}(\hat{V}) = \operatorname{Ker} g^{(1)} \oplus \Big(\operatorname{Im}(f^{(1)}) \otimes \mathbb{T}(1)\Big).$$

Proof We put $X = \Delta^l$, $D_i = \{z_i = 0\}$ and $D = \bigcup_{i=1}^l D_i$. Let $(E, \overline{\partial}_E, h, \theta)$ and $(\hat{E}, \overline{\partial}_{\hat{E}}, \hat{h}, \hat{\theta})$ be the harmonic bundles over $X - D$, corresponding to $(V, W, \boldsymbol{N}, S)$ and $(\hat{V}, \hat{W}, \hat{\boldsymbol{N}}, \hat{S})$ respectively. Since we discuss in the category of mixed twistor structures, such decomposition holds at $\lambda = 1$ if and only if such decomposition holds over $\mathbb{P}^1$. Let $(\mathcal{E}, \mathbb{D})$ and $(\hat{\mathcal{E}}, \hat{\mathbb{D}})$ be the associated deformed holomorphic bundles, and $^\diamond\mathcal{E}$ and $^\diamond\hat{\mathcal{E}}$ be the canonical prolongment.

The morphisms f and g induce the morphisms $F : {}^\diamond\mathcal{E} \longrightarrow {}^\diamond\hat{\mathcal{E}}$ and $G : {}^\diamond\hat{\mathcal{E}} \longrightarrow {}^\diamond\mathcal{E}$, which are flat with respect to the λ-connections $\mathbb{D}$ and $\hat{\mathbb{D}}$.

Pick a point $Q \in D_1 \cap D_2 - \bigcup_{j>2}(D_j \cap D_1 \cap D_2)$. By using the normalizing frame, we obtain the isomorphism:

$$\big({}^\diamond\mathcal{E}_{|\,(1,O)}, \operatorname{Res}_1(\mathbb{D}), \operatorname{Res}_2(\mathbb{D})\big) \simeq \big({}^\diamond\hat{\mathcal{E}}_{|\,(1,Q)}, \operatorname{Res}_1(\mathbb{D}), \operatorname{Res}_2(\mathbb{D})\big).$$

We have a similar isomorphism for $\hat{\mathcal{E}}$. Hence we have only to show the following decomposition:

$$\operatorname{Gr}^{W(N)}({}^\diamond\hat{\mathcal{E}}_{|(1,Q)}) = \operatorname{Im}(F_{|(1,Q)}) \oplus \operatorname{Ker}(G_{|(1,Q)}). \tag{3.39}$$

Let $\pi : X \longrightarrow D_1 \cap D_2$ denote the projection onto the first two components. We obtain the surface $\pi^{-1}(Q) \simeq \Delta^2 \subset X$. The restrictions of $(E_i, \overline{\partial}_{E_i}, h_i, \theta_i)$ to $\pi^{-1}(Q)$ are nilpotent orbits (Lemma 3.80, for example). Thus we obtain the decomposition (3.39) from Lemma 3.133. Thus the proof of Proposition 3.134 is accomplished. □

3.9.4. The preliminary decomposition. Let us consider a pair $\boldsymbol{I} = (J, K)$ of subsets J and K of $\underline{l}$, such that $J \cap K = \emptyset$ and $J \neq \emptyset$. For such a pair, we put $|\boldsymbol{I}| := |J| + |K|$. For two pairs $\boldsymbol{I}_i = (J_i, K_i)$ $(i = 1, 2)$, we mean $J_1 \subset J_2$ and $K_1 \subset K_2$ by $\boldsymbol{I}_1 \subset \boldsymbol{I}_2$. In the case $\boldsymbol{I}_1 \subset \boldsymbol{I}_2$, we put $\boldsymbol{I}_1 - \boldsymbol{I}_2 := (J_2 - J_1) \cup (K_2 - K_1)$.

Let $(V, W, \boldsymbol{N}, S)$ be a Pol-MTS of type (n, l). We put $N_K := \prod_{k \in K} N_k$. We have the naturally induced filtration W, the tuple of the nilpotent maps $\boldsymbol{N}$, and the pairing S on $\mathrm{Im}(N_K)$. Due to Proposition 3.126, the tuple $(\mathrm{Im}(N_K), W, \boldsymbol{N}, S)$ is a Pol-MTS of type $(n + |K|, l)$. By the construction given in the subsection 3.9.2, we obtain the Pol-MTS $\mathcal{V}_J(\mathrm{Im}(N_K))$ of type $(n + |K| - |J| + 1, l + 1)$ (See Corollary 3.132).

Let us pick an element i of K, and we put $K' := K - \{i\}$. The inclusion $\mathrm{Im}(N_K) \subset \mathrm{Im}(N_{K'}) \otimes \mathbb{T}(-1)$ is denoted by var_i. The morphism N_i induces the morphism $\mathrm{Im}(N_{K'}) \longrightarrow \mathrm{Im}(N_K)$, which is denoted by can_i. The morphism var_i and can_i induce the morphisms $\mathrm{var}_i : \mathcal{V}_J(\mathrm{Im}(N_K)) \longrightarrow \mathcal{V}_J(\mathrm{Im}(N_{K'})) \otimes \mathbb{T}(-1)$ and $\mathrm{can}_i : \mathcal{V}_J(\mathrm{Im}(N_{K'})) \longrightarrow \mathcal{V}_J(\mathrm{Im}(N_K))$.

Let us pick an element $i \in J$, and we put $J' := J - \{i\}$. The identity of $\mathrm{Im}(N_K)[N]$ induces the morphism $\mathrm{var}_i : \mathcal{V}_J(\mathrm{Im}(N_K)) \longrightarrow \mathcal{V}_{J'}(\mathrm{Im}(N_K))$. The morphism $(N - N_i) : \mathrm{Im}(N_K)[N] \longrightarrow \mathrm{Im}(N_K)[N] \otimes \mathbb{T}(-1)$ induces the morphism $\mathrm{can}_i : \mathcal{V}_{J'}(\mathrm{Im}(N_K)) \otimes \mathbb{T}(1) \longrightarrow \mathcal{V}_J(\mathrm{Im}(N_K))$.

We introduce the following notation. For a pair $\boldsymbol{I} = (J, K)$ as above, we put $\mathcal{U}_{\boldsymbol{I}} := \mathcal{V}_J(\mathrm{Im}(N_K)) \otimes \mathbb{T}(|K|)$. Then we obtain the morphisms $\mathrm{can}_i : \mathcal{U}_{\boldsymbol{I}'} \longrightarrow \mathcal{U}_{\boldsymbol{I}} \otimes \mathbb{T}(-1)$ and $\mathrm{var}_i : \mathcal{U}_{\boldsymbol{I}} \longrightarrow \mathcal{U}_{\boldsymbol{I}'}$ for $\boldsymbol{I}' \subset \boldsymbol{I}$ such that $\boldsymbol{I} - \boldsymbol{I}' = \{i\}$. They induce the morphisms $\mathrm{can}_i : P_h \,\mathrm{Gr}_h^{W(N)} \mathcal{U}_{\boldsymbol{I}'} \longrightarrow P_h \,\mathrm{Gr}_h^{W(N)} \mathcal{U}_{\boldsymbol{I}} \otimes \mathbb{T}(-1)$ and $\mathrm{var}_i : P_h \,\mathrm{Gr}_h^{W(N)} \mathcal{U}_{\boldsymbol{I}} \longrightarrow P_h \,\mathrm{Gr}_h^{W(N)} \mathcal{U}_{\boldsymbol{I}'}$.

PROPOSITION 3.135. *Let $\boldsymbol{I} = (J, K)$ be a pair of subsets J and K of $\underline{l}$ such that $J \cap K = \emptyset$ and $J \neq \emptyset$. Let $\boldsymbol{I}'$ be a pair of subsets such that $\boldsymbol{I}' \subset \boldsymbol{I}$ and $\boldsymbol{I} - \boldsymbol{I}' = \{i\}$. Then we have the following decomposition:*

$$P_h \,\mathrm{Gr}_h^{W(N)} \big(\mathcal{U}_{\boldsymbol{I}}\big) = \mathrm{Ker}(\mathrm{var}_i) \oplus \mathrm{Im}\big(\mathrm{can}_i \otimes \mathbb{T}(1)\big).$$

Proof We can reduce the problem to the case of equivariant $\boldsymbol{R}$-equivariant nilpotent orbit, due to Corollary 3.110. Then it follows from Proposition 3.134. □

3.9.5. The decomposition. We will often omit to denote the Tate twistor in the following argument, for simplicity. For pairs $\boldsymbol{I}' \subset \boldsymbol{I}$, we have the naturally defined morphisms $\mathrm{can}(\boldsymbol{I}', \boldsymbol{I}) : \mathcal{U}_{\boldsymbol{I}'} \longrightarrow \mathcal{U}_{\boldsymbol{I}}$ and $\mathrm{var}(\boldsymbol{I}', \boldsymbol{I}) : \mathcal{U}_{\boldsymbol{I}} \longrightarrow \mathcal{U}_{\boldsymbol{I}'}$:

$$\mathrm{can}(\boldsymbol{I}', \boldsymbol{I}) := \prod_{i \in \boldsymbol{I} - \boldsymbol{I}'} \mathrm{can}_i, \qquad \mathrm{var}(\boldsymbol{I}', \boldsymbol{I}) := \prod_{i \in \boldsymbol{I} - \boldsymbol{I}'} \mathrm{var}_i\,.$$

Let $\boldsymbol{I}_i = (J_i, K_i)$ $(i = 0, 1, 2, 3)$ be pairs of subsets of $\underline{l}$. We assume $\boldsymbol{I}_1 \cap \boldsymbol{I}_2 := (J_1 \cap J_2, K_1 \cap K_2) = \boldsymbol{I}_0$, and $\boldsymbol{I}_1 \cup \boldsymbol{I}_2 := (J_1 \cup J_2, K_1 \cup K_2) = \boldsymbol{I}_3$.

LEMMA 3.136. *The following diagrams are commutative:*

$$\begin{array}{ccc}
\mathcal{U}_{\boldsymbol{I}_0} \xrightarrow{\mathrm{can}(\boldsymbol{I}_0,\boldsymbol{I}_1)} \mathcal{U}_{\boldsymbol{I}_1} & & \mathcal{U}_{\boldsymbol{I}_0} \xleftarrow{\mathrm{var}(\boldsymbol{I}_0,\boldsymbol{I}_1)} \mathcal{U}_{\boldsymbol{I}_1} \\
\mathrm{can}(\boldsymbol{I}_0,\boldsymbol{I}_2) \downarrow \quad \mathrm{can}(\boldsymbol{I}_1,\boldsymbol{I}_3) \downarrow & & \mathrm{var}(\boldsymbol{I}_0,\boldsymbol{I}_2) \uparrow \quad \mathrm{var}(\boldsymbol{I}_1,\boldsymbol{I}_3) \uparrow \\
\mathcal{U}_{\boldsymbol{I}_2} \xrightarrow{\mathrm{can}(\boldsymbol{I}_2,\boldsymbol{I}_3)} \mathcal{U}_{\boldsymbol{I}_3}, & & \mathcal{U}_{\boldsymbol{I}_2} \xleftarrow{\mathrm{var}(\boldsymbol{I}_2,\boldsymbol{I}_3)} \mathcal{U}_{\boldsymbol{I}_3},
\end{array}$$

$$\begin{array}{c}
\mathcal{U}_{\boldsymbol{I}_0} \xleftarrow{\mathrm{var}(\boldsymbol{I}_0,\boldsymbol{I}_1)} \mathcal{U}_{\boldsymbol{I}_1} \\
\mathrm{can}(\boldsymbol{I}_0,\boldsymbol{I}_2) \downarrow \quad \mathrm{can}(\boldsymbol{I}_1,\boldsymbol{I}_3) \downarrow \\
\mathcal{U}_{\boldsymbol{I}_2} \xleftarrow{\mathrm{var}(\boldsymbol{I}_2,\boldsymbol{I}_3)} \mathcal{U}_{\boldsymbol{I}_3}.
\end{array}$$

Proof It can be directly checked from the definition. □

For $\boldsymbol{I}' \subset \boldsymbol{I}$, the morphisms $\mathrm{can}(\boldsymbol{I}',\boldsymbol{I}) : P_h \mathrm{Gr}_h^{W(N)}(\mathcal{U}_{\boldsymbol{I}'}) \longrightarrow P_h \mathrm{Gr}_h^{W(N)}(\mathcal{U}_{\boldsymbol{I}})$ and $\mathrm{var}(\boldsymbol{I}',\boldsymbol{I}) : P_h \mathrm{Gr}_h^{W(N)}(\mathcal{U}_{\boldsymbol{I}}) \longrightarrow P_h \mathrm{Gr}_h^{W(N)}(\mathcal{U}_{\boldsymbol{I}'})$ are given.

PROPOSITION 3.137. *For $\boldsymbol{I}' \subset \boldsymbol{I}$, we have the following decomposition:*

$$(3.40) \quad P_h \mathrm{Gr}_h^{W(N)}(\mathcal{U}_{\boldsymbol{I}}) = \mathrm{Ker}\big(\mathrm{var}(\boldsymbol{I}',\boldsymbol{I})\big) \oplus \mathrm{Im}\big(\mathrm{can}(\boldsymbol{I}',\boldsymbol{I}) \otimes \mathbb{T}(|\boldsymbol{I}| - |\boldsymbol{I}'|)\big).$$

Proof We use an induction on $|\boldsymbol{I}| - |\boldsymbol{I}'|$. The case $|\boldsymbol{I}| - |\boldsymbol{I}'| = 1$ is shown in Proposition 3.135. We assume that the claim holds in the case $|\boldsymbol{I}| - |\boldsymbol{I}'| < a$, and we will prove that the claim also holds in the case $|\boldsymbol{I}| - |\boldsymbol{I}'| = a$.

We take $\boldsymbol{I}_i$ $(i = 0,1,2,3)$ as in Lemma 3.136 satisfying the following:

$$\boldsymbol{I}_0 = \boldsymbol{I}', \quad \boldsymbol{I}_3 = \boldsymbol{I}, \quad \boldsymbol{I}_0 \subsetneq \boldsymbol{I}_i \subsetneq \boldsymbol{I}_3, \ (i=1,2).$$

Due to the hypothesis of the induction, we have the following decompositions:

$$(3.41) \quad \begin{array}{l} P_h \mathrm{Gr}_h^{W(N)}(\mathcal{U}_{\boldsymbol{I}_3}) = \mathrm{Im}\big(\mathrm{can}(\boldsymbol{I}_2,\boldsymbol{I}_3)\big) \oplus \mathrm{Ker}\big(\mathrm{var}(\boldsymbol{I}_2,\boldsymbol{I}_3)\big), \\ P_h \mathrm{Gr}_h^{W(N)}(\mathcal{U}_{\boldsymbol{I}_1}) = \mathrm{Im}\big(\mathrm{can}(\boldsymbol{I}_0,\boldsymbol{I}_1)\big) \oplus \mathrm{Ker}\big(\mathrm{var}(\boldsymbol{I}_0,\boldsymbol{I}_1)\big). \end{array}$$

Here we have omitted to denote the Tate twist, for simplicity.

Due to the commutativity in Lemma 3.136, the morphism $\mathrm{can}(\boldsymbol{I}_1,\boldsymbol{I}_3)$ and $\mathrm{var}(\boldsymbol{I}_1,\boldsymbol{I}_3)$ are compatible with the decompositions in (3.41). Then it is easy to derive the decomposition (3.40) for $\boldsymbol{I} = \boldsymbol{I}_3$ and $\boldsymbol{I}' = \boldsymbol{I}_0$. □

We put as follows:

$$\mathcal{C}_{\boldsymbol{I}} := \bigcap_{\boldsymbol{I}' \subsetneq \boldsymbol{I}} \mathrm{Ker}\big(\mathrm{var}(\boldsymbol{I}',\boldsymbol{I})\big) \subset P_h \mathrm{Gr}_h^{W(N)}(\mathcal{U}_{\boldsymbol{I}}),$$

$$\mathcal{C}_{\boldsymbol{I},\boldsymbol{I}'} := \mathrm{can}(\boldsymbol{I}',\boldsymbol{I})\big(\mathcal{C}_{\boldsymbol{I}'}\big) \subset P_h \mathrm{Gr}_h^{W(N)}(\mathcal{U}_{\boldsymbol{I}}).$$

LEMMA 3.138. *We have the following:*

$$\mathcal{C}_{\boldsymbol{I}} = \bigcap_{\substack{\boldsymbol{I}' \subset \boldsymbol{I}, \\ |\boldsymbol{I}|-|\boldsymbol{I}'|=1}} \mathrm{Ker}\big(\mathrm{var}(\boldsymbol{I}',\boldsymbol{I})\big) \subset P_h \mathrm{Gr}_h^{W(N)}(\mathcal{U}_{\boldsymbol{I}}).$$

Proof It can be shown by an argument similar to the proof of Proposition 3.137. □

LEMMA 3.139. *Let us consider pairs $\boldsymbol{I}_i \subset \boldsymbol{I}$ $(i=1,2)$. Assume $\boldsymbol{I}_1 \not\subset \boldsymbol{I}_2$. Then we have $\mathrm{var}(\boldsymbol{I}_2,\boldsymbol{I})\big(\mathcal{C}_{\boldsymbol{I}_1,\boldsymbol{I}}\big) = 0$.*

Proof It immediately follows from the commutativity (Lemma 3.136) as follows:

$$\begin{aligned}(3.42)\quad \operatorname{var}(\boldsymbol{I}_2,\boldsymbol{I})\big(\mathcal{C}_{\boldsymbol{I}_1,\boldsymbol{I}}\big) &= \operatorname{var}(\boldsymbol{I}_2,\boldsymbol{I})\circ\operatorname{can}(\boldsymbol{I}_1,\boldsymbol{I})\big(\mathcal{C}_{\boldsymbol{I}_1}\big)\\ &= \operatorname{can}(\boldsymbol{I}_1\cap\boldsymbol{I}_2,\boldsymbol{I}_2)\circ\operatorname{var}(\boldsymbol{I}_1\cap\boldsymbol{I}_2,\boldsymbol{I}_1)\big(\mathcal{C}_{\boldsymbol{I}_1}\big)=0.\end{aligned}$$

□

LEMMA 3.140. *Let us consider sub-pairs $\boldsymbol{I}_1\subset\boldsymbol{I}_2\subset\boldsymbol{I}$. Then the restriction of the morphism $\operatorname{var}(\boldsymbol{I}_2,\boldsymbol{I})_{|\mathcal{C}_{\boldsymbol{I},\boldsymbol{I}_1}}$ is injective.*

Proof It follows from $\mathcal{C}_{\boldsymbol{I},\boldsymbol{I}_1}\subset\operatorname{Im}\big(\operatorname{can}(\boldsymbol{I}_2,\boldsymbol{I})\big)$ and the decomposition in Proposition 3.137. □

LEMMA 3.141. *The natural morphism $\phi:\bigoplus_{\boldsymbol{I}'\subset\boldsymbol{I}}\mathcal{C}_{\boldsymbol{I},\boldsymbol{I}'}\longrightarrow P_h\operatorname{Gr}_h^{W(N)}\mathcal{U}_{\boldsymbol{I}}$ is onto.*

Proof It follows from the surjectivity of $\sum\operatorname{Im}\big(\operatorname{can}(\boldsymbol{I}',\boldsymbol{I})\big)\oplus\mathcal{C}_{\boldsymbol{I}}\longrightarrow P_h\operatorname{Gr}_h^{W(N)}(\mathcal{U}_{\boldsymbol{I}})$. □

PROPOSITION 3.142. *We have the following decomposition:*

$$P_h\operatorname{Gr}_h^{W(N)}(\mathcal{U}_{\boldsymbol{I}})\simeq\bigoplus_{\boldsymbol{I}'\subset\boldsymbol{I}}\mathcal{C}_{\boldsymbol{I},\boldsymbol{I}'}.$$

Proof We have only to show the injectivity of ϕ in Lemma 3.141. Assume $\sum v_{\boldsymbol{I}'}=0$, where $v_{\boldsymbol{I}'}$ are elements of $\mathcal{C}_{\boldsymbol{I},\boldsymbol{I}'}$. We put $\mathcal{S}=\{\boldsymbol{I}'\,|\,v_{\boldsymbol{I}'}\neq 0\}$. Assume $\mathcal{S}\neq\emptyset$, and we will derive a contradiction. Let $\boldsymbol{I}_0'$ be the minimal element. Then we obtain the following:

$$0=\operatorname{var}(\boldsymbol{I}_0',\boldsymbol{I})\Big(\sum v_{\boldsymbol{I}'}\Big)=\operatorname{var}(\boldsymbol{I}_0',\boldsymbol{I})(v_{\boldsymbol{I}_0'}).$$

It implies $v_{\boldsymbol{I}_0'}=0$ due to Lemma 3.140. But it contradicts our choice of $\boldsymbol{I}_0'$. Thus we obtain Proposition 3.142. □

LEMMA 3.143. *For a pair $\boldsymbol{I}=(J,K)$ such that $J\cup K=\underline{l}$, $\mathcal{C}_{\boldsymbol{I}}$ is pure twistor of weight $n-l+1+h$. The pairing of $P_h\operatorname{Gr}_h^{W(N)}(\mathcal{U}_{\boldsymbol{I}})$ induce the polarization on $\mathcal{C}_{\boldsymbol{I}}$.*

Proof We denote $J\sqcup K$ also by $\boldsymbol{I}$, for simplicity of notation. $\mathcal{C}_{\boldsymbol{I}}$ is a direct summand of $P_h\operatorname{Gr}_h^{W(N)}(\mathcal{U}_{\boldsymbol{I}})$. On $\mathcal{C}_{\boldsymbol{I}}$, the nilpotent maps $\mathcal{N}_i$ $(i\in\boldsymbol{I})$ are 0, by definition of $\mathcal{C}_{\boldsymbol{I}}$. Thus the weight filtration $\mathcal{W}(\boldsymbol{I})$ induced by $\sum_{i\in\boldsymbol{I}}\mathcal{N}_i$ is trivial on $\mathcal{C}_{\boldsymbol{I}}$. Recall that the filtration $\mathcal{W}(\boldsymbol{I})$ is same as the induced mixed twistor structure of $P_h\operatorname{Gr}_h^{W(N)}\mathcal{U}_{\boldsymbol{I}}$, up to shift of the degree. Thus we obtain the first claim. The second claim can be shown similarly. □

3.10. $\mathcal{R}$-triple in dimension 0 and twistor structure

3.10.1. Perfect strict $\mathcal{R}$-triple and vector bundle over $\mathbb{P}^1$. We would like to compare the twistor structures of Simpson and Sabbah. Let $\sigma:\boldsymbol{C}_\lambda^*\longrightarrow\boldsymbol{C}_\lambda^*$ be the anti-holomorphic map given by $\sigma(\lambda)=-\overline{\lambda}^{-1}$. For an $\mathcal{O}_{\boldsymbol{C}_\lambda^*}$-module $\mathcal{M}$, we have the natural $\mathcal{O}_{\boldsymbol{C}_\lambda^*}$-module structure of $\sigma^*\mathcal{M}$, which is given by $f\cdot\sigma^*(m)=\sigma^*\big(\overline{\sigma^*(f)}\cdot m\big)$. Let $\mathcal{M}'$ and $\mathcal{M}''$ are $\mathcal{O}_{\boldsymbol{C}_\lambda}$-modules. A sesqui-linear pairing of $\mathcal{M}'$ and $\mathcal{M}''$ is an $\mathcal{O}_{\boldsymbol{C}_\lambda^*}$-homomorphism:

$$\mathcal{M}'_{|\boldsymbol{C}_\lambda^*}\otimes\sigma^*\mathcal{M}''_{|\boldsymbol{C}_\lambda^*}\longrightarrow\mathcal{O}_{\boldsymbol{C}_\lambda^*}.$$

Such a tuple $(\mathcal{M}',\mathcal{M}'',C)$ is called an $\mathcal{R}$-triple in dimension 0. (See [**72**] or the section 14.3 of this paper for more detail on $\mathcal{R}$-triple.)

REMARK 3.144. Following Sabbah, we often use the notation $\overline{\mathcal{M}}$ to denote $\sigma^*\mathcal{M}$. A section $\sigma^* f$ of $\sigma^*\mathcal{M}$ is often denoted by $\overline{f}$. □

DEFINITION 3.145. $\mathcal{T} = (\mathcal{M}', \mathcal{M}'', C)$ be an $\mathcal{R}$-triple in 0-dimension. It is called perfect and strict if the following holds:

- The $\mathcal{O}_{\boldsymbol{C}_\lambda}$-modules $\mathcal{M}'$ and $\mathcal{M}''$ are locally free.
- The sesqui-linear pairing $C : \mathcal{M}'_{|\boldsymbol{C}^*_\lambda} \otimes \sigma^*\mathcal{M}''_{|\boldsymbol{C}^*_\lambda} \longrightarrow \mathcal{O}_{\boldsymbol{C}^*_\lambda}$ is perfect. □

Let V be a vector bundle over $\mathbb{P}^1$. We put $V_0 := V_{|\boldsymbol{C}_\lambda}$ and $V_\infty := V_{|\boldsymbol{C}_\mu}$. We have the natural perfect pairing $C_V : V_0^\vee{}_{|\boldsymbol{C}^*_\lambda} \otimes \sigma V_{\infty\,|\boldsymbol{C}^*_\lambda} \longrightarrow \mathcal{O}_{\boldsymbol{C}^*}$ given as follows: Let U be an open set of $\boldsymbol{C}^*_\lambda$. Let f be a section of V_0 on U, and let g be a section of V_∞ on $\sigma(U)$. Then $\sigma^* g$ gives the section of $\sigma^* V_\infty$ on U. The pairing $C_V(f, \sigma^* g)$ is given by $\langle f, g\rangle$, where $\langle\cdot,\cdot\rangle$ denote the natural pairing of V and $V^\vee$. We put $\Theta(V) := \big(V_0^\vee, \sigma(V_\infty), C_V\big)$, which is clearly a strict perfect $\mathcal{R}$-triple.

On the other hand, if $\mathcal{T} = (\mathcal{M}', \mathcal{M}'', C)$ is a strict perfect $\mathcal{R}$-triple, then C induces the isomorphism $\mathcal{M}'^\vee_{|\boldsymbol{C}^*} \simeq \sigma^{-1}(\mathcal{M}'')_{|\boldsymbol{C}^*}$.

Let $V^{(i)}$ $(i = 1, 2)$ be a locally free $\mathcal{O}_{\mathbb{P}^1}$-modules, and let $f : V^{(1)} \longrightarrow V^{(2)}$ be a morphism of $\mathcal{O}$-sheaves. Then we obtain the morphism $\big(f_0^\vee, \sigma(f_\infty)\big) : \Theta(V^{(1)}) \longrightarrow \Theta(V^{(2)})$.

LEMMA 3.146. *By the correspondence Θ above, we obtain the equivalence of the categories of vector bundles over $\mathbb{P}^1$ and strict perfect $\mathcal{R}$-triples in dimension 0.* □

For perfect strict $\mathcal{R}$-triples $\mathcal{T}_i = (\mathcal{M}'_i, \mathcal{M}''_i, C_i)$ $(i = 1, 2)$, the tensor product $\mathcal{T}_1 \otimes \mathcal{T}_2$ is given by $(\mathcal{M}'_1 \otimes \mathcal{M}'_2, \mathcal{M}''_1 \otimes \mathcal{M}''_2, C_1 \otimes C_2)$. The direct sum is also naturally defined.

Let $\mathcal{T} = (\mathcal{M}', \mathcal{M}'', C)$ be a perfect strict $\mathcal{R}$-triple. The pairing C induces the perfect pairing $C' : \mathcal{M}'^\vee_{|\boldsymbol{C}^*} \otimes \mathcal{M}''^\vee_{|\boldsymbol{C}^*} \longrightarrow \mathcal{O}_{\boldsymbol{C}^*}$. Then we obtain the dual $\mathcal{T}^\vee = (\mathcal{M}'^\vee, \mathcal{M}''^\vee, C')$. The following lemma is easy.

LEMMA 3.147. *The functor Θ preserves tensor products, duals and direct sums.* □

We have Sabbah's hermitian adjoint $\mathcal{T}^* = (\mathcal{M}'', \mathcal{M}', C^*)$ in the category of $\mathcal{R}$-triples ([**72**]), where C^* is given by $C^*(f, \sigma^* g) = \overline{\sigma^* C(g, \sigma^* f)}$. The hermitian adjoint in the category of vector bundles over $\mathbb{P}^1$ is given by $V^* := \sigma(V)^\vee$.

3.10.2. Tate objects. The $\mathcal{R}$-triple corresponding to $\mathcal{O}_{\mathbb{P}^1}(p, q)$ is as follows:

$$\Theta(\mathcal{O}_{\mathbb{P}^1}(p,q)) = \Big(\mathcal{O}_{\boldsymbol{C}_\lambda}\cdot f_0^{(-p,-q)},\ \ \mathcal{O}_{\boldsymbol{C}_\lambda}\cdot\sigma(f_\infty^{(p,q)}),\ \ C\Big).$$

Here the pairing C is given as follows:

$$C\big(f_0^{(-p,-q)}, \sigma(f_\infty^{(p,q)})\big) = \big(\sqrt{-1}\lambda\big)^{p+q}.$$

For simplicity, we denote it by $\big(\sqrt{-1}\lambda\big)^{p+q}$. In particular, if we forget the torus action, $\Theta(\mathcal{O}(n))$ is given as follows:

$$\Theta(\mathcal{O}_{\mathbb{P}^1}(n)) = \Big(\mathcal{O}_{\boldsymbol{C}_\lambda}\cdot f_0^{(-n)},\ \ \mathcal{O}_{\boldsymbol{C}_\lambda}\cdot\sigma(f_\infty^{(n)}),\ \ \big(\sqrt{-1}\lambda\big)^n\Big).$$

The $\mathcal{R}$-triple $\Theta(\sigma(\mathcal{O}_{\mathbb{P}^1}(-n)))$ is given by $\Big(\mathcal{O}_{\boldsymbol{C}_\lambda}\cdot\sigma(f_\infty^{(n)}),\ \mathcal{O}_{\boldsymbol{C}_\lambda}\cdot f_0^{(-n)},\ (-1)^n\cdot(\sqrt{-1}\lambda)^{-n}\Big)$. The latter is same as the hermitian adjoint $\Theta\big(\mathcal{O}_{\mathbb{P}^1}(n)\big)^*$ in the category of $\mathcal{R}$-triples. Hence the isomorphism $\sigma^*\mathcal{O}(-n)\longrightarrow\mathcal{O}(-n)$ (the subsection 3.3.2) induces the isomorphism $\Theta\big(\mathcal{O}_{\mathbb{P}^1}(n)\big)^*\longrightarrow\Theta\big(\mathcal{O}_{\mathbb{P}^1}(-n)\big)$.

LEMMA 3.148. *The induced isomorphism $\Theta\big(\mathcal{O}_{\mathbb{P}^1}(n)\big)^*\longrightarrow\Theta\big(\mathcal{O}_{\mathbb{P}^1}(-n)\big)$ is given by the pair of the maps (φ_1,φ_2):*

$$\mathcal{O}_{\boldsymbol{C}_\lambda}\cdot\sigma\big(f_\infty^{(n)}\big)\xleftarrow{\varphi_1}\mathcal{O}_{\boldsymbol{C}_\lambda}\cdot f_0^{(n)},\qquad \varphi_1(f_0^{(n)})=\sqrt{-1}^n\sigma(f_\infty^{(n)}),$$

$$\mathcal{O}_{\boldsymbol{C}_\lambda}\cdot f_0^{(-n)}\xrightarrow{\varphi_2}\mathcal{O}_{\boldsymbol{C}_\lambda}\cdot\sigma\big(f_\infty^{(-n)}\big),\quad \varphi_2\big(f_0^{(-n)}\big)=\sqrt{-1}^n\sigma\big(f_\infty^{(-n)}\big).$$

Proof The morphism φ_1 is the dual of the isomorphism $\iota_{-n}:\sigma\big(\mathcal{O}(-n)_\infty\big)\longrightarrow\mathcal{O}(-n)_0$. The morphism φ_2 is given by $\sigma(\iota_{-n})$. Then the claim can be checked by a direct calculation. □

For any half integers $k\in\frac{1}{2}\mathbb{Z}$, we put as follows:

$$\mathbb{T}^S(k):=\Big(\mathcal{O}_{\boldsymbol{C}_\lambda}\cdot e_0^{(2k)},\ \mathcal{O}_{\boldsymbol{C}_\lambda}\cdot e_\infty^{(2k)},\ (\sqrt{-1}\lambda)^{-2k}\Big).$$

It is the Tate object in the category of $\mathcal{R}$-triples. Namely, the Tate twist in the category of $\mathcal{R}$-triple is given by the tensor product with $\mathbb{T}^S(k)$. Recall that the canonical isomorphism $\mathbb{T}^S(k)\longrightarrow\mathbb{T}^S(-k)^*$ is given by the pair of maps $\big((-1)^{2k},1\big)$ (see [**72**]).

We fix the complex number a such that $a^2=-\sqrt{-1}$. Then we take the isomorphism $\Phi^{(n)}=(\Phi_1^{(n)},\Phi_2^{(n)}):\Theta\big(\mathcal{O}_{\mathbb{P}^1}(n)\big)\longrightarrow\mathbb{T}^S(-n/2)$ given as follows:

$$\mathcal{O}_{\boldsymbol{C}_\lambda}\cdot f_0^{(-n)}\xleftarrow{\Phi_1^{(n)}}\mathcal{O}_{\boldsymbol{C}_\lambda}\cdot e_0^{(-n)},\qquad \Phi_1(e_0^{(-n)})=a^n\cdot f_0^{(-n)},$$

$$\mathcal{O}_{\boldsymbol{C}_\lambda}\cdot\sigma(f_\infty^{(n)})\xrightarrow{\Phi_2^{(n)}}\mathcal{O}_{\boldsymbol{C}_\lambda}\cdot e_\infty^{(-n)},\quad \Phi_2\big(\sigma(f_\infty^{(n)})\big)=a^{-n}\cdot e_\infty^{(-n)}.$$

In particular, the isomorphism $\Theta(\mathbb{T}(n))\longrightarrow\mathbb{T}^S(n)$ is given by $\big(\Phi_1^{(-2n)},\Phi_2^{(-2n)}\big)$:

$$\Phi_1^{(-2n)}(e_0^{(2n)})=(\sqrt{-1})^n\cdot t_0^{(-n)},\qquad \Phi_2^{(-2n)}(\sigma(t_\infty^{(n)}))=(\sqrt{-1})^{-n}\cdot e_\infty^{(2n)}.$$

LEMMA 3.149. *The following diagramm is commutative:*

$$\begin{array}{ccc}\Theta\big(\mathcal{O}_{\mathbb{P}^1}(n)\big)^* & \longrightarrow & \Theta\big(\mathcal{O}_{\mathbb{P}^1}(-n)\big)\\ \uparrow & & \downarrow\\ \mathbb{T}^S(-n/2)^* & \longrightarrow & \mathbb{T}^S(n/2).\end{array}\tag{3.43}$$

Proof We have the following diagramms:

$$\begin{array}{ccc}\mathcal{O}\cdot\sigma(f_\infty^{(n)}) & \xleftarrow{(\sqrt{-1})^n} & \mathcal{O}\cdot f_0^{(n)}\\ \big\downarrow a^{-n} & & a^{-n}\big\uparrow\\ \mathcal{O}\cdot e_\infty^{(-n)} & \xleftarrow{(-1)^n} & \mathcal{O}\cdot e_0^{(n)},\end{array}\qquad \begin{array}{ccc}\mathcal{O}\cdot f_0^{(-n)} & \xrightarrow{(\sqrt{-1})^n} & \mathcal{O}\cdot\sigma\big(f_\infty^{(-n)}\big)\\ \big\uparrow a^n & & \big\downarrow a^n\\ \mathcal{O}\cdot e_0^{(-n)} & \xrightarrow{1} & \mathcal{O}\cdot e_\infty^{(n)}.\end{array}\tag{3.44}$$

Since we have $a^{-2n}\cdot(\sqrt{-1})^n=(-1)^n$ and $a^{2n}\cdot(\sqrt{-1})^n=1$ due to our choice of a, we obtain the commutativity of the diagramms in (3.44) and thus the commutativity of the diagramm (3.43). □

LEMMA 3.150. *The functor Θ essentially commutes with the Tate twists of the both categories. The functor Θ preserves the compatibility of the Tate twist and the adjunction.*

Proof It follows from Lemma 3.149. □

Recall that we have the canonical isomorphism $\Theta\big(\sigma(V^\vee)\big) \simeq \Theta(V)^*$.

COROLLARY 3.151. *The pairing $S : V \otimes \sigma^*(V) \longrightarrow \mathbb{T}(-n)$ is a polarization of pure twistor structure of weight n if and only if the induced morphism $S : \Theta(V) \longrightarrow \Theta(V)^* \otimes \mathbb{T}^S(-n)$ is a polarization of pure twistor $\mathcal{R}$-triple of weight n in 0-dimension.*

Proof In the case $n = 0$, it is clear. Since Θ is compatible with the Tate twist, we can reduce the problem to the case $n = 0$, as follows:
$S : V \otimes \sigma^*(V) \longrightarrow \mathbb{T}(-n)$ is a polarization,

$$\iff S(-n) : \big(V \otimes \mathcal{O}_{\mathbb{P}^1}(-n)\big) \otimes \sigma^*\big(V \otimes \mathcal{O}(-n)\big) \longrightarrow \mathbb{T}(0) \text{ is a polarization,}$$

$$\iff \Theta\big(V \otimes \mathcal{O}_{\mathbb{P}^1}(-n)\big) \longrightarrow \Theta\big(V \otimes \mathcal{O}_{\mathbb{P}^1}(-n)\big)^* \text{ is a polarization,}$$

$$\iff \Theta(V) \otimes \mathbb{T}(n/2) \longrightarrow \big(\Theta(V) \otimes \mathbb{T}(n/2)\big)^* \text{ is a polarization,}$$

$$\iff \Theta(V) \longrightarrow \Theta(V)^* \otimes \mathbb{T}^S(-n) \text{ is a polarization.}$$

□

3.10.3. The induced polarized pure twistor. Let V be a pure twistor of weight n and $S : V \otimes \sigma(V) \longrightarrow \mathbb{T}(-n)$ be a polarization, which induces the pairing $\sigma(S) : \sigma(V) \otimes V \longrightarrow \mathbb{T}(-n)$, which is also the polarization. Then we obtain the isomorphism $\rho_S : \sigma(V) \longrightarrow V^\vee \otimes \mathbb{T}(-n)$. The induced morphism (ρ'_S, ρ''_S) : $\Theta\big(\sigma(V)\big) \longrightarrow \Theta\big(\sigma(V)\big)^* \otimes \mathbb{T}^S(-n)$ gives the polarization of $\Theta\big(\sigma(V)\big)$. We can naturally regarded ρ'_S and ρ''_S as morphisms $V_0 \longrightarrow \sigma\big(V_\infty^\vee\big)$. Let $C' : V_{0\,|\,\boldsymbol{C}^*} \otimes \sigma^* V_{0\,|\,\boldsymbol{C}^*} \longrightarrow \mathcal{O}_{\boldsymbol{C}^*}$ be the sesqui-linear pairing given as follows:

$$C'(u, \bar{v}) := C_{\sigma(V)}\big(\rho''_S(u), \bar{v}\big).$$

Then we obtain the pure twistor structure (V_0, V_0, C') in Sabbah's sense,

LEMMA 3.152. *The morphism $\big((-1)^n, 1\big) : (V_0, V_0, C') \longrightarrow (V_0, V_0, C') \otimes \mathbb{T}^S(-n)$ gives the polarization.*

Proof We have only to check the claim in the case $V = \mathcal{O}(n)$ with the canonical polarization by a direct calculation. □

Let us calculate the pairing C'. First we consider the case $V = \mathcal{O}(n)$. We denote the pairing of V by S. The pairing $\sigma(S)$ gives the correspondence $\sigma(f_0^{(n)}) \otimes f_\infty^{(n)} \longmapsto (\sqrt{-1})^{-n} \cdot t_\infty^{(-n)}$. Thus we have $\rho_S(\sigma(f_0^{(n)})) = (\sqrt{-1})^{-n} \cdot f_\infty^{(-n)} \otimes t_\infty^{(-n)}$. Hence we have the following:

$$\begin{aligned}(3.45)\quad \rho''_S(f_0^{(n)}) &= \sigma\big(\rho_S(\sigma(f_0^{(n)}))\big) = (\sqrt{-1})^n \cdot \sigma(f_\infty^{(-n)}) \otimes \sigma(t_\infty^{(-n)}) \\ &\longmapsto (\sqrt{-1})^n \cdot \sigma(f_\infty^{(-n)}) \cdot (\sqrt{-1})^n \cdot e_\infty^{(-2n)} = (-1)^n \cdot \sigma(f_\infty^{(-n)}) \cdot e_\infty^{(-2n)}.\end{aligned}$$

Then the pairing $C' : \mathcal{O}(n)_{0\,|\,\boldsymbol{C}^*} \otimes \sigma^*\mathcal{O}(n)_{0\,|\,\boldsymbol{C}^*} \longrightarrow \mathcal{O}_{\boldsymbol{C}^*}$ is given as follows:

(3.46)
$$C'\big(f_0^{(n)} \otimes \overline{f_0^{(n)}}\big) = C_{\sigma(\mathcal{O}(n))}\big(\sigma(f_\infty^{(-n)}) \cdot (-1)^n, \overline{f_0^{(n)}}\big) = (-1)^n \cdot \big\langle \sigma(f_\infty^{(-n)}), \sigma(f_0^{(n)})\big\rangle$$
$$= (-1)^n \cdot \sigma^* \overline{\big\langle f_\infty^{(-n)}, f_0^{(n)}\big\rangle} = (-1)^n \cdot \sigma^* \overline{\big((\sqrt{-1}\lambda)^{-n}\big)} = (\sqrt{-1}\lambda)^n.$$

On the other hand, we have the following:

$$(3.47)\quad S\big(f_0^{(n)}, \sigma(f_0^{(n)})\big) = S\big(f_0^{(n)}, \sigma((\sqrt{-1}\lambda)^{-n} \cdot f_\infty^{(n)})\big)$$
$$= (\sqrt{-1}\lambda)^n \cdot (-1)^n \cdot (\sqrt{-1})^{-n} \cdot t_0^{(-n)} = (\sqrt{-1}\lambda)^n \cdot (\sqrt{-1})^n \cdot t_0^{(-n)}.$$

Hence we have the following formula:

$$C'\big(f_0^{(n)}, \overline{f_0^{(n)}}\big) \cdot t_0^{(-n)} = (\sqrt{-1})^{-n} \cdot S\big(f_0^{(n)}, \sigma^*(f_0^{(n)})\big).$$

LEMMA 3.153. *Let (V, S) be a polarized pure twistor of weight n. Then the $\mathcal{R}$-triple (V_0, V_0, C') with the hermitian sesqui-linear duality $\big((-1)^h, 1\big)$ is a pure twistor of weight n in Sabbah's sense, where the sesqui-linear pairing C' is given as follows:*
$$C'(u, \overline{v}) \cdot t_0^{(-n)} = (\sqrt{-1})^{-n} \cdot S(u \otimes \sigma^*(v)).$$

Proof Since any polarized pure twistor of weight n is isomorphic to a direct sum of $(\mathcal{O}(n), S)$, we can reduce the lemma to the case $V = \mathcal{O}(n)$. It has been already shown above. □

3.10.4. A lemma. First we recall the result in the subsection 3.9.5 with a slightly different notation. Let $(V, W, \boldsymbol{N}, S)$ be a polarized mixed twistor structure of weight 0 in n-variables. Let I and J be subsets of $\underline{n}$ such that $I \cap J = \emptyset$ and $I \neq \emptyset$. We put $N_J := \prod_{j \in J} N_j$. We have the naturally defined filtration W, the tuple of filtrations $\boldsymbol{N}$ and the pairing S on $\mathrm{Im}(N_J) \subset V \otimes \mathbb{T}(-|J|)$. (See the subsection 3.9.1 for the induced pairing.) The tuple $\big(\mathrm{Im}(N_J), W, \boldsymbol{N}, S\big)$ is a polarized mixed twistor structure with weight $|J|$ in n-variables, due to Proposition 3.126. By applying Kashiwara's construction in the subsection 3.9.2, we obtain the vector bundle $\mathcal{V}_I(\mathrm{Im}(N_J))$, on which the filtration W, the nilpotent maps $\boldsymbol{N}$ and the pairing $\mathcal{V}_I(S)$ are naturally defined. The multiplication of N naturally gives the nilpotent map, which is also denoted by N. The tuple of the nilpotent maps N and N_i ($i \in \underline{n}$) are denoted by $\widehat{\boldsymbol{N}}$. Due to Corollary 3.132, the tuple $\big(\mathcal{V}_I(\mathrm{Im}\, N_J), W, \widehat{\boldsymbol{N}}, \mathcal{V}_I(S)\big)$ is the polarized mixed twistor structure with weight $|J| - |I| + 1$ in $(n+1)$-variables.

Let i be an element of I. We put $I' = I - \{i\}$. The identity of $\mathrm{Im}(N_J)$ naturally induces the map $\mathrm{var}_i : \mathcal{V}_I(\mathrm{Im}(N_J)) \longrightarrow \mathcal{V}_{I'}(\mathrm{Im}(N_J))$. Let i be an element of J. We put $J' = J - \{j\}$. We have the natural map $\mathrm{Im}(N_J) \longrightarrow \mathrm{Im}(N_{J'}) \otimes \mathbb{T}(-1)$. It naturally induces the map $\mathrm{var}_i : \mathcal{V}_I\big(\mathrm{Im}(N_J)\big) \longrightarrow \mathcal{V}_I\big(\mathrm{Im}(N_{J'})\big)$. They induce the maps $P_h \mathrm{Gr}_h^{W(N)} \mathcal{V}_I\big(\mathrm{Im}(N_J)\big) \longrightarrow P_h \mathrm{Gr}_h^{W(N)} \mathcal{V}_I\big(\mathrm{Im}(N_{J'})\big)$, which are also denoted by var_i.

Let us consider the case I and J gives the decomposition of $\underline{n}$, i.e., $\underline{n} = I \sqcup J$. We obtain the subbundle $\mathcal{C}_h(I, J)$ of $P_h \mathrm{Gr}_h^{W(N)} \mathcal{V}_I(\mathrm{Im}\, N_J)$, which is given as follows:

$$(3.48)\qquad \mathcal{C}_h(I, J) := \bigcap_{i \in \underline{n}} \mathrm{Ker}\, \mathrm{var}_i\,.$$

Due to Lemma 3.143, $\mathcal{C}_h(I,J)$ is a pure twistor structure of weight $h+|J|-|I|+1$, and the pairing $\mathcal{V}_I(S)(N^h\otimes\mathrm{id})$ gives the polarization. As a result, we obtain the pure twistor structure $\mathcal{C}_h(I,J)\otimes\mathcal{O}(|I|-1-|J|)$ of weight h with the polarization.

In general, we put $V_0:=V_{|\boldsymbol{C}_\lambda}$ for a vector bundle over $\mathbb{P}^1$. We put $Y=\mathcal{C}_h(I,J)\otimes\mathcal{O}(|I|-1-|J|)$. Then we obtain the pure twistor structure, (Y_0,Y_0,C') due to the result in the subsection 3.10.3. Let us calculate the sesqui-linear pairing C'. In the following, $\overline{N}_a$ denote the endomorphism of V_0 determined by $N_a=\overline{N}_a\cdot t_0^{(-1)}$.

We have the following:

$$\mathcal{V}_I(\operatorname{Im}N_J)_0=\bigoplus_{i=0}^{\infty}(\operatorname{Im}N_J)_0\otimes t_0^{(i)}\cdot N^i\Big/\prod_{a\in I}(N-N_a).$$

Let u_i and v_j be sections of $\operatorname{Im}(N_J)_0$ on appropriate open subsets of $\boldsymbol{C}^*$, and let k be any non-negative integer. We put $\tilde{N}_I:=\prod_{i\in I}(N-N_i)$. We have the following equalities:

$$\begin{aligned}(3.49)\quad & S'\big(\tilde{N}_I^{-1}(u_i\cdot N^{i+k}\cdot t_0^{(i)}),\sigma(v_j\cdot t_0^{(j)}\cdot N^j)\big)\\ &=\sum_{\boldsymbol{n}}(-1)^j\cdot N^{-|I|-|\boldsymbol{n}|+i+j+k}S\Big(\prod_{a\in I}N_a^{n_a}u_i\otimes t_0^{(i)},\sigma\big(v_j\cdot t_0^{(j)}\big)\Big)\\ &=\sum_{\boldsymbol{n}}(\sqrt{-1}\lambda)^{-2j}\cdot N^{-|I|-|\boldsymbol{n}|+i+j+k}\cdot S\Big(\prod_{a\in I}\overline{N}_a^{n_a}\cdot u_i,\sigma(v_j)\Big)\cdot t_0^{(-|\boldsymbol{n}|+i+j)}.\end{aligned}$$

We can take sections $\widetilde{u}_i$ and $\widetilde{v}_j$ of V_0 on appropriate open sets satisfying the following:

$$(3.50)\qquad u_i=\Big(\prod_{a\in J}N_a\Big)\widetilde{u}_i,\qquad v_j=\Big(\prod_{a\in J}N_a\Big)\widetilde{v}_j.$$

We put $\overline{N}_J:=\prod_{j\in J}\overline{N}_j$. Due to the construction of the pairing on $\operatorname{Im}(N_J)$, the right hand side of (3.49) can be rewritten as follows:

$$\sum_{\boldsymbol{n}}(\sqrt{-1}\lambda)^{-2j}\cdot N^{-|I|-|\boldsymbol{n}|+i+j+k}\cdot S\Big(\prod_{a\in I}\overline{N}_a^{n_a}\cdot\overline{N}_J\cdot\widetilde{u}_i,\sigma(\widetilde{v}_j)\Big)\cdot t_0^{(-|\boldsymbol{n}|+i+j-|J|)}.$$

Hence we obtain the following:

$$\begin{aligned}(3.51)\quad & \mathcal{V}_I(S)\big(u_i\cdot t_0^{(i)}\cdot N^{i+k},\ \sigma(v_j\cdot t_0^{(j)}\cdot N^j)\big)\\ &=\sum_{\boldsymbol{n}\in S(k)}(\sqrt{-1}\lambda)^{-2j}\cdot S\Big(\prod_{a\in I}\overline{N}_a^{n_a}\cdot\overline{N}_J\cdot\widetilde{u}_i,\ \sigma(\widetilde{v}_j)\Big)\cdot t_0^{(|I|-1-k-|J|)}.\end{aligned}$$

Here we put $S(k):=\big\{\boldsymbol{n}\in\mathbb{Z}_{\geq 0}^I\ \big|\ -|\boldsymbol{n}|-|I|+i+j+k=-1\big\}$.

LEMMA 3.154. *When we twist them by $\mathcal{O}(|I|-1-|J|)$, we have the following:*

$$\begin{aligned}(3.52)\quad & \mathcal{V}_{\underline{I}}(S)\Big(u_i\cdot t_0^{(i)}\cdot N^{i+k}\cdot f_0^{(|I|-1-|J|)},\ \sigma\big(v_j\cdot t_0^{(j)}\cdot N^j\cdot f_0^{(|I|-1-|J|)}\big)\Big)\\ &=\sum_{\boldsymbol{n}\in S(k)}(\sqrt{-1}\lambda)^{|I|-1-|J|-2j}\cdot(\sqrt{-1})^{|I|-1-|J|}\cdot S\Big(\prod_{a\in I}\overline{N}_a^{n_a}\cdot\overline{N}_J\cdot\widetilde{u}_i,\ \sigma(\widetilde{v}_j)\Big)\cdot t_0^{(-k)}.\end{aligned}$$

Proof We have the following correspondence for the canonical polarization of $\mathcal{O}(l)$:

$$(3.53)\quad f_0^{(l)} \otimes \sigma\big(f_0^{(l)}\big) = f_0^{(l)} \otimes \sigma\big((\sqrt{-1}\lambda)^{-l} \cdot f_\infty^{(l)}\big) \longmapsto \\ (-1)^{-l} \cdot (\sqrt{-1}\lambda)^l \cdot (\sqrt{-1})^{-l} \cdot t_0^{(-l)} = (\sqrt{-1})^l \cdot (\sqrt{-1}\lambda)^l \cdot t_0^{(-l)}.$$

Then the claim is clear. □

Then we obtain the following formula:

$$(3.54)\quad \mathcal{V}_I(S)\Big(u_i \cdot t_0^{(i)} \cdot (-\sqrt{-1}N)^i \cdot N^k \cdot f_0^{(l)},\ \sigma\big(v_j \cdot t_0^{(j)} \cdot (\sqrt{-1}N)^j \cdot f_0^{(l)}\big)\Big) \\ = \sqrt{-1}^k \sum_{\boldsymbol{n}\in S(k)} (\sqrt{-1}\lambda)^{l-2j} S\Big(\prod_{a\in I}(-\sqrt{-1}N_a) \cdot \prod_{b\in J}(-\sqrt{-1}N_b) \cdot \widetilde{u}_i, \sigma(\widetilde{v}_j)\Big) \cdot t_0^{(-k)}.$$

Here we put $l = |I| - |J| - 1$. Hence we obtain the following.

LEMMA 3.155. *We put $Y := \mathcal{C}_h(I,J) \otimes \mathcal{O}(|I| - |J| - 1)$. Then the $\mathcal{R}$-triple (Y_0, Y_0, C') with the Hermitian sesqui-linear duality $\big((-1)^h, 1\big)$ is polarized pure twistor of weight h. The sesqui-linear pairing C' is given as follows:*

$$(3.55)\quad C'\big(u_i \cdot (-\sqrt{-1}N)^i, \overline{v \cdot (\sqrt{-1}N)^j}\big) \\ = \sum_{\boldsymbol{n}\in S(k)} (\sqrt{-1}\lambda)^{|I|-|J|-1-2j} \cdot S\Big(\prod_{a\in I}(-\sqrt{-1}N_a)^{n_a} \cdot \prod_{b\in J}(-\sqrt{-1}N_b) \cdot \widetilde{u}_i, \sigma(\widetilde{v}_j)\Big).$$

Here $\widetilde{u}_i$ and $\widetilde{v}_j$ are sections of V_0 such that $\Big(\prod_{j\in J} N_j\Big)\widetilde{u}_i = u_i$ and $\Big(\prod_{j\in J} N_j\Big)\widetilde{v}_j = v_j$.

Proof It follows from Lemma 3.153 and the formula (3.54). □

CHAPTER 4

Preliminary for Filtrations

4.1. Filtrations and decompositions on a vector space

4.1.1. Filtration. Let V be a finite dimensional vector space over a field k.

DEFINITION 4.1. An increasing filtration F of V indexed by $\boldsymbol{R}$ is defined to be a family of subspaces $\{F_\eta \subset V \mid \eta \in \boldsymbol{R}\}$ satisfying the conditions $F_\eta \subset F_{\eta'}$ $(\eta \leq \eta')$ and $F_\eta = V$ for any sufficiently large η. In this paper, we mainly use the increasing filtration. So 'filtration' will mean 'increasing filtration' if we do not mention. □

When we consider a tuple of filtrations, we often use the notation $\boldsymbol{F} = ({}^iF \mid i \in I)$. We also use the notation ${}^i\mathrm{Gr}^F$ to denote the associated graded vector space $\mathrm{Gr}^{{}^iF}$.

Let $\boldsymbol{F} := \bigl({}^iF \bigm| i \in I\bigr)$ be a tuple of filtrations indexed by $\boldsymbol{R}$. For any subset $J \subset I$ and for any $\boldsymbol{\eta} \in \boldsymbol{R}^J$, we put as follows:

$$ {}^JF_{\boldsymbol{\eta}} = \bigcap_{j \in J} {}^jF_{\eta_j}. $$

DEFINITION 4.2. $\boldsymbol{F} := \bigl({}^iF \bigm| i \in I\bigr)$ be a tuple of filtrations indexed by $\boldsymbol{R}$. It is called compatible, if we have a decomposition of $V = \bigoplus_{\boldsymbol{\eta} \in \boldsymbol{R}^I} U_{\boldsymbol{\eta}}$ satisfying the following:

$$ {}^IF_{\boldsymbol{\rho}} = \bigoplus_{\boldsymbol{\eta} \in \boldsymbol{R}^I, \boldsymbol{\eta} \leq \boldsymbol{\rho}} U_{\boldsymbol{\eta}}. \tag{4.1} $$

If $\boldsymbol{F}$ is compatible, a decomposition of V satisfying (4.1) is called a splitting of $\boldsymbol{F}$. □

Recall that $\boldsymbol{F} = ({}^iF \mid i \in I)$ is called sequentially compatible if the following is satisfied:

- We have the induced filtrations ${}^iF^{(1)}$ $(i = 2, \ldots, l)$ on ${}^1\mathrm{Gr}^F$. Then ${}^iF^{(1)}$ are sequentially compatible.
- The following map is surjective:

$$ \bigcap_{i=1}^{l} {}^iF_{h_i} \longrightarrow {}^1\mathrm{Gr}^F_{h_1} \cap \bigcap_{i=2}^{l} {}^iF^{(1)}_{h_i}. $$

LEMMA 4.3. *Compatibility and sequential compatibility are equivalent.*

Proof We saw that sequential compatibility implies compatibility in our previous paper [**65**]. It is easy to see that compatibility implies sequential compatibility. □

Notation Let $\boldsymbol{F} = ({}^iF \mid i \in I)$ be a tuple of filtrations. For any subset $J \subset I$ and $\boldsymbol{\eta} \in \boldsymbol{R}^J$, we put as follows:

$$ {}^J \mathrm{Gr}^{F}_{\boldsymbol{\eta}}(V) := \frac{{}^J F_{\boldsymbol{\eta}}}{\sum_{\boldsymbol{\eta}' \lneq \boldsymbol{\eta}} {}^I F_{\boldsymbol{\eta}'}}. $$

LEMMA 4.4. *Let $\boldsymbol{F} = ({}^iF \bigm| i \in I)$ be a tuple of filtrations on V. Then we have the following inequality:*

$$ \sum_{\boldsymbol{\eta}} \dim {}^I \mathrm{Gr}^F_{\boldsymbol{\eta}} \geq \dim V. $$

If the equality holds, then the filtrations ${}^iF$ $(i = 1, \ldots, l)$ are compatible.

Proof We have the surjection $\pi_{\boldsymbol{\eta}} : {}^I F_{\boldsymbol{\eta}} \longrightarrow {}^I \mathrm{Gr}^F_{\boldsymbol{\eta}}$. We pick a subspace $U_{\boldsymbol{\eta}} \subset {}^I F_{\boldsymbol{\eta}}$ such that the restriction of $\pi_{\boldsymbol{\eta}}$ to $U_{\boldsymbol{\eta}}$ gives the isomorphism of $U_{\boldsymbol{\eta}}$ and ${}^I \mathrm{Gr}^F_{\boldsymbol{\eta}}$. Then we have the naturally defined surjection $f : \bigoplus_{\boldsymbol{\eta}} U_{\boldsymbol{\eta}} \longrightarrow V$. It implies the inequality. If the equality holds, then f is isomorphic, and thus the decomposition $V = \bigoplus_{\boldsymbol{\eta}} U_{\boldsymbol{\eta}}$ gives a splitting of the filtrations. □

4.1.2. Decomposition. Let k be a field. Let V be a finite dimensional vector space over k.

DEFINITION 4.5. A decomposition $\boldsymbol{E}$ of V indexed by a set S is defined to be a family of subspaces $\{\mathbb{E}_\alpha \subset V \bigm| \alpha \in S\}$ such that $V = \bigoplus_{\alpha \in S} \mathbb{E}_\alpha$. □

When we consider a tuple of decompositions, we often use the notation $\boldsymbol{E} = ({}^i\mathbb{E} \bigm| i \in I)$. For any subset $J \subset I$ and for any $\boldsymbol{\alpha} \in \boldsymbol{C}^J$, we put as follows:

$$ {}^J\mathbb{E}_{\boldsymbol{\alpha}} := \bigcap_{j \in J} {}^j\mathbb{E}_{\alpha_j}. $$

DEFINITION 4.6. Let $\boldsymbol{E} = ({}^i\mathbb{E} \bigm| i \in I)$ be a tuple of decompositions indexed by a set S. It is called compatible, if the following holds:

$$ V = \bigoplus_{\boldsymbol{\alpha} \in S^I} {}^I\mathbb{E}_{\boldsymbol{\alpha}}. $$

□

4.1.3. Filtrations and decompositions. Let V be a vector space over k.

DEFINITION 4.7. Let $\boldsymbol{E} = ({}^i\mathbb{E} \bigm| i \in I)$ be a tuple of decompositions of V. Let $\boldsymbol{F} = ({}^jF \bigm| j \in J)$ be a tuple of filtrations of F. Then the tuple $(\boldsymbol{E}, \boldsymbol{F})$ is called compatible, if the following holds:

- The tuple $\boldsymbol{E}$ of decompositions is compatible.
- The tuple $\boldsymbol{F}$ of the filtrations is compatible, in the sense of Definition 4.2.
- We have ${}^jF_b = \bigoplus \left({}^jF_b \cap {}^I\mathbb{E}_{\boldsymbol{\alpha}}\right)$. □

Let J_1 be a subset of J and $\boldsymbol{\eta}$ be an element of $\boldsymbol{R}^{J_1}$. The filtrations ${}^iF$ $(i \notin J_1)$ induce the filtration on ${}^{J_1}\mathrm{Gr}^F_{\boldsymbol{\eta}}$. We denote the induced filtration by ${}^iF^{J_1}$. Similarly, we use the notation ${}^{J_2}F^{J_1}$ and ${}^I\mathbb{E}^{J_1}$. We use the notation $\boldsymbol{F}^{J_1}$ and $\boldsymbol{E}^{J_1}$ to denote the induced tuples $\left({}^jF^{J_1} \bigm| j \in J - J_1\right)$ and $\left({}^i\mathbb{E}^{J_1} \bigm| i \in I\right)$.

4.2. Filtrations and decompositions on a vector bundle

4.2.1. Filtrations on a vector bundle. Let X be a complex manifold or scheme equipped with an action of a finite group G. Let V be a G-equivariant vector bundle.

DEFINITION 4.8.

- A G-filtration F of V indexed by $\boldsymbol{R}$ in the category of G-equivariant vector bundles is defined to be a family of G-equivariant subbundles $\{F_\eta \subset V \,|\, \eta \in \boldsymbol{R}\}$ satisfying the conditions $F_\eta \subset F_{\eta'}$ $(\eta \leq \eta')$ and $F_\eta = V$ for any sufficiently large η.
- A G-decomposition $\mathbb{E}$ of V indexed by $\boldsymbol{C}$ in the category of G-equivariant vector bundles is defined to be a family of G-equivariant subbundles $\{\mathbb{E}_\alpha \,|\, \alpha \in \boldsymbol{C}\}$ such that $V = \bigoplus_{\alpha \in \boldsymbol{C}} \mathbb{E}_\alpha$. □

We often omit to denote "G-", if there are any confusion.

Let $\boldsymbol{F} = ({}^iF \,|\, i \in I)$ be a tuple of G-filtrations of V. Let P be a point of X. Then we obtain the tuple of filtrations $({}^iF_{|P} \,|\, i \in I)$ of the vector space $V_{|P}$. We denote the tuple by $\boldsymbol{F}_{|P}$. Similarly, we use the notation $\boldsymbol{E}_{|P}$ to denote the restriction of $\boldsymbol{E}$ to the fibers $V_{|P}$.

DEFINITION 4.9. Let $\boldsymbol{E} = \{{}^i\mathbb{E} \,|\, i \in I\}$ be a tuple of G-decompositions of V. Let $\boldsymbol{F} = \{{}^jF \,|\, j \in J\}$ be a tuple of G-filtrations of V. The tuple $(\boldsymbol{E}, \boldsymbol{F})$ is called compatible, if the following holds:

- For any point $P \in X$, the tuple $(\boldsymbol{E}_{|P}, \boldsymbol{F}_{|P})$ is compatible in the sense of Definition 4.7.
- Let I_1 and J_1 be subsets of I and J respectively. Let $\boldsymbol{\alpha}$ and $\boldsymbol{\eta}$ be elements of $\boldsymbol{R}^{J_1}$ and $\boldsymbol{C}^{I_1}$. Then $\left\{{}^{J_1}F_{\boldsymbol{\eta}\,|\,P} \cap {}^{I_1}\mathbb{E}_{\boldsymbol{\alpha}\,|\,P} \,\middle|\, P \in X\right\}$ forms a vector bundle over X. □

The second condition can be reworded as follows:

LEMMA 4.10. *We put as follows:*

$$\bar{d}(\boldsymbol{\alpha}, \boldsymbol{\eta}, P) := \dim\Big({}^{J_1}F_{\boldsymbol{\eta}\,|\,P} \cap {}^{I_1}\mathbb{E}_{\boldsymbol{\alpha}\,|\,P}\Big).$$

Then $\left\{{}^{J_1}F_{\boldsymbol{\eta}\,|\,P} \cap {}^{I_1}\mathbb{E}_{\boldsymbol{\alpha}\,|\,P} \,\middle|\, P \in X\right\}$ *forms a vector bundle over* X, *if and only if the numbers* $\bar{d}(\boldsymbol{\alpha}, \boldsymbol{\eta}, P)$ *are independent of a choice of* $P \in X$. □

REMARK 4.11. We put as follows:

$$d(\boldsymbol{\alpha}, \boldsymbol{\eta}, P) := \dim\Big({}^J\operatorname{Gr}^F_{\boldsymbol{\eta}\,|\,P} \cap {}^I\mathbb{E}^J_{\boldsymbol{\alpha}\,|\,P}\Big).$$

If the tuple $(\boldsymbol{E}_{|P}, \boldsymbol{F}_{|P})$ is compatible, the family of the numbers $\{d(\boldsymbol{\alpha}, \boldsymbol{\eta}, P) \,|\, \boldsymbol{\alpha} \in \boldsymbol{C}^I,\ \boldsymbol{\eta} \in \boldsymbol{R}^J\}$ can be reconstructed from the family of the numbers $\{\bar{d}(\boldsymbol{\alpha}, \boldsymbol{\eta}, P) \,|\, \boldsymbol{\alpha} \in \boldsymbol{C}^I,\ \boldsymbol{\eta} \in \boldsymbol{R}^J\}$. Hence, if the tuple $(\boldsymbol{E}, \boldsymbol{F})$ is compatible, we obtain the vector bundle ${}^J\operatorname{Gr}^F_{\boldsymbol{\eta}} \cap {}^I\mathbb{E}^J_{\boldsymbol{\alpha}}$ on X.

Such an argument will be used in many times without mention. □

Let $\boldsymbol{E} = ({}^i\mathbb{E} \,|\, i \in I)$ be a tuple of G-decompositions of V, and let $\boldsymbol{F} = ({}^jF \,|\, j \in J)$ be a tuple of G-filtrations of V. For any subset $I_1 \subset I$, we denote the tuple $({}^i\mathbb{E} \,|\, i \in I_1)$ by $q_{I_1}(\boldsymbol{E})$. We also use the notation $q_{J_1}(\boldsymbol{F})$ for any subset $J_1 \subset J$.

LEMMA 4.12. *Let $\boldsymbol{E}$, $\boldsymbol{F}$ be as above. Assume that $(\boldsymbol{E},\boldsymbol{F})$ is compatible. Then $(q_{I_1}(\boldsymbol{E}), q_{J_1}(\boldsymbol{F}))$ is compatible for any subsets $I_1 \subset I$ and $J_1 \subset J$.* □

DEFINITION 4.13. Let $\boldsymbol{E} = \big({}^{i}\mathbb{E} \,\big|\, i \in I\big)$ be a tuple of G-decompositions of V, and let $\boldsymbol{F} = \big({}^{j}F \,\big|\, j \in J\big)$ be a tuple of G-filtrations of V. Assume that $(\boldsymbol{E},\boldsymbol{F})$ is compatible. A decomposition $V = \bigoplus_{(\boldsymbol{a},\boldsymbol{\alpha})\in \boldsymbol{R}^J\times \boldsymbol{C}^I} U_{(\boldsymbol{a},\boldsymbol{\alpha})}$ is called a splitting of $(\boldsymbol{E},\boldsymbol{F})$, if the following holds for any $(\boldsymbol{a},\boldsymbol{\alpha}) \in \boldsymbol{R}^J \times \boldsymbol{C}^I$:

$$ {}^{J}F_{\boldsymbol{a}} \cap {}^{I}\mathbb{E}_{\boldsymbol{\alpha}} = \bigoplus_{\boldsymbol{b}\le \boldsymbol{a}} U_{(\boldsymbol{b},\boldsymbol{\alpha})}. $$

□

4.2.2. Compatible tuple $(\boldsymbol{E},\boldsymbol{F},\boldsymbol{W})$. Let V be a G-equivariant vector bundle over X. Let $\boldsymbol{E} = \big({}^{s}\mathbb{E} \,\big|\, s \in S\big)$ be a compatible tuple of G-decompositions of V. Let $\boldsymbol{F} = \big({}^{i}F \,\big|\, i \in \underline{l}\big)$ be a compatible tuple of G-filtrations of V. Let m be an integer such that $1 \le m \le l$. For any element $(\boldsymbol{\alpha},\boldsymbol{\eta}) \in \boldsymbol{C}^S \times \boldsymbol{R}^m$, we have the G-vector bundle ${}^{\underline{m}}\mathrm{Gr}^{F}_{\boldsymbol{\eta}} \cap {}^{S}\mathbb{E}^{\underline{m}}_{\boldsymbol{\alpha}}$.

Let l_1 be an integer such that $1 \le l_1 \le l$. Let us consider a tuple $\boldsymbol{W} = \big(W(\underline{m}) \,\big|\, m \in \underline{l_1}\big)$ such that each $W(\underline{m})$ is a G-filtration of ${}^{\underline{m}}\mathrm{Gr}^{F}_{\boldsymbol{\eta}}$ indexed by $\mathbb{Z}$ which is compatible with the tuple $(\boldsymbol{E}^{\underline{m}}, \boldsymbol{F}^{\underline{m}})$. (See the subsection 4.1.3 for $\boldsymbol{E}^{\underline{m}}$ and $\boldsymbol{F}^{\underline{m}}$). In that case, the filtrations $W(\underline{i})$ $(i \le m)$ induce the filtrations on ${}^{\underline{m}}\mathrm{Gr}^{F}_{\boldsymbol{\eta}}$ for any $\boldsymbol{\eta} \in \boldsymbol{R}^m$. We denote the induced filtrations by $W^{\underline{m}}(\underline{i})$. We denote the tuple $\big(W^{\underline{m}}(\underline{i}) \,\big|\, i \le m\big)$ by $\boldsymbol{W}^{\underline{m}}$.

DEFINITION 4.14. Let V, $\boldsymbol{E}$, $\boldsymbol{F}$, l_1 and $\boldsymbol{W}$ be as above. The tuple $(\boldsymbol{E},\boldsymbol{F},\boldsymbol{W})$ is called compatible if $(\boldsymbol{E}^{\underline{m}}, \boldsymbol{F}^{\underline{m}}, \boldsymbol{W}^{\underline{m}})$ is compatible for any $m \le l$, in the sense of Definition 4.9. □

Let J be a subset of $\underline{l}$. Let $m(J)$ denote the number determined by the conditions $\underline{m(J)} \subset J \cap \underline{l_1}$ and $m(J)+1 \notin J \cap \underline{l_1}$. We put $q_J(\boldsymbol{W}) := \big(W(\underline{i}) \,\big|\, i = 1,\ldots,m(J)\big)$. Then the following lemma is easy to see.

LEMMA 4.15. *Let S' and J be subsets of S and I respectively. Then the induced tuple $\big(q_{S'}(\boldsymbol{E}), q_J(\boldsymbol{F}), q_J(\boldsymbol{W})\big)$ is compatible.* □

4.2.3. Splitting of $(\boldsymbol{E},\boldsymbol{F},\boldsymbol{W})$. Let l' be any integer such that $1 \le l' \le l$, and we put $l_1' := \min\{l', l_1\}$. We have the projection $\boldsymbol{R}^l \longrightarrow \boldsymbol{R}^{l'}$, taking the first l'-components. Then we obtain the projection $\pi_1 : \boldsymbol{C}^S \times \boldsymbol{R}^l \times \mathbb{Z}^{l_1} \longrightarrow \boldsymbol{C}^S \times \boldsymbol{R}^{l'}$. We also have the projection $\pi_3 : \boldsymbol{C}^S \times \boldsymbol{R}^l \times \mathbb{Z}^{l_1'} \longrightarrow \boldsymbol{C}^S \times \boldsymbol{R}^{l'}$.

On the other hand, we have the projection $\mathbb{Z}^{l_1} \longrightarrow \mathbb{Z}^{l_1'}$, taking the first l_1'-components. Then we obtain the projection $\pi_2 : \boldsymbol{C}^S \times \boldsymbol{R}^l \times \mathbb{Z}^{l_1} \longrightarrow \boldsymbol{C}^S \times \boldsymbol{R}^l \times \mathbb{Z}^{l_1'}$. Note we have $\pi_1 = \pi_3 \circ \pi_2$.

Let us consider a G-equivariant decomposition of V:

$$ V = \bigoplus_{\boldsymbol{u}\in \boldsymbol{C}^S\times \boldsymbol{R}^l\times \mathbb{Z}^{l_1}} U_{\boldsymbol{u}}. \tag{4.2} $$

For elements $\boldsymbol{v} \in \boldsymbol{C}^S \times \boldsymbol{R}^{l'}$ and $\boldsymbol{u}_1 \in \boldsymbol{C}^S \times \boldsymbol{R}^l \times \mathbb{Z}^{l_1'}$, we put as follows:

$$ {}^{l'}C_{\boldsymbol{v}} := \bigoplus_{\pi_1(\boldsymbol{u})=\boldsymbol{v}} U_{\boldsymbol{u}}, \quad {}^{l_1'}B_{\boldsymbol{u}_1} := \bigoplus_{\pi_2(\boldsymbol{u})=\boldsymbol{u}_1} U_{\boldsymbol{u}}. $$

Then we have the following:

$$ {}^{l'}C_{\boldsymbol{v}} = \bigoplus_{\pi_3(\boldsymbol{u}_1)=\boldsymbol{v}} {}^{l'_1}B_{\boldsymbol{u}_1}. $$

DEFINITION 4.16. Assume that $(\boldsymbol{E}, \boldsymbol{F}, \boldsymbol{W})$ is compatible. The decomposition (4.2) is called a splitting of $(\boldsymbol{E}, \boldsymbol{F}, \boldsymbol{W})$, if the following holds:

(1) The decomposition $V = \bigoplus_{\boldsymbol{v}\in \boldsymbol{C}^S\times \boldsymbol{R}^{l'}} {}^{l'}C_{\boldsymbol{v}}$ is a splitting of $(\boldsymbol{E}, q_{\underline{l}'}(\boldsymbol{F}))$ for any l', in the sense of Definition 4.13.

(2) For any element $\boldsymbol{v} = (\boldsymbol{\alpha}, \boldsymbol{a}) \in \boldsymbol{C}^S \times \boldsymbol{R}^{l'}$, we have the induced decomposition via the isomorphism ${}^{\underline{l}'}\operatorname{Gr}^F_{\boldsymbol{a}} \cap^S\mathbb{E}^{\underline{l}'}_{\boldsymbol{\alpha}} \simeq {}^{l'}C_{\boldsymbol{v}}$ given by the previous condition:

$$ {}^{\underline{l}'}\operatorname{Gr}^F_{\boldsymbol{a}} \cap^S\mathbb{E}^{\underline{l}'}_{\boldsymbol{\alpha}} \simeq \bigoplus_{\pi_3(\boldsymbol{u}_1)=\boldsymbol{v}} {}^{l'_1}B_{\boldsymbol{u}_1}. \tag{4.3} $$

Then the decomposition (4.3) is a splitting of the tuple $(\boldsymbol{F}^{\underline{l}'}, \boldsymbol{W}^{\underline{l}'})$ in the sense of Definition 4.13.

We also say that $\big(U_{\boldsymbol{u}} \,\big|\, \boldsymbol{u} \in \boldsymbol{C}^S \times \boldsymbol{R}^l \times \mathbb{Z}^{l_1}\big)$ is a splitting. □

Let $(\boldsymbol{E}, \boldsymbol{F}, \boldsymbol{W})$ be a compatible tuple, and let $\big(U_{\boldsymbol{u}} \,\big|\, \boldsymbol{u} \in \boldsymbol{C}^S \times \boldsymbol{R}^l \times \mathbb{Z}^{l_1}\big)$ be a splitting of $(\boldsymbol{E}, \boldsymbol{F}, \boldsymbol{W})$. Let S' and J be subsets of S and I respectively. We have the naturally defined projection:

$$ q_{S',J} : \boldsymbol{C}^S \times \boldsymbol{R}^l \times \mathbb{Z}^{l_1} \longrightarrow \boldsymbol{C}^{S'} \times \boldsymbol{R}^J \times \mathbb{Z}^{m(J)}. $$

Then we put as follows, for any element $\boldsymbol{u}_1 \in \boldsymbol{C}^{S'} \times \boldsymbol{R}^J \times \mathbb{Z}^{m(J)}$:

$$ U'_{\boldsymbol{u}_1} := \bigoplus_{q_{S',J}(\boldsymbol{u})=\boldsymbol{u}_1} U_{\boldsymbol{u}}. $$

The following lemma is easy to see.

LEMMA 4.17. *The decomposition* $\big(U'_{\boldsymbol{u}_1} \,\big|\, \boldsymbol{u}_1 \in \boldsymbol{C}^{S'} \times \boldsymbol{R}^J \times \mathbb{Z}^{m(J)}\big)$ *gives a splitting of the compatible tuple* $\big(q_S(\boldsymbol{E}), q_J(\boldsymbol{F}), q_J(\boldsymbol{W})\big)$. □

4.3. Compatibility of the filtrations and nilpotent maps

4.3.1. A lemma. Let V be a vector space over k.

LEMMA 4.18. *Let F be a filtration on V, and N be a nilpotent map on V preserving the filtration F. On the associated graded vector space $\operatorname{Gr}^F(V)$, we have the induced filtration $W^F(N)$ and the induced nilpotent map N^F. Assume $W(N^F) = W^F(N)$.*

- *The isomorphism $N^h : \operatorname{Gr}^{W(N)}_h \longrightarrow \operatorname{Gr}^{W(N)}_{-h}$ is strict with respect to the induced filtration $F^{(1)}$.*
- *The induced filtration $F^{(1)}$ and the primitive decomposition of $\operatorname{Gr}^{W(N)}$ are compatible.*

Proof Let us consider the induced isomorphism:

$$ N^h : \operatorname{Gr}^{W(N)}_h \longrightarrow \operatorname{Gr}^{W(N)}_{-h}. \tag{4.4} $$

It preserves the filtration $F^{(1)}$. First of all, we would like to show that the morphism (4.4) is strict with respect to the filtration $F^{(1)}$.

Since the isomorphism (4.4) preserves the filtration, the equality $\dim F_a^{(1)} \cap \operatorname{Gr}_h^{W(N)} = \dim F_a^{(1)} \cap \operatorname{Gr}_{-h}^{W(N)}$ implies the strictness of (4.4). So we have only to show the equality. For that purpose, we have only to show the following equality for any a and any h:

$$\dim \operatorname{Gr}_a^{F^{(1)}}\big(\operatorname{Gr}_h^{W(N)}\big) = \dim \operatorname{Gr}_a^{F^{(1)}}\big(\operatorname{Gr}_{-h}^{W(N)}\big). \tag{4.5}$$

Note we have the following equalities for any a and h:

$$\dim \operatorname{Gr}_a^{F^{(1)}}\big(\operatorname{Gr}_h^{W(N)}\big) = \dim \operatorname{Gr}_h^{W^F(N)}\big(\operatorname{Gr}_a^F\big) = \dim \operatorname{Gr}_h^{W(N^F)}\big(\operatorname{Gr}_a^F\big). \tag{4.6}$$

By definition of the weight filtrations, we have the following equality:

$$\dim \operatorname{Gr}_h^{W(N^F)}\big(\operatorname{Gr}_a^F\big) = \dim \operatorname{Gr}_{-h}^{W(N^F)}\big(\operatorname{Gr}_a^F\big). \tag{4.7}$$

From (4.6) and (4.7), we obtain the equality (4.5). Thus we obtain the strictness of the morphism of (4.4).

Recall that the primitive part is given as follows:

$$P_h \operatorname{Gr}_h^{W(N)} = \operatorname{Ker}\big(N^{h+1} : \operatorname{Gr}_h^{W(N)} \longrightarrow \operatorname{Gr}_{-h-2}^{W(N)}\big),$$

We put $P_h \operatorname{Gr}_{h-2a}^{W(N)} = N^a P_h \operatorname{Gr}_h^{W(N)}$ for any $0 \leq a \leq h$, and then we have the decomposition.

$$\operatorname{Gr}_h^{W(N)} = \bigoplus_{a \geq 0} P_{|h|+2a} \operatorname{Gr}_h^{W(N)}.$$

Let x be an element of $F_b^{(1)} \cap \operatorname{Gr}_h^{W(N)}$. We have the primitive decomposition $x = \sum x_a$, where $x_a \in P_{|h|+2a} \operatorname{Gr}_h^{W(N)}$. We would like to show each element x_a is contained in $F_b^{(1)}$.

Since the isomorphism (4.4) is strict, we have only to consider the case $h \geq 0$. We assume that $\{a \mid x_a \notin F_b^{(1)}\}$, and we will derive the contradiction. We put $a_0 := \max\{a \mid x_a \notin F_b^{(1)}\}$. We may assume that $x = \sum_{a \leq a_0} x_a$. Then we have the following:

$$N^{h+a_0} x = N^{h+a_0} x_{a_0} \in F_b^{(1)}.$$

Due to the strictness of (4.4), there exists the element $y \in F_b^{(1)} \cap \operatorname{Gr}_{h+2a_0}^{W(N)}$ such that $N^{h+2a_0} y = N^{h+a_0} x_{a_0}$. Due to the property of the primitive decomposition, we have the equality $N^{a_0} y = x_{a_0} \in F_b^{(1)}$, which contradicts $x_{a_0} \notin F_b^{(1)}$. Thus we are done. □

COROLLARY 4.19. *Under the assumption of Lemma* 4.18, *we have the induced filtration* $F^{(1)}$ *on the primitive part* $P_h \operatorname{Gr}_h^{W(N)}$. *We have the natural isomorphism* $\operatorname{Gr}^{F^{(1)}} P_h \operatorname{Gr}_h^{W(N)} \simeq P_h \operatorname{Gr}^{W(N^F)} \operatorname{Gr}^F$. □

4.3.2. Sequential compatibility.

DEFINITION 4.20. Let V be a finite dimensional vector space. Let $\mathcal{N}_1, \ldots, \mathcal{N}_m$ be commuting tuple of nilpotent maps of V. Let ${}^1F, \ldots, {}^lF$ be filtrations on V. The tuple $(\mathcal{N}_1, \ldots, \mathcal{N}_m; {}^1F, \ldots, {}^lF)$ is called sequentially compatible, if the following conditions hold:

(1) Each $\mathcal{N}_j$ preserves the filtrations ${}^iF$.

(2) We put $\mathcal{N}(\underline{j}) := \sum_{i\leq j} \mathcal{N}_i$. We denote the weight filtration of $\mathcal{N}(\underline{j})$ by $W(\underline{j})$. Then the filtrations $(W(\underline{1}), W(\underline{2}), \ldots, W(\underline{m}), {}^1F, \ldots, {}^lF)$ are compatible, in the sense of Definition 4.2.

(3) On the associated graded vector spaces ${}^I\operatorname{Gr}^F_{\boldsymbol{a}}$ $(\boldsymbol{a} \in \boldsymbol{R}^I)$, we have the induced filtrations ${}^IW(\underline{j})$ and the induced nilpotent maps ${}^I\mathcal{N}(\underline{j})$. Then we have $W({}^I\mathcal{N}(\underline{j})) = {}^IW(\underline{j})$.

(4) On the associated graded vector spaces ${}^I\operatorname{Gr}^F_{\boldsymbol{a}}$ $(\boldsymbol{a} \in \boldsymbol{R}^I)$, the tuple of nilpotent maps ${}^I\mathcal{N}_1, \ldots, {}^I\mathcal{N}_m$ are sequentially compatible (see Definition 2.7 in [**65**]). □

LEMMA 4.21. *Assume a tuple $\mathcal{S} = (\mathcal{N}_1, \ldots, \mathcal{N}_m, {}^1F, \ldots, {}^lF)$ is sequentially compatible.*

(A): *The induced tuple $\mathcal{S}(I) := \big({}^I\mathcal{N}_1, \ldots, {}^I\mathcal{N}_m, {}^iF\ (i \in I^c)\big)$ on ${}^I\operatorname{Gr}^F$ is sequentially compatible.*

(B): *The induced tuple $\mathcal{S}^{(1)} := \big(\mathcal{N}_2^{(1)}, \ldots, \mathcal{N}_m^{(1)}, {}^iF\ (i \in \underline{l})\big)$ on $\operatorname{Gr}^{W(\underline{1})}$ is sequentially compatible. Moreover, it is compatible with the primitive decomposition of $\operatorname{Gr}^{W(\underline{1})}$.*

Proof First we see the claim (A). The conditions (1), (3) and (4) in Definition 4.20 are clear. The conditions (2) and (3) for $\mathcal{S}$ imply the condition (2) for $\mathcal{S}(I)$.

Let us show the claim (B). The condition (1) is clear. Note that we have $W(\mathcal{N}^{(1)}(\underline{j}))_h = W^{(1)}(\underline{j})_{h+a}$ on $\operatorname{Gr}^{W(\underline{1})}_a$ due to the condition (4) for $\mathcal{S}$. Then the condition (2) for $\mathcal{S}^{(1)}$ follows from the same condition for $\mathcal{S}$. We have the following on $\operatorname{Gr}^{W^F(\underline{1})}_a {}^I\operatorname{Gr}^F \simeq {}^I\operatorname{Gr}^{F^{(1)}} \operatorname{Gr}^{W(\underline{1})}_a$:

$$ {}^IW(\underline{j})^{(1)}_{h+a} = W\big({}^I\mathcal{N}^{(1)}(\underline{j})\big)_h. $$

Thus the conditions (3) and (4) for $\mathcal{S}^{(1)}$ follows from the conditions (2), (3) and (4) for $\mathcal{S}$.

The compatibility with the primitive decomposition follows from Lemma 4.18. □

COROLLARY 4.22. *Assume that a tuple $(\mathcal{N}_1, \ldots, \mathcal{N}_m, {}^1F, \ldots, {}^lF)$ is sequentially compatible. Then the induced tuple $(\mathcal{N}_2^{(1)}, \ldots, \mathcal{N}_m^{(1)}, {}^1F^{(1)}, \ldots, {}^lF^{(1)})$ on the primitive part $P\operatorname{Gr}^{W(\underline{1})}$ is sequentially compatible.*

Proof It follows from the claim (B) in Lemma 4.21. □

DEFINITION 4.23. Let X be a complex manifold or scheme. Let V be a vector bundle over X. Let $\mathcal{N}_1, \ldots, \mathcal{N}_m$ be nilpotent maps on V. Let ${}^1F, \ldots, {}^lF$ be filtrations of V in the category of vector bundles.

A tuple $(\mathcal{N}_1, \ldots, \mathcal{N}_m, {}^1F, \ldots, {}^lF)$ is called sequentially compatible, if the following conditions hold:

(1) For any point $P \in X$, the tuple $(\mathcal{N}_{1\,|\,P}, \ldots, \mathcal{N}_{m\,|\,P}, {}^1F_{|P}, \ldots, {}^lF_{|P})$ is sequentially compatible, in the sense of Definition 4.20.

(2) The family $\Big\{\bigcap_{j=1}^m W(\underline{j})_{h_j\,|\,P} \cap \bigcap_{i=1}^l F_{a_i\,|\,P} \,\Big|\, P \in X\Big\}$ forms a vector bundle over X for each $(\boldsymbol{h}, \boldsymbol{a})$. □

4.3.3. Compatibility of the nilpotent maps defined on the associated graded bundles. Let V be a vector bundle over X. Let $\boldsymbol{E}=\big({}^s\mathbb{E}\,\big|\,s\in S\big)$ be a compatible tuple of the decompositions. Let $\boldsymbol{F}=\big({}^iF\,\big|\,i\in\underline{l}\big)$ be a compatible tuple of the filtrations. Let us consider a tuple of nilpotent maps $\boldsymbol{N}=(N_i)$, where N_i are defined on ${}^{\underline{i}}\mathrm{Gr}^F(V)$. In that case, we have the induced nilpotent maps $N_i^{\underline{m}}$ on ${}^{\underline{m}}\mathrm{Gr}^F$ for any $i\leq m$. We denote the tuple $(N_1^{\underline{m}},\dots,N_m^{\underline{m}})$ by $\boldsymbol{N}^{\underline{m}}$.

DEFINITION 4.24. The tuple $(\boldsymbol{E},\boldsymbol{F},\boldsymbol{N})$ is called sequentially compatible, if the following holds:

- N_i preserves the decomposition $\boldsymbol{E}^{\underline{i}}$.
- The tuple $(\boldsymbol{N}^{\underline{m}},\boldsymbol{F}^{\underline{m}})$ on ${}^{\underline{m}}\mathrm{Gr}^F$ is sequentially compatible for any $m\leq l$, in the sense of Definition 4.23. (See the subsection 4.1.3 for $\boldsymbol{F}^{\underline{m}}$.) □

When we are given a compatible tuple $(\boldsymbol{E},\boldsymbol{F},\boldsymbol{N})$, we obtain the weight filtration $W(\underline{m})$ on ${}^{\underline{m}}\mathrm{Gr}^F$, and thus the tuple $\boldsymbol{W}=\big(W(\underline{m})\,\big|\,m\in\underline{l}\big)$. Hence we obtain the induced tuple $(\boldsymbol{E},\boldsymbol{F},\boldsymbol{W})$ as in the subsection 4.2.2.

LEMMA 4.25. *If $(\boldsymbol{E},\boldsymbol{F},\boldsymbol{N})$ is compatible in the sense of Definition* 4.24, *then the induced tuple $(\boldsymbol{E},\boldsymbol{F},\boldsymbol{W})$ is compatible in the sense of Definition* 4.14.

Proof It immediately follows from the definition. □

DEFINITION 4.26. A splitting of a sequentially compatible tuple $(\boldsymbol{E},\boldsymbol{F},\boldsymbol{N})$ is defined to be a splitting of the induced tuple $(\boldsymbol{E},\boldsymbol{F},\boldsymbol{W})$, in the sense of Definition 4.16. □

4.3.4. Strongly sequential compatibility of nilpotent maps. Let V be a vector bundle on X. Let $\boldsymbol{E}=\big({}^s\mathbb{E}\,\big|\,s\in S\big)$ be a compatible tuple of G-equivariant decompositions. Let $\boldsymbol{N}=\big(N_i\,|\,i=1,\dots,l\big)$ be a tuple of nilpotent maps of V.

DEFINITION 4.27. The tuple $(\boldsymbol{E},\boldsymbol{N})$ is called strongly sequentially compatible, if the following holds:

- Each nilpotent map N_i preserves the decompositions $\boldsymbol{E}$.
- The tuple of nilpotent maps $\boldsymbol{N}$ is strongly sequentially compatible (see Definition 2.9 in [**65**]). □

DEFINITION 4.28. Let $(\boldsymbol{E},\boldsymbol{N})$ be a strongly sequentially compatible tuple. A splitting of $(\boldsymbol{E},\boldsymbol{N})$ is defined to be decompositions:

$$V=\bigoplus_{\substack{\boldsymbol{\alpha}\in\boldsymbol{C}^S,\\ \boldsymbol{h}\in\mathbb{Z}^m}}U_{\boldsymbol{\alpha},\boldsymbol{h}},\qquad U_{\boldsymbol{\alpha},\boldsymbol{h}}=\bigoplus_{a\geq 0}P_{|q_1(\boldsymbol{h})|+2a}U_{\boldsymbol{\alpha},\boldsymbol{h}}.$$

They are assumed to satisfy the following conditions:

- $\mathcal{N}(\underline{1})\big(P_hU_{\boldsymbol{\alpha},\boldsymbol{h}}\big)=P_hU_{\boldsymbol{\alpha},\boldsymbol{h}-2\boldsymbol{\delta}}$.
- $\mathcal{N}(\underline{1})^{h+1}\big(P_hU_{\boldsymbol{\alpha},\boldsymbol{h}}\big)=0$.
- $\mathcal{N}(\underline{1})^h:P_hU_{\boldsymbol{\alpha},\boldsymbol{h}}\longrightarrow P_hU_{\boldsymbol{\alpha},\boldsymbol{h}-2h\boldsymbol{\delta}}$ is isomorphic if $q_1(\boldsymbol{h})=h\geq 0$.

Here we put $\boldsymbol{\delta}:=(1,\dots,1)\in\mathbb{Z}^m$. □

For any subset $S'\subset S$, the tuple $\big({}^s\mathbb{E}\,\big|\,s\in S'\big)$ is denoted by $q_{S'}(\boldsymbol{E})$. For any integer m' such that $1\leq m'\leq m$, the tuple $(N_i\,|\,i=1,\dots,m')$ is denoted by $q_{\underline{m'}}(\boldsymbol{N})$.

The following lemma is easy to see.

LEMMA 4.29.

- *For any subset $S' \subset S$ and for any integer m' such that $1 \leq m' \leq m$, the induced tuple $\big(q_{S'}(\boldsymbol{E}), q_{\underline{m}'}(\boldsymbol{N})\big)$ is compatible.*
- *Assume $(U_{\boldsymbol{\alpha},\boldsymbol{h}}, P_h U_{\boldsymbol{\alpha},\boldsymbol{h}} \,|\, \boldsymbol{\alpha} \in \boldsymbol{C}^S,\ \boldsymbol{h} \in \mathbb{Z}^m)$ is a splitting of $(\boldsymbol{E}, \boldsymbol{N})$. For any elements $\boldsymbol{\beta} \in \boldsymbol{C}^{S'}$ and $\boldsymbol{k} \in \mathbb{Z}^{m'}$, we put as follows:*

$$U'_{\boldsymbol{\beta},\boldsymbol{k}} := \bigoplus_{\substack{q_{S'}(\boldsymbol{\alpha})=\boldsymbol{\beta}\\ q_{\underline{m}'}(\boldsymbol{h})=\boldsymbol{k}}} U_{\boldsymbol{\alpha},\boldsymbol{h}}, \qquad P_h U'_{\boldsymbol{\beta},\boldsymbol{k}} := \bigoplus_{\substack{q_{S'}(\boldsymbol{\alpha})=\boldsymbol{\beta},\\ q_{\underline{m}'}(\boldsymbol{h})=\boldsymbol{k}}} P_h U_{\boldsymbol{\alpha},\boldsymbol{h}}.$$

Then $(U'_{\boldsymbol{\beta},\boldsymbol{k}}, P_h U'_{\boldsymbol{\beta},\boldsymbol{k}})$ is a splitting. □

REMARK 4.30. We can consider the strongly sequential compatibility of the tuple $\big(\boldsymbol{E}, \boldsymbol{F}, \boldsymbol{W}\big)$. However it is rather complicated, and it is useless for our purpose. □

4.4. Extension of splittings

4.4.1. A lift of a splitting of a compatible filtrations. We put $X := \Delta^n$, $D_i := \{z_i = 0\}$ and $D := \bigcup_{i=1}^p D_i$ for some $p \leq n$. Let $\boldsymbol{c}$ be an element of $\mathbb{Z}^m_{>0}$, and ρ be a $\mu_{\boldsymbol{c}}$-action on X. Let V be a $\mu_{\boldsymbol{c}}$-bundle, and let $\boldsymbol{F} := \big({}^iF \,\big|\, i \in I\big)$ be a compatible tuple of equivariant filtrations of V.

LEMMA 4.31. *Assume that we have an equivariant splitting $\big(U^D_{\boldsymbol{\eta}} \,\big|\, \boldsymbol{\eta} \in \boldsymbol{R}^I\big)$ of $\boldsymbol{F}_{|D}$, i.e., we have a decomposition of $V_{|D}$ as follows:*

$$V_{|D} = \bigoplus_{\boldsymbol{\eta}\in\boldsymbol{R}^I} U^D_{\boldsymbol{\eta}}, \qquad {}^IF_{\rho|D} = \bigoplus_{\boldsymbol{\eta}\leq\rho} U_{\boldsymbol{\eta}}.$$

Then there exists an equivariant splitting $\big(U_{\boldsymbol{\eta}} \,\big|\, \boldsymbol{\eta} \in \boldsymbol{R}^I\big)$ of $\boldsymbol{F}$ on a neighbourhood of O such that $U_{\boldsymbol{\eta}|D} = U^D_{\boldsymbol{\eta}}$.

Proof We have the equivariant surjection ${}^IF_{\boldsymbol{\eta}} \longrightarrow {}^I\mathrm{Gr}^F_{\boldsymbol{\eta}}$ over X. On the divisor D, we have the subbundle $U^D_{\boldsymbol{\eta}} \subset {}^IF_{\boldsymbol{\eta}\,|\,D}$ is given. Then we may extend it to the subbundle $\tilde{U}_{\boldsymbol{\eta}} \subset {}^IF_{\boldsymbol{\eta}}$ on a neighbourhood of O, by using Lemma 2.14. □

4.4.2. A lift of equivariant splitting. We put $X = \Delta^n$, $D_i := \{z_i = 0\}$ and $D = \bigcup_{i=1}^p D_i$. We take a $\mu_{\boldsymbol{c}}$-action on X. Let V be a $\mu_{\boldsymbol{c}}$-equivariant bundle over X. Let $\boldsymbol{F} = \big({}^iF \,\big|\, i \in \underline{l}\big)$ be a compatible tuple of equivariant filtrations of a vector bundle V.

We have the vector bundle ${}^l\,\mathrm{Gr}^F$, and we have the tuple of the induced filtrations $\boldsymbol{F}^{(1)}_a := \big({}^l\,\mathrm{Gr}^F_a \cap {}^iF^{(1)} \,\big|\, i = 1, \ldots, l-1\big)$ on ${}^l\,\mathrm{Gr}^F_a$ for any $a \in \boldsymbol{R}$.

We have the natural isomorphism $\boldsymbol{R}^l \simeq \boldsymbol{R}^{l-1} \times \boldsymbol{R}$. We use the notation $(\boldsymbol{\eta}, a)$ to denote an element of $\boldsymbol{R}^l$, where $\boldsymbol{\eta}$ and a denote elements of $\boldsymbol{R}^{l-1}$ and $\boldsymbol{R}$ respectively.

Let $\big(U^D_{(\boldsymbol{\eta},a)} \,\big|\, (\boldsymbol{\eta}, a) \in \boldsymbol{R}^l\big)$ be a splitting of the restriction $\boldsymbol{F}_{|D}$. For any $a \in \boldsymbol{R}$, let $\big(U^{(1)}_{(\boldsymbol{\eta},a)} \,\big|\, \boldsymbol{\eta} \in \boldsymbol{R}^{l-1}\big)$ be a splitting of the filtration $\boldsymbol{F}^{(1)}_a$ of ${}^l\,\mathrm{Gr}^F_a$.

We have the naturally defined surjection:

$$\pi_a : {}^{\underline{l}}F_{(\boldsymbol{\eta},a)} \longrightarrow {}^l\,\mathrm{Gr}^F_a \cap {}^{\underline{l-1}}F^{(1)}_{\boldsymbol{\eta}}.$$

We assume $\pi_{a|D}\big(U^D_{(\boldsymbol{\eta},a)}\big) = U^{(1)}_{(\boldsymbol{\eta},a)}$.

LEMMA 4.32. *We have a splitting* $\big(U_{(\boldsymbol{\eta},a)}\,\big|\,\boldsymbol{\eta}\in\boldsymbol{R}^{l-1}\big)$ *of* $\boldsymbol{F}$ *satisfying the following:*

$$U_{(\boldsymbol{\eta},a)\,|\,D}=U^D_{(\boldsymbol{\eta},a)},\qquad \pi_a\big(U_{(\boldsymbol{\eta},a)}\big)=U^{(1)}_{(\boldsymbol{\eta},a)}.$$

Such a splitting $\big(U_{(\boldsymbol{\eta},a)}\,\big|\,\boldsymbol{\eta}\in\boldsymbol{R}^{l-1}\big)$ *is called a lift of* $\big(U^{(1)}_{(\boldsymbol{\eta},a)}\,\big|\,(\boldsymbol{\eta},a)\in\boldsymbol{R}^l\big)$ *extending* $\big(U^D_{(\boldsymbol{\eta},a)}\,\big|\,\boldsymbol{\eta}\in\boldsymbol{R}^{l-1}\big)$.

Proof We have only to take an equivariant lift of $U^{(1)}_{(\boldsymbol{\eta},a)}$ extending $U^D_{(\boldsymbol{\eta},a)}$ by applying Lemma 2.14. □

4.4.3. Extension of a splitting of a compatible tuple $(\boldsymbol{E},\boldsymbol{F},\boldsymbol{W})$. We put $X=\Delta^n$, $D_i:=\{z_i=0\}$ and $D=\bigcup_{i=1}^p D_i$ for some $p\leq n$. We take a $\mu_{\boldsymbol{c}}$-action on X. Let V be an equivariant holomorphic bundle on X. Let $(\boldsymbol{E},\boldsymbol{F},\boldsymbol{W})$ be a compatible tuple as in Definition 4.14. Assume that we are given a splitting $\big(U^D_{\boldsymbol{u}}\,\big|\,\boldsymbol{u}\in\boldsymbol{C}^S\times\boldsymbol{R}^l\times\mathbb{Z}^{l_1}\big)$ of the restriction $(\boldsymbol{E},\boldsymbol{F},\boldsymbol{W})_{|D}$ (see Definition 4.16). We would like to extend it to a splitting of $(\boldsymbol{E},\boldsymbol{F},\boldsymbol{W})$ on a neighbourhood of O.

Let l' be integer such that $1\leq l'\leq l_1$. We have the projection $\boldsymbol{R}^l\longrightarrow\boldsymbol{R}^{l'}$, taking the first l'-components. Then we obtain the projections $\pi_{1,l'}:\boldsymbol{C}^S\times\boldsymbol{R}^l\times\mathbb{Z}^{l_1}\longrightarrow\boldsymbol{C}^S\times\boldsymbol{R}^{l'}$ and $\pi_{3,l'}:\boldsymbol{C}^S\times\boldsymbol{R}^l\times\mathbb{Z}^{l'}\longrightarrow\boldsymbol{C}^S\times\boldsymbol{R}^{l'}$. We have the projection $\mathbb{Z}^{l_1}\longrightarrow\mathbb{Z}^{l'}$, taking the first l'-components. Then we obtain the projection $\pi_{2,l'}:\boldsymbol{C}^S\times\boldsymbol{R}^l\times\mathbb{Z}^{l_1}\longrightarrow\boldsymbol{C}^S\times\boldsymbol{R}^l\times\mathbb{Z}^{l'}$.

For any element $\boldsymbol{u}\in\boldsymbol{C}^S\times\boldsymbol{R}^l\times\mathbb{Z}^{l_1}$, we put $\pi_{1,l'}(\boldsymbol{u})=(\boldsymbol{\alpha},\boldsymbol{a})$, and we put as follows:

$${}^{\underline{l'}}U^D_{\boldsymbol{u}}:=\pi_{\boldsymbol{a}}\big(U^D_{\boldsymbol{u}}\big)\subset\big({}^{\underline{l'}}\operatorname{Gr}^F_{\boldsymbol{a}}\cap{}^S\mathbb{E}_{\boldsymbol{\alpha}}\big)_{|D}.$$

For an element $\boldsymbol{v}\in\boldsymbol{C}^S\times\boldsymbol{R}^l\times\mathbb{Z}^{l'}$, we put as follows:

$${}^{\underline{l'}}B^D_{\boldsymbol{v}}:=\bigoplus_{\pi_{2,l'}(\boldsymbol{u})=\boldsymbol{v}}{}^{\underline{l'}}U^D_{\boldsymbol{u}}.$$

Then we have the decomposition

$$\big({}^{\underline{l'}}\operatorname{Gr}^F_{\boldsymbol{a}}\cap{}^S\mathbb{E}_{\boldsymbol{\alpha}}\big)_{|D}=\bigoplus_{\pi_{3,l'}(\boldsymbol{v})=(\boldsymbol{a},\boldsymbol{\alpha})}{}^{\underline{l'}}B^D_{\boldsymbol{v}}.\tag{4.8}$$

Then the decomposition (4.8) gives a splitting of $(\boldsymbol{F}^{\underline{l'}},\boldsymbol{W}^{\underline{l'}})_{|D}$, by definition.

LEMMA 4.33. *Let l' be any integer such that $0\leq l'\leq l_1$, and let $\boldsymbol{u}$ be any element of $\boldsymbol{C}^S\times\boldsymbol{R}^l\times\mathbb{Z}^{l_1}$. We can take subbundles ${}^{\underline{l'}}U'_{\boldsymbol{u}}$ of ${}^{\underline{l'}}\operatorname{Gr}^F_{\pi_{1,l'}(\boldsymbol{u})}$ satisfying the following:*

(1) *We have the following decomposition, for any element $(\boldsymbol{\alpha},\boldsymbol{a})\in\boldsymbol{C}^S\times\boldsymbol{R}^{l'}$:*

$${}^S\mathbb{E}_{\boldsymbol{\alpha}}{}^{\underline{l'}}\operatorname{Gr}^F_{\boldsymbol{a}}=\bigoplus_{\pi_{1,l'}(\boldsymbol{u})=(\boldsymbol{\alpha},\boldsymbol{a})}{}^{\underline{l'}}U'_{\boldsymbol{u}}.$$

(2) *For any element $\boldsymbol{v}\in\boldsymbol{C}^S\times\boldsymbol{R}^{l'}\times\mathbb{Z}^{l'}$, we put as follows:*

$${}^{\underline{l'}}B'_{\boldsymbol{v}}:=\bigoplus_{\pi_{2,l'}(\boldsymbol{u})=\boldsymbol{v}}{}^{\underline{l'}}U'_{\boldsymbol{u}}.\tag{4.9}$$

We have the decomposition ${}^S\mathbb{E}_{\boldsymbol{\alpha}}{}^{\underline{l'}}\operatorname{Gr}^F_{\boldsymbol{a}}=\bigoplus_{\pi_{3,l'}(\boldsymbol{v})=(\boldsymbol{\alpha},\boldsymbol{a})}{}^{\underline{l'}}B'_{\boldsymbol{v}}$ for any $\boldsymbol{a}\in\boldsymbol{R}^{l'}$, due to the first condition. Then the decomposition gives a splitting of $(\boldsymbol{F}^{\underline{l'}},\boldsymbol{W}^{\underline{l'}})$.

(3) *We have* ${}^{\underline{l'}}U_{\boldsymbol{u}\,|\,D} = {}^{\underline{l'}}U^D_{\boldsymbol{u}}$. *In particular, we have* ${}^{\underline{l'}}B_{\boldsymbol{v}\,|\,D} = {}^{\underline{l'}}B^D_{\boldsymbol{v}}$.

Proof We can restrict our attention to each component ${}^S\mathbb{E}_{\boldsymbol{\alpha}}$ ($\boldsymbol{\alpha} \in \boldsymbol{C}^S$). Thus we may assume that $|S| = 1$. In the following, we omit to denote ${}^S\mathbb{E}_{\boldsymbol{\alpha}}$. We use a descending induction on l'.
1. First we construct such decomposition in the case $l' = l_1$. We have the following decomposition, for any $\boldsymbol{a} \in \boldsymbol{R}^{l_1}$:

$$
{}^{\underline{l_1}}\mathrm{Gr}^F_{\boldsymbol{a}\,|\,D} = \bigoplus_{\pi_{1,l_1}(\boldsymbol{u})=\boldsymbol{a}} {}^{\underline{l_1}}U^D_{\boldsymbol{u}}. \tag{4.10}
$$

Then it is a splitting of the compatible tuple $(\boldsymbol{F}^{\underline{l_1}}, \boldsymbol{W}^{\underline{l_1}})_{|D}$. By using Lemma 4.31, we can obtain a splitting $\big({}^{\underline{l_1}}U'_{\boldsymbol{u}} \,\big|\, \boldsymbol{u} \in \boldsymbol{R}^l \times \mathbb{Z}^{l_1},\ \pi_{1,l_1}(\boldsymbol{u}) = \boldsymbol{a}\big)$ extending the splitting $\big({}^{\underline{l_1}}U^D_{\boldsymbol{u}} \,\big|\, \boldsymbol{u} \in \boldsymbol{R}^l \times \mathbb{Z}^{l_1}\ \pi_{1,l_1}(\boldsymbol{u}) = \boldsymbol{a}\big)$. Thus the claim has been shown in the case $l' = l_1$.

2. We assume that we have already obtained the vector subbundles ${}^{\underline{l'+1}}U'_{\boldsymbol{u}}$, and we will construct ${}^{\underline{l'}}U'_{\boldsymbol{u}}$.

We have the projection $\mathbb{Z}^{l'+1} \longrightarrow \mathbb{Z}^{l'}$ by taking the first l'-components. We denote the induced projection $\boldsymbol{R}^l \times \mathbb{Z}^{l'+1} \longrightarrow \boldsymbol{R}^l \times \mathbb{Z}^{l'}$ by $\pi_{4,l'}$. Similarly the projection $\boldsymbol{R}^l \longrightarrow \boldsymbol{R}^{l'+1}$ induces the morphism $\pi_{5,l'} : \boldsymbol{R}^l \times \mathbb{Z}^{l'} \longrightarrow \boldsymbol{R}^{l'+1}$.

We take ${}^{\underline{l'+1}}B'_{\boldsymbol{v}}$ ($\boldsymbol{v} \in \boldsymbol{R}^l \times \mathbb{Z}^{l'+1}$) as in (4.9). Then we put as follows, for any element $\boldsymbol{v} \in \boldsymbol{R}^l \times \mathbb{Z}^{l'}$:

$$
{}^{\underline{l'+1}}B''_{\boldsymbol{v}} = \bigoplus_{\pi_{4,l'}(\boldsymbol{v}_1)=\boldsymbol{v}} {}^{\underline{l'+1}}B'_{\boldsymbol{v}_1} \subset {}^{\underline{l'+1}}\mathrm{Gr}^F_{\pi_{5,l'}(\boldsymbol{v})}.
$$

Then we have the decomposition, for any $(\boldsymbol{a}, b) \in \boldsymbol{R}^{l'+1} = \boldsymbol{R}^{l'} \times \boldsymbol{R}$:

$$
{}^{\underline{l'+1}}\mathrm{Gr}^F_{(\boldsymbol{a},b)} = \bigoplus_{\pi_{5,l'}(\boldsymbol{v})=(\boldsymbol{a},b)} {}^{\underline{l'+1}}B''_{\boldsymbol{v}}.
$$

Recall that $\big({}^{\underline{l'+1}}B'_{\boldsymbol{v}} \,\big|\, \pi_{2,l'+1}(\boldsymbol{v}) = (\boldsymbol{a}, b)\big)$ gives the splitting of the following tuple of the filtrations on ${}^{\underline{l'+1}}\mathrm{Gr}^F_{(\boldsymbol{a},b)}$:

$$
\big(\boldsymbol{F}^{\underline{l'+1}}, \boldsymbol{W}^{\underline{l'+1}}\big) = \big({}^{l'+2}F^{\underline{l'+1}}, \ldots, {}^{l}F^{\underline{l'+1}}; W^{\underline{l'+1}}(\underline{1}), \ldots, W^{\underline{l'+1}}(\underline{l'+1})\big).
$$

Hence the decomposition $\big({}^{\underline{l'+1}}B''_{\boldsymbol{v}} \,\big|\, \pi_{5,l'+1}(\boldsymbol{v}) = (\boldsymbol{a}, b)\big)$ gives the splitting of the following compatible tuple of the filtrations on ${}^{\underline{l'+1}}\mathrm{Gr}^F_{(\boldsymbol{a},b)}$:

$$
\big({}^{l'+2}F^{\underline{l'+1}}, \ldots, {}^{l}F^{\underline{l'+1}}; W^{\underline{l'+1}}(\underline{1}), \ldots, W^{\underline{l'+1}}(\underline{l'})\big).
$$

We have the compatible tuple of filtrations $(\boldsymbol{F}^{\underline{l'}}, \boldsymbol{W}^{\underline{l'}})$ on ${}^{\underline{l'}}\mathrm{Gr}^F_{\boldsymbol{a}}$. We have the isomorphism ${}^{l'+1}\mathrm{Gr}_b\, {}^{\underline{l'}}\mathrm{Gr}^F_{\boldsymbol{a}} \simeq {}^{\underline{l'+1}}\mathrm{Gr}^F_{(\boldsymbol{a},b)}$.

Then we can take a lift $\big({}^{\underline{l'}}B'_{\boldsymbol{v}} \,\big|\, \boldsymbol{v} \in \boldsymbol{R}^l \times \boldsymbol{R}^{l'_1}\big)$ of $\big({}^{\underline{l'+1}}B''_{\boldsymbol{v}} \,\big|\, \boldsymbol{v} \in \boldsymbol{R}^l \times \boldsymbol{R}^{l'_1}\big)$ extending $\big({}^{\underline{l'}}B^D_{\boldsymbol{v}} \,\big|\, \boldsymbol{v} \in \boldsymbol{R}^l \times \boldsymbol{R}^{l'_1}\big)$, by applying Lemma 4.32.

Then we have the naturally defined isomorphism ${}^{\underline{l'}}B'_{\boldsymbol{v}} \longrightarrow {}^{\underline{l'+1}}B''_{\boldsymbol{v}}$ for any $\boldsymbol{v} \in \boldsymbol{R}^l \times \mathbb{Z}^{l'}$. For any $\boldsymbol{u} \in \boldsymbol{R}^l \times \mathbb{Z}^{l_1}$ such that $\pi_{2,l'}(\boldsymbol{u}) = \boldsymbol{v}$, we can lift ${}^{\underline{l'}}U'_{\boldsymbol{u}}$ of ${}^{\underline{l'+1}}U'_{\boldsymbol{u}}$ extending ${}^{\underline{l'}}U^D_{\boldsymbol{u}}$, by applying Lemma 2.14. Thus the induction can proceed. □

COROLLARY 4.34. *Assume that we are given a splitting* $\big(U_{\boldsymbol{u}}^D \,\big|\, \boldsymbol{u} \in \boldsymbol{C}^S \times \boldsymbol{R}^l \times \mathbb{Z}^{l_1}\big)$ *of the restriction* $(\boldsymbol{E},\boldsymbol{F},\boldsymbol{W})_{|D}$. *Then we have a splitting* $\big(U_{\boldsymbol{u}} \,\big|\, \boldsymbol{u} \in \boldsymbol{C}^S \times \boldsymbol{R}^l \times \mathbb{Z}^{l_1}\big)$ *of* $(\boldsymbol{E},\boldsymbol{F},\boldsymbol{W})$ *defined around the origin* O, *such that* $U_{\boldsymbol{u}\,|\,D} = U_{\boldsymbol{u}}^D$.

Proof We have only to put $U_{\boldsymbol{u}} := {}^{\underline{0}}U_{\boldsymbol{u}}$. □

4.4.4. Extension of a splitting of a sequentially compatible tuple $(\boldsymbol{E},\boldsymbol{F},\boldsymbol{N})$**.** We put $X = \Delta^n$, $D_i := \{z_i = 0\}$ and $D = \bigcup_{i=1}^{p} D_i$ for some $p \leq n$. We take a $\mu_{\boldsymbol{c}}$-action on X. Let V be an equivariant holomorphic bundle on X. Let $(\boldsymbol{E},\boldsymbol{F},\boldsymbol{N})$ be a sequentially compatible tuple as in Definition 4.24.

LEMMA 4.35. *Assume that we are given a splitting* $\big(U_{\boldsymbol{u}}^D \,\big|\, \boldsymbol{u} \in \boldsymbol{C}^S \times \boldsymbol{R}^l \times \mathbb{Z}^{l_1}\big)$ *of the restriction* $(\boldsymbol{E},\boldsymbol{F},\boldsymbol{N})_{|D}$. *Then we have a splitting* $\big(U_{\boldsymbol{u}} \,\big|\, \boldsymbol{u} \in \boldsymbol{C}^S \times \boldsymbol{R}^l \times \mathbb{Z}^{l_1}\big)$ *of* $(\boldsymbol{E},\boldsymbol{F},\boldsymbol{N})$, *such that* $U_{\boldsymbol{u}\,|\,D} = U_{\boldsymbol{u}}^D$.

Proof It immediately follows from Definition 4.26 and Corollary 4.34. □

4.4.5. Extension of splitting of strongly sequentially compatible tuple $(\boldsymbol{E},\boldsymbol{N})$**.** We put $X = \Delta^n$, $D_i := \{z_i = 0\}$ and $D = \bigcup_{i=1}^{p} D_i$ for some $p \leq n$. We take a $\mu_{\boldsymbol{c}}$-action on X. Let V be an equivariant holomorphic bundle on X. Let $(\boldsymbol{E},\boldsymbol{N})$ be a strongly sequentially compatible tuple as in Definition 4.28.

LEMMA 4.36. *Assume that we are given a splitting* $\big(U_{\boldsymbol{u}}^D,\, P_h U_{\boldsymbol{u}}^D \,\big|\, \boldsymbol{u} \in \boldsymbol{C}^S \times \mathbb{Z}^{l_1}, h \in \mathbb{Z}_{\geq 0}\big)$ *of the restriction* $(\boldsymbol{E},\boldsymbol{N})_{|D}$. *Then we have a splitting* $\big(U_{\boldsymbol{u}},\, P_h U_{\boldsymbol{u}} \,\big|\, \boldsymbol{u} \in \boldsymbol{C}^S \times \mathbb{Z}^{l_1}, h \in \mathbb{Z}_{\geq 0}\big)$ *of* $(\boldsymbol{E},\boldsymbol{N})$ *such that* $U_{\boldsymbol{u}\,|\,D} = U_{\boldsymbol{u}}^D$ *and* $P_h U_{\boldsymbol{u}\,|\,D} = P_h U_{\boldsymbol{u}}^D$.

Proof We have only to consider each component ${}^S\mathbb{E}_{\boldsymbol{\alpha}}$. Thus we may assume $|S| = 1$. In the following, we omit to denote $\boldsymbol{\alpha}$.

We put $V^{(1)} := P_h \operatorname{Gr}_h^{W(N_1)}(V)$. We have the surjection $\pi_h : \operatorname{Ker}(N_1^{h+1}) \longrightarrow P_h \operatorname{Gr}_h^{W(N_1)}(V)$. The image of $P_h U_{\boldsymbol{h}}^D$ via the restriction of the morphism $\pi_{h\,|\,D}$ is denoted by $P_h U_{\boldsymbol{h}}^{\prime D}$.

We obtain the induced sequentially compatible tuple $\boldsymbol{N}^{(1)} := (N_2^{(1)},\ldots,N_{l_1}^{(1)})$ on $V^{(1)}$, and we obtain the splitting $\big(P_h U_{\boldsymbol{h}}^{\prime D} \,\big|\, \boldsymbol{h} \in \mathbb{Z}^{l_1}, q_1(\boldsymbol{h}) = h\big)$. We can extend it to a splitting $\big(P_h U'_{\boldsymbol{h}} \,\big|\, \boldsymbol{h} \in \mathbb{Z}^{l_1}, q_1(\boldsymbol{h}) = h\big)$ by using Lemma 4.31. We can take an equivariant lift $P_h U_{\boldsymbol{h}}$ of $P_h U'_{\boldsymbol{h}}$ extending $P_h U_{\boldsymbol{h}}^D$. We have only to put as follows, for any $a \in \mathbb{Z}_{\geq 0}$ and for any $\boldsymbol{h} \in \mathbb{Z}^{l_1}$:

$$P_{|q_1(\boldsymbol{h})|+2a} U_{\boldsymbol{h}} = \begin{cases} N_1^a\big(P_{q_1(\boldsymbol{h})+2a} U_{\boldsymbol{h}+2a\boldsymbol{\delta}}\big), & (q_1(\boldsymbol{h}) \geq 0), \\ N_1^{a+|q_1(\boldsymbol{h})|}\big(P_{q_1(\boldsymbol{h})+2a} U_{\boldsymbol{h}+2a\boldsymbol{\delta}}\big), & (q_1(\boldsymbol{h}) < 0). \end{cases}$$

Then $\big(P_h U_{\boldsymbol{h}} \,\big|\, h \in \mathbb{Z}_{\geq 0},\, \boldsymbol{h} \in \mathbb{Z}^{l_1}\big)$ gives a desired decomposition. □

4.5. Compatibility of the filtrations and nilpotent maps on the divisors

4.5.1. Compatibility of filtrations and decompositions. We put $X = \Delta^n$ and $D = \bigcup_{i=1}^{l} D_i$. We take a $\mu_{\boldsymbol{c}}$-action on X. Let V be an equivariant vector bundle over X. Let $\boldsymbol{E} = \big({}^i\mathbb{E} \,\big|\, i = 1,\ldots,l\big)$ be a tuple of equivariant decompositions ${}^i\mathbb{E}$ of $V_{|D_i}$. Let $\boldsymbol{F} = \big({}^iF \,\big|\, i = 1,\ldots,l\big)$ be a tuple of equivariant filtrations of $V_{|D_i}$.

Let I be a subset of $\underline{l}$. Then we obtain the tuple ${}^I\boldsymbol{E} := \big({}^i\mathbb{E} \,\big|\, i \in I\big)$ of the equivariant decompositions of $V_{|D_I}$ and the tuple ${}^I\boldsymbol{F} := \big({}^iF \,\big|\, i \in I\big)$ of the equivariant filtrations of $V_{|D_I}$.

DEFINITION 4.37. The tuple $(\boldsymbol{E},\boldsymbol{F})$ is called compatible if $({}^I\boldsymbol{E},{}^I\boldsymbol{F})$ are compatible for any $I\subset \underline{l}$, in the sense of Definition 4.9. □

Let $\boldsymbol{E}=\big({}^i\mathbb{E}\,\big|\,i\in\underline{l}\big)$ and $\boldsymbol{F}=\big({}^iF\,\big|\,i\in\underline{l}\big)$ be as above. Assume that $(\boldsymbol{E},\boldsymbol{F})$ is compatible. Let us consider splittings ${}^I\boldsymbol{U}=\big({}^IU_{\boldsymbol{u}}\,\big|\,\boldsymbol{u}\in\boldsymbol{C}^I\times\boldsymbol{R}^I\big)$ for any subset $I\subset\underline{l}$ of $({}^I\boldsymbol{E},{}^I\boldsymbol{F})$, (Definition 4.9).

For any subset $I\subset I'$, let q_I denote the projection $\boldsymbol{C}^{I'}\times\boldsymbol{R}^{I'}\longrightarrow\boldsymbol{C}^I\times\boldsymbol{R}^I$.

DEFINITION 4.38. A tuple of splittings $\big({}^I\boldsymbol{U}\,\big|\,I\subset\underline{l}\big)$ is called a splitting of the tuple $\big(\boldsymbol{E},\boldsymbol{F}\big)$, if ${}^IU_{\boldsymbol{u}\,|\,D_{I'}}=\bigoplus_{q_I(\boldsymbol{u}_1)=\boldsymbol{u}}{}^{I'}U_{\boldsymbol{u}_1}$ hold for any $I\subset I'$. □

4.5.2. Compatibility of $(\boldsymbol{E},\boldsymbol{F},\boldsymbol{W})$. We put $X=\Delta^n$, $D_i=\{z_i=0\}$ and $D=\bigcup_{i=1}^{l}D_i$ for some $l\leq n$. Let $\boldsymbol{E}=\big({}^i\mathbb{E}\,\big|\,i\in\underline{l}\big)$ be a tuple of decompositions of $V_{|D_i}$ $(i\in\underline{l})$. Let $\boldsymbol{F}=\big({}^iF\,\big|\,i\in\underline{l}\big)$ be a tuple of filtrations of $V_{|D_i}$ $(i\in\underline{l})$. We assume that $(\boldsymbol{E},\boldsymbol{F})$ is compatible.

Let us consider a tuple $\boldsymbol{W}=\big(W(\underline{i})\,\big|\,i\in\underline{l_1}\big)$, where $W(\underline{i})$ are filtrations of the vector bundle ${}^{\underline{i}}\operatorname{Gr}^F(V)$ on $D_{\underline{i}}$. We have the induced filtrations $W^J(\underline{i})$ of ${}^J\operatorname{Gr}^F(V)$ on D_J for any subset J such that $\underline{i}\subset J$.

For any subset $J\subset\underline{l}$, we have the number $m(J)$ determined by the condition $\underline{m(J)}\subset J$ and $m(J)+1\not\in J$. We denote the tuple $\big(W^J(\underline{i})\,\big|\,i\in\underline{m(J)}\big)$ by ${}^J\boldsymbol{W}$. Then we obtain the tuple $({}^J\boldsymbol{E},{}^J\boldsymbol{F},{}^J\boldsymbol{W})$.

DEFINITION 4.39. The tuple $(\boldsymbol{E},\boldsymbol{F},\boldsymbol{W})$ is called compatible, if the induced tuples $({}^J\boldsymbol{E},{}^J\boldsymbol{F},{}^J\boldsymbol{W})$ on D_J are compatible for any subset $J\subset\underline{l}$, in the sense of Definition 4.14. □

Let us consider splittings ${}^J\boldsymbol{U}=\big({}^JU_{\boldsymbol{u}}\,\big|\,\boldsymbol{u}\in\boldsymbol{C}^J\times\boldsymbol{R}^J\times\mathbb{Z}^{m(J)}\big)$ of the tuple $\big({}^J\boldsymbol{E},{}^J\boldsymbol{F},{}^J\boldsymbol{W}\big)$, in the sense of Definition 4.16. For any pair of subsets $I\subset I'$ of $\underline{l}$, let q_I denote the projection $\boldsymbol{C}^{I'}\times\boldsymbol{R}^{I'}\times\mathbb{Z}^{m(I')}\longrightarrow\boldsymbol{C}^I\times\boldsymbol{R}^I\times\mathbb{Z}^{m(I)}$.

DEFINITION 4.40. A tuple of splittings $\boldsymbol{U}=\big({}^J\boldsymbol{U}\,\big|\,J\subset\underline{l}\big)$ is called a splitting of $(\boldsymbol{E},\boldsymbol{F},\boldsymbol{W})$, if ${}^IU_{\boldsymbol{u}\,|\,D_{I'}}=\bigoplus_{q_I(\boldsymbol{u}')=\boldsymbol{u}}{}^{I'}U_{\boldsymbol{u}'}$ hold for any subset $I\subset I'$ and for any $\boldsymbol{u}\in\boldsymbol{C}^I\times\boldsymbol{R}^I\times\mathbb{Z}^{m(I)}$. □

PROPOSITION 4.41. *Let $(\boldsymbol{E},\boldsymbol{F},\boldsymbol{W})$ be as above, and we assume that it is compatible. Then there exists a splitting of $(\boldsymbol{E},\boldsymbol{F},\boldsymbol{W})$, in the sense of Definition* 4.40.

Proof On $D_{\underline{l}}$, we take a splitting ${}^{\underline{l}}\boldsymbol{U}$ of the compatible tuple $({}^{\underline{l}}\boldsymbol{E},{}^{\underline{l}}F,{}^{\underline{l}}W)$. Then we construct the splittings ${}^I\boldsymbol{U}$ descending inductively on $|I|$. We assume that we have the splittings ${}^{I'}\boldsymbol{U}$ for any $I'\supsetneq I$. Then we will construct the splitting ${}^I\boldsymbol{U}$.

We put $\partial D_I:=\bigcup_{I'\supsetneq I}D_{I'}$. From the given splittings ${}^{I'}\boldsymbol{U}$ $(I'\supsetneq I)$, we obtain the splitting ${}^I\boldsymbol{U}^{\partial D_I}$ of the restriction $({}^I\boldsymbol{E},{}^I\boldsymbol{F},{}^I\boldsymbol{W})_{|\partial D_I}$. Then we can extend ${}^I\boldsymbol{U}^{\partial D_I}$ to ${}^I\boldsymbol{U}$, due to Lemma 4.33. Thus the inductive construction can proceed, and hence we are done. □

COROLLARY 4.42. *Let $(\boldsymbol{E},\boldsymbol{F})$ be a compatible tuple as in Definition* 4.37. *Then we have a splitting of $(\boldsymbol{E},\boldsymbol{F})$ in the sense of Definition* 4.38.

Proof We have only to consider Proposition 4.41 in the case where the filtrations $W(\underline{m})$ are trivial. □

4.5.3. Sequential compatibility. We put $X = \Delta^n$, $D_i = \{z_i = 0\}$ and $D = \bigcup_{i=1}^{l} D_i$ for some $l \leq n$. Let $\boldsymbol{E} = \big({}^i\mathbb{E} \,\big|\, i \in \underline{l}\big)$ be a tuple of decompositions of $V_{|D_i}$ $(i \in \underline{l})$. Let $\boldsymbol{F} = \big({}^iF \,\big|\, i \in \underline{l}\big)$ be a tuple of filtrations of $V_{|D_i}$ $(i \in \underline{l})$. We assume that $(\boldsymbol{E}, \boldsymbol{F})$ is compatible.

Let us consider a tuple $\boldsymbol{N} = \big(N_i \,\big|\, i \in \underline{l_1}\big)$, where N_i are nilpotent maps of the vector bundle ${}^{\underline{i}}\operatorname{Gr}^F(V)$ on $D_{\underline{i}}$. We have the induced filtrations N_i^J of ${}^J\operatorname{Gr}^F(V)$ on D_J for any subset J such that $\underline{i} \subset J$.

For any subset $J \subset \underline{l}$, we have the number $m(J)$ determined by the condition $\underline{m(J)} \subset J$ and $m(J)+1 \notin J$. We denote the tuple $\big(N_i \,\big|\, i \in \underline{m(J)}\big)$ by ${}^J\boldsymbol{N}$. Then we obtain the tuple $({}^J\boldsymbol{E}, {}^J\boldsymbol{F}, {}^J\boldsymbol{N})$.

DEFINITION 4.43. The tuple $(\boldsymbol{E}, \boldsymbol{F}, \boldsymbol{N})$ is called sequentially compatible, if the induced tuples $\big({}^J\boldsymbol{E}, {}^J\boldsymbol{F}, {}^J\boldsymbol{N}\big)$ are sequentially compatible for any subset $J \subset \underline{l}$, in the sense of Definition 4.24. □

The nilpotent endomorphism $N(\underline{i})$ induces the filtration $W(\underline{i})$ of the vector bundle ${}^{\underline{i}}\operatorname{Gr}^F(V)$ over $D_{\underline{i}}$. Thus we obtain the tuple $\boldsymbol{W} = \big(W(\underline{i}) \,\big|\, i \in \underline{l_1}\big)$.

LEMMA 4.44. *The tuple $(\boldsymbol{E}, \boldsymbol{F}, \boldsymbol{W})$ is compatible, in the sense of Definition 4.39.*

Proof It immediately follows from the definition of compatibility. □

Let us consider splittings ${}^J\boldsymbol{U} = \big({}^JU_{\boldsymbol{u}} \,\big|\, \boldsymbol{u} \in \boldsymbol{C}^J \times \boldsymbol{R}^J \times \mathbb{Z}^{m(J)}\big)$ of the tuple $\big({}^J\boldsymbol{E}, {}^J\boldsymbol{F}, {}^J\boldsymbol{N}\big)$, in the sense of Definition 4.26. For any pair of subsets $I \subset I'$ of $\underline{l}$, let q_I denote the projection $\boldsymbol{C}^{I'} \times \boldsymbol{R}^{I'} \times \mathbb{Z}^{m(I')} \longrightarrow \boldsymbol{C}^I \times \boldsymbol{R}^I \times \mathbb{Z}^{m(I)}$.

DEFINITION 4.45. A tuple of splittings $\boldsymbol{U} = \big({}^J\boldsymbol{U} \,\big|\, J \subset \underline{l}\big)$ is called a splitting of $(\boldsymbol{E}, \boldsymbol{F}, \boldsymbol{N})$, if ${}^IU_{\boldsymbol{u}\,|\,D_{I'}} = \bigoplus_{q_I(\boldsymbol{u}')=\boldsymbol{u}} {}^{I'}U_{\boldsymbol{u}'}$ hold for any subset $I \subset I'$ and for any $\boldsymbol{u} \in \boldsymbol{C}^I \times \boldsymbol{R}^I \times \mathbb{Z}^{m(I)}$. □

PROPOSITION 4.46. *Let $(\boldsymbol{E}, \boldsymbol{F}, \boldsymbol{N})$ be sequentially compatible in the sense of Definition 4.43. and we assume that it is compatible. Then there exists a splitting of $(\boldsymbol{E}, \boldsymbol{F}, \boldsymbol{N})$, in the sense of Definition 4.45.*

Proof It immediately follows from Lemma 4.44 and Proposition 4.41. □

COROLLARY 4.47. *Assume that $(\boldsymbol{E}, \boldsymbol{F}, \boldsymbol{N})$ is sequentially compatible. Then there exists a frame $\boldsymbol{v}$ compatible with the decompositions ${}^i\mathbb{E}$, the filtrations ${}^iF$ on D_i, and the filtrations $W(\underline{m})$ on $D_{\underline{m}}$.*

Proof We have a splitting of $(\boldsymbol{E}, \boldsymbol{F}, \boldsymbol{N})$, due to Proposition 4.46. Then we have only to take a frame compatible with splittings (Lemma 2.16). □

For any frame $\boldsymbol{v}$ compatible with $(\boldsymbol{E}, \boldsymbol{F}, \boldsymbol{N})$, we have the decomposition $\boldsymbol{v} = \bigcup \boldsymbol{v}_{\boldsymbol{a},\boldsymbol{\alpha},\boldsymbol{h}}$:

$$\boldsymbol{v}_{\boldsymbol{a},\boldsymbol{\alpha},\boldsymbol{h}} = \Big(v_i \,\Big|\, \mathbf{deg}^{\mathbb{E},F}(v_i) = (\boldsymbol{\alpha}, \boldsymbol{a}),\ \deg^{W(\underline{m})}(v_i) = h_m\Big).$$

Let us consider the special case $N_i = 0$ $(i = 1, \ldots, l)$. Then we obtain the following corollary.

COROLLARY 4.48. *Let $(\boldsymbol{E}, \boldsymbol{F})$ be a compatible tuple, in the sense of Definition 4.37. Then there exists an equivariant frame $\boldsymbol{v}$ compatible with the decompositions ${}^i\mathbb{E}$ and the filtrations ${}^iF$.* □

4.5.4. Strongly sequential compatibility. We put $X = \Delta^n$, $D_i = \{z_i = 0\}$ and $D = \bigcup_{i=1}^{l} D_i$ for some $l \leq n$. Let $\boldsymbol{E} = ({}^i\mathbb{E} \,|\, i \in \underline{l})$ be a tuple of decompositions of $V_{|D_i}$ $(i \in \underline{l})$. Let $\boldsymbol{N} = (N_i \,|\, i \in \underline{l})$ be a tuple of nilpotent maps of $V_{|D_i}$ $(i \in \underline{l})$.

Let I be a subset of $\underline{l}$. Then we obtain the tuple ${}^I\boldsymbol{E}$ of the decompositions of $V_{|D_I}$. We also obtain the tuple ${}^I\boldsymbol{N} := (N_{i\,|\,D_I} \,|\, i \in I)$ of nilpotent maps of $V_{|D_I}$.

DEFINITION 4.49. The tuple $(\boldsymbol{E}, \boldsymbol{N})$ is called strongly sequentially compatible, if $({}^I\boldsymbol{E}, {}^I\boldsymbol{N})$ are strongly sequentially compatible for any subset $I \subset \underline{l}$, in the sense of Definition 4.27 □

Let $\boldsymbol{E}$ and $\boldsymbol{N}$ be as above. We assume that they are strongly sequentially compatible. Let us consider splittings ${}^I\boldsymbol{U}$ of $({}^I\boldsymbol{E}, {}^I\boldsymbol{N})$, in the sense of Definition 4.28. For any subset $I \subset I'$, let q_I denote the projection $\boldsymbol{C}^{I'} \times \mathbb{Z}^{I'} \longrightarrow \boldsymbol{C}^I \times \mathbb{Z}^I$.

DEFINITION 4.50. A tuple of splittings $({}^I\boldsymbol{U} \,|\, I \subset \underline{l})$ is called a splitting of the tuple $(\boldsymbol{E}, \boldsymbol{N})$, if ${}^IU_{\boldsymbol{u}\,|\,D_{I'}} = \bigoplus_{q_I(\boldsymbol{u}')=\boldsymbol{u}} {}^{I'}U_{\boldsymbol{u}'}$ hold for any $I \subset I'$. □

PROPOSITION 4.51. *Let $(\boldsymbol{E}, \boldsymbol{N})$ be as above, and we assume that it is compatible. Then there exists a splitting of $(\boldsymbol{E}, \boldsymbol{N})$.*

Proof It can be shown by an argument similar to the proof of Proposition 4.41. □

Let $(\boldsymbol{E}, \boldsymbol{N})$ be strongly sequentially compatible tuple. From the nilpotent maps $N(\underline{m}) = \sum_{i=1}^{m} N_i$ of $V_{|D_{\underline{m}}}$, we have the weight filtrations $W(\underline{m})$ of $V_{|D_{\underline{m}}}$. We denote the tuple by $\boldsymbol{W}$.

COROLLARY 4.52. *We can take a frame $\boldsymbol{v}$ compatible with $(\boldsymbol{E}, \boldsymbol{W})$.*

Proof Similar to Corollary 4.47. □

We have the decomposition $\boldsymbol{v} = \bigcup_{(\boldsymbol{a},\boldsymbol{h}) \in \boldsymbol{C}^l \times \mathbb{Z}^l} \boldsymbol{v}_{(\boldsymbol{a},\boldsymbol{h})}$.

COROLLARY 4.53. *We can take a compatible frame $\boldsymbol{v}$ satisfying the following:*

- *We have the decomposition $\boldsymbol{v}_{(\boldsymbol{a},\boldsymbol{h})} = \bigcup_{a \geq 0} P\boldsymbol{v}_{(\boldsymbol{a},\boldsymbol{h}),h(a)}$, where we put $h(a) = |q_1(\boldsymbol{h})| + 2a$.*
- *$P\boldsymbol{v}_{(\boldsymbol{a},\boldsymbol{h}),h}$ consists of sections $v_{(\boldsymbol{a},\boldsymbol{h}),h}, i$ $\big(i = 1, \dots, d(\boldsymbol{a}, \boldsymbol{h}, h)\big)$, and the following holds:*

$$N(\underline{1})\big(v_{(\boldsymbol{a},\boldsymbol{h}),h,i}\big) = \begin{cases} v_{(\boldsymbol{a},\boldsymbol{h}-2\boldsymbol{\delta}),h,i}, & (-h+2 \leq q_1(\boldsymbol{h}) \leq h), \\ 0, & (\text{otherwise}). \end{cases}$$

Such $\boldsymbol{v}$ is called strongly compatible with $(\boldsymbol{E}, \boldsymbol{N})$. □

CHAPTER 5

Some Lemmas for Generically Splitted Case

5.1. Filtrations

5.1.1. One nilpotent map. Let R be a discrete valuation ring, K be the quotient field, and k be the residue field. Let V be a free R-module of finite rank. Assume that we are given the following data, in this subsection.

CONDITION 5.1.

(1) Let ${}^{i}F$ $(i = 1, \dots, l)$ be compatible filtrations of V in the category of R-free modules. We denote the tuple of the filtrations $({}^{i}F \mid i \in \underline{l})$ by $\boldsymbol{F}$.
(2) We have the decomposition $V_{|K} = \bigoplus_{a} U_{a}$, which gives a splitting of $\boldsymbol{F}_{|K}$.
(3) Let N be a nilpotent endomorphism of V preserving $\boldsymbol{F}$.
(4) The restriction $N_{|K}$ preserves $U_{\boldsymbol{b}}$ for any $\boldsymbol{b}$.
(5) The endomorphism N induces the endomorphism ${}^{\underline{l}}N^{F}_{\boldsymbol{a}}$ of ${}^{\underline{l}}\mathrm{Gr}^{F}_{\boldsymbol{a}}(V)$. Then the conjugacy classes of ${}^{\underline{l}}N^{F}_{\boldsymbol{a}\,|\,x}$ $(x = k, K)$ are same. □

LEMMA 5.2. *Let m be an integer such that $1 \leq m \leq l$.*

- *The conjugacy classes of ${}^{\underline{m}^c}N^{F}$ on ${}^{\underline{m}^c}\mathrm{Gr}^{F}$ are constant, i.e., the conjugacy classes of ${}^{\underline{m}^c}N^{F}_{|x}$ $(x = k, K)$ are same. Here we put $\underline{m}^c = \underline{l} - \underline{m}$.*
- *We put $H_{m,\boldsymbol{a},h} := W_h({}^{\underline{m}^c}N^{F}) \cap {}^{\underline{m}}F_{\rho_m(\boldsymbol{a})} \cap {}^{\underline{m}^c}\mathrm{Gr}^{F}_{\eta_m(\boldsymbol{a})}$. Here we put $\rho_m(\boldsymbol{a}) = (a_1, \dots, a_m)$ and $\eta_m(\boldsymbol{a}) = (a_{m+1}, \dots, a_l)$. Then $H_{m,\boldsymbol{a},h}$ forms a vector subbundle of ${}^{\underline{m}^c}\mathrm{Gr}^{F}_{\eta_m(\boldsymbol{a})}$, and the image of $\phi_{m,\boldsymbol{a},h} : H_{m,\boldsymbol{a},h} \longrightarrow {}^{\underline{m-1}^c}\mathrm{Gr}^{F}_{\eta_{m-1}(\boldsymbol{a})}$ is same as $H_{m-1,\boldsymbol{a},h}$.*

Proof We have the sequence of degeneration ${}^{\underline{m}^c}N^{F}_{|K} \Longrightarrow {}^{\underline{m}^c}N^{F}_{|k} \Longrightarrow {}^{\underline{l}}N^{F}_{|k}$. Since the conjugacy classes of the first one and the last one are same, we obtain the first claim. In particular, $W({}^{\underline{m}^c}N^{F}) \cap {}^{\underline{m}^c}\mathrm{Gr}^{F}_{\boldsymbol{a}}$ gives a filtration of ${}^{\underline{m}^c}\mathrm{Gr}^{F}_{\boldsymbol{a}}$ in the category of the vector bundles. We also remark that the tuple of ${}^{i}F$ $(i \in I)$ and N satisfy Condition 5.1 for any $I \subset \underline{l}$.

As a preparation of the proof of the second claim, we put as follows for $x = k, K$:

$$H_{m\,\boldsymbol{a},h\,|\,x} := W_h({}^{\underline{m}^c}N^{F}_{\eta_m(\boldsymbol{a})})_{|x} \cap {}^{\underline{m}}F_{\rho_m(\boldsymbol{a})\,|\,x} \cap {}^{\underline{m}^c}\mathrm{Gr}^{F}_{\eta_m(\boldsymbol{a})\,|\,x}\,.$$

To see that $H_{m,\boldsymbol{a},h}$ forms a vector subbundle, we have only to show the following equality:

$$\dim H_{m,\boldsymbol{a},h\,|\,k} = \dim H_{m,\boldsymbol{a},h\,|\,K}. \tag{5.1}$$

Let us show the second claim in Lemma 5.2 by an induction on (l, m). Let $P(l, m)$ denote the second claim for (l, m). Note $0 \leq m \leq l$. The claim $P(0, 0)$ is clear. The claim $P(l, 0)$ follows from the condition (5) in Condition 5.1. Thus we have only to show that $P(l, m-1) + P(l-1, m-1)$ implies $P(l, m)$.

See the following naturally defined morphism:

$$\varphi_{m,\boldsymbol{a},h} : H_{m,\boldsymbol{a},h} \longrightarrow {}^{m}\operatorname{Gr}^{F}_{a_m}\,{}^{\underline{m}^c}\operatorname{Gr}^{F}_{\eta_m(\boldsymbol{a})} = {}^{\underline{m-1}^c}\operatorname{Gr}^{F}_{\eta_{m-1}(\boldsymbol{a})}\,.$$

On a generic point K, we have the splitting of the filtration F compatible with N. Hence we have $\operatorname{Im}(\varphi_{m,\boldsymbol{a},h\,|\,K}) = H_{m-1,\boldsymbol{a},h\,|\,K}$. In particular, we have the equality $\dim \operatorname{Im}(\varphi_{m,\boldsymbol{a},h\,|\,K}) = \dim H_{m-1\,\boldsymbol{a}\,h\,|\,K}$. Since $H_{m-1,\boldsymbol{a},h}$ is a subbundle of ${}^{\underline{m-1}^c}\operatorname{Gr}^{F}_{\eta_{m-1}(\boldsymbol{a})}$ due to the hypothesis $P(l,m-1)$ of the induction. We also obtain the following:

$$\operatorname{Im}\bigl(\varphi_{m,\boldsymbol{a},h}\bigr) \subset H_{m-1,\boldsymbol{a},h}. \tag{5.2}$$

We have the morphisms $\varphi_{m,\boldsymbol{a},h\,|\,k} : H_{m,\boldsymbol{a},h\,|\,k} \longrightarrow {}^{\underline{m-1}^c}\operatorname{Gr}^{F}_{\eta_m(\boldsymbol{a})\,|\,k}$. Due to (5.2), we have the implication $\operatorname{Im}(\varphi_{m,\boldsymbol{a},h\,|\,k}) \subset H_{m-1,\boldsymbol{a},h\,|\,k}$. Thus we obtain the following inequality:

$$\dim \operatorname{Im}(\varphi_{m,\boldsymbol{a},h\,|\,k}) \le \dim H_{m-1,\boldsymbol{a},h\,|\,k}. \tag{5.3}$$

Due to the hypothesis $P(l,m-1)$ of the induction, we have $\dim H_{m-1,\boldsymbol{a},h\,|\,K} = \dim H_{m-1\,\boldsymbol{a},h\,|\,k}$. Hence we obtain the following:

$$\dim \operatorname{Im}(\varphi_{m,\boldsymbol{a},h\,|\,k}) \le \dim \operatorname{Im}(\varphi_{m,\boldsymbol{a},h\,|\,K}). \tag{5.4}$$

We have the following for some $\epsilon > 0$:

$$\operatorname{Ker} \varphi_{m,\boldsymbol{a},h\,|\,x} = H_{m,\boldsymbol{a},h\,|\,x} \cap {}^{m}F_{<a_m\,|\,x} = H_{m,\boldsymbol{a}-\epsilon\boldsymbol{\delta}_m,h\,|\,x}.$$

For any elements $\boldsymbol{b} \in \boldsymbol{R}^{m-1}$ and $\boldsymbol{c} \in \boldsymbol{R}^{\underline{m}^c}$, we put as follows:

$$I_{\boldsymbol{b},\boldsymbol{c},h} := {}^{\underline{m-1}}F_{\boldsymbol{b}} \cap {}^{\underline{m}^c}\operatorname{Gr}^{F}_{\boldsymbol{c}} \cap W_h({}^{\underline{m}^c}N^F).$$

Note that the tuple $\bigl(N, {}^{1}F, \ldots, {}^{m-1}F, {}^{m+1}F, \ldots {}^{l}F\bigr)$ also satisfies Condition 5.1. Due to the hypothesis $P(l-1,m-1)$ of the induction, $I_{\boldsymbol{b},\boldsymbol{c},h}$ is a vector subbundle, and we have the following:

$$\begin{aligned} &I_{\boldsymbol{b},\boldsymbol{c},h\,|\,x} = {}^{\underline{m-1}}F_{\boldsymbol{b}\,|\,x} \cap {}^{\underline{m}^c}\operatorname{Gr}^{F}_{\boldsymbol{c}\,|\,x} \cap W_h({}^{\underline{m}^c}N^F)_{|x}, \quad (x = k, K) \\ &\dim(I_{\boldsymbol{b},\boldsymbol{c},h\,|\,k}) = \dim(I_{\boldsymbol{b},\boldsymbol{c},h\,|\,K}). \end{aligned} \tag{5.5}$$

We have the induced filtration ${}^{m}F(I_{\boldsymbol{b},\boldsymbol{c},h\,|\,x})$ on $I_{\boldsymbol{b},\boldsymbol{c},h\,|\,x}$, and we have the following:

$$\dim \operatorname{Im} \varphi_{m,\boldsymbol{a},h\,|\,x} = \dim \operatorname{Gr}^{{}^{m}F}_{a_m}\bigl(I_{\rho_{m-1}(\boldsymbol{a}),\eta_m(\boldsymbol{a}),h\,|\,x}\bigr). \tag{5.6}$$

Hence we have the following equality for $x = k, K$:

$$\sum_{\substack{\rho_{m-1}(\boldsymbol{a})=\boldsymbol{b},\\ \eta_m(\boldsymbol{a})=\boldsymbol{c}}} \dim \operatorname{Im}(\varphi_{m,\boldsymbol{a},h\,|\,x}) = \dim(I_{\boldsymbol{b},\boldsymbol{c}\,|\,x}). \tag{5.7}$$

Thus we obtain the following equality, from (5.4), (5.5) and (5.7):

$$\dim \operatorname{Im}(\varphi_{m\,\boldsymbol{a}\,h\,|\,k}) = \dim \operatorname{Im}(\varphi_{m\,\boldsymbol{a}\,h\,|\,K}). \tag{5.8}$$

We have the following:

$$\dim H_{m,\boldsymbol{a},h\,|\,x} = \sum_{\substack{\pi_m(\boldsymbol{b})=\boldsymbol{a},\\ q_m(\boldsymbol{b})\le q_m(\boldsymbol{a})}} \dim \operatorname{Im}(\varphi_{m,\boldsymbol{b},h\,|\,x}). \tag{5.9}$$

Here π_m denote the projection $\boldsymbol{R}^l \longrightarrow \boldsymbol{R}^{l-1}$, omitting the m-th component.

Thus we obtain (5.1) from (5.8) and (5.9). Due to (5.8) and $\dim \varphi_{m,\boldsymbol{a},h\,|\,K} = \dim H_{m-1,\boldsymbol{a},h\,|\,K}$, we obtain $\dim \mathrm{Im}(\varphi_{m,\boldsymbol{a},h\,|\,k}) = \dim H_{m-1,\boldsymbol{a},h\,|\,k}$, which implies $\mathrm{Im}(\varphi_{m,\boldsymbol{a},h}) = H_{m-1,\boldsymbol{a},h}$. Thus the induction can proceed. □

5.1.2. A tuple of nilpotent maps. Let V be a free R-module of finite rank. Assume that we are given the following datum.

CONDITION 5.3.

(1) Let ${}^iF$ $(i = 1, \dots, l)$ be compatible filtrations, and N_j $(j = 1, \dots, \alpha)$ be a commuting tuple of nilpotent maps. We denote the tuple of filtrations by $\boldsymbol{F}$, and we put $N(\underline{i}) = \sum_{j \leq i} N_j$.
(2) We have the splitting $V_{|K} = \bigoplus U_{\boldsymbol{a}}$ of $\boldsymbol{F}$.
(3) $N_{j\,|\,K}$ preserves $U_{\boldsymbol{a}}$, and N_j preserves $\boldsymbol{F}$.
(4) We have the induced morphisms ${}^{\underline{l}}N_j^F$ $(j = 1, \dots, \alpha)$ on ${}^{\underline{l}}\mathrm{Gr}^F$. Then $({}^{\underline{l}}N_1^F, \dots, {}^{\underline{l}}N_\alpha^F)$ is sequentially compatible. □

We put as follows:

$$J_{\boldsymbol{h},\boldsymbol{a},m} := \bigcap_{j=1}^{\alpha} W(\underline{j})_{h_j} \cap {}^{\underline{m}}F_{\rho_m(\boldsymbol{a})} \cap {}^{\underline{m}^c}\mathrm{Gr}_{\eta_m(\boldsymbol{a})}.$$

Let us consider the following morphism:

$$\psi_{\boldsymbol{h},\boldsymbol{a},m} : J_{\boldsymbol{h},\boldsymbol{a},m} \longrightarrow {}^{\underline{m-1}^c}\mathrm{Gr}_{\eta_{m-1}(\boldsymbol{a})}.$$

LEMMA 5.4. *$J_{\boldsymbol{h},\boldsymbol{a},m}$ is a subbundle of ${}^{\underline{m}^c}\mathrm{Gr}_{\eta_m(\boldsymbol{a})}$, and we have $\mathrm{Im}(\psi_{\boldsymbol{h},\boldsymbol{a},m}) = J_{\boldsymbol{h},\boldsymbol{a},m-1}$.*

Proof The argument is essentially same as the proof of the previous one. We put as follows:

$$J_{\boldsymbol{h},\boldsymbol{a},m\,|\,x} := \bigcap_{j=1}^{\alpha} W(\underline{j})_{h_j\,|\,x} \cap {}^{\underline{m}}F_{\rho_m(\boldsymbol{a})\,|\,x} \cap {}^{\underline{m}^c}\mathrm{Gr}_{\eta_m(\boldsymbol{a})\,|\,x}.$$

Then we have only to show the following equalities:

(A): $\dim J_{\boldsymbol{h},\boldsymbol{a},m\,|\,k} = \dim J_{\boldsymbol{h},\boldsymbol{a},m\,|\,K}$.
(B): $\dim \mathrm{Im}\, \psi_{\boldsymbol{h},\boldsymbol{a},m\,|\,k} = \dim J_{\boldsymbol{h},\boldsymbol{a},m-1\,|\,k}$.

We denote the claims for m by $P(m)$. We show $P(m)$ by an induction on m. The claim (A) in $P(0)$ follows from the condition (4) in Condition 5.3. The claim (B) in $P(0)$ is trivial.

Let us consider the following morphisms:

$$\psi_1 : W(\underline{j})_{h_j} \cap {}^{\underline{m}^c}\mathrm{Gr}_{\eta_m(\boldsymbol{a})} \longrightarrow {}^{\underline{m-1}^c}\mathrm{Gr}_{\eta_{m-1}(\boldsymbol{a})}.$$

LEMMA 5.5. *We have the following:*

$$\mathrm{Im}(\psi_1) = W(\underline{j})_{h_j} \cap {}^{\underline{m-1}^c}\mathrm{Gr}_{\eta_{m-1}(\boldsymbol{a})}.$$

Proof It immediately follows from Lemma 5.2. □

Hence we obtain $\mathrm{Im}(\psi_{\boldsymbol{h},\boldsymbol{a},m\,|\,x}) \subset J_{\boldsymbol{h},\boldsymbol{a},m-1\,|\,x}$. Since we are given the splitting of the filtrations $\boldsymbol{F}$ compatible with the nilpotent maps on the generic point K, we have the following:

$$\mathrm{Im}(\psi_{\boldsymbol{h},\boldsymbol{a},m\,|\,K}) = J_{\boldsymbol{h},\boldsymbol{a},m-1\,|\,K}.$$

Due to the hypothesis $P(m-1)$ of the induction, we have the following equality:

$$\dim J_{\boldsymbol{h},\boldsymbol{a},m-1\,|\,k} = \dim J_{\boldsymbol{h},\boldsymbol{a},m-1\,|\,K}. \tag{5.10}$$

Hence we obtain the following inequality:

$$\dim \operatorname{Im} \psi_{\boldsymbol{h},\boldsymbol{a},m\,|\,k} \leq \dim \operatorname{Im} \psi_{\boldsymbol{h},\boldsymbol{a},m\,|\,K}. \tag{5.11}$$

On the other hand, we have $\operatorname{Ker}(\psi_{\boldsymbol{h},\boldsymbol{a},m\,|\,x}) = J_{\boldsymbol{h},\boldsymbol{a}-\epsilon\boldsymbol{\delta}_m,m\,|\,x}$ for some small positive number ϵ.

For any elements $\boldsymbol{b} \in \boldsymbol{R}^{m-1}$ and $\boldsymbol{c} \in \boldsymbol{R}^{\underline{m}^c}$, we put as follows:

$$I'_{\boldsymbol{h},\boldsymbol{b},\boldsymbol{c}\,|\,x} := \bigcap_{j=1}^{\alpha} W(\underline{j})_{h_j\,|\,x} \cap {}^{\underline{m-1}}F_{\boldsymbol{b}\,|\,x} \cap {}^{\underline{m}^c}\operatorname{Gr}_{\boldsymbol{c}\,|\,x}.$$

We have the following inequality:

$$\dim(I'_{\boldsymbol{h},\boldsymbol{b},\boldsymbol{c}\,|\,k}) = \dim(I'_{\boldsymbol{h},\boldsymbol{b},\boldsymbol{c}\,|\,K}). \tag{5.12}$$

We have the induced filtration ${}^mF\big(I'_{\boldsymbol{h},\boldsymbol{b},\boldsymbol{c}\,|\,x}\big)$ and the following equality:

$$\dim \operatorname{Im}(\psi_{\boldsymbol{h},\boldsymbol{a},m\,|\,x}) = \dim \operatorname{Gr}^{{}^mF}_{a_m}\big(I'_{\boldsymbol{h},\rho_{m-1}(\boldsymbol{a}),\eta_m(\boldsymbol{a})\,|\,x}\big).$$

Thus we have the following:

$$\sum_{\substack{\rho_{m-1}(\boldsymbol{a})=\boldsymbol{b},\\ \eta_m(\boldsymbol{a})=\boldsymbol{c}}} \dim \operatorname{Im}(\psi_{\boldsymbol{h},\boldsymbol{a},m\,|\,x}) = \dim(I'_{\boldsymbol{h},\boldsymbol{b},\boldsymbol{c}\,|\,x}). \tag{5.13}$$

We obtain the equality $\dim \operatorname{Im} \psi_{\boldsymbol{h},\boldsymbol{a},m\,|\,k} = \dim \operatorname{Im} \psi_{\boldsymbol{h},\boldsymbol{a},m\,|\,K}$, due to (5.10), (5.12) and (5.13). It implies (B) in $P(m)$.

We have the following:

$$\dim J_{\boldsymbol{h},\boldsymbol{a},m\,|\,x} = \sum_{\substack{\pi_m(\boldsymbol{b})=\pi_m(\boldsymbol{a}),\\ q_m(\boldsymbol{b})\leq q_m(\boldsymbol{a})}} \dim \operatorname{Im}(\psi_{\boldsymbol{h},\boldsymbol{b},m\,|\,x}).$$

Here π_m denotes the projection $\boldsymbol{R}^l \longrightarrow \boldsymbol{R}^{l-1}$, forgetting the m-th component. Then we obtain the claim (A) in $P(m)$. □

COROLLARY 5.6. *The tuple of the filtrations* $\big(W(\underline{1}),\ldots,W(\underline{\alpha}),{}^1F,\ldots,{}^lF\big)$ *are compatible, in the sense of Definition* 4.2.

Proof We have the following morphisms:

$$\bigcap_{j=1}^{\alpha} W(\underline{j})_{h_j} \cap {}^{\underline{l}}F_{\boldsymbol{a}} \xrightarrow{\psi_{\boldsymbol{h},\boldsymbol{a}}} \bigcap_{j=1}^{\alpha} W(\underline{j})_{h_j} \cap {}^{\underline{l}}\operatorname{Gr}^F_{\boldsymbol{a}} \xrightarrow{\phi_{\boldsymbol{h},\boldsymbol{a}}} \operatorname{Gr}^W_{\boldsymbol{h}}\,{}^{\underline{l}}\operatorname{Gr}^F_{\boldsymbol{a}}.$$

The morphism $\psi_{\boldsymbol{h},\boldsymbol{a}}$ is surjective due to Lemma 5.2, and the morphism $\phi_{\boldsymbol{h},\boldsymbol{a}}$ is surjective due to the condition (4) in Condition 5.3.

Let us pick subbundles $C_{\boldsymbol{h},\boldsymbol{a}} \subset \bigcap_{j=1}^{\alpha} W(\underline{j})_{h_j} \cap {}^{\underline{l}}F_{\boldsymbol{a}}$ such that the restriction of $\phi_{\boldsymbol{h},\boldsymbol{a}} \circ \psi_{\boldsymbol{h},\boldsymbol{a}}$ to $C_{\boldsymbol{h},\boldsymbol{a}}$ is isomorphic.

Since the filtrations $W^{\underline{l}}(\underline{1}),\ldots W^{\underline{l}}(\underline{\alpha})$ are compatible, the following holds:

$${}^{\underline{l}}\operatorname{Gr}^F_{\boldsymbol{a}} = \bigoplus_{\boldsymbol{h}} \psi_{\boldsymbol{h},\boldsymbol{a}}(C_{\boldsymbol{h},\boldsymbol{a}}).$$

Since ${}^iF$ $(i=1,\ldots,l)$ are compatible, the following holds:

$$V = \bigoplus_{\boldsymbol{a}} \bigoplus_{\boldsymbol{h}} C_{\boldsymbol{h},\boldsymbol{a}}.$$

We have only to show the following:

$$\bigcap_{j=1}^{\alpha} W(\underline{j})_{h_j} \cap {}^{\underline{l}}F_{\boldsymbol{a}} = \bigoplus_{(\boldsymbol{k},\boldsymbol{b})\leq(\boldsymbol{h},\boldsymbol{a})} C_{\boldsymbol{h},\boldsymbol{a}}. \tag{5.14}$$

Since we have the splitting on the generic point K, it is easy to see that the restriction of (5.14) holds over the generic point K. We have already known that the both sides of (5.14) are subbundles of V (Lemma 5.2). Then we can conclude that (5.14) holds on R. □

COROLLARY 5.7. *The conjugacy classes of $\mathcal{N}(\underline{j})$ are constant over R, and $\bigcap_{j=1}^{\alpha} W(\underline{j})_{h_j} \cap {}^{\underline{l}}F_{\boldsymbol{a}}$ are vector bundles for any $\boldsymbol{a}$ and any $\boldsymbol{h}$.* □

PROPOSITION 5.8. *$(\mathcal{N}_1,\ldots,\mathcal{N}_\alpha,{}^1F,\ldots,{}^lF)$ are sequentially compatible, in the sense of Definition 4.23.*

Proof We have only to show that $(\mathcal{N}_{1\,|\,k},\ldots,\mathcal{N}_{\alpha\,|\,k},{}^1F_{|k},\ldots,{}^lF_{|k})$ are sequentially compatible in the sense of Definition 4.20.

The condition (1) in Definition 4.20 follows from (3) in Condition 5.3. The condition (2) in Definition 4.20 follows from Corollary 5.6. The condition (3) follows Lemma 5.2.

Let us see the condition (4). We use an induction on α. In the case $\alpha = 1$, there remain nothing to prove. We assume that the claim holds in the case $\alpha - 1$, and we will prove the claim also holds in the case α.

We have the induced filtrations ${}^1F^{(1)},\ldots {}^lF^{(1)}$ and the induced nilpotent maps $\mathcal{N}_2^{(1)},\ldots,\mathcal{N}_\alpha^{(1)}$ on the vector bundle $\mathrm{Gr}^{W(N_1)}(V)$. Due to the hypothesis of the induction, the tuple $\big(\mathcal{N}_2^{(1)},\ldots,\mathcal{N}_\alpha^{(2)},{}^1F^{(1)},\ldots,{}^lF^{(1)}\big)$ is sequentially compatible.

We put $\mathcal{N}^{(1)}(\underline{i}) = \sum_{j\leq i}\mathcal{N}_j^{(1)}$, which is same as the induced morphism by $\mathcal{N}(\underline{i})$ on $\mathrm{Gr}^{W(N_1)}$. Let $W^{(1)}(\underline{i})$ denote the induced filtration on $\mathrm{Gr}^{W(N_1)}$ by $W(\underline{i})$. We have only to show the following:

$$W^{(1)}(\underline{i})_{h+a} \cap \mathrm{Gr}_a^{W(N_1)}\,{}^I\mathrm{Gr}^F = W(\mathcal{N}^{(1)}(\underline{i}))_h \cap \mathrm{Gr}_a^{W(N_1)}\,{}^I\mathrm{Gr}^F. \tag{5.15}$$

Since we have the splitting on the generic point K, it is easy to see that (5.15) holds when it is restricted to the generic point K. We have already known that the both sides are vector subbundles of $\mathrm{Gr}_a^{W(N_1)}\,{}^I\mathrm{Gr}^F$, we can conclude that (5.15) holds on R. □

5.2. Compatibility of morphisms and filtrations

5.2.1. A compatible tuple of filtrations and a morphism. Let R be a discrete valuation ring, K be the quotient field, and k be the residue field. Let $V^{(a)}$ be free R-modules $(a = 1, 2)$, and let $\boldsymbol{F} := \big({}^iF(V^{(a)})\,\big|\, i\in I\big)$ be a compatible tuple of filtrations of $V^{(a)}$ in the category of free R-modules. Let $f : V^{(1)} \longrightarrow V^{(2)}$ be the morphism preserving the filtrations. We have the induced morphism ${}^I\mathrm{Gr}_{\boldsymbol{\eta}}^F(f)$: ${}^I\mathrm{Gr}_{\boldsymbol{\eta}}^F\big(V^{(1)}\big) \longrightarrow {}^I\mathrm{Gr}_{\boldsymbol{\eta}}^F\big(V^{(2)}\big)$ for any element $\boldsymbol{\eta}\in\boldsymbol{R}^I$.

Let S be a finite subset of $\boldsymbol{R}^I$. For simplicity, we use the following notation:

$${}^IF_S(V^{(a)}) := \sum_{\boldsymbol{\eta}\in S}{}^IF_{\boldsymbol{\eta}}(V^{(a)}),\quad {}^I\mathrm{Gr}_S^F(V^{(a)}) := \bigoplus_{\boldsymbol{\eta}\in S}{}^I\mathrm{Gr}_{\boldsymbol{\eta}}(V^{(a)}).$$

We have the naturally defined projection $\pi_S : {}^IF_S\big(V^{(a)}\big) \longrightarrow {}^I\mathrm{Gr}_S^F\big(V^{(a)}\big)$.

PROPOSITION 5.9. *Assume the following:*

- *We have a splitting* $V^{(a)}_{|K} = \bigoplus_{\boldsymbol{\eta}\in \boldsymbol{R}^I} U^{(a)}_{\boldsymbol{\eta}}$ *of the tuple of filtrations* $\boldsymbol{F}_{|K}$ *satisfying* $f_K(U^{(1)}_{\boldsymbol{\eta}}) \subset U^{(2)}_{\boldsymbol{\eta}}$.
- *The image* $\operatorname{Im}{}^I\operatorname{Gr}^F_{\boldsymbol{\eta}}(f)$ *is the vector subbundle of* ${}^I\operatorname{Gr}^F_{\boldsymbol{\eta}}(V^{(2)})$, *for any* $\boldsymbol{\eta}\in\boldsymbol{R}^I$.

Then the following claims hold.

(1) *For any finite subset* $S\subset \boldsymbol{R}^I$, *the image* $f\big({}^IF_S(V^{(1)})\big)$ *is a vector subbundle of* $V^{(2)}$.

(2) *We have the following:*

$$f\big({}^IF_S(V^{(1)})\big) = \operatorname{Im}(f)\cap {}^IF_S(V^{(2)}).$$

Proof For any finite subset S of $\boldsymbol{R}^I$, we put as follows:

$$L(S) := \max\Big\{\sum_{i\in I} q_i(\boldsymbol{a})\,\Big|\,\boldsymbol{a}\in S\Big\}.$$

The following claim is denoted by $P(r)$:

$P(r)$**:** The claim of Proposition 5.9 holds in the case $L(S)\leq r$.

LEMMA 5.10.

- *The claim* $P(r)$ *holds for any sufficiently negative* r.
- *If the claim* $P(r)$ *holds for some* $r\in\boldsymbol{R}$, *then there exists a positive number such that* $P(r')$ *holds for any* r' *such that* $r\leq r'\leq r+\epsilon$.

Proof It follows from the finiteness of the set $\big\{b\in\boldsymbol{R}\,\big|\,\exists i,\,\exists a,\,{}^i\operatorname{Gr}^F_b\big(V^{(a)}\big)\neq 0\big\}$. □

Due to Lemma 5.10, we have only to show that $P(r)$ holds under the assumption $P(r')$ holds for any real numbers $r'<r$, which we will show in the following.

LEMMA 5.11. *Let S be a finite subset of $\boldsymbol{R}^I$. Let us consider the projection* $\pi_S : {}^IF_S\big(V^{(a)}\big)\longrightarrow {}^I\operatorname{Gr}^F_S\big(V^{(a)}\big)$. *The kernel of π_S is described as the form* ${}^IF_{S'}\big(V^{(a)}\big)$ *for some finite subset* $S'\subset\boldsymbol{R}^I$ *such that* $L(S')<L(S)$.

Proof We have the following:

$$\operatorname{Ker}\pi_S = \sum_{\boldsymbol{a}\in S}\sum_{\boldsymbol{\eta}\lneq\boldsymbol{a}} {}^IF_{\boldsymbol{\eta}}\big(V^{(a)}\big).$$

Then it is clear from the compatibility of the tuple $\boldsymbol{F}$. □

LEMMA 5.12. *Let* $\pi'_S : f\big({}^IF_S(V^{(1)})\big)\longrightarrow {}^I\operatorname{Gr}^F_S(V^{(2)})$ *be the morphism induced by the projection π_S. Then we have the following:*

$$\operatorname{Im}\pi'_S = \operatorname{Im}{}^I\operatorname{Gr}^F_S(f)\subset {}^I\operatorname{Gr}^F_S(V^{(2)}).$$

Proof We have the following commutative diagramm:

$$\begin{array}{ccc}
{}^IF_S(V^{(1)}) & \longrightarrow & {}^IF_S(V^{(2)})\\
{\scriptstyle\pi_S}\big\downarrow & & {\scriptstyle\pi_S}\big\downarrow\\
{}^I\operatorname{Gr}^F_S(V^{(1)}) & \xrightarrow{{}^I\operatorname{Gr}^F_S(f)} & {}^I\operatorname{Gr}^F_S(V^{(2)}).
\end{array}$$

Then Lemma 5.12 immediately follows. □

We always have $f\big({}^I F_S(V^{(1)})\big) \subset \operatorname{Im}(f) \cap {}^I F_S(V^{(2)})$. Hence we always have the following:

$$\pi_S\Big(f\big({}^I F_S(V^{(1)})\big)\Big) \subset \pi_S\Big(\operatorname{Im}(f) \cap {}^I F_S(V^{(2)})\Big). \tag{5.16}$$

LEMMA 5.13. *In* (5.16), *the equality holds.*

Proof Since we have the splitting on the generic point K, it is easy to see that the equality holds, when we restrict (5.16) to the generic point. Since the left hand side is a vector subbundle in ${}^I\operatorname{Gr}^F_S(V^{(2)})$ due to Lemma 5.12, we obtain the equality on R. □

Let us pick an appropriate finite subset $S' \subset \boldsymbol{R}^I$ such that $L(S') < L(S)$ and $\operatorname{Ker}\pi_S = {}^I F_{S'}\big(V^{(a)}\big)$ $(a = 1, 2)$. We have the following:

$$\operatorname{Im}(f) \cap {}^I F_S(V^{(2)}) \cap \operatorname{Ker}\pi_S = \operatorname{Im}(f) \cap {}^I F_{S'}\big(V^{(2)}\big). \tag{5.17}$$

We have the following implication:

$$f\big({}^I F_{S'}\big(V^{(1)}\big)\big) \subset f\big({}^I F_S\big(V^{(1)}\big)\big) \cap \operatorname{Ker}\pi_S \subset \operatorname{Im}(f) \cap {}^I F_S\big(V^{(1)}\big) \cap \operatorname{Ker}\pi_S. \tag{5.18}$$

Due to the assumption $P(r')$ $(r' < r)$, we have $f\big({}^I F_{S'}\big(V^{(1)}\big)\big) = \operatorname{Im}(f) \cap {}^I F_{S'}\big(V^{(2)}\big)$. Then we obtain the following equality from (5.17) and (5.18):

$$f\big({}^I F_S\big(V^{(1)}\big)\big) \cap \operatorname{Ker}\pi_S = f\big({}^I F_{S'}\big(V^{(1)}\big)\big) = \operatorname{Im}(f) \cap {}^I F_{S'}\big(V^{(2)}\big) = \operatorname{Im}(f) \cap \operatorname{Ker}\pi_S. \tag{5.19}$$

Hence we obtain the equality $f\big({}^I F_S\big(V^{(1)}\big)\big) = \operatorname{Im}(f) \cap {}^I F_S\big(V^{(2)}\big)$, from Lemma 5.13 and (5.19).

Due to the assumption $P(r')$ $(r' < r)$, $f\big({}^I F_S\big(V^{(1)}\big)\big) \cap \operatorname{Ker}\pi_S$ is a vector subbundle of $V^{(2)}$. It follows that $f\big({}^I F_S\big(V^{(1)}\big)\big)$ is a vector subbundle of $V^{(2)}$. Thus the proof of Proposition 5.9 is accomplished. □

COROLLARY 5.14. *Under the assumption of Proposition* 5.9, *we have the equality* ${}^I\operatorname{Gr}^F_{\boldsymbol{\eta}}(\operatorname{Im}(f)) = \operatorname{Im}{}^I\operatorname{Gr}^F_{\boldsymbol{\eta}}(f)$, *for any element* $\boldsymbol{\eta} \in \boldsymbol{R}^I$. □

5.2.2. Decomposition. Let R, K, k, $V^{(a)}$ and $\boldsymbol{F} = \big({}^i F \,\big|\, i \in I\big)$ be as in the subsection 5.2.1. Let $f : V^{(1)} \longrightarrow V^{(2)}$ and $g : V^{(2)} \longrightarrow V^{(1)}$ be morphisms preserving the filtrations.

LEMMA 5.15. *Assume the following:*

- *We have splittings* $V^{(a)}_{|K} = \bigoplus U^{(a)}_{\boldsymbol{\eta}}$ *of the filtrations* $\boldsymbol{F}\big(V^{(a)}\big)_{|K}$ *on the generic point* K.
- *We have* $f_{|K}\big(U^{(1)}_{\boldsymbol{\eta}}\big) \subset U^{(2)}_{\boldsymbol{\eta}}$, *and* $g_{|K}\big(U^{(2)}_{\boldsymbol{\eta}}\big) \subset U^{(1)}_{\boldsymbol{\eta}}$ *on the generic point* K.
- *We have* ${}^I\operatorname{Gr}_{\boldsymbol{\eta}}\big(V^{(2)}\big) = \operatorname{Im}\big({}^I\operatorname{Gr}_{\boldsymbol{\eta}}(f)\big) \oplus \operatorname{Ker}\big({}^I\operatorname{Gr}_{\boldsymbol{\eta}}(g)\big)$. *In particular,* $\operatorname{Im}\big({}^I\operatorname{Gr}_{\boldsymbol{\eta}}(f)\big)$, $\operatorname{Ker}\big({}^I\operatorname{Gr}_{\boldsymbol{\eta}}(g)\big)$ *and* $\operatorname{Im}\big({}^I\operatorname{Gr}_{\boldsymbol{\eta}}(g)\big)$ *are the vector subbundles.*

Then we have the decomposition $V^{(2)} = \operatorname{Im}(f) \oplus \operatorname{Ker}(g)$.

Proof $\operatorname{Im}(f)$ and $\operatorname{Ker}(g)$ are vector subbundles of $V^{(2)}$, due to Proposition 5.9. The tuple of filtrations $\boldsymbol{F}\big(V^{(2)}\big)$ induce the tuple of filtrations $\boldsymbol{F}\big(\operatorname{Im}(f)\big)$ and $\boldsymbol{F}\big(\operatorname{Ker}(g)\big)$ on R. The decomposition $\big(U^{(a)}_{\boldsymbol{\eta}} \,\big|\, \boldsymbol{\eta} \in \boldsymbol{R}^I\big)$ of $V^{(2)}_{|K}$ induces the decomposition of $\operatorname{Im}(f)$ and $\operatorname{Ker}(g)$ on the generic point. Let us consider the naturally defined morphism $\Phi : \mathrm{Im}(f) \oplus \mathrm{Ker}(g) \longrightarrow V^{(2)}$. Then it is easy to see that the assumption of Proposition 5.9 is satisfied. Thus the image $\mathrm{Im}\,\Phi$ is a vector subbundle of $V^{(2)}$. Since $\mathrm{Im}(\Phi_{|K})$ and $V^{(2)}_{|K}$ are same, we obtain $\mathrm{Im}(\Phi) = V^{(2)}$ on R. □

CHAPTER 6

Model Bundles

We give a basic example of tame harmonic bundles. They are completely calculable. We will use the notation in Part 2–3.

6.1. Basic example I

6.1.1. Preliminary. We put $X := \Delta$ and $D = \{O\}$. Let $u = (a, \alpha)$ be an element of $\boldsymbol{R} \times \boldsymbol{C}$. We put $E := \mathcal{O} \cdot e$. We consider the hermitian metric h given by $h(e, e) = |z|^{-2a}$. We consider the Higgs field θ given by $\theta(e) = e \cdot \alpha \cdot dz/z$. Then we have the following:

$$\overline{\partial}_E e = 0, \quad \partial_E e = e \cdot \Big(-a \cdot \frac{dz}{z} \Big), \quad \theta^\dagger(e) = e \cdot \overline{\alpha} \cdot \frac{d\bar{z}}{\bar{z}}.$$

It is easy to check that the above tuple $(E, \overline{\partial}_E, \theta, h)$ is a harmonic bundle on Δ^*. It will be often denoted by $L(u)$ and called a model bundle of rank one.

We put $v = \exp\big(-\bar{\alpha} \cdot \lambda \cdot \log |z|^2\big) \cdot e$. Then we obtain the following:

$$\big(\overline{\partial}_E + \lambda \theta^\dagger\big) v = 0, \quad \mathbb{D} v = \big(\lambda \cdot \partial_E + \theta\big) \cdot v = \big(-\bar{\alpha} \cdot \lambda^2 - a \cdot \lambda + \alpha\big) v \cdot \frac{dz}{z}.$$

Then v gives a holomorphic frame of $\mathcal{E}$ over $\mathcal{X} - \mathcal{D}$. We put as follows:

$$s := \exp\big(\big(\overline{\alpha} \cdot \lambda + a - \alpha \cdot \lambda^{-1}\big) \cdot \log z\big) \cdot v = \exp\big(-\alpha \cdot \lambda^{-1} \log z + a \cdot \log z - \overline{\alpha} \cdot \lambda \cdot \log \overline{z}\big) \cdot e.$$

Then s is a frame of the holomorphic bundle $\mathcal{H}(E)$ over $\boldsymbol{C}^*$.

We put $e^\dagger := |z|^{2a} \cdot e$. Then we have $\partial_E e^\dagger = 0$. Namely $e^\dagger$ gives the frame of (E, ∂_E). We also have $h(e^\dagger, e^\dagger) = |z|^{2a}$, and $\overline{\partial}_E e^\dagger = e^\dagger \cdot a \cdot d\bar{z}/\bar{z}$.

We put $v^\dagger := \exp\big(-\alpha \cdot \mu \cdot \log |z|^2\big) \cdot e^\dagger$. Then we obtain the following:

$$\big(\partial_E + \mu \cdot \theta\big) \cdot v^\dagger = 0, \quad \mathbb{D}^\dagger v^\dagger = \big(\mu \cdot \overline{\partial}_E + \theta^\dagger\big) \cdot v^\dagger = \big(-\alpha \cdot \mu^2 + a \cdot \mu + \overline{\alpha}\big) \cdot v^\dagger \cdot \frac{d\bar{z}}{\bar{z}}.$$

Namely $v^\dagger$ is a frame of $\mathcal{E}^\dagger$ over $\mathcal{X}^\dagger - \mathcal{D}^\dagger$. We put as follows:

$$s^\dagger = \exp\big(\big(\mu \cdot \alpha - a - \overline{\alpha} \cdot \mu^{-1}\big) \log \bar{z}\big) \cdot v^\dagger.$$

Then $s^\dagger$ is a frame of $\mathcal{H}^\dagger(E)$.

LEMMA 6.1. *We have* $s = s^\dagger$.

Proof It can be shown by a direct calculation, as follows:

$$(6.1) \quad \begin{aligned} s^\dagger &= \exp\big(-\overline{\alpha} \cdot \mu^{-1} \cdot \log \bar{z} - a \cdot \log \bar{z} - \alpha \cdot \mu \cdot \log z\big) \cdot e^\dagger \\ &= \exp\big(-\overline{\alpha} \cdot \mu^{-1} \cdot \log \bar{z} + a \cdot \log z - \alpha \cdot \mu \cdot \log z\big) e = s. \end{aligned}$$

Thus we are done. □

We put $w := v_{|\boldsymbol{C}_\lambda \times \{O\}}$ and $w^\dagger := v^\dagger_{|\boldsymbol{C}_\mu \times \{O\}}$. The induced objects $\mathcal{G}_u$ and $\mathcal{G}^\dagger_{u^\dagger}$ are generated by w and $w^\dagger$ respectively.

6.1.2. The gluings of $S_u^{\mathrm{can}}(E)$ and $S_u(E,P)$. See the sections 11.3 for the bundles $S_u^{\mathrm{can}}(E)$ and $S_u(E,P)$. By definition, we have $\Phi_u^{\mathrm{can}}(s) = w_{|\boldsymbol{C}_\lambda^*}$ and $\Phi_{u^\dagger}^{\dagger\,\mathrm{can}}(s^\dagger) = w^\dagger_{|\boldsymbol{C}_\mu^*}$. We also have $s = s^\dagger$. Then the gluing of $S_u^{\mathrm{can}}(E)$ is given by the relation $w_{|\boldsymbol{C}_\lambda^*} = w^\dagger_{\boldsymbol{C}_\mu^*}$. In particular, $S_u^{\mathrm{can}}(E)$ is isomorphic to $\mathcal{O}_{\mathbb{P}^1}$.

Let us see the gluing of $S_u(E,P)$. We denote $v_{|\boldsymbol{C}_\lambda^*\times\{P\}}$ by $v_{|P}$ for simplicity of notation. We will also use a similar convention. We have the following:

$$v_{|P} = \exp\bigl(-\overline{\alpha}\cdot\lambda\cdot\log|z(P)|^2\bigr)\cdot e_{|P}.$$

Then we have the following:

$$\begin{aligned}(6.2)\qquad v^\dagger_{|P} &= \exp(-\alpha\cdot\mu\cdot\log|z(P)|^2)\cdot e^\dagger_{|P} = \exp\bigl(-\alpha\cdot\mu\cdot\log|z(P)|^2 + a\cdot\log|z(P)|^2\bigr)\cdot e_{|P}\\ &= \exp\bigl((-\alpha\cdot\mu + a + \overline{\alpha}\cdot\lambda)\cdot\log|z(P)|^2\bigr)\cdot v_{|P}.\end{aligned}$$

We put as follows:

$$\tilde{w} := \exp\bigl(\overline{\alpha}\cdot\lambda\cdot\log|z(P)|^2\bigr)\cdot w,\quad \tilde{w}^\dagger := \exp\bigl(\alpha\cdot\mu\cdot\log|z(P)|^2\bigr)\cdot w^\dagger.$$

Then $\mathcal{G}_u$ and $\mathcal{G}_u^\dagger$ are generated by w and $w^\dagger$ respectively, and the gluing of $S_u(E,P)$ is given by the relation $\tilde{u}^\dagger = \tilde{u}\cdot|z(P)|^{2a}$. In particular, $S_u(E,P)$ is isomorphic to $\mathcal{O}_{\mathbb{P}^1}$.

COROLLARY 6.2. *Let $(E,\overline{\partial}_E,\theta,h)$ be a model bundle $L(u)$ for some $u\in\boldsymbol{R}\times\boldsymbol{C}$.*

- *The vector bundles $S_u^{\mathrm{can}}(E)$ and $S_u(E,P)$ are pure twistors of weight 0.*
- *A frame of $H^0\bigl(\mathbb{P}^1, S_u^{\mathrm{can}}(E)\bigr)$ is given by $w = w^\dagger$.*
- *A frame of $H^0\bigl(\mathbb{P}^1, S_u(E,P)\bigr)$ is given by $\tilde{u} = |z(P)|^{-2a}\cdot\tilde{u}^\dagger$.* □

6.1.3. Pairing. The pairing $S:\mathcal{E}\otimes\sigma^*\mathcal{E}\longrightarrow\mathcal{O}_{\mathcal{X}-\mathcal{D}}$ is given as follows:

$$\begin{aligned}(6.3)\qquad S\bigl(v,\sigma^*(v^\dagger)\bigr) &= h\bigl(v(\lambda,x), v^\dagger(-\overline{\lambda},x)\bigr)\\ &= \exp\bigl(-\overline{\alpha}\cdot\lambda\cdot\log|z|^2\bigr)\cdot\overline{\exp\bigl(-\alpha\cdot(-\overline{\lambda})\cdot\log|z|^2\bigr)} = 1.\end{aligned}$$

Thus the induced pairing $S:\mathcal{G}_u\otimes\sigma^*\mathcal{G}^\dagger_{u^\dagger}\longrightarrow\mathcal{O}_{\boldsymbol{C}_\lambda}$ is given by $S\bigl(w,\sigma^\dagger w^\dagger\bigr) = 1$.

For the global section $v_1 = w = w^\dagger$ of $S^{\mathrm{can}}(E)$, we have $S(v_1,\sigma^*v_1) = S(w,\sigma^*w^\dagger) = 1 > 0$. In particular, we obtain the following.

LEMMA 6.3. *The pairing $S: S_u^{\mathrm{can}}(E)\otimes\sigma^*S_u^{\mathrm{can}}(E)\longrightarrow\mathbb{T}(0)$ induces the polarization of $S_u^{\mathrm{can}}(E)$.* □

For the global section $v_2 = \tilde{u} = |z(P)|^{-2a}\cdot\tilde{u}^\dagger$, we have the following:

$$\begin{aligned}(6.4)\qquad S\bigl(v_2,\sigma^*v_2\bigr) &= S\bigl(\tilde{u},\sigma^*|z|^{-2a}\tilde{u}^\dagger\bigr)\\ &= |z|^{-2a}\cdot\exp\bigl(\overline{\alpha}\cdot\lambda\cdot\log|z(P)|^2\bigr)\cdot\overline{\exp\bigl(-\alpha\cdot\overline{\lambda}\cdot\log|z(P)|^2\bigr)}\cdot S(w,\sigma^*w^\dagger)\\ &= |z(P)|^{-2a} > 0.\end{aligned}$$

In particular, the pairing $S: S_u(E,P)\otimes\sigma^*S_u(E,P)\longrightarrow\mathbb{T}(0)$ is the polarization. As a result, we obtain the following lemma.

LEMMA 6.4. *Let $(E,\overline{\partial}_E,\theta,h)$ be a model bundle $L(u)$ for some $u=(a,\alpha)$. Then the induced tuples $\bigl(S_u^{\mathrm{can}}(E),W,\mathcal{N}^\triangle,S\bigr)$ and $\bigl(S_u(E,P),W,\mathcal{N}^\triangle,S\bigr)$ are polarized mixed twistor structures.* □

6.1.4. The canonical frame. Let λ_0 be an element of $\boldsymbol{C}_\lambda$. Let $u=(a,\alpha)$ be an element of $\boldsymbol{R}\times\boldsymbol{C}$. Let $\mathcal{L}(u)$ be the deformed holomorphic bundle of $L(u)$.

The case $\alpha\neq 0$ Let b be a real number such that $b\not\in\mathfrak{p}(\lambda_0,u)+\mathbb{Z}$. In the case we have the integer ν determined by the condition $b-1<\nu+\mathfrak{p}(\lambda_0,u)<b$. Let v be as in the subsection 6.1.1. Then the section $z^{-\nu}v$ is a frame of ${}_b\mathcal{L}(u)$, which we call the canonical frame at λ_0.

The case $\alpha=0$. In the case, we recall $\mathfrak{p}(\lambda,u)=a$ for any λ, and we have $v=e$. Let b be any real number. We have the integer ν determined by the condition $b-1<\nu+a\leq b$. Then the frame $z^{-\nu}\cdot v$ is called the canonical frame of ${}_b\mathcal{L}(u)$ at λ_0.

6.1.5. The higher dimensional case. Let $\boldsymbol{u}=(u_1,\ldots,u_n)$ be an element of $(\boldsymbol{R}\times\boldsymbol{C})^n$, where $u_i=(a_i,\alpha_i)$. We put $X=\Delta^n$ and $D=\bigcup_{i=1}D_i$, where $D_i=\{z_i=0\}$. Then we put $L(\boldsymbol{u}):=\mathcal{O}_{X-D}\cdot e$. We have the Higgs field θ and the metric h determined as follows:

$$\theta\cdot e=e\cdot\sum\alpha_i\cdot\frac{dz_i}{z_i},\quad h(e,e)=\prod_{i=1}^{n}|z_i|^{-2a_i}.$$

Let q_i denote the projection of X onto the i-th component. Then we have the isomorphism $L(\boldsymbol{u})\simeq\bigoplus_{i=1}^{n}q_i^{*}L(u_i)$ compatible with the metric and the Higgs field. Hence $(L(\boldsymbol{u}),\theta,h)$ is a harmonic bundle. We have the holomorphic frame of the deformed holomorphic bundle $\mathcal{L}(\boldsymbol{u})$ over $\mathcal{X}-\mathcal{D}$:

$$v=\exp\Bigl(-\sum\overline{\alpha_i}\cdot\log|z_i|^2\Bigr)\cdot e.$$

Let λ_0 be an element of $\boldsymbol{C}_\lambda$. We have the canonical frame f_i of $\mathcal{L}(u_i)$ around λ_0. Then the section $\prod_{i=1}^{n}q_i^{*}f_i$ gives the canonical frame of ${}_{\boldsymbol{b}}\mathcal{L}(\boldsymbol{u})$ around λ_0.

In the case $\boldsymbol{u}=(\boldsymbol{b},\boldsymbol{0})\in\boldsymbol{R}^l\times\boldsymbol{C}^l$, $L(\boldsymbol{u})$ is also denoted by $L(\boldsymbol{b})$.

6.2. Basic example II

6.2.1. Preliminary. We put $X:=\Delta$ and $D:=\{O\}$. We put $y:=-\log|z|^2$. We put $E:=\mathcal{O}_{X-D}\cdot e_1\oplus\mathcal{O}_{X-D}\cdot e_{-1}$. The metric h is given as follows:

$$H(h,\boldsymbol{e})=\begin{pmatrix}y&0\\0&y^{-1}\end{pmatrix}.$$

The Higgs field θ is given as follows:

$$\theta\cdot\boldsymbol{e}=\boldsymbol{e}\cdot\begin{pmatrix}0&0\\1&0\end{pmatrix}\frac{dz}{z}.$$

Then $Mod(2)=(E,\overline{\partial}_E,\theta,h)$ is a harmonic bundle. (See our previous paper [**65**], for example). We take a frame $\boldsymbol{v}=(v^{1,0},v^{0,1})$ of the deformed holomorphic bundle $\mathcal{E}$ given as follows:

$$\boldsymbol{v}=\boldsymbol{e}\cdot\begin{pmatrix}1&-\lambda\cdot y^{-1}\\0&1\end{pmatrix}.\tag{6.5}$$

Then $\boldsymbol{v}$ gives a normalizing frame of ${}^{\diamond}\mathcal{E}$, i.e., it satisfies the following condition:

$$\mathbb{D}\boldsymbol{v}=\boldsymbol{v}\cdot\begin{pmatrix}0&0\\1&0\end{pmatrix}\frac{dz}{z}.$$

We also put as follows:

$$\boldsymbol{e}^{\dagger}=\boldsymbol{e}\cdot\begin{pmatrix} y^{-1} & 0\\ 0 & y\end{pmatrix},\quad \boldsymbol{v}^{\dagger}=\boldsymbol{e}^{\dagger}\cdot\begin{pmatrix} 1 & 0\\ -\mu\cdot y^{-1} & 1\end{pmatrix}.$$

Then $\boldsymbol{v}^{\dagger}$ gives a normalizing frame of ${}^{\diamond}\mathcal{E}^{\dagger}$:

$$\mathbb{D}^{\dagger}\boldsymbol{v}^{\dagger}=\boldsymbol{v}^{\dagger}\cdot\begin{pmatrix} 0 & 1\\ 0 & 0\end{pmatrix}\frac{d\bar{z}}{\bar{z}}.$$

6.2.2. The induced objects. It is easy to see that $\mathcal{KMS}(\mathcal{E}^0)=\{(n,0)\,|\,n\in\mathbb{Z}\}\subset\boldsymbol{R}\times\boldsymbol{C}$. We have the natural isomorphisms $S^{\mathrm{can}}_{(n,0)}(E)\simeq S^{\mathrm{can}}_{(0,0)}(E)$ and $S_{(n,0)}(E,P)\simeq S_{(0,0)}(E,P)$. Hence we only consider the case $\boldsymbol{u}=(0,0)$. In the following, we omit to denote the subscript $\boldsymbol{u}$. We also omit to denote '$\underline{1}$'. We also use the notation "$|P$" instead of "$|\{P\}\times\boldsymbol{C}^*_{\lambda}$" for simplicity.

We put $u^{1,0}:=v^{1,0}_{|O\times\boldsymbol{C}_{\lambda}}$ and $u^{0,1}=v^{0,1}_{|O\times\boldsymbol{C}_{\lambda}}$. We have $\mathcal{G}=\mathcal{O}_{\boldsymbol{C}_{\lambda}}\cdot u^{1,0}\oplus\mathcal{O}_{\boldsymbol{C}_{\lambda}}\cdot u^{0,1}$, by definition. We also have $\mathcal{G}(\mathcal{E})={}^{\diamond}\mathcal{E}$. Thus we have the following:

$$\mathcal{G}(\mathcal{E})_{|P}=\mathcal{E}_{|P}=\mathcal{O}_{\boldsymbol{C}_{\lambda}}\cdot v^{1,0}_{|P}\oplus\mathcal{O}_{\boldsymbol{C}_{\lambda}}\cdot v^{0,1}_{|P}=\mathcal{O}_{\boldsymbol{C}_{\lambda}}\cdot e_{1\,|\,P}\oplus\mathcal{O}_{\boldsymbol{C}_{\lambda}}\cdot e_{-1\,|\,P}.$$

We have the following equality on the plane $\{P\}\times\boldsymbol{C}_{\lambda}$:

$$\boldsymbol{v}_{|P}=\boldsymbol{e}_{|P}\cdot\begin{pmatrix} 1 & -\lambda\cdot y(P)^{-1}\\ 0 & 1\end{pmatrix}.$$

On the conjugate side, we have the following:

$$\mathcal{G}^{\dagger}=\mathcal{O}_{\boldsymbol{C}_{\mu}}\cdot u^{\dagger\,1,0}\oplus\mathcal{O}_{\boldsymbol{C}_{\mu}}\cdot u^{\dagger\,0,1},\qquad u^{\dagger\,1,0}:=v^{\dagger\,1,0}_{|O\times\boldsymbol{C}_{\mu}},\qquad u^{\dagger\,0,1}:=v^{\dagger\,0,1}_{|O\times\boldsymbol{C}_{\mu}}.$$

We also have the following:

$$\mathcal{G}^{\dagger}(\mathcal{E}^{\dagger})_{|P}=\mathcal{O}_{\boldsymbol{C}_{\mu}}\cdot v^{\dagger\,1,0}_{|P}\oplus\mathcal{O}_{\boldsymbol{C}_{\mu}}\cdot v^{\dagger\,0,1}_{|P}=\mathcal{O}_{\boldsymbol{C}_{\mu}}\cdot e^{\dagger}_{1\,|\,P}\oplus\mathcal{O}_{\boldsymbol{C}_{\mu}}\cdot e^{\dagger}_{-1\,|\,P}.$$

Then we have the following:

$$\boldsymbol{v}^{\dagger}_{|P}=\boldsymbol{e}^{\dagger}_{|P}\cdot\begin{pmatrix} 1 & 0\\ -\mu\cdot y(P)^{-1} & 1\end{pmatrix},\quad \boldsymbol{e}^{\dagger}_{|P}=\boldsymbol{e}_{|P}\begin{pmatrix} y(P)^{-1} & 0\\ 0 & y(P)\end{pmatrix}. \tag{6.6}$$

6.2.3. The gluing of $S(P)$ and the nilpotent maps. Then we have the following:

(6.7)

$$\Phi_{(P,O)}(\boldsymbol{e}_{|P})=\boldsymbol{u}\cdot\begin{pmatrix} 1 & \lambda\cdot y(P)^{-1}\\ 0 & 1\end{pmatrix},\qquad \Phi^{\dagger}_{(P,O)}(\boldsymbol{e}^{\dagger}_{|P})=\boldsymbol{u}^{\dagger}\cdot\begin{pmatrix} 1 & 0\\ \mu\cdot y(P)^{-1} & 1\end{pmatrix}.$$

From (6.7) and the first equation in (6.6), the vector bundle $S(E,P)$ is obtained by the following gluing:

$$\begin{aligned}\boldsymbol{u}=\boldsymbol{u}^{\dagger}&\begin{pmatrix} 1 & 0\\ \mu\cdot y(P)^{-1} & 1\end{pmatrix}\cdot\begin{pmatrix} y(P) & 0\\ 0 & y(P)^{-1}\end{pmatrix}\cdot\begin{pmatrix} 1 & -y(P)^{-1}\cdot\lambda\\ 0 & 1\end{pmatrix}\\ &=\boldsymbol{u}^{\dagger}\begin{pmatrix} y(P) & -\lambda\\ \mu & 0\end{pmatrix}.\end{aligned} \tag{6.8}$$

The nilpotent map $\mathcal{N}^{\triangle}$ is given as follows:

$$\mathcal{N}^{\triangle}_{|\boldsymbol{C}_{\lambda}}:\quad u^{1,0}\longmapsto u^{0,1}\otimes t_0^{(-1)},\quad u^{0,1}\longmapsto 0,\quad\text{on the plane }\boldsymbol{C}_{\lambda},$$

$$\mathcal{N}^{\triangle}_{|\boldsymbol{C}_{\mu}}:\quad u^{\dagger\,1,0}\longmapsto 0,\quad u^{\dagger\,0,1}\longmapsto -u^{\dagger\,1,0}\otimes t_{\infty}^{(-1)},\quad\text{on the plane }\boldsymbol{C}_{\mu}.$$

Hence the vector bundle $\mathrm{Gr}_1^{W^{\triangle}}$ is given by the gluing $u^{1,0} = \mu \cdot u^{\dagger\,0,1}$. The vector bundle $\mathrm{Gr}_{-1}^{W^{\triangle}}$ is given by $u^{0,1} = -\lambda \cdot u^{\dagger\,1,0}$. Therefore $\mathrm{Gr}_a^{W^{\triangle}}$ is a pure twistor of weight a, and $(S(E,P), W^{\triangle})$ is a mixed twistor.

6.2.4. The pairing. The induced pairing $S : \mathrm{Gr}_1^{W^{\triangle}} \otimes \sigma^* \mathrm{Gr}_{-1}^{W^{\triangle}} \longrightarrow \mathbb{T}(0)$ is given as follows:

$$u^{1,0} \otimes \sigma^*(u^{\dagger\,1,0}) \longmapsto 1, \quad u^{\dagger\,0,1} \otimes \sigma^*(u^{0,1}) \longmapsto 1.$$

REMARK 6.5. Note that we have the following equality on $\boldsymbol{C}_\lambda^*$:

$$(6.9) \quad u^{1,0} \otimes \sigma^*(u^{\dagger\,1,0}) = \mu \cdot u^{\dagger\,0,1} \otimes \sigma^*(-\lambda \cdot u^{0,1})$$
$$= \mu \cdot \overline{\left(-(-\overline{\lambda})\right)} \cdot u^{\dagger\,0,1} \otimes \sigma^*(u^{0,1}) = u^{\dagger\,0,1} \otimes \sigma^* u^{0,1}.$$

Thus the pairing S is well defined, of course. □

We have induced the pairing $S(\mathcal{N}^{\triangle} \otimes \mathrm{id}) : \mathrm{Gr}_1^{W^{\triangle}} \otimes \sigma^* \mathrm{Gr}_1^{W^{\triangle}} \longrightarrow \mathbb{T}(-1)$.

LEMMA 6.6. *It gives a polarization.*

Proof We consider the induced pairing $\mathrm{Gr}_1^{W^{\triangle}} \otimes \mathcal{O}(-1) \otimes \sigma^*\big(\mathrm{Gr}_1^{W^{\triangle}} \otimes \mathcal{O}(-1)\big) \longrightarrow \mathbb{T}(0)$. A base s of the one dimensional vector space $H^0\big(\mathbb{P}^1, \mathrm{Gr}_1^{W^{\triangle}} \otimes \mathcal{O}(-1)\big)$ is given by the following:

$$s = -\sqrt{-1} \cdot u^{1,0} \otimes f_0^{(-1)} = u^{\dagger\,0,1} \otimes f_\infty^{(-1)}.$$

Via the pairing $S\big(\mathcal{N}^{\triangle} \otimes \mathrm{id}\big)$, we have the composite of the following correspondences:

$$(6.10) \quad \big(-\sqrt{-1} \cdot u^{1,0} \otimes f_0^{(-1)}\big) \otimes \sigma^*\big(u^{\dagger\,0,1} \otimes f_\infty^{(-1)}\big) \longmapsto$$
$$-\sqrt{-1}\big(u^{0,1} \otimes t_0^{(-1)} \otimes f_0^{(-1)}\big) \otimes \sigma^*\big(u^{\dagger\,0,1} \otimes f_\infty^{(-1)}\big) \longmapsto -\sqrt{-1} \cdot f_0^{(1)} \otimes \sigma^*(f_\infty^{(-1)}) \longmapsto 1.$$

Thus we are done. □

In all, we obtain the following.

COROLLARY 6.7. *The tuple* $\big(S(E,P), W, \mathcal{N}^{\triangle}, S\big)$ *is a polarized mixed twistor structure.* □

6.2.5. The case $S^{\mathrm{can}}(E)$. Next we consider the vector bundle $S^{\mathrm{can}}(E)$. Let $\pi : \mathbb{H} \longrightarrow \Delta^*$ be the universal covering given by $\zeta \longmapsto \exp\big(\sqrt{-1}\zeta\big) = z$. Pick a point $\tilde{P}$ such that $\pi(\tilde{P}) = P$. We put as follows:

$$\boldsymbol{s} = \boldsymbol{v} \cdot \exp\Big(-\lambda^{-1} \cdot \big(\log z - \log z(\tilde{P})\big) \begin{pmatrix} 0 & 0 \\ 1 & 0 \end{pmatrix}\Big).$$

Then $\boldsymbol{s}$ is a frame of the space of multi-valued flat sections, such that $\boldsymbol{s}(\tilde{P}) = \boldsymbol{v}(P)$.

We put as follows:

$$\boldsymbol{s}^\dagger = \boldsymbol{v}^\dagger \cdot \exp\Big(-\mu^{-1} \cdot \big(\log \bar{z} - \log \bar{z}(\tilde{P})\big) \begin{pmatrix} 0 & 1 \\ 0 & 0 \end{pmatrix}\Big).$$

Then $\boldsymbol{s}^\dagger$ be a frame of the space of multi-valued sections such that $\boldsymbol{s}^\dagger(\tilde{P}) = \boldsymbol{v}^\dagger(P)$. Then we have the following relation:

$$\boldsymbol{s} = \boldsymbol{s}^\dagger \begin{pmatrix} y & -\lambda \\ \mu & 0 \end{pmatrix}.$$

To see the morphism Φ^{can}, we develop $\boldsymbol{s}$ as a polynomial of $\log z$, and we take the degree 0-part. Therefore the morphism $\mathcal{H} \longrightarrow \mathcal{G}_{|\boldsymbol{C}_\lambda^*}$ is given as follows:

$$\boldsymbol{s} \longmapsto \boldsymbol{u} \cdot \exp\Bigl(\lambda^{-1} \log z(\tilde{P}) \begin{pmatrix} 0 & 0 \\ 1 & 0 \end{pmatrix}\Bigr).$$

Similarly, we obtain the morphism $\mathcal{H} \longrightarrow \mathcal{G}^{\dagger}_{|\boldsymbol{C}_\lambda^*}$:

$$\boldsymbol{s}^{\dagger} \longmapsto \boldsymbol{u}^{\dagger} \cdot \exp\Bigl(\mu^{-1} \log \bar{z}(\tilde{P}) \begin{pmatrix} 0 & 1 \\ 0 & 0 \end{pmatrix}\Bigr).$$

Then the gluing of $S^{\mathrm{can}}(E)$ is given as follows:

(6.11) $\boldsymbol{u} =$

$$\begin{aligned} &\boldsymbol{u}^{\dagger} \cdot \exp\left(\mu^{-1} \log \bar{z}(\tilde{P}) \begin{pmatrix} 0 & 1 \\ 0 & 0 \end{pmatrix}\right) \cdot \begin{pmatrix} y & -\lambda \\ \mu & 0 \end{pmatrix} \cdot \exp\left(-\lambda^{-1} \log z(\tilde{P}) \begin{pmatrix} 0 & 0 \\ 1 & 0 \end{pmatrix}\right) \\ &= \boldsymbol{u}^{\dagger} \cdot \exp\left(\mu^{-1} \log |z(P)|^2 \begin{pmatrix} 0 & 1 \\ 0 & 0 \end{pmatrix}\right) \cdot \begin{pmatrix} y & -\lambda \\ \mu & 0 \end{pmatrix} \\ &= \boldsymbol{u}^{\dagger} \cdot \begin{pmatrix} 1 & -\mu^{-1} y \\ 0 & 1 \end{pmatrix} \cdot \begin{pmatrix} y & -\lambda \\ \mu & 0 \end{pmatrix} = \boldsymbol{u}^{\dagger} \begin{pmatrix} 0 & -\lambda \\ \mu & 0 \end{pmatrix}. \end{aligned}$$

Thus $S^{\mathrm{can}}(E)$ is naturally isomorphic to $\mathrm{Gr}_1^{W^{\triangle}} \oplus \mathrm{Gr}_{-1}^{W^{\triangle}}$.

The filtration $W^{\triangle}$ gives the mixed twistor structure, and the pairing S gives the polarization, which has already been shown. In all, we obtain the following.

LEMMA 6.8. *The tuples $\bigl(S^{\mathrm{can}}(E), W, \mathcal{N}^{\triangle}, S\bigr)$ and $\bigl(S(E,P), W, \mathcal{N}^{\triangle}, S\bigr)$ are polarized mixed twistor structures.* □

6.2.6. The rank l case. By taking the $(l-1)$-th symmetric product of the harmonic bundle $Mod(2)$ given in the subsection 6.2.1, we obtain the harmonic bundle $Mod(l)$, whose rank is l. Let $\mathcal{M}od(l)$ denote the deformed holomorphic bundle of $Mod(l)$.

The frame $\boldsymbol{w}$ of $\mathcal{M}od(l)$ is called canonical, if the λ-connection form with respect to $\boldsymbol{w}$ is of the form $A \cdot dz/z$ for some constant nilpotent matrix A. The frame $\boldsymbol{v}$ of $\mathcal{M}od(2)$ given in (6.5) induces a canonical frame of $\mathcal{M}od(l)$.

The following lemma is clear.

LEMMA 6.9. *Let V be a vector space over $\boldsymbol{C}$, and let N be a nilpotent map of V. We have a harmonic bundle of the form $\bigoplus_i Mod(l_i)$ such that the residue is isomorphic to (V, N). Such a harmonic bundle is denoted by $E(V, N)$, although it is not uniquely determined.* □

6.2.7. General model bundles. In general, the harmonic bundle of the following form is called a model bundle:

$$\bigoplus_i Mod(l_i) \otimes L(u_i).$$

Let λ_0 be a point of $\boldsymbol{C}_\lambda$. We have canonical frames of $\mathcal{L}(u_i)$ and $\mathcal{M}od(l_i)$ around λ_0. The tensor products of them induce the frame of $\bigoplus \mathcal{M}od(l_i) \otimes \mathcal{L}(u_i)$. The induced frame is called a canonical frame.

LEMMA 6.10. *Let $(E, \overline{\partial}_E, \theta, h)$ be a model bundle on Δ^*. The induced tuples $\bigl(S^{\mathrm{can}}(E), W, \mathcal{N}^{\triangle}, S\bigr)$ and $\bigl(S(E,P), W, \mathcal{N}^{\triangle}\bigr)$ are polarized mixed twistor structures.*

Proof It follows from Lemma 6.4, Lemma 6.8 and the functoriality of the construction of the induced tuples (the subsection 11.3.8). □

Part 2

Prolongation of Deformed Holomorphic Bundles

CHAPTER 7

Harmonic Bundles on a Punctured Disc

7.1. Simpson's main estimate

7.1.1. Statement. Let $R > 0$ be a positive number. We put $X = \overline{\Delta}(R)$, and $D = \{O\}$. Let $(E, \overline{\partial}_E, \theta, h)$ be a tame harmonic bundle over $X - D$. We denote $\theta = f_0 \cdot dz/z$. Assume that we have the decomposition $E = \bigoplus_{a \in S_0} E_a$ satisfying the following conditions:

CONDITION 7.1.

(1) The endomorphism f_0 preserves the decomposition, that is, $f_0 = \bigoplus_a f_{0\,a}$ for $f_{0\,a} \in End(E_a)$.
(2) There exist $C_0 > 0$ and $\epsilon_0 > 0$, such that $|b - a| < C_0 \cdot |z|^{\epsilon_0}$ for any eigenvalue b of $f_{0\,a}(z)$.
(3) We put $\xi := \sum_{a \in \mathcal{S}p(\mathcal{E}^0)} \operatorname{rank}(E_a) \cdot |a|^2$. Then we have $\xi \neq 0$. $\square$

We take a total order $\leq_1$ of $S_0 := \mathcal{S}p(\mathcal{E}^0)$. Then we obtain the filtration F defined as follows:

$$F_a E := \bigoplus_{b \leq_1 a} E_b.$$

Let E'_a denote the orthogonal complement of $F_{<a}(E)$ in $F_a(E)$. We have $\dim(E_a) = \dim(E'_a)$, but $E_a \neq E'_a$ in general. We put as follows:

$$\rho := \bigoplus_{a \in S_0} a \cdot id_{E_a} \in \bigoplus_{a \in S_0} \operatorname{End}(E_a), \quad \rho' := \bigoplus_{a \in S_0} a \cdot id_{E'_a} \in \bigoplus_{a \in S_0} \operatorname{End}(E'_a). \tag{7.1}$$

Note we have $|\rho'|_h^2 = \xi$.

The following proposition is our main purpose in the section 7.1. It will be proved in the subsections 7.1.2–7.1.6.

PROPOSITION 7.2.

(I): *Let R_1 be a positive number such that $R_1 < R$. There exists $C_1 > 0$ satisfying the following inequality on $\Delta^*(R_1)$:*

$$|f_0 - \rho'|_h \leq C_1 \cdot \left(-\log \frac{|z|}{R}\right)^{-1}. \tag{7.2}$$

(II): *There exist positive constants $C_2 > 0$, $\epsilon_2 > 0$ and $R_2 > 0$, satisfying the following inequality on $\Delta^*(R_2)$:*

$$\left|\rho - \rho'\right|_h \leq C_2 \cdot |z|^{\epsilon_2}. \tag{7.3}$$

Here C_1, C_2, ϵ_2, R_2 depends only on the constants R, C_0, ϵ_0, R_1, S_0, $\operatorname{rank}(E)$ *and the set* $\{\operatorname{rank} E_a \mid a \in S_0\}$.

For simplicity, we introduce the following terminology which is used only in this section.

DEFINITION 7.3. A constant is called good, if it depends only on the constants R, C_0, ϵ_0, R_1, S_0, $\mathrm{rank}(E)$ and the set $\{\mathrm{rank}\, E_a \,|\, a \in S_0\}$. □

REMARK 7.4. Proposition 7.2 was proved by Simpson (Theorem 1 in [**81**]). In fact, we closely follow his argument. However we would like to care the dependence of constants, to use the proposition in the higher dimensional case. It is the reason why we give the detail. □

7.1.2. Preliminary. We put $f := f_0 \cdot z^{-1}$. Recall we have the following fundamental inequality due to Simpson.

LEMMA 7.5 (Simpson, the page 729 in [**81**]). *The following inequality holds:*

$$\Delta \log |f|_h^2 \leq \frac{-\big|[f, f^{\dagger}]\big|_h^2}{|f|_h^2}.$$

□

Due to (1) in Condition 7.1, we have the endomorphisms $f_{0\,a\,|\,Q} \in End(E_{a\,|\,Q})$. We have the $\mathbb{E}$-decomposition of $E_{a\,|\,Q}$:

$$E_{a\,|\,Q} = \bigoplus_{\alpha \in \mathcal{S}p(f_{0\,a\,|\,Q})} \mathbb{E}(f_{0\,a\,|\,Q}, \alpha).$$

We have the natural bijection $\mathcal{S}p(f_{0\,|\,Q}) \simeq \{(a,\alpha) \,|\, a \in S_0,\ \alpha \in \mathcal{S}p(f_{0\,a\,|\,Q})\}$. We pick a total order $\leq_2$ on $\mathcal{S}p(f_{0\,a\,|\,Q})$ on each a. Then we obtain the total order $\leq_3$ on $\mathcal{S}p(f_{0\,|\,Q})$, which is given by the lexicographic order of $\leq_1$ and $\leq_2$.

We obtain the filtration $F^{(1)}$ on $E_{|Q}$ defined as follows:

$$F^{(1)}_{(a,\alpha)}(E_{|Q}) = \bigoplus_{(b,\beta) \leq_3 (a,\alpha)} \mathbb{E}(f_{0\,b\,|\,Q}, \beta).$$

Let $H_{(a,\alpha)}$ denote the orthogonal complement of $F^{(1)}_{<(a,\alpha)}$ in $F^{(1)}_{(a,\alpha)}$. We have the following:

$$E'_{a\,|\,Q} = \bigoplus_{\alpha \in \mathcal{S}p(f_{0\,a\,|\,Q})} H_{(a,\alpha)}.$$

We put as follows:

$$\tilde{\rho}_Q := \bigoplus_{(a,\alpha) \in \mathcal{S}p(f_{0\,|\,Q})} \alpha \cdot id_{H_{(a,\alpha)}}.$$

LEMMA 7.6. *There exists a good constant C and ϵ_0 satisfying the following:*

$$\big|\tilde{\rho}_Q - \rho'_{|Q}\big|_h \leq C \cdot |z(Q)|^{\epsilon_0}.$$

Proof It follows from (2) in Condition 7.1. □

COROLLARY 7.7. *There exists a good constant A_2 satisfying the following:*

$$|\tilde{\rho}_Q|_h^2 \leq A_2.$$

Proof It immediately follows from Lemma 7.6. □

We decompose as $f_{|Q} := \tilde{\rho}_Q \cdot z(Q)^{-1} + g_Q$. Then we have the following equality:

$$|f_{|Q}|_h^2 = |\tilde{\rho}_Q|_h^2 \cdot |z(Q)|^{-2} + |g_Q|_h^2.$$

LEMMA 7.8. *There exists a good constant A_1 satisfying the following:*

$$\big|[f_{|Q},\, f^{\dagger}_{|Q}]\big|_h \geq A_1 \cdot |g_Q|^2. \tag{7.4}$$

Proof It can be shown by an elementary linear algebraic argument. □

7.1.3. Step 1 for the proof of (I).

LEMMA 7.9. *There exist good constants A_3 and A_4 satisfying the following: For any point $Q \in \Delta^*(R)$, one of the following holds:*

- $|f(Q)|_h^2 \leq A_3 \cdot |z(Q)|^{-2}$.
- $\Delta \log |f|_h^2(Q) \leq -A_4 \cdot |f|_h^2(Q)$.

Proof Assume $|f_{|Q}|_h^2 \geq 2 \cdot A_2 \cdot |z(Q)|^{-2}$. Then we obtain $|g_Q|_h^2 \geq 2^{-1} \cdot |f_{|Q}|_h^2$. Hence we obtain the following:

$$(\Delta \log |f|_h^2)(Q) \leq \frac{-\big|[f_{|Q},\, f^{\dagger}_{|Q}]\big|_h^2}{|f_{|Q}|_h^2} \leq \frac{-A_1^2 \cdot |g_Q|_h^4}{|f_{|Q}|_h^2} \leq \frac{-A_1^2}{4} \cdot |f_{|Q}|_h^2.$$

Thus we may take $A_3 = 2A_2$ and $A_4 = 4^{-1}A_1^2$. □

Let η and B be positive numbers. We put as follows:

$$m_{\eta,B} := \frac{B}{(|z|-\eta)^2 \cdot (|z|-R)^2}.$$

It is a C^∞-function on the region $\{z \,|\, \eta < |z| < R\}$. Note that $m_{\eta,B}$ is ∞ on the boundary $\{z \,|\, |z| = \eta\} \cup \{z \,|\, |z| = R\}$.

LEMMA 7.10. *Let us take a positive number B satisfying the following inequalities:*

$$B \geq \frac{4R^2}{A_4}, \qquad B > A_3 \cdot R^2.$$

In particular, B is a good constant. Then we obtain the following inequalities:

$$\Delta \log m_{\eta,B} \geq -A_4 \cdot m_{\eta,B}, \qquad m_{\eta,B} \geq A_3 \cdot |z|^{-2}.$$

Proof We have the following formula:

$$\begin{aligned} \Delta \log m_{\eta,B} &= \frac{-2\eta}{|z| \cdot \big(|z|-\eta\big)^2} + \frac{-2R}{|z| \cdot \big(|z|-R\big)^2} \\ &= \frac{-2}{B}\Big(\frac{\big(|z|-R\big)^2}{|z|} \cdot \eta + \frac{\big(|z|-\eta\big)^2}{|z|} \cdot R\Big) \cdot m_{\eta,B}. \end{aligned} \tag{7.5}$$

From $\eta < |z| < R$, we have the following inequalities:

$$\frac{\eta}{|z|} < 1, \quad (|z|-R)^2 \leq R^2, \quad \frac{(|z|-\eta)^2}{|z|} \leq |z| - \eta \leq R.$$

Thus we have the following inequality:

$$\frac{\eta}{|z|} \cdot \big(|z|-R\big)^2 + \frac{\big(|z|-\eta\big)^2}{|z|} \cdot R \leq 2R^2. \tag{7.6}$$

From (7.5) and (7.6), we obtain the following inequality on the region $\{z \,|\, \eta < |z| < R\}$:

$$\Delta \log m_{\eta,B} \geq -\frac{4}{B} \cdot R^2 \cdot m_{\eta,B}.$$

For any $B \geq \frac{4R^2}{A_4}$ and for any $\eta > 0$, we obtain the first inequality on the region $\{\eta < |z| < R\}$.

We also have the following inequalities:

$$m_{\eta,B} = \frac{B}{(|z|-\eta)^2 \cdot (|z|-R)^2} \geq \frac{A_3 \cdot R^2}{(|z|-\eta)^2 \cdot (|z|-R)^2} \geq \frac{A_3}{(|z|-\eta)^2} \geq \frac{A_3}{|z|^2}.$$

Hence we obtain the second inequality. □

We put $S_1 := \{z \in \Delta^*(R) \,|\, |f(z)|_h^2 > m_{\eta,B}(z)\}$.

LEMMA 7.11. *The set S_1 is empty. In other words, we have the inequality $|f(z)|_h^2 \leq m_{\eta,B}(z)$ for any point $z \in \Delta^*(R)$ such that $|z| > \eta$.*

Proof Assume that S_1 is not empty, and we will derive a contradiction. On the region S_1, we have the inequality $|f|_h^2 > m_{\eta,B} \geq A_3 \cdot |z|^{-2}$. Hence we have $\Delta \log |f|_h^2 \leq -A_4 \cdot |f|_h^2$ due to Lemma 7.9. Then we obtain the following inequality on S_1:

$$\Delta \log\big(|f|_h^2 / m_{\eta,B}\big) \leq -A_4 \cdot \big(|f|_h^2 - m_{\eta,B}\big) < 0.$$

It implies that the function $\log\big(|f|_h^2/m_{\eta,B}\big)$ cannot take the maximal in the region S_1.

Let $\bar{S}_1$ denote the closure of S_1 in $\boldsymbol{C}$. Since we have $m_\eta = \infty$ on the boundary $\{|z| = \eta\} \cup \{|z| = R\}$, the intersection of the sets $\bar{S}_1$ and $\{|z| = \eta\} \cup \{|z| = R\}$ is empty. Hence we have $|f|_h^2 = m_{\eta,B}$ on the boundary of $\bar{S}_1$. Therefore, we obtain $|f|_h^2 \leq m_{\eta,B}$ on S_1, but it contradicts the definition of S_1. Hence we obtain $S_1 = \emptyset$. □

When we take a limit $\eta \to 0$, we obtain the inequality on $\Delta^*(R)$:

$$|f(z)|_h^2 \leq \frac{B}{|z|^2 \cdot \big(|z| - R\big)^2}.$$

Hence, we have arrived at the following:

LEMMA 7.12. *For any R_3 such that $R_1 < R_3 < R$, there exists a good constant A_5 such that the following inequality holds on $\Delta^*(R_3)$:*

$$|f(z)|_h^2 \leq \frac{A_5}{|z|^2}.$$

□

7.1.4. Step 2 for the proof of (I).

We put as follows: $k := \log |f|_h^2 - \log\big(\xi \cdot |z|^{-2}\big)$. Then we have the following:

$$k(Q) = \log\Big(\frac{|z(Q)|^2}{\xi} \cdot |f_{|Q}|^2\Big) = \log\Big(\frac{|z(Q)|^2}{\xi} \cdot \big(|\tilde{\rho}_Q|_h^2 \cdot |z(Q)|^{-2} + |g_Q|_h^2\big)\Big).$$

We put $b_Q := |\tilde{\rho}_Q|^2 - \xi$.

LEMMA 7.13. *We have a good constant A_6 such that $|b_Q| \leq A_6 \cdot |z(Q)|^{\epsilon_0}$.*

Proof This is a reformulation of Lemma 7.6. □

LEMMA 7.14. *There exists a good constant A_7 satisfying the following for any point $Q \in \Delta^*(R_3)$:*

$$A_7 \cdot \Big(\xi^{-1} \cdot b_Q + \frac{|z(Q)|^2}{\xi} \cdot |g_Q|_h^2\Big) \leq k(Q) \leq \xi^{-1} \cdot b_Q + \frac{|z(Q)|^2}{\xi} \cdot |g_Q|_h^2. \tag{7.7}$$

Proof We have the following description:

$$k(Q)=\log\Bigl(1+\xi^{-1}\cdot b_Q+\frac{|z(Q)|^2}{\xi}\cdot|g_Q|_h^2\Bigr).$$

Then the right inequality is obvious. We have only to obtain the left inequality.

Recall we have obtained the following estimate on $\Delta^*(R_3)$ in Step 1:

$$|g_Q|_h^2\leq|f_{|Q}|_h^2\leq A_5\cdot|z(Q)|^{-2}.$$

Hence we have the following inequality:

$$0\leq\frac{|z(Q)|^2}{\xi}\cdot|g_Q|_h^2\leq\frac{A_5}{\xi}.$$

Then we obtain the left inequality for some good constant, for example, by using the convexity of the logarithmic function. □

We will show Lemma 7.15 later.

LEMMA 7.15. *There exists a good constant A_8 satisfying the following:*

$$k\leq A_8\cdot\Bigl(-\log\frac{|z|}{R_3}\Bigr)^{-2}.$$

LEMMA 7.16. *Lemma* 7.15 *implies the claim (I) of Proposition* 7.2.

Proof Assume that we have shown Lemma 7.15, then we obtain the following inequality on $\Delta^*(R_1)$:

$$|f_0-\rho'|_h^2=b_Q+|z(Q)|^2\cdot|g_Q|_h^2\leq\xi\cdot A_7^{-1}\cdot k\leq\xi\cdot A_7^{-1}\cdot A_8\cdot\Bigl(-\log\frac{|z|}{R_3}\Bigr)^{-2}.$$

It implies the desired inequality (7.2) in (I). Thus we have reduced the proof of the claim (I) to Lemma 7.15. □

7.1.5. Proof of Lemma 7.15. Let us prove Lemma 7.15. There exists a good constant A_{10} satisfying the following inequality on $\Delta^*(R_3)$:

$$\xi^{-1}\cdot A_6\cdot|z|^{\epsilon_0}\leq\frac{1}{2}A_{10}\Bigl(-\log\frac{|z|}{R_3}\Bigr)^{-2}.$$

Here A_6 appeared in Lemma 7.13. Take a good constant A_{11} satisfying the following:

$$A_{11}<\frac{A_1^2\cdot\xi^2}{4\cdot A_5},\quad A_{11}<\frac{6}{A_{10}}. \tag{7.8}$$

The first condition will be used in Lemma 7.17 and the second condition will be used to obtain the inequality (7.11).

LEMMA 7.17. *One of the following holds:*

- $k(Q)<A_{10}\cdot\Bigl(\log\frac{|z(Q)|}{R_3}\Bigr)^{-2}$.
- $(\Delta k)(Q)<-A_{11}\cdot k(Q)^2\cdot|z(Q)|^{-2}$.

Proof Assume $k(Q) \geq A_{10} \cdot \big(-\log \frac{|z(Q)|}{R_3}\big)^{-2}$. Then we obtain the following:

$$\frac{1}{2}k(Q) \geq \xi^{-1} \cdot A_6 \cdot |z|^{\epsilon_0} \geq \xi^{-1} \cdot |b_Q|.$$

Namely we obtain the following:

$$k(Q) - \xi^{-1} \cdot b_Q \geq \frac{k(Q)}{2}.$$

By using the right inequality in (7.7), we obtain the following inequality:

$$\frac{|z(Q)|^2}{\xi} \cdot |g_Q|_h^2 \geq \frac{k(Q)}{2}.$$

Thus we obtain the following:

$$-|g_Q|_h^4 \leq \frac{-\xi^2}{4 \cdot |z(Q)|^4} k(Q)^2. \tag{7.9}$$

Note the following:

$$\Delta k = \Delta \log |f|_h^2 \leq \frac{-\big|[f, f^\dagger]\big|_h^2}{|f|_h^2}.$$

Then we obtain the following inequality by using Lemma 7.12, the inequality (7.4), the left inequality in (7.8), and the inequality (7.9):

$$(\Delta k)(Q) \leq -A_1^2 \cdot \frac{\xi^2}{|z(Q)|^4 \cdot 4} \cdot k(Q)^2 \cdot \frac{|z(Q)|^2}{A_5} = \frac{-A_1^2 \cdot \xi^2 \cdot k(Q)^2}{4 \cdot A_5 \cdot |z(Q)|^2} < -A_{11} \cdot \frac{k(Q)^2}{|z(Q)|^2}.$$

Thus we are done. □

For any positive numbers ϵ and B, we put as follows:

$$p_{B,\epsilon} = B \cdot \left(-\log \frac{|z|}{R_3} \right)^{-2} + \epsilon \cdot \left(-\log \frac{|z|}{R_3} \right).$$

LEMMA 7.18. *We put $B = 6 \cdot A_{11}^{-1}$. We have the following inequalities:*

$$\Delta p_{B,\epsilon} \geq -A_{11} \cdot \frac{p_{B,\epsilon}^2}{|z|^2}, \qquad p_{B,\epsilon} > A_{10} \cdot \left(-\log \frac{|z|}{R_3} \right)^{-2}. \tag{7.10}$$

Proof We have the following formula, where we use the real coordinate $z = r \cdot e^{\sqrt{-1}\theta}$:

$$\Delta p_{B,\epsilon} = -\left(\frac{\partial}{\partial r} + \frac{1}{r} \right) \left(\frac{2B}{(-\log r/R_3)^3 \cdot r} \right) = \frac{-6B}{(-\log r/R_3)^4 \cdot r^2} \geq \frac{-6 p_{B,\epsilon}^2}{B \cdot r^2}.$$

Here we have used the inequality:

$$\frac{p_{B,\epsilon}^2}{r^2} \geq \frac{B^2}{(-\log r/R_3)^2 \cdot r^2}.$$

Since we put $B = 6 \cdot A_{11}^{-1}$, we obtain the first inequality in (7.10).

Note that we have the following, due to the second inequality in (7.8):

$$B = 6 \cdot A_{11}^{-1} > A_{10}. \tag{7.11}$$

Thus we obtain the second equality in (7.10). □

We put $S_2 := \big\{ Q \,\big|\, k(Q) > p_\epsilon(Q) \big\}$.

LEMMA 7.19. *The set S_2 is empty. In other words, we have $k(Q) \leq p_\epsilon(Q)$ for any point $Q \in \Delta^*$.*

Proof We assume that S_2 is not empty, and we will derive a contradiction. Let Q be a point of S_2, and then we have the inequality:

$$k(Q) > p_\epsilon(Q) \geq A_{10} \cdot \Bigl(-\log \frac{|z(Q)|}{R_3}\Bigr)^{-2}.$$

Then we obtain the following inequality due to Lemma 7.17:

$$\Delta k(Q) < -A_{11} \frac{|k(Q)|^2}{|z(Q)|^2}.$$

Hence we have the following inequality:

$$\Delta(k - p_\epsilon)(Q) < -A_{11} \frac{k(Q)^2 - p_\epsilon(Q)^2}{|z(Q)|^2} < 0.$$

It means that the function $k - p_\epsilon$ does not have any maximal point in the region S_2. Since we have $p_\epsilon = \infty$ on the boundary $\{|z| = 0\} \cup \{|z| = R_3\}$, the intersection of the sets S_2 and $\{|z| = 0\} \cup \{|z| = R_3\}$ is empty. Hence we have $k(Q) = p_\epsilon(Q)$ on the boundary of S_2. Thus we obtain the inequality $k \leq p_\epsilon$ on the region S_2, which contradicts the definition of S_2. □

Let us return to the proof of Lemma 7.15. We obtain $k(Q) \leq p_{B,\epsilon}$ for any positive number ϵ due to Lemma 7.19. Thus we obtain the following inequality for a good constant:

$$k \leq B \cdot \Bigl(-\log \frac{|z|}{R_3}\Bigr)^{-2}.$$

Therefore the proof of Lemma 7.15 and the proof of the claim (I) are accomplished. □

7.1.6. The proof of (II). For any element $l \in \bigoplus_{a \leq_1 b} Hom(E'_b, E'_a)$, the adjoint $\mathrm{ad}(l)$ induces the endomorphism of $\bigoplus_{a <_1 b} Hom(E'_b, E'_a)$, which we denote by F_l. Recall that we put $f_0 = f \cdot z$.

LEMMA 7.20. *There exist good constants R_4 and A_{14} such that F_{f_0} is invertible on $\Delta^*(R_4)$, and the norms of F_{f_0} and $F_{f_0}^{-1}$ are dominated by A_{14}.*

Proof We put $\tilde{g} := f_0 - \rho'$, which is a section of $\bigoplus_{a \leq_1 b} Hom(E'_b, E'_a)$. Then we have the following inequality on $\Delta^*(R_1)$ due to the claim (I):

$$|\tilde{g}|_h \leq C_1 \cdot \Bigl(-\log \frac{|z|}{R}\Bigr)^{-1}. \tag{7.12}$$

Hence the norm of the endomorphism $F_{\tilde{g}}$ is dominated by $A_{13} \cdot \bigl(-\log(|z| \cdot R^{-1})\bigr)^{-1}$ for a good constant A_{13}. On the other hand, the endomorphism $F_{\rho'}$ is invertible and the norms of $F_{\rho'}$ and $F_{\rho'}^{-1}$ are dominated by a good constant A_{12}. Thus we obtain Lemma 7.20. □

We put $q := \rho - \rho'$, which is an element of $\bigoplus_{a <_1 b} Hom(E'_b, E'_a)$.

LEMMA 7.21. *There exists a good constant A_{15} such that $|q|_h \leq A_{15} \cdot \bigl(-\log(|z| \cdot R^{-1})\bigr)^{-1}$ on the region $\Delta^*(R_4)$.*

Proof We have the following equality by using $[\rho',\rho']=0$:

$$0=[f_0,\rho]=[f_0,\rho'+q]=F_{f_0}(q)+[\rho'+\tilde{g},\rho']=F_{f_0}(q)+[\tilde{g},\rho'].$$

Then we obtain Lemma 7.21 by using Lemma 7.20 and (7.12). □

LEMMA 7.22. *There exist good constants R_5 and A_{16} such that the following inequality holds on the region $\Delta^*(R_5)$:*

$$\big|[\rho,f^\dagger]\big|_h^2\geq A_{16}\cdot|q|_h^2\cdot|z|^{-2}.$$

Proof We have the following equality by a direct calculation:

$$[\rho,f^\dagger]=[\rho',f^\dagger]+[q,\,\bar{z}^{-1}\cdot\bar{\rho}']+[q,\,\bar{z}^{-1}\cdot\tilde{g}^\dagger].$$

Here we put $\bar{\rho}'=\bigoplus_{a\in S_0}\bar{a}\cdot id_{E'_a}$, and we have used the relation $\rho'^\dagger=\bar{\rho}'$. Note the following:

$$[\rho',f^\dagger]\in\bigoplus_{a>_1b}Hom(E'_b,E'_a),\qquad[q,\,\bar{z}^{-1}\cdot\bar{\rho}']\in\bigoplus_{a<_1b}Hom(E'_b,E'_a).$$

There exist good constants $C>0$ and $C'>0$ satisfying the following:

$$\big|[q,\bar{z}^{-1}\cdot\tilde{g}^\dagger]\big|_h\leq C'\cdot|z|^{-1}\cdot|\operatorname{ad}(\tilde{g})|_h\cdot|q|_h\leq C\cdot\left(-|z|\log\tfrac{|z|}{R}\right)^{-1}\cdot|q|_h,$$

$$\big|[q,\bar{z}^{-1}\bar{\rho}']\big|_h\geq C\cdot|z|^{-1}|q|_h.$$

In all, we obtain Lemma 7.22. □

Recall the following inequality due to Simpson (in the page 731 of [**81**]):

$$\Delta\log|\rho|_h^2\leq\frac{-\big|[\rho,f^\dagger]\big|_h^2}{|\rho|_h^2}. \tag{7.13}$$

LEMMA 7.23. *There exist good constants R_6 and A_{17} such that the following inequality holds on $\Delta^*(R_6)$:*

$$\Delta\log|\rho|_h^2\leq-A_{17}\cdot|q|_h^2\cdot|z|^{-2}.$$

Proof It follows from (7.13) and Lemma 7.22. □

Since we have $|\rho|_h^2=\xi+|q|_h^2$, we obtain the following inequality on $\Delta^*(R_6)$:

$$\Delta\log(1+\xi^{-1}\cdot|q|_h^2)\leq-A_{17}\cdot|q|_h^2\cdot|z|^{-2}. \tag{7.14}$$

LEMMA 7.24. *There exists a good constant A_{18} such that the following holds:*

$$A_{18}\cdot\xi^{-1}\cdot|q|_h^2\leq\log(1+\xi^{-1}\cdot|q|_h^2)\leq\xi^{-1}\cdot|q|_h^2. \tag{7.15}$$

Proof The right inequality in (7.15) is clear. Since we have $|q|_h\leq A_{15}\cdot\big(-\log(|z|\cdot R^{-1})\big)^{-1}$ on $\Delta^*(R_6)$ due to Lemma 7.21, we have the following:

$$|q|_h^2\cdot\xi^{-1}\leq\xi^{-1}\cdot A_{15}^2\cdot\left(-\log\frac{|z|}{R}\right)^{-2}. \tag{7.16}$$

In particular, we have a good constant $C>0$ such that $0\leq|q|_h^2\cdot\xi^{-1}\leq C$ on $\Delta^*(R_6)$. Thus there exists a good constant for the left inequality in (7.15). □

We put $k:=\log(1+\xi^{-1}\cdot|q|_h^2)$.

LEMMA 7.25. *There exists a good constant A_{19} such that the following holds:*

$$\Delta k\leq-A_{19}\cdot k\cdot|z|^{-2}. \tag{7.17}$$

Proof It follows from (7.14) and the right inequality in (7.15). □

LEMMA 7.26. *We have the following inequality on* $\Delta^*(R_6)$*:*

$$k \leq \xi^{-1} \cdot |q|_h^2 \leq A_{15} \cdot \xi^{-1} \cdot \big(-\log(|z| \cdot R^{-1})\big)^{-2}.$$

Proof It immediately follows from the right inequality in (7.15) and (7.16). □

COROLLARY 7.27. *There exists a good constant* A_{20} *such that* $k \leq A_{20}$. □

For positive numbers B, ϵ and u, we put as follows:

$$p_{B,\epsilon,u} := B \cdot \Big(|z|^u + \epsilon \cdot \Big(-\log \frac{|z|}{R}\Big)\Big).$$

It is easy to check $\Delta p_{B,\epsilon,u} = -u^2 \cdot |z|^{u-2} \cdot B$.

LEMMA 7.28.

- *Take* $u > 0$ *satisfying* $u^2 < A_{19}$. *Then we have the following inequality:*

$$\Delta p_{B,\epsilon,u} > -A_{19} \cdot |z|^{-2} \cdot |z|^u \cdot B > -A_{19} \cdot |z|^{-2} \cdot p_{B,\epsilon,u}. \tag{7.18}$$

- *Fix* $u > 0$. *Take* $B > 0$ *as* $B \cdot R_6^u > A_{20}$. *Then we obtain the following:*

$$p_{B,\epsilon,u}(R_6) = B \cdot \Big(R_6^u + \epsilon \cdot \Big(-\log \frac{|z|}{R_6}\Big)\Big) > A_{20} > k(R_6). \tag{7.19}$$

□

Let us fix good constants u and B as in Lemma 7.28. We use the notation p_ϵ instead of $p_{B,\epsilon,u}$. Then we obtain the following inequality from (7.17) and (7.18):

$$\Delta(k - p_\epsilon) < -\frac{A_{19}}{|z|^2} \cdot (k - p_\epsilon). \tag{7.20}$$

We put $S_3 := \big\{Q \in \Delta^*(R_0) \,\big|\, k(Q) - p_\epsilon > 0\big\}$.

LEMMA 7.29. *The set* S_3 *is empty. In other words, we have* $k(Q) \leq p_\epsilon(Q)$ *for any point* $Q \in \Delta^*$.

Proof Assume that S_3 is not empty, and we will derive a contradiction. The function $k - p_\epsilon$ has no maximal point in the region S_3 due to the inequality (7.20). Since we have $p_\epsilon(R_6) > k(R_6)$ due to the inequality (7.19), and since we have $p_\epsilon(0) = \infty$ by definition, the intersection of the sets S_3 and $\{|z| = 0\} \cup \{|z| = R_6\}$ is empty. Hence we have $k = p_\epsilon$ on the boundary of the closure $\bar{S}_3$. Hence we obtain $k \leq p_\epsilon$ on S_3, which is a contradiction. Hence the set S_3 is empty. □

When we take limit $\epsilon \to 0$, we obtain the inequality $k \leq B \cdot r^u$ on $\Delta^*(R_6)$. Then there exists a good constant A_{21} such that the following inequality holds on $\Delta^*(R_6)$:

$$|q|_h \leq A_{21} \cdot |z|^u.$$

Thus the proof of the claim (II) is accomplished. □

7.1.7. Some consequences and the asymptotic orthogonality.

COROLLARY 7.30. *For any $R_1 < R$, there exists a good constant C such that the following holds on $\Delta^*(R_1)$:*

$$|f_0 - \rho|_h \leq C \cdot \Big(- \log \frac{|z|}{R}\Big)^{-1}.$$

Proof It follows from the estimate of $q = \rho - \rho'$ and (I) in Proposition 7.2. □

We have the decomposition $f_0 = \sum_{a \geq_1 b} \tilde{f}_{0,a,b}$, where $\tilde{f}_{0,a,b} \in Hom(E'_a, E'_b)$.

COROLLARY 7.31. *There exist good constants $C > 0$ and $\epsilon > 0$ such that the following holds:*

- $|\tilde{f}_{0\,a\,b}|_h \leq C \cdot |z|^{\epsilon}$ *in the case $a \neq b$.*
- $|\tilde{f}_{0\,a\,a} - a \cdot id_{E'_a}|_h \leq C \cdot \Big(-\log \dfrac{|z|}{R}\Big)^{-1}$.

Proof The second inequality immediately follows from (I) in Proposition 7.2. Due to the commutativity $[f_0, \rho] = 0$, we have $[f_0, \rho'] + [f_0, q] = 0$. We have the following:

$$[f_0, \rho'] = \sum_{a >_1 b} (b - a) \cdot \tilde{f}_{0\,a\,b}.$$

On the other hand, we have the estimate

$$\big|[f_0, q]\big|_h \leq C \cdot |f_0|_h \cdot |q|_h \leq C' |z|^{\epsilon_2}.$$

Hence we obtain the estimates for $\tilde{f}_{0\,a\,b}$. □

We have the decomposition $f_0^{\dagger} = \sum_{a,b} \tilde{f}^{\dagger}_{0\,a\,b}$, where $\tilde{f}^{\dagger}_{0\,a\,b} \in Hom(E'_a, E'_b)$.

COROLLARY 7.32. *There exist good constants C and ϵ such that the following holds:*

- $|\tilde{f}^{\dagger}_{0\,a\,b}|_h \leq C \cdot |z|^{\epsilon}$ *in the case $a \neq b$.*
- $|\tilde{f}^{\dagger}_{0\,a\,a} - \bar{a} \cdot id_{E'_a}|_h \leq C \cdot (-\log|z|)^{-1}$.

Proof It immediately follows from Corollary 7.31. □

We put $\bar{\rho} := \bigoplus_{a \in S_0} \bar{a} \cdot id_{E_a}$. We also put $\bar{\rho}' := \bigoplus_{a \in S_0} \bar{a} \cdot id_{E'_a}$. Note that $\bar{\rho}'$ is adjoint of ρ', and that $\bar{\rho} - \bar{\rho}' \in \bigoplus_{a <_1 b} Hom(E'_b, E'_a)$.

LEMMA 7.33. *There exists good constants C', R' and ϵ' such that the following holds on $\Delta^*(R')$:*

$$\big|\bar{\rho} - \bar{\rho}'\big|_h \leq C' \cdot |z|^{\epsilon'}.$$

Proof The argument is similar to the proof of Corollary 7.31. We have the following formula:

$$0 = [\bar{\rho}, \rho] = [\bar{\rho}, \rho'] + [\bar{\rho}, q] = \big[\bar{\rho} - \bar{\rho}', \rho'\big] + [\bar{\rho}, q].$$

There exist good positive constants R'', C'', C''' and ϵ'' satisfying the following inequalities on $\Delta^*(R'')$:

$$\big|[\bar{\rho}, q]\big|_h \leq C'' \cdot |z|^{\epsilon''}, \qquad \big|\big[\bar{\rho} - \bar{\rho}', \rho'\big]\big|_h \geq C'' \cdot \big|\bar{\rho} - \bar{\rho}'\big|_h.$$

Thus we are done. □

COROLLARY 7.34. *There exist good positive constants C' and R' such that the following holds on $\Delta^*(R')$:*

$$\left|f_0^{\dagger} - \bar{\rho}\right|_h \leq C \cdot \Big(-\log \frac{|z|}{R}\Big)^{-1}.$$

Proof We have $f_0^{\dagger} - \bar{\rho} = (f_0^{\dagger} - \bar{\rho}') + (\bar{\rho}' - \bar{\rho}) = (f_0^{\dagger} - \rho'^{\dagger}) + (\bar{\rho}' - \bar{\rho})$. Then we obtain the result from (I) in Proposition 7.2 and Lemma 7.33. □

In general, let us consider an element $g \in \bigoplus_a End(E_a)$. Then we have the decomposition:

$$g = \sum_{a \geq_1 b} g_{a\,b}, \qquad g_{a\,b} \in Hom(E'_a, E'_b).$$

LEMMA 7.35. *We have the estimate $|g_{a\,b}|_h \leq C \cdot |z|^{\epsilon} \cdot |g|_h$ for $a \neq b$.*

Proof We have $0 = [g, \rho] = [g, \rho'] + [g, q]$. We have an estimate $\big|[g, q]\big|_h \leq C' \cdot |g|_h \cdot |z|^{\epsilon'}$ on $\Delta^*(R')$ for some good positive constants C', ϵ' and R'. On the other hand, we have the following:

$$[g, \rho'] = \sum_{a >_1 b} (b - a) \cdot g_{a,b}.$$

Then the claim immediately follows. □

For the endomorphism g above, we also obtain the adjoint $g^{\dagger} \in End(E)$, and we have the decomposition:

$$g^{\dagger} = \sum_{a \leq_1 b} (g^{\dagger})_{a\,b}, \qquad (g^{\dagger})_{a\,b} \in Hom(E'_a, E'_b).$$

LEMMA 7.36. *We have $|(g^{\dagger})_{a\,b}|_h \leq C \cdot |z|^{\epsilon} \cdot |g|_h$ if $a \neq b$.*

Proof It immediately follows from Lemma 7.35. □

We have the following asymptotic orthogonality.

PROPOSITION 7.37. *There exist good constants $C_3 > 0$, $\epsilon_3 > 0$ and $R_{10} > 0$. Let a_1 and a_2 be elements of S_0 such that $a_1 \neq a_2$. Then E_{a_1} and E_{a_2} are $|z|^{\epsilon_3}$-asymptotically orthogonal. More precisely, let v_i be C^{∞}-sections of E_i. Then it holds $\big|(v_1, v_2)_h\big| \leq C_3 \cdot |z|^{\epsilon_3} \cdot |v_1|_h \cdot |v_2|_h$ on $\Delta^*(R_{10})$.*

Proof Let v be a C^{∞}-section of E_a. We have the following decomposition:

$$v = \sum_{b \leq_1 a} v_b, \qquad v_b \in C^{\infty}\big(X - D, E'_b\big).$$

We have the equalities $\rho(v) = a \cdot v = \sum_{b \leq a} a \cdot v_b$. On the other hand, we have the following equalities:

$$\rho(v) = \rho'(v) + q(v) = \sum_{b \leq_1 a} b \cdot v_b + q(v).$$

Hence we obtain the following:

$$\sum_{b <_1 a} (a - b) \cdot v_b = q(v).$$

Therefore there exists a good constant A_{22} such that the following holds for any v and for any $b <_1 a$:

$$|v_b|_h \leq A_{22} \cdot |z|^u \cdot |v|_h.$$

Let v be a C^∞-section of E_a and w be a C^∞-section of E_c such that $c < a$. Then we have the following:

$$|(v,w)_h| = \Big|\sum_{b \leq_1 a} (v_b, w)_h\Big| = \Big|\sum_{b \leq_1 c} (v_b, w)_h\Big| \leq \sum_{b \leq_1 c} |v_b|_h \cdot |w|_h \leq A_{22} \cdot |z|^u \cdot \sum_{b \leq_1 c} |v|_h |w|_h.$$

Hence there exists a good constant $A_{23} > 0$ such that the following holds:

$$|(v,w)_h| \leq A_{23} \cdot |z|^u \cdot |v| \cdot |w|.$$

Thus we are done. □

7.2. The KMS-structure of tame harmonic bundles on a punctured disc

7.2.1. Prolongment of $\mathcal{E}^\lambda$. We put $X = \Delta$ and $D = \{O\}$. Let $(E, \overline{\partial}_E, \theta, h)$ be a tame harmonic bundle over $X - D$. We have the deformed holomorphic bundle $(\mathcal{E}^\lambda, \mathbb{D}^\lambda)$ over $\mathcal{X}^\lambda - \mathcal{D}^\lambda$ with the metric h. Let us recall some of the results on the prolongment of $\mathcal{E}^\lambda$. (See the section 10 of [**80**] and the section 3 of [**81**]. See also the subsection 4.3.1–4.3.3 in [**65**].)

(1) For any real number $b \in \boldsymbol{R}$, the $\mathcal{O}_X$-module ${}_b\mathcal{E}^\lambda$ is locally free.

(2) For any real numbers $a < b$, we have the canonical inclusion ${}_a\mathcal{E}^\lambda \longrightarrow {}_b\mathcal{E}^\lambda$ of $\mathcal{O}_X$-modules. Then we obtain the parabolic filtration of ${}_b\mathcal{E}^\lambda{}_{|O}$. Namely we put $F_a({}_b\mathcal{E}^\lambda) := \mathrm{Im}({}_a\mathcal{E}^\lambda{}_{|O} \longrightarrow {}_b\mathcal{E}^\lambda{}_{|O})$. Then we have the following inclusions for any $b - 1 \leq a \leq b$:

$$0 = F_{b-1}({}_b\mathcal{E}^\lambda) \subset F_a({}_b\mathcal{E}^\lambda) \subset F_b({}_b\mathcal{E}^\lambda) = {}_b\mathcal{E}^\lambda{}_{|O}.$$

(3) We put as follows:

$$F_{<a}(\mathcal{E}^\lambda) = \sum_{c<a} F_c(\mathcal{E}^\lambda) = \bigcup_{c<a} F_c(\mathcal{E}^\lambda), \quad \mathrm{Gr}_a^F({}_b\mathcal{E}^\lambda) := \frac{F_a({}_b\mathcal{E}^\lambda)}{F_{<a}({}_b\mathcal{E}^\lambda)}.$$

If $a \leq b < a+1$, we have the canonical isomorphism $\mathrm{Gr}_a^F({}_a\mathcal{E}^\lambda) \longrightarrow \mathrm{Gr}_a^F({}_b\mathcal{E}^\lambda)$. Hence we omit to denote b in this case.

On the contrary, if $b < a$ or $b \geq a+1$, then we have $\mathrm{Gr}_a^F({}_b\mathcal{E}^\lambda) = 0$ by definition.

(4) Let $\boldsymbol{v} = (v_i)$ be a holomorphic frame of ${}_b\mathcal{E}^\lambda$ over X compatible with the parabolic filtration at D. We put $b_i := \deg^F(v_i)$. We put $v_i' := |z|^{b_i} \cdot v_i$, and then we obtain the C^∞-frame $\boldsymbol{v}' := (v_i')$ of $\mathcal{E}^\lambda$ over $\mathcal{X}^\lambda - \mathcal{D}^\lambda$. Then $\boldsymbol{v}'$ is adapted up to log order

(5) $\mathbb{D}^\lambda$ is logarithmic in the following sense: if f is a holomorphic section of ${}_b\mathcal{E}^\lambda$, then $\mathbb{D}^\lambda f$ is a holomorphic section of ${}_{b+1}\mathcal{E}^\lambda \otimes \Omega^{1,0} = {}_b\mathcal{E}^\lambda \otimes \Omega^{1,0}(\log O)$. In particular, we obtain the residue $\mathrm{Res}(\mathbb{D}^\lambda) \in \mathrm{End}({}_b\mathcal{E}^\lambda{}_{|O})$, which preserves the parabolic filtration F.

We have the $\mathbb{E}$-decomposition of ${}_b\mathcal{E}^\lambda{}_{|O}$ for $\mathrm{Res}(\mathbb{D})$:

$${}_b\mathcal{E}^\lambda{}_{|O} = \bigoplus_{\alpha \in \boldsymbol{C}} \mathbb{E}({}_b\mathcal{E}^\lambda{}_{|O}, \alpha).$$

LEMMA 7.38. *The decomposition $\mathbb{E}$ and the parabolic filtration F is compatible.*

Proof Since $F_a({}_b\mathcal{E}^\lambda)$ is stable under the action of $\mathrm{Res}(\mathbb{D}^\lambda)$, we have only to apply Lemma 2.8. □

For any $u = (a, \alpha) \in \boldsymbol{R} \times \boldsymbol{C}$ and $b \in]a, a+1]$, we put as follows:

$$\mathrm{Gr}_u^{F,\mathbb{E}}(\mathcal{E}^\lambda) := \mathbb{E}(\mathrm{Gr}_a^F({}_b\mathcal{E}^\lambda), \alpha) = \mathrm{Gr}_a^F\,\mathbb{E}\big({}_b\mathcal{E}^\lambda{}_{|O}, \alpha\big).$$

It is independent of a choice of $b \in]a, a+1]$. And we have the induced morphism $\mathrm{Gr}_u^{F,\mathbb{E}}(\mathrm{Res}(\mathbb{D}^\lambda))$ on $\mathrm{Gr}_u^{F,\mathbb{E}}(\mathcal{E}^\lambda)$. The nilpotent part is denoted by $\mathcal{N}_u$. Then we obtain the weight filtration W of $\mathcal{N}_u$.

The estimate in (5) is strengthened as follows: Let $\boldsymbol{v} = (v_i)$ be a frame of ${}_b\mathcal{E}^\lambda$ which is compatible with the parabolic filtration. Moreover, we assume that the induced frame $\boldsymbol{v}^{(1)} = \big(v_i^{(1)}\big)$ of $\mathrm{Gr}^F({}_b\mathcal{E}^\lambda)$ is compatible with the weight filtration. We put $b_i := \deg^F(v_i)$ and $k_i := \deg^W(v_i^{(1)})$. We put $v_i' := |z|^{b_i} \cdot \big(-\log|z|\big)^{-k_i/2} \cdot v_i$, and we obtain the C^∞-frame $\boldsymbol{v}' = (v_i')$ of $\mathcal{E}^\lambda$ on $X - D$.

LEMMA 7.39 (Simpson [**82**], see also the section 4.3.3 in [**65**]). *$\boldsymbol{v}'$ is adapted.* □

LEMMA 7.40. *Let f be a section of ${}_b\mathcal{E}^\lambda$. Then there exists a real numbers a, k, and C_i $(i = 1, 2)$ such that the following holds:*

$$0 < C_1 < |f|_h \cdot |z|^a \cdot \big(-\log|z|\big)^{-k/2} \le C_2.$$

Proof Let $\boldsymbol{v}$ and $\boldsymbol{v}'$ be as above. We have the expression $f = \sum f_i \cdot v_i$. Due to the adaptedness of $\boldsymbol{v}'$, the claim follows. □

7.2.2. KMS-spectrum.

DEFINITION 7.41. For a harmonic bundle $(E, \overline{\partial}_E, \theta, h)$, the set $\mathcal{KMS}(\mathcal{E}^\lambda)$ is defined as follows:

$$\mathcal{KMS}(\mathcal{E}^\lambda) := \big\{u \in \boldsymbol{R} \times \boldsymbol{C} \,\big|\, \dim \mathrm{Gr}_u^{F,\mathbb{E}}(\mathcal{E}) \neq 0\big\}.$$

It is called the KMS-spectrum set of $\mathcal{E}$ at (λ, O). For any $u \in \mathcal{KMS}(\mathcal{E}^\lambda)$, the number $\mathfrak{m}(\lambda, u)$ is defined as follows:

$$\mathfrak{m}(\lambda, u) := \dim \mathrm{Gr}_u^{F,\mathbb{E}}(\mathcal{E}^\lambda).$$

It is called the multiplicity of the KMS-spectrum u. □

The natural morphisms $\mathcal{KMS}(\mathcal{E}^\lambda) \longrightarrow \boldsymbol{R}$ and $\mathcal{KMS}(\mathcal{E}^\lambda) \longrightarrow \boldsymbol{C}$ are denoted by $\pi^{\mathfrak{p}}$ and $\pi^{\mathfrak{e}}$. We put $\mathcal{P}ar(\mathcal{E}^\lambda) := \mathrm{Im}(\pi^{\mathfrak{p}})$ and $\mathcal{S}p(\mathcal{E}^\lambda) := \mathrm{Im}(\pi^{\mathfrak{e}})$.

PROPOSITION 7.42. *We have the isomorphism $\mathrm{Gr}_u^{F,\mathbb{E}}(\mathcal{E}^\lambda) \simeq \mathrm{Gr}_{u+(1,-\lambda)}^{F,\mathbb{E}}(\mathcal{E}^\lambda)$.*

Proof Let $\boldsymbol{v} = (v_i)$ be a holomorphic frame of ${}_a\mathcal{E}^\lambda$ compatible with $\mathbb{E}$ and F. We put $b_i := \deg^F(v_i)$. We put $\tilde{v}_i := z^{-1} \cdot v_i$ and $\widetilde{\boldsymbol{v}} := (\tilde{v}_i)$.

LEMMA 7.43. *$\widetilde{\boldsymbol{v}}$ gives a holomorphic frame of ${}_{a+1}\mathcal{E}^\lambda$.*

Proof We put $\tilde{v}_i' := \tilde{v}_i \cdot |z|^{b_i - 1}$, and $\tilde{\boldsymbol{v}}' = (\tilde{v}_i)$. Then it is C^∞-frame of $\mathcal{E}^\lambda$ over $\mathcal{X}^\lambda - \mathcal{D}^\lambda$, and it is adapted up to log order. Then $\widetilde{\boldsymbol{v}}$ gives a frame of ${}_a\mathcal{E}^\lambda$ compatible with the parabolic filtration due to Lemma 2.8 and Lemma 2.9. □

Let us return to the proof of Proposition 7.42. Let $\mathcal{A}$ be the λ-connection form of $\mathbb{D}^\lambda$ with respect to the frame $\boldsymbol{v}$, i.e. $\mathbb{D}^\lambda \boldsymbol{v} = \boldsymbol{v} \cdot \mathcal{A}$ holds. Then we have the following:

$$\mathbb{D}^\lambda \widetilde{\boldsymbol{v}} = \mathbb{D}^\lambda\big(z^{-1} \cdot \boldsymbol{v}\big) = \widetilde{\boldsymbol{v}} \cdot \Big(\mathcal{A} - \lambda \frac{dz}{z}\Big).$$

We obtain the isomorphism ${}_a\mathcal{E}^\lambda{}_{|O} \longrightarrow {}_{a+1}\mathcal{E}^\lambda{}_{|O}$ defined by the correspondence $v_i(0) \longmapsto \tilde{v}_i(0)$. Then it induces the isomorphism:

$$\mathrm{Gr}_u^{\mathbb{E},F}(\mathcal{E}^\lambda) \longrightarrow \mathrm{Gr}_{u+(1,-\lambda)}^{\mathbb{E},F}(\mathcal{E}^\lambda).$$

Thus we are done. □

COROLLARY 7.44. *We have the equality* $\mathfrak{m}(\lambda, u) = \mathfrak{m}\big(\lambda, u + (1, -\lambda)\big)$. □

We have the free $\mathbb{Z}$-action on $\mathcal{KMS}(\mathcal{E}^\lambda)$:

$$\mathbb{Z} \times \mathcal{KMS}(\mathcal{E}^\lambda) \longrightarrow \mathcal{KMS}(\mathcal{E}^\lambda), \quad (n, u) \longmapsto u + n \cdot (1, -\lambda).$$

It preserves the multiplicities.

DEFINITION 7.45. We put $\overline{\mathcal{KMS}}(\mathcal{E}^\lambda) := \mathcal{KMS}(\mathcal{E}^\lambda)/\mathbb{Z}$. Note that the multiplicity of any element $u \in \overline{\mathcal{KMS}}(\mathcal{E}^\lambda)$ is naturally defined, due to Corollary 7.44. □

DEFINITION 7.46. We put as follows:

$$\begin{aligned} \mathcal{KMS}({}_b\mathcal{E}^\lambda) &:= \big\{u \in \mathcal{KMS}(\mathcal{E}^\lambda) \,\big|\, b - 1 < \pi^{\mathfrak{p}}(u) \leq b\big\} \\ &= \big\{u \in \boldsymbol{R} \times \boldsymbol{C} \,\big|\, \mathrm{Gr}_u^{\mathbb{E},F}({}_b\mathcal{E}^\lambda) \neq 0\big\}. \end{aligned} \tag{7.21}$$

□

We have the natural morphism $\pi : \mathcal{KMS}({}_b\mathcal{E}^\lambda) \longrightarrow \overline{\mathcal{KMS}}(\mathcal{E}^\lambda)$. The following lemma is clear.

LEMMA 7.47. *The morphism* π *is bijective.* □

The restriction of $\pi^{\mathfrak{p}}$ to $\mathcal{KMS}({}_b\mathcal{E}^\lambda)$ gives the morphisms $\mathcal{KMS}({}_b\mathcal{E}^\lambda) \longrightarrow \boldsymbol{R}$. The image $\pi^{\mathfrak{p}}\big(\mathcal{KMS}({}_b\mathcal{E}^\lambda)\big)$ is denoted by $\mathcal{P}ar({}_b\mathcal{E}^\lambda)$. The following lemma is clear.

LEMMA 7.48. *We have* $\mathcal{P}ar({}_b\mathcal{E}^\lambda) = \big\{a \in \boldsymbol{R} \,\big|\, \mathrm{Gr}_a^F({}_b\mathcal{E}^\lambda) \neq 0\big\}$. □

The restriction of the morphism $\pi^{\mathfrak{e}}$ to $\mathcal{KMS}({}_b\mathcal{E}^\lambda)$ gives the morphism $\pi^{\mathfrak{e}} : \mathcal{KMS}({}_b\mathcal{E}^\lambda) \longrightarrow \boldsymbol{C}$. The image $\pi^{\mathfrak{e}}\big(\mathcal{KMS}({}_b\mathcal{E}^\lambda)\big)$ is denoted by $\mathcal{S}p({}_b\mathcal{E}^\lambda)$. The following lemma is clear.

LEMMA 7.49. *We have* $\mathcal{S}p({}_b\mathcal{E}^\lambda) = \big\{\alpha \in \boldsymbol{C} \,\big|\, \mathbb{E}\big({}_b\mathcal{E}^\lambda{}_{|O}, \alpha\big) \neq 0\big\}$. □

7.2.3. The functoriality of the KMS structure for pull back. We will use the notation κ and ν given in the subsection 2.1.5. Let c be a positive integer, and $\psi_c : X \longrightarrow X$ be the morphism given by $z \longmapsto z^c$. Let f be a holomorphic section of ${}^\diamond\mathcal{E}^\lambda$ over X. We put $b = -\operatorname{ord}(f)$. Assume $-1 < b \leq 0$, i.e., $f_{|O} \neq 0$ in ${}^\diamond\mathcal{E}^\lambda{}_{|O}$. We have the holomorphic section $\psi_c^{-1}(f)$ of $\psi_c^{-1}(\mathcal{E}^\lambda)$ over $\mathcal{X}^\lambda - \mathcal{D}^\lambda$. Then we have $\operatorname{ord}(\psi_c^{-1}(f)) = -c \cdot b$. We put as follows:

$$\tilde{f} := z^{\nu(c \cdot b)} \cdot \psi_c^{-1}(f).$$

Then we have the following:

$$-\operatorname{ord}(\tilde{f}) = -\nu(c\cdot b) - \operatorname{ord}(\psi_c^{-1}(f)) = -\nu(c\cdot b) + c\cdot b = \kappa(c\cdot b).$$

Hence $\tilde{f}$ is a section of ${}_{\kappa(cb)}\psi_c^{-1}\mathcal{E}^\lambda$. In particular, it gives a section of ${}^\diamond\psi_c^{-1}\mathcal{E}^\lambda$.

Let $\boldsymbol{v}$ be the holomorphic frame of ${}^\diamond\mathcal{E}^\lambda$ compatible with F. We put $b_i := \deg^F(v_i)$. We put $v_i' := |z|^{b_i}\cdot v_i$ and $\boldsymbol{v}' := (v_i')$. Recall that $\boldsymbol{v}'$ is a C^∞-frame of ${}^\diamond\mathcal{E}^\lambda$ over $\mathcal{X}^\lambda - \mathcal{D}^\lambda$, which is adapted up to log order. For each v_i, take the holomorphic section $\tilde{v}_i$ of ${}^\diamond\psi_c^{-1}\mathcal{E}^\lambda$ as above. Then we obtain the tuple of sections $\widetilde{\boldsymbol{v}} = (\tilde{v}_i)$ of ${}^\diamond\psi_c^{-1}\mathcal{E}^\lambda$.

LEMMA 7.50. *$\widetilde{\boldsymbol{v}}$ is a holomorphic frame of ${}^\diamond\psi_c^{-1}\mathcal{E}^\lambda$.*

Proof We put as follows:

$$\tilde{v}_i' := |z|^{\kappa(cb_i)}\tilde{v}_i = C_i(z)\cdot\psi_c^{-1}(v_i)'.$$

Here we have $|C_i(z)| = 1$. We obtain the tuple of the C^∞-sections $\tilde{\boldsymbol{v}}' = (\tilde{v}_i')$ of $\psi_c^{-1}\mathcal{E}^\lambda$. It is a C^∞-frame of $\psi_c^{-1}\mathcal{E}^\lambda$ over $\mathcal{X}^\lambda - \mathcal{D}^\lambda$, and it is adapted up to log order. Hence we have only to apply Lemma 2.8. □

COROLLARY 7.51. *We have the surjective morphism $\psi_c^* : \mathcal{KMS}({}^\diamond\mathcal{E}^\lambda) \longrightarrow \mathcal{KMS}({}^\diamond\psi_c^{-1}\mathcal{E}^\lambda)$ given as follows:*

$$(b,\beta) \longmapsto \big(\kappa(c\cdot b),\ c\cdot\beta + \nu(c\cdot b)\cdot\lambda\big) = c\cdot(b,\beta) + \nu(c\cdot b)\cdot(-1,\lambda).$$

We have isomorphisms:

$$\operatorname{Gr}_u^{\mathbb{E},F}({}^\diamond\psi_c^{-1}\mathcal{E}^\lambda) \simeq \bigoplus_{\psi_c^*(u)=u'} \operatorname{Gr}_{u'}^{\mathbb{E},F}({}^\diamond\mathcal{E}^\lambda), \qquad \operatorname{Gr}_a^F({}^\diamond\psi_c^{-1}\mathcal{E}^\lambda) \simeq \bigoplus_{\kappa(cb)=a} \operatorname{Gr}_b({}^\diamond\mathcal{E}^\lambda).$$

The isomorphisms are given by $\boldsymbol{v}$ and $\widetilde{\boldsymbol{v}}$.

Proof We have only to note that $\tilde{\boldsymbol{v}}$ is compatible with $\mathbb{E}$ and F. The compatibility of $\tilde{\boldsymbol{v}}$ and F follows from the fact that $\tilde{\boldsymbol{v}}'$ is adapted up to log order (Lemma 2.9).

Let $\mathcal{A}$ be the λ-connection form of $\mathbb{D}$ with respect to the frame $\boldsymbol{v}$, namely we have $\mathbb{D}^\lambda\boldsymbol{v} = \boldsymbol{v}\cdot\mathcal{A}$. Then we obtain $\tilde{\mathbb{D}}^\lambda\tilde{\boldsymbol{v}} = \tilde{\boldsymbol{v}}\cdot\tilde{\mathcal{A}}$, where $\tilde{\mathcal{A}} = \psi_c^{-1}\mathcal{A} + B\cdot dz/z$, and B denotes the diagonal matrix such that $B_{j\,j} = \nu(c\cdot b_j)$. Thus we obtain the compatibility of $\tilde{\boldsymbol{v}}$ and the decomposition $\mathbb{E}$. □

COROLLARY 7.52. *The following holds:*

$$\mathcal{P}ar(\psi_c^{-1}(\mathcal{E})) = \bigcup_{a\in\mathcal{P}ar({}^\diamond\mathcal{E}^\lambda)} (c\cdot a + \mathbb{Z}).$$

□

7.2.4. The action. Assume that c is a positive integer which is sufficiently large with respect to $\mathcal{P}ar({}^\diamond\mathcal{E}^\lambda)$. (See Definition 2.1). We have the action of μ_c on X, given by $\omega^*z = \omega\cdot z$. It can be naturally lifted to the action on $\psi_c^{-1}\mathcal{E}^\lambda$. Since v_i are invariant under the action of μ_c, we have the following:

$$\omega^*\tilde{v}_i = \omega^{\nu(cb_i)}\cdot\tilde{v}_i.$$

We have the weight decomposition:

$${}^\diamond\psi_c^{-1}\mathcal{E}_{|O} = \bigoplus_h U_h.$$

Here $\omega^* = \omega^h$ on U_h for $-c+1 \le h \le 0$. The following lemma is clear.

LEMMA 7.53. *$U_h = \langle \tilde{v}_i \,|\, \nu(c\cdot b_i) = h \rangle$.* □

We put $S := \{h \,|\, -c+1 \leq h \leq 0,\ U_h \neq 0\}$. Then we have $S = \{\nu(c\cdot b) \,|\, b \in \mathcal{P}ar({}^{\diamond}\mathcal{E}^{\lambda})\}$. Since c is sufficiently large with respect to $\mathcal{P}ar({}^{\diamond}\mathcal{E}^{\lambda})$, any element $b \in \mathcal{P}ar({}^{\diamond}\mathcal{E}^{\lambda})$ is uniquely determined by the number $\nu(c\cdot b) \in S$. Thus we have the map $\varphi : S \longrightarrow \mathcal{P}ar({}^{\diamond}\psi_c^{-1}\mathcal{E}^{\lambda})$ given by the following correspondence:

$$\nu(c\cdot b) \longmapsto \kappa(c\cdot b).$$

Let us consider the filtration F' given as follows:

$$F'_{\tilde{b}} := \bigoplus_{\substack{h\in S,\\ \varphi(h)\leq \tilde{b}}} U_h.$$

LEMMA 7.54. *We have $F'_b({}^{\diamond}\psi_c^{-1}\mathcal{E}^{\lambda}) = F_b({}^{\diamond}\psi_c^{-1}\mathcal{E}^{\lambda})$.*

Proof It follows from the following equalities:

$$F_b({}^{\diamond}\psi_c^{-1}\mathcal{E}^{\lambda}) = \langle \tilde{v}_i \,|\, \kappa(c\cdot b_i) \leq b \rangle = F'_b({}^{\diamond}\psi_c^{-1}\mathcal{E}^{\lambda}).$$

Thus we are done. □

COROLLARY 7.55. *The decomposition ${}^{\diamond}\psi_c^{-1}\mathcal{E}^{\lambda}{}_{|O} = \bigoplus_h U_h$ gives a splitting of the parabolic filtration in the following sense:*

$$F_b = \bigoplus_{\substack{h\in S\\ \varphi(h)\leq b}} U_h.$$

In particular, U_h is naturally isomorphic to $\mathrm{Gr}_b({}^{\diamond}\psi_c^{-1}\mathcal{E}^{\lambda}{}_{|O})$ $(\varphi(h) = b)$. □

7.2.5. Descent of the frame. On the other hand, we can descend the equivariant frame. Let f be a holomorphic section of ${}^{\diamond}\psi_c^{-1}\mathcal{E}^{\lambda}$, such that $\omega^*(f) = \omega^h \cdot f$ for some integer h such that $-c < h \leq 0$. We put $f_1 := z^{-h}\cdot f$, and then we have $\omega^*(f_1) = f_1$. Hence f_1 induces the holomorphic section $\tilde{f}$ of $\mathcal{E}^{\lambda}$ over $X - D$. We have the following:

$$-\operatorname{ord}(\tilde{f}) = -c^{-1}\cdot \operatorname{ord}(f_1) = c^{-1}\cdot \big(h - \operatorname{ord}(f)\big) \leq 0.$$

Hence $\tilde{f}$ gives a holomorphic section of ${}^{\diamond}\mathcal{E}^{\lambda}$.

Let $\boldsymbol{v} = (v_i)$ be a holomorphic frame of ${}^{\diamond}\psi_c^{-1}\mathcal{E}^{\lambda}$ satisfying the following:

- It is equivariant in the sense $\omega^*\tilde{v}_i = \omega^{h_i}\cdot \tilde{v}_i$ for $-c < h_i \leq 0$.
- It is compatible with the parabolic filtration F.

We put $b_i := \deg^F(v_i)$, and then we have $-1 < b_i \leq 0$.

LEMMA 7.56. *We have $-1 < c^{-1}\cdot(h + b_i) \leq 0$.*

Proof Since we have $-c+1 \leq h \leq 0$ and $-1 < b_i \leq 0$, we obtain $-c < h + b_i \leq 0$. □

Let us take the section $\tilde{v}_i$ of ${}^{\diamond}\mathcal{E}^{\lambda}$ for each v_i as above. Then we obtain the tuple of sections $\tilde{\boldsymbol{v}} = (\tilde{v}_i)$.

LEMMA 7.57. *$\tilde{\boldsymbol{v}}$ is a holomorphic frame of ${}^{\diamond}\mathcal{E}^{\lambda}$, compatible with the parabolic filtration.*

Proof We put $\tilde{b}_i := c^{-1} \cdot (b_i + h - c)$. We put as follows:

$$\tilde{v}_i' := |z|^{\tilde{b}_i} \cdot \tilde{v}_i, \quad \tilde{\boldsymbol{v}} = (\tilde{v}_i).$$

Then it can be checked that $\tilde{\boldsymbol{v}}$ is adapted up to log order. Thus we have only to apply Lemma 2.8 and Lemma 2.9. □

7.2.6. Functoriality for tensor product. Let $(E^{(a)}, \overline{\partial}_{E^{(a)}}, \theta^{(a)}, h^{(a)})$ $(a = 1,2)$ be tame harmonic bundles over $X - D$. We obtain the prolonged deformed holomorphic bundles ${}^{\diamond}\mathcal{E}^{(a)\,\lambda}$. Let $\boldsymbol{v}^{(a)}$ be holomorphic frames of ${}^{\diamond}\mathcal{E}^{(a)\,\lambda}$ compatible with the parabolic filtration F. We put $b_i^{(a)} := \deg^F(v_i^{(a)})$, and $v_i^{(a)\,\prime} := |z|^{b_i^{(a)}} v_i^{(a)}$. The tuple of sections $\boldsymbol{v}^{(a)\,\prime} = (v_i^{(a)\,\prime})$ gives a C^∞-frame of $\mathcal{E}^{\lambda\,(a)}$ over $X - D$, which is adapted up to log order.

Then we obtain the C^∞-frame $\boldsymbol{v}^{(1)\,\prime} \otimes \boldsymbol{v}^{(2)\,\prime}$ of $\mathcal{E}^{(1)\,\lambda} \otimes \mathcal{E}^{(2)\,\lambda}$, given as follows:

$$\boldsymbol{v}^{(1)\,\prime} \otimes \boldsymbol{v}^{(2)\,\prime} = \left(v_i^{(1)\,\prime} \otimes v_j^{(2)\,\prime} \,\middle|\, 1 \le i \le \operatorname{rank} E^{(1)},\ 1 \le j \le \operatorname{rank} E^{(2)} \right).$$

It is adapted up to log order. Hence we put $w_{i\,j} := z^{-\epsilon(i,j)} \cdot v_i^{(1)} \otimes v_j^{(2)}$, where $\epsilon(i,j)$ are given as follows:

$$\epsilon(i,j) := \begin{cases} 1, & (b_i^{(1)} + b_j^{(2)} \le -1), \\ 0, & (\text{otherwise, i.e., } -1 < b_i^{(1)} + b_j^{(2)} \le 0). \end{cases}$$

Then we obtain the tuple of holomorphic sections $\boldsymbol{w} = \left(w_{i\,j} \,\middle|\, 1 \le i \le \operatorname{rank} E^{(1)},\ 1 \le j \le \operatorname{rank} E^{(2)} \right)$, and it gives the holomorphic frame of ${}^{\diamond}\left(\mathcal{E}^{(1)\,\lambda} \otimes \mathcal{E}^{(2)\,\lambda}\right)$, compatible with the filtration.

COROLLARY 7.58. *We have the surjective morphism:*

$$\psi : \mathcal{KMS}({}^{\diamond}\mathcal{E}^{(1)\,\lambda}) \times \mathcal{KMS}({}^{\diamond}\mathcal{E}^{(2)\,\lambda}) \longrightarrow \mathcal{KMS}\left({}^{\diamond}\left(\mathcal{E}^{(1)\,\lambda} \otimes \mathcal{E}^{(2)\,\lambda}\right)\right).$$

Here $\psi(u_1, u_2) \in \boldsymbol{R} \times \boldsymbol{C}$ is given as follows, for elements $u_i = (b_i, \beta_i) \in \boldsymbol{R} \times \boldsymbol{C}$ $(i = 1,2)$:

$$\psi(u_1, u_2) = \left(\kappa(b_1 + b_2),\ \beta_1 + \beta_2 - \nu(b_1 + b_2) \cdot \lambda \right).$$

We have the equality of the multiplicities:

$$\mathfrak{m}(\lambda, u) = \sum_{\psi(u_1,u_2)=u} \mathfrak{m}(\lambda, u_1) \cdot \mathfrak{m}(\lambda, u_2).$$

□

COROLLARY 7.59. *We have the isomorphisms:*

$$\operatorname{Gr}_b^F\left({}^{\diamond}\left(\mathcal{E}^{(1)\,\lambda} \otimes \mathcal{E}^{(2)\,\lambda}\right)\right) \simeq \bigoplus_{\kappa(b_1+b_2)=b} \operatorname{Gr}_{b_1}^F({}^{\diamond}\mathcal{E}^{(1)\,\lambda}) \otimes \operatorname{Gr}_{b_2}^F({}^{\diamond}\mathcal{E}^{(2)\,\lambda}),$$

$$\operatorname{Gr}_u^{F,\mathbb{E}}\left({}^{\diamond}\left(\mathcal{E}^{(1)\,\lambda} \otimes \mathcal{E}^{(2)\,\lambda}\right)\right) \simeq \bigoplus_{\psi(u_1,u_2)=u} \operatorname{Gr}_{u_1}^{F,\mathbb{E}}({}^{\diamond}\mathcal{E}^{(1)\,\lambda}) \otimes \operatorname{Gr}_{u_2}^{F,\mathbb{E}}({}^{\diamond}\mathcal{E}^{(2)\,\lambda}).$$

□

COROLLARY 7.60. *We have the isomorphism:*

$$\operatorname{Gr}_b^F({}^{\diamond}\operatorname{Sym}^h \mathcal{E}^\lambda) \simeq \bigoplus_{(\boldsymbol{b},\boldsymbol{m}) \in \mathcal{S}(b,h)} \bigotimes_i \operatorname{Sym}^{m_i}(\operatorname{Gr}_{b_i}^F {}^{\diamond}\mathcal{E}^\lambda).$$

Here we put as follows:

$$\mathcal{S}(b,h) := \Big\{(\boldsymbol{b},\boldsymbol{m})\,\Big|\, \sum m_i = h,\quad \kappa\Big(\sum m_i b_i\Big) = b\Big\}.$$

In all, we have an isomorphism $\mathrm{Gr}^F({}^{\diamond}\mathrm{Sym}^{\cdot}\,\mathcal{E}^{\lambda}) \simeq \mathrm{Sym}^{\cdot}(\mathrm{Gr}^F\,{}^{\diamond}\mathcal{E}^{\lambda})$.

We also have an isomorphism:

$$\mathrm{Gr}^F_b\Big({}^{\diamond}\bigwedge^{h}\mathcal{E}^{\lambda}\Big) \simeq \bigoplus_{(\boldsymbol{b},\boldsymbol{m})\in\mathcal{S}(b,h)} \bigotimes_i \bigwedge^{m_i}(\mathrm{Gr}^F_{b_i}\,{}^{\diamond}\mathcal{E}^{\lambda}).$$

We have $\mathrm{Gr}^F({}^{\diamond}\bigwedge^{\cdot}\mathcal{E}^{\lambda}) \simeq \bigwedge^{\cdot}(\mathrm{Gr}^F\,{}^{\diamond}\mathcal{E}^{\lambda})$. □

7.2.7. Functoriality for dual. Let $\boldsymbol{v} = (v_i)$ be a holomorphic frame of ${}^{\diamond}\mathcal{E}^{\lambda}$ compatible with F. We put $b_i := \deg^F(v_i)$, $v_i' := |z|^{b_i}\cdot v_i$, and $\boldsymbol{v}' = (v_i')$. Then we obtain the dual frame $\boldsymbol{v}'^{\vee}$ of $\mathcal{E}^{\lambda\,\vee}$ over $X - D$. It is adapted up to log order.

Let $\boldsymbol{v}^{\vee} = (v_i^{\vee})$ be the dual frame of $\boldsymbol{v}$ over $X - D$. Then it gives a holomorphic frame of ${}_{1-\epsilon}\mathcal{E}^{\lambda\,\vee}$ for some $\epsilon > 0$.

Then we put $w_i := z^{\epsilon(i)}\cdot v_i^{\vee}$ and $\boldsymbol{w} = (w_i)$, where $\epsilon(i)$ is given as follows:

$$\epsilon(i) := \begin{cases} 1, & (b_i \neq 0),\\ 0, & (b_i = 0).\end{cases}$$

Then $\boldsymbol{w}$ is a holomorphic frame of ${}^{\diamond}\mathcal{E}^{\vee\,\lambda}$ compatible with F.

COROLLARY 7.61. *We have the bijection* $\psi : \mathcal{KMS}({}^{\diamond}\mathcal{E}^{\lambda}) \longrightarrow \mathcal{KMS}({}^{\diamond}\mathcal{E}^{\lambda\,\vee})$. *For any* $u = (b,\beta)$, $\psi(u)$ *is given by* $\big(\kappa(-b), -\beta - \nu(-\beta)\big)$. *We also have the isomorphism* $\mathrm{Gr}^{F\,\mathbb{E}}_u({}^{\diamond}\mathcal{E}^{\lambda})^{\vee} \simeq \mathrm{Gr}^{F\,\mathbb{E}}_{\psi(u)}({}^{\diamond}\mathcal{E}^{\lambda\,\vee})$. □

7.2.8. L^2-property and growth estimate. We put $\overline{X} = \overline{\Delta}$, $X = \Delta$ and $D = \{O\}$. Let $(E, \overline{\partial}_E, \theta, h)$ be a tame harmonic bundle on $\overline{X} - D$. We use the Poincaré metric of $X - D = \Delta^*$. Let b be a real number. For a measurable section f of $\mathcal{E}^{\lambda}$ on $X - D$, we put as follows:

$$\|f\|_{b,N}^2 := \int_{X-D} |f|_h^2\cdot|z|^{2b}\cdot\big(-\log|z|\big)^{-N}\cdot \mathrm{dvol}.$$

Since $(\mathcal{E}^{\lambda}, h)$ is acceptable, we can apply the results in the section 2.8.

LEMMA 7.62. *Let f be a section of ${}_b\mathcal{E}^{\lambda}$ on $\overline{X}$. Let ϵ be any positive number, and N be a negative number. Then* $\|f\|_{b+\epsilon,N} < \infty$.

Proof Since f is given over $\overline{X}$, we have only to care the behaviour of f around O. Let $\boldsymbol{v} = (v_i)$ be a holomorphic frame of ${}_b\mathcal{E}$ on a neighbourhood U of O, which is compatible with the filtrations F and W. We put $b_i := \deg^F(v_i)$ and $k_i := \deg^W(v_i^{(1)})$. We put $v_i' := |z|^{b_i}\cdot\big(-\log|z|\big)^{-k_i/2}\cdot v_i$, and we obtain the C^{∞}-tuple $\boldsymbol{v}' = (v_i')$ of $\mathcal{E}^{\lambda}$ on $X - D$. Recall that $\boldsymbol{v}'$ is adapted.

We have the expression $f = \sum f_i\cdot v_i$, where f_i are holomorphic function on $U - D$. Due to the adaptedness of $\boldsymbol{v}'$, there exist positive numbers C_1 and C_2 such that the following holds on $U - D$:

$$C_1|f|_h^2\cdot|z|^{2(b+\epsilon)}\cdot\big(-\log|z|\big)^{-N} \leq \sum |f_i|^2\cdot|z|^{2(b-b_i+\epsilon)}\cdot\big(-\log|z|\big)^{-N+k_i} \leq C_2|f|_h^2\cdot|z|^{2(b+\epsilon)}\cdot\big(-\log|z|\big)^{-N}. \tag{7.22}$$

Since we have $b - b_i + \epsilon > 0$, we obtain the desired integrability around O. □

Let us take a positive number ϵ such that $]b, b+\epsilon] \cap \mathcal{P}ar({}_b\mathcal{E}^\lambda) = \emptyset$.

LEMMA 7.63. *Let f be a holomorphic section of $\mathcal{E}^\lambda$ on $X - D$. Assume that there exists a number N such that $\|f\|_{b+\epsilon,N} < \infty$. Then f is a holomorphic section of ${}_b\mathcal{E}^\lambda$ on X.*

Proof Let us take a holomorphic frame $\boldsymbol{v}$ of ${}_b\mathcal{E}^\lambda$ as in the proof of Lemma 7.62. We have the estimate (7.22). Hence $\|f\|_{b+\epsilon,N} < \infty$ implies the following finiteness:

$$\int |f_i|^2 \cdot |z|^{2(b-b_i+\epsilon)} \cdot \bigl(-\log|z|\bigr)^{-N+k_i} \cdot \frac{dz \cdot d\bar{z}}{|z|^2 \cdot \bigl(-\log|z|\bigr)^2} < \infty.$$

Due to the assumption $]b, b+\epsilon] \cap \mathcal{P}ar\bigl({}_b\mathcal{E}^\lambda\bigr) = \emptyset$, we have $b - b_i + \epsilon < 1$. Hence f_i are, in fact, holomorphic on X. Thus we are done. □

7.3. Basic comparison due to Simpson

7.3.1. The statement. We put $X = \Delta$ and $D = \{O\}$. Let $(E, \overline{\partial}_E, \theta, h)$ be a tame harmonic bundle over $X - D$. As is already seen, we obtain the vector space $V_u := \mathrm{Gr}_u^{\mathbb{E},F}({}^\diamond E_{|O})$ and N_u. We have the model bundle $E(V, N)$ as in Lemma 6.9:

$$(E_0, \overline{\partial}_{E_0}, h_0, \theta_0) = \bigoplus_{u \in \mathcal{KMS}({}^\diamond E)} E(V_u, N_u) \otimes L(u).$$

We can pick an isomorphism $\Phi : {}^\diamond E_0 \longrightarrow {}^\diamond E$ satisfying the following:

- Φ preserves the parabolic filtrations.
- The induced morphism $\mathrm{Gr}_a^F(\Phi) \in Hom\bigl(\mathrm{Gr}_a^F({}^\diamond E_0), \mathrm{Gr}_a^F({}^\diamond E)\bigr)$ is compatible with the morphisms $\mathrm{Gr}_a^F(\mathrm{Res}(\theta_0))$ and $\mathrm{Gr}_a^F(\mathrm{Res}(\theta))$.

PROPOSITION 7.64 (Simpson). *Φ and Φ^{-1} are bounded.*

Proof See the subsubsection 4.3.3 in the previous paper [**65**], for example. □

Since we have $\mathcal{E}^\lambda = E$ and $\mathcal{E}_0^\lambda = E_0$ as C^∞-bundles, we obtain the C^∞-isomorphism $\Phi : \mathcal{E}_0^\lambda \longrightarrow \mathcal{E}^\lambda$ on $\mathcal{X}^\lambda - \mathcal{D}^\lambda$. Let us take holomorphic frames $\boldsymbol{v}_0$ and $\boldsymbol{v}$ of ${}^\diamond\mathcal{E}_0^\lambda$ and ${}^\diamond\mathcal{E}^\lambda$, which are compatible with generalized eigen decompositions $\mathbb{E}$, parabolic filtrations F and the weight filtrations W. We put as follows:

$$\deg^F(v_j) = b_j, \qquad \deg^{\mathbb{E}}(v_j) = \beta_j, \qquad \frac{\deg^W(v_j)}{2} = k_j,$$

$$\deg^F(v_{0\,i}) = b_{0\,i}, \quad \deg^{\mathbb{E}}(v_{0\,i}) = \beta_{0\,i}, \qquad \frac{\deg^W(v_{0\,i})}{2} = k_{0\,i}.$$

We also put as follows:

$$v_j' := v_j \cdot |z|^{b_j} \cdot (-\log|z|)^{-k_j}, \qquad \boldsymbol{v}' = (v_j'),$$

$$v_{0\,i}' := v_{0\,i} \cdot |z|^{b_{0\,i}} \cdot (-\log|z|)^{-k_{0\,i}}, \quad \boldsymbol{v}_0' = (v_{0\,i}').$$

Then we obtain the C^∞-functions I and I' of $X - D$ to $M(r)$ defined as follows:

$$\Phi(v_{0\,i}) = \sum I_{j,i} \cdot v_j, \quad \Phi(v_{0\,i}') = \sum I_{j,i}' \cdot v_j'.$$

The following lemma can be checked by a direct calculation.

LEMMA 7.65. *We have the equality $I_{j,i}' = I_{j,i} \cdot |z|^{b_{0,i}-b_j} \cdot \bigl(-\log|z|\bigr)^{-k_{0,i}+k_j}$.* □

By the isomorphism Φ, we identify E and E_0. Let $\theta^\dagger$ be the conjugate of θ with respect to h, and $\theta_0^\dagger$ be the conjugate of θ_0 with respect to h_0. Recall the following lemma.

LEMMA 7.66 (Simpson, Lemma 7.3, Lemma 7.7 [**81**]). *We have the following inequalities:*

$$|\theta^\dagger - \theta_0^\dagger|_h \leq C\cdot|z|^{-1}(-\log|z|)^{-1}, \qquad \int |\theta^\dagger - \theta_0^\dagger|_h\cdot|z|\cdot\big(-\log|z|\big)\frac{|dz\cdot d\bar{z}|}{|z|^2\cdot\big(-\log|z|\big)} < \infty.$$

Here C denotes some positive constant, and the metric of $\mathrm{End}(E)\otimes\Omega^{0,1}_{X-D}$ *is given by h of E and the Euclidean metric $dz\cdot d\bar{z}$ of $X-D$.* □

We also recall the following proposition.

PROPOSITION 7.67 (Simpson).

(1) *I' and I'^{-1} are bounded.*
(2) *We have the inequality $|I'_{j\,i}| \leq C\cdot(-\log|z|)^{-1}$ in the case $(b_j,\beta_j)\neq(b_{0\,i},\beta_{0\,i})$.*
(3) *We have the inequality $||I'_{j\,i}||_W < \infty$ in the case $k_j\neq k_{0\,i}$. (See the page* 764 *of* [**81**] *or the subsubsection* 4.3.4 *of* [**65**] *for the norm $||\cdot||_W$.)*

In the case $\lambda = 1$, the proposition is given by Simpson (the section 7 in [**81**]). His argument clearly works in general case. Hence we only indicate an outline. See loc.cit. for more detail.

7.3.2. Outline of the proof of Proposition 7.67. In this subsection, we will use the Euclidean metric $dz\cdot d\bar{z}$ on $X-D$. The claim (1) immediately follows from the boundedness of Φ and Φ^{-1}, and the adaptedness of $\boldsymbol{v}'$ and $\boldsymbol{v}'_0$. Note that $\bar{\partial}\Phi = \lambda\cdot(\theta_0^\dagger - \theta^\dagger)$. We can apply the argument of Simpson, and we obtain the claim (3). (See also the section 4.3.4 of [**65**].)

Let us see the outline of the proof of the claim (2). The holomorphic sections $\psi, \psi^{(1)}$ of $End(\mathcal{E}^\lambda)\otimes\Omega^{1,0}$ are given as follows:

$$\psi(v_j) := v_j\cdot(\beta_j + \lambda\cdot b_j)\cdot\frac{dz}{z}, \qquad \psi^{(1)}(v_j) := v_j\cdot(-\lambda\cdot b_j)\frac{dz}{z}. \tag{7.23}$$

For any holomorphic section f of $\mathcal{E}^\lambda$ such that $|f|_h \leq C_1\cdot|z|^{-b}\cdot(-\log|z|)^k$, it is easy to see that there exists a positive constant C_2 satisfying the following:

$$\big|\mathbb{D}^\lambda f - \psi(f) - \psi_1^{(1)}(f)\big|_h \leq C_2\cdot|z|^{-b-1}(-\log|z|)^{k-1}.$$

LEMMA 7.68 (Lemma 7.2 of [**81**]). *We have the following finiteness:*

$$\int \big|\psi - (1+|\lambda|^2)\theta\big|_h^2\cdot(-\log|z|)^{1-\epsilon}\cdot|dz\cdot d\bar{z}| < \infty.$$

Proof We have only to show the following claim: If $|f|_h \leq C_3\cdot|z|^{-b}\cdot(-\log|z|)^k$, then we have the finiteness:

$$\int \big|\psi(f) - (1+|\lambda|^2)\theta(f)\big|_h\cdot|z|^{2b}\cdot\big(-\log|z|\big)^{1-\epsilon-2k}\cdot|dz\cdot d\bar{z}| < \infty. \tag{7.24}$$

We have only to show the inequality (7.24) in the case $b=0$. We may assume that $|f|_h \sim \big(-\log|z|\big)^k$ for some half integer k (see Lemma 7.40). Note the following inequality:

(7.25)
$$\big|\psi(f)-(1+|\lambda|^2)\cdot\theta(f)\big|_h^2 \leq 2\cdot\big|\mathbb{D}^{\lambda}(f)-(1+|\lambda|^2)\cdot\theta(f)\big|_h^2+C\cdot|z|^{-2}\big(-\log|z|\big)^{2(k-1)}$$
$$= 2\cdot|\lambda|^2\cdot|\partial_{\mathcal{E}^\lambda}(f)|_h^2+C\cdot|z|^{-2}\cdot\big(-\log|z|\big)^{2(k-1)}.$$

We have the Weitzenbeck formula:

$$\Delta|f|_h^2 = -\big|\partial_{\mathcal{E}^\lambda}f\big|_h^2+(1+|\lambda|^2)\cdot\big(|\theta f|_h^2-|\theta^\dagger f|_h^2\big). \tag{7.26}$$

Hence we obtain the following, for some positive constant C:

$$\Delta\big(|f|_h^2\cdot(-\log|z|)^{1-\epsilon-2k}\big)\leq -\frac{1}{2}\big|\partial_{\mathcal{E}^\lambda}f\big|_h^2\cdot(-\log|z|)^{1-\epsilon-2k} + C\cdot|f|_h^2\cdot|z|^{-2}\cdot\big(-\log|z|\big)^{-1-\epsilon-2k}. \tag{7.27}$$

By using (7.27) and the equivalence of the norms $|f|_h\sim(-\log|z|)^k$, we obtain the following:

$$\Delta\Big(\big(|f|_h^2\cdot(-\log|z|)^{-2k}-C\big)\cdot(-\log|z|)^{1-\epsilon}\Big)\leq -\frac{1}{2}|\partial_{\mathcal{E}^\lambda}f|_h^2\cdot(-\log|z|)^{1-\epsilon-2k}.$$

We put as follows:

$$F:=\Big(|f|_h^2\cdot(-\log|z|)^{-2k}-C\Big)\cdot(-\log|z|)^{1-\epsilon}.$$

Then we obtain the following:

$$\lim_{|z|\to 0}\frac{F}{-\log|z|}=0.$$

Hence we obtain the inequality due to Lemma 2.2 of [**81**] (see Proposition 2.24):

$$\int|\partial_{\mathcal{E}^\lambda}f|_h^2\cdot(-\log|z|)^{1-\epsilon-2k}\cdot|dz\cdot d\bar{z}|<\infty. \tag{7.28}$$

From (7.25) and (7.28), we obtain Lemma 7.68. □

Since Simpson's proof of Lemma 7.4 and Lemma 7.5 in [**81**] (for the case $\lambda=1$) can be also easily applied to the general case $\lambda\neq 1$, we omit to give a proof of the following lemma.

LEMMA 7.69 (Lemma 7.4 and Lemma 7.5 in [**81**]). *We have the following inequalities:*

$$|(1+|\lambda|^2)\theta-\psi|_h\leq|z|^{-1}(-\log|z|)^{-1}, \tag{7.29}$$

$$|\psi-\psi_0|_h\leq|z|^{-1}(-\log|z|)^{-1}. \tag{7.30}$$

□

COROLLARY 7.70 (Lemma 7.11 in [**81**]). *In the case $\beta_j+\lambda\cdot b_j\neq\beta_{0\,i}+\lambda\cdot b_i$, we have the inequality $|I'_{i,j}|\leq(-\log|z|)^{-1}$.*

Proof It is not difficult to derive the claim from (7.30). See the proof of Lemma 7.11 of [**81**]. □

Note the relation $\overline{\partial}\Phi=\lambda\cdot(\theta_0^\dagger-\theta^\dagger)$. Then we obtain the following:

$$|\overline{\partial}I_{i\,j}|\cdot|z|^{b_{0\,i}-b_j}\cdot(-\log|z|)^{-k_{0,i}+k_j}\leq C\cdot|z|^{-1}\cdot(-\log|z|)^{-1}.$$

In the case $b_{0,i} - b_j \neq 0$, we can pick the C^∞-function $K(\overline{\partial} I_{i\,j})$ on Δ^* satisfying the following:

- $\overline{\partial} K(\overline{\partial} I_{i\,j}) = \overline{\partial} I_{i\,j}$.
- $|K(\overline{\partial} I_{i\,j})| \cdot |z|^{b_{0,i}-b_j} \cdot (-\log|z|)^{-k_{0,i}+k_j} \leq C \cdot \bigl(-\log|z|\bigr)^{-1}$.

Then we obtain the following:

- $K(\overline{\partial} I_{i\,j}) - I_{i\,j}$ is holomorphic on $X - D$.
- $\bigl|K(\overline{\partial} I_{i\,j}) - I_{i\,j}\bigr| \cdot |z|^{b_{0\,i}-b_j} \cdot (-\log|z|)^{-k_{0\,i}+k_j}$ is bounded.

Hence we obtain the following inequality in the case $b_{0\,i} - b_j \notin \mathbb{Z}$:

$$\bigl|K(\overline{\partial} I_{i\,j}) - I_{i\,j}\bigr| \cdot |z|^{b_{0\,i}-b_j} \cdot (-\log|z|)^{-k_{0\,i}+k_j} \leq C \cdot (-\log|z|)^{-1}. \tag{7.31}$$

Hence we have the inequality $|I_{i\,j}| \leq C \cdot (-\log|z|)^{-1}$ in the case $b_{0\,i} \neq b_j$. Thus we obtain the claim (2) from Corollary 7.70 and (7.31), and the outline of the proof of Proposition 7.67 is finished. □

7.3.3. Some consequences. Recall the bijection $\mathfrak{k}(\lambda) : \boldsymbol{C} \times \boldsymbol{R} \longrightarrow \boldsymbol{C} \times \boldsymbol{R}$ is defined in the subsection 2.1.7.

COROLLARY 7.71. *Let u be an element of $\mathcal{KMS}(\mathcal{E}^0)$. Then $\mathfrak{k}(\lambda, u)$ is contained in $\mathcal{KMS}(\mathcal{E}^\lambda)$, and we have the equality:*

$$\dim \operatorname{Gr}^W_k\bigl(\operatorname{Gr}^{\mathbb{E},F}_u(\mathcal{E}^0)\bigr) = \dim \operatorname{Gr}^W_k\bigl(\operatorname{Gr}^{\mathbb{E},F}_{\mathfrak{k}(\lambda,u)}(\mathcal{E}^\lambda)\bigr).$$

In particular, we obtain the bijective morphism $\mathfrak{k}(\lambda) : \mathcal{KMS}(\mathcal{E}^0) \longrightarrow \mathcal{KMS}(\mathcal{E}^\lambda)$ and the following equality:

$$\dim\bigl(\operatorname{Gr}^{\mathbb{E},F}_u(\mathcal{E}^0)\bigr) = \dim\bigl(\operatorname{Gr}^{\mathbb{E},F}_{\mathfrak{k}(\lambda,u)}(\mathcal{E}^\lambda)\bigr), \quad \mathfrak{m}(0,u) = \mathfrak{m}\bigl(\lambda, \mathfrak{k}(\lambda,u)\bigr).$$

Proof The claims for the model bundles can be checked by direct calculations. Then the claims for general tame harmonic bundles follow from Proposition 7.67. □

COROLLARY 7.72. *Let $\boldsymbol{e}$ be a holomorphic frame of ${}^\diamond E$ compatible with $\mathbb{E}$, F and W. Let $\boldsymbol{v}$ be a holomorphic frame of ${}^\diamond\mathcal{E}^\lambda$ compatible with $\mathbb{E}$, F and W. We put as follows:*

$$b(e_j) := \deg^F(e_j), \quad k(e_j) := \frac{\deg^W(e_j)}{2}, \quad b(v_i) := \deg^F(v_i), \quad k(v_i) := \frac{\deg^W(v_i)}{2}.$$

We put as follows:

$$e'_j := e_j \cdot |z|^{b(e_j)} \cdot (-\log|z|)^{-k(e_j)}, \quad v'_i := v_i \cdot |z|^{b(v_i)} \cdot (-\log|z|)^{-k(v_i)}.$$

The C^∞-function $B : X - D \longrightarrow M(r)$ is given as follows:

$$v'_i = \sum B'_{j\,i} \cdot e'_j.$$

Then we have the following:

- *B' and B'^{-1} are bounded.*
- *$|B'_{j\,i}| \leq C \cdot (-\log|z|)^{-1}$ in the case $\deg^{F,\mathbb{E}}(v_i) \neq \mathfrak{k}(\lambda, \deg^{F,\mathbb{E}}(e_j))$.*

Proof For the case of model bundles, we pick $\boldsymbol{e}_0$, $\boldsymbol{e}'_0$, $\boldsymbol{v}_0$ and $\boldsymbol{v}'_0$ similarly, and then we obtain $B'_{0,j\,i}$. In this case, we may assume the following, due to the construction of model bundles:

(A): $B'_{0,j\,i} = 0$ if $\deg^{F,\mathbb{E}}(v_{0\,i}) \neq \mathfrak{k}(\lambda, \deg^{F,\mathbb{E}}(e_{0\,j}))$.

From our construction of Φ, we may assume $\Phi(\boldsymbol{e}_0) = \boldsymbol{e}$. Then we obtain our claims due to the assumption (A) above and the claim (2) in Proposition 7.67. □

7.4. Multi-valued flat sections

7.4.1. Order of multi-valued flat sections. We put $X = \Delta$ and $D = \{O\}$. Let $(E, \overline{\partial}_E, \theta, h)$ be a tame harmonic bundle over $X - D$. Let us fix $\lambda \in \boldsymbol{C}^*$. We have the λ-connection $(\mathcal{E}^\lambda, \mathbb{D}^\lambda)$. We have the associated flat connection $\mathbb{D}^{\lambda,f}$. Then we obtain the space of multi-valued flat sections, which we denote by $H(\mathcal{E}^\lambda)$.

LEMMA 7.73. *For any $s \in H(\mathcal{E}^\lambda)$ and any positive number C_0, there exist positive constants C_1 and b satisfying the following:*

$$|s|_h \leq C_1 \cdot |z|^{-b} \qquad \text{on } \big\{z \,\big|\, |\arg z| < C_0\big\}.$$

Proof It follows from tameness of our harmonic bundles. (See Remark in the page 732 of [**81**], for example. Or it is not difficult to show directly.) □

We have the universal covering map $\pi : \mathbb{H} \longrightarrow \Delta^*$, given by $\zeta = x + \sqrt{-1}y \longmapsto \exp(\sqrt{-1}\zeta) = z$. We may regard $s \in H(\mathcal{E}^\lambda)$ as a flat section of $\pi^*\mathcal{E}^\lambda$. We have the following equality:

$$\frac{\partial h(s,s)}{\partial x} = 2\operatorname{Re}\Big(h(\nabla^{\lambda,u}_{\partial_x} s, s)\Big) = 2\operatorname{Re}\Big(h((\nabla^{\lambda,u}_{\partial_x} - \mathbb{D}^{\lambda,f}_{\partial_x})s,\ s)\Big).$$

Here $\nabla^{\lambda,u}$ denote the unitary connection for $(\mathcal{E}^\lambda, h)$, and ∂_x denote the vector field $\partial/\partial x$. The difference $\mathbb{D}^{\lambda,f} - \nabla^{\lambda,u}$ is given by $a \cdot \theta + b \cdot \theta^\dagger$ $(a, b \in \boldsymbol{C})$. We have the description $\theta = \theta^\zeta \cdot d\zeta$ and $\theta^\dagger = \theta^{\zeta\,\dagger} d\bar{\zeta}$. Due to Simpson's main estimate (Proposition 7.2 (I)), we have the boundedness $|\theta^\zeta|_h \leq C$ and $|\theta^{\zeta\,\dagger}|_h \leq C$. Hence we obtain the following inequality for some positive constant C:

$$\left|\frac{\partial}{\partial x}\log|s|_h^2\right| = \left|\frac{\partial h(s,s)}{\partial x}\cdot h(s,s)^{-1}\right| \leq 2|a|\cdot\big|\theta^\zeta\big|_h + 2|b|\cdot\big|\theta^{\zeta\,\dagger}\big|_h \leq C. \tag{7.32}$$

LEMMA 7.74. *For any positive number C_1, there exists a positive constants C_2 such that the following holds:*

- *For any $x_i \in \boldsymbol{R}$ $(i = 1, 2)$ such that $|x_i| < C_1$, and for any $y > 1$, the following inequality holds:*

$$\Big|\log\big|s(x_1, y)\big|_h^2 - \log\big|s(x_2, y)\big|_h^2\Big| \leq C_2.$$

Proof It immediately follows from the inequality (7.32). □

DEFINITION 7.75. Let s be an element of $H(\mathcal{E}^\lambda)$, b be any real number. Then '$-\operatorname{ord}(s) \leq b$' means the following:

- Pick any real number $x_1 \in \boldsymbol{R}$. For any positive number $\epsilon > 0$, there exists a positive constant C such that $|s(x_1, y)| \leq C \cdot e^{(b+\epsilon)\cdot y}$.

Note that such property does not depend on a choice of x_1, due to Lemma 7.74. □

7.4.2. KMS-structure of $H(\mathcal{E}^\lambda)$.

DEFINITION 7.76. We put as follows:

$$\mathcal{F}_b(H(\mathcal{E}^\lambda)) := \big\{s \in H(\mathcal{E}^\lambda) \,\big|\, -\operatorname{ord}(s) \leq b\big\}.$$

Thus we obtain the filtration $\mathcal{F}$ on $H(\mathcal{E}^\lambda)$. □

Let M^{λ} be the monodromy on $H(\mathcal{E}^{\lambda})$.

LEMMA 7.77. *The filtration $\mathcal{F}$ is preserved by M^{λ}. In particular, $\mathcal{F}$ is compatible with the generalized eigen decomposition of M^{λ}.*

Proof It is clear from our definition of the filtration $\mathcal{F}$. □

For $u = (a, \alpha) \in \boldsymbol{R} \times \boldsymbol{C}^*$, we put as follows:

$$\mathrm{Gr}_u^{\mathcal{F},\mathbb{E}} H(\mathcal{E}^{\lambda}) := \mathrm{Gr}_a^{\mathcal{F}} \mathbb{E}\big(H(\mathcal{E}^{\lambda}), \alpha\big).$$

We put as follows:

$$\overline{\mathcal{KMS}}^f(\mathcal{E}^{\lambda}) := \Big\{ u \in \boldsymbol{R} \times \boldsymbol{C}^* \,\Big|\, \mathrm{Gr}_u^{\mathcal{F},\mathbb{E}} H(\mathcal{E}^{\lambda}) \neq 0 \Big\},$$

$$\mathcal{P}ar^f(\mathcal{E}^{\lambda}) := \Big\{ a \in \boldsymbol{R} \,\Big|\, \mathrm{Gr}_a^{\mathcal{F}} H(\mathcal{E}^{\lambda}) \neq 0 \Big\},$$

$$\mathcal{S}p^f(\mathcal{E}^{\lambda}) := \mathcal{S}p(M^{\lambda}) = \Big\{ \alpha \in \boldsymbol{C}^* \,\Big|\, \mathbb{E}\big(H(\mathcal{E}^{\lambda}), \alpha\big) \neq 0 \Big\}$$

We put $\mathfrak{m}^f(\lambda, u) := \dim \mathrm{Gr}_u^{\mathcal{F},\mathbb{E}} H(\mathcal{E}^{\lambda})$ for $u \in \boldsymbol{R} \times \boldsymbol{C}$, which is called the multiplicity. Similarly, $\mathfrak{m}^f(\lambda, a)$ and $\mathfrak{m}^f(\lambda, \alpha)$ are given for $a \in \boldsymbol{R}$ and $\alpha \in \boldsymbol{C}^*$.

7.4.3. Compatibility of the orders for holomorphic sections and flat sections. Let us consider the generalized eigen decomposition of $H(\mathcal{E}^{\lambda})$ with respect to the monodromy M^{λ}:

$$H(\mathcal{E}^{\lambda}) = \bigoplus_{\omega \in Sp(M^{\lambda})} \mathbb{E}\big(H(\mathcal{E}^{\lambda}), \omega\big).$$

Let M_{ω}^{λ} denote the restriction of M^{λ} to $\mathbb{E}\big(H(\mathcal{E}^{\lambda}), \omega\big)$.

Pick a real number $b \in \boldsymbol{R}$. Then there exists the unique complex number $\alpha = \alpha(b, \omega)$ satisfying the following:

- $\exp\big(-2\pi\sqrt{-1} \cdot \alpha\big) = \omega$.
- $b \leq \mathrm{Re}(\alpha) < b + 1$.

Let $M_{\omega}^{\lambda,u}$ denote the unipotent part of M_{ω}^{λ}, and we put as follows:

$$N_{\omega}^{\lambda} := \frac{-1}{2\pi\sqrt{-1}} \log M_{\omega}^{\lambda,u} = \frac{-1}{2\pi\sqrt{-1}} \sum_{n=1}^{\infty} \frac{(-1)^{n-1}}{n} (M_{\omega}^{\lambda,u} - 1)^n.$$

Then we have the following:

$$\exp\Big(-2\pi\sqrt{-1}\big(\alpha(b, \omega) + N_{\omega}^{\lambda}\big)\Big) = M_{\omega}^{\lambda}.$$

Let s be an element of $\mathbb{E}\big(H(\mathcal{E}^{\lambda}), \omega\big)$. We put as follows:

$$F(s, b) := \exp\Big(\sqrt{-1}\zeta \cdot \big(\alpha(b, \omega) + N_{\omega}^{\lambda}\big)\Big) \cdot s = \exp\Big(\log z \cdot \big(\alpha(b, \omega) + N_{\omega}^{\lambda}\big)\Big) \cdot s. \tag{7.33}$$

Then $F(s, b)$ induces the holomorphic section of $X - D$.

LEMMA 7.78. *Let s be an element of $\mathbb{E}\big(H(\mathcal{E}^{\lambda}), \omega\big)$. We have the following equality:*

$$-\operatorname{ord}\big(F(s, b)\big) = -\operatorname{ord}(s) - \mathrm{Re}\big(\alpha(b, \omega)\big). \tag{7.34}$$

Proof We take the multi-valued flat sections s_1 $(i = 1, 2, \ldots)$ as $s_h = (N_\omega^\lambda)^h \cdot s$. We have the following equality, for some l:

$$F(s,b) = z^{\alpha(b,\omega)} \cdot \Big(s + \sum_{i=1}^{l} \frac{1}{i!} s_i\Big). \tag{7.35}$$

Hence we obtain $-\operatorname{ord}(F(s,b)) \leq -\operatorname{ord}(s) - \operatorname{Re}(\alpha(b,\omega))$.

Let us consider the case that $s \in \mathcal{F}_{b_1}$ and $N_\omega s \in \mathcal{F}_{<b_1}$. Then s_i $(i = 1, \ldots, l)$ above are contained in $\mathcal{F}_{<b_1}$, and thus we have $-\operatorname{ord}(F(s,b)) = -\operatorname{ord}(s) - \operatorname{Re}\big(\alpha(b,\omega)\big)$.

Assume $s \in \mathcal{F}_{b_1}$. The number $i(s)$ is determined for s by the following condition:

$$s_{i(s)} \notin \mathcal{F}_{<b_1}, \quad s_{i(s)+1} \in \mathcal{F}_{<b_1}.$$

To show the equality (7.34), we use an induction on $i(s)$. If $i(s) = 0$, then we have already shown the claim. We assume that the claim holds for any s such that $i(s) < i_0$, and we will show the claim for s such that $i(s) = i_0$. Note that $i(N_\omega s) = i_0 - 1$, and we have the equality $-\operatorname{ord}(F(N_\omega s)) = -\operatorname{ord}(N_\omega s) - \operatorname{Re}(\alpha(b,\omega)) = b_3$ due to the hypothesis of the induction. Note the following equality:

$$\mathbb{D}^{\lambda,f}(F(s,b)) = \alpha(b,\omega) \cdot F(s,b) \cdot \frac{dz}{z} + F(N_\omega \cdot s, b) \cdot \frac{dz}{z}.$$

Assume $b_2 := -\operatorname{ord}(F(s,b)) < -\operatorname{ord}(s) - \operatorname{Re}(\alpha(b,\omega)) =: b_3$, and we will derive the contradiction. Note that $\mathbb{D}^{\lambda,f} F(s,b)$ and $\alpha(b,\omega) \cdot F(s,b)\frac{dz}{z}$ are sections of ${}_{b_2+1}\mathcal{E} \otimes \Omega_X^{1,0}$. On the other hand, $F(N_\omega \cdot s, b)\frac{dz}{z}$ is a section of ${}_{b_3+1}\mathcal{E} \otimes \Omega_X^{1,0}$ such that it is not 0 in $\operatorname{Gr}^F_{b_3+1}$. Hence we have arrived at the contradiction. It implies $-\operatorname{ord}(F(s,b)) = -\operatorname{ord}(s) - \operatorname{Re}(\alpha(b,\omega))$, and thus the induction on $i(s)$ can proceed. Hence we are done. □

7.4.4. The compatibility of the KMS-structures at λ. Let us pick real numbers c and a, and a complex number $\omega \in \boldsymbol{C}$. Recall we have the following inequality, by definition:

$$c - a - 1 < -\operatorname{Re}\big(\alpha(a-c,\omega)\big) \leq c - a.$$

In the case $-\operatorname{ord}(s) \leq a$, we obtain the following:

$$-\operatorname{ord}\big(F(s, a-c)\big) = -\operatorname{ord}(s) - \operatorname{Re}\big(\alpha(a-c,\omega)\big) \leq a - \operatorname{Re}\big(\alpha(a-c,\omega)\big) \leq c.$$

Hence $F(s, a-c)$ gives a section of ${}_c\mathcal{E}^\lambda$.

We put $d(a,\omega) := a - \operatorname{Re}\big(\alpha(a,\omega)\big) < 0$. Then the morphism $\mathcal{F}_a\mathbb{E}\big(H(\mathcal{E}^\lambda),\omega\big) \longrightarrow {}_{d(a,\omega)}\mathcal{E}^\lambda$ is given by the correspondence $s \longmapsto F(s,a)$, and we obtain the following induced morphism:

$$\varphi_{(a,\omega)} : \operatorname{Gr}^{\mathcal{F}}_a \mathbb{E}\big(H(\mathcal{E}^\lambda),\omega\big) \longrightarrow \operatorname{Gr}^F_{d(a,\omega)}({}^\diamond\mathcal{E}^\lambda{}_{|O}).$$

LEMMA 7.79. *The morphism $\varphi_{(a,\omega)}$ is injective.*

Proof It follows from Lemma 7.78. □

We have the action of $\operatorname{Gr}^{\mathcal{F}}_a(N_\omega^\lambda)$ on $\operatorname{Gr}^{\mathcal{F}}_a \mathbb{E}(H(\mathcal{E}^\lambda),\omega)$ for each a. Then we obtain the following endomorphism:

$$\alpha(a,\omega) + \operatorname{Gr}^{\mathcal{F}}_a(N_\omega^\lambda) \in \operatorname{End}\Big(\operatorname{Gr}^{\mathcal{F}}_a \mathbb{E}\big(H(\mathcal{E}^\lambda),\omega\big)\Big).$$

On the other hand, the endomorphism $\mathrm{Res}(\mathbb{D}^{\lambda,f})$ on ${}^{\diamond}\mathcal{E}^{\lambda}{}_{|O}$ induces the following endomorphism:

$$\mathrm{Gr}^F_{d(a,\omega)}\big(\mathrm{Res}(\mathbb{D}^{\lambda,f})\big) \in \mathrm{End}\Big(\mathrm{Gr}^F_{d(a,\omega)}\big({}^{\diamond}\mathcal{E}^{\lambda}{}_{|O}\big)\Big).$$

LEMMA 7.80. *We have the following equality:*

$$\varphi_{(a,\omega)} \circ \Big(\alpha(a,\omega) + \mathrm{Gr}^{\mathcal{F}}_a(N^{\lambda}_{\omega})\Big) = \mathrm{Gr}^F_{d(a,\omega)}(\mathrm{Res}(\mathbb{D}^{\lambda,f})) \circ \varphi_{(a,\omega)}.$$

Proof We have the following morphism due to $F(\cdot,a)$:

$$\Big(\mathcal{F}_a\mathbb{E}(H(\mathcal{E}^{\lambda}),\omega),\ \alpha(a,\omega) + \mathrm{Gr}^{\mathcal{F}}_a(N^{\lambda}_{\omega})\Big) \longrightarrow \Big(\Gamma(X, {}_d\mathcal{E}^{\lambda}),\ \mathbb{D}^{\lambda,f}\Big).$$

Here we put $d := d(a,\omega)$. We also have the following morphisms:

$$\Gamma(X, {}_d\mathcal{E}^{\lambda}) \longrightarrow F_d({}^{\diamond}\mathcal{E}^{\lambda}{}_{|O}) \longrightarrow \mathrm{Gr}^F_d({}^{\diamond}\mathcal{E}^{\lambda}{}_{|O}).$$

These morphisms are equivariant with respect to the operators $\mathbb{D}^{\lambda,f}$, $\mathrm{Res}(\mathbb{D}^{\lambda,f})$ and $\mathrm{Gr}^F_d(\mathrm{Res}(\mathbb{D}^{\lambda,f}))$ respectively. Thus we are done. □

COROLLARY 7.81. *We have the following implication in* $\mathrm{Gr}^F_d({}^{\diamond}\mathcal{E}^{\lambda}{}_{|O})$*:*

$$\mathrm{Im}(\varphi_{(a,\omega)}) \subset \mathbb{E}\Big(\mathrm{Gr}^F_d(\mathrm{Res}(\mathbb{D}^{\lambda,f})),\ \alpha(a,\omega)\Big) = \mathbb{E}\big(\mathrm{Gr}^F_d(\mathrm{Res}(\mathbb{D}^{\lambda})),\ \lambda\cdot\alpha(a,\omega)\big).$$

Here we put $d := d(a,\omega) = a - \mathrm{Re}(\alpha(a,\omega)) < 0$.

Proof It immediately follows from Lemma 7.80. □

For any $u_1 = (a,\omega) \in \boldsymbol{R}\times\boldsymbol{C}^*$, we put as follows:

$$\mathfrak{t}(u_1) = \big(d(a,\omega),\ \lambda\cdot\alpha(a,\omega)\big).$$

Thus we obtain the bijective map $\mathfrak{t}: \boldsymbol{R}\times\boldsymbol{C}^* \longrightarrow]-1,0]\times\boldsymbol{C}$.

Recall that we put $\mathrm{Gr}^{\mathcal{F},\mathbb{E}}_{u_1}(H(\mathcal{E}^{\lambda})) := \mathrm{Gr}^{\mathcal{F}}_a\,\mathbb{E}\big(H(\mathcal{E}^{\lambda}),\omega\big)$ for any $u_1 = (a,\omega) \in \boldsymbol{R}\times\boldsymbol{C}^*$. Then we obtain the injection, due to Lemma 7.79 and Corollary 7.81:

$$\varphi_{u_1} : \mathrm{Gr}^{\mathcal{F},\mathbb{E}}_{u_1}(H(\mathcal{E}^{\lambda})) \longrightarrow \mathrm{Gr}^{F,\mathbb{E}}_{\mathfrak{t}(u_1)}({}^{\diamond}\mathcal{E}^{\lambda}).$$

Thus we obtain the following injection:

$$\bigoplus_{u_1\in\boldsymbol{R}\times\boldsymbol{C}^*}\varphi_{u_1} : \bigoplus_{u_1\in\boldsymbol{R}\times\boldsymbol{C}^*}\mathrm{Gr}^{\mathcal{F},\mathbb{E}}_{u_1}(H(\mathcal{E}^{\lambda})) \longrightarrow \bigoplus_{u\in\mathcal{KMS}({}^{\diamond}\mathcal{E}^{\lambda})}\mathrm{Gr}^{F,\mathbb{E}}_u({}^{\diamond}\mathcal{E}^{\lambda}).$$

PROPOSITION 7.82. *The morphism* $\bigoplus_{u_1\in\boldsymbol{R}\times\boldsymbol{C}^*}\varphi_{u_1}$ *is isomorphic. Each* φ_{u_1} *is isomorphic.*

Proof We have already known that φ_{u_1} are injective (Lemma 7.79). We have the following equalities:

$$\sum_{u_1}\dim\mathrm{Gr}^{\mathcal{F},\mathbb{E}}_{u_1}H(\mathcal{E}^{\lambda}) = \mathrm{rank}\,\mathcal{E}^{\lambda} = \sum_{u\in\mathcal{KMS}({}^{\diamond}\mathcal{E}^{\lambda})}\dim\mathrm{Gr}^{F,\mathbb{E}}_u({}^{\diamond}\mathcal{E}^{\lambda}).$$

The claims follow from the equalities above. □

We have the weight filtration W on $\mathrm{Gr}^{\mathcal{F},\mathbb{E}}_u H(\mathcal{E}^{\lambda})$ induced by the nilpotent map $\mathrm{Gr}^{\mathcal{F}}_a(N_{\omega})$, where we have $u = (a,\omega)$.

LEMMA 7.83. *The morphism* φ_u *preserves the weight filtrations* W.

Proof It immediately follows from Lemma 7.80. □

COROLLARY 7.84. *The map* $\mathfrak{t}$ *gives the bijection* $\overline{\mathcal{KMS}}^f(\mathcal{E}^\lambda) \longrightarrow \mathcal{KMS}({}^\diamond\mathcal{E}^\lambda)$. *It preserves the multiplicity, i.e.,* $\mathfrak{m}^f(\lambda, u) = \mathfrak{m}(\lambda, \mathfrak{t}(u))$. □

7.4.5. The correspondence of KMS-spectrum. We will use the maps $\mathfrak{p}^f(\lambda)$, $\mathfrak{e}^f(\lambda)$ and $\mathfrak{k}^f(\lambda)$ given in the subsection 2.1.7. We put as follows for any real number c:

$$\mathcal{K}(\mathcal{E}, \lambda, c) := \{u \in \mathcal{KMS}(\mathcal{E}^0) \mid c - 1 \le \mathfrak{p}(\lambda, u) < c\}. \tag{7.36}$$

Then we have the isomorphism $\mathfrak{k}(\lambda) : \mathcal{K}(\mathcal{E}, \lambda, c) \longrightarrow \mathcal{KMS}({}_c\mathcal{E}^\lambda)$. Let us consider the case $c = 0$.

LEMMA 7.85. *Let* u *be an element of* $\mathcal{K}(\mathcal{E}, \lambda, 0)$. *We have the relation* $\mathfrak{k}(\lambda, u) = \mathfrak{t}(\mathfrak{k}^f(\lambda, u))$. *(See the subsection 7.4.4.)*

Proof From the formula (2.2) and the inequality $-1 < \mathfrak{p}(\lambda, u) \le 0$, we have the following:

$$\mathfrak{e}^f(\lambda, u) = \exp\left(-2\pi\sqrt{-1}\lambda^{-1} \cdot \mathfrak{e}(\lambda, u)\right), \quad \mathfrak{p}^f(\lambda, u) \le \mathrm{Re}\left(\lambda^{-1} \cdot \mathfrak{e}(\lambda, u)\right) < \mathfrak{p}^f(\lambda, u) + 1.$$

Thus we obtain the following, by definition of $\alpha(b, \omega)$:

$$\lambda^{-1} \cdot \mathfrak{e}(\lambda, u) = \alpha\left(\mathfrak{p}^f(\lambda, u), \mathfrak{e}^f(\lambda, u)\right).$$

We also obtain the following:

$$\mathfrak{p}(\lambda, u) = \mathfrak{p}^f(\lambda, u) - \mathrm{Re}\left(\alpha(\mathfrak{p}^f(\lambda, u)), \mathfrak{e}^f(\lambda, u)\right) = d\left(\mathfrak{p}^f(\lambda, u), \mathfrak{e}^f(\lambda, u)\right).$$

It means $\mathfrak{k}(\lambda, u) = \mathfrak{t}\left(\mathfrak{k}^f(\lambda, u)\right)$. □

LEMMA 7.86. *The image of* $\mathcal{KMS}(\mathcal{E}^0)$ *via the morphism* $\mathfrak{k}^f(\lambda)$ *is* $\overline{\mathcal{KMS}}^f(\mathcal{E}^\lambda)$, *and we have the equality:* $\mathfrak{m}^f(\lambda, \mathfrak{k}^f(\lambda, u)) = \mathfrak{m}(\lambda, \mathfrak{k}(\lambda, u))$.

Proof From Proposition 7.82 and Lemma 7.85, the image of $\mathcal{K}(\mathcal{E}, \lambda, 0)$ via the morphism $\mathfrak{k}^f(\lambda)$ is same as $\overline{\mathcal{KMS}}^f(\mathcal{E}^\lambda)$, and we have the equality of the multiplicity. Note that we have the equalities $\mathfrak{k}^f(\lambda, u + (1, 0)) = \mathfrak{k}^f(\lambda, u)$ and $\mathfrak{m}(\lambda, \mathfrak{k}(\lambda, u)) = \mathfrak{m}\left(\lambda, \mathfrak{k}(\lambda, u + (1, 0))\right)$. Thus we are done. □

LEMMA 7.87.

(1) *We have the* $\mathbb{Z}$*-action on* $\mathcal{KMS}(\mathcal{E}^0)$. *The map* $\mathfrak{k}^f(\lambda)$ *induces the isomorphism* $\overline{\mathcal{KMS}}(\mathcal{E}^0) \simeq \overline{\mathcal{KMS}}^f(\mathcal{E}^\lambda)$, *which is also denoted by* $\mathfrak{k}^f(\lambda)$.
(2) *We have* $\mathfrak{m}(0, u) = \mathfrak{m}^f(\lambda, \mathfrak{k}^f(\lambda, u))$.

Proof It follows from Corollary 7.71 and Lemma 7.86. □

7.4.6. Norm estimate for the multi-valued flat sections. Let $\boldsymbol{s} = (s_i)$ be a frame of $H(\mathcal{E}^\lambda)$ satisfying the following conditions:

CONDITION 7.88.

(1) It is compatible with $\mathbb{E}$. We put $\deg^{\mathbb{E}}(s_i) = \omega_i$.
(2) It is compatible with $\mathcal{F}$ on $\mathbb{E}\left(H(\mathcal{E}^\lambda), \omega\right)$ for any ω. We put $\deg^{\mathcal{F}}(s_i) = a_i$.
(3) From the conditions (1) and (2) above, we obtain the induced frame $\boldsymbol{s}^{(1)}$ on $\mathrm{Gr}^{\mathcal{F},\mathbb{E}} H(\mathcal{E}^\lambda)$. The frame $\boldsymbol{s}^{(1)}$ is compatible with the weight filtration W. We put $\deg^W(s_i^{(1)}) = k_i$.

If the conditions above are satisfied, we say that $\boldsymbol{s}$ is compatible with $\mathcal{F}$, $\mathbb{E}$ and W. □

The matrix $(b^{(n)}_{j\,i})$ is determined by the following condition:

$$(N^{\lambda})^n s_i = \sum_j b^{(n)}_{j\,i} \cdot s_j.$$

LEMMA 7.89. *We have* $b^{(n)}_{j\,i} = 0$ *in the following cases:*

- $\omega_i \neq \omega_j$.
- $\omega_i = \omega_j$ *and* $a_i < a_j$.
- $(\omega_i, a_i) = (\omega_j, a_j)$ *and* $k_i - n < k_j$.

Proof It is clear from our choice of $\boldsymbol{s}$. □

We put $v_i := F(s_i, a_i)$ and $\boldsymbol{v} = (v_i)$. It gives a tuple of sections of ${}^{\diamond}\mathcal{E}^{\lambda}$. We put $\alpha_i = \alpha(a_i, \omega_i)$.

LEMMA 7.90. *The tuple of sections* $\boldsymbol{v}$ *is compatible with the generalized eigen decomposition* $\mathbb{E}$*, the parabolic filtration* F *and the weight filtration* W *of* ${}^{\diamond}\mathcal{E}_{|O}$.

Proof It immediately follows from Proposition 7.82 and Lemma 7.80. □

LEMMA 7.91. *We have the following equality:*

$$(N^{\lambda})^n v_i = \sum_{\substack{\omega_i=\omega_j,\\ a_i \geq a_j}} b^{(n)}_{j\,i} \cdot z^{\alpha_i-\alpha_j} \cdot v_j.$$

Note that if $\omega_i = \omega_j$ *and* $a_i \geq a_j$*, then* $\alpha_i - \alpha_j$ *is a non-negative integer.*

Proof We have the following equalities:

$$(7.37)\quad (N^{\lambda})^n v_i = F\big((N^{\lambda})^n s_i, -a_i\big) = \sum_{\substack{\omega_i=\omega_j,\\ a_i \geq a_j}} b^{(n)}_{j\,i} \cdot F(s_j, -a_i) = \sum_{\substack{\omega_i=\omega_j,\\ a_i \geq a_j}} b^{(n)}_{j\,i} \cdot z^{\alpha_i-\alpha_j} \cdot F(s_j, -a_j).$$

Thus we are done. □

Let C be any positive number. We put $\alpha_i := \alpha(a_i, \omega_i)$. On the region $\big\{z \,\big|\, |\arg z| < C\big\}$, we have the following equalities:

$$(7.38)\qquad s_i = z^{-\alpha_i} \cdot \exp(-\log z \cdot N^{\lambda}) \cdot v_i = z^{-\alpha_i} \sum_{n=0}^{\infty} \frac{1}{n!}(-\log z)^n \cdot (N^{\lambda})^n \cdot v_i.$$

It can be described as $s_i = z^{-\alpha_i} \sum f_j \cdot v_j$ for some multi-valued holomorphic functions f_j.

LEMMA 7.92. *Let* C *be a positive number. We have the following, on the region* $\big\{|\arg z| < C\big\}$*:*

- $f_i = 1$.
- $|f_j| \leq C' \cdot (-\log|z|)^{(k_i-k_j)/2}$ *for some positive constant* C'*, in the case* $(a_i, \omega_i) = (a_j, \omega_j)$ *and* $k_i > k_j$.
- $|f_j| \leq C' \cdot (-\log|z|)^{M}$ *for some positive constants* C' *and* M*, in the case* $\omega_i = \omega_j$ *and* $a_i > a_j$.
- *Otherwise,* f_j *vanishes identically.*

Proof It immediately follows from Lemma 7.91 and (7.38). □

We put $s_i' := s_i \cdot |z|^{a_i} \cdot (-\log|z|)^{-k_i/2}$, and $\boldsymbol{s}' := (s_i')$.

PROPOSITION 7.93. *$\boldsymbol{s}'$ is adapted on the region $\{s \,\big|\, |\arg z| < C\}$ for any positive constant C.*

Proof It is a direct corollary of Lemma 7.92 and the adaptedness of $\boldsymbol{v}'$. □

7.4.7. The decomposition and the filtration of the flat bundle $\mathcal{E}^\lambda$. We have the generalized eigen decomposition over $X - D$ for the monodromy M^λ:

$$\mathcal{E}^\lambda = \bigoplus_{\omega \in \mathcal{S}p(M^\lambda)} \mathbb{E}(\mathcal{E}^\lambda, \omega). \tag{7.39}$$

COROLLARY 7.94. *The decomposition is quasi adapted (Definition 2.6).*

Proof Let $\boldsymbol{v}$ be the frame of ${}^\diamond\mathcal{E}^\lambda$ obtained from $\boldsymbol{s}$ as in the subsection 7.4.6. Then $\boldsymbol{v}$ is compatible with the generalized eigen decomposition above, and $\boldsymbol{v}'$ is adapted. Thus the decomposition is quasi adapted. □

Let c be any real number. Let $\boldsymbol{s}$ be a frame of $H(\mathcal{E}^\lambda)$, compatible with $\mathbb{E}$, $\mathcal{F}$ and W. We put ${}_cv_i := F(s_i, a_i - c)$, and ${}_c\boldsymbol{v} = ({}_cv_i)$. We put ${}_c\alpha_i := \alpha(a_i - c_i, \omega_i)$.

LEMMA 7.95. *The tuple of sections ${}_c\boldsymbol{v}$ is a frame of ${}_c\mathcal{E}^\lambda$ compatible with F, $\mathbb{E}$ and W. We have the following:*

$$\mathbb{D}^{\lambda,f}({}_cv_i) = \Big({}_c\alpha_i \cdot {}_cv_i + \sum_{\substack{\omega_i = \omega_j,\\ a_i \geq a_j}} b_{j\,i}^{(1)} \cdot z^{{}_c\alpha_i - {}_c\alpha_j} \cdot {}_cv_j\Big) \cdot \frac{dz}{z}. \tag{7.40}$$

Proof Since $\boldsymbol{v}$ is a frame of ${}^\diamond\mathcal{E}^\lambda$ compatible with the parabolic filtration, it is easy to check ${}_c\boldsymbol{v}$ is a frame of ${}_c\mathcal{E}$, compatible with the parabolic filtration. The equality (7.40) follows from the following:

$$\begin{aligned} \mathbb{D}^{\lambda,f}({}_cv_i) = \mathbb{D}^{\lambda,f}\big(z^{{}_c\alpha_i} \cdot \exp(\log z \cdot N_{\omega_i}^\lambda) \cdot s_i\big) &= \big({}_c\alpha_i + N_{\omega_i}^\lambda\big) \cdot {}_cv_i \cdot \frac{dz}{z} \\ &= \Big({}_c\alpha_i \cdot {}_cv_i + \sum_{\substack{\omega_i = \omega_j,\\ a_i \geq a_j}} b_{j\,i}^{(1)} \cdot z^{{}_c\alpha_i - {}_c\alpha_j} \cdot {}_cv_j\Big) \frac{dz}{z}. \end{aligned} \tag{7.41}$$

The compatibility of ${}_c\boldsymbol{v}$ with $\mathbb{E}$ and W follows from the formula (7.40). □

COROLLARY 7.96. *The decomposition (7.39) is prolonged to the following:*

$${}_c\mathcal{E}^\lambda = \bigoplus_{\omega \in \mathcal{S}p(M^\lambda)} {}_c\mathbb{E}(\mathcal{E}^\lambda, \omega).$$

In particular, ${}_c\mathbb{E}(\mathcal{E}^\lambda, \omega)$ is locally free.

Proof It is easy to check the claim by using the frame ${}_c\boldsymbol{v}$. □

We obtain two decomposition of ${}_c\mathcal{E}^\lambda{}_{|O}$:

$${}_c\mathcal{E}^\lambda{}_{|O} = \bigoplus_\beta \mathbb{E}\big(\operatorname{Res}\mathbb{D}^\lambda, \beta\big) = \bigoplus_{\omega \in \mathcal{S}p(M^\lambda)} {}_c\mathbb{E}(\mathcal{E}^\lambda, \omega)_{|O}.$$

COROLLARY 7.97. *The following holds:*

$$ {}_c\mathbb{E}(\mathcal{E}^{\lambda},\omega)_{|O} = \bigoplus_{\exp(-2\pi\sqrt{-1}\lambda^{-1}\cdot\beta)=\omega} \mathbb{E}\big(\operatorname{Res}\mathbb{D}^{\lambda},\beta\big). $$

Proof It immediately follows from the formula (7.40). □

The filtration $\mathcal{F}$ on $\mathbb{E}(H(\mathcal{E}^{\lambda}),\omega)$ induces the filtration $\mathcal{F}\big(\mathbb{E}(\mathcal{E}^{\lambda},\omega)\big)$ on $\mathbb{E}(\mathcal{E}^{\lambda},\omega)$.

LEMMA 7.98. *The filtration $\mathcal{F}\big(\mathbb{E}(\mathcal{E}^{\lambda},\omega)\big)$ can be prolonged to the filtration ${}_c\mathcal{F}\big(\mathbb{E}(\mathcal{E}^{\lambda},\omega)\big)$. We have the following equality:*

$$ {}_c\mathcal{F}_a\big(\mathbb{E}(\mathcal{E}^{\lambda},\omega)\big) = \big\langle {}_cv_i \,\big|\, \omega_i=\omega, a_i\leq a\big\rangle. $$

We also have ${}_c\mathcal{F}_a(\mathcal{E}^{\lambda}) = \bigoplus_{\omega} {}_c\mathcal{F}_a\big(\mathbb{E}(\mathcal{E}^{\lambda},\omega)\big)$.

Proof It is clear from our definition. □

Then we obtain the two filtrations on ${}_c\mathcal{E}^{\lambda}{}_{|O}$, one is $F({}_c\mathcal{E}^{\lambda}{}_{|O})$ and the other is ${}_c\mathcal{F}(\mathcal{E}^{\lambda})_{|O}$.

LEMMA 7.99. *We have the following relation:*

$$ {}_c\mathcal{F}_a\big(\mathbb{E}(\mathcal{E}^{\lambda},\omega)\big)_{|O} = F_{d(a,\omega)}\big(\mathbb{E}({}_c\mathcal{E}^{\lambda}{}_{|O},\lambda\cdot\alpha(a-c,\omega))\big) \oplus \bigoplus_{\substack{\exp(-2\pi\sqrt{-1}\lambda^{-1}\beta)=\omega,\\ \operatorname{Re}(\lambda^{-1}\beta)<\operatorname{Re}(\alpha(a-c,\omega))}} \mathbb{E}({}_c\mathcal{E}^{\lambda}{}_{|O},\beta). $$

Proof It can be shown by using the frame ${}_c\boldsymbol{v}$. □

The nilpotent morphism N^{λ}_{ω} on $\mathbb{E}(H(\mathcal{E}^{\lambda}),\omega)$ induces the endomorphism N^{λ}_{ω} on $\mathbb{E}(\mathcal{E}^{\lambda},\omega)$ over $X-D$. It is prolonged to the endomorphism of ${}_c\mathbb{E}(\mathcal{E}^{\lambda},\omega)$ over X. It preserves the filtration $\mathcal{F}\big(\mathbb{E}(\mathcal{E}^{\lambda},\omega)\big)$. Then we obtain the nilpotent morphism $\operatorname{Gr}^{\mathcal{F},\mathbb{E}}_u(N^{\lambda})$ on $\operatorname{Gr}^{\mathcal{F},\mathbb{E}}_u(\mathcal{E}^{\lambda})$ over X.

LEMMA 7.100. *The conjugacy class of $\operatorname{Gr}^{\mathcal{F},\mathbb{E}}_u(N^{\lambda})_{|P}$ is independent of $P\in X$. Hence we obtain the weight filtration W of $\operatorname{Gr}^{\mathcal{F},\mathbb{E}}_u(\mathcal{E}^{\lambda})$ in the category of the vector bundles.* □

7.4.8. Functoriality for tensor products. Let $(E_i,\overline{\partial}_{E_i},h_i,\theta_i)$ $(i=1,2)$ be harmonic bundles over $X-D$. We denote the deformed holomorphic bundle by $\mathcal{E}_i^{\lambda}$ $(i=1,2)$. We have the natural following isomorphism:

$$ H(\mathcal{E}_1^{\lambda}\otimes\mathcal{E}_2^{\lambda}) \simeq H(\mathcal{E}_1^{\lambda})\otimes H(\mathcal{E}_2^{\lambda}). $$

LEMMA 7.101. *We have the natural isomorphisms:*

$$ \mathbb{E}\big(H(\mathcal{E}_1^{\lambda}\otimes\mathcal{E}_2^{\lambda}),\omega\big) = \bigoplus_{\omega_1\times\omega_2=\omega} \mathbb{E}\big(H(\mathcal{E}_1^{\lambda}),\omega_1\big)\otimes\mathbb{E}\big(H(\mathcal{E}_2^{\lambda}),\omega_2\big). $$

□

We obtain the corresponding decomposition:

$$ \mathbb{E}(\mathcal{E}^{\lambda},\omega) = \bigoplus_{\omega_1\times\omega_2=\omega} \mathbb{E}(\mathcal{E}_1^{\lambda},\omega_1)\otimes\mathbb{E}(\mathcal{E}_2^{\lambda},\omega_2). \tag{7.42} $$

LEMMA 7.102.

(1) *The decomposition* (7.42) *is quasi adapted (Definition* 2.6*).*

(2) *We have the following:*

$$\mathcal{F}_a\mathbb{E}(H(\mathcal{E}_1^\lambda\otimes\mathcal{E}_2^\lambda),\omega)=\bigoplus_{\omega_1\times\omega_2=\omega}\sum_{a_1+a_2\leq a}\mathcal{F}_{a_1}\mathbb{E}(H(\mathcal{E}_1^\lambda),\omega_1)\otimes\mathcal{F}_{a_2}\mathbb{E}(H(\mathcal{E}_2^\lambda),\omega_2).$$

Proof Let $\boldsymbol{s}_i$ be a frame of $H(\mathcal{E}_i^\lambda)$ compatible with $\mathbb{E}$, $\mathcal{F}$ and W. Then we obtain the adapted frame $\boldsymbol{s}_i'$. By using $\boldsymbol{s}_1'$ and $\boldsymbol{s}_2'$, we obtain the first claim. The second claim follows from the first claim. □

We have the following product of the abelian group $\boldsymbol{R}\times\boldsymbol{C}^*$: For $u_i=(a_i,\alpha_i)$, we put $u_1\cdot u_2=(a_1+a_2,\alpha_1\times\alpha_2)$.

COROLLARY 7.103. *We have the isomorphism:*

$$\mathrm{Gr}_u^{\mathcal{F},\mathbb{E}}(H(\mathcal{E}^\lambda))\simeq\bigoplus_{u_1\cdot u_2=u}\mathrm{Gr}_{u_1}^{\mathcal{F},\mathbb{E}}(H(\mathcal{E}_1^\lambda))\otimes\mathrm{Gr}_{u_2}^{\mathcal{F},\mathbb{E}}(H(\mathcal{E}_2^\lambda)).$$

□

We have the natural isomorphisms:

$$\mathbb{E}(\bigwedge^a H(\mathcal{E}^\lambda),\omega)\simeq\bigoplus_{f\in\mathcal{S}(a,\omega)}\bigotimes_{\omega'\in\mathcal{S}p(M^\lambda)}\bigwedge^{f(\omega')}\mathbb{E}(H(\mathcal{E}^\lambda),\omega'),\tag{7.43}$$

$$\mathbb{E}(\mathrm{Sym}^a H(\mathcal{E}^\lambda),\omega)\simeq\bigoplus_{f\in\mathcal{S}(a,\omega)}\bigotimes_{\omega'\in\mathcal{S}p(M^\lambda)}\mathrm{Sym}^{f(\omega')}\mathbb{E}(H(\mathcal{E}^\lambda),\omega'),\tag{7.44}$$

$$\mathcal{S}(a,\omega):=\Big\{f:\mathcal{S}p(M^\lambda)\longrightarrow\mathbb{Z}_{\geq 0}\,\Big|\,\prod\omega'^{f(\omega')}=\omega,\ \sum f(\omega')=a\Big\}.\tag{7.45}$$

COROLLARY 7.104.

- *The decompositions* (7.43), (7.44) *and* (7.45) *are quasi adapted.*
- *The parabolic filtrations on the left hand sides of* (7.43), (7.44) *and* (7.45) *are isomorphic to the induced filtrations on the right hand side.*
- *The weight filtrations are also isomorphic.* □

7.4.9. Functoriality for dual. We have the natural isomorphism: $H(\mathcal{E}^{\vee\,\lambda})\simeq H(\mathcal{E}^\lambda)^\vee$.

LEMMA 7.105. *Under the isomorphism, $\mathcal{F}_aH(\mathcal{E}^{\vee\,\lambda})$ is same as the following:*

$$\Big\{f\in H(\mathcal{E}^\lambda)^\vee\,\Big|\,f(\mathcal{F}_bH(\mathcal{E}^\lambda))\subset\mathcal{F}_{b+a}H(\mathcal{E}^\lambda)\Big\}.$$

Proof Let $\boldsymbol{s}=(s_i)$ be a base of $H(\mathcal{E}^\lambda)$, compatible with $\mathbb{E}$, $\mathcal{F}$, W. Let $\boldsymbol{s}^\vee=(s_i^\vee)$ denote the dual base.

We put $\deg^{\mathcal{F}}(s_i)=a_i$. We put $s_i':=s_i\cdot|z|^{a_i}$, and $\boldsymbol{s}':=(s_i')$. Then $\boldsymbol{s}'$ is adapted up to log order. We put $s_i^{\vee\,\prime}:=s_i^\vee\cdot|z|^{-a_i}$, and $\boldsymbol{s}^{\vee\,\prime}=(s_i^{\vee\,\prime})$. Then $\boldsymbol{s}^{\vee\,\prime}$ is the dual base of $\boldsymbol{s}'$, and $\boldsymbol{s}^{\vee\,\prime}$ is adapted up to log order. Then the claim follows easily. □

7.4.10. Functoriality for pull back. Let $\psi_c:X\longrightarrow X$ given by $z\longmapsto z^c$. We have the natural isomorphism:

$$\psi_c^{-1}:H(\mathcal{E}^\lambda)\simeq H(\psi_c^{-1}\mathcal{E}^\lambda).$$

Let M_1^{λ} denote the monodromy of $\psi_c^{-1}\mathcal{E}^{\lambda}$. We obtain the following isomorphism for any $\omega_1 \in \mathcal{S}p(M^{\lambda_1})$:

$$\mathbb{E}\big(H(\psi_c^{-1}\mathcal{E}^{\lambda}),\omega_1\big) \simeq \bigoplus_{\substack{\omega\in\mathcal{S}p(M^{\lambda})\\ \omega^c=\omega_1}} \mathbb{E}(H(\mathcal{E}^{\lambda}),\omega).$$

LEMMA 7.106. *We have the following, for any element $\omega_1 \in \mathcal{S}p(M_1^{\lambda})$:*

$$\mathcal{F}_{c\cdot a}\big(\mathbb{E}\big(H(\psi_c^{-1}\mathcal{E}^{\lambda}),\omega_1\big)\big) = \bigoplus_{\substack{\omega\in\mathcal{S}p(M^{\lambda})\\ \omega^c=\omega_1}} \mathcal{F}_a\big(\mathbb{E}\big(H(\mathcal{E}^{\lambda}),\omega\big)\big). \tag{7.46}$$

The weight filtrations are compatible.

Proof The compatibility for the weight filtration is clear. Let $\boldsymbol{s}$ be a frame of $H(\mathcal{E}^{\lambda})$ compatible with $\mathbb{E}$, $\mathcal{F}$ and W. We put $a_i := \deg^{\mathcal{F}}(s_i)$, and $s_i' := s_i \cdot |z|^{a_i}$. Then $\boldsymbol{s}' = (s_i')$ is adapted up to log order.

Consider $\psi_c^{-1}(s_i') = \psi_c^{-1}(s_i)\cdot|z|^{c\cdot a_i}$. The frame $\psi_c^{-1}(\boldsymbol{s}') = \big(\psi_c^{-1}(s_i')\big)$ is adapted up to log order. Hence the equality (7.46) follows. □

7.5. The case where λ is generic

7.5.1. Genericity. We have the induced morphisms $\mathfrak{p}^f(\lambda) : \overline{\mathcal{KMS}}(\mathcal{E}^0) \longrightarrow \mathcal{P}ar^f(\mathcal{E}^{\lambda})$ and $\mathfrak{e}^f(\lambda) : \overline{\mathcal{KMS}}(\mathcal{E}^0) \longrightarrow \mathcal{S}p(M^{\lambda})$.

DEFINITION 7.107. λ is called generic with respect to $(E,\overline{\partial}_E,\theta,h)$, if the map $\mathfrak{e}^f(\lambda) : \overline{\mathcal{KMS}}(\mathcal{E}^0) \longrightarrow \mathcal{S}p(M^{\lambda})$ is bijective. In other words, the map $\mathfrak{e}(\lambda) : \mathcal{KMS}(\mathcal{E}^0) \longrightarrow \mathcal{S}p(\mathcal{E}^{\lambda})$ is bijective. □

REMARK 7.108. We can consider $\mathcal{KMS}({}^{\diamond}\mathcal{E}^0)$ instead of $\overline{\mathcal{KMS}}(\mathcal{E}^0)$. □

LEMMA 7.109. *Let S be the set of $\lambda \in \boldsymbol{C}$, which are generic with respect to $(E,\overline{\partial}_E,\theta,h)$. Then the set $\boldsymbol{C}^* - S$ is discrete in $\boldsymbol{C}^*$. In particular, it is countable.*

Proof Pick $u = (a,\alpha)$ and $v = (b,\beta)$ be different elements of $\mathcal{KMS}({}^{\diamond}\mathcal{E}^0)$. Let us consider the following condition for λ:

$$\mathfrak{e}^f(\lambda,u) = \mathfrak{e}^f(\lambda,v).$$

It is equivalent to the following:

$$\lambda^{-1}\cdot(\alpha-\beta) - (a-b) - \lambda\cdot(\bar{\alpha}-\bar{\beta}) \in \mathbb{Z}.$$

Let n be an integer. Consider the following equation:

$$\lambda^{-1}\cdot(\alpha-\beta) - (a-b) - \lambda\cdot(\bar{\alpha}-\bar{\beta}) = n. \tag{7.47}$$

Let $\lambda_i(n)$ $(i = 1,2)$ be the solutions of the equation (7.47). Then we have the following relation:

$$|\lambda_1(n)\cdot\lambda_2(n)| = \left|-\frac{\alpha-\beta}{\bar{\alpha}-\bar{\beta}}\right| = 1. \tag{7.48}$$

We also have the following:

$$\lambda_1(n)+\lambda_2(n) = \frac{-(a-b+n)}{\bar{\alpha}-\bar{\beta}}. \tag{7.49}$$

Then we obtain the following:

$$\lim_{|n|\to\infty} n^{-1}\cdot\big|\lambda_1(n)+\lambda_2(n)\big| = \big|\bar{\alpha}-\bar{\beta}\big|^{-1} \neq 0.$$

Hence the set of solutions of (7.48) is discrete in $\boldsymbol{C}^*$. Since $\mathcal{KMS}({}^{\diamond}\mathcal{E}^0)$ is finite, the claim follows. □

Assume λ is generic. Then for any $\omega \in \mathcal{S}p(M^\lambda)$, there exists the unique element $u_0 \in \overline{\mathcal{KMS}}(\mathcal{E}^0)$ satisfying $\mathfrak{e}^f(\lambda, u_0) = \omega$. Note that the parabolic structure of $\mathbb{E}(H(\mathcal{E}^\lambda), \omega)$ is trivial in the following sense: $\mathrm{Gr}_b^{\mathcal{F}}\, \mathbb{E}(H(\mathcal{E}^\lambda), \omega) \neq 0$ if and only if $b = \mathfrak{p}^f(\lambda, u_0)$.

LEMMA 7.110. *Let Z be a countable subset of $\boldsymbol{C}^*_\lambda$. Assume the following:*

For any $\lambda \in \boldsymbol{C}^* - Z$, we know the set $\mathcal{S}p(M^\lambda)$ and the multiplicity of each element $\alpha \in \mathcal{S}p(M^\lambda)$.

Then we know the set $\mathcal{KMS}(\mathcal{E}^\lambda)$ and the multiplicities $\mathfrak{m}(\lambda, u)$ ($\lambda \in \boldsymbol{C}$, $u \in \mathcal{KMS}(\mathcal{E}^\lambda)$).

Proof It follows from Lemma 7.109. □

7.5.2. Quasi canonical prolongment. Let b be a real number. We have the quasi canonical prolongment $QC_b(\mathcal{E}^\lambda)$ of $\mathcal{E}^\lambda$, namely, a holomorphic vector bundle over X satisfying the following:

- The restriction $QC_b(\mathcal{E}^\lambda)_{|X-D}$ is isomorphic to $\mathcal{E}^\lambda$.
- Let g be a holomorphic section of $QC_b(\mathcal{E}^\lambda)$. Then $\mathbb{D}^{\lambda,f}$ gives a holomorphic section of $QC_b(\mathcal{E}^\lambda) \otimes \Omega_X(\log D)$.
- Let β be an eigenvalue of the residue $\mathrm{Res}(\mathbb{D}^{\lambda,f})$ on $QC_b(\mathcal{E}^\lambda)_D$. Then the inequality $b \leq \mathrm{Re}(\beta) < b+1$ holds.

Recall that $QC_b(\mathcal{E}^\lambda)$ is uniquely determined as the subsheaf of $j_*\mathcal{E}^\lambda$, where j denotes the inclusion $X - D \longrightarrow X$.

We have the decomposition:

$$QC_b(\mathcal{E}^\lambda) = \bigoplus_{\omega} QC_b\big(\mathbb{E}(\mathcal{E}^\lambda, \omega)\big).$$

We have the natural filtration given by the following:

$$QC_b\big(\mathcal{F}_a\mathbb{E}(\mathcal{E}^\lambda, \omega)\big).$$

Let $\boldsymbol{s}$ be a base of $H(\mathcal{E}^\lambda)$. We put $v_i := F(s_i, b)$ and $\boldsymbol{v} = (v_i)$. Then $\boldsymbol{v}$ gives a holomorphic frame of $\mathcal{E}^\lambda$ over $X - D$. Let us consider the prolongment of $\mathcal{E}^\lambda$ by $\boldsymbol{v}$. Then it satisfies the conditions above. Hence $\boldsymbol{v}$ gives the holomorphic frame of $QC_b(\mathcal{E}^\lambda)$.

Assume that λ is generic. Let us consider the following:

$$QC_0(\mathcal{E}^\lambda) := \bigoplus_{u \in \overline{\mathcal{KMS}}(\mathcal{E}^0)} QC_0\big(\mathbb{E}\big(\mathcal{E}^\lambda, \mathfrak{e}^f(\lambda, u)\big)\big).$$

LEMMA 7.111. $QC_0\big(\mathbb{E}\big(\mathcal{E}^\lambda, \mathfrak{e}^f(\lambda, u)\big)\big) = {}_d\mathbb{E}\big(\mathcal{E}^\lambda, \mathfrak{e}^f(\lambda, u)\big)$. *Here we put $d := \mathfrak{p}^f(\lambda, u) - \mathrm{Re}\big(\alpha\big(0, \mathfrak{e}^f(\lambda, u)\big)\big)$.*

Proof Let s be a non-zero element of $\mathbb{E}\big(H(\mathcal{E}^\lambda), \mathfrak{e}^f(\lambda, u)\big)$. Then we have the following:

$$-\mathrm{ord}(F(s, \mathfrak{e}^f(\lambda, u))) = -\mathrm{ord}(s) - \mathrm{Re}(\alpha(0, \mathfrak{e}^f(\lambda, u))).$$

Then the claim follows, from the uniqueness of the quasi canonical prolongment [**19**]. □

LEMMA 7.112. *We have the following relation:*

$$ {}_c\mathcal{E}^\lambda = \bigoplus_{u \in \overline{\mathcal{KMS}}(\mathcal{E}^0)} QC_0\bigl(\mathbb{E}(\mathcal{E}^\lambda, \mathfrak{e}^f(\lambda, u))\bigr) \cdot z^{N(u)}. $$

Here we put $N(u) := \nu_c\Bigl(\mathfrak{p}^f(\lambda, u) - \mathrm{Re}\bigl(\alpha\bigl(0, \mathfrak{e}^f(\lambda, u)\bigr)\bigr)\Bigr)$. *(See the subsection* 2.1.5 *for* ν_c*).*

Proof It follows from $c - 1 < -N(u) + d \leq c$. □

REMARK 7.113. The lemma says that quasi canonical prolongment is essentially same as the prolongment by an increasing order of the norms, in the case that λ is generic. □

REMARK 7.114. For any point $\lambda \in \boldsymbol{C}^*$, not necessarily generic, the vector bundle ${}_c\mathcal{E}^\lambda$ is obtained from $QC_0(\mathcal{E}^\lambda)$ by a sequence of elementary transformations. □

7.5.3. Normalizing frame. Let $\boldsymbol{s}$ be a frame of $H(\mathcal{E}^\lambda)$, which is compatible with $\mathbb{E}$ and $\mathcal{F}$. We put $\deg^{\mathcal{F}}(s_i) = a_i$ and $\deg^{\mathbb{E}}(s_i) = \omega_i$. Pick any real number c.

Then we have the elements $u_i \in \mathcal{K}(\mathcal{E}, \lambda, c)$ such that $\mathfrak{k}^f(\lambda, u_i) = (a_i, \omega_i)$. We put ${}_cv_i := F(s_i, a_i - c)$, which is the section of ${}_c\mathcal{E}^\lambda$.

LEMMA 7.115. ${}_c\boldsymbol{v} = ({}_cv_i) \in {}_c\mathcal{E}^\lambda$ *gives the frame of* ${}_c\mathcal{E}^\lambda$*, which is compatible with the parabolic filtration* F *and the generalized eigen decomposition* $\mathbb{E}$*. We have* $\deg^{F,\mathbb{E}}(v_i) = \mathfrak{k}(\lambda, u_i)$*. We also have the following:*

$$ \mathbb{D}\boldsymbol{v} = \boldsymbol{v} \cdot \bigl(C + N\bigr) \cdot \frac{dz}{z}. $$

Here C *denotes the diagonal matrix whose* (i,i)*-component is* $\mathfrak{e}(\lambda, u_i)$*, and* N *denotes the nilpotent matrix.*

Proof In the case $c = 0$, the claim follows from Proposition 7.82. The general case can be shown similarly. □

7.6. Family of multi-valued sections

7.6.1. The structure of holomorphic bundle. Let $(E, \overline{\partial}_E, \theta, h)$ be a tame harmonic bundle over $X - D$, and $\mathcal{E}$ be the deformed holomorphic bundle with λ-connection $\mathbb{D}$ over $\mathcal{X} - \mathcal{D} = (X - D) \times \boldsymbol{C}_\lambda$. Let us consider the family of the multi-valued flat sections $\bigl\{H(\mathcal{E}^\lambda) \bigm| \lambda \in \boldsymbol{C}^*_\lambda\bigr\}$.

Let $\pi : \mathbb{H} \longrightarrow X - D$ is the universal covering, and let P be a point of $\mathbb{H}$. We have the holomorphic vector bundle $\pi^{-1}\mathcal{E}_{|\boldsymbol{C}^*_\lambda \times \{P\}}$ over $\boldsymbol{C}^*_\lambda$. Since we can pick the isomorphism $\mathcal{E}_{|(\lambda, P)} \simeq H(\mathcal{E}^\lambda)$, we obtain the structure of holomorphic vector bundle on the family $\{H(\mathcal{E}^\lambda) \,|\, \lambda \in \boldsymbol{C}^*_\lambda\}$. The holomorphic vector bundle is denoted by $\mathcal{H}(\mathcal{E})$ or simply by $\mathcal{H}$. Clearly the holomorphic structure does not depend on a choice of P.

7.6.2. The $\mathbb{E}^{(\lambda_0)}$-decomposition. Pick $\lambda_0 \in \boldsymbol{C}^*_\lambda$. We have the monodromy M^{λ_0} on $H(\mathcal{E}^{\lambda_0})$. We put $S_0 := \mathcal{S}p(M^{\lambda_0})$. Pick a positive number ϵ_1 satisfying the following:

$$ \epsilon_1 < \min\bigl\{|a - b| \bigm| a \neq b \in S_0\bigr\}. $$

Pick sufficiently small $\epsilon_0 > 0$ such that we have the following decomposition on $\Delta(\lambda_0, \epsilon_0)$:

$$(7.50) \qquad \mathcal{H}_{|\Delta(\lambda_0,\epsilon_0)} = \bigoplus_{\omega \in S_0} \mathbb{E}^{(\lambda_0)}(\mathcal{H}, \omega), \quad \mathbb{E}^{(\lambda_0)}(\mathcal{H}, \omega)_{|\lambda} := \mathbb{E}_{\epsilon_1}\big(H(\mathcal{E}^\lambda), \omega\big).$$

See (2.21) for the notation $\mathbb{E}_{\epsilon_1}$. The subset $\mathcal{S}(\omega) \subset \mathcal{KMS}({}^\circ\mathcal{E}^0)$ is given as follows:

$$\mathcal{S}(\omega) := \mathfrak{e}^f(\lambda_0)^{-1}(\omega) = \Big\{ u \in \overline{\mathcal{KMS}}(\mathcal{E}^0) \;\Big|\; \mathfrak{e}^f(\lambda_0, u) = \omega \Big\}.$$

We may assume that any point $\lambda \in \Delta^*(\lambda_0, \epsilon_0)$ are generic, due to Lemma 7.109. Then we have the following decomposition on the punctured disc $\Delta^*(\lambda_0, \epsilon_0)$:

$$(7.51) \qquad \mathcal{H}_{|\Delta^*(\lambda_0,\epsilon_0)} = \bigoplus_{u \in \overline{\mathcal{KMS}}(\mathcal{E}^0)} \mathcal{H}_u, \quad \mathcal{H}_{u\,|\,\lambda} := \mathbb{E}\big(H(\mathcal{E}^\lambda), \mathfrak{e}^f(\lambda, u)\big).$$

LEMMA 7.116. *We have the following decomposition:*

$$\mathbb{E}^{(\lambda_0)}(\mathcal{H}, \omega)_{|\Delta^*(\lambda_0,\epsilon_0)} = \bigoplus_{u \in \mathcal{S}(\omega)} \mathcal{H}_u.$$

Proof It immediately follows from the definition of $\mathbb{E}^{(\lambda_0)}$ in (7.50) and the decomposition (7.51). □

7.6.3. The filtration $\mathcal{F}^{(\lambda_0)}$.

We remark the following.

LEMMA 7.117. *The map $\mathfrak{p}^f(\lambda_0) : \mathcal{S}(\omega) \longrightarrow \boldsymbol{R}$ is injective.*

Proof It follows from Lemma 2.2. □

On the vector bundle $\mathbb{E}^{(\lambda_0)}(\mathcal{H}, \omega)_{|\Delta^*(\lambda_0,\epsilon_0)}$, we have the filtration $\mathcal{F}^{(\lambda_0)}$ defined as follows:

$$\mathcal{F}^{(\lambda_0)}_d \mathbb{E}^{(\lambda_0)}(\mathcal{H}, \omega)_{|\Delta^*(\lambda_0,\epsilon_0)} := \bigoplus_{\substack{u \in \mathcal{S}(\omega), \\ \mathfrak{p}^f(\lambda_0, u) \leq d}} \mathcal{H}_u.$$

LEMMA 7.118. *The above filtration $\mathcal{F}^{(\lambda_0)}$ of $\mathcal{H}_{\omega\,|\,\Delta^*(\lambda_0,\epsilon_0)}$ can be prolonged to the filtration of $\mathcal{H}_\omega$ over $\Delta(\lambda_0, \epsilon_0)$.*

Proof Since $\mathcal{F}^{(\lambda_0)}$ is defined by using the generalized eigen-decomposition of holomorphic endomorphisms M^λ, the claim holds. □

We denote the prolonged filtration also by $\mathcal{F}^{(\lambda_0)}$.

PROPOSITION 7.119. *We have the following equality:*

$$(7.52) \qquad \Big(\mathcal{F}^{(\lambda_0)}_d \mathbb{E}^{(\lambda_0)}(\mathcal{H}, \omega)\Big)_{|\lambda_0} = \mathcal{F}_d\Big(\mathbb{E}(H(\mathcal{E}^{\lambda_0}), \omega)\Big).$$

Proof We use the induction as is explained in the following. For any sufficiently small d, the both sides in (7.52) are 0. Hence the equality (7.52) holds trivially. If the equality (7.52) holds for d, then (7.52) holds for $d + \eta$ for any sufficiently small $\eta > 0$. Hence we have only to show the following claim:

(C): Assume that (7.52) holds for any $d < d_0$. Then the equality (7.52) holds for d_0.

We assume that $\mathfrak{p}^f(\lambda_0, u_0) = d_0$ for $u_0 \in \mathcal{S}(\omega)$. Note that the element u_0 is uniquely determined due to Lemma 7.117. We put $R := \operatorname{rank} \mathcal{F}_d^{(\lambda_0)}$. We have the natural isomorphism:

$$\bigwedge^R \mathcal{H}(\mathcal{E}) \simeq \mathcal{H}(\bigwedge^R \mathcal{E}). \tag{7.53}$$

We do not distinguish them in the following argument. We put as follows:

$$u_1 = \sum_{\substack{u \in \mathcal{S}(\omega) \\ \mathfrak{p}^f(\lambda_0, u) \leq d_0}} \mathfrak{m}(u, 0) \cdot u.$$

We put $d_1 := \mathfrak{p}^f(\lambda_0, u_1)$.

LEMMA 7.120. *Let $a : \mathcal{S}(\omega) \longrightarrow \mathbb{Z}_{\geq 0}$ be a map satisfying the following conditions:*

$$a(u) \leq \mathfrak{m}(u, 0), \qquad \sum a(u) = R.$$

Then we have the following inequality:

$$\sum_{u \in \mathcal{S}(\omega)} a(u) \cdot \mathfrak{p}^f(\lambda_0, u) \geq d_1. \tag{7.54}$$

The equality in (7.54) holds if and only if $\{a(u) \mid u \in \mathcal{S}(\omega)\}$ satisfies the following:

$$a(u) = \begin{cases} \mathfrak{m}(u, 0), & \big(\mathfrak{p}^f(\lambda_0, u) \leq \mathfrak{p}^f(\lambda_0, u_0)\big), \\ 0, & \big(\mathfrak{p}^f(\lambda_0, u) > \mathfrak{p}^f(\lambda_0, u_0)\big). \end{cases}$$

Proof It immediately follows from our choice of d_1. □

LEMMA 7.121. *We have the following equality:*

$$\begin{aligned} &\mathcal{F}_{d_1} \cap \bigwedge^R \mathbb{E}\big(H(\mathcal{E}^{\lambda_0}), \omega\big) = \bigwedge^R \big(\mathcal{F}_{d_0} \mathbb{E}(H(\mathcal{E}^{\lambda_0}), \omega)\big), \\ &\mathcal{F}_{<d_1} \cap \bigwedge^R \mathbb{E}\big(H(\mathcal{E}^{\lambda_0}), \omega\big) = 0. \end{aligned} \tag{7.55}$$

In particular, the rank of $\mathcal{F}_{d_1} \cap \bigwedge^R \mathbb{E}\big(H(\mathcal{E}^{\lambda_0}), \omega\big)$ is one.

Proof We put $R' := \operatorname{rank} \mathbb{E}\big(H(\mathcal{E}^{\lambda_0}), \omega\big)$. Let $\boldsymbol{s} = (s_1, \ldots, s_{R'})$ be a frame of $\mathbb{E}\big(H(\mathcal{E}^{\lambda_0}), \omega\big)$ which is compatible with $\mathcal{F}$ and $\mathbb{E}$. We may assume that $(s_1, \ldots, s_R)$ be a frame of $\mathcal{F}_{d_0}\mathbb{E}\big(H(\mathcal{E}^{\lambda_0}), \omega\big)$. For any subset $I \subset \{1, \ldots, R'\}$, we put $s_I := \bigwedge_{i \in I} s_i$. Due to the norm estimate for the multi-valued sections (Proposition 7.93), the tuple $\{s_I \,\big|\, |I| = R\}$ gives the frame of $\bigwedge^R \mathbb{E}\big(H(\mathcal{E}^{\lambda_0}), \omega\big)$, which is also compatible with the filtration $\mathcal{F}$, and we have the inequality $-\operatorname{ord}(s_I) = -\sum_{i \in I} \operatorname{ord}(s_i)$. Thus we obtain $-\operatorname{ord}(s_I) \geq d_1$, and the equality holds if and only if I is same as the set $\{1, \ldots, R\}$ due to Lemma 7.120. Then we obtain the equalities (7.55). □

We have the line subbundle $\mathcal{L} := \bigwedge^R \Big(\mathcal{F}_{d_0}^{(\lambda_0)} \mathbb{E}^{(\lambda_0)}(\mathcal{H}(\mathcal{E}), \omega)\Big)$ of $\bigwedge^R \mathcal{H}(\mathcal{E})$. Let us pick a section s of $\mathcal{L}$ such that $s_{|\lambda_0} \neq 0$.

LEMMA 7.122. *We have $-\operatorname{ord}(s_{|\lambda_0}) \leq d_1$.*

Proof For any $\lambda \in \Delta(\lambda_0, \epsilon_0)$, the element $s_{|\lambda} \in H(\mathcal{E}^\lambda)$ is an eigenvector of M^λ, and the eigenvalue is $\bigwedge^R \mathfrak{e}^f(\lambda, u_1)$. We put as follows:

$$v := \exp\Big(\log z \cdot \mathfrak{e}^f(\lambda, u)\Big) \cdot s.$$

Then it gives a holomorphic section of $\bigwedge^R \mathcal{E}$ defined over $\Delta(\lambda_0,\epsilon_0)\times(X-D)$.

We may assume that any $\lambda\in\Delta^*(\lambda_0,\epsilon_0)$ is generic. Then we obtain the following equality for any point $\lambda\in\Delta^*(\lambda_0,\epsilon_0)$:

$$-\operatorname{ord}\bigl(v_{|\{\lambda\}\times(X-D)}\bigr)=\mathfrak{p}(\lambda,u_1).$$

We also note that $|v|$ is bounded over the compact set $\Delta(\lambda_0,\epsilon_0)\times\{|z|=1/2\}$. Since $\bigwedge^R(\mathcal{E}^\lambda,h)$ is acceptable for any λ (Theorem 1 in [**81**], or Corollary 8.19 in this paper), there exists a positive constant $M>0$ satisfying the following, due to Corollary 2.53:

- For any $\epsilon>0$, there exists $C_\epsilon>0$ such that the following inequality holds for any $\lambda\in\Delta^*(\lambda_0,\epsilon_0)$:

$$\bigl|v_{|\{\lambda\}\times(X-D)}\bigr|_h\leq C_\epsilon\cdot|z|^{-\mathfrak{p}(\lambda,u_1)}\cdot\bigl(-\log|z|\bigr)^M.$$

Then we obtain the inequality for $\lambda=\lambda_0$:

$$\bigl|v_{|\{\lambda_0\}\times(X-D)}\bigr|_h\leq C_\epsilon\cdot|z|^{-\mathfrak{p}(\lambda_0,u_1)}\cdot\bigl(-\log|z|\bigr)^M.$$

Hence we obtain the inequality $-\operatorname{ord}(v_{|\{\lambda_0\}\times(X-D)})\leq\mathfrak{p}(\lambda_0,u_1)$ by continuity. It implies $-\operatorname{ord}(s_{|\lambda_0})\leq\mathfrak{p}^f(\lambda_0,u_1)=d_1$ due to Lemma 7.78. □

Let us return to the proof of Proposition 7.119. Due to Lemma 7.121 and 7.122, we obtain the following equality:

$$\bigwedge^R\Bigl(\mathcal{F}^{(\lambda_0)}_{d_0}\mathbb{E}^{(\lambda_0)}\bigl(\mathcal{H}(\mathcal{E}^\lambda),\omega\bigr)\Bigr)_{|\lambda_0}=\bigwedge^R\Bigl(\mathcal{F}_{d_0}\mathbb{E}\bigl(H(\mathcal{E}^{\lambda_0}),\omega\bigr)\Bigr).$$

It implies the equality (7.52) for d_0. Thus the proof of Proposition 7.119 is accomplished. □

7.6.4. The filtration $\mathcal{F}^{(\lambda_0)}$ and the decomposition $\mathbb{E}^{(\lambda_0)}$ on $\mathcal{E}$. The filtration and the decomposition for $\mathcal{H}$ on $\Delta(\lambda_0,\epsilon_0)$ induce those for $\mathcal{E}$ on $\Delta(\lambda_0,\epsilon_0)\times(X-D)$. We only summarize the result.

We have the following decomposition of the family of the λ-connections on $\Delta(\lambda_0,\epsilon_0)\times(X-D)$:

$$\mathcal{E}=\bigoplus_{\omega\in S_0}\mathbb{E}^{(\lambda_0)}(\mathcal{E},\omega),\qquad\mathbb{E}^{(\lambda_0)}(\mathcal{E},\omega):=\mathbb{E}_{\epsilon_2}(\mathcal{E},\omega).$$

Moreover we have the following decomposition on $\Delta^*(\lambda_0,\epsilon_0)\times(X-D)$:

$$\mathbb{E}^{(\lambda_0)}(\mathcal{E},\omega)_{|\Delta^*(\lambda_0,\epsilon_0)}=\bigoplus_{u\in\mathcal{S}(\omega)}\mathbb{E}(\mathcal{E},\mathfrak{e}^f(\lambda,u)).$$

On the vector bundle $\mathbb{E}^{(\lambda_0)}(\mathcal{E},\omega)$, we have the filtration $\mathcal{F}^{(\lambda_0)}$ satisfying the following conditions:

- On $\Delta^*(\lambda_0,\epsilon_0)\times(X-D)$, we have the following splitting:

$$\mathcal{F}^{(\lambda_0)}_d\bigl(\mathbb{E}^{(\lambda_0)}(\mathcal{E},\omega)\bigr)_{|\Delta^*(\lambda_0,\epsilon_0)\times(X-D)}=\bigoplus_{\substack{u\in\mathcal{S}(\omega)\\ \mathfrak{p}^f(\lambda_0,u)\leq d}}\mathbb{E}\bigl(\mathcal{E},\mathfrak{e}^f(\lambda,u)\bigr).$$

- On $\{\lambda_0\}\times(X-D)$, we have the following:

$$\mathcal{F}^{(\lambda_0)}_d\bigl(\mathbb{E}^{(\lambda_0)}(\mathcal{E},\omega)\bigr)_{|\{\lambda_0\}\times(X-D)}=\mathcal{F}_d(\mathbb{E}(\mathcal{E}^{\lambda_0},\omega)).$$

7.7. Asymptotic orthogonality

7.7.1. Asymptotic orthogonality for $\mathbb{E}$-decomposition of $\mathcal{E}^0$.

We put $X = \Delta$ and $D = \{O\}$. Let $(E, \overline{\partial}_E, \theta, h)$ be a tame harmonic bundle over $X - D$. Let ϵ_1 be a sufficiently small number, and C be a sufficiently small number. Then we have the following decomposition over $\Delta^*(C)$:

$$\diamond\mathcal{E}^0 = \bigoplus_{\alpha \in \mathcal{S}p(\mathrm{Res}(\theta))} \mathbb{E}_{\epsilon_1}(\diamond\mathcal{E}^0, \alpha). \tag{7.56}$$

See (2.21) for the notation $\mathbb{E}_{\epsilon_1}$.

LEMMA 7.123. *There exists a positive constant ϵ_2 such that the decomposition is $|z|^{\epsilon_2}$-asymptotically orthogonal.*

Proof It is shown in Proposition 7.37. □

7.7.2. An asymptotically orthogonal decomposition of $\mathcal{E}^\lambda$. Let $\lambda \in \boldsymbol{C}^*_\lambda$ be generic. Let $\boldsymbol{v} = (v_i)$ be a normalizing frame of $\diamond\mathcal{E}$. We have the decomposition:

$$\diamond\mathcal{E}^\lambda = \bigoplus_{-1<b\leq 0} \Big(\bigoplus_{\substack{u \in \mathcal{KMS}(\mathcal{E}^0) \\ \kappa(\mathfrak{p}(\lambda,u))=b}} \diamond\mathbb{E}(\mathcal{E}^\lambda, \mathfrak{e}^f(\lambda, u)) \Big). \tag{7.57}$$

We would like to show that the decomposition above is $(-\log|z|)^{-1}$-asymptotically orthogonal.

The flat connection $\tilde{\nabla}^\lambda$ is given by $\mathbb{D}^{\lambda,f} - \lambda^{-1} \cdot \psi$, where $\psi = \psi_0 dz/z$ is defined as in the formula (7.23). On the other hand, we have the unitary connection $\nabla^\lambda = \overline{\partial}_E + \lambda\theta^\dagger + \partial_E - \bar{\lambda}\theta$.

We put $\Phi^\lambda := \tilde{\nabla}^\lambda - \nabla^\lambda$. Then we have the following formula:

(7.58)
$$\Phi^\lambda = \Big(\overline{\partial}_E + \lambda\theta^\dagger + \partial_E + \lambda^{-1}\theta - \lambda^{-1}\psi\Big) - \Big(\overline{\partial}_E + \lambda\theta^\dagger + \partial_E - \bar{\lambda}\theta\Big) = \lambda^{-1}\Big((1+|\lambda|^2)\cdot\theta - \psi\Big).$$

Let η be a positive number. Let $M_\eta(\tilde{\nabla}^\lambda)$ denote the monodromy of $\tilde{\nabla}^\lambda$ along the circle $|z| = \eta$.

LEMMA 7.124. *There exists $C > 0$ satisfying the following:*

$$\Big|\big(M(\tilde{\nabla}^\lambda)u, M(\tilde{\nabla}^\lambda)v\big)_h - (u,v)_h\Big| \leq C \cdot (-\log|z|)^{-1} \cdot |u|_h \cdot |v|_h. \tag{7.59}$$

Proof We use the real coordinate (x, y) $(0 \leq x < 2\pi,\ 0 < y)$ given by $z = \exp(\sqrt{-1}x - y)$. Let V denote the vector field given by $\partial/\partial x$. Let s be a flat section with respect to $\tilde{\nabla}^\lambda$ over $|z| = \eta$. It satisfies the following:

$$\frac{d}{dx} h(s,s) = 2\,\mathrm{Re}\big(h(\nabla^\lambda_V(s), s)\big) = -2\,\mathrm{Re}\big(h(\Phi^\lambda_V(s), s)\big).$$

Hence we obtain the following:

$$\frac{d}{dx} \log|s|_h^2 = -2\frac{\mathrm{Re}\big(h(\Phi^\lambda_V(s), s)\big)}{|s|_h^2}.$$

Due to the estimate (7.29), we have $|\Phi^\lambda_V|_h \leq C_0 \cdot y^{-1}$ for some $C_0 > 0$. Thus there exists a positive constant $C_1 > 0$, which is independent of x and y, and satisfying the following:

$$\left|\frac{d}{dx} \log|s(x,y)|_h^2\right| \leq C_1 \cdot y^{-1}.$$

Hence there exists $C_2 > 0$ satisfying the following for any $0 \leq x \leq 2\pi$ and $y > 0$:

$$\big|s(x,y)\big|_h \leq C_2 \cdot \big|s(0,y)\big|_h. \tag{7.60}$$

On the other hand, we also have the following equality:

$$\frac{d}{dx}(s_1, s_2)_h = -(\Phi_V^{\lambda} s_1, s_2)_h - (s_1, \Phi_V^{\lambda} s_2)_h.$$

Then we obtain the following inequality from (7.29), (7.58) and (7.60):

$$\left|\frac{d}{dx}(s_1, s_2)_h\right| \leq C_3 \cdot y^{-1} \cdot |s_1|_h \cdot |s_2|_h \leq C_4 \cdot y^{-1} \cdot |s_1(0)|_h \cdot |s_2(0)|_h.$$

The inequality (7.59) immediately follows. □

LEMMA 7.125. *The decomposition* (7.57) *is the generalized eigen decomposition with respect to the monodromy* $M_\eta(\tilde{\nabla}^\lambda)$. *Namely we have the following equality:*

$$\bigoplus_{\substack{u \in \mathcal{KMS}({}^\diamond\mathcal{E}^0) \\ \mathfrak{p}(\lambda,u)=b}} \mathbb{E}\big(\mathcal{E}^\lambda, \mathfrak{e}^f(\lambda,u)\big) = \mathbb{E}\Big(M_\eta(\tilde{\nabla}^\lambda),\ \exp(2\pi\sqrt{-1}b)\Big).$$

Proof Since the endomorphism ψ_0 is flat with respect to $\mathbb{D}^{\lambda,f}$, we have the following:

$$M(\tilde{\nabla}^\lambda) = M(\mathbb{D}^{\lambda,f}) \circ \exp\big(2\pi\lambda^{-1}\sqrt{-1}\psi_0\big).$$

Hence the eigenvalue of $M(\tilde{\nabla}^\lambda)$ corresponding to v_i is as follows:

$$\exp\big(-2\pi\sqrt{-1}\lambda^{-1} \cdot \beta_i\big) \cdot \exp\big(2\pi\sqrt{-1}(\lambda^{-1} \cdot \beta_i + b_i)\big) = \exp\big(2\pi\sqrt{-1}b_i\big).$$

It implies the lemma. □

LEMMA 7.126. *There exists a positive number $C > 0$ such that the following holds for any i:*

$$\Big|M_\eta(\tilde{\nabla}^\lambda)v_i - \exp(2\pi\sqrt{-1}b_i)v_i\Big|_h \leq C \cdot |v_i| \cdot (-\log|z|)^{-1}.$$

Proof We have the following equality:

$$\begin{aligned} &M_\eta(\tilde{\nabla}^\lambda)v_i - \exp(2\pi\sqrt{-1}b_i) \cdot v_i \\ &= \exp\big(2\pi\sqrt{-1}(b_i + \lambda^{-1} \cdot \beta_i)\big) \cdot \Big(M(\mathbb{D}^{\lambda,f}) \cdot v_i - \exp(-2\pi\sqrt{-1}\lambda^{-1} \cdot \beta_i) \cdot v_i\Big) \\ &= \exp(2\pi\sqrt{-1}b_i) \cdot \sum_{n=1}^{\infty} \frac{(-2\pi)^n}{n!}(N_{\omega_i}^\lambda)^n v_i. \end{aligned} \tag{7.61}$$

Since we have $|(N_{\omega_i}^\lambda)^n v_i| \leq C \cdot |v_i| \cdot (-\log|z|)^{-1}$, we obtain the result. □

LEMMA 7.127. *There exists a positive number C satisfying the following:*

Let γ_i $(i=1,2)$ be elements of $\mathcal{S}p(M_\eta(\tilde{\nabla}^\lambda))$, and u_i be elements of $\mathbb{E}(M_\eta(\tilde{\nabla}^\lambda), \gamma_i)$. If $\gamma_1 \neq \gamma_2$, then the following holds:

$$\big|(u_1, u_2)_h\big| \leq C \cdot (-\log|z|)^{-1} \cdot |u_1|_h \cdot |u_2|_h. \tag{7.62}$$

Namely the generalized eigen-decomposition of the monodromy endomorphism of the connection $\tilde{\nabla}^\lambda$ is $(-\log|z|)^{-1}$-asymptotically orthogonal.

Proof Due to Lemma 7.124, we have the following inequality:

$$(7.63)\quad \Big|\big(M(\tilde{\nabla}^{\lambda})\cdot u_1, M(\tilde{\nabla}^{\lambda})\cdot u_2\big)_h - (u_1,u_2)_h\Big| \le C\cdot(-\log|z|)^{-1}\cdot|u_1|_h\cdot|u_2|_h.$$

On the other hand, we have the following inequality due to Lemma 7.126:

$$(7.64)\quad \Big|\big(M(\tilde{\nabla}^{\lambda})u,\ M(\tilde{\nabla}^{\lambda})v\big)_h - \gamma_1\cdot\bar{\gamma}_2\cdot(u,v)_h\Big|$$
$$= \Big|\big((M(\tilde{\nabla}^{\lambda})-\gamma_1)u,\ M(\tilde{\nabla}^{\lambda})v\big)_h + \big(\gamma_1\cdot u,\ (M(\tilde{\nabla}^{\lambda})-\gamma_2)v\big)_h\Big|$$
$$\le C'\cdot(-\log|z|)^{-1}\cdot|u|_h\cdot|v|_h.$$

Since we have $|\gamma_i|=1$, the condition $\gamma_1\neq\gamma_2$ implies $\gamma_1\cdot\bar{\gamma}_2\neq 1$. Hence we obtain the inequality (7.62) from (7.63) and (7.64). □

7.7.3. Asymptotic orthogonality for the parabolic filtration of $\mathcal{E}^0$.

LEMMA 7.128. *We pick $\lambda\in \boldsymbol{C}^*_{\lambda}$ satisfying the following:*

(1) *λ is generic (Definition 7.107).*
(2) *$u\neq u' \Longrightarrow \mathfrak{p}(\lambda,u)\neq\mathfrak{p}(\lambda,u')$.*

Proof Note $|\lambda|$ is sufficiently small, we may always assume the second condition holds. □

Let us pick λ as in Lemma 7.128. Let $\boldsymbol{e}$ be a frame of ${}^{\diamond}\mathcal{E}^0$ compatible with $\mathbb{E}$, F and W. Let $\boldsymbol{v}$ be a frame of ${}^{\diamond}\mathcal{E}^{\lambda}$ compatible with $\mathbb{E}$, F and W. We put $b(e_i)=\deg^F(e_i)$ and $b(v_i)=\deg^F(v_i)$. We put $k(e_i)=\deg^W(e_i)$ and $k(v_i)=\deg^W(v_i)$. We put $u(e_i)=\deg^{F,\mathbb{E}}(e_i)$ and $u(v_i)=\deg^{F,\mathbb{E}}(v_i)$.

We put $e'_i := e_i\cdot|z|^{b(e_i)}\cdot(-\log|z|)^{-k(e_i)/2}$ and $v'_i := v_i\cdot|z|^{b(v_i)}\cdot(-\log|z|)^{-k(v_i)/2}$. We have the bounded C^{∞}-function $B: X-D\longrightarrow M(r)$ determined by the following:

$$e'_i=\sum_j B'_{j\,i}\cdot v'_j.$$

Recall that $|B'_{j\,i}|\le C\cdot(-\log|z|)^{-1}$ unless $\mathfrak{k}\big(\lambda,u(e_i)\big)=u(v_j)$ (Corollary 7.72).

LEMMA 7.129. *There exists a positive constant $C>0$ satisfying the following condition:*

> *The inequality $\big|(e_i,e_j)_h\big|\le C\cdot(-\log|z|)^{-1}\cdot|e_i|_h\cdot|e_j|_h$ holds for some positive constant C in the case $u(e_i)\neq u(e_j)$.*

Proof We have already obtained stronger estimate in the case $\deg^{\mathbb{E}}(e_i)\neq\deg^{\mathbb{E}}(e_j)$ (Lemma 7.123). We have only to prove the following estimate, for some positive number C_1 in the case $b(e_i)\neq b(e_j)$:

$$\big|(e'_i,e'_j)_h\big|\le C_1\cdot(-\log|z|)^{-1}.$$

We have the following formula:

$$(7.65)\qquad (e'_i,e'_j)_h=\sum B'_{k\,i}\cdot\bar{B}'_{l\,j}\cdot(v'_k,v'_l)_h.$$

We have the inequality $|B'_{k\,i}|\cdot|\bar{B}'_{l\,j}|<C\cdot(-\log|z|)^{-1}$, unless the following condition holds:

$$(7.66)\qquad \mathfrak{k}\big(\lambda,u(e_i)\big)=u(v_k),\quad \mathfrak{k}\big(\lambda,u(e_j)\big)=u(v_l).$$

Assume that $b(e_i) \neq b(e_j)$ and the equalities (7.66) hold. Then we have $b(v_k) \neq b(v_l)$ due to our choice of λ. Thus we obtain the inequality $(v'_k, v'_l)_h \leq C \cdot (-\log|z|)^{-1}$ for some positive constant C. Hence the right hand side in (7.65) is dominated by $(-\log|z|)^{-1}$. □

Let us take a decomposition of the vector bundle $\mathbb{E}_{\epsilon_2}({}^{\diamond}\mathcal{E}^0, \alpha)$:

$$\mathbb{E}_{\epsilon_2}({}^{\diamond}\mathcal{E}^0, \alpha) = \bigoplus_{a \in \mathcal{P}ar({}^{\diamond}\mathcal{E}^0)} V_{(a,\alpha)}. \tag{7.67}$$

We assume that the decomposition gives a splitting of the parabolic filtration F, in the following sense:

$$F_a({}^{\diamond}\mathcal{E}^0{}_{|O}) = \bigoplus_{\alpha} \bigoplus_{b \leq a} V_{b,\alpha\,|\,O}. \tag{7.68}$$

For example, we can pick V_a as the vector subbundle of $\mathbb{E}_{\epsilon_2}({}^{\diamond}\mathcal{E}, \alpha)$ generated by $\{e_i \,|\, u(e_i) = (a, \alpha)\}$. On the other hand, if we are given such decomposition, then we can pick the frame $\boldsymbol{e}$ compatible with $\mathbb{E}$, F and W such that V_a is generated by $\{e_i \,|\, u(e_i) = (a, \alpha)\}$.

PROPOSITION 7.130. *If the condition* (7.68) *is satisfied, then the decomposition* (7.67) *is* $(-\log|z|)^{-1}$*-asymptotically orthogonal.*

Proof It follows from Lemma 7.129. □

7.7.4. Asymptotic orthogonality for the weight filtration. Let us take a model bundle $(E_0, \overline{\partial}_{E_0}, \theta_0, h_0)$ for $(E, \overline{\partial}_E, \theta, h)$ as in the subsection 7.3.1. We use the notation in the section 7.3. We also put $\Delta^*(C) = \{z \in \Delta, \,|\, 0 < |z| < C\}$ for some $C < 1$. Recall the finiteness due to Simpson (Lemma 7.66):

$$\int |\theta^{\dagger} - \theta_0^{\dagger}|_h^2 \cdot \bigl(-|z| \cdot \log|z|\bigr)^2 \cdot \frac{|dz \cdot d\bar{z}|}{|z|^2(-\log|z|)} < \infty.$$

Hence, for any $\epsilon > 0$ there exists a subset $Z_\epsilon \subset \Delta^*(C)$ satisfying the following:

- The volume of $\Delta^*(C) - Z_\epsilon$ is finite with respect to the measure $|dz \cdot d\bar{z}| \cdot |z|^{-2}(-\log|z|)^{-1}$.
- We have the estimate $|\theta^{\dagger} - \theta_0^{\dagger}|_h \cdot \bigl(-|z|\log|z|\bigr) \leq \epsilon$ on Z_ϵ.

The endomorphisms A_0 and A are determined by the following:

$$A_0 \cdot \frac{dz \cdot d\bar{z}}{|z|^2(-\log|z|)^2} = \theta_0 \cdot \theta_0^{\dagger} + \theta_0^{\dagger} \cdot \theta_0, \qquad A \cdot \frac{dz \cdot d\bar{z}}{|z|^2(-\log|z|)^2} = \theta \cdot \theta^{\dagger} + \theta^{\dagger} \cdot \theta.$$

Then A_0 is self dual with respect to h_0, and A is self dual with respect to h. By a direct calculation, it can be checked that the eigenvalues of A_0 are integers.

We have the decomposition $E_0 = \bigoplus_{k \in \mathbb{Z}} U_k$ satisfying the following:

- $W_k = \bigoplus_{h \leq k} U_h$.
- U_k is eigen space of A_0 corresponding to the integer k.

The following lemma is clear.

LEMMA 7.131.

- *There exist* $C_4 > 0$ *and* $\delta > 0$ *such that* $|\theta - \theta_0|_h \cdot (-|z|\log|z|) \leq C_4 \cdot |z|^{\delta}$.
- *There exists* $C_5 > 0$ *the inequality* $|A_0 - A|_h \leq C_5 \cdot \epsilon$ *holds on* Z_ϵ *for any* $\epsilon > 0$. □

Let ϵ be a sufficiently small positive number. Let v_i be a C^∞-section of U_{k_i} for $k_1 \neq k_2$. Since A is self dual with respect to h, we have $(Av_1, v_2)_h - (v_1, Av_2)_h = 0$. On the other hand, we have the following on Z_ϵ:

$$|A \cdot v_i - k_i \cdot v_i|_h = |A \cdot v_i - A_0 \cdot v_i|_h \leq C_5 \cdot \epsilon \cdot |v|_h.$$

Then we obtain the following:

$$(7.69) \quad |k_1 - k_2| \cdot \big|(v_1, v_2)_h\big| = \Big|\big(k_1 \cdot v_1, v_2\big)_h - \big(v_1, k_2 \cdot v_2\big)_h\Big|$$
$$= \Big|\big((k_1 - A)v_1, v_2\big)_h - \big(v_1, (k_2 - A)v_2\big)_h\Big| \leq 2C_5 \cdot \epsilon \cdot |v_1|_h \cdot |v_2|_h.$$

Here we have used the self-duality of A with respect to h. Hence we obtain the inequality $\big|(v_1, v_2)_h\big| \leq C_6 \cdot \epsilon \cdot |v_1|_h \cdot |v_2|_h$ for some $C_6 > 0$, which is independent of v_i $(i = 1, 2)$. In all, we obtain the following proposition.

PROPOSITION 7.132. *For any $\epsilon > 0$, there exists a subset $Z_\epsilon \subset \Delta^*(C)$ satisfying the following conditions:*

- *The measure of $\Delta^*(C) - Z_\epsilon$ with respect to $|dz \cdot d\bar{z}| \cdot |z|^{-2}(\log|z|)^{-1}$ is finite.*
- *Let v, w be sections of ${}^\diamond\mathcal{E}^0$ such that $\deg^{F,\mathbb{E}}(v) = \deg^{F,\mathbb{E}}(w)$ and $\deg^W(v) \neq \deg^W(w)$. On Z_ϵ, we have the estimate $|(v, w)| \leq \epsilon \cdot |v|_h \cdot |w|_h$.* □

REMARK 7.133. We remark that the volumes of $Z_\epsilon \cap \Delta^*(e^{-2})$ with respect the measure $|z|^{-2}\big(-\log|z|\big)^{-1} \cdot \log\big(-\log|z|\big)^{-1} \cdot |dz \cdot d\bar{z}|$ are also infinite. □

7.8. Maximum principle for the distance of the harmonic metrics

Let $(E^{(i)}, \overline{\partial}_{E^{(i)}}, h^{(i)}, \theta^{(i)})$ $(i = 1, 2)$ be tame harmonic bundles on $\overline{\Delta}^*$ of rank r. We denote the corresponding λ-connections by $\big(\mathcal{E}^{(i)\,\lambda}, \mathbb{D}^\lambda\big)$. Assume the following:

- The set of KMS-spectra are same, i.e., $\mathcal{KMS}(\mathcal{E}^{(1)\,0}) = \mathcal{KMS}(\mathcal{E}^{(2)\,0})$.
- Let λ be a complex number which is generic with respect to $\mathcal{KMS}(\mathcal{E}^{(i)\,0})$, and we are given a flat isomorphism $\Phi : \big(\mathcal{E}^{(1)\,\lambda}, \mathbb{D}^\lambda\big) \longrightarrow \big(\mathcal{E}^{(2)\,\lambda}, \mathbb{D}^\lambda\big)$.

LEMMA 7.134. *The functions $\log|\Phi|^2$ and $\log|\Phi^{-1}|^2$ are subharmonic on the disc Δ.*

Proof First, let us see the boundedness of Φ and Φ^{-1}. Since λ is generic, the harmonic metrics h_i of $\big(\mathcal{E}^{(i)\,\lambda}, \mathbb{D}^\lambda\big)$ are determined by the monodromy, up to boundedness. Use the norm estimate (Lemma 7.39) and the normalizing frame. given in the subsection 7.5.3.

We have the inequality $\Delta \log|\Phi|^2 \leq 0$ and $\Delta \log|\Phi^{-1}|^2 \leq 0$ on the punctured disc Δ^*, due to the Weitzenbeck formula of Simpson (Lemma 4.1 in [**81**]). Since $\log|\Phi|^2$ and $\log|\Phi^{-1}|^2$ are bounded, the inequalities hold on the disc Δ. (See Lemma 2.2 in [**81**] or Proposition 2.24.) Thus we are done. □

We identify $\big(\mathcal{E}^{(1)}, \mathbb{D}^\lambda\big)$ and $\big(\mathcal{E}^{(2)}, \mathbb{D}^\lambda\big)$ via the morphism Φ. For any point $P \in \overline{\Delta}^*$, we can consider the distance of $h_{1\,|\,P}$ and $h_{2\,|\,P}$ in the space $\mathcal{PH}(r)$ (the subsection 21.2.1).

LEMMA 7.135. *Let R be a real number such that the following inequalities hold for any point $Q \in \partial\overline{\Delta}$:*

$$d_{\mathcal{PH}(r)}\big(h_{1\,|\,Q}, h_{2\,|\,Q}\big) \leq R.$$

Then the following inequalities hold for any point $P \in \overline{\Delta}^*$*:*

$$(7.70)\quad d_{\mathcal{PH}(r)}\bigl(h_{1\,|\,P},h_{2\,|\,P}\bigr)\leq\Bigl(\frac{e^R-e^{-R}}{2R}\Bigr)\cdot\max\Bigl\{d_{\mathcal{PH}(r)}\bigl(h_{1\,|\,Q},h_{2\,|\,Q}\bigr)\,\Big|\,Q\in\partial\overline{\Delta}\Bigr\}.$$

Proof We always have the following, due to Lemma 21.21:

$$(7.71)\qquad d_{\mathcal{PH}(r)}\bigl(h_{1\,|\,P},h_{2\,|\,P}\bigr)\leq\frac{|\Phi_{|P}|^2+|\Phi_{|P}^{-1}|^2-2r}{2}.$$

For any point $Q\in\partial X$, we have the following, due to Lemma 21.21 again:

$$(7.72)\qquad \frac{|\Phi_{|Q}|^2+|\Phi_{|Q}^{-1}|^2-2r}{2}\leq\frac{e^R-e^{-R}}{2R}\cdot d_{\mathcal{PH}(r)}\bigl(h_{1\,|\,Q},h_{2\,|\,Q}\bigr).$$

Then (7.70) follows from the inequalities (7.71), (7.72) and Lemma 7.134. □

CHAPTER 8

Harmonic Bundles on a Product of Punctured Discs

8.1. Preliminary

8.1.1. KMS-Spectrum. We put $X := \Delta^n$, $D_i := \{z_i = 0\}$ and $D = \bigcup_{i=1}^{l} D_i$. Let π_i denote the projection $X \longrightarrow D_i$, forgetting the i-th component. Let $(E, \overline{\partial}_E, \theta, h)$ be a tame harmonic bundle over $X - D$ (see Definition 4.4 in [**65**]). We have the deformed holomorphic bundle $(\mathcal{E}^\lambda, \mathbb{D}^\lambda)$. Let i be an element of $\underline{l}$, and P be a point of $D_i^\circ := D_i - \bigcup_{j\neq i,\ 1\leq j\leq l} D_i \cap D_j$. We put as follows:

$$\mathcal{KMS}(\mathcal{E}^\lambda, i, P) := \mathcal{KMS}\big(\mathcal{E}^\lambda_{|\pi_i^{-1}(P)}\big).$$

LEMMA 8.1. *The set $\mathcal{KMS}(\mathcal{E}^\lambda, i, P)$ and the multiplicities of any element $u \in \mathcal{KMS}(\mathcal{E}^\lambda, i, P)$ are independent of $P \in D_i^\circ$.*

Proof Let P_i $(i = 1, 2)$ be points of D_i°. For any $\lambda \in \boldsymbol{C}_\lambda^*$, we have $\mathcal{S}p(\mathcal{E}^\lambda, i, P_1) = \mathcal{S}p(\mathcal{E}^\lambda, i, P_2)$, and the multiplicities are same. Then we obtain the result due to Lemma 7.110. □

In the following, we use the notation $\mathcal{KMS}(\mathcal{E}^\lambda, i)$ instead of $\mathcal{KMS}(\mathcal{E}^\lambda, i, P)$. Similarly, the sets $\overline{\mathcal{KMS}}(\mathcal{E}^\lambda, i)$, $\mathcal{S}p(\mathcal{E}^\lambda, i)$ and $\mathcal{P}ar(\mathcal{E}^\lambda, i)$ are obtained.

For any element $\boldsymbol{b} \in \boldsymbol{R}^l$, the sets $\mathcal{KMS}({}_{\boldsymbol{b}}\mathcal{E}^\lambda, i) := \mathcal{KMS}\big({}_{b_i}\big(\mathcal{E}^\lambda_{|\pi_i^{-1}(P)}\big)\big)$ are also obtained, where b_i denotes the i-th component of $\boldsymbol{b}$. Similarly, $\mathcal{P}ar\big({}_{\boldsymbol{b}}\mathcal{E}^\lambda, i\big)$ and $\mathcal{S}p\big({}_{\boldsymbol{b}}\mathcal{E}^\lambda, i\big)$ are obtained.

Let us consider the Higgs field:

$$\theta = \sum_{i=1}^{l} f_i \cdot \frac{dz_i}{z_i} + \sum_{j=l+1}^{n} g_j \cdot dz_j.$$

We have the characteristic polynomials $\det(t - f_i)$ and $\det(t - g_j)$, whose coefficients are holomorphic functions defined over $X - D$. Since our harmonic bundle is tame, the coefficients are prolonged to the holomorphic functions defined over X, by definition. We denote them by the same notation.

LEMMA 8.2. *We have* $\det(t - f_i)_{|P_1} = \det(t - f_i)_{|P_2}$ *if* $P_a \in D_i$ $(a = 1, 2)$.

Proof If $P \in D_i^\circ$, then we have $\det(t - f_i)_{|P} = \det\big(t - f_{i\,|\pi_i^{-1}(P)}\big)_{|P}$. Due to Lemma 8.1, the right hand side is independent of a choice of a point $P \in D_i^\circ$. Then we obtain the result for D_i. □

LEMMA 8.3. *The norms $\big|f_i\big|_h$ and $\big|g_j\big|_h$ are bounded.*

Proof The boundedness of $|f_i|_h$ follows from the bounded in the one dimensional case (the claim (I) of Proposition 7.2). We remark that the eigenvalues of g_j are bounded. Then the boundedness $|g_j|_h$ is shown in [**82**], for example. □

8.1.2. Rank 1. Let $(E, \overline{\partial}_E, \theta, h)$ be a tame harmonic bundle of rank 1 over $X - D$. In this case, the set $\mathcal{KMS}({}^\diamond\mathcal{E}^0, i)$ consists of only one element u_i for $i = 1, \ldots, l$. We have the model bundle $L(\boldsymbol{u})$ for $\boldsymbol{u} = (u_1, \ldots, u_l)$ (see the section 6.1 for $L(\boldsymbol{u})$). Then $(E, \overline{\partial}_E, \theta, h) \otimes L(\boldsymbol{u})$ is tame and nilpotent. Moreover the parabolic structure is trivial. Hence it is the restriction of the harmonic bundle $(E_0, \overline{\partial}_{E_0}, \theta_0, h_0)$ of rank 1 over X, due to Corollary 4.10 in [**65**]. Namely we obtain the following.

PROPOSITION 8.4. *Let $(E, \overline{\partial}_E, \theta, h)$ be a tame harmonic bundle over $X - D$ of rank 1. Then it is isomorphic to the tensor product of a model bundle $L(\boldsymbol{u})$ and a harmonic bundle $(E_0, \overline{\partial}_{E_0}, h_0, \theta_0)$ over X.* □

8.1.3. A characterization of tameness.

LEMMA 8.5. *Let X be a simply connected compact region of $\boldsymbol{C}^n$. Let $\pi : X \times \overline{\Delta}^* \longrightarrow X$ denote the projection. Let $(E, \overline{\partial}_E, \theta, h)$ be a harmonic bundle on $X \times \overline{\Delta}^*$. Assume the following:*

- *For any point $P \in X$, the restriction $(E, \overline{\partial}_E, \theta, h)_{|\pi^{-1}(P)}$ is tame with respect to $(P, O) \in \pi^{-1}(P)$.*

Then the harmonic bundle $(E, \overline{\partial}_E, \theta, h)$ is tame with respect to the divisor $X \times \{O\}$.

Proof Let $(z_1, \ldots, z_n)$ be the coordinate of X, and w be the coordinate of $\overline{\Delta}$. We have the description $\theta = \sum_{i=1}^n g_i \cdot dz_i + f \cdot dw/w$. We would like to show that the coefficients of $\det(t - f)$ and $\det(t - g_i)$ are holomorphic on $X \times \overline{\Delta}$. Due to our assumption, the coefficients of $\det(t - f)_{|\pi^{-1}(P)}$ are holomorphic for any $P \in X$. Then it is easy to derive that the coefficients of $\det(t - f)$ are holomorphic on $X \times \overline{\Delta}$. So we have only to see $\det(t - g_i)$.

We remark that $\mathcal{KMS}(\mathcal{E}^0_{|\pi^{-1}(P)})$ is independent of a choice of $P \in X$, which can be shown by the same argument as the proof of Lemma 8.1. Let us take a complex number λ, which is generic with respect to $\mathcal{KMS}(\mathcal{E}^0_{|\pi^{-1}(P)})$. We have the flat bundle $(\mathcal{E}^\lambda, \mathbb{D}^{\lambda,f})$ on $X \times \overline{\Delta}^*$. In particular, we have the following flat isomorphism for any points $P_1, P_2 \in X$:

$$(\mathcal{E}^\lambda, \mathbb{D}^{\lambda,f})_{|\pi^{-1}(P_1)} \simeq (\mathcal{E}^\lambda, \mathbb{D}^{\lambda,f})_{|\pi^{-1}(P_2)}. \tag{8.1}$$

We put $h_P := h_{|\pi^{-1}(P)}$. For any point $Q \in \overline{\Delta}^*$, we can consider the distance $d_{\mathcal{PH}(r)}(h_{P_1\,|\,Q}, h_{P_2\,|\,Q})$ for any points $P_i \in X$, via the isomorphism (8.1).

Let P be a point of X, and v be a real tangent vector of X at P. It naturally gives the tangent vector v_Q of $X \times \overline{\Delta}^*$ at (P, Q) for any point $Q \in \overline{\Delta}^*$. From Lemma 7.135 and the formula (2.54), we obtain the following maximum principle:

$$\left|\overline{\lambda} \cdot \theta(v_Q) + \lambda \cdot \theta^\dagger(v_Q)\right|^2 \leq \max\Big\{\left|\overline{\lambda} \cdot \theta(v_{Q'}) + \lambda \cdot \theta^\dagger(v_{Q'})\right|^2 \,\Big|\, Q' \in \partial\overline{\Delta}\Big\}. \tag{8.2}$$

We remark that $(2R)^{-1} \cdot (e^R - e^{-R})$ goes to 1 when R goes to 0.

Let us use the real coordinate $z_i = x_i + \sqrt{-1}y_i$. Let ∂_{x_i} and ∂_{y_i} denote the vector fields $\partial/\partial x_i$ and $\partial/\partial y_i$. Since $X \times \partial\overline{\Delta}$ is compact, we obtain the boundedness of the following, from (8.2):

$$\overline{\lambda}\cdot\theta(\partial_{x_i})+\lambda\cdot\theta^{\dagger}(\partial_{x_i})=\overline{\lambda}\cdot g_i+\lambda\cdot g_i^{\dagger}.$$

$$\overline{\lambda}\cdot\theta(\partial_{y_i})+\lambda\cdot\theta^{\dagger}(\partial_{y_i})=\overline{\lambda}\cdot g_i-\lambda\cdot g_i^{\dagger}.$$

Hence we obtain the boundedness of g_i.

Then we obtain the boundedness of the coefficients of $\det(t-g_i)$, which implies that they are holomorphic on $X\times\overline{\Delta}$. □

LEMMA 8.6. *We put $X=\Delta^n$, $D_i:=\{z_i=0\}$ and $D=\bigcup_{i=1}^{l}D_i$. Let π_i denote the naturally defined projection $X\longrightarrow D_i$. Let $(E,\overline{\partial}_E,\theta,h)$ be a harmonic bundle on $X-D$. Assume the following:*

- *For any point $Q\in D_i^{\circ}=D_i\setminus\bigcup_{j\neq i}D_j$, the restriction $(E,\overline{\partial}_E,\theta,h)_{|\pi^{-1}(Q)}$ is tame.*

Then the harmonic bundle is tame with respect to the divisor D.

Proof We put $D^{[2]}:=\bigcup_{i\neq j}D_i\cap D_j$, which is of codimension 2 in X. We have the description:

$$\theta=\sum_{i=1}^{l}f_i\cdot\frac{dz_i}{z_i}+\sum_{i=l+1}^{n}g_i\cdot dz_i.$$

Due to Lemma 8.5, the coefficients of $\det(t-f_i)$ and $\det(t-g_i)$ are holomorphic on $X-D^{[2]}$. Then we can conclude that they are holomorphic on X due to the theorem of Hartogs. □

We have a straightforward corollary.

COROLLARY 8.7. *Let X be a complex manifold, and D be a normal crossing divisor of X. Let $(E,\overline{\partial}_E,\theta,h)$ be a harmonic bundle on $X-D$. Assume the following:*

- *Let C be any smooth curve contained in X, which intersects with the smooth part of D transversely. Then the restriction $(E,\overline{\partial}_E,\theta,h)_{|C}$ is tame.*

Then the harmonic bundle $(E,\overline{\partial}_E,\theta,h)$ is tame. □

8.2. Simpson's Main estimate in the higher dimensional case

8.2.1. Preliminary decomposition. We put $X=\Delta^n$, $D_i=\{z_i=0\}$ and $D=\bigcup_{i=1}^{l}D_i$. Let $(E,\overline{\partial}_E,\theta,h)$ be a tame harmonic bundle on $X-D$. If we replace X by a sufficiently small neighbourhood of the origin O, and if we replace the coordinate appropriately, then we may assume that there exist positive constants ϵ, ϵ_1 and C, satisfying the following conditions:

- We have the decomposition $E=\bigoplus_{a\in\mathcal{S}p(\mathcal{E}^0,i)}\mathbb{E}_{\epsilon_1}(f_i,a)$ over $X-D$, and the decomposition $f_i=\bigoplus_{a\in\mathcal{S}p(\mathcal{E}^0,i)}f_{i\,a}$. (See the subsection 2.7.3 for $\mathbb{E}_{\epsilon_1}$.)
- We have the inequality $|\alpha-a|<C\cdot|z|^{\epsilon}$ for any eigenvalue of $f_{i\,a}$.

We may also assume $\sum_{a\in\mathcal{S}p(\mathcal{E}^0,i)}|a|^2\neq 0$ for any i, by tensoring some model bundle $L(\boldsymbol{u})$ of rank 1.

Since f_k $(k=1,\dots,l)$ and g_j $(j=l+1,\dots,n)$ are commutative, $\mathbb{E}_{\epsilon_1}(f_i,a)$ are preserved by f_k and g_j. By an inductive procedure, we may assume the following:

CONDITION 8.8.

- The subset $\mathcal{S}p(\theta) \subset \prod_{i=1}^{l} \mathcal{S}p(\mathcal{E}^0, i)$ is given.
- The holomorphic decomposition $E = \bigoplus_{\boldsymbol{a} \in \mathcal{S}p(\theta)} E_{\boldsymbol{a}}$ is given.
- Each $E_{\boldsymbol{a}}$ is preserved by f_k $(k = 1, \ldots, l)$ and g_j $(j = l+1, \ldots, n)$. Hence we have the decompositions:

$$f_i = \bigoplus_{\boldsymbol{a} \in \mathcal{S}p(\theta)} f_{i\,\boldsymbol{a}}, \quad f_{i\,\boldsymbol{a}} \in End(E_{\boldsymbol{a}}),$$

$$g_j = \bigoplus_{\boldsymbol{a} \in \mathcal{S}p(\theta)} g_{j\,\boldsymbol{a}}, \quad g_{j\,\boldsymbol{a}} \in End(E_{\boldsymbol{a}}).$$

- There exist positive constants ϵ and C such that the inequality $|\alpha - q_i(\boldsymbol{a})| \leq C \cdot |z|^{\epsilon}$ holds for any eigenvalue α of $f_{i\,\boldsymbol{a}}$. Here q_i denotes the projection onto the i-th component.
- We have $\sum_{a \in \mathcal{S}p(\mathcal{E}^0, i)} |a|^2 \neq 0$. □

REMARK 8.9. The condition $\sum_{a \in \mathcal{S}p(\mathcal{E}^0, i)} |a|^2 \neq 0$ is not essential. To see it, we have only to consider the tensor product of $(E, \overline{\partial}_E, \theta, h)$ and an appropriate model bundle $L(\boldsymbol{b})$. □

LEMMA 8.10.

- *We put* $h_0 := \bigoplus_{\boldsymbol{a} \in \mathcal{S}p(\theta)} h_{|E_{\boldsymbol{a}}}$. *Then* h_0 *and* h *are mutually bounded.*
- *Namely there exist positive constants* C_1 *and* C_2 *such that the following inequalities hold for any section* $v = \sum_{\boldsymbol{a} \in \mathcal{S}p(\theta)} v_{\boldsymbol{a}}$:

$$C_1 \sum_{\boldsymbol{a} \in \mathcal{S}p(\theta)} |v_{\boldsymbol{a}}|_h \leq |v|_h \leq C_2 \sum_{\boldsymbol{a} \in \mathcal{S}p(\theta)} |v_{\boldsymbol{a}}|_h.$$

Proof Since we know the boundedness of $\left|f_i\right|_h$ for $i = 1, \ldots, l$ (Lemma 8.3), we have only to apply Lemma 2.37 inductively. □

LEMMA 8.11. *Let* $v_{\boldsymbol{a}}$ *and* $v_{\boldsymbol{b}}$ *be* C^{∞}*-sections of* $E_{\boldsymbol{a}}$ *and* $E_{\boldsymbol{b}}$ *respectively. We put* $\mathrm{Diff}(\boldsymbol{a}, \boldsymbol{b}) := \left\{ i \,\middle|\, q_i(\boldsymbol{a}) \neq q_i(\boldsymbol{b}) \right\}$. *Then there is a positive constant* ϵ *such that the following holds:*

$$\left|(v_{\boldsymbol{a}}, v_{\boldsymbol{b}})\right| \leq C \cdot \prod_{i \in \mathrm{Diff}(\boldsymbol{a}, \boldsymbol{b})} |z_i|^{\epsilon} \cdot \left|v_{\boldsymbol{a}}\right|_h \cdot \left|v_{\boldsymbol{b}}\right|_h.$$

Proof Due to the asymptotic orthogonality in the one dimensional case (Proposition 7.37), we have the following, for some $\epsilon' > 0$:

$$\left|(v_{\boldsymbol{a}}, v_{\boldsymbol{b}})\right| \leq C \cdot \left(\min_{i \in \mathrm{Diff}(\boldsymbol{a}, \boldsymbol{b})} |z_i|^{\epsilon'} \right) \cdot \left|v_{\boldsymbol{a}}\right|_h \cdot \left|v_{\boldsymbol{b}}\right|_h.$$

Then the claim immediately follows. □

We put ${}^iE_a = \bigoplus_{q_i(\boldsymbol{a})=a} E_{\boldsymbol{a}}$, on which we have two metrics $\bigoplus_{q_i(\boldsymbol{a})=a} h_{|E_{\boldsymbol{a}}}$ and $h_{|{}^iE_a}$.

LEMMA 8.12. *The metrics* $h_{|{}^iE_a}$ *and* $\bigoplus_{q_i(\boldsymbol{a})=a} h_{|E_{\boldsymbol{a}}}$ *are mutually bounded.*

Proof Similar to Lemma 8.10 □

8.2.2. Inequalities. The restriction of f_i to ${}^jE_a$ is denoted by ${}^jf_{i\,a}$.

LEMMA 8.13. *We have the following inequalities for a positive constant C and for any a:*

$$|{}^if_{i\,a} - a\cdot id_{{}^iE_a}|_h \le C\cdot(-\log|z_i|)^{-1}. \tag{8.3}$$

Proof It follows from Corollary 7.31. □

LEMMA 8.14. *We have the following inequality for some positive constant C':*

$$|f_{i\,\boldsymbol{a}} - q_i(\boldsymbol{a})\cdot id_{E_{\boldsymbol{a}}}|_h \le C'\cdot(-\log|z_i|)^{-1}. \tag{8.4}$$

Proof The estimate (8.4) follows from (8.3) and Lemma 8.12. □

We have the decomposition $g_j = \bigoplus g_{j\,\boldsymbol{a}}$, where $g_{j\,\boldsymbol{a}} \in End(E_{\boldsymbol{a}})$. The following lemma is easy.

LEMMA 8.15. *We have $|g_{j\,\boldsymbol{a}}|_h \le C$ for some positive constant C.* □

We have the adjoint maps $f_i^{\dagger}$ and the decomposition $f_i^{\dagger} = \sum (f_i^{\dagger})_{\boldsymbol{a}\,\boldsymbol{b}}$, where $(f_i^{\dagger})_{\boldsymbol{a}\,\boldsymbol{b}} \in Hom(E_{\boldsymbol{a}}, E_{\boldsymbol{b}})$. We also have the adjoint $g_j^{\dagger}$ and the decomposition $g_j^{\dagger} = \sum (g_j^{\dagger})_{\boldsymbol{a}\,\boldsymbol{b}}$, where $(g_j^{\dagger})_{\boldsymbol{a}\,\boldsymbol{b}} \in Hom(E_{\boldsymbol{a}}, E_{\boldsymbol{b}})$.

For any elements $\boldsymbol{a}$ and $\boldsymbol{b}$ of $\boldsymbol{C}^l$, we put $\mathrm{Diff}(\boldsymbol{a},\boldsymbol{b}) := \{j \,|\, q_j(\boldsymbol{a}) \ne q_j(\boldsymbol{b})\}$.

LEMMA 8.16. *There exist positive constants C and ϵ such that the following holds:*

(A): $\big|(f_i^{\dagger})_{\boldsymbol{a}\,\boldsymbol{a}} - \overline{q_i(\boldsymbol{a})}\cdot id_{E_{\boldsymbol{a}}}\big|_h \le C\cdot(-\log|z_i|)^{-1}$.

(B): *If $q_i(\boldsymbol{a}) = q_i(\boldsymbol{b})$ and $\boldsymbol{a}\ne\boldsymbol{b}$, we have the following inequality:*

$$\big|(f_i^{\dagger})_{\boldsymbol{a}\,\boldsymbol{b}}\big|_h \le C\cdot(-\log|z_i|)^{-1}\cdot \prod_{k\in\mathrm{Diff}(\boldsymbol{a},\boldsymbol{b})} |z_k|^{\epsilon}.$$

(C): *If $q_i(\boldsymbol{a}) \ne q_i(\boldsymbol{b})$, then we have the following inequality:*

$$\big|(f_i^{\dagger})_{\boldsymbol{a}\,\boldsymbol{b}}\big|_h \le C\cdot \prod_{k\in\mathrm{Diff}(\boldsymbol{a},\boldsymbol{b})} |z_k|^{\epsilon}.$$

(D): *We have the following inequality:*

$$\big|(g_j^{\dagger})_{\boldsymbol{a}\,\boldsymbol{b}}\big|_h \le C\cdot \prod_{k\in\mathrm{Diff}(\boldsymbol{a},\boldsymbol{b})} |z_k|^{\epsilon}.$$

Proof We put $\rho_i := \sum_a a\cdot id_{{}^iE_a}$, and $\bar\rho_i := \sum_a \bar a\cdot id_{{}^iE_a}$. We denote the adjoint of ρ_i by $\rho_i^{\dagger}$. Due to the result in the case of curves, we obtain the following:

- $|\rho_i^{\dagger} - \bar\rho_i|_h \le C\cdot|z_i|^{\epsilon}$, (Lemma 7.33).
- $|f_i - \rho_i|_h \le C\cdot(-\log|z_i|)^{-1}$, (Corollary 7.30).

In particular, we have $|f_i^{\dagger} - \bar\rho_i|_h \le C\cdot(-\log|z_i|)^{-1}$. It implies (A) due to Lemma 8.10.

We also have the following decomposition for $j\ne i$:

$$(f_i-\rho_i)^{\dagger} = \sum {}^j(f_i-\rho_i)^{\dagger}_{a,b}, \qquad {}^j(f_i-\rho_i)^{\dagger}_{a,b} \in Hom({}^jE_a, {}^jE_b).$$

We have the following inequalities due to Lemma 2.37 and Lemma 2.39:

$$\big|{}^j(f_i-\rho_i)^{\dagger}_{a\,b}\big|_h \le C\cdot|z_j|^{\epsilon}\cdot|f_i-\rho_i|_h \le C\cdot|z_j|^{\epsilon}\cdot(-\log|z_i|)^{-1}, \quad (a\ne b),$$

$$\big|{}^j(f_i-\rho_i)^{\dagger}_{a\,a}\big|_h \le C\cdot|f_i-\rho_i|_h \le C\cdot(-\log|z_i|)^{-1}, \qquad (a=b),$$

Due to Lemma 8.12, there exist positive constants C and ϵ such that the following holds:

$$(8.5)\qquad \left|(f_i-\rho_i)^{\dagger}_{\boldsymbol{a}\,\boldsymbol{b}}\right|_h \leq \begin{cases} C\cdot|z_j|^{\epsilon}\cdot(-\log|z_i|)^{-1}, & (q_j(\boldsymbol{a})\neq q_j(\boldsymbol{b})),\\ C\cdot(-\log|z_i|)^{-1}, & (q_j(\boldsymbol{a})=q_j(\boldsymbol{b})).\end{cases}$$

For any subset $I\subset \underline{l}$ and for any positive number ϵ, there exists a positive number ϵ' such that the following inequality holds:

$$(8.6)\qquad \min\bigl\{|z_i|^{\epsilon}\,\big|\, i\in I\bigr\}\leq \prod_{i\in I}|z_i|^{\epsilon'}.$$

From (8.5) and (8.6), we obtain the following inequality, for some sufficiently small $\epsilon>0$:

$$(8.7)\qquad \left|(f_i-\rho_i)^{\dagger}_{\boldsymbol{a}\,\boldsymbol{b}}\right|_h \leq C\cdot(-\log|z_i|)^{-1}\cdot \prod_{\substack{j\neq i,\\ j\in \mathrm{Diff}(\boldsymbol{a},\boldsymbol{b})}}|z_j|^{\epsilon}.$$

Similarly we have the following:

$$|(\rho_i^{\dagger})_{\boldsymbol{a},\boldsymbol{b}}|_h\leq \begin{cases} C\cdot|z_j|^{\epsilon}, & \text{if } q_j(\boldsymbol{a})\neq q_j(\boldsymbol{b}),\\ C, & \text{if } q_j(\boldsymbol{a})=q_j(\boldsymbol{b}).\end{cases}$$

By definition, we have $\bar{\rho}_{i\,\boldsymbol{a}\,\boldsymbol{b}}=0$ if $\boldsymbol{a}\neq\boldsymbol{b}$. We also have the following for any $\boldsymbol{a}$ and $\boldsymbol{b}$ from Lemma 7.33:

$$\left|(\rho_i^{\dagger}-\bar{\rho}_i)_{\boldsymbol{a}\,\boldsymbol{b}}\right|_h\leq C\cdot|z_i|^{\epsilon}.$$

In all, we obtain the following for some positive constants C and ϵ:

$$(8.8)\qquad |(\rho_i^{\dagger}-\bar{\rho}_i)_{\boldsymbol{a}\,\boldsymbol{b}}|_h\leq C\cdot|z_i|^{\epsilon}\cdot \prod_{\substack{j\neq i\\ j\in \mathrm{Diff}(\boldsymbol{a},\boldsymbol{b})}}|z_j|^{\epsilon}.$$

Hence, in the case $\boldsymbol{a}\neq\boldsymbol{b}$, we obtain the following inequality from (8.7), (8.8) and $\bar{\rho}_{i\,\boldsymbol{a},\boldsymbol{b}}=0$ $(\boldsymbol{a}\neq\boldsymbol{b})$:

$$|(f_i^{\dagger})_{\boldsymbol{a}\,\boldsymbol{b}}|_h\leq C\cdot(-\log|z_j|)^{-1}\cdot \prod_{\substack{j\neq i\\ j\in \mathrm{Diff}(\boldsymbol{a},\boldsymbol{b})}}|z_j|^{\epsilon}.$$

It implies the claim (B).

In the case $j\neq i$, we have the following for some positive constants C and ϵ:

$$(8.9)\qquad \left|{}^j(f_i^{\dagger})_{a,b}\right|_h\leq \begin{cases} C\cdot|z_j|^{\epsilon}, & (a\neq b),\\ C, & (a=b).\end{cases}$$

From (8.9) and (8.6), we obtain the following estimate, for some positive constants C and ϵ:

$$\left|(f_i^{\dagger})_{\boldsymbol{a}\,\boldsymbol{b}}\right|_h\leq \begin{cases} C\cdot|z_j|^{\epsilon}, & \bigl(q_j(\boldsymbol{a})\neq q_j(\boldsymbol{b})\bigr),\\ C, & \bigl(q_j(\boldsymbol{a})=q_j(\boldsymbol{b})\bigr).\end{cases}$$

On the other hand, in the case $q_i(\boldsymbol{a})\neq q_i(\boldsymbol{b})$, we have the following for some positive constants C and ϵ:

$$\left|(f_i^{\dagger})_{\boldsymbol{a}\,\boldsymbol{b}}\right|_h\leq C\cdot|z_i|^{\epsilon}.$$

In all, we obtain the following inequality for some positive constants C and ϵ, if $q_i(\boldsymbol{a}) \neq q_i(\boldsymbol{b})$:

$$|(f_i^{\dagger})_{\boldsymbol{a}\,\boldsymbol{b}}|_h \leq C \cdot \prod_{j \in \mathrm{Diff}(\boldsymbol{a},\boldsymbol{b})} |z_j|^{\epsilon}.$$

Hence we obtain the claim (C).

The claim (D) can be obtained similarly. □

8.2.3. Some consequences. Let $\boldsymbol{a}$ and $\boldsymbol{b}$ be elements of $\mathcal{S}p(\theta)$. We put as follows:

$$F_{(i,\boldsymbol{a}),(j,\boldsymbol{b})} := f_{i\,\boldsymbol{a}} \circ (f_j^{\dagger})_{\boldsymbol{b}\,\boldsymbol{a}} - (f_j^{\dagger})_{\boldsymbol{b}\,\boldsymbol{a}} \circ f_{i\,\boldsymbol{b}}.$$

LEMMA 8.17. *There exist positive constants C and ϵ such that the following inequalities hold:*
The case $q_i(\boldsymbol{a}) = q_i(\boldsymbol{b})$ and $q_j(\boldsymbol{a}) = q_j(\boldsymbol{b})$:

$$|F_{(i,\boldsymbol{a}),(j,\boldsymbol{b})}|_h \leq C \cdot (-\log|z_i|)^{-1} \cdot (-\log|z_j|)^{-1} \cdot \prod_{k \in \mathrm{Diff}(\boldsymbol{a},\boldsymbol{b})} |z_k|^{\epsilon}.$$

The case $q_i(\boldsymbol{a}) = q_i(\boldsymbol{b})$ and $q_j(\boldsymbol{a}) \neq q_j(\boldsymbol{b})$:

$$|F_{(i,\boldsymbol{a}),(j,\boldsymbol{b})}|_h \leq C \cdot (-\log|z_i|)^{-1} \cdot \prod_{k \in \mathrm{Diff}(\boldsymbol{a},\boldsymbol{b})} |z_k|^{\epsilon}.$$

The case $q_i(\boldsymbol{a}) \neq q_i(\boldsymbol{b})$ and $q_j(\boldsymbol{a}) \neq q_j(\boldsymbol{b})$:

$$|F_{(i,\boldsymbol{a}),(j,\boldsymbol{b})}|_h \leq C \cdot \prod_{k \in \mathrm{Diff}(\boldsymbol{a},\boldsymbol{b})} |z_k|^{\epsilon}.$$

Proof Let us consider the case $q_i(\boldsymbol{a}) = q_i(\boldsymbol{b})$ and $q_j(\boldsymbol{a}) = q_j(\boldsymbol{b}) = d$. We put $\delta_{\boldsymbol{a},\boldsymbol{b}} := 1$ $(\boldsymbol{a} \neq \boldsymbol{b})$, or $\delta_{\boldsymbol{a},\boldsymbol{b}} := 0$ $(\boldsymbol{a} \neq \boldsymbol{b})$. Then we have the following:

$$\begin{aligned}(8.10)\quad F_{(i,\boldsymbol{a}),(j,\boldsymbol{b})} = &\Big(f_{i\,\boldsymbol{a}} - c \cdot \delta_{\boldsymbol{a}\,\boldsymbol{b}} \cdot id_{E_{\boldsymbol{a}\,\boldsymbol{b}}}\Big) \circ \Big((f_j^{\dagger})_{\boldsymbol{b},\boldsymbol{a}} - \bar{d} \cdot \delta_{\boldsymbol{a}\,\boldsymbol{b}} \cdot id_{E_{\boldsymbol{a}\,\boldsymbol{b}}}\Big)\\ &- \Big((f_j^{\dagger})_{\boldsymbol{b},\boldsymbol{a}} - \bar{d} \cdot \delta_{\boldsymbol{a}\,\boldsymbol{b}} \cdot id_{E_{\boldsymbol{a}\,\boldsymbol{b}}}\Big) \circ \Big(f_{i\,\boldsymbol{b}} - c \cdot \delta_{\boldsymbol{a}\,\boldsymbol{b}} \cdot id_{E_{\boldsymbol{a}\,\boldsymbol{b}}}\Big).\end{aligned}$$

Therefore we obtain the following inequality:

$$|F_{(i,\boldsymbol{a}),(j,\boldsymbol{b})}|_h \leq C \cdot (-\log|z_i|)^{-1} \cdot (-\log|z_j|)^{-1} \cdot \prod_{k \in \mathrm{Diff}(\boldsymbol{a},\boldsymbol{b})} |z_k|^{\epsilon}.$$

Thus we are done in the case $q_i(\boldsymbol{a}) = q_i(\boldsymbol{b})$ and $q_j(\boldsymbol{a}) = q_j(\boldsymbol{b}) = d$. The rest cases can be shown similarly. □

PROPOSITION 8.18. *The two form $\theta \wedge \theta^{\dagger} + \theta^{\dagger} \wedge \theta$ is dominated by the Kahler form of the Poincaré metric of $X - D$:*

$$\omega_{X-D} = \omega_{\mathbf{P}} = \sum_{i=1}^{l} \frac{\sqrt{-1} \cdot dz_i \cdot d\bar{z}_i}{|z_i|^2(-\log|z_i|)^2} + \sum_{i=l+1}^{n} \frac{\sqrt{-1} \cdot dz_i \cdot d\bar{z}_i}{(1-|z_i|^2)^2}.$$

Proof The contributions of $f_i \cdot dz_i/z_i$ and $f_j^{\dagger} \cdot d\bar{z}_j/\bar{z}_j$ to the form $\theta \wedge \theta^{\dagger} + \theta^{\dagger} \wedge \theta$ is a sum of the following forms:

$$F_{(i,\boldsymbol{a}),(j,\boldsymbol{b})} \cdot \frac{dz_i}{z_i} \wedge \frac{d\bar{z}_j}{\bar{z}_j}.$$

Then it is dominated by the form $q_i^* \omega_{\Delta^*} + q_j^* \omega_{\Delta^*}$. The rest terms can be dominated similarly. □

We have the unitary connection of $\mathcal{E}^\lambda$ given by the following:

$$\overline{\partial}_{\mathcal{E}^\lambda} + \partial_{\mathcal{E}^\lambda} = \overline{\partial}_E + \partial_E + \lambda \cdot \theta^\dagger - \bar{\lambda} \cdot \theta.$$

The curvature $R(\overline{\partial}_{\mathcal{E}^\lambda} + \partial_{\mathcal{E}^\lambda})$ is as follows:

$$\big(1 + |\lambda|^2\big) \cdot \big(\theta \wedge \theta^\dagger + \theta^\dagger \wedge \theta\big).$$

COROLLARY 8.19. *The curvature $R(\overline{\partial}_{\mathcal{E}^\lambda} + \partial_{\mathcal{E}^\lambda})$ is dominated by $\omega_{\mathbf{p}}$. Namely, the hermitian holomorphic bundle $(\mathcal{E}^\lambda, h)$ is acceptable.*

Proof It follows from Proposition 8.18. □

REMARK 8.20. The curvature of the hermitian holomorphic bundle $(\mathcal{E}, h)$ over $\mathcal{X} - \mathcal{D} = (X - D) \times \boldsymbol{C}_\lambda$ contains the terms of the following form:

$$-d\overline{\lambda} \wedge \theta + d\lambda \wedge \theta^\dagger.$$

Hence $(\mathcal{E}, h)$ is not acceptable, unless the residue of θ is nilpotent. □

8.3. Prolongation in the case that λ is generic

8.3.1. The quasi canonical prolongation. We continue to use the setting in the section 8.2. Definition 7.107 is generalized as follows.

DEFINITION 8.21. λ is called generic with respect to $(E, \overline{\partial}_E, \theta, h)$ if the maps $\mathfrak{e}(\lambda) : \mathcal{KMS}(\mathcal{E}^0, i) \longrightarrow \mathcal{S}p(\mathcal{E}^\lambda, i)$ are bijective for any i. □

Assume that λ is generic. Let P be a point of $X - D$. Then we have the endomorphisms M_i^λ on $\mathcal{E}^\lambda_{|P}$. Here M_i^λ is the monodromy with respect to the loop around D_i with the anti-clockwise direction. We have the set of eigenvalues $\mathcal{S}p^f(\mathcal{E}^\lambda, i)$ of M_i^λ. Then we have the set $\mathcal{S}p^f(\mathcal{E}^\lambda, \underline{l}) \subset \prod_{i=1}^l \mathcal{S}p^f(\mathcal{E}^\lambda, i)$, and the generalized eigen decomposition of $\mathcal{E}^\lambda_{|P}$ with respect to the tuple of the endomorphisms $\boldsymbol{M}^\lambda = (M_1^\lambda, \ldots, M_l^\lambda)$ (the subsection 2.7.2):

$$\mathcal{E}^\lambda_{|P} = \bigoplus_{\boldsymbol{\omega} \in \mathcal{S}p^f(\mathcal{E}^\lambda, \underline{l})} \mathbb{E}(\boldsymbol{M}^\lambda, \boldsymbol{\omega}).$$

Thus we obtain the decomposition $\mathcal{E}^\lambda = \bigoplus_{\boldsymbol{\omega} \in \mathcal{S}p^f(\mathcal{E}^\lambda, \underline{l})} \mathbb{E}(\mathcal{E}^\lambda, \boldsymbol{\omega})$. It is compatible with the flat connection $\mathbb{D}^{\lambda f}$. We put $q_i(\boldsymbol{\omega}) = \omega_i$. We put as follows:

$$\mathcal{K}(\mathcal{E}, \lambda, \boldsymbol{0}, i) := \big\{u \in \mathcal{KMS}(\mathcal{E}^0, i) \,\big|\, -1 < \mathfrak{p}(\lambda, u) \leq 0\big\}.$$

Then we have the unique element $u_i = u_i(\boldsymbol{\omega}) \in \mathcal{K}(\mathcal{E}, \lambda, \boldsymbol{0}, i)$ such that $\mathfrak{e}^f(\lambda, u_i(\boldsymbol{\omega})) = \omega_i$. Then we obtain the number $b_i = b_i(\boldsymbol{\omega}) := \mathfrak{p}^f(\lambda, u_i(\boldsymbol{\omega}))$. We put $\boldsymbol{b}(\boldsymbol{\omega}) := (b_1, \ldots, b_l)$.

For $s \in \mathbb{E}\big(H(\mathcal{E}^\lambda), \boldsymbol{\omega}\big)$, we put as follows (see the page 169 for $\alpha(b, \omega)$):

$$F\big(s, \boldsymbol{b}(\boldsymbol{\omega})\big) := \exp\Big(\sum \log z_i \cdot \big(\alpha(b_i, \omega_i) + N_{i\,\boldsymbol{\omega}}\big)\Big) \cdot s.$$

Note that we have the equality $-\operatorname{ord}(s_{|\pi_i^{-1}(P)}) = b_i$ for any point $P \in D_i^\circ$.

LEMMA 8.22. *We have the following, for some positive constants C and M:*

$$|F(s, \boldsymbol{b}(\boldsymbol{\omega}))|_h \leq C \cdot \prod_{i=1}^l |z_i|^{-\mathfrak{p}(\lambda, u_i)} \cdot \Big(-\sum_{i=1}^l \log |z_i|\Big)^M.$$

Proof For each $P \in D_i^{\circ}$, we have the equality $-\operatorname{ord}\big(F(s,\boldsymbol{b}(\boldsymbol{\omega}))_{|\pi_i^{-1}(P)}\big) = \mathfrak{p}(\lambda, u_i)$ (Lemma 7.115). Then we obtain the result due to Corollary 2.53 and Corollary 8.19. □

COROLLARY 8.23. *$F(s, \boldsymbol{b}(\boldsymbol{\omega}))$ are sections of ${}^{\diamond}\mathcal{E}^{\lambda}$.* □

Let $\boldsymbol{s} = (s_j)$ be a base of $H(\mathcal{E}^{\lambda})$ compatible with $\mathbb{E}$. If $s_j \in \mathbb{E}\big(H(\mathcal{E}^{\lambda}), \boldsymbol{\omega}_j\big)$, we put ${}^{\underline{l}}\deg^{\mathbb{E}}(s_j) = \boldsymbol{\omega}_j$. Then we put $v_j := F(s_j, \boldsymbol{b}(\boldsymbol{\omega}_j))$ and $\boldsymbol{v} := (v_j)$. It is a tuple of sections of ${}^{\diamond}\mathcal{E}^{\lambda}$.

LEMMA 8.24. *If λ is generic, the $\mathcal{O}_X$-module ${}^{\diamond}\mathcal{E}^{\lambda}$ is coherent and locally free, and $\boldsymbol{v}$ is a frame of ${}^{\diamond}\mathcal{E}^{\lambda}$.*

Proof For any point $P \in D_i^{\circ}$, the tuple $\boldsymbol{v}_{|\pi_i^{-1}(P)}$ is a frame of ${}^{\diamond}(\mathcal{E}^{\lambda}_{|\pi_i^{-1}(P)})$, due to Lemma 7.90. We put as follows:

$$c_i := \sum_{a \in \mathcal{P}ar({}^{\diamond}\mathcal{E}^{\lambda}, i)} \mathfrak{m}(\lambda, a) \cdot a.$$

We have the line bundle ${}_{\boldsymbol{c}}\det(\mathcal{E}^{\lambda})$ over X. We have the following on $\pi_i^{-1}(P)$:

$$\big({}_{\boldsymbol{c}}\det(\mathcal{E}^{\lambda})\big)_{|\pi_i^{-1}(P)} = {}_{c_i}\big(\det(\mathcal{E}^{\lambda})_{|\pi_i^{-1}(P)}\big) = \det\big({}^{\diamond}\big(\mathcal{E}^{\lambda}_{|\pi_i^{-1}(P)}\big)\big).$$

We put $\Omega(\boldsymbol{v}) = v_1 \wedge \cdots \wedge v_r$. Then $\Omega(\boldsymbol{v})_{|\pi_i^{-1}(P)}$ gives a frame of $\det\big({}^{\diamond}\big(\mathcal{E}^{\lambda}_{|\pi_i^{-1}(P)}\big)\big)$. Hence $\Omega(\boldsymbol{v})$ is a frame of ${}_{\boldsymbol{c}}\det(\mathcal{E}^{\lambda})$ over X.

Let f be a holomorphic section of ${}^{\diamond}\mathcal{E}^{\lambda}$. Then f is described as $\sum f_j \cdot v_j$ for holomorphic functions f_i on $X - D$. Let us consider $f \wedge v_2 \wedge \cdots \wedge v_r$. On the curve $\pi_i^{-1}(P)$, it gives a section of ${}_{\boldsymbol{c}}\det(\mathcal{E}^{\lambda})_{|\pi_i^{-1}(P)}$. Hence $f \wedge v_2 \wedge \cdots \wedge v_r$ is a holomorphic section of ${}_{\boldsymbol{c}}\det(\mathcal{E}^{\lambda})$ over X. We have $f \wedge v_2 \wedge \cdots \wedge v_r = f_1 \cdot \Omega(\boldsymbol{v})$. Hence f_1 is holomorphic over X. Similarly f_j is holomorphic for each j. □

REMARK 8.25. The last argument can be found in [**17**]. □

8.3.2. Weak norm estimate. In the case $s_j \in \mathbb{E}\big(H(\mathcal{E}^{\lambda}), \boldsymbol{\omega}\big)$, we put $u_i(v_j) := u_i(\boldsymbol{\omega})$. We put as follows:

$$v_j' := v_j \cdot \prod_{i=1}^{l} |z_i|^{\mathfrak{p}(\lambda, u_i(v_j))}.$$

We put $\boldsymbol{v}' := (v_j')$. Then there exist positive constants M and C satisfying the following, due to Lemma 8.22:

$$H(h, \boldsymbol{v}') \leq C \cdot \Big(-\sum_{i=1}^{l} \log |z_i|\Big)^{M}. \tag{8.11}$$

Let $\boldsymbol{v}^{\vee} = (v_j^{\vee})$ be the dual frame of $\boldsymbol{v}$ over $X - D$.

LEMMA 8.26. *There exits positive constants C and M satisfying the following:*

$$|v_j^{\vee}|_h \leq C \cdot \prod_{i=1}^{l} |z_i|^{\mathfrak{p}(\lambda, u_i(v_j))} \cdot \Big(-\sum_{i=1}^{l} \log |z_i|\Big)^{M}.$$

Proof We have the estimates for the restrictions $v^{\vee}_{j\,|\,\pi_i^{-1}(P)}$ due to the result in the subsection 7.4.9. Then we can derive the estimate desired, due to Corollary 2.53. □

We put as follows:

$$v_j^{\vee\prime} := v_j^{\vee} \cdot \prod_i |z_i|^{-\mathfrak{p}(\lambda, u_i(v_j))}.$$

We put $\boldsymbol{v}^{\vee\prime} := (v_j^{\vee\prime})$. Then $\boldsymbol{v}^{\vee\prime}$ is the dual frame of $\boldsymbol{v}'$ over $X - D$, and there exist positive constants C and M satisfying the following, due to Lemma 8.26:

$$H(h^{\vee}, \boldsymbol{v}^{\vee\prime}) \leq C \cdot \Bigl(-\sum_{i=1}^{l} \log |z_i|\Bigr)^{M}. \tag{8.12}$$

Hence we obtain the following.

LEMMA 8.27. *The frame $\boldsymbol{v}'$ is adapted up to log order. Namely, there exist positive constants C_1, C_2 and M satisfying the following:*

$$C_1 \cdot \Bigl(-\sum_{i=1}^{l} \log |z_i|\Bigr)^{-M} \leq H(h, \boldsymbol{v}') \leq C_2 \cdot \Bigl(-\sum_{i=1}^{l} \log |z_i|\Bigr)^{M}.$$

Proof The right inequality is given in (8.11). The left inequality immediately follows from (8.12). □

8.3.3. Minor generalization and a λ-connection form. Let $\boldsymbol{c}$ be an element of $\boldsymbol{R}^l$. We put $\mathcal{K}(\mathcal{E}, \lambda, \boldsymbol{c}, i) := \bigl\{u \in \mathcal{KMS}(\mathcal{E}^0, i) \,\big|\, c_i - 1 < \mathfrak{p}(\lambda, u) \leq c_i\bigr\}$. Let $\boldsymbol{s}$ be a frame of $H(\mathcal{E}^{\lambda})$ as in the subsection 8.3.1. We put ${}_{\boldsymbol{c}}v_j := F\bigl(s_j, \boldsymbol{b}(\boldsymbol{\omega}_j) - \boldsymbol{c}\bigr)$, and we put ${}_{\boldsymbol{c}}\boldsymbol{v} := \bigl({}_{\boldsymbol{c}}v_j\bigr)$.

LEMMA 8.28. *If λ is generic, the sheaf ${}_{\boldsymbol{c}}\mathcal{E}^{\lambda}$ is locally free, and ${}_{\boldsymbol{c}}\boldsymbol{v}$ gives a frame of ${}_{\boldsymbol{c}}\mathcal{E}^{\lambda}$.*

Proof It can be shown by an argument similar to the proof of Lemma 8.24. □

In the case $s_j \in \mathbb{E}\bigl(H(\mathcal{E}^{\lambda}), \boldsymbol{\omega}\bigr)$, $u_i({}_{\boldsymbol{c}}v_j)$ denotes the unique element of $\mathcal{K}(\mathcal{E}, \lambda, \boldsymbol{c}, i)$ such that $\mathfrak{e}^f(\lambda, u_i({}_{\boldsymbol{c}}v_j)) = \omega_i$. We put as follows:

$${}_{\boldsymbol{c}}v_j{}' := {}_{\boldsymbol{c}}v_j \cdot \prod_{i=1}^{l} |z_i|^{\mathfrak{p}(\lambda, u_i({}_{\boldsymbol{c}}v_j))}.$$

We put ${}_{\boldsymbol{c}}\boldsymbol{v}' = ({}_{\boldsymbol{c}}v_j{}')$.

LEMMA 8.29. *The frame ${}_{\boldsymbol{c}}\boldsymbol{v}'$ is adapted up to log order.*

Proof It can be shown by an argument similar to Lemma 8.27. □

LEMMA 8.30. *We have the following relation:*

$$\mathbb{D}^{\lambda}\,{}_{\boldsymbol{c}}\boldsymbol{v} = {}_{\boldsymbol{c}}\boldsymbol{v} \cdot \sum_i (C_i + N_i) \cdot \frac{dz_i}{z_i}.$$

Here C_i denote the diagonal matrices whose (j,j)-components are $\mathfrak{e}\bigl(\lambda, u_i({}_{\boldsymbol{c}}v_j)\bigr)$, and N_i denote the nilpotent matrices.

Proof It follows from Lemma 7.115. □

8.3.4. Restriction to a diagonal curve. Let ϵ_i $(i=1,2)$ be positive numbers such that $\epsilon_1+\epsilon_2<1/2$. Assume the following for $i=1,2$:

$$\mathcal{P}ar\big({}^{\diamond}\mathcal{E}^{\lambda},i\big)\subset]-\epsilon_i,0[.$$

Pick an element $({}^0z_3,\ldots,{}^0z_n)\in(\Delta^*)^{n-2}$, we put $C_0:=\{(z,z,{}^0z_3,\ldots,{}^0z_n)\in X\}$. Let us discuss the KMS-structure of the restriction $(E,\overline{\partial}_E,\theta,h)_{|C_0}$.

LEMMA 8.31.

(1) *We have the following implication:*

$$\mathcal{KMS}\big({}^{\diamond}(\mathcal{E}^{\lambda}_{|C_0})\big)\subset\big\{u_1+u_2\,\big|\,u_i\in\,\mathcal{KMS}({}^{\diamond}\mathcal{E}^{\lambda},i)\big\}.$$

(2) *We have the following equality:*

$$\sum_{b\in\mathcal{P}ar\left({}^{\diamond}\left(\mathcal{E}^{\lambda}_{|C_0}\right)\right)} b\cdot\mathfrak{m}(\lambda,b)=\sum_{i=1,2}\sum_{b_i\in\mathcal{P}ar\left({}^{\diamond}\mathcal{E}^{\lambda},i\right)} b_i\cdot\mathfrak{m}(\lambda,b_i). \tag{8.13}$$

Proof Let us take the frame $\boldsymbol{v}$ as in the subsection 8.3.2. For each v_j, we have the elements $u_i(v_j)\in\mathcal{K}(\mathcal{E},\lambda,0,j)$. Let us consider the restriction $\widetilde{v}_j:=v_{j\,|\,C_0}$. We put as follows:

$$\widetilde{v}'_j:=\widetilde{v}_j\cdot|z|^{\mathfrak{p}\left(\lambda,u_1(v_j)+u_2(v_j)\right)},\quad\widetilde{\boldsymbol{v}}':=\big(\widetilde{v}'_j\big).$$

Then the frame $\widetilde{\boldsymbol{v}}'$ is adapted up to log order, due to Lemma 8.29. Since we have $-1<-\epsilon_1-\epsilon_2<\mathfrak{p}\big(\lambda,u_1(v_j)+u_2(v_j)\big)\leq0$ due to the assumption, $\widetilde{\boldsymbol{v}}$ gives a frame of ${}^{\diamond}\big(\mathcal{E}^{\lambda}_{|C_0}\big)$ due to Lemma 2.8. We also have the following relation, due to Lemma 8.30:

$$\mathbb{D}\widetilde{\boldsymbol{v}}=\widetilde{\boldsymbol{v}}\cdot\big(C_1+C_2+N_1+N_2\big)\frac{dz}{z}. \tag{8.14}$$

Here C_i and N_j denote the matrices given as in Lemma 8.30 for $\boldsymbol{v}$. From the adaptedness of $\widetilde{\boldsymbol{v}}'$ up to log order and (8.14), we obtain the following:

$$\deg^{F,\mathbb{E}}\big(\widetilde{v}_j\big)=\mathfrak{k}\big(\lambda,u_1(v_j)\big)+\mathfrak{k}\big(\lambda,u_2(v_j)\big).$$

Thus we obtain the first claim. We also obtain the following equality:

$$\sum_j-\operatorname{ord}\big(\widetilde{v}_j\big)=\sum_j\sum_{i=1,2}-{}^{i}\operatorname{ord}\big(v_j\big). \tag{8.15}$$

The left (resp. right) hand side of (8.15) are same as the left (resp. right) hand side of (8.13). Thus we obtain the second claim. □

8.4. Extension of holomorphic sections on a hyperplane

8.4.1. Preliminary I, Estimates of Higgs fields. We put $X=\Delta_\zeta\times\Delta_z^{n-1}$ and $X^{(0)}=\{0\}\times\Delta_z^{n-1}$. Let D'_i denote the divisor of Δ_z^{n-1} defined by $z_i=0$. We put $D_i=\Delta_\zeta\times D'_i$ and $D=\bigcup_{i=1}^{l}D_i$. We put $D_i^{(0)}=\{0\}\times D'_i$ and $D^{(0)}=\bigcup_{i=1}^{l}D_i^{(0)}$. Let $(E,\overline{\partial}_E,\theta,h)$ be a tame harmonic bundle on $X-D$.

Let $\lambda_0\in\boldsymbol{C}_\lambda$ be *generic* with respect to $(E,\overline{\partial}_E,\theta,h)$. Let $\boldsymbol{v}$ be a frame of ${}^{\diamond}\mathcal{E}^{\lambda_0}$ as in the subsection 8.3.1. Let ${}^{i}\deg^{F}(v_j)$ denote the degree of $v_{j\,|\,D_i}$ with respect to the filtration ${}^{i}F$. For simplicity of the notation, we put $b_h(v_j):={}^{h}\deg^{F}(v_j)$. Then we put as follows:

$$v'_i:=v_i\cdot\prod_{h=1}^{l}|z_h|^{b_h(v_i)},\quad\boldsymbol{v}':=(v'_i).$$

We define the function $B : X - D \longrightarrow M(r)$ as follows:

$$B_{ij} := \begin{cases} \prod_{h=1}^{l} |z_h|^{b_h(v_i)}, & (i = j), \\ 0, & (i \neq j). \end{cases}$$

Then we have $\boldsymbol{v}' = \boldsymbol{v} \cdot B$.

We put $\overline{\partial}_{\mathcal{E}^{\lambda_0}} = \overline{\partial}_E + \lambda_0 \cdot \theta^\dagger$, which is the holomorphic structure of $\mathcal{E}^{\lambda_0}$. We define $K \in C^\infty\big(X - D, \Omega_X^{1,0} \otimes M(r)\big)$ by $K = B^{-1} \cdot \overline{\partial} B = \overline{\partial} B \cdot B^{-1}$. The following lemma is clear.

LEMMA 8.32. *We have the following formula:*

$$K_{ij} := \begin{cases} \displaystyle\sum_{h=1}^{l} \frac{b_h(v_i)}{2} \frac{d\bar{z}_h}{\bar{z}_h}, & (i = j), \\ 0, & (i \neq j). \end{cases}$$

We have $\overline{\partial}_{\mathcal{E}^{\lambda_0}} \boldsymbol{v}' = \boldsymbol{v}' \cdot K$. □

The one forms $\Theta \in C^\infty\big(X - D, \Omega^{1,0} \otimes M(r)\big)$ and $\Theta^\dagger \in C^\infty\big(X - D, \Omega_X^{0,1} \otimes M(r)\big)$ are given as follows:

$$\theta \cdot \boldsymbol{v}' = \boldsymbol{v}' \cdot \Theta, \qquad \theta^\dagger \cdot \boldsymbol{v}' = \boldsymbol{v}' \cdot \Theta^\dagger.$$

The functions Θ^ζ, Θ^k, $\Theta^{\dagger\,\zeta}$ and $\Theta^{\dagger\,k}$ are defined as follows:

$$\Theta = \Theta^\zeta \cdot d\zeta + \sum_{k=1}^{l} \Theta^k \cdot \frac{dz_k}{z_k} + \sum_{k=l+1}^{n} \Theta^k \cdot dz_k,$$

$$\Theta^\dagger = \Theta^{\dagger\zeta} \cdot d\bar{\zeta} + \sum_{k=1}^{l} \Theta^{\dagger\,k} \cdot \frac{d\bar{z}_k}{\bar{z}_k} + \sum_{k=l+1}^{n} \Theta^{\dagger\,k} \cdot d\bar{z}_k.$$

LEMMA 8.33. *The functions* Θ^ζ, Θ^k, $\Theta^{\dagger\,\zeta}$ *and* $\Theta^{\dagger\,k}$ *are dominated by polynomials of* $-\log|z_i|$ $(i = 1, \dots, l)$.

Proof For the decompositions $\theta = \sum_{i=1}^{l} f_i \cdot dz_i / z_i + \sum_{i=l+1}^{n} g_i \cdot dz_i$, the norms of f_i and g_j with respect to h are bounded (see the subsection 8.2.2). We also know that $\boldsymbol{v}'$ is adapted up to log order (Lemma 8.27). Then the result follows. □

We have the λ_0-connection $\mathbb{D}^{\lambda_0}$. Let A denote the λ_0-connection one form of $\mathbb{D}^{\lambda_0}$ with respect to $\boldsymbol{v}$, i.e., A is a holomorphic section of $M(r) \otimes \Omega_X^{1,0}(\log D)$ determined by the condition:

$$\mathbb{D}^{\lambda_0} \boldsymbol{v} = \boldsymbol{v} \cdot A.$$

Due to our choice of $\boldsymbol{v}$, we have the decomposition $A = \bigoplus A_{\boldsymbol{\omega}}$ which corresponds to the decomposition $\mathcal{E}^{\lambda_0} = \bigoplus_{\boldsymbol{\omega}} \mathbb{E}(\mathcal{E}^{\lambda_0}, \boldsymbol{\omega})$, namely, $A_{\boldsymbol{\omega}}$ is a holomorphic section of $\mathrm{End}\big(\mathbb{E}(\mathcal{E}^{\lambda_0}, \boldsymbol{\omega})\big) \otimes \Omega^{1,0}(\log D)$.

LEMMA 8.34. *A and B are commutative.*

Proof It follows from the decomposition $B = \bigoplus_{\boldsymbol{\omega}} B(\boldsymbol{\omega}) \cdot Id_{\mathbb{E}(\mathcal{E}^{\lambda_0}, \boldsymbol{\omega})}$ for $B(\boldsymbol{\omega}) \in C^\infty(X - D)$. □

We have the following formula:

$$\begin{aligned}(8.16)\quad \mathbb{D}^{\lambda_0}\boldsymbol{v}' &= \mathbb{D}^{\lambda_0}(\boldsymbol{v}\cdot B) = \boldsymbol{v}\cdot A\cdot B + \boldsymbol{v}\cdot\lambda_0\partial(B) + \boldsymbol{v}\cdot\overline{\partial}B \\ &= \boldsymbol{v}'\cdot(B^{-1}AB + \lambda_0 B^{-1}\partial B + B^{-1}\overline{\partial}B) = \boldsymbol{v}'(A + \lambda_0\overline{K} + K).\end{aligned}$$

We have the description:

$$A + \lambda_0\cdot\overline{K} = \sum_{i=1}^{l} g_i\cdot\frac{dz_i}{z_i}.$$

LEMMA 8.35. *The functions g_i are bounded.*

Proof Since A is a holomorphic section of $M(r)\otimes\Omega^{1,0}(\log D)$, the lemma follows from Lemma 8.32. □

LEMMA 8.36. *We have the following equalities:*

$$(8.17)\qquad \partial\Theta^{\dagger} + \Big[\lambda_0^{-1}\cdot\big(A + \lambda_0\overline{K} - \Theta\big), \Theta^{\dagger}\Big] = 0, \quad \overline{\partial}\Theta + \Big[K - \lambda_0\Theta^{\dagger}, \Theta\Big] = 0.$$

Proof Recall the equalities $\partial_E(\theta^{\dagger}) = 0$ and $\partial_E = \lambda_0^{-1}\big(\mathbb{D}^{\lambda_0} - \theta - \overline{\partial}_{\mathcal{E}^{\lambda_0}}\big)$. We obtain the first equality in (8.17) from (8.16) and the definition of Θ.

As for the second equality in (8.17), we have only to use $\overline{\partial}_E(\theta) = 0$, $\overline{\partial}_E = \overline{\partial}_{\mathcal{E}^{\lambda_0}} - \lambda_0\theta^{\dagger}$ and $\overline{\partial}_{\mathcal{E}^{\lambda_0}}\boldsymbol{v}' = \boldsymbol{v}'\cdot K$. □

LEMMA 8.37. *We regard $\Theta^{\dagger}(\zeta, z)$ and $\Theta(\zeta, z)$ as functions of ζ. Then the following holds:*

- *They are L_k^p for any $k\in\mathbb{Z}_{\geq 0}$ and for any p.*
- *The L_k^p-norms of $\Theta^{\dagger}(\zeta, z)$ and $\Theta(\zeta, z)$ are dominated by polynomials of $(-\log|z_j|)$ $(j = 1,\ldots,l)$.*

Proof It follows from Lemma 8.33, the formula (8.17), the estimate for A and K, and Corollary 2.84. □

LEMMA 8.38. *We have the descriptions as follows:*

$$\partial_k\Theta^{\dagger\,\zeta} = C_k\cdot\frac{dz_k}{z_k}\cdot d\bar{\zeta}.$$

Here, C_k denotes an element of $C^{\infty}\big(X - D, M(r)\big)$. Their sup-norms are dominated by polynomials of $-\log|z_j|$ $(j = 1,\ldots,l)$.

Proof It follows from the estimates for Θ, A and K, due to Lemma 2.86. □

8.4.2. Preliminary II, Construction and estimate of a cocycle. We continue to use the setting in the subsection 8.4.1. Let λ be any element of $\boldsymbol{C}_{\lambda}$, and λ_0 be generic as in the subsection 8.4.1. Let f be a holomorphic section of the sheaf ${}^{\diamond}\big(\mathcal{E}^{\lambda}_{|X^{(0)}-D^{(0)}}\big)$ over $X^{(0)}$. Before proceeding, let us recall the standard method to extend f on $X^{(0)} - D^{(0)}$ to a holomorphic section on $X - D$.

We have the line bundle $\mathcal{O}_X(-X^{(0)})$. It can naturally be seen as the subsheaf of $\mathcal{O}_X$. We put $\mathcal{E}^{\lambda}(-X^{(0)}) := \mathcal{E}^{\lambda}\otimes\mathcal{O}(-X^{(0)})$, which gives the subsheaf of $\mathcal{E}^{\lambda}$. Hence a C^{∞}-section of $\mathcal{E}^{\lambda}(-X^{(0)})$ induces the C^{∞}-section of $\mathcal{E}^{\lambda}$. Or, a measurable section of $\mathcal{E}^{\lambda}(-X^{(0)})$ induces the measurable section of $\mathcal{E}^{\lambda}$. In the following, we use such a view point without mention.

Assume that we have a C^∞-section ρ of $\mathcal{E}^\lambda$ such that $\rho_{|X^{(0)}-D^{(0)}} = f$, and that we have a section G of $\mathcal{E}^\lambda(-X^{(0)})$ such that $\overline{\partial} G = \overline{\partial}\rho$. Then we put $\widetilde{f} := \rho - G$. It is holomorphic and $\widetilde{f}_{|X^{(0)}-D^{(0)}} = f$. Namely $\widetilde{f}$ is a desired holomorphic section. It is our purpose in this subsection to construct an appropriate cocycle ρ from f (Proposition 8.40).

As a preparation, we give some estimate of f and the derivative. Since the tuple of sections $\boldsymbol{v}'$ gives C^∞-frame of C^∞-bundle $\mathcal{E}^\lambda = \mathcal{E}^{\lambda_0}$ over $X - D$, we have the following description:

$$f = \sum_i f_i(z) \cdot v_i'(0,z), \qquad f_i \in C^\infty(X^{(0)} - D^{(0)}).$$

LEMMA 8.39. *There exist positive constants C and M satisfying the following:*

$$|f|_h \leq C \cdot \Bigl(-\sum_{i=1}^{l} \log|z_i|\Bigr)^M.$$

Proof We obtain such estimate for the restrictions $f_{|\pi_i^{-1}(P)}$ for any point $P \in D_i^{(0)} \setminus \bigcup_{j\neq i, j\leq i} D_j^{(0)} \cap D_i^{(0)}$. Then we obtain the result due to Lemma 7.40 and Corollary 2.53. □

Due to the adaptedness of $\boldsymbol{v}'$ up to log order, we obtain the following inequalities for some positive constants C and M and for any $i = 1, \ldots, l$:

$$|f_i| \leq C \cdot \Bigl(-\sum_{j=1}^{l} \log|z_j|\Bigr)^M. \tag{8.18}$$

Note that we have the following relation:

$$\overline{\partial}_{\mathcal{E}^\lambda}\boldsymbol{v}' = \Bigl(\overline{\partial}_{\mathcal{E}^{\lambda_0}} + (\lambda - \lambda_0)\cdot\theta^\dagger\Bigr)\boldsymbol{v}' = \boldsymbol{v}' \cdot \Bigl(K + (\lambda-\lambda_0)\cdot\Theta^\dagger\Bigr). \tag{8.19}$$

Hence we obtain the following equality on $X^{(0)} - D^{(0)}$:

$$(8.20)\quad 0 = \overline{\partial} f(z) =$$
$$\sum_i \Bigl(\overline{\partial} f_i(z) + \sum_{k,j}\bigl((\lambda-\lambda_0)\cdot\Theta^{k\,\dagger}_{i\,j}(0,z) + \frac{b_k(v_i)}{2}\cdot\delta_{i\,j}\bigr)\cdot\bar{\eta}_k\cdot f_j\Bigr)\cdot v_i(0,z).$$

Here we put as follows:

$$\delta_{i\,j} := \begin{cases} 1, & (i=j),\\ 0, & (i\neq j),\end{cases} \qquad \bar{\eta}_k = \begin{cases} d\bar{z}_k/\bar{z}_k, & (k\leq l),\\ d\bar{z}_k, & (k\geq l+1).\end{cases}$$

Note that we obtain the estimate of $\overline{\partial} f_i$ from (8.20) and the estimates of f_i and the Higgs field.

We use the notation in the subsection 2.8.5. The metric $h_{\boldsymbol{a},N}$ of $\mathcal{E}^\lambda$ is given as in (2.27). Let us fix a C^∞-hermitian metric of $h_{\mathcal{O}(-X^{(0)})}$. The metrics $h_{\mathcal{O}(-X^{(0)})}$ and $h_{\boldsymbol{a},N}$ induce the hermitian metric of $\mathcal{E}^\lambda(-X^{(0)})$, which is denoted by $h'_{\boldsymbol{a},N}$.

Let $\boldsymbol{\delta}$ and $\boldsymbol{0}$ denote the elements $(\overbrace{1,\ldots,1}^{l})$ and $(\overbrace{0,\ldots,0}^{l})$ respectively. We use the notation in the section 2.8.

PROPOSITION 8.40. *There exists an element* $\rho \in C^{\infty}(X-D, \mathcal{E}^{\lambda})$ *satisfying the following:*

(1) *Let* ϵ *be any positive number, and* N *be any negative number. Then* ρ *is bounded as a section of* $\mathcal{E}^{\lambda}$, *up to the polynomial order of* $-\log|z_i|$ $(i=1,\ldots,l)$, *with respect to* $h_{\mathbf{0},N}$. *In particular,* ρ *is an* L^2*-section of* $\mathcal{E}^{\lambda}$ *with respect to* $h_{\epsilon\cdot\boldsymbol{\delta},N}$.

(2) *Let* ϵ *be any positive number, and* N *be any negative number. Then* $\overline{\partial}_{\mathcal{E}^{\lambda}}\rho$ *is bounded as a section of* $\mathcal{E}^{\lambda}(-X^{(0)})\otimes\Omega_X^{0,1}$ *up to the polynomial order of* $-\log|z_i|$ $(i=1,\ldots,l)$, *with respect to the metric* $h'_{\mathbf{0},N}$ *of* $\mathcal{E}^{\lambda}(-X^{(0)})$ *and the Poincaré metric of* $\Omega^{0,1}_{X-D}$. *The restriction to* $X-(X^{(0)}\cup D)$ *is* C^{∞}. *In particular,* $\overline{\partial}_{\mathcal{E}^{\lambda}}\rho \in A^{0,1}_{\epsilon\boldsymbol{\delta},N}\big(\mathcal{E}^{\lambda}(-X^{(0)})\big)$.

(3) $\rho_{|X^{(0)}-D^{(0)}} = f$.

(4) *The support of* ρ *is contained in the region* $\big\{\zeta \,\big|\, |\zeta|<1/3\big\}\times\big(X^{(0)}-D^{(0)}\big)$.

Proof Let ϵ be any positive number, and N be any negative number. Let χ be any function over Δ_{ζ} satisfying the following:

$$\chi(\zeta)=\begin{cases} 1, & (|\zeta|\leq 1/3),\\ 0, & (|\zeta|>2/3).\end{cases}$$

We put as follows:

$$F^1 := \sum_i f_i(z)\cdot v_i(\zeta,z),\quad F^2 := \overline{\zeta}\cdot(\lambda-\lambda_0)\cdot g^2,\quad g^2 := \sum_{i,j}\Theta^{\zeta\,\dagger}_{i\,j}(0,z)\cdot f_j(z)\cdot v_i(\zeta,z).$$

We put $\rho := \chi\cdot(F^1-F^2)$. Then the claims (3) and (4) are clear. Let us show the claims (1) and (2).

LEMMA 8.41. ρ *is bounded up to the polynomial order of* $-\log|z_i|$ $(i=1,\ldots,l)$ *with respect to* $h_{\mathbf{0},N}$.

Proof It follows from (8.18) and Lemma 8.33. □

We have the following formula, where we put $\nu(\lambda):=\lambda-\lambda_0$:

$$(8.21)\quad \overline{\partial}_{\mathcal{E}^{\lambda}}F^1 =$$

$$\sum_i\Big(\overline{\partial}f_i+\sum_j\Big(\sum_k\Big(\nu(\lambda)\cdot\Theta^{k\,\dagger}_{i\,j}(\zeta,z)+\frac{b_k(v_i)}{2}\delta_{i\,j}\Big)\cdot\overline{\eta}_k+\nu(\lambda)\cdot\Theta^{\zeta\,\dagger}_{i\,j}(\zeta,z)\cdot d\overline{\zeta}\Big)\cdot f_j\Big)\cdot v_i(\zeta,z)$$

$$=\nu(\lambda)\cdot\sum_{i,j}\left(\sum_k\Big(\Theta^{k\,\dagger}_{i\,j}(\zeta,z)-\Theta^{k\,\dagger}_{i\,j}(0,z)\Big)\cdot\overline{\eta}_k+\Theta^{\zeta\,\dagger}_{i\,j}(\zeta,z)\cdot d\overline{\zeta}\right)\cdot f_j\cdot v_i(\zeta,z).$$

We also have the following:

$$\overline{\partial}_{\mathcal{E}^{\lambda}}F^2=(\lambda-\lambda_0)\cdot\left(\sum_{i,j}\Theta^{\zeta\,\dagger}_{i\,j}(0,z)\cdot f_j(z)\cdot v_i(\zeta,z)\cdot d\overline{\zeta}+\overline{\zeta}\cdot\overline{\partial}g^2\right).$$

Hence we obtain the following:

$$(8.22)\quad \overline{\partial}_{\mathcal{E}^\lambda}\rho = \overline{\partial}\chi\cdot(F^1-F^2)+\chi\cdot\overline{\partial}(F^1-F^2) = \overline{\partial}\chi\cdot(F^1-F^2) + \chi\cdot(\lambda-\lambda_0)\times\Bigg(\sum_{i,k,j}\big(\Theta^{k\,\dagger}_{i\,j}(\zeta,z)-\Theta^{k\,\dagger}_{i\,j}(0,z)\big)\cdot\overline{\eta}_k\cdot f_j\cdot v_i(\zeta,z) + \sum_{i\,j}\big(\Theta^{\zeta\,\dagger}_{i\,j}(\zeta,z)-\Theta^{\zeta\,\dagger}_{i\,j}(0,z)\big)\cdot f_j(z)\cdot v_i(\zeta,z)\cdot d\overline{\zeta}+\overline{\zeta}\cdot\overline{\partial}g^2\Bigg).$$

It is easy to see that the restriction of $\overline{\partial}_{\mathcal{E}^\lambda}\rho$ to $X-(X^{(0)}\cup D)$ is C^∞.

LEMMA 8.42. *We regard $\overline{\partial}\chi\cdot(F^1-F^2)$ as a section of $\mathcal{E}^\lambda(-X^{(0)})\otimes\Omega^{0,1}$. Then it is bounded up to the polynomial order of $-\log|z_i|$ $(i=1,\dots,l)$ with respect to the metric $h'_{\mathbf{0},N}$ of $\mathcal{E}^\lambda(-X^{(0)})$ and the Poincaré metric of $\Omega^{0,1}_{X-D}$.*

Proof Note that $\overline{\partial}\chi$ vanishes on $|\zeta|<1/3$. Then the claim follows from Lemma 8.33 and Lemma 8.39. □

LEMMA 8.43. *We regard the following as a section of $\mathcal{E}^\lambda(-X^{(0)})\otimes\Omega^{0,1}$:*

$$(8.23)\quad \chi\times\Big(\sum_{i,k,j}\big(\Theta^{k\,\dagger}_{i\,j}(\zeta,z)-\Theta^{k\,\dagger}_{i\,j}(0,z)\big)\cdot\overline{\eta}_k\cdot f_j\cdot v_i(\zeta,z) + \sum_{i\,j}\big(\Theta^{\zeta\,\dagger}_{i\,j}(\zeta,z)-\Theta^{\zeta\,\dagger}_{i\,j}(0,z)\big)\cdot f_j(z)\cdot v_i(\zeta,z)\cdot d\overline{\zeta}\Big).$$

Then it is bounded up to the polynomial order of $-\log|z_i|$ $(i=1,\dots,l)$, with respect to the metrics $h'_{\mathbf{0},N}$ of $\mathcal{E}^\lambda(-X^{(0)})$ and the Poincaré metric of $\Omega^{0,1}_{X-D}$.

Proof We put $K_0=\{\zeta\,|\,|\zeta|<2/3\}$. On $K_0\times(X^{(0)}-D^{(0)})$, we have the following for $a=1,\dots,n-1$, or ζ:

$$\Big|\Theta^{a\,\dagger}_{i\,j}(\zeta,z)-\Theta^{a\,\dagger}_{i\,j}(0,z)\Big|\leq|\zeta|\times\big|\big|\Theta^{a\,\dagger}_{i\,j}(\cdot,z)\big|\big|_{C^1(K_0\times\{z\})}.$$

Here $\big|\big|\Theta^{a\,\dagger}_{i\,j}(\cdot,z)\big|\big|_{C^1(K_0\times\{z\})}$ denotes the C^1-norm of the restriction of $\Theta^{a\,\dagger}_{i\,j}$ to $K_0\times\{z\}$. Then we obtain the result from Lemma 8.37. □

LEMMA 8.44. *We regard $\chi\cdot\overline{\zeta}\cdot\overline{\partial}g^2$ as a section of $\mathcal{E}(-X^{(0)})\otimes\Omega^{0,1}$. It is bounded up to the polynomial order of $-\log|z_i|$ $(i=1,\dots,l)$, with respect to the metric $h'_{\mathbf{0},N}$ of $\mathcal{E}^\lambda(-X^{(0)})$ and the Poincaré metric of $\Omega^{0,1}_{X-D}$.*

Proof We have the following:

$$(8.24)\quad \chi\cdot\overline{\zeta}\cdot\overline{\partial}g^2=\chi\cdot\overline{\zeta}\times \sum_{i\,j}\Big(\overline{\partial}\Theta^{\zeta\,\dagger}_{i\,j}(0,z)\cdot f_j(z)\cdot v_i'(\zeta,z)+\Theta^{\zeta\,\dagger}_{i\,j}(0,z)\overline{\partial}f_j(z)\cdot v_i(\zeta,z)+\Theta^{\zeta\,\dagger}_{i\,j}(0,z)\cdot f_j(z)\cdot\overline{\partial}v_i(\zeta,z)\Big).$$

The first summand can be estimated due to Lemma 8.38. The second summand can be estimated due to (8.20). The third summand can be estimated due to (8.19). □

Hence the claims (2) and (1) are obtained, and the proof of Proposition 8.40 is accomplished. □

8.4.3. Extension of holomorphic sections. We put $X = \Delta^n$ and $D = \bigcup_{i=1}^{l} D_i$. We put as follows:

$$X^{(1)} := \big\{(z_1,\ldots,z_n) \in X \,\big|\, z_1 = z_2\big\}, \quad D^{(1)} := X^{(1)} \cap D,$$

$$X_0 := \big\{(z_1,\ldots,z_n) \in X \,\big|\, z_1 = z_2 = 0\big\}.$$

CONDITION 8.45. Let $(E, \overline{\partial}_E, \theta, h)$ be a tame harmonic bundle over $X - D$. Let ϵ_1 and ϵ_2 be positive numbers such that $\epsilon_1 + \epsilon_2 < 1$. Assume that $\mathcal{P}ar({}^{\diamond}\mathcal{E}^{\lambda}, i)$ is contained in $]-\epsilon_i, 0]$ for $i = 1, 2$. □

We remark $\mathcal{P}ar(\mathcal{E}^{\lambda}, i) \cap]0, 1 - \epsilon_i] = \emptyset$ under the condition.

PROPOSITION 8.46. *Let $(E, \overline{\partial}_E, \theta, h)$ be a tame harmonic bundle satisfying Condition 8.45. Let f be a holomorphic section of ${}^{\diamond}(\mathcal{E}^{\lambda}_{|X^{(1)}-D^{(1)}})$ on $X^{(1)}$. Then there exists a neighbourhood U of X_0 in X, and there exists a holomorphic section $\widetilde{f} \in \Gamma(U, {}^{\diamond}\mathcal{E}^{\lambda})$, such that $\widetilde{f}_{|X^{(1)}\cap U} = f_{|X^{(1)}\cap U}$.*

Proof We use the setting in the subsections 2.8.8–2.8.10. We impose the additional condition $\psi_1(\overline{X}^{(1)}) = X^{(1)}$. We remark that $\psi_1(\overline{X})$ gives a neighbourhood of X_0. We take the numbers ϵ, a and b be as in Lemma 2.61. We remark that $\mathcal{P}ar(\mathcal{E}^{\lambda}, 1) \cap]0, a + 2\epsilon] \subset \mathcal{P}ar(\mathcal{E}^{\lambda}, 1) \cap]0, 1 - \epsilon_1] = \emptyset$. Similarly, we have $\mathcal{P}ar(\mathcal{E}^{\lambda}, 2) \cap]0, b + 2\epsilon] = \emptyset$. Moreover, we may assume that $\mathcal{P}ar(\mathcal{E}^{\lambda}, i) \cap]0, \epsilon] = \emptyset$ for $i = 3, \ldots, l$, if ϵ is sufficiently small.

Let us take a sufficiently negative number N such that (2.30) holds. We take the metric $\widetilde{h}$ of the bundle $\psi_1^{-1}\mathcal{E}^{\lambda}(-\overline{X}^{(1)})$ as in (2.40):

$$\widetilde{h} := \widetilde{h}_{N,\epsilon,a,b} = \psi_1^{-1} h_{\epsilon\cdot\boldsymbol{\delta},N} \cdot h_{-1,a,b} \cdot H_1^{-1} \cdot \tau_1^{2+\epsilon} \big(\tau_2 \cdot \tau_3\big)^{\epsilon}.$$

We also put $\widetilde{h}_0 := \psi_1^{-1} h_{\boldsymbol{0},N} \cdot h_{-1,a,b} \cdot H_1^{-1} \cdot \tau_1^{2+\epsilon} \big(\tau_2 \cdot \tau_3\big)^{\epsilon}$. We remark that we can use Corollary 2.63 in this setting. We take the metrics $\widehat{h}$ and $\widehat{h}_0$ of $\psi_1^{-1}\mathcal{E}^{\lambda}$:

$$\widehat{h} := \psi_1^{*} h_{\epsilon\cdot\boldsymbol{\delta},N} \cdot h_{0,a,b} \cdot H_1^{-1} \cdot \tau_1^{2+\epsilon} \big(\tau_2 \cdot \tau_3\big)^{\epsilon},$$
$$\widehat{h}_0 := \psi_1^{*} h_{\boldsymbol{0},N} \cdot h_{0,a,b} \cdot H_1^{-1} \cdot \tau_1^{2+\epsilon} \big(\tau_2 \cdot \tau_3\big)^{\epsilon}.$$

Take an embedding $\kappa : \Delta_{\zeta} \longrightarrow \mathbb{P}^1 - \{0, \infty\}$ such that $\kappa(0) = [1:1] \in \mathbb{P}^1$. We take a holomorphic function η on $\pi^{-1}(\kappa(\Delta_{\zeta}))$ such that $\eta^{-1}(0)$ is the intersection of the exceptional divisor $\varphi^{-1}(O)$ and $\pi^{-1}(\kappa(\Delta_{\zeta}))$.

The section f induces a holomorphic section of $\big(\psi_1^{*}\mathcal{E}^{\lambda}\big)_{|\overline{X}^{(1)}}$ over $\overline{X}^{(1)} \setminus \overline{D}^{(1)}$, which we also denote by f. By using Proposition 8.40, we can take a C^{∞}-section ρ of $\psi_1^{-1}\mathcal{E}^{\lambda}$ over $\overline{X} - \overline{D}$ satisfying the following:

- The support of ρ is contained in $\pi^{-1}(\kappa(\Delta_{\zeta})) \times \Delta_w^{n-2}$.
- ρ is bounded as a section of $\psi_1^{-1}\mathcal{E}^{\lambda}$ up to the polynomial order of $-\log|z_i|$ $(i = 3, \ldots, l)$ and $-\log|\eta|$, with respect to $\widehat{h}_0$. In particular, ρ is L^2-section of $\psi_1^{-1}\mathcal{E}^{\lambda}$ with respect to $\widehat{h}$.
- $\overline{\partial}(\rho)$ is bounded as a section of $\psi_1^{-1}\mathcal{E}^{\lambda}(-\overline{X}^{(1)}) \otimes \Omega^{1,0}_{\overline{X}-\overline{D}}$ up to the polynomial order of $-\log|z_i|$ $(i = 3, \ldots, l)$ and $-\log|\eta|$, with respect to the metric $\widetilde{h}_0$ of $\psi_1^{-1}\mathcal{E}^{\lambda}(-\overline{X}^{(1)}) \otimes \Omega^{1,0}$ and the metric $g_{\overline{X}-\overline{D}}$ of $\Omega^{0,1}_{\overline{X}-\overline{D}}$. The restriction of $\overline{\partial}\rho$ to $\overline{X} - (\overline{D} \cup \overline{X}^{(1)})$ are C^{∞}. In particular, $\overline{\partial}\rho$ is contained in $A^{0,1}_{\widetilde{h}}\big(\psi_1^{*}\mathcal{E}^{\lambda}(-\overline{X}^{(1)})\big)$.

- We have $\rho_{|\overline{X}^{(1)}-\overline{D}^{(1)}} = f$.

Due to Corollary 2.63, we can pick an element G of $A^{0,0}_{\widetilde{h}}\big(\psi_1^{-1}\mathcal{E}^0(-\overline{X}^{(1)})\big)$ such that $\overline{\partial}G = \overline{\partial}\rho$. Then we put $\widetilde{f} := \rho - G$. Then it satisfies $\overline{\partial}\widetilde{f} = 0$, $\widetilde{f} \in A^{0,0}_{\widetilde{h}}(\psi_1^{*}\mathcal{E}^0)$.

LEMMA 8.47. *We have* $\widetilde{f}_{|\overline{X}^{(1)}-\overline{D}^{(1)}} = f$.

Proof If we take a sufficiently small positive number ϵ_3, we may have the natural inclusion $\kappa(\Delta_\varsigma) \times \Delta_\eta(\epsilon_3) \subset \pi^{-1}\big(\kappa(\Delta_\varsigma)\big)$. We naturally regard $\kappa(\Delta_\varsigma) \times \Delta_\eta(\epsilon_3) \times \Delta_w^{n-2}$ as an open subset of $\overline{X}$. Let P be any point of $W = \big(\overline{X}^{(1)} - \overline{D}^{(1)}\big) \cap \{[1 : 1]\} \times \Delta_\eta(\epsilon_3) \times \Delta_w^{n-2}$. We have the natural inclusion of $\Delta_\varsigma \times \{P\} \subset \overline{X} - \overline{D}$. We remark that the restrictions of $g_{\overline{X}-\overline{D}}$ and $\widetilde{h}$ to $\Delta_\varsigma \times \{P\}$ give the C^∞-metrics of $\Delta_\varsigma \times \{P\}$ and $\psi_1^{-1}\mathcal{E}^\lambda \otimes \mathcal{O}(-\overline{X}^{(1)})_{|\Delta_\varsigma\times\{P\}}$ respectively.

For almost all points P of W, the following holds:

- $G_{|\Delta_\varsigma\times\{P\}}$ is L^2 with respect to the restrictions of $g_{\overline{X}-\overline{D}}$ and $\widetilde{h}$ to $\Delta_\varsigma\times\{P\}$.
- $\overline{\partial}G_{|\Delta_\varsigma\times\{P\}} = \overline{\partial}\rho_{|\Delta_\varsigma\times\{P\}}$.

Due to Corollary 2.27, G is bounded as a section of $\psi_1^{-1}\mathcal{E}^\lambda \otimes \mathcal{O}(-\overline{X}^{(1)})_{|\Delta_\varsigma\times\{P\}}$. Therefore, if we regard $G_{|\Delta_\varsigma\times\{P\}}$ as a section of $\psi^{-1}\mathcal{E}^\lambda_{|\Delta_\varsigma\times\{P\}}$, then the value $G_{|(0,P)}$ is 0 in $\psi^{-1}\mathcal{E}^\lambda_{|(0,P)}$. Hence we have $\widetilde{f}_{|(0,P)} = \rho_{|(0,P)} = f_{|(0,P)}$ for almost all $P \in \overline{X}^{(1)} - \overline{D}^{(1)}$. Hence we obtain $\widetilde{f}_{|\overline{X}^{(1)}-\overline{D}^{(1)}} = f$. □

We have an open subset V of X such that $\psi_1\big(\overline{X} - \overline{D}\big) = V \cap \big(X - D\big) = V \setminus D$. By the identification, we can regard $\widetilde{f}$ as a holomorphic section of $\mathcal{E}^\lambda_{|V\setminus D}$. We would like to show that $\widetilde{f}$ gives the section of ${}^\diamond\mathcal{E}^\lambda$ on V.

We put $D_i^\circ := D_i \setminus \bigcup_{j\leq l, j\neq i} D_j$.

LEMMA 8.48. *Let P be a point of $V \cap D_1^\circ$. We have* $-\operatorname{ord}\big(\widetilde{f}_{|\pi_1^{-1}(P)\cap V}\big) \leq 0$.

Proof We put $C_P := \pi_1^{-1}(P) \cap V$. We may assume that the closure of C_P in $\boldsymbol{C}^n$ is contained in X, by shrinking V. Let us take a small neighbourhood U of P in D_1°. We may assume $C_P \times U \subset V$. The metric $\widehat{h}_{|C_P\times U}$ and the metric $h \cdot |z_1|^{2(a+2\epsilon)}$ are mutually bounded up to the polynomial order of $-\log|z_1|$. Therefore $\widetilde{f}_{|C_P\times U}$ is L^2, with respect to the metric $h \cdot |z_1|^{2(a+2\epsilon)} \cdot \big(-\log|z_1|\big)^{-M}$ for a some positive M and the metric $g_{X-D\,|\,C_P\times U}$. Due to Corollary 2.51, the restriction $\widetilde{f}_{|C_P}$ is L^2 with respect to $h \cdot |z_1|^{2(a+2\epsilon)} \cdot \big(-\log|z_1|\big)^{-M}$ and the metric $g_{X-D\,|\,C_P}$. We remark $\mathcal{P}ar({}^\diamond\mathcal{E}^\lambda, 1) \cap]0, a+2\epsilon] = \emptyset$. Due to Lemma 7.63, we obtain that $-\operatorname{ord}(\widetilde{f}_{|C_P}) \leq 0$. □

Similarly, we can show the following lemma.

LEMMA 8.49. *We have* $-\operatorname{ord}(\widetilde{f}_{|\pi_i^{-1}(P)\cap V}) \leq 0$ *for any* $P \in D_i^\circ \cap V$ *and any* $i = 1, 2, , \dots, l$. □

LEMMA 8.50. $\widetilde{f}$ *is a section of* ${}^\diamond\mathcal{E}^\lambda$ *over* V.

Proof It follows from Lemma 8.49, Corollary 2.53 and Corollary 8.19. □

As a result, we obtain the holomorphic section $\widetilde{f}$ of ${}^{\diamond}\mathcal{E}^{\lambda}$ over U such that $\widetilde{f}_{|X^{(1)}} = f$. Thus the proof of Proposition 8.46 is accomplished. □

8.4.4. Extension property in the codimension one case. We put $X := \Delta_z^n \times \Delta_w$, $D_i := \{z_i = 0\}$ and $D = \bigcup_{i=1}^{l} D_i$. We put $X^{(2)} := \Delta_z^n \times \{0\}$ and $D^{(2)} = D \cap X^{(2)}$. We have the origin $(O,0) \in X^{(2)} \subset X$. Let $(E, \overline{\partial}_E, \theta, h)$ be any tame harmonic bundle over $X - D$. Let us consider the restriction of $\mathcal{E}^{\lambda}$ to $X^{(2)} - D^{(2)}$.

LEMMA 8.51. *Let f be a holomorphic section of ${}_{\boldsymbol{b}}\big(\mathcal{E}^{\lambda}_{|X^{(2)}-D^{(2)}}\big)$ defined on a neighbourhood of $(O,0)$ in $X^{(2)}$. Then there exists a holomorphic section $\tilde{f}$ of ${}_{\boldsymbol{b}}\mathcal{E}^{\lambda}$ defined on a neighbourhood of $(O,0)$, which satisfies $\tilde{f}_{|X^{(2)}} = f$.*

Proof The argument is essentially same as the proof of Proposition 8.46. In fact, we can show the claim more simply, by using the results in the subsections 8.4.1–8.4.2 and the vanishing in Lemma 2.48. □

8.4.5. Local freeness in codimension one. We put $X = \Delta \times \Delta_w^n$, $D = \{0\} \times \Delta_w^n$. Let $(E, \overline{\partial}_E, \theta, h)$ be any tame harmonic bundle on $X - D$. Let π denote the projection of X to D. Let P be a point of D. Then we obtain the smooth curve $\pi^{-1}(P)$.

COROLLARY 8.52. *Let b be any real number. Let f be a holomorphic section of ${}_b\big(\mathcal{E}^{\lambda}_{|\pi^{-1}(P)}\big)$ defined on a neighbourhood of $(0,P)$ in $\pi^{-1}(P)$. Then there exists a holomorphic section $\tilde{f}$ of ${}_b\mathcal{E}^{\lambda}$ defined on a neighbourhood of $(0,P)$ in X, such that $\tilde{f}_{|\pi^{-1}(P)} = f$.*

Proof We have only to use Lemma 8.51 inductively. □

We use the following corollary without mention.

COROLLARY 8.53. *The sheaf ${}_b\mathcal{E}^{\lambda}$ is locally free. The restriction ${}_b\mathcal{E}^{\lambda} \longrightarrow {}_b\mathcal{E}^{\lambda}{}_{|\pi^{-1}(P)}$ is surjective.*

Proof The second claim is shown in Corollary 8.52. Let us show the first claim. We have only to prove the case $b = 0$. We have the set $\mathcal{P}ar({}^{\diamond}\mathcal{E}^{\lambda}, 1)$. Let $\boldsymbol{v}$ be a frame of ${}^{\diamond}\mathcal{E}^{\lambda}_{|\pi_i^{-1}(P)}$, which is compatible with the parabolic filtration F. For each v_i, we have the number $b_i := \deg^F(v_i)$.

Then we can pick sections $\tilde{v}_i$ of ${}_{b_i}\mathcal{E}^{\lambda}$ such that $\tilde{v}_{i\,|\,\pi_i^{-1}(P)} = v_i$, by using Corollary 8.52. Thus we obtain the tuple $\tilde{\boldsymbol{v}} := (\tilde{v}_i)$ of sections of ${}^{\diamond}\mathcal{E}^{\lambda}$. We would like to show that $\tilde{\boldsymbol{v}}$ gives a frame of ${}^{\diamond}\mathcal{E}^{\lambda}$.

We put $\widetilde{b} := \sum_{b \in \mathcal{P}ar({}^{\diamond}\mathcal{E}^{\lambda},1)} \mathfrak{m}(\lambda, b) \cdot b$. Then we have the natural isomorphism ${}_{\tilde{b}}\big(\det \mathcal{E}^{\lambda}\big)_{|\pi^{-1}(P)} \simeq \det\big({}^{\diamond}\mathcal{E}^{\lambda}_{|\pi^{-1}(P)}\big)$. Let us consider the section $\Omega(\tilde{\boldsymbol{v}}) := \tilde{v}_1 \wedge \cdots \wedge \tilde{v}_{\operatorname{rank} E}$. Since $\boldsymbol{v}$ is a frame of ${}^{\diamond}\mathcal{E}^{\lambda}_{|\pi^{-1}(P)}$, we have $\Omega(\tilde{\boldsymbol{v}})_{|(0,P)} \neq 0$. Hence $\Omega(\tilde{\boldsymbol{v}})$ gives a frame of ${}_{\tilde{b}}\big(\det \mathcal{E}^{\lambda}\big)_{|\pi^{-1}(P)}$ around $(0,P)$. Then we can conclude that $\tilde{\boldsymbol{v}}$ gives a frame of ${}^{\diamond}\mathcal{E}^{\lambda}$ on a neighbourhood of $(0,P)$, due to the last argument in the proof of Lemma 8.24. □

8.5. Preliminary prolongation of $\mathcal{E}^{\lambda}$ (Special case)

8.5.1. Preliminary. We use the setting in the section 8.2. Moreover we impose the following assumption in this section.

- Let ϵ_i $(i = 1, \ldots, l)$ be small positive numbers such that $\sum_i \epsilon_i < 1/2$. We assume $\mathcal{P}ar({}^{\diamond}\mathcal{E}^{\lambda}, i) \subset] - \epsilon_i, 0[$ for each i.

Pick an element $({}^{0}z_3, \ldots, {}^{0}z_n) \in (\Delta^*)^{n-2}$, and let us consider the restriction of $(E, \overline{\partial}_E, \theta, h)$ to the curve $C_0 := \{(z, z, {}^{0}z_3, \ldots, {}^{0}z_n) \in X\}$ as in the subsection 8.3.4.

LEMMA 8.54. *If η is sufficiently small, the set $\mathcal{KMS}({}^{\diamond}\mathcal{E}^{\lambda'}, i)$ depends on $\lambda' \in \Delta(\lambda, \eta)$ continuously, and we have $\mathcal{P}ar\big({}^{\diamond}\mathcal{E}^{\lambda'}, i\big) \subset] - \epsilon_i, 0[$ for $i = 1, 2$.*

Proof We put $\mathcal{K}(\mathcal{E}, \lambda, \mathbf{0}, i) := \big\{u \in \mathcal{KMS}(\mathcal{E}^0, i) \,\big|\, \mathfrak{k}(\lambda, u) \in \mathcal{KMS}({}^{\diamond}\mathcal{E}^{\lambda}, i)\big\}$. Recall that we have assumed $\big\{\mathfrak{p}(\lambda, u) \,\big|\, u \in \mathcal{K}(\mathcal{E}, \lambda, \mathbf{0}, i)\big\} = \mathcal{P}ar\big({}^{\diamond}\mathcal{E}^{\lambda}, i\big) \subset] - \epsilon_i, 0[$. If η is sufficiently small, we have $\big\{\mathfrak{p}(\lambda', u) \,\big|\, u \in \mathcal{K}(\mathcal{E}, \lambda, \mathbf{0}, i)\big\} \subset] - \epsilon_i, 0[$ for any $\lambda' \in \Delta(\lambda, \eta)$. We also have the following:

$$\sum_{u \in \mathcal{K}(\mathcal{E}, \lambda, \mathbf{0}, i)} \mathfrak{m}(\lambda', \mathfrak{k}(\lambda', u)) = \operatorname{rank} \mathcal{E}^{\lambda'}.$$

Hence we obtain $\mathcal{KMS}({}^{\diamond}\mathcal{E}^{\lambda'}, i) = \big\{\mathfrak{k}(\lambda', u) \,\big|\, u \in \mathcal{K}(\mathcal{E}, \lambda, \mathbf{0}, i)\big\}$. Thus we are done. □

LEMMA 8.55. *We have the following claims for the KMS-structure of ${}^{\diamond}\mathcal{E}_{|C_0}$.*

- $\mathcal{P}ar({}^{\diamond}\mathcal{E}^{\lambda}{}_{|C_0}) \subset] - \epsilon_1 - \epsilon_1, 0[$.
- *We have the following equality:*

$$\sum_{b \in \mathcal{P}ar({}^{\diamond}\mathcal{E}^{\lambda}_{|C_0})} b \cdot \mathfrak{m}(\lambda, b) = \sum_{i=1,2} \sum_{b_i \in \mathcal{P}ar({}^{\diamond}\mathcal{E}^{\lambda}, i)} b_i \cdot \mathfrak{m}(\lambda, b_i).$$

Proof Let η be sufficiently small positive number as in Lemma 8.54. Let $\lambda' \in \Delta(\lambda, \eta)$ be generic. Due to Lemma 8.31, the following holds:

$$\mathcal{KMS}\big({}^{\diamond}(\mathcal{E}^{\lambda'}_{|C_0})\big) \subset \big\{u_1 + u_2 \,\big|\, u_i \in \mathcal{KMS}({}^{\diamond}\mathcal{E}^{\lambda'}, i)\big\}.$$

Then the following holds for any $\lambda' \in \Delta(\lambda, \eta)$, due to Lemma 8.54:

$$\mathcal{KMS}\big({}^{\diamond}(\mathcal{E}^{\lambda'}_{|C_0})\big) \subset \big\{u_1 + u_2 \,\big|\, u_i \in \mathcal{KMS}({}^{\diamond}\mathcal{E}^{\lambda'}, i)\big\},$$
$$\mathcal{P}ar\big({}^{\diamond}(\mathcal{E}^{\lambda'}_{|C_0})\big) \subset \big\{a_1 + a_2 \,\big|\, a_i \in \mathcal{P}ar({}^{\diamond}\mathcal{E}^{\lambda'}, i)\big\} \subset] - \epsilon_1 - \epsilon_2, 0[.$$

Hence we obtain the first claim. We can show the second claim similarly by using Lemma 8.31. □

8.5.2. Preliminary prolongation.

LEMMA 8.56. *Let $\epsilon_1, \ldots, \epsilon_l$ be positive numbers such that $\sum_{i=1}^{l} \epsilon_i < 1/2$. We assume $\mathcal{P}ar({}^{\diamond}\mathcal{E}^{\lambda}, i) \subset] - \epsilon_i, 0[$ for each i. Then the $\mathcal{O}_X$-sheaf ${}^{\diamond}\mathcal{E}^{\lambda}$ is locally free.*

Proof We use an induction on the dimension of X. As the hypothesis of the induction, we assume the following:

The $\mathcal{O}_X$-sheaf ${}^{\diamond}\mathcal{E}^{\lambda}$ is locally free, if the following holds:

- $\dim(X) \leq n - 1$.
- $\mathcal{P}ar({}^{\diamond}\mathcal{E}^{\lambda}, i) \subset] - \epsilon_i, 0[$ and $\sum \epsilon_i < 1$.

We use the notation in the subsection 8.4.3. Due to the hypothesis of the induction and the first claim of Lemma 8.55, we have the local freeness of ${}^\diamond(\mathcal{E}^\lambda_{|X^{(1)}-D^{(1)}})$. Pick a frame $\boldsymbol{v}=(v_i)$ of ${}^\diamond(\mathcal{E}^\lambda_{|X^{(1)}-D^{(1)}})$ over $X^{(1)}$. For each v_i, we pick a section $\tilde{v}_i$ of ${}^\diamond\mathcal{E}^\lambda$ over a neighbourhood U of X_0 such that $\tilde{v}_{i\,|\,U\cap X^{(1)}}=v_{i\,|\,U\cap X^{(1)}}$, by using Proposition 8.46. We may assume that v_i are defined over X by shrinking X. Clearly Lemma 8.56 can be reduced to the following lemma.

LEMMA 8.57. *$\tilde{\boldsymbol{v}}$ gives a frame of ${}^\diamond\mathcal{E}^\lambda$ around X_0.*

Proof We put as follows:

$$\tilde{\boldsymbol{b}}:=(\tilde{b}_1,\ldots,\tilde{b}_l),\quad \tilde{b}_i:=\sum_{b\in\mathcal{P}ar({}^\diamond\mathcal{E}^\lambda,i)} b\cdot\mathfrak{m}(\lambda,b).$$

The restriction $\tilde{\boldsymbol{v}}_{|\pi_i^{-1}(P)}$ gives a tuple of holomorphic sections of ${}^\diamond(\mathcal{E}^\lambda_{|\pi_i^{-1}(P)})$ for any $P\in D_i^\circ$. Hence $\Omega(\tilde{\boldsymbol{v}})_{|\pi_i^{-1}(P)}$ is a holomorphic section of $\det\big({}^\diamond(\mathcal{E}^\lambda_{|\pi_i^{-1}(P)})\big)={}_{\tilde{b}_i}\big(\det(\mathcal{E}^\lambda)_{|\pi_i^{-1}(P)}\big)$. It implies $\Omega(\tilde{\boldsymbol{v}})$ is a holomorphic section of ${}_{\tilde{\boldsymbol{b}}}\det(\mathcal{E}^\lambda)$.

We have the natural isomorphism $({}_{\tilde{\boldsymbol{b}}}\det(\mathcal{E}^\lambda))_{|X^{(1)}}\simeq\det\big({}^\diamond(\mathcal{E}^\lambda_{|X^{(1)}})\big)$. Since $\Omega(\tilde{\boldsymbol{v}})_{|X^{(1)}}$ gives a frame of $\det\big({}^\diamond(\mathcal{E}^\lambda_{|X^{(1)}})\big)$, we obtain $\Omega(\tilde{\boldsymbol{v}})_{|O}\neq 0$. It is a standard argument to conclude that $\boldsymbol{v}$ gives a frame around the origin O. (See the proof of Lemma 8.24). Hence we obtain Lemma 8.57, and thus Lemma 8.56. □

8.6. Prolongation of $\mathcal{E}^\lambda$ and the compatibility of the parabolic filtrations

8.6.1. Statements of the theorems. We put $X=\Delta^n$, $D_i=\{z_i=0\}$ and $D=\bigcup_{i=1}^l D_i$. Let $(E,\overline{\partial}_E,\theta,h)$ be a tame harmonic bundle on $X-D$. The following theorems will be proved in the subsections 8.6.2–8.6.3.

THEOREM 8.58. *For any $\boldsymbol{b}\in\boldsymbol{R}^l$, the $\mathcal{O}_X$-module ${}_{\boldsymbol{b}}\mathcal{E}^\lambda$ are coherent and locally free. In particular, ${}^\diamond\mathcal{E}^\lambda$ is locally free.*

Let us pick an element $\boldsymbol{b}=(b_1,\ldots,b_l)\in\boldsymbol{R}^l$. Let $\boldsymbol{\delta}_i$ denote the element $(\overbrace{0,\ldots,0}^{i-1},1,0,\ldots,0)$. For any $\boldsymbol{b}'\leq\boldsymbol{b}$, we have the naturally defined morphism ${}_{\boldsymbol{b}'}\mathcal{E}^\lambda\longrightarrow{}_{\boldsymbol{b}}\mathcal{E}^\lambda$. For $b_i-1\leq b\leq b_i$, we put $\boldsymbol{b}'=\boldsymbol{b}+(b-b_i)\boldsymbol{\delta}_i$ and as follows:

$${}^iF_b\big({}_{\boldsymbol{b}}\mathcal{E}^\lambda\big):=\mathrm{Im}\big({}_{\boldsymbol{b}'}\mathcal{E}^\lambda{}_{|D_i}\longrightarrow{}_{\boldsymbol{b}}\mathcal{E}^\lambda{}_{|D_i}\big).$$

Then we obtain the filtration ${}^iF({}_{\boldsymbol{b}}\mathcal{E}^\lambda):=\big\{{}^iF_b({}_{\boldsymbol{b}}\mathcal{E}^\lambda)\,\big|\,b_i-1\leq b\leq b_i\big\}$ of $\mathcal{O}_{D_i}$-modules.

THEOREM 8.59.

- *${}^iF({}_{\boldsymbol{b}}\mathcal{E}^\lambda)$ is a filtration in the category of vector bundles on D_i.*
- *The tuple of the filtrations $\big({}^iF\,\big|\,i=1,\ldots,l\big)$ on the divisors are compatible. (Definition 4.37).*
- *Let $\boldsymbol{\eta}$ be an element of $\prod_{i\in I}[b_i-1,b_i]$. We put $\tilde{\boldsymbol{\eta}}:=\boldsymbol{b}+\boldsymbol{\eta}-q_I(\boldsymbol{b})$. Then we have the following:*

$${}^IF_{\boldsymbol{\eta}}\big({}_{\boldsymbol{b}}\mathcal{E}^\lambda\big):=\bigcap_{i\in I}{}^iF_{\eta_i\,|\,D_I}\big({}_{\boldsymbol{b}}\mathcal{E}^\lambda\big)=\mathrm{Im}\big({}_{\tilde{\boldsymbol{\eta}}}\mathcal{E}^\lambda{}_{|D_I}\longrightarrow{}_{\boldsymbol{b}}\mathcal{E}^\lambda{}_{|D_I}\big).$$

Before entering the proof of the theorems, we remark the following.

LEMMA 8.60. *We have only to show the claims for* ${}^{\diamond}\mathcal{E}^{\lambda}$ *of any tame harmonic bundle* $(E, \overline{\partial}_E, h, \theta)$ *to prove Theorem* 8.58 *and Theorem* 8.59.

Proof For a tame harmonic bundle $(E, \overline{\partial}_E, h, \theta)$, and $\boldsymbol{b} \in \boldsymbol{R}^l$, we have the harmonic bundle $(E', \overline{\partial}_{E'}, h', \theta') = (E, \overline{\partial}_E, h, \theta) \otimes L(-\boldsymbol{b})$ (see the section 6.1.5 for $L(-\boldsymbol{b})$). We denote the deformed holomorphic bundle of $(E', \overline{\partial}_{E'}, h', \theta')$ by $\mathcal{E}'^{\lambda}$. Then we have the natural isomorphism ${}_{\boldsymbol{b}}\mathcal{E}^{\lambda} \simeq {}^{\diamond}\mathcal{E}'^{\lambda}$ by definition. Thus we are done. □

We also remark that we use Corollary 8.53 without mention.

8.6.2. Step 1.

CONDITION 8.61. Let us take an element $\boldsymbol{c} \in \mathbb{Z}^l_{>0}$ as follows for any i:

(1) c_i are sufficiently large with respect to $\mathcal{KMS}({}^{\diamond}\mathcal{E}^{\lambda}, i)$ (Definition 2.1).
(2) There exist numbers $b_i \in]-1, (4l)^{-1}[$ satisfying the following:

$$\{-b_i + \kappa(c_i \cdot a_i) \,|\, a_i \in \mathcal{P}ar(\mathcal{E}^{\lambda}, i)\} \subset]-(2l)^{-1}, 0[.$$

See the subsection 2.1.5 for κ. We put $\boldsymbol{b} = (b_1, \ldots, b_l)$. □

We put $(E_1, \overline{\partial}_{E_1}, h_1, \theta_1) := \psi_{\boldsymbol{c}}^{-1}(E, \overline{\partial}_E, h, \theta) \otimes L(-\boldsymbol{b})$. We denote the deformed holomorphic bundle of $(E_1, \overline{\partial}_{E_1}, h_1, \theta_1)$ by $\mathcal{E}_1^{\lambda}$. Then we have the natural isomorphism ${}^{\diamond}(\psi_{\boldsymbol{c}}^{-1}\mathcal{E}^{\lambda}) = {}_{\boldsymbol{b}}(\psi_{\boldsymbol{c}}^{-1}\mathcal{E}^{\lambda}) \simeq {}^{\diamond}\mathcal{E}_1^{\lambda}$. We also have the following, due to the result in the one dimensional case (see the subsection 7.2.3):

$$\mathcal{KMS}({}^{\diamond}\mathcal{E}_1^{\lambda}, i) = \{-b_i + \kappa(c_i \cdot a_i) \,|\, a_i \in \mathcal{P}ar(\mathcal{E}^{\lambda}, i)\} \subset]-(2l)^{-1}, 0[. \tag{8.25}$$

LEMMA 8.62. ${}^{\diamond}\mathcal{E}_1^{\lambda}$ *is locally free.*

Proof It follows from (8.25) and the preliminary prolongation (Lemma 8.56). □

COROLLARY 8.63. *The sheaf* ${}^{\diamond}(\psi_{\boldsymbol{c}}^{-1}\mathcal{E}^{\lambda})$ *is a locally free* $\mathcal{O}_X$*-module.* □

We have the natural $\mu_{\boldsymbol{c}}$-action on $\psi_{\boldsymbol{c}}^{-1}\mathcal{E}^{\lambda}$, which is prolonged to the action on ${}^{\diamond}(\psi_{\boldsymbol{c}}^{-1}\mathcal{E}^{\lambda})$. In particular, we obtain the μ_{c_i}-action on ${}^{\diamond}(\psi_{\boldsymbol{c}}^{-1}\mathcal{E}^{\lambda})_{|D_i}$. Since the action of μ_{c_i} on D_i is trivial, we have the decomposition:

$${}^{\diamond}\psi_{\boldsymbol{c}}^{-1}\mathcal{E}^{\lambda}{}_{|D_i} = \bigoplus_{c_i-1 \leq h \leq 0} V_h.$$

Here the generator ω of μ_{c_i} acts as ω^h on V_h.

Let us pick a point P of $D_i^{\circ} = D_i \setminus \bigcup_{j \neq i, j \leq l} D_j \cap D_i$. We have the following morphism due to our choice of $\boldsymbol{c}$ ((1) in Condition 8.61) and the result in the subsection 7.2.4:

$$\varphi : \{h \,|\, -c_i + 1 \leq h \leq 0,\ V_h \neq 0\} \longrightarrow \{\tilde{b} \,|\, -1 < \tilde{b} \leq 0,\ \mathrm{Gr}^F_{\tilde{b}}\big({}^{\diamond}\psi_{\boldsymbol{c}}^{-1}\mathcal{E}^{\lambda}_{|\pi_i^{-1}(P)}\big) \neq 0\}.$$

We consider the filtration ${}^iF'$ of ${}^{\diamond}(\psi_{\boldsymbol{c}}^{-1}\mathcal{E}^{\lambda})_{|D_i}$ in the category of vector bundles on D_i, given as follows for any $-1 < b < 0$:

$${}^iF'_b := \bigoplus_{\varphi(h) \leq b} V_h.$$

Due to the construction, it is easy to see that the filtrations $({}^iF' \,|\, i = 1, \ldots, l)$ are compatible in the sense of Definition 4.37.

We consider the subsheaf ${}_{b\delta_i}\big(\psi_{\boldsymbol{c}}^{-1}\mathcal{E}^\lambda\big)'$ of ${}^\diamond\psi_{\boldsymbol{c}}^{-1}(\mathcal{E}^\lambda)$, given as follows:

$$ {}_{b\delta_i}\big(\psi_{\boldsymbol{c}}^{-1}\mathcal{E}^\lambda\big)' = \mathrm{Ker}\Big(\pi : {}^\diamond\big(\psi_{\boldsymbol{c}}^{-1}\mathcal{E}^\lambda\big) \longrightarrow \frac{{}^\diamond\big(\psi_{\boldsymbol{c}}^{-1}\mathcal{E}^\lambda\big)_{|D_i}}{{}^iF'_b}\Big). $$

Here π denotes the naturally defined morphism.

LEMMA 8.64. *We have the following, for any $-1 < b < 0$:*

$$ {}_{b\delta_i}\big(\psi_{\boldsymbol{c}}^{-1}\mathcal{E}^\lambda\big)' = {}_{b\delta_i}\big(\psi_{\boldsymbol{c}}^{-1}\mathcal{E}^\lambda\big), \qquad {}^iF'_b = {}^iF_b $$

Proof Let f be a holomorphic section of ${}_{b\delta_i}\big(\psi_{\boldsymbol{c}}^{-1}\mathcal{E}^\lambda\big)$. It can be also regarded as a section of ${}^\diamond\psi_{\boldsymbol{c}}^{-1}\mathcal{E}^\lambda$. Let P be a point of D_i°. We have the element $f(P)$ of ${}^\diamond\psi_{\boldsymbol{c}}^{-1}\mathcal{E}^\lambda{}_{|P} = {}^\diamond\big(\psi_{\boldsymbol{c}}^{-1}\mathcal{E}^\lambda{}_{|\pi_i^{-1}(P)}\big)_{|P}$. Due to Lemma 7.54, we obtain $f(P) \in {}^iF'_{b\,|\,P}$.

Let $\bar{f}$ denote the image of f via the projection π. Then $\bar{f}(P) = 0$ for any $P \in D_i^\circ$. It implies $\bar{f} = 0$ on D_i. Hence we obtain $f \in {}_{b\delta_i}\big(\psi_{\boldsymbol{c}}^{-1}\mathcal{E}^\lambda\big)'$.

On the other hand, pick a section $f \in {}_{b\delta_i}\big(\psi_{\boldsymbol{c}}^{-1}\mathcal{E}^\lambda\big)'$. Due to Lemma 7.54, we obtain the following inequality for any $P \in D_i^\circ$:

$$ -\mathrm{ord}\big(f_{|\pi_i^{-1}(P)}\big) \leq b. $$

Then we obtain $f \in {}_{b\delta_i}\big(\psi_{\boldsymbol{c}}^{-1}\mathcal{E}^\lambda\big)$ due to Corollary 2.53 and Corollary 8.19.

In all, we obtain ${}_{b\delta_i}\big(\psi_{\boldsymbol{c}}^{-1}\mathcal{E}^\lambda\big) = {}_{b\delta_i}\big(\psi_{\boldsymbol{c}}^{-1}\mathcal{E}^\lambda\big)'$. It implies that ${}_{b\delta_i}\big(\psi_{\boldsymbol{c}}^{-1}\mathcal{E}^\lambda\big)$ is locally free, and ${}^iF_b = {}^iF'_b$. □

LEMMA 8.65. *We have the following:*

$$ {}^IF_{\boldsymbol{b}} = \mathrm{Im}\big({}_{\boldsymbol{b}}\psi_{\boldsymbol{c}}^{-1}\mathcal{E}^\lambda{}_{|D_I} \longrightarrow {}^\diamond\psi_{\boldsymbol{c}}^{-1}\mathcal{E}^\lambda{}_{|D_I}\big). $$

Proof It can be shown by an argument similar to the proof of Lemma 8.64. □

8.6.3. Step 2. Let f be a section of ${}^\diamond\psi_{\boldsymbol{c}}^{-1}\mathcal{E}^\lambda$. Assume the following:

- $f(0) \neq 0$, and f is compatible with the filtrations ${}^iF$ $(i = 1, \dots, l)$, i.e., there exists a splitting of ${}^iF$ $(i = 1, \dots, l)$ which is compatible with f.
- f is equivariant, i.e., $g^*(f) = \prod \omega_i^{h_i} \cdot f$ for some $-c_i + 1 \leq h_i \leq 0$. Here $g = (\omega_1, \dots, \omega_n) \in \mu_{\boldsymbol{c}}$.

We put $f_1 := \prod z_i^{-h_i} \cdot f$. Then it is a section of $\psi_{\boldsymbol{c}}^{-1}\mathcal{E}^\lambda$, and it is $\mu_{\boldsymbol{c}}$-invariant, i.e., $g^*(f_1) = f_1$ for any $g \in \mu_{\boldsymbol{c}}$. Hence there exists the unique section $\bar{f}$ of $\mathcal{E}^\lambda$ on $X - D$, such that $\psi_{\boldsymbol{c}}^{-1}\bar{f} = f_{1\,|\,X-D}$. Note the following:

$$ -{}^i\mathrm{ord}(\bar{f}) = c_i^{-1}(h_i - {}^i\mathrm{ord}(f)) \leq 0. $$

Hence $\bar{f}$ gives the section of ${}^\diamond\mathcal{E}^\lambda$.

Let us take a frame $\boldsymbol{v} = (v_i)$ of ${}^\diamond\psi_{\boldsymbol{c}}^{-1}\mathcal{E}^\lambda$ satisfying the following conditions (Corollary 4.48):

- It is equivariant.
- It is compatible with the filtrations $({}^1F, \dots, {}^lF)$.

Then we obtain a tuple $\bar{\boldsymbol{v}} = (\bar{v}_1, \dots, \bar{v}_r)$ of sections of ${}^\diamond\mathcal{E}^\lambda$ by the procedure above.

LEMMA 8.66. *${}^\diamond\mathcal{E}^\lambda$ is locally free, and $\bar{\boldsymbol{v}}$ gives a local frame on a neighbourhood of the origin O.*

Proof Recall that we have already obtained the result in the case $\dim(X) = 1$ (Lemma 7.57). Let us consider the element $\tilde{\boldsymbol{b}} = (\tilde{b}_1, \ldots, \tilde{b}_l)$ of $\boldsymbol{R}^l$, given as follows:

$$\tilde{b}_i := \sum_{b \in \mathcal{P}ar({}^{\diamond}\mathcal{E}^{\lambda}, i)} b \cdot \mathfrak{m}(\lambda, b).$$

Let P be a point of D_i°. Then $\Omega(\bar{\boldsymbol{v}})_{|\pi_i^{-1}(P)}$ is a frame of $\det\left({}^{\diamond}\left(\mathcal{E}^{\lambda}_{|\pi_i^{-1}(P)}\right)\right) = {}_{\tilde{b}_i}\det\left(\mathcal{E}^{\lambda}_{|\pi_i^{-1}(P)}\right)$. Thus we obtain that $\Omega(\bar{\boldsymbol{v}})$ is a frame of ${}_{\tilde{\boldsymbol{b}}}\det(\mathcal{E}^{\lambda})$.

Let $f = \sum f_i \cdot \bar{v}_i$ be a holomorphic section of ${}^{\diamond}\mathcal{E}^{\lambda}$. As usual, we can show that f_i are holomorphic over X, and thus $\bar{\boldsymbol{v}}$ gives a frame of ${}^{\diamond}\mathcal{E}^{\lambda}$ (see the proof of Lemma 8.24). In particular, the sheaf ${}^{\diamond}\mathcal{E}^{\lambda}$ is locally free. □

We consider the filtration ${}^{i}F'_b$ of ${}^{\diamond}\mathcal{E}^{\lambda}{}_{|D_i}$ in the category of the vector bundles over D_i, given as follows:

$${}^{i}F'_b := \left\langle \bar{v}_{j\,|\,D_i} \;\middle|\; -{}^{i}\mathrm{ord}(\bar{v}_j) \leq b \right\rangle.$$

For any $-1 < b \leq 0$, we consider the subsheaf ${}_{b \cdot \boldsymbol{\delta}_i}\mathcal{E}^{\lambda\,\prime}$ of ${}^{\diamond}\mathcal{E}^{\lambda}$ given as follows:

$${}_{b \cdot \boldsymbol{\delta}_i}\left(\mathcal{E}^{\lambda}\right)' := \mathrm{Ker}\left(\pi : {}^{\diamond}\mathcal{E}^{\lambda} \longrightarrow \frac{{}^{\diamond}\mathcal{E}^{\lambda}{}_{|D_i}}{{}^{i}F'_b}\right).$$

Here π denotes the naturally defined morphism. Then ${}_{b \cdot \boldsymbol{\delta}_i}\left(\mathcal{E}^{\lambda}\right)'$ is locally free.

LEMMA 8.67. *We have* ${}_{b\boldsymbol{\delta}_i}\mathcal{E}^{\lambda} = {}_{b\boldsymbol{\delta}_i}\left(\mathcal{E}^{\lambda}\right)'$ *and* ${}^{i}F'_b = {}^{i}F_b$.

Proof We have already known that the claim holds if $\dim X = 1$ (Lemma 7.57).

Let f be a holomorphic section of ${}_{b\boldsymbol{\delta}_i}\mathcal{E}^{\lambda}$. We can also regard it as a section of ${}^{\diamond}\mathcal{E}^{\lambda}$. By applying Lemma 7.57 to $f_{|\pi_i^{-1}(P)} \in {}^{\diamond}\left(\mathcal{E}^{\lambda}_{|\pi_i^{-1}(P)}\right)$, we obtain that $f(P) \in {}^{i}F'_{|P}$ for any $P \in D_i^{\circ}$. Then it is easy to derive that f is contained in ${}_{b\boldsymbol{\delta}_i}\left(\mathcal{E}^{\lambda}\right)'$.

On the other hand, let f be a holomorphic section of ${}_{b\boldsymbol{\delta}_i}\left(\mathcal{E}^{\lambda}\right)'$. Applying Lemma 7.57 to $f_{|\pi_i^{-1}(P)}$, we obtain $-\mathrm{ord}(f_{|\pi_i^{-1}(P)}) \leq b$ for any $P \in D_i^{\circ}$. Then we obtain $f \in {}_{b\boldsymbol{\delta}_i}\mathcal{E}^{\lambda}$ due to Corollary 2.53. Therefore we obtain ${}_{b\boldsymbol{\delta}_i}\mathcal{E}^{\lambda} = {}_{b\boldsymbol{\delta}_i}\left(\mathcal{E}^{\lambda}\right)'$, and thus ${}^{i}F_b = {}^{i}F'_b$. □

LEMMA 8.68. *The filtration ${}^{i}F$ is a filtration in the category of the vector bundles over D_i. The filtrations $({}^{i}F \,|\, i = 1, \ldots, l)$ are compatible.*

Proof By our construction, ${}^{i}F'$ is the filtration in the category of the vector bundles over D_i, and the tuple $({}^{i}F' \,|\, i = 1, \ldots, l)$ is compatible. Then the lemma follows from Lemma 8.67. □

LEMMA 8.69. *We have the following equality:*

$${}^{I}F_{\boldsymbol{b}}({}^{\diamond}\mathcal{E}^{\lambda}) = \mathrm{Im}({}_{\boldsymbol{b}}\mathcal{E}^{\lambda}{}_{|D_I} \longrightarrow {}^{\diamond}\mathcal{E}^{\lambda}{}_{|D_I}).$$

Proof It can be shown by an argument similar to the proof of Lemma 8.67. □

Then Theorem 8.58 follows from Lemma 8.66 and Lemma 8.60, and Theorem 8.59 follows from Lemma 8.68, Lemma 8.69 and Lemma 8.60. □

8.6.4. Weak norm estimate of holomorphic sections. Let $\boldsymbol{v}$ be a frame of ${}_{\boldsymbol{b}}\mathcal{E}^{\lambda}$ compatible with the parabolic filtrations $({}^{i}F\,|\,i=1,\ldots,l)$. We obtain the numbers ${}^{i}b(v_j):={}^{i}\deg^{F}(v_j)$. We put as follows:

$$v'_j := v_j\cdot\prod_{i=1}^{l}|z_i|^{{}^{i}b(v_j)},\qquad \boldsymbol{v}'=(v'_j).$$

Then $\boldsymbol{v}'$ is a C^{∞}-frame of $\mathcal{E}^{\lambda}$ over $X-D$.

PROPOSITION 8.70. *$\boldsymbol{v}'$ is adapted up to log order.*

Proof The argument is essentially same as the proof of Lemma 8.27. By our construction of $\boldsymbol{v}'$, the following is clear:

$$H(h,\boldsymbol{v}')\leq C_1\cdot\big(-\sum\log|z_i|\big)^{M}.$$

Let $\boldsymbol{v}^{\vee}$ denote the dual frame of $\boldsymbol{v}$. Then $\boldsymbol{v}^{\vee}$ gives a tuple of sections of ${}_{-\boldsymbol{b}+(1-\epsilon)\boldsymbol{\delta}}\mathcal{E}^{\vee\,\lambda}$ for some $\epsilon>0$. Let P be a point of D_i°. Due to the result in the case of curves, $\boldsymbol{v}^{\vee}_{|\pi_i^{-1}(P)}$ gives a frame of ${}_{-b_i+(1-\epsilon)}\mathcal{E}^{\vee\,\lambda}{}_{|\pi_i^{-1}(P)}$, which is compatible with the parabolic filtration. We have ${}^{i}\deg^{F}(v_j^{\vee})=\deg^{F}\big(v^{\vee}_{j\,|\,\pi_i^{-1}(P)}\big)=-{}^{i}b(v_j)$ for any point $P\in D_i^{\circ}$. We put as follows:

$$\boldsymbol{v}^{\vee\,\prime}=(v_j^{\vee\,\prime}),\qquad v_j^{\vee\,\prime}:=v_j^{\vee}\cdot\prod_{i=1}^{l}|z_i|^{-{}^{i}b(v_j)}.$$

Due to Corollary 2.53, we obtain the following (see the subsection 8.3.2):

$$H(h^{\vee},\boldsymbol{v}^{\vee\,\prime})\leq C_2\cdot\Big(-\sum_{i=1}^{l}\log|z_i|\Big)^{M}.$$

It implies the following:

$$C_3\cdot\Big(-\sum_{i=1}^{l}\log|z_i|\Big)^{-M}\leq H(h,\boldsymbol{v}').$$

Thus we are done. □

8.7. Prolongation of $\mathcal{E}_{|\Delta(\lambda_0,\epsilon_0)\times(X-D)}$

8.7.1. Preliminary. We continue to use the setting in the section 8.6. Recall that we may assume to have the following decomposition (Condition 8.8 and Remark 8.9):

$$\mathcal{E}^{0}=\bigoplus_{\boldsymbol{a}\in\mathcal{S}p(\theta)}E_{\boldsymbol{a}}.$$

Due to Lemma 8.11, there exists a positive constant ϵ_1 such that the subbundles $E_{\boldsymbol{a}}$ and $E_{\boldsymbol{b}}$ ($\boldsymbol{a}\neq\boldsymbol{b}$) are $\prod_{j\in\mathrm{Diff}(\boldsymbol{a},\boldsymbol{b})}|z_j|^{\epsilon_1}$-asymptotically orthogonal.

Consider the λ-dependent section $g(\lambda)$ of $End(E)$ over $X-D$, given as follows:

$$g(\lambda):=\bigoplus_{\boldsymbol{a}}\exp\Big(\lambda\sum_{i=1}^{l}\bar{a}_i\cdot\log|z_i|^2\Big)\cdot id_{E_{\boldsymbol{a}}}.\tag{8.26}$$

LEMMA 8.71. *We have the following equality:*

$$g(\lambda-\lambda_0)\cdot\overline{\partial}_E g(\lambda-\lambda_0)^{-1}=-(\lambda-\lambda_0)\cdot\sum_{\boldsymbol{a}}\Big(\sum_i \bar{a}_i\frac{d\bar{z}_i}{\bar{z}_i}\Big)\cdot id_{E_{\boldsymbol{a}}}. \tag{8.27}$$

Proof It can be checked by a direct calculation. □

We have the decomposition $\theta^\dagger=\phi_1+\phi_2+\phi_3$ satisfying the following:

- $\phi_1=\sum_{\boldsymbol{a}}\big(\sum_{i=1}^{l}\bar{a}_i\cdot\bar{z}_i^{-1}d\bar{z}_i\big)\cdot id_{E_{\boldsymbol{a}}}$.
- $\phi_2=\sum_{\boldsymbol{a}}\phi_{2\,\boldsymbol{a}}$ where $\phi_{2\,\boldsymbol{a}}\in End(E_{\boldsymbol{a}})\otimes\Omega^{0,1}_{X-D}$, and $|\phi_{2\,\boldsymbol{a}}|_{h,\mathbf{p}}$ is bounded. Here $|\cdot|_{h,\mathbf{p}}$ denotes the norm with respect to h and the Poincaré metric of $X-D$.
- $\phi_3=\sum_{\boldsymbol{a}\neq\boldsymbol{b}}\phi_{3\,\boldsymbol{a},\boldsymbol{b}}$, where $\phi_{3\,\boldsymbol{a}\,\boldsymbol{b}}$ are sections of $Hom(E_{\boldsymbol{a}},E_{\boldsymbol{b}})\otimes\Omega^{0,1}_{X-D}$. We have the following estimate for some positive constants ϵ_2 and C:
$$\big|\phi_{3\,\boldsymbol{a},\boldsymbol{b}}\big|_{h,\mathbf{p}}\leq C\cdot\prod_{i\in\mathrm{Diff}(\boldsymbol{a},\boldsymbol{b})}|z_i|^{\epsilon_2}.$$

The following lemma is clear

LEMMA 8.72. *$g(\lambda)$ and ϕ_i $(i=1,2)$ are commutative.* □

LEMMA 8.73. *We have the following formula:*

$$\begin{aligned}g(\lambda-\lambda_0)\cdot(\overline{\partial}_E+\lambda\theta^\dagger)\cdot g(\lambda-\lambda_0)^{-1}&=\\ \overline{\partial}_E+\lambda_0\cdot\theta^\dagger+(\lambda-\lambda_0)\cdot\big(\phi_2+\phi_3\big)&+\lambda\cdot\Big(g(\lambda-\lambda_0)\cdot\phi_3\cdot g(\lambda-\lambda_0)^{-1}-\phi_3\Big).\end{aligned}\tag{8.28}$$

Proof We have the following equality:

$$\begin{aligned}&g(\lambda-\lambda_0)\cdot(\overline{\partial}_E+\lambda\theta^\dagger)\cdot g(\lambda-\lambda_0)^{-1}=\\ &\quad\overline{\partial}_E+g(\lambda-\lambda_0)\cdot\overline{\partial}g(\lambda-\lambda_0)^{-1}+\lambda\cdot g(\lambda-\lambda_0)\cdot\theta^\dagger\cdot g(\lambda-\lambda_0)^{-1}=\\ &\overline{\partial}_E+\lambda_0\cdot\theta^\dagger+g(\lambda-\lambda_0)\cdot\overline{\partial}g(\lambda-\lambda_0)^{-1}+(\lambda-\lambda_0)\cdot\theta^\dagger+\lambda\cdot\Big(g(\lambda-\lambda_0)\cdot\theta^\dagger\cdot g(\lambda-\lambda_0)^{-1}-\theta^\dagger\Big).\end{aligned}\tag{8.29}$$

We obtain the following from (8.27):

$$g(\lambda-\lambda_0)\cdot\overline{\partial}g(\lambda-\lambda_0)^{-1}+(\lambda-\lambda_0)\cdot\theta^\dagger=(\lambda-\lambda_0)\cdot\big(\phi_2+\phi_3\big).\tag{8.30}$$

We also obtain the following from Lemma 8.72:

$$g(\lambda-\lambda_0)\cdot\theta^\dagger\cdot g(\lambda-\lambda_0)^{-1}-\theta^\dagger=g(\lambda-\lambda_0)\cdot\phi_3\cdot g(\lambda-\lambda_0)^{-1}-\phi_3.\tag{8.31}$$

Then we obtain (8.28) from (8.29), (8.30) and (8.31). □

LEMMA 8.74. *We put $\psi(\lambda):=g(\lambda-\lambda_0)\cdot\phi_3\cdot g(\lambda-\lambda_0)^{-1}-\phi_3$. There exist positive constants η, ϵ' and C such that the following holds for any $\lambda,\lambda'\in\Delta(\lambda_0,\eta)$:*

$$\big|\big(\psi(\lambda)-\psi(\lambda')\big)_{\boldsymbol{a},\boldsymbol{b}}\big|_{h,\mathbf{p}}\leq C\cdot|\lambda-\lambda'|\cdot\prod_{i\in\mathrm{Diff}(\boldsymbol{a},\boldsymbol{b})}|z_i|^{\epsilon'}.$$

Here $\big(A\big)_{\boldsymbol{a},\boldsymbol{b}}\in Hom\big(E_{\boldsymbol{a}},E_{\boldsymbol{b}}\big)$ denotes the $(\boldsymbol{a},\boldsymbol{b})$-component of A, and $|\cdot|_{h,\mathbf{p}}$ denotes the norm with respect to h and the Poincaré metric of $X-D$.

Proof By definition, the $(\boldsymbol{a},\boldsymbol{b})$-component of $\psi(\lambda)-\psi(\lambda')$ is as follows:

$$\phi_{3\,\boldsymbol{a},\boldsymbol{b}}\cdot\Big(\prod_{i=1}^{l}|z_i|^{2\lambda(a_i-b_i)}-\prod_{i=1}^{l}|z_i|^{2\lambda'(a_i-b_i)}\Big).$$

Hence the norm is dominated by the following:

$$(8.32)\quad |\phi_{3\,\boldsymbol{a}\,\boldsymbol{b}}|_{h,\mathbf{p}}\cdot\prod_{i=1}^{l}|z_i|^{2\,\mathrm{Re}(\lambda'(a_i-b_i))}\cdot\Big(\prod_{i=1}^{l}|z_i|^{2(\lambda-\lambda')(a_i-b_i)}-1\Big)$$
$$\leq C\cdot\prod_{i\in\mathrm{Diff}(\boldsymbol{a},\boldsymbol{b})}|z_i|^{\epsilon-\eta|a_i-b_i|}(-\log|z_i|)\cdot|\lambda-\lambda'|.$$

Thus we are done. □

Let us pick a point $\lambda_0\in\boldsymbol{C}_\lambda$. We put as follows for any $\lambda\in\boldsymbol{C}_\lambda$:

$$(8.33)\qquad d''(\lambda):=g(\lambda-\lambda_0)\cdot(\overline{\partial}_E+\lambda\cdot\theta^\dagger)\cdot g(\lambda-\lambda_0)^{-1}.$$

Then we have the following equality due to (8.28):

$$d''(\lambda)=\overline{\partial}_E+\lambda_0\cdot\theta^\dagger+(\lambda-\lambda_0)\cdot(\phi_2+\phi_3)+\lambda\cdot\psi(\lambda).$$

It gives the holomorphic structure of C^∞-bundle E over $X-D$, and it is equivalent to $\overline{\partial}_{\mathcal{E}^\lambda}=\overline{\partial}_E+\lambda\cdot\theta^\dagger$ up to the (not unitary) gauge transformation.

LEMMA 8.75. *If $\eta>0$ is sufficiently small, then there exists a positive constant C such that the following holds for any $\lambda,\lambda'\in\Delta(\lambda_0,\eta)$:*

$$(8.34)\qquad |d''(\lambda)-d''(\lambda')|_{h,\mathbf{p}}\leq|\lambda-\lambda'|\cdot C.$$

Note that $d''(\lambda)-d''(\lambda')$ are $(0,1)$-forms.

Proof We have the following:

$$d''(\lambda)-d''(\lambda')=(\lambda-\lambda')\cdot(\phi_2+\phi_3)+\lambda\cdot(\psi(\lambda)-\psi(\lambda')).$$

Thus we obtain the result from Lemma 8.74 and the estimates for ϕ_i $(i=2,3)$. □

Let p_λ denote the projection $\Delta(\lambda_0,\eta)\times(X-D)\longrightarrow X-D$, and then we have the C^∞-bundle $p_\lambda^{-1}E$ on $\Delta(\lambda_0,\eta)\times(X-D)$. We have the naturally defined operator:

$$\overline{\partial}_\lambda+d''(\lambda):C^\infty(p_\lambda^{-1}(E))\longrightarrow C^\infty\Big(p_\lambda^{-1}(E)\otimes\Omega^{0,1}_{(X-D)\times\Delta(\lambda_0,\eta)}\Big).$$

LEMMA 8.76. *The operator $\overline{\partial}_\lambda+d''(\lambda)$ gives a holomorphic structure, i.e., $(\overline{\partial}_\lambda+d''(\lambda))^2=0$.*

Proof Note that $g(\lambda-\lambda_0)$ is holomorphic with respect to λ. Then the claim can be checked by a direct calculation. □

We use the notation in the subsection 2.8.5. We also put as follows, for each $\lambda\in\Delta(\lambda_0,\eta)$:

$$\langle f_1,f_2\rangle_{\lambda,\boldsymbol{c},N}:=\int(f_1,f_2)_{\boldsymbol{c},N}\,\mathrm{dvol}+\int\big(d''(\lambda)f_1,\ d''(\lambda)f_2\big)_{\boldsymbol{c},N}\,\mathrm{dvol},$$

$$||f||^2_{\lambda,\boldsymbol{c},N}:=\langle f,f\rangle_{\lambda,\boldsymbol{c},N}.$$

LEMMA 8.77. *For any $\lambda\in\Delta(\lambda_0,\eta)$, the norms $||\cdot||_{\lambda,\boldsymbol{c},N}$ and $||\cdot||_{\lambda_0,\boldsymbol{c},N}$ are equivalent.*

Proof It follows from the inequality (8.34) □

Hence the completions with respect to the norms $||\cdot||_{\lambda,\mathbf{c},N}$ are independent of a choice of $\lambda \in \Delta(\lambda_0,\eta)$ for some sufficiently small positive number η. Let $A^{0,q}_{\mathbf{c},N}(\mathcal{E}^{\lambda_0})$ be the completion of the space $A^{0,q}_{c}(\mathcal{E}^{\lambda_0})$ with respect to the norm $||\cdot||_{\mathbf{c},N}$. Then we have the family of complexes $\big(A^{0,\cdot}_{\mathbf{c},N}(\mathcal{E}^{\lambda_0}), d''(\lambda)\big)$ $(\lambda \in \Delta(\lambda_0,\eta))$.

8.7.2. Extension of holomorphic sections.

LEMMA 8.78. *There exists a positive number $\eta > 0$, a sufficiently negative number N, and the family of linear morphisms $G(\lambda) : \mathrm{Ker}(d''(\lambda_0)) \longrightarrow A^{0,0}_{\mathbf{c},N}(\mathcal{E}^{\lambda_0})$ holomorphically depending $\lambda \in \Delta(\lambda_0,\eta)$, which satisfies the following conditions:*

- *The vanishing $H^i(A^{0,\cdot}_{\mathbf{c},N}(\mathcal{E}^{\lambda_0}), d''(\lambda)) = 0$ holds for any $i > 0$ and for any $\lambda \in \Delta(\lambda_0,\eta)$.*
- *The morphism $G(\lambda)$ satisfies the conditions (1), (2) and (3) in Lemma 2.65. It gives the trivialization of the family $\big\{\mathrm{Ker}\, d''(\lambda) \,|\, \lambda \in \Delta(\lambda_0,\eta)\big\}$, namely $G(\lambda)$ gives the homeomorphism of $\mathrm{Ker}\, d''(\lambda_0)$ and $\mathrm{Ker}\, d''(\lambda)$ for any point $\lambda \in \Delta(\lambda_0,\eta)$.*

Proof Note that $d''(\lambda_0) = \overline{\partial}_{\mathcal{E}^{\lambda_0}}$, and hence the conditions in Lemma 2.68 is satisfied due to Lemma 2.48. Then we obtain the result due to Lemma 2.68. □

Recall that we have the C^∞-bundle $p_\lambda^{-1}(E)$ with the hermitian metric $h_{\mathbf{c},N} = h\cdot\prod_{i=1}^{l}|z_i|^{2c_i}\cdot\big(-\log|z_i|^2\big)^{-N}\cdot\prod_{i=l+1}^{n}(1-|z_i|^2)^{-N}$. We also have the holomorphic structure $\overline{\partial}_\lambda + d''(\lambda)$ (Lemma 8.76).

COROLLARY 8.79. *For any holomorphic section f of $\mathcal{E}^{\lambda_0}$ over $\mathcal{X}^{\lambda_0}$, we have a holomorphic section $\tilde{f}$ of the holomorphic bundle $\big(p_\lambda^{-1}E, \overline{\partial}_\lambda + d''(\lambda)\big)$ over $\Delta(\lambda_0,\eta)\times(X-D)$ such that $\tilde{f}_{|\{\lambda\}\times(X-D)} \in A^{0,0}_{\mathbf{c},N}(\mathcal{E}^{\lambda_0})$ for each $\lambda \in \Delta(\lambda_0,\eta)$.*

Proof Let $G(\lambda)$ be the family of the morphism given in Lemma 8.78. We put $\tilde{f}_{|\{\lambda\}\times(X-D)} := G(\lambda)(f)$. We have the absolute convergent series in $A^{0,0}_{\mathbf{c},N}(\mathcal{E}^{\lambda_0})$:

$$\tilde{f} = \sum(\lambda-\lambda_0)^i\cdot f_i, \qquad f_i \in A^{0,0}_{\mathbf{c},N}(\mathcal{E}^{\lambda_0}). \tag{8.35}$$

By our construction, it is clear that the restrictions $\tilde{f}_{|\{\lambda\}\times(X-D)}$ are contained in $A^{0,0}_{\mathbf{c},N}(\mathcal{E}^{\lambda_0})$. Since (8.35) is absolute convergent, we also have the following finiteness:

$$\int_{\Delta(\lambda_0,\epsilon)\times(X-D)} |\tilde{f}|_h^2\cdot\prod_{i=1}^{l}|z_i|^{2c_i}(-\log|z_i|^2)^{-N}\,\mathrm{dvol} < \sum\frac{\pi}{h+1}\epsilon^{2(h+2)}||f_i|| < \infty. \tag{8.36}$$

The finiteness (8.36) implies that $\tilde{f}$ can be regarded as an L^2-section on $\Delta(\lambda_0,\eta)\times(X-D)$ with respect to the metric $h_{\mathbf{c},N}$.

We clearly have $\overline{\partial}_\lambda\tilde{f} = 0$. We have $d''(\lambda)(\tilde{f}_{|\lambda}) = 0$ for any point $\lambda \in \Delta(\lambda_0,\eta)$, by our construction. Then we obtain $d''(\lambda)(\tilde{f}) = 0$ in the distribution sense, due to Fubini's theorem. Hence we obtain $\big(\overline{\partial}_\lambda + d''(\lambda)\big)\tilde{f} = 0$ in the distribution sense. Thus we can conclude that $\tilde{f}$ is holomorphic section with respect to $\overline{\partial}_\lambda + d''(\lambda)$. □

We put $F(\lambda) := g(\lambda-\lambda_0)^{-1}\cdot\tilde{f}$. Then F gives a section of the C^∞-bundle $p_\lambda^{-1}(E)$ over $\Delta(\lambda_0,\eta)\times X$.

LEMMA 8.80. *For any positive number ϵ, there exists a positive number η satisfying the following:*

- *F is holomorphic with respect to the holomorphic structure $\overline{\partial}_{\mathcal{E}} = \overline{\partial}_\lambda + \overline{\partial}_E + \lambda \cdot \theta^\dagger$, i.e., F gives a holomorphic section of $\mathcal{E}$ over $\Delta(\lambda_0, \eta) \times (X - D)$.*
- *For any $\lambda \in \Delta(\lambda_0, \eta)$ and for any $\epsilon' > 0$ there exists a positive constant C satisfying the following inequality:*

$$\left|F(\lambda)\right|_h \leq C \cdot \Big(\prod |z_i|^{-c_i - \epsilon - \epsilon'}\Big).$$

Proof Since g is holomorphic with respect to λ, we have the following relation from (8.33):

$$\overline{\partial}_\lambda + d''(\lambda) = g(\lambda - \lambda_0) \cdot \Big(\overline{\partial}_\lambda + \overline{\partial}_E + \lambda \cdot \theta^\dagger\Big) \cdot g(\lambda - \lambda)^{-1}.$$

Then the holomorphic property of F with respect to $\overline{\partial}_{\mathcal{E}}$ follows from the holomorphic property of $\tilde{f}$ with respect to $\overline{\partial}_\lambda + d''(\lambda)$.

For any positive number ϵ, there exist positive constants $\eta > 0$ and C_1 such that the following holds for any $\lambda \in \Delta(\lambda_0, \eta)$:

$$\left|g(\lambda - \lambda_0)\right|_h \leq C_1 \cdot \Big(\prod |z_i|^{-\epsilon}\Big). \tag{8.37}$$

Since $\tilde{f}_{|\{\lambda\}\times(X-D)}$ is an element of $A^{0,0}_{\boldsymbol{c},N}(\mathcal{E}^{\lambda_0})$, we obtain the following finiteness for any $\lambda \in \Delta(\lambda_0, \eta)$ from (8.37):

$$\int_{\{\lambda\}\times(X-D)} |F(\lambda)|_h^2 \cdot \prod |z_i|^{c_i + \epsilon} \cdot \big(-\log|z_i|\big)^{-N} \mathrm{dvol} < \infty.$$

It implies the second claim. See the subsection 2.8.7 (in particular Remark 2.54) to derive the growth estimate from the L^2-estimate. See also Lemma 7.63 in the one dimensional case. □

We reformulate the result in this subsection.

PROPOSITION 8.81. *Let $\boldsymbol{b} = (b_1, \ldots, b_l)$ be an element of $\boldsymbol{R}^l$. Assume $b_i \notin \mathcal{P}ar(\mathcal{E}^{\lambda_0}, i)$. Then there exists a positive number η satisfying the following:*

- *For any holomorphic section f of ${}_{\boldsymbol{b}}\mathcal{E}^{\lambda_0}$, there exists a holomorphic section F of ${}_{\boldsymbol{b}}\mathcal{E}$ over $X \times \Delta(\lambda_0, \eta)$ such that $F_{|\{\lambda\}\times X} = f$.*

Proof It immediately follows from Lemma 8.80. □

8.7.3. Prolongation of $\mathcal{E}$ around λ_0.

CONDITION 8.82. Let $\boldsymbol{b} = (b_1, \ldots, b_l)$ be an element of $\boldsymbol{R}^l$ such that $b_i \notin \mathcal{P}ar(\mathcal{E}^{\lambda_0}, i)$ for any i. □

THEOREM 8.83. *Let $\boldsymbol{b}$ be an element of $\boldsymbol{R}^l$ as in Condition 8.82. Then there exists a positive number η such that ${}_{\boldsymbol{b}}\mathcal{E}$ is locally free over $\Delta(\lambda_0, \eta) \times X$.*

Proof By considering the tensor product of $(E, \overline{\partial}_E, \theta, h)$ and the model bundle $L(\boldsymbol{u})$, we may assume that the residue of $\mathrm{tr}(\theta)$ is trivial. We can also assume that $\boldsymbol{b} = 0$. Note we have $0 \notin \mathcal{P}ar(\mathcal{E}^{\lambda_0}, i)$ from our assumption, in that case.

Let $\boldsymbol{v}$ be a frame of ${}^{\diamond}\mathcal{E}^{\lambda_0}$, compatible with ${}^{i}F$ $(i=1,\ldots,l)$. We put ${}^{i}a_j := {}^{i}\deg^{F}(v_j)$. We have $-1 < {}^{i}a_j < 0$. We put as follows:

$$\boldsymbol{a}_j := \big({}^{1}a_j, {}^{2}a_j, \ldots, {}^{l}a_j\big), \qquad \boldsymbol{c} := \sum_j \boldsymbol{a}_j, \qquad {}^{i}c = \sum_j^{\operatorname{rank} E} {}^{i}a_j.$$

Then v_j is a holomorphic section of ${}_{\boldsymbol{a}_j}\mathcal{E}^{\lambda_0}$, and we have $\det({}^{\diamond}\mathcal{E}^{\lambda_0}) = {}_{\boldsymbol{c}}\det(\mathcal{E}^{\lambda_0})$.

By using Proposition 8.81, we can take $\epsilon > 0$ and $\eta > 0$, and holomorphic sections $\tilde{v}_j$ of ${}_{\boldsymbol{a}_j+\epsilon\cdot\boldsymbol{\delta}}\mathcal{E}^{\lambda_0}$ over $\Delta(\lambda_0,\eta)\times X$ such that $\tilde{v}_{j\,|\,\{\lambda_0\}\times X} = v_j$. In particular, $\tilde{v}_j$ give holomorphic sections of ${}^{\diamond}\mathcal{E}$ over $\Delta(\lambda_0,\eta)\times X$. Hence we obtain the tuple of sections $\tilde{\boldsymbol{v}} := (\tilde{v}_i)$ of ${}^{\diamond}\mathcal{E}$ over $\Delta(\lambda_0,\eta)\times X$.

Then we have the following inequality for any sufficiently small $\epsilon > 0$ and for any $\lambda \in \Delta(\lambda_0,\eta)$:

$${}^{i}\deg\big(\Omega(\tilde{\boldsymbol{v}})_{|\{\lambda\}\times X}\big) < \sum_j {}^{i}a_j + \operatorname{rank}\mathcal{E}\cdot\epsilon < {}^{i}c + 1.$$

Hence $\Omega(\tilde{\boldsymbol{v}})_{|\{\lambda\}\times X}$ is a holomorphic section of ${}_{\boldsymbol{c}}\det(\mathcal{E}^{\lambda}) = {}_{\boldsymbol{c}}\det(\mathcal{E})_{|\{\lambda\}\times X}$. Since we have assumed that the residue of $\operatorname{tr}(\theta)$ is trivial, we have ${}_{\boldsymbol{c}}\det(\mathcal{E})_{|\lambda} = \det({}^{\diamond}\mathcal{E}^{\lambda})$. Since we have $\Omega(\tilde{\boldsymbol{v}})_{|(\lambda_0,O)} \neq 0$, the section $\Omega(\tilde{\boldsymbol{v}})$ gives a frame around (λ_0,O).

Then we can use the standard argument as follows (the last argument in the proof of Lemma 8.24). Let f be a holomorphic section of ${}^{\diamond}\mathcal{E}$ on a neighbourhood U of (λ_0,O). On $\big(U\cap(X-D)\big)\times\Delta(\lambda_0,\epsilon)$, we have the following description:

$$f = \sum f_i\cdot\tilde{v}_i.$$

As usual, we consider the section $f\wedge\tilde{v}_2\wedge\cdots\wedge\tilde{v}_r = f_1\cdot\Omega(\tilde{\boldsymbol{v}})$ of $\det({}^{\diamond}\mathcal{E}) = {}_{\boldsymbol{c}}\det(\mathcal{E})$ over U, and we can derive that f_1 is holomorphic. Similarly, we can show that f_i are holomorphic for any i. It implies that $\tilde{\boldsymbol{v}}$ gives a frame of ${}^{\diamond}\mathcal{E}$ on a neighbourhood of (λ_0,O). □

8.8. The KMS-structure of ${}_{\boldsymbol{b}}\mathcal{E}$

8.8.1. The parabolic filtrations ${}^{i}F^{(\lambda_0)}({}_{\boldsymbol{b}}\mathcal{E})$. We continue to use the setting in the section 8.7. Let $\boldsymbol{b}$ be an element of $\boldsymbol{R}^l$ as in Condition 8.82. Let $\boldsymbol{v} = (v_j)$ be a frame of ${}_{\boldsymbol{b}}\mathcal{E}^{\lambda_0}$ on X compatible with the parabolic structure. For each v_j, we obtain the element $\boldsymbol{a}(v_j) = (a_1(v_j),\ldots,a_l(v_j)) \in \boldsymbol{R}^l$, where we put $a_i(v_j) := {}^{i}\deg^{F}(v_j)$.

Let us pick a positive number ϵ as follows:

$$0 < \epsilon < \frac{1}{3}\min\Big(\bigcup_{i=1}^{l}\big\{|a_1-a_2|\,\big|\,a_1,a_2\in\mathcal{P}ar(\mathcal{E}^{\lambda_0},i),\ a_1\neq a_2\big\}\Big). \tag{8.38}$$

We put $\mathcal{Y}(\lambda_0,\epsilon_0) := Y\times\Delta(\lambda_0,\epsilon_0)$ for a complex variety. We also use the notation $\mathcal{Y}^{\lambda} = Y\times\{\lambda\}$. We can pick a positive number ϵ_0 such that there exist sections $\tilde{v}_j$ of ${}_{\boldsymbol{a}(v_j)+\epsilon\cdot\boldsymbol{\delta}}\mathcal{E}$ on $\mathcal{X}(\lambda_0,\epsilon_0)$, such that $\tilde{v}_{j|\mathcal{X}^{\lambda_0}} = v_j$. If ϵ_0 is sufficiently small, we may assume the following:

CONDITION 8.84.

- Let u_1 and u_2 be elements of $\mathcal{KMS}(\mathcal{E}^0,i)$ such that $\mathfrak{p}(\lambda_0,u_1) < \mathfrak{p}(\lambda_0,u_2)$. Note we have $\mathfrak{p}(\lambda_0,u_1)+\epsilon < \mathfrak{p}(\lambda_0,u_2)$, due to (8.38). Then the inequality $\mathfrak{p}(\lambda,u_1)+\epsilon < \mathfrak{p}(\lambda',u_2)$ holds for any $\lambda,\lambda'\in\Delta(\lambda_0,\epsilon_0)$.

- Let u be an element of $\mathcal{KMS}(\mathcal{E}^0, i)$, and let λ be a point of $\Delta(\lambda_0, \epsilon_0)$. Then $\mathfrak{p}(\lambda_0, u) > b_i - 1$ if and only if $\mathfrak{p}(\lambda, u) > b_i - 1$. □

On $\mathcal{D}_i(\lambda_0, \epsilon_0)$, the filtration ${}^iF^{(\lambda_0)}$ is obtained as follows:

$$ {}^iF_a^{(\lambda_0)}\bigl({}_{\boldsymbol{b}}\mathcal{E}_{|\mathcal{D}_i(\lambda_0,\epsilon_0)}\bigr) := \bigl\langle \tilde{v}_{j\,|\,\mathcal{D}_i(\lambda_0,\epsilon_0)} \bigm| {}^i\deg(v_j) \leq a' \bigr\rangle, $$

$$ a' := \max\bigl\{x \in \mathcal{P}ar({}_{\boldsymbol{b}}\mathcal{E}^{\lambda_0}, i) \bigm| x \leq a\bigr\}. $$

For any $\lambda \in \Delta(\lambda_0, \eta)$ and $c \in \mathcal{P}ar({}_{\boldsymbol{b}}\mathcal{E}^{\lambda_0}, i)$, we put as follows:

$$ d(\lambda, \lambda_0, c) := \max\bigl\{\mathfrak{p}(\lambda, u) \bigm| u \in \mathcal{KMS}(\mathcal{E}^0, i),\ \mathfrak{p}(\lambda_0, u) = c\bigr\}. $$

Recall that we have the parabolic filtration ${}^iF$ of ${}_{\boldsymbol{b}}\mathcal{E}^{\lambda}$. So we compare the two vector subbundles ${}^iF^{(\lambda_0)}_{c\,|\,\mathcal{D}_i^\lambda}$ and ${}^iF_{d(\lambda,\lambda_0,c)}$ of ${}_{\boldsymbol{b}}\mathcal{E}^{\lambda}{}_{|\mathcal{D}_i^\lambda}$.

LEMMA 8.85. *For any $\lambda \in \Delta(\lambda_0, \epsilon_0)$, we have the equality ${}^iF^{(\lambda_0)}_{c\,|\,\mathcal{D}_i^\lambda} = {}^iF_{d(\lambda,\lambda_0,c)}$. In particular, the filtration ${}^iF^{(\lambda_0)}$ is independent of choices of a compatible frame $\boldsymbol{v}$ and an extension $\tilde{\boldsymbol{v}}$.*

Proof Since both of ${}^iF^{(\lambda_0)}_{c\,|\,\{\lambda\}\times D_i}$ and ${}^iF_{d(\lambda,\lambda_0,c)}$ are vector subbundles, we have only to show ${}^iF^{(\lambda_0)}_{c\,|\,(\lambda,P)} = {}^iF_{d(\lambda,\lambda_0,c)\,|\,(\lambda,P)}$ for any $P \in D_i^\circ$. For that purpose, we have only to consider the restriction of $\mathcal{E}$ to the curve $\pi_i^{-1}(P)$. Thus we can restrict ourselves to the case $\dim(X) = 1$, which is assumed in the following of the proof.

In the case $\deg^F(v_i) = c$, we have the following inequality for any $\lambda \in \Delta(\lambda_0, \epsilon_0)$, due to Condition 8.84:

$$ \deg^F\bigl(\tilde{v}_{i\,|\,\mathcal{X}^\lambda}\bigr) \leq c + \epsilon < \min\bigl\{d \in \mathcal{P}ar(\mathcal{E}^\lambda, i) \bigm| d > d(\lambda, \lambda_0, c)\bigr\}. $$

It implies $\tilde{v}_{i\,|\,(\lambda,O)}$ is contained in the space $F_{d(\lambda,\lambda_0,c)}({}_b\mathcal{E}^\lambda{}_{|O})$. Hence we obtain $F^{(\lambda_0)}_{c\,|\,(\lambda,O)} \subset F_{d(\lambda,\lambda_0,c)}({}_b\mathcal{E}^\lambda{}_{|O})$.

Due to our construction of the filtration $F^{(\lambda_0)}$, we have the following equality (Corollary 7.71):

$$ \operatorname{rank} F_c^{(\lambda_0)} = \operatorname{rank} F_c\bigl({}_b\mathcal{E}^\lambda{}_{|O}\bigr) = \sum_{b-1<c'\leq c} \mathfrak{m}(\lambda_0, c') = \sum_{\substack{u\in\mathcal{KMS}(\mathcal{E}^0),\\ b-1<\mathfrak{p}(\lambda_0,u)\leq c}} \mathfrak{m}(0, u). $$

On the other hand, we have the following equality:

(8.39)

$$ \begin{aligned} \operatorname{rank} F_{d(\lambda,\lambda_0,c)}({}_b\mathcal{E}^\lambda{}_{|O}) = \sum_{b-1<c'\leq d(\lambda,\lambda_0,c)} \mathfrak{m}(\lambda, c') &= \sum_{\substack{u\in\mathcal{KMS}(\mathcal{E}^0),\\ b-1<\mathfrak{p}(\lambda,u)\leq d(\lambda,\lambda_0,c)}} \mathfrak{m}(0, u) \\ &= \sum_{\substack{u\in\mathcal{KMS}(\mathcal{E}^0),\\ b-1<\mathfrak{p}(\lambda_0,u)\leq c}} \mathfrak{m}(0, u). \end{aligned} $$

Note $\mathfrak{p}(\lambda_0, u) > b - 1$ if and only if $\mathfrak{p}(\lambda, u) > b - 1$ due to our assumption. Thus we obtain $F_c^{(\lambda_0)}({}_b\mathcal{E})_{|(\lambda,O)} = F_{d(\lambda,\lambda_0,c)}({}_b\mathcal{E}^\lambda{}_{|O})$. □

On $\mathcal{D}_i(\lambda_0,\epsilon_0)$, we obtain the following vector bundle:

$$ {}^i\mathrm{Gr}_c^{F^{(\lambda_0)}}\bigl({}_{\boldsymbol{b}}\mathcal{E}_{|\mathcal{D}_i(\lambda_0,\epsilon_0)}\bigr) := \frac{{}^iF_c^{(\lambda_0)}\bigl({}_{\boldsymbol{b}}\mathcal{E}_{|\mathcal{D}_i(\lambda_0,\epsilon_0)}\bigr)}{{}^iF_{<c}^{(\lambda_0)}\bigl({}_{\boldsymbol{b}}\mathcal{E}_{|\mathcal{D}_i(\lambda_0,\epsilon_0)}\bigr)}. $$

COROLLARY 8.86. *We have the following isomorphism on $\mathcal{D}_i^\lambda$ for any $\lambda \in \Delta(\lambda_0,\eta)$:*

$$ {}^i\mathrm{Gr}_c^{F^{(\lambda_0)}}\bigl({}_{\boldsymbol{b}}\mathcal{E}_{|\mathcal{D}_i(\lambda_0,\epsilon_0)}\bigr)_{|\mathcal{D}_i^\lambda} \simeq \frac{{}^iF_{d(\lambda,\lambda_0,c)}\bigl({}_{\boldsymbol{b}}\mathcal{E}^\lambda{}_{|\mathcal{D}_i^\lambda}\bigr)}{\sum_{b<c}{}^iF_{d(\lambda,\lambda_0,b)}\bigl({}_{\boldsymbol{b}}\mathcal{E}^\lambda{}_{|\mathcal{D}_i^\lambda}\bigr)}. \tag{8.40}$$

Proof It immediately follows from Lemma 8.85. □

REMARK 8.87. For simplicity of the notation, ${}^i\mathrm{Gr}_c^{F^{(\lambda_0)}}\bigl({}_{\boldsymbol{b}}\mathcal{E}^\lambda{}_{|\mathcal{D}_i(\lambda_0,\epsilon_0)}\bigr)$ is often denoted by ${}^i\mathrm{Gr}_c^{F^{(\lambda_0)}}\bigl({}_{\boldsymbol{b}}\mathcal{E}^\lambda\bigr)$. □

8.8.2. Regularity of $\mathbb{D}$ and $\mathbb{D}^\lambda$.

LEMMA 8.88. *$\mathbb{D}$ is a family of the regular λ-connection, namely, if f is a holomorphic section of ${}_{\boldsymbol{b}}\mathcal{E}$, then $\mathbb{D}f$ is a section of ${}_{\boldsymbol{b}}\mathcal{E}\otimes p_\lambda^{-1}\Omega_X^{1,0}(\log D)$ over $\mathcal{X}(\lambda_0,\epsilon_0)$. Here p_λ denotes the natural projection $\mathcal{X}(\lambda_0,\epsilon_0)\longrightarrow X$.*

Proof Let us consider the case $\lambda_0\neq 0$. We put $\mathcal{X}^*(\lambda_0,\epsilon_0) := X\times\Delta^*(\lambda_0,\epsilon_0)$. We may assume that any $\lambda\in\Delta^*(\lambda_0,\epsilon_0)$ is generic. In this case, the prolongment ${}_{\boldsymbol{b}}\mathcal{E}_{|\mathcal{X}^*(\lambda_0,\epsilon_0)}$ can essentially be regarded as a quasi canonical prolongment. Thus $\mathbb{D}f_{|\mathcal{X}^*(\lambda_0,\epsilon_0)}$ gives a section of the bundle ${}_{\boldsymbol{b}}\mathcal{E}\otimes p_\lambda^{-1}\Omega_X^{1,0}(\log D)_{|\mathcal{X}^*(\lambda_0,\epsilon_0)}$. Hence $\mathbb{D}f$ gives a section of ${}_{\boldsymbol{b}}\mathcal{E}\otimes p_\lambda^{-1}\Omega_X^{1,0}(\log D)$ over $\mathcal{X}(\lambda_0,\epsilon_0)-\mathcal{D}^{\lambda_0}$. Note that the codimension of $\mathcal{D}^{\lambda_0}$ in $\mathcal{X}(\lambda_0,\epsilon_0)$ is two. Thus $\mathbb{D}f$ gives a section of ${}_{\boldsymbol{b}}\mathcal{E}\otimes\Omega_X^{1,0}(\log D)$ over $\mathcal{X}(\lambda_0,\epsilon_0)$.

Then we can show the claim in the case $\lambda=0$ by the same argument. □

COROLLARY 8.89. *$\mathbb{D}^\lambda$ is the regular λ-connection for any λ.*

Proof Let $\boldsymbol{b}=(b_1,\ldots,b_l)$ be an element of $\boldsymbol{R}^l$ and let f be a section of ${}_{\boldsymbol{b}}\mathcal{E}^\lambda$. We may assume that $b_i\notin \mathcal{P}ar(\mathcal{E}^\lambda,i)$ for any i. We can take a holomorphic section F of ${}_{\boldsymbol{b}}\mathcal{E}$ over $\mathcal{X}(\lambda,\eta)$ such that $F_{|\mathcal{X}^\lambda}=f$ for sufficiently small $\eta>0$ (Proposition 8.81). Due to Lemma 8.88, $\mathbb{D}F$ is a section of ${}_{\boldsymbol{b}}\mathcal{E}\otimes p_\lambda^{-1}\Omega_X^{1,0}(\log D)$. Since we have $\mathbb{D}F_{|\mathcal{X}^\lambda}=\mathbb{D}^\lambda f$, we obtain the result. □

8.8.3. The residue and the λ-connection on the divisors. On $\mathcal{D}_i(\lambda_0,\epsilon_0)$, we have ${}_{\boldsymbol{b}}\mathcal{E}_{|\mathcal{D}_i(\lambda_0,\epsilon_0)}$. Then we have the endomorphisms $\mathrm{Res}_i(\mathbb{D})$, which preserve the parabolic filtrations due to Lemma 8.88. We also have the induced λ-connection ${}^i\mathbb{D}$ of ${}_{\boldsymbol{b}}\mathcal{E}_{|\mathcal{D}_i(\lambda_0,\epsilon_0)}$, which is defined as follows: For any $f\in{}_{\boldsymbol{b}}\mathcal{E}_{|\mathcal{D}_i(\lambda_0,\epsilon_0)}$, pick $F\in{}_{\boldsymbol{b}}\mathcal{E}$ such that $F_{|\mathcal{D}_i(\lambda_0,\epsilon_0)}=f$. Then we put ${}^i\mathbb{D}(f):=\mathbb{D}F_{|\mathcal{D}_i(\lambda_0,\epsilon_0)}$.

LEMMA 8.90. *It is well defined.*

Proof Assume $F_{|\mathcal{D}_i(\lambda_0,\epsilon_0)}=0$. Then we have the description $F=z_i\cdot G$ for some $G\in{}_{\boldsymbol{b}}\mathcal{E}$. We have the following:

$$\mathbb{D}(z_i\cdot G)=\lambda\cdot dz_i\cdot G+z_i\cdot\mathbb{D}G.$$

Thus $\mathbb{D}(z_i\cdot G)_{|\mathcal{D}_i(\lambda_0,\epsilon_0)}=0$ in ${}_{\boldsymbol{b}}\mathcal{E}_{\mathcal{D}_i(\lambda_0,\epsilon_0)}\otimes p_\lambda^{-1}\Omega_{D_i}(\log D\cdot D_i)$, where $D\cdot D_i$ denotes $\bigcup_{j\neq i,j\leq l}D_i\cap D_j$. Hence we are done. □

Let $\boldsymbol{v}$ be a frame of ${}_{\boldsymbol{b}}\mathcal{E}$, which is compatible with the parabolic filtrations $\bigl({}^{i}F \,|\, i=1,\ldots,l\bigr)$. Then we obtain the λ-connection form $\mathcal{A}$ determined by $\mathbb{D}\boldsymbol{v} = \boldsymbol{v}\cdot\mathcal{A}$. We develop $\mathcal{A}$ as $\mathcal{A} = \sum \mathcal{A}^{k}\frac{dz_k}{z_k}$.

LEMMA 8.91. *Then we have the following formula:*

$$\mathrm{Res}_i(\mathbb{D})\boldsymbol{v}_{|\mathcal{D}_i(\lambda_0,\epsilon_0)} = \boldsymbol{v}_{|\mathcal{D}_i(\lambda_0,\epsilon_0)}\cdot\mathcal{A}^{i}_{|\mathcal{D}_i(\lambda_0,\epsilon_0)},$$

$${}^{i}\mathbb{D}\boldsymbol{v}_{|\mathcal{D}_i(\lambda_0,\epsilon_0)} = \boldsymbol{v}_{|\mathcal{D}_i(\lambda_0,\epsilon_0)}\cdot\sum_{k\neq i}\mathcal{A}^{k}_{|\mathcal{D}_i(\lambda_0,\epsilon_0)}\frac{dz_k}{z_k}.$$

Proof It immediately follows from the definitions. □

LEMMA 8.92. ${}^{i}\mathbb{D}$ *and* $\mathrm{Res}_i(\mathbb{D})$ *preserve the filtration* ${}^{i}F^{(\lambda_0)}$.

Proof We remark that we may replace ϵ_0 with a smaller positive number, due to Lemma 8.85. Let f be a section of ${}^{i}F^{(\lambda_0)}_{c}\bigl({}_{\boldsymbol{b}}\mathcal{E}_{|\mathcal{D}_i(\lambda_0,\epsilon_0)}\bigr)$. Take a small positive number ϵ', and let $\boldsymbol{b}'$ be the element of $\boldsymbol{R}^{l}$ as follows:

$$q_j(\boldsymbol{b}') = \begin{cases} b_j, & (j\neq i),\\ c+\epsilon', & (j=i).\end{cases}$$

Then by definition of the filtration $F^{(\lambda_0)}$, we can take a holomorphic section $G\in {}_{\boldsymbol{b}'}\mathcal{E}$ on $\mathcal{X}(\lambda_0,\epsilon_0')$, such that $\pi(G_{|\mathcal{D}_i(\lambda_0,\epsilon_0')}) = f_{|\mathcal{D}_i(\lambda_0,\epsilon_0')}$ for some $0<\epsilon_0'<\epsilon_0$, where π denotes the natural morphism ${}_{\boldsymbol{b}'}\mathcal{E}_{|\mathcal{D}_i(\lambda_0,\epsilon_0')}\longrightarrow {}_{\boldsymbol{b}}\mathcal{E}_{|\mathcal{D}_i(\lambda_0,\epsilon_0')}$. By definition, ${}^{i}\mathbb{D}(f)$ is the image of $\mathbb{D}G_{|\mathcal{D}_i(\lambda_0,\epsilon_0)}$. Then it is easy to see that ${}^{i}\mathbb{D}$ preserves the filtration ${}^{i}F^{(\lambda_0)}$, due to the regularity of $\mathbb{D}$. By a similar argument, we can show that $\mathrm{Res}_i(\mathbb{D})$ preserves the filtration ${}^{i}F^{(\lambda_0)}$. □

We put $\mathcal{D}_{i,j}(\lambda_0,\epsilon_0) := \mathcal{D}_i(\lambda_0,\epsilon_0)\cap\mathcal{D}_j(\lambda_0,\epsilon_0)$.

COROLLARY 8.93. $\mathrm{Res}_i(\mathbb{D})_{|\mathcal{D}_{i,j}(\lambda_0,\epsilon_0)}$ *preserves the filtration* ${}^{j}F^{(\lambda_0)}$.

Proof Since we have $\mathrm{Res}_i(\mathbb{D})_{|\mathcal{D}_{i,j}(\lambda_0,\epsilon_0)} = \mathrm{Res}_i({}^{j}\mathbb{D})$, the claim immediately follows from Lemma 8.92. □

COROLLARY 8.94. *Assume* ${}^{k}\deg^{F^{(\lambda_0)}}(v_i) > {}^{k}\deg^{F^{(\lambda_0)}}(v_j)$. *Then we have* $\mathcal{A}^{h}_{i\,j\,|\,\mathcal{D}_k} = 0$ *for any* $h=1,\ldots,l$. □

LEMMA 8.95. *The endomorphism* $\mathrm{Res}_i(\mathbb{D})$ *is flat with respect to* ${}^{i}\mathbb{D}$.

Proof It immediately follows from the flatness of $\mathbb{D}$. □

We have the λ-connection ${}^{i}\mathbb{D}^{\lambda}$ of ${}_{\boldsymbol{b}}\mathcal{E}^{\lambda}{}_{|\mathcal{D}_i^{\lambda}}$, which is similar to ${}^{i}\mathbb{D}$. Since $\mathrm{Res}_i(\mathbb{D}^{\lambda})$ and ${}^{i}\mathbb{D}^{\lambda}$ are obtained as the specializations of $\mathrm{Res}_i(\mathbb{D})$ and ${}^{i}\mathbb{D}$, similar results are obtained.

LEMMA 8.96.

- $\mathrm{Res}_i(\mathbb{D}^{\lambda})$ *and* ${}^{i}\mathbb{D}^{\lambda}$ *preserve the filtration* ${}^{i}F$.
- $\mathrm{Res}_i(\mathbb{D}^{\lambda})$ *preserves the filtration* ${}^{j}F$, *on* $D_i\cap D_j$. □

8.8.4. The $\mathbb{E}$-decomposition and KMS-spectrum for $\mathcal{E}^{\lambda}$. We recall that the eigenvalues of $\mathrm{Res}_i(\mathbb{D}^{\lambda})$ on ${}_{\boldsymbol{b}}\mathcal{E}^{\lambda}$ are constant, and the set of them is given by $\mathcal{S}p\bigl({}_{\boldsymbol{b}}\mathcal{E}^{\lambda},i\bigr)$. Hence we have the generalized eigen-decomposition ${}^{i}\mathbb{E}$ of ${}_{\boldsymbol{b}}\mathcal{E}^{\lambda}{}_{|\mathcal{D}_i^{\lambda}}$:

$${}_{\boldsymbol{b}}\mathcal{E}^{\lambda}{}_{|\mathcal{D}_i^{\lambda}} = \bigoplus_{\beta\in\mathcal{S}p({}_{\boldsymbol{b}}\mathcal{E}^{\lambda},i)}\mathbb{E}\bigl({}_{\boldsymbol{b}}\mathcal{E}^{\lambda}{}_{|\mathcal{D}_i^{\lambda}},\beta\bigr).$$

LEMMA 8.97. *The tuple of the decompositions and the filtrations* $\bigl({}^iF, {}^i\mathbb{E} \,\big|\, i \in \underline{l}\bigr)$ *is compatible, in the sense of Definition* 4.9.

Proof Since $\mathrm{Res}_i(\mathbb{D}^\lambda)$ preserves the filtrations ${}^jF$ $(j = 1, \ldots, l)$ on $D_i \cap D_j$, the claim is easily checked. □

For $u = (a, \alpha)$, we put ${}^i\,\mathrm{Gr}^{F,\mathbb{E}}_u\bigl({}_{\boldsymbol{b}}\mathcal{E}^\lambda\bigr) := {}^i\mathbb{E}\bigl({}^i\,\mathrm{Gr}^F_a\bigl({}_{\boldsymbol{b}}\mathcal{E}^\lambda{}_{|\mathcal{D}_i^\lambda}\bigr), \alpha\bigr)$. More generally, let I be a subset of $\underline{l}$ and $\boldsymbol{u} = (\boldsymbol{a}, \boldsymbol{\alpha})$ be an element of $\boldsymbol{R}^I \times \boldsymbol{C}^I$. Then we put ${}^I\,\mathrm{Gr}^{F,\mathbb{E}}_{\boldsymbol{u}}\bigl({}_{\boldsymbol{b}}\mathcal{E}^\lambda\bigr) := {}^I\mathbb{E}\bigl({}^I\,\mathrm{Gr}^F_{\boldsymbol{a}}\bigl({}_{\boldsymbol{b}}\mathcal{E}^\lambda{}_{|\mathcal{D}_I^\lambda}\bigr), \boldsymbol{\alpha}\bigr)$.

We put as follows:

$$\mathcal{KMS}\bigl({}_{\boldsymbol{b}}\mathcal{E}^\lambda, I\bigr) := \bigl\{\boldsymbol{u} \in \boldsymbol{R}^I \times \boldsymbol{C}^I \,\big|\, {}^I\,\mathrm{Gr}^{F,\mathbb{E}}_{\boldsymbol{u}}\bigl({}_{\boldsymbol{b}}\mathcal{E}^\lambda\bigr) \neq 0\bigr\}.$$

Then we put as follows:

$$\mathcal{KMS}\bigl(\mathcal{E}^\lambda, I\bigr) := \bigcup_{\boldsymbol{b} \in \boldsymbol{R}^l} \mathcal{KMS}\bigl({}_{\boldsymbol{b}}\mathcal{E}^\lambda, I\bigr).$$

We put $D_I^\circ := D_I \setminus \bigcup_{j \notin I, j \leq l} D_I \cap D_j$. The following lemma is clear from our construction.

LEMMA 8.98. *Let* $\boldsymbol{b}_i$ *be elements of* $\boldsymbol{R}^l$ $(i = 1, 2)$. *Let* $\boldsymbol{u}$ *be an element of* $\boldsymbol{R}^I \times \boldsymbol{C}^I$. *If* ${}^I\,\mathrm{Gr}^{F,\mathbb{E}}_{\boldsymbol{u}}\bigl({}_{\boldsymbol{b}_i}\mathcal{E}^\lambda\bigr)$ $(i = 1, 2)$ *are not* 0, *the restrictions of them to* D_I° *are canonically isomorphic. In particular, we have* $\mathrm{rank}\,{}^I\,\mathrm{Gr}^{F,\mathbb{E}}_{\boldsymbol{u}}\bigl({}_{\boldsymbol{b}_1}\mathcal{E}^\lambda\bigr) = \mathrm{rank}\,{}^I\,\mathrm{Gr}^{F,\mathbb{E}}_{\boldsymbol{u}}\bigl({}_{\boldsymbol{b}_2}\mathcal{E}^\lambda\bigr)$. □

Notation For $\boldsymbol{u} \in \mathcal{KMS}\bigl(\mathcal{E}^\lambda, I\bigr)$, we take $\boldsymbol{b} \in \boldsymbol{R}^l$ such that ${}^I\,\mathrm{Gr}^{F,\mathbb{E}}_{\boldsymbol{u}}\bigl({}_{\boldsymbol{b}}\mathcal{E}^\lambda\bigr) \neq 0$, and we put $\mathfrak{m}(\lambda, \boldsymbol{u}, I) := \mathrm{rank}\,{}^I\,\mathrm{Gr}^{F,\mathbb{E}}_{\boldsymbol{u}}\bigl({}_{\boldsymbol{b}}\mathcal{E}^\lambda\bigr)$. If I is fixed, we use $\mathfrak{m}(\lambda, \boldsymbol{u})$.

REMARK 8.99. In particular, ${}^{\underline{l}}\,\mathrm{Gr}^{F,\mathbb{E}}_{(\boldsymbol{a},\boldsymbol{\alpha})}\bigl({}_{\boldsymbol{b}}\mathcal{E}^\lambda\bigr)$ is independent of a choice of $\boldsymbol{b}$ such that $\boldsymbol{a} \leq \boldsymbol{b} < \boldsymbol{a} + \boldsymbol{\delta}$ up to the canonical isomorphisms, where $\boldsymbol{\delta} = (\overbrace{1, \ldots, 1}^{l})$. Hence we often use the notation ${}^{\underline{l}}\,\mathrm{Gr}^{F,\mathbb{E}}_{(\boldsymbol{a},\boldsymbol{\alpha})}\bigl(\mathcal{E}^\lambda\bigr)$ instead of ${}^{\underline{l}}\,\mathrm{Gr}^{F,\mathbb{E}}_{(\boldsymbol{a},\boldsymbol{\alpha})}\bigl({}_{\boldsymbol{b}}\mathcal{E}^\lambda\bigr)$. □

LEMMA 8.100. *For any element* $u \in \boldsymbol{C} \times \boldsymbol{R}$, $\mathfrak{m}(0, u, i) := \mathfrak{m}(\lambda, \mathfrak{k}(\lambda, u), i)$.

Proof It follows from the result in the one dimensional case (Corollary 7.71). □

For later use, we put as follows:

$$\begin{aligned} \mathcal{K}(\mathcal{E}, \lambda, \boldsymbol{b}, i) &:= \bigl\{u \in \mathcal{KMS}\bigl(\mathcal{E}^0, i\bigr) \,\big|\, b_i - 1 < \mathfrak{p}(\lambda, u) \leq b_i\bigr\}, \\ \mathcal{T}(\mathcal{E}, \lambda, c, i) &:= \bigl\{u \in \mathcal{KMS}\bigl(\mathcal{E}^0, i\bigr) \,\big|\, \mathfrak{p}(\lambda, u) = c\bigr\}. \end{aligned} \tag{8.41}$$

REMARK 8.101. In Part 4, the notation $\mathcal{K}(\mathcal{E}, \lambda, \boldsymbol{b}, i)$ will be used in the different meaning. □

8.8.5. The $\mathbb{E}^{(\lambda_0)}$-decomposition for $\mathcal{E}$. Let us pick any point $\lambda_0 \in \boldsymbol{C}_\lambda$. Then there exist small positive numbers η_2 and ϵ_2 such that we have the following decomposition into vector bundles on $\mathcal{D}_i(\lambda_0, \eta_2)$:

$${}_{\boldsymbol{b}}\mathcal{E}_{|\mathcal{D}_i(\lambda_0,\eta_2)} = \bigoplus_{\beta \in \mathcal{S}p({}_{\boldsymbol{b}}\mathcal{E}^{\lambda_0}, i)} {}^i\mathbb{E}^{(\lambda_0)}\bigl({}_{\boldsymbol{b}}\mathcal{E}_{|\mathcal{D}_i(\lambda_0,\eta_2)}, \beta\bigr).$$

Here we put ${}^i\mathbb{E}^{(\lambda_0)}\big({}_{\boldsymbol{b}}\mathcal{E},\beta\big) := \mathbb{E}_{\epsilon_2}\big(\mathrm{Res}_i(\mathbb{D}),\beta\big)$. (See the subsection 2.7.3 for $\mathbb{E}_{\epsilon_2}$.) Namely, we put as follows:

$$ {}^i\mathbb{E}^{(\lambda_0)}\big({}_{\boldsymbol{b}}\mathcal{E}_{|\mathcal{D}_i(\lambda_0,\eta_2)},\beta\big)_{|(\lambda,P)} := \bigoplus_{\substack{\alpha\in\mathcal{S}p\left(\mathrm{Res}_i(\mathbb{D}^\lambda)\right)\\ |\beta-\alpha|\leq\epsilon_2}} \mathbb{E}\big(\mathrm{Res}_i(\mathbb{D})_{|(\lambda,P)},\alpha\big) \subset {}_{\boldsymbol{b}}\mathcal{E}^\lambda{}_{|(\lambda,P)}. $$

LEMMA 8.102. *Assume η_2 and ϵ_2 are sufficiently small. Let λ be any element of $\Delta^*(\lambda_0,\eta_2)$. Then we have the following generalized eigen-decomposition:*

$$ {}^i\mathbb{E}^{(\lambda_0)}\big({}_{\boldsymbol{b}}\mathcal{E}_{|\mathcal{D}_i(\lambda_0,\eta_2)},\beta\big)_{|\mathcal{D}_i^\lambda} = \bigoplus_{\substack{u\in\mathcal{K}(\mathcal{E},\lambda_0,\boldsymbol{b},i),\\ \mathfrak{e}(\lambda_0,u)=\beta}} {}^i\mathbb{E}\big({}_{\boldsymbol{b}}\mathcal{E}^\lambda{}_{|\mathcal{D}_i^\lambda},\mathfrak{e}(\lambda,u)\big). $$

Proof Since we have assumed $b_i\notin\mathcal{P}ar(\mathcal{E}^\lambda,i)$, we have $\mathcal{K}(\mathcal{E},\lambda,\boldsymbol{b},i)=\mathcal{K}(\mathcal{E},\lambda_0,\boldsymbol{b},i)$ if η_2 is sufficiently small. If ϵ_2 is sufficiently small, $|\mathfrak{e}(\lambda,u)-\beta|<\epsilon_2$ implies $\mathfrak{e}(\lambda_0,u)=\beta$. If η_2 is sufficiently small, $u\neq u'$ and $\mathfrak{e}(\lambda_0,u)=\mathfrak{e}(\lambda_0,u')=\beta$ implies $\mathfrak{e}(\lambda,u)\neq\mathfrak{e}(\lambda,u')$ for $u,u'\in\mathcal{K}(\mathcal{E},\lambda_0,\boldsymbol{b},i)$. Thus we are done. □

LEMMA 8.103. *The tuple $\big({}^iF^{(\lambda_0)},{}^i\mathbb{E}^{(\lambda_0)}\,\big|\,i=1,\dots,l\big)$ of the filtrations and the decompositions is compatible in the sense of Definition 4.9. Moreover, we have the following for any $\lambda\in\Delta^*(\lambda_0,\epsilon_0)$:*

$$ {}^iF_c^{(\lambda_0)}\,{}^i\mathbb{E}^{(\lambda_0)}\big({}_{\boldsymbol{b}}\mathcal{E}_{|\mathcal{D}_i(\lambda_0,\eta_2)},\beta\big)_{|\mathcal{D}_i^\lambda} = \bigoplus_{\substack{u\in\mathcal{K}(\mathcal{E},\lambda_0,\boldsymbol{b},i),\\ \mathfrak{e}(\lambda_0,u)=\beta,\\ \mathfrak{p}(\lambda_0,u)\leq c}} {}^i\mathbb{E}\big({}_{\boldsymbol{b}}\mathcal{E}^\lambda{}_{|\mathcal{D}_i^\lambda},\mathfrak{e}(\lambda,u)\big). $$

Proof Due to Lemma 8.97, the tuple of the decompositions and the filtrations $\big({}^iF,{}^i\mathbb{E}\,\big|\,i\in\underline{l}\big)$ is compatible for each ${}_{\boldsymbol{b}}\mathcal{E}^\lambda$. Then Lemma 8.103 follows from Lemma 8.102 and Lemma 8.85. □

LEMMA 8.104. *${}^i\mathbb{D}$ and $\mathrm{Res}_i(\mathbb{D})$ preserve the decomposition ${}^i\mathbb{E}^{(\lambda_0)}$.*

Proof As for $\mathrm{Res}_i(\mathbb{D})$, it is clear. Since $\mathrm{Res}_i(\mathbb{D})$ is flat with respect to ${}^i\mathbb{D}$, the generalized eigen decomposition is preserved by ${}^i\mathbb{D}$. □

COROLLARY 8.105. *$\mathrm{Res}_i(\mathbb{D})$ preserves ${}^j\mathbb{E}$ on $\mathcal{D}_i(\lambda_0,\epsilon_0)\cap\mathcal{D}_j(\lambda_0,\epsilon_0)$.*

Proof Similar to Corollary 8.93. □

8.8.6. The associated graded bundles. The decomposition ${}^i\mathbb{E}^{(\lambda_0)}$ induces the decomposition of ${}^i\,\mathrm{Gr}^{F^{(\lambda_0)}}\big({}_{\boldsymbol{b}}\mathcal{E}\big)$.

Notation For any element $u=(a,\alpha)\in\boldsymbol{R}\times\boldsymbol{C}$, let ${}^i\,\mathrm{Gr}_u^{F^{(\lambda_0)},\mathbb{E}^{(\lambda_0)}}\big({}_{\boldsymbol{b}}\mathcal{E}_{|\mathcal{D}_i(\lambda_0,\eta_2)}\big)$ denote the bundle ${}^i\mathbb{E}^{(\lambda_0)}\Big({}^i\,\mathrm{Gr}_a^{F^{(\lambda_0)}}\big({}_{\boldsymbol{b}}\mathcal{E}_{|\mathcal{D}_i(\lambda,\eta_2)}\big),\alpha\Big)$ over $\mathcal{D}_i(\lambda_0,\epsilon_0)$. We omit to denote $\mathcal{D}_i(\lambda_0,\eta_2)$ or even ${}_{\boldsymbol{b}}\mathcal{E}$, if there are no confusion. □

LEMMA 8.106. *If η_2 is sufficiently small, we have the following equality on $\mathcal{D}_i^\lambda$, for any $\lambda\in\Delta(\lambda_0,\eta_2)$ and for any $u\in\mathcal{K}(\mathcal{E},\lambda_0,\boldsymbol{b},i)$:*

$$ {}^i\,\mathrm{Gr}^{F^{(\lambda_0)},\mathbb{E}^{(\lambda_0)}}_{\mathfrak{k}(\lambda_0,u)}\big({}_{\boldsymbol{b}}\mathcal{E}\big)_{|\mathcal{D}_i^\lambda} = {}^i\,\mathrm{Gr}^{F,\mathbb{E}}_{\mathfrak{k}(\lambda,u)}\big({}_{\boldsymbol{b}}\mathcal{E}^\lambda\big). $$

Proof We put $(c,\beta)=\mathfrak{k}(\lambda_0,u)\in\boldsymbol{R}\times\boldsymbol{C}$. Since $\mathrm{Res}_i(\mathbb{D})$ preserves the filtration ${}^iF^{(\lambda_0)}$ due to Lemma 8.92, we have the induced action of $\mathrm{Res}_i(\mathbb{D})$ on ${}^i\,\mathrm{Gr}_c^{F^{(\lambda_0)}}\big({}_{\boldsymbol{b}}\mathcal{E}\big)$.

Then there exist small positive numbers η_2 and ϵ_2 such that we have the following decomposition on $\Delta(\lambda_0, \eta_2) \times D_i$:

$$ {}^i\operatorname{Gr}_c^{F^{(\lambda_0)}}({}_{\boldsymbol{b}}\mathcal{E}) = \bigoplus_{u \in \mathcal{T}(\mathcal{E},\lambda_0,c,i)} \mathbb{E}_{\epsilon_2}\big({}^i\operatorname{Gr}_c^{F^{(\lambda_0)}}({}_{\boldsymbol{b}}\mathcal{E}),\ \mathfrak{e}(\lambda_0,u)\big). $$

It is easy to see $\mathbb{E}_{\epsilon_2}\big({}^i\operatorname{Gr}_c^{F^{(\lambda_0)}}({}_{\boldsymbol{b}}\mathcal{E}),\ \mathfrak{e}(\lambda_0,u)\big) \simeq {}^i\operatorname{Gr}_{\mathfrak{k}(\lambda_0,u)}^{F^{(\lambda_0)},\mathbb{E}^{(\lambda_0)}}({}_{\boldsymbol{b}}\mathcal{E})$.

We put $b := \max\{d' \in \mathcal{P}ar(\mathcal{E}^{(\lambda_0)}, i), \,|\, d' < c\}$. Since we have the isomorphism (8.40), the set of the eigenvalues of the induced morphism $\operatorname{Res}_i(\mathbb{D}^\lambda)$ on ${}^i\operatorname{Gr}_c^{F^{(\lambda_0)}}({}_{\boldsymbol{b}}\mathcal{E})_{|(\lambda,P)}$ is as follows:

$$ (8.42) \quad \big\{\mathfrak{e}(\lambda,u') \,\big|\, u' \in \mathcal{K}(\mathcal{E},\lambda,\boldsymbol{b},i),\ d(\lambda,\lambda_0,b) < \mathfrak{p}(\lambda,u') \leq d(\lambda,\lambda_0,c)\big\} $$
$$ = \big\{\mathfrak{e}(\lambda,u') \,\big|\, u' \in \mathcal{K}(\mathcal{E},\lambda,\boldsymbol{b},i),\ \mathfrak{p}(\lambda_0,u') = c\big\} = \big\{\mathfrak{e}(\lambda,u) \,\big|\, u \in \mathcal{T}(\mathcal{E},\lambda_0,c,i)\big\}. $$

Let u and u' be elements of $\mathcal{T}(\mathcal{E},\lambda_0,c,i)$. If η_2 is sufficiently small, $|\mathfrak{e}(\lambda,u) - \mathfrak{e}(\lambda,u')| < \epsilon_2$ if and only if $\mathfrak{e}(\lambda_0,u) = \mathfrak{e}(\lambda_0,u')$. Since we have $\mathfrak{p}(\lambda_0,u) = \mathfrak{p}(\lambda_0,u') = c$, the condition $\mathfrak{e}(\lambda_0,u) = \mathfrak{e}(\lambda_0,u')$ implies $u = u'$ (Lemma 2.2). Hence the condition $|\mathfrak{e}(\lambda,u) - \mathfrak{e}(\lambda,u')| < \epsilon_2$ implies $u = u'$, and thus $\mathfrak{e}(\lambda,u) = \mathfrak{e}(\lambda,u')$. Thus we obtain the result. □

More generally, we obtain the vector bundle on $\mathcal{D}_I(\lambda_0,\eta_2)$ for $\boldsymbol{u} = (\boldsymbol{a},\boldsymbol{\alpha}) \in \boldsymbol{R}^I \times \boldsymbol{C}^I$:

$$ {}^I\operatorname{Gr}_{\boldsymbol{u}}^{F^{(\lambda_0)},\mathbb{E}^{(\lambda_0)}}\big({}_{\boldsymbol{b}}\mathcal{E}_{|\mathcal{D}_I(\lambda_0,\eta_2)}\big) = {}^I\mathbb{E}^{(\lambda_0)}\Big({}^I\operatorname{Gr}_{\boldsymbol{a}}^{F^{(\lambda_0)}}\big({}_{\boldsymbol{b}}\mathcal{E}_{|\mathcal{D}_I(\lambda_0,\eta_2)}\big), \boldsymbol{\alpha}\Big). $$

We often omit to denote "$|\mathcal{D}_I(\lambda_0,\eta_2)$" if there are no confusion.

LEMMA 8.107. *If η_2 is sufficiently small, we have the following equality for any $\lambda \in \Delta(\lambda_0,\eta_2)$ and for any $\boldsymbol{u} \in \prod_{i\in I}\mathcal{K}(\mathcal{E},\lambda_0,\boldsymbol{b},i)$:*

$$ {}^I\operatorname{Gr}_{\mathfrak{k}(\lambda_0,\boldsymbol{u})}^{F^{(\lambda_0)},\mathbb{E}^{(\lambda_0)}}\big({}_{\boldsymbol{b}}\mathcal{E}\big)_{|\mathcal{D}_I^\lambda} = {}^I\operatorname{Gr}_{\mathfrak{k}(\lambda,\boldsymbol{u})}^{F,\mathbb{E}}\big({}_{\boldsymbol{b}}\mathcal{E}^\lambda{}_{|\mathcal{D}_I^\lambda}\big). $$

Proof Since we have the compatibilities of the tuples $\big({}^iF^{(\lambda_0)}, {}^i\mathbb{E}^{(\lambda_0)} \,\big|\, i \in \underline{l}\big)$ and $\big({}^iF, {}^i\mathbb{E} \,\big|\, i \in \underline{l}\big)$, it follows from Lemma 8.106. □

We have the bijection $\mathfrak{k}(\lambda) : \prod_{i\in I}\mathcal{KMS}(\mathcal{E}^0,i) \longrightarrow \prod_{i\in I}\mathcal{KMS}(\mathcal{E}^\lambda,i)$ (Corollary 7.71).

PROPOSITION 8.108. *The bijection $\mathfrak{k}(\lambda) : \mathcal{KMS}(\mathcal{E}^0,I) \longrightarrow \mathcal{KMS}(\mathcal{E}^\lambda,I)$ is induced, and we have the equality of the multiplicities $\mathfrak{m}(0,\boldsymbol{u},I) = \mathfrak{m}\big(\lambda,\mathfrak{k}(\lambda,\boldsymbol{u}),I\big)$.*

Proof Let $\boldsymbol{u}$ be an element of $\prod_{i\in I}\mathcal{KMS}(\mathcal{E}^0,i)$, then we obtain the element $\mathfrak{k}(\lambda_0,\boldsymbol{u}) \in \prod_{i\in I}\mathcal{KMS}(\mathcal{E}^{\lambda_0},i)$. Assume $\mathfrak{k}(\lambda_0,\boldsymbol{u}) \in \mathcal{KMS}({}_{\boldsymbol{b}}\mathcal{E}^{\lambda_0},I)$. Then we have the vector bundle ${}^I\operatorname{Gr}_{\mathfrak{k}(\lambda_0,\boldsymbol{u})}^{F^{(\lambda_0)},\mathbb{E}^{(\lambda_0)}}({}_{\boldsymbol{b}}\mathcal{E}) \neq 0$ over $\mathcal{D}_I(\lambda_0,\epsilon_0)$ for some sufficiently small ϵ_0. We have ${}^I\operatorname{Gr}_{\mathfrak{k}(\lambda_0,\boldsymbol{u}_0)}^{F^{(\lambda_0)},\mathbb{E}^{(\lambda_0)}}\big({}_{\boldsymbol{b}}\mathcal{E}\big)_{|\mathcal{D}_I^\lambda} = {}^I\operatorname{Gr}_{\mathfrak{k}(\lambda,\boldsymbol{u})}^{F,\mathbb{E}}\big({}_{\boldsymbol{b}}\mathcal{E}^\lambda\big)$ for $\lambda \in \Delta(\lambda_0,\epsilon_0)$, due to Lemma 8.107.

To show Proposition 8.108, let us consider the following condition for λ and $\boldsymbol{u} \in \prod_{i\in I}\mathcal{KMS}(\mathcal{E}^0,i)$:

$P(\lambda,\boldsymbol{u})$: $\mathfrak{k}(\lambda,\boldsymbol{u}) \in \mathcal{KMS}(\mathcal{E}^\lambda,I)$.

For any λ, there exists a positive number $\eta(\lambda)$ such that $P(\lambda,\boldsymbol{u})$ holds if and only if $P(\lambda',\boldsymbol{u})$ holds for some $\lambda' \in \Delta(\lambda,\eta(\lambda))$. Then the proposition can be shown by an elementary continuity method. □

8.8.7. Weak norm estimate. Pick a frame $\boldsymbol{v}$ of ${}_{\boldsymbol{b}}\mathcal{E}$ over $\mathcal{X}(\lambda_0,\eta_2)$, which is compatible with $F^{(\lambda_0)}$ and $\mathbb{E}^{(\lambda_0)}$. For each v_j and for each i, we have the unique element $u_i(v_j)\in\mathcal{KMS}(\mathcal{E}^0,i)$ satisfying the following:

$$ {}^{i}\deg^{F^{(\lambda_0)},\mathbb{E}^{(\lambda_0)}}(v_j)=\mathfrak{k}(\lambda_0,u_i(v_j)). $$

LEMMA 8.109. *The frame $\boldsymbol{v}_{|\mathcal{X}^\lambda}$ of ${}_{\boldsymbol{b}}\mathcal{E}^\lambda$ is compatible with $\big({}^iF\,\big|\,i=1,\dots,l\big)$, for any $\lambda\in\Delta(\lambda_0,\eta_0)$. Moreover, we have ${}^i\deg^F(v_j)=\mathfrak{p}(\lambda,u_i(v_j))$.*

Proof The decomposition ${}^i\mathbb{E}^{(\lambda_0)}$ induces that of ${}^i\operatorname{Gr}_c^{F^{(\lambda_0)}}\big({}_{\boldsymbol{b}}\mathcal{E}^\lambda\big)$, which is also denoted by ${}^i\mathbb{E}^{(\lambda_0)}$. Recall the isomorphism (8.40). Hence we have the naturally defined morphism $\varphi:{}^iF_a\big({}_{\boldsymbol{b}}\mathcal{E}^\lambda{}_{|\mathcal{D}_i^\lambda}\big)\longrightarrow{}^i\operatorname{Gr}^F_{d(\lambda,\lambda_0,c)}\big({}_{\boldsymbol{b}}\mathcal{E}^\lambda\big)$ for any $a\le d(\lambda,\lambda_0,c)$. It is easy to see the following:

$$ \operatorname{Im}(\varphi)=\bigoplus_{\substack{\mathfrak{p}(\lambda_0,u)=c\\ \mathfrak{p}(\lambda,u)\le a}}{}^i\mathbb{E}^{(\lambda_0)}\big({}^i\operatorname{Gr}_c^{F^{(\lambda_0)}}\big({}_{\boldsymbol{b}}\mathcal{E}^\lambda\big),\mathfrak{e}(\lambda_0,u)\big). $$

Since $\boldsymbol{v}$ is compatible with the ${}^iF^{(\lambda_0)}$ and ${}^i\mathbb{E}^{(\lambda_0)}$, the induced frame on ${}^i\operatorname{Gr}^{F^{(\lambda_0)}}$ is compatible with the decomposition ${}^i\mathbb{E}^{(\lambda_0)}$. Hence it is compatible with the filtration ${}^iF$. □

Let us consider the C^∞-frame $\boldsymbol{v}'$ of $\mathcal{E}$ over $\Delta(\lambda_0,\eta_2)\times(X-D)$, given as follows:

$$ v_j':=v_j\cdot\prod_{i=1}^{l}|z_i|^{\mathfrak{p}(\lambda,u_i(v_j))},\qquad \boldsymbol{v}':=(v_j'). $$

By a standard argument, we obtain the following.

PROPOSITION 8.110. *The frame $\boldsymbol{v}'$ is adapted up to log order.*

Proof We recall that the holomorphic bundle $\mathcal{E}^\lambda$ with the hermitian metric h is acceptable, i.e., the curvature of $R(h,\mathcal{E}^\lambda)\in End(\mathcal{E}^\lambda)\otimes\Omega^{1,1}_{X-D}$ is bounded with respect the metric h and the Poincaré metric of $X-D$ (Corollary 8.19). Moreover, the estimate is uniform for $\lambda\in\Delta(\lambda_0,\epsilon_0)$. Therefore we remark that the constant M in Corollary 2.53 can be taken independently of $\lambda\in\Delta(\lambda_0,\epsilon_0)$.

Then it is easy to see that $H(h,\boldsymbol{v}')\le C\cdot\big(-\sum_{i=1}^{l}\log|z_i|\big)^M$ for some positive constants M and C, which follows from Lemma 8.109 and Corollary 2.53. By considering the dual frame $\boldsymbol{v}^{\vee\prime}$, we obtain $H(h,\boldsymbol{v}')\ge C'\cdot\big(-\sum_{i=1}^{l}\log|z_i|\big)^{-M'}$, as usual (see the proof of Proposition 8.70, for example). Thus we are done. □

8.8.8. Some functoriality. Let us consider the functoriality for tensor product. Let $(E^{(a)},\overline{\partial}_{E^{(a)}},h^{(a)},\theta^{(a)})$ $(a=1,2)$ be tame harmonic bundles on $X-D$. We obtain the deformed holomorphic bundles $\mathcal{E}^{(a)}$. Pick $\lambda_0\in\boldsymbol{C}$. For simplicity, we consider the following situation:

- There are small positive numbers $\eta_{a\,i}$ for $a=1,2$ and for $i=1,\dots,l$ such that $\eta_{1\,i}+\eta_{2\,i}<1$.
- For any $\lambda\in\Delta(\lambda_0,\epsilon_0)$ and for any $i=1,\dots,l$, the sets $\mathcal{P}ar({}^\diamond\mathcal{E}^{(a)\,\lambda},i)$ $(a=1,2)$ are contained in the open intervals $]-\eta_{a\,i},0[$.

Take $\boldsymbol{v}^{(a)}$ be frames of ${}^\diamond\mathcal{E}^{(a)}$ over $\Delta(\lambda_0,\epsilon_0)$, which are compatible with the filtration ${}^iF^{(\lambda_0)}$ and the decomposition ${}^i\mathbb{E}^{(\lambda_0)}$ $(i=1,\dots,l)$.

LEMMA 8.111. *$\boldsymbol{v}^{(1)}\otimes\boldsymbol{v}^{(2)}$ is a frame of ${}^{\diamond}\big(\mathcal{E}^{(1)}\otimes\mathcal{E}^{(2)}\big)$, which is compatible with the filtration $F^{(\lambda_0)}$ and the decomposition $\mathbb{E}^{(\lambda_0)}$.*

Proof We obtain the frame $\boldsymbol{v}^{(a)\prime}$ from $\boldsymbol{v}^{(a)}$ as in the subsection 8.8.7, which is adapted up to log order (Proposition 8.110). Then $\boldsymbol{v}^{(1)\prime}\otimes\boldsymbol{v}^{(2)\prime}$ is adapted up to log order, as in the subsection 8.8.7. Due to our assumption, we obtain that $\boldsymbol{v}^{(1)}\otimes\boldsymbol{v}^{(2)}$ gives a frame of ${}^{\diamond}\big(\mathcal{E}^{(1)}\otimes\mathcal{E}^{(2)}\big)$. Then the compatibilities with the filtration F and the decomposition $\mathbb{E}$ are clear. □

COROLLARY 8.112. *Let η be a positive number and R be positive integer such that $R\cdot\eta<1$. Assume $\mathcal{P}ar({}^{\diamond}\mathcal{E}^{\lambda},i)\subset]-\eta,0[$ for any $\lambda\in\Delta(\lambda_0,\epsilon_0)$, for simplicity. Then we have the following canonical isomorphism:*

$$
{}^{\diamond}\bigwedge^{R}\mathcal{E}\simeq\bigwedge^{R}{}^{\diamond}\mathcal{E},\qquad {}^{\diamond}\mathrm{Sym}^{R}\mathcal{E}\simeq\mathrm{Sym}^{R}({}^{\diamond}\mathcal{E}).
$$

□

8.9. The induced vector bundle

8.9.1. The induced vector bundles ${}^{\underline{l}}\mathcal{G}_{\boldsymbol{u}}$ on $\mathcal{D}_{\underline{l}}$. We continue to use the setting in the section 8.8. Let $\boldsymbol{u}$ be an element of $\mathcal{KMS}(\mathcal{E}^0,\underline{l})$. Pick $\lambda_0\in\boldsymbol{C}_{\lambda}$, and we put $(\boldsymbol{a},\boldsymbol{\alpha}):=\mathfrak{k}(\lambda_0,\boldsymbol{u})$. Take $\boldsymbol{b}\in\boldsymbol{R}^l$ such that ${}^{\underline{l}}\mathrm{Gr}^{F}_{\boldsymbol{a}}\big({}_{\boldsymbol{b}}\mathcal{E}\big)\neq 0$ and $b_i\not\in\mathcal{KMS}(\mathcal{E}^{\lambda_0},i)$ for each i. Then we obtain the vector bundle over $\mathcal{D}_{\underline{l}}(\lambda_0,\epsilon_0)$, for some small positive number ϵ_0:

$$
{}^{\underline{l}}\mathcal{G}^{(\lambda_0)}_{\boldsymbol{u}}:={}^{\underline{l}}\mathrm{Gr}^{F^{(\lambda_0)},\mathbb{E}^{(\lambda_0)}}_{\mathfrak{k}(\lambda_0,\boldsymbol{u})}\big({}_{\boldsymbol{b}}\mathcal{E}\big).
$$

Clearly, it does not depend on a choice of $\boldsymbol{b}$ on a neighbourhood of λ_0.

Pick $\lambda_1\in\Delta(\lambda_0,\epsilon_0)$ and ϵ_0' as $\Delta^*(\lambda_0,\epsilon_0)\supset\Delta(\lambda_1,\epsilon_0')$. We obtain the following vector bundles on $\mathcal{D}_{\underline{l}}(\lambda_1,\epsilon_0')$:

$$
{}^{\underline{l}}\mathcal{G}^{(\lambda_1)}_{\boldsymbol{u}},\qquad {}^{\underline{l}}\mathcal{G}^{(\lambda_0)}_{\boldsymbol{u}\,|\,\mathcal{D}_{\underline{l}}(\lambda_1,\epsilon_0')}.
$$

LEMMA 8.113. *The vector bundles ${}^{\underline{l}}\mathcal{G}^{(\lambda_1)}_{\boldsymbol{u}}$ and ${}^{\underline{l}}\mathcal{G}^{(\lambda_0)}_{\boldsymbol{u}\,|\,\mathcal{D}_{\underline{l}}(\lambda_1,\epsilon_0')}$ are naturally isomorphic.*

Proof It follows from Lemma 8.107. □

Therefore we obtain the vector bundle ${}^{\underline{l}}\mathcal{G}_{\boldsymbol{u}}$ over $\mathcal{D}_{\underline{l}}$. When we distinguish the dependence of ${}^{\underline{l}}\mathcal{G}_{\boldsymbol{u}}$ on the harmonic bundle $(E,\overline{\partial}_E,\theta,h)$, we use the notation ${}^{\underline{l}}\mathcal{G}_{\boldsymbol{u}}(E,\overline{\partial}_E,\theta,h)$, or simply ${}^{\underline{l}}\mathcal{G}_{\boldsymbol{u}}(E)$.

8.9.2. Shift. Let $\boldsymbol{\delta}_0$ denote the element $(1,0)\in\boldsymbol{R}\times\boldsymbol{C}$, and let $\boldsymbol{\delta}_{0,i}$ denote the element of $(\boldsymbol{R}\times\boldsymbol{C})^l$ such that $q_i(\boldsymbol{\delta}_{0,i})=\boldsymbol{\delta}_0$ and $q_j(\boldsymbol{\delta}_{0,i})=(0,0)$ for $j\neq i$. Here q_j denote the projection onto the j-th component.

In the setting of the subsection 8.9.1, let f be a section of ${}^{\underline{l}}\mathrm{Gr}^{F^{(\lambda_0)},\mathbb{E}^{(\lambda_0)}}_{\mathfrak{k}(\lambda_0,\boldsymbol{u})}\big({}_{\boldsymbol{b}}\mathcal{E}\big)$. Then we can take a holomorphic section G of ${}_{\boldsymbol{b}}\mathcal{E}$ such that $G_{|\mathcal{D}_{\underline{l}}(\lambda_0,\epsilon_0)}$ is contained in ${}^{\underline{l}}F^{(\lambda_0)}_{\boldsymbol{a}}{}^{\underline{l}}\mathbb{E}^{(\lambda_0)}\big({}_{\boldsymbol{b}}\mathcal{E},\boldsymbol{\alpha}\big)$, which naturally induces $f\in{}^{\underline{l}}\mathcal{G}^{(\lambda_0)}_{\boldsymbol{u}}$ in the associated graded level. Then it is easy to see that $z_i^{-1}G$ naturally induces a holomorphic section of ${}^{\underline{l}}\mathcal{G}^{(\lambda_0)}_{\boldsymbol{u}+\boldsymbol{\delta}_{0,i}}$, which is independent of a choice of G. Since it can be glued, we obtain the isomorphism ${}^{\underline{l}}\mathcal{G}_{\boldsymbol{u}}\longrightarrow{}^{\underline{l}}\mathcal{G}_{\boldsymbol{u}+\boldsymbol{\delta}_{0,i}}$. Namely, we obtain the following lemma.

LEMMA 8.114. *We have the isomorphism* sh : ${}^{\underline{l}}\mathcal{G}_{\boldsymbol{u}} \simeq {}^{\underline{l}}\mathcal{G}_{\boldsymbol{u}+\boldsymbol{\delta}_{0,i}}$ *for any* $\boldsymbol{u} \in \mathcal{KMS}(\mathcal{E}^0, \underline{l})$ *and any* $i \in \underline{l}$. *The isomorphism is given as above.* □

Therefore the induced bundle can be defined for the elements of $\overline{\mathcal{KMS}}(\mathcal{E}^0, \underline{l})$. For any elements $\boldsymbol{u} \in \overline{\mathcal{KMS}}(\mathcal{E}^0, \underline{l})$, the induced bundle is denoted by ${}^{\underline{l}}\mathcal{G}_{\boldsymbol{u}}$, for simplicity of the notation.

8.9.3. The induced frame. Let $\boldsymbol{v} = (v_i)$ be a compatible frame of ${}_{\boldsymbol{b}}\mathcal{E}$ over $\mathcal{X}(\lambda_0, \epsilon_0)$. For each section v_i, we have the element $\boldsymbol{u}(v_i) \in \mathcal{KMS}(\mathcal{E}^0, \underline{l})$ such that ${}^{\underline{l}}\deg^{\mathbb{E}^{(\lambda_0)}, F^{(\lambda_0)}}(v_i) = \mathfrak{k}(\lambda_0, \boldsymbol{u}(v_i))$. For any element $\boldsymbol{u} \in \mathcal{KMS}(\mathcal{E}^0, \underline{l})$, we put as follows:

$$\boldsymbol{v}_{\boldsymbol{u}} := \left(z^{-\boldsymbol{n}} \cdot v_i \,\middle|\, \boldsymbol{u}(v_i) + \boldsymbol{n} = \boldsymbol{u}\right).$$

LEMMA 8.115. *The tuple of sections* $\boldsymbol{v}_{\boldsymbol{u}}$ *induces the frame of* ${}^{\underline{l}}\mathcal{G}_{\boldsymbol{u}}^{(\lambda_0)}$.

Proof First we remark the following: Let $\boldsymbol{n}$ be an element of $\mathbb{Z}^l$. We put $\tilde{\boldsymbol{v}} := (z^{-\boldsymbol{n}} v_i)$. Then $\tilde{\boldsymbol{v}}$ gives the frame of ${}_{\boldsymbol{b}+\boldsymbol{n}}\mathcal{E}$ over $\mathcal{X}(\lambda_0, \epsilon_0)$.

It is easy to see that $\boldsymbol{v}_{\boldsymbol{u}}$ induces the tuple of sections of ${}^{\underline{l}}\mathcal{G}_{\boldsymbol{u}}^{(\lambda_0)}$. By using the remark above, we can show that $\boldsymbol{v}_{\boldsymbol{u}}$ is a frame. □

8.9.4. The nilpotent map $\mathcal{N}_{i,\boldsymbol{u}}$ and the pairing with the dual. Let $\boldsymbol{u}$ be an element of $\mathcal{KMS}(\mathcal{E}^0, \underline{l})$. We have the endomorphism $\mathrm{Res}_i(\mathbb{D}^\lambda)$ on ${}^{\underline{l}}\mathcal{G}_{\boldsymbol{u}}$. The unique eigenvalue of $\mathrm{Res}_i(\mathbb{D}^\lambda)$ on ${}^{\underline{l}}\mathcal{G}_{\boldsymbol{u}\,|\,\lambda}$ is $\mathfrak{e}(\lambda, q_i(\boldsymbol{u}))$. Hence the nilpotent part $\mathcal{N}_{i\,\boldsymbol{u}} := \mathrm{Res}_i(\mathbb{D}^\lambda) - \mathfrak{e}(\lambda, q_i(\boldsymbol{u}))$ gives the holomorphic endomorphism.

Let λ_0 be an element of $\boldsymbol{C}_\lambda$ and $\epsilon_0 > 0$. We have the naturally defined morphism ${}_{\boldsymbol{b}}\mathcal{E} \otimes \left({}_{-\boldsymbol{b}+\boldsymbol{\delta}}\mathcal{E}^\vee\right) \longrightarrow \mathcal{O}$ over $\Delta(\lambda_0, \epsilon_0)$, which is the morphism of λ-connections. (We remark $b_i \notin \mathcal{KMS}(\mathcal{E}^{\lambda_0}, i)$.) It is easy to see that ${}^{i}\mathbb{E}^{(\lambda_0)}$ and ${}^{i}F^{(\lambda_0)}$ are preserved over $\mathcal{D}_i(\lambda_0, \epsilon_0)$. Thus we obtain the following morphism over $\mathcal{D}_{\underline{l}}(\lambda_0, \epsilon_0)$:

$${}^{\underline{l}}\mathrm{Gr}^{F^{(\lambda_0)}, \mathbb{E}^{(\lambda_0)}}_{\mathfrak{k}(\lambda_0, \boldsymbol{u})}(\mathcal{E}) \otimes {}^{\underline{l}}\mathrm{Gr}^{F^{(\lambda_0)}, \mathbb{E}^{(\lambda_0)}}_{\mathfrak{k}(\lambda_0, -\boldsymbol{u})}(\mathcal{E}^\vee) \longrightarrow \mathcal{O}_{\mathcal{D}_{\underline{l}}(\lambda_0, \epsilon_0)}.$$

Then we obtain the morphism $S : {}^{\underline{l}}\mathcal{G}_{\boldsymbol{u}}(E) \otimes {}^{\underline{l}}\mathcal{G}_{-\boldsymbol{u}}(E^\vee) \longrightarrow \mathcal{O}_{\mathcal{D}_{\underline{l}}}$.

LEMMA 8.116. *Under the isomorphisms* ${}^{\underline{l}}\mathcal{G}_{\boldsymbol{u}}(E) \simeq {}^{\underline{l}}\mathcal{G}_{\boldsymbol{u}+\boldsymbol{\delta}_{0,i}}(E)$ *and* ${}^{\underline{l}}\mathcal{G}_{-\boldsymbol{u}}(E^\vee) \simeq {}^{\underline{l}}\mathcal{G}_{-\boldsymbol{u}-\boldsymbol{\delta}_{0,i}}(E^\vee)$ *given in the subsection* 8.9.2, *the pairings* S *are same.* □

LEMMA 8.117. *We have the equality* $S\left(\mathcal{N}_{i\,\boldsymbol{u}} \otimes \mathrm{id}\right) + S\left(\mathrm{id} \otimes \mathcal{N}_{i,-\boldsymbol{u}}\right) = 0$.

Proof It follows from $S\left(\mathbb{D} \otimes \mathrm{id}\right) + S\left(\mathrm{id} \otimes \mathbb{D}\right) = \mathbb{D} \circ S$. □

LEMMA 8.118. *Under the isomorphism* ${}^{\underline{l}}\mathcal{G}_{\boldsymbol{u}} \simeq {}^{\underline{l}}\mathcal{G}_{\boldsymbol{u}+\boldsymbol{\delta}_{0,i}}$ *given in the subsection* 8.9.2, *we have the coincidence* $\mathcal{N}_{j,\boldsymbol{u}} \simeq \mathcal{N}_{j,\boldsymbol{u}+\boldsymbol{\delta}_{0,i}}$.

Proof We give an only sketch. Let $\boldsymbol{v} = (v_k)$ be a holomorphic frame of ${}_{\boldsymbol{b}}\mathcal{E}$ over $\mathcal{X}(\lambda_0, \epsilon_0)$. We put $\widetilde{v}_k = z_i^{-1} \cdot v_k$. It gives the holomorphic frame of $z_i^{-1}{}_{\boldsymbol{b}}\mathcal{E}$, and it is easy to check the coincidence of the nilpotent parts of the connection forms with respect to $\widetilde{\boldsymbol{v}}$ and $\boldsymbol{v}$. Then the claim immediately follows. □

LEMMA 8.119. *Let* $\boldsymbol{u}$ *be an element of* $\overline{\mathcal{KMS}}(\mathcal{E}^0, \underline{l})$. *The induced bundle* ${}^{\underline{l}}\mathcal{G}_{\boldsymbol{u}}(E)$ *with the nilpotent maps* $\mathcal{N}_{i,\boldsymbol{u}}$ *are well defined. The pairing* ${}^{\underline{l}}\mathcal{G}_{\boldsymbol{u}}(E) \otimes {}^{\underline{l}}\mathcal{G}_{-\boldsymbol{u}}(E^\vee) \longrightarrow \mathcal{O}_{\mathcal{D}_{\underline{l}}}$ *is also well defined.* □

8.9.5. Functoriality for dual. Due to the results in the subsection 8.9.4, we obtain the naturally defined morphism ${}^{\underline{l}}\mathcal{G}_{-\boldsymbol{u}}(E^{\vee}) \longrightarrow {}^{\underline{l}}\mathcal{G}_{\boldsymbol{u}}(E)^{\vee}$.

LEMMA 8.120. *The naturally defined morphism* ${}^{\underline{l}}\mathcal{G}_{-\boldsymbol{u}}(E^{\vee}) \longrightarrow {}^{\underline{l}}\mathcal{G}_{\boldsymbol{u}}(E)^{\vee}$ *is isomorphic.*

Proof Let $\boldsymbol{v}$ be a frame of ${}_{\boldsymbol{b}}\mathcal{E}$, which is compatible with $\mathbb{E}^{(\lambda_0)}$ and $F^{(\lambda_0)}$. The dual frame $\boldsymbol{v}^{\vee}$ gives the frame of ${}_{-\boldsymbol{b}+\boldsymbol{\delta}}\mathcal{E}$ on $\mathcal{X}(\lambda_0,\epsilon_0)$, due to the weak norm estimate (Proposition 8.110). It is also compatible with $\mathbb{E}^{(\lambda_0)}$ and $F^{(\lambda_0)}$. By using Lemma 8.115, we obtain the induced frames of ${}^{\underline{l}}\mathcal{G}_{\boldsymbol{u}}(E)$ and ${}^{\underline{l}}\mathcal{G}_{-\boldsymbol{u}}(E^{\vee})$ from $\boldsymbol{v}_{\boldsymbol{u}}$ and $\left(\boldsymbol{v}^{\vee}\right)_{-\boldsymbol{u}}$ respectively. Thus we are done. □

8.9.6. Functoriality for tensor products. Let $(E_i,\overline{\partial}_{E_i},\theta_i,h_i)$ $(i=1,2)$ be tame harmonic bundles over $X-D$. Let $\boldsymbol{u}_i$ $(i=1,2)$ be elements of $\mathcal{KMS}(\mathcal{E}_i^0,\underline{l})$. We have the induced morphisms on $\mathcal{X}(\lambda_0,\epsilon_0)$:

$$ {}^{\underline{l}}\mathrm{Gr}^{F^{(\lambda_0)}}_{\mathfrak{p}(\lambda_0,\boldsymbol{u}_1)}(\mathcal{E}_1)\otimes {}^{\underline{l}}\mathrm{Gr}^{F^{(\lambda_0)}}_{\mathfrak{p}(\lambda_0,\boldsymbol{u}_2)}(\mathcal{E}_2)\longrightarrow {}^{\underline{l}}\mathrm{Gr}^{F^{(\lambda_0)}}_{\mathfrak{p}(\lambda_0,\boldsymbol{u}_1+\boldsymbol{u}_2)}(\mathcal{E}_1\otimes\mathcal{E}_2). $$

Since the morphism is compatible with the residues of the λ-connections, we obtain the following induced morphisms:

$$ F_{\boldsymbol{u}_1,\boldsymbol{u}_2}: {}^{\underline{l}}\mathcal{G}^{(\lambda_0)}_{\boldsymbol{u}_1}(E_1)\otimes {}^{\underline{l}}\mathcal{G}^{(\lambda_0)}_{\boldsymbol{u}_2}(E_2)\longrightarrow {}^{\underline{l}}\mathcal{G}^{(\lambda_0)}_{\boldsymbol{u}_1+\boldsymbol{u}_2}(E_1\otimes E_2). $$

The nilpotent maps $\mathcal{N}_i$ of ${}^{\underline{l}}\mathcal{G}_{\boldsymbol{u}_i}(E_i)$ induces the nilpotent map $\mathcal{N}_1\otimes 1+1\otimes\mathcal{N}_2$ on the left hand side. On the other hand, we have the nilpotent maps on the right hand side.

LEMMA 8.121. *The morphism* $F_{\boldsymbol{u}_1,\boldsymbol{u}_2}$ *is compatible with the nilpotent maps of the induced vector bundles.*

Proof It is clear from our construction. □

LEMMA 8.122. *Let* $\boldsymbol{u}$ *be an element of* $\overline{\mathcal{KMS}}(\mathcal{E}_1^0\otimes\mathcal{E}_2^0,\underline{l})$. *We have the isomorphism:*

$$ \bigoplus_{\substack{\boldsymbol{u}_i\in\overline{\mathcal{KMS}}(\mathcal{E}_i^0,\underline{l}),\\ \boldsymbol{u}_1+\boldsymbol{u}_2=\boldsymbol{u}}} {}^{\underline{l}}\mathcal{G}_{\boldsymbol{u}_1}(\mathcal{E}_1)\otimes {}^{\underline{l}}\mathcal{G}_{\boldsymbol{u}_2}(\mathcal{E}_2)\simeq {}^{\underline{l}}\mathcal{G}_{\boldsymbol{u}}(\mathcal{E}_1\otimes\mathcal{E}_2). $$

It is compatible with the induced nilpotent maps.

Proof We have only to show that the morphism is isomorphic locally. We can show it by using Lemma 8.115. □

8.9.7. Functoriality for pull backs. We put $X^{(1)}:=\Delta_z^n$ and $X^{(2)}:=\Delta_\zeta^m$. We put $D_i^{(1)}=\{z_i=0\}$ and $D^{(1)}=\bigcup_{i=1}^n D_i^{(1)}$. We put $D_i^{(2)}:=\{\zeta_i=0\}$ and $D^{(2)}:=\bigcup_{i=1}^m D_i^{(2)}$. Let $(E,\overline{\partial}_E,\theta,h)$ be a tame harmonic bundle over $X^{(2)}-D^{(2)}$.

Let $\boldsymbol{c}=\left(c_{j\,i}\,\middle|\,1\leq j\leq n,\ 1\leq i\leq m\right)$ be an element of $\mathbb{Z}_{\geq 0}^{n\cdot m}$. Let us consider the morphism $\psi: X^{(1)}\longrightarrow X^{(2)}$ given as follows:

$$ \psi^*(\zeta_i):=\prod_{j=1}^{n} z_j^{c_{j\,i}}. $$

We assume that the origin of $X^{(1)}$ is mapped to the origin of $X^{(2)}$. We obtain the harmonic bundle $\psi^{-1}(E,\overline{\partial}_E,\theta,h)$ over $X^{(1)}-D^{(1)}$. We would like to compare the induced vector bundles for $(E,\overline{\partial}_E,\theta,h)$ and $\psi^{-1}(E,\overline{\partial}_E,\theta,h)$.

We pick any point $\lambda_0 \in \boldsymbol{C}_\lambda$. Let $\boldsymbol{b}^{(2)} = (b_i^{(2)})$ be an element of $\boldsymbol{R}^l$ such that $b_i^{(2)} \notin \mathcal{KMS}(\mathcal{E}^{\lambda_0}, i)$ for each i. Let $\boldsymbol{b}^{(1)}$ be an element of $\boldsymbol{R}^n$ such that $b_i^{(1)} \notin \mathcal{KMS}(\psi^{-1}\mathcal{E}^{\lambda_0}, i)$ for each i.

Take a sufficiently small $\epsilon_0 > 0$ such that ${}_{\boldsymbol{b}^{(2)}}\mathcal{E}$ is locally free on $\mathcal{X}^{(2)}(\lambda_0, \epsilon_0)$ and that we have the filtrations ${}^iF^{(\lambda_0)}$ and the decompositions ${}^i\mathbb{E}^{(\lambda_0)}$. Let $\boldsymbol{v} = (v_i)$ be a frame of ${}_{\boldsymbol{b}^{(2)}}\mathcal{E}$ which is compatible with the filtrations ${}^iF^{(\lambda_0)}$ and the decompositions ${}^i\mathbb{E}^{(\lambda_0)}$ $(i = 1, \ldots, m)$. We have the elements $\boldsymbol{u}(v_i) \in \mathcal{KMS}(\mathcal{E}^0, \underline{m})$ for each v_i satisfying the following:

$$\deg^{\mathbb{E}^{(\lambda_0)}, F^{(\lambda_0)}}(v_i) = \mathfrak{k}(\lambda_0, \boldsymbol{u}(v_i)).$$

We put as follows:

$$v_i' := v_i \cdot \prod_{k=1}^{l} |\zeta_k|^{\mathfrak{p}(\lambda, u_k(v_i))}, \quad \boldsymbol{v}' = (v_i').$$

We have already seen that C^∞-frame $\boldsymbol{v}'$ on $\mathcal{X}^{(2)}(\lambda_0, \epsilon_0) - \mathcal{D}^{(2)}(\lambda_0, \epsilon_0)$ is adapted up to log order.

We obtain the holomorphic frame $\psi^{-1}\boldsymbol{v}$ and the C^∞-frame $\psi^{-1}\boldsymbol{v}'$ of $\psi^{-1}\mathcal{E}$ over $\mathcal{X}^{(1)}(\lambda_0, \epsilon_0) - \mathcal{D}^{(1)}(\lambda_0, \epsilon_0)$. Note that $\psi^{-1}\boldsymbol{v}'$ is adapted up to log order.

We would like to modify $\psi^{-1}\boldsymbol{v}$ to obtain a holomorphic frame of ${}_{\boldsymbol{b}^{(1)}}\psi^{-1}\mathcal{E}$. We put as follows:

$$\boldsymbol{c} \cdot \boldsymbol{u}(v_i) := \Big(\sum_k c_{1\,k} \cdot u_k(v_i), \ldots, \sum_k c_{n\,k} \cdot u_k(v_i)\Big) \in (\boldsymbol{R} \times \boldsymbol{C})^n.$$

The elements $\boldsymbol{n}(v_i) \in \mathbb{Z}^n$ are determined by the following conditions:

$$\boldsymbol{b}^{(1)} - \boldsymbol{\delta} < \mathfrak{p}\big(\lambda_0, \boldsymbol{c} \cdot \boldsymbol{u}(v_i)\big) + \boldsymbol{n}(v_i) < \boldsymbol{b}^{(1)}.$$

We remark our choice $b_i^{(1)} \notin \mathcal{KMS}(\psi^{-1}\mathcal{E}^{\lambda_0}, i)$. We put as follows:

$$w_i := \psi^* v_i \cdot \prod z_j^{-n_j(v_i)}, \quad \boldsymbol{w} = (w_i), \quad \boldsymbol{d}(w_i) := \mathfrak{p}\big(\lambda_0, \boldsymbol{c} \cdot \boldsymbol{u}(v_i)\big) + \boldsymbol{n}(v_i).$$

We remark that $\boldsymbol{b}^{(1)} - \boldsymbol{\delta} < \boldsymbol{d}(w_i) < \boldsymbol{b}^{(1)}$ from our choice. Then we put as follows:

$$w_i' := w_i \cdot \prod_{j=1}^{n} |z_j|^{d_j(w_i)}, \quad \boldsymbol{w}' := (w_i').$$

Thus we obtain C^∞-frame $\boldsymbol{w}'$ of $\mathcal{E}$ over $\mathcal{X}^{(1)}(\lambda_0, \epsilon_1) - \mathcal{D}^{(1)}(\lambda_0, \epsilon_1)$.

LEMMA 8.123. *The tuple $\boldsymbol{w}'$ is adapted up to log order.*

Proof It is easy to see that we have some C^∞-functions f_i such that $w_i' = f_i \cdot \psi^{-1} v_i'$ and $|f_i| = 1$ hold. Since $\psi^{-1}\boldsymbol{v}$ is adapted up to log order, we obtain the lemma. □

LEMMA 8.124. *If ϵ_1 is sufficiently small, the tuple $\boldsymbol{w}$ gives the frame of ${}_{\boldsymbol{b}^{(1)}}\psi^*\mathcal{E}$ over $\mathcal{X}^{(1)}(\lambda_0, \epsilon_1)$. It is compatible with $F^{(\lambda_0)}$ and $\mathbb{E}^{(\lambda_0)}$.*

Proof The first claim and the compatibility with $\mathbb{F}^{(\lambda_0)}$ follow from Lemma 2.8 and Lemma 2.9.

Let $A = \sum A_k \cdot d\zeta_k/\zeta_k$ denote the λ-connection form of $\mathbb{D}$ with respect to $\boldsymbol{v}$, i.e., $\mathbb{D}\boldsymbol{v} = \boldsymbol{v} \cdot A$. Then we have $\psi^*\mathbb{D}\psi^*\boldsymbol{v} = \psi^*\boldsymbol{v} \cdot \psi^* A$. We have the following:

$$\psi^* A = \sum_i \sum_j \frac{dz_j}{z_j} \cdot c_{j\,i} \cdot \psi^* A_i.$$

Thus we obtain the following:

$$\psi^*\mathbb{D} \cdot \boldsymbol{w} = \boldsymbol{w} \cdot \sum_j \Bigl(\sum_k c_{j\,k} \cdot \psi^* A_k + C_j \Bigr) \cdot \frac{dz_j}{z_j}. \tag{8.43}$$

Here C_j denotes the diagonal matrix whose i-th components are $-n_j(v_i) \cdot \lambda$. Then it is easy to see that $\boldsymbol{w}$ is compatible with the decompositions ${}^i\mathbb{E}^{(\lambda_0)}$. □

For $\boldsymbol{u}_1 \in \mathcal{KMS}\bigl({}_{\boldsymbol{b}^{(1)}}\psi^{-1}\mathcal{E}^{\lambda_0}, \underline{n}\bigr)$, we put as follows:

$$S(\boldsymbol{u}_1) = \bigl\{ \boldsymbol{u} \in \mathcal{KMS}(\mathcal{E}^0, \underline{m}) \,\big|\, \boldsymbol{b}^{(2)} - \boldsymbol{\delta} < \mathfrak{p}(\lambda_0, \boldsymbol{u}) < \boldsymbol{b}^{(2)},\ \boldsymbol{c} \cdot \boldsymbol{u} \equiv \boldsymbol{u}_1 \bigr\}.$$

Here "$\boldsymbol{u}_1 \equiv \boldsymbol{u}_2$" means $\boldsymbol{u}_1 - \boldsymbol{u}_2 \in (\mathbb{Z} \times \{0\})^n \subset (\boldsymbol{R} \times \boldsymbol{C})^n$ for $\boldsymbol{u}_1, \boldsymbol{u}_2 \in (\boldsymbol{R} \times \boldsymbol{C})^n$. We obtain the following isomorphism via the frames $\boldsymbol{v}$ and $\boldsymbol{w}$:

$$\bigoplus_{\boldsymbol{u} \in S(\boldsymbol{u}_1)} \psi_0^{-1} {}^{\underline{m}}\mathrm{Gr}^{F^{(\lambda_0)}, \mathbb{E}^{(\lambda_0)}}_{\mathfrak{k}(\lambda_0, \boldsymbol{u})}(\mathcal{E}) \xrightarrow{\simeq} {}^{\underline{n}}\mathrm{Gr}^{F^{(\lambda_0)}, \mathbb{E}^{(\lambda_0)}}_{\mathfrak{k}(\lambda_0, \boldsymbol{u}_1)}(\psi^{-1}\mathcal{E}). \tag{8.44}$$

Here ψ_0 denotes the morphism $D^{(1)}_{\underline{n}} \longrightarrow D^{(2)}_{\underline{m}}$ induced by ψ.

LEMMA 8.125. *The isomorphism is independent of a choice of the frame $\boldsymbol{v}$.*

Proof The isomorphism (8.44) induces the decomposition of the right hand side. We have only to check that the decomposition is independent of a choice of the frame, which is checked as follows: Let $\widetilde{\boldsymbol{v}}$ be other frame of ${}_{\boldsymbol{b}^{(2)}}\mathcal{E}$ compatible with the filtrations ${}^iF^{(\lambda_0)}$ and the decompositions ${}^i\mathbb{E}^{(\lambda_0)}$ $(i = 1, \ldots, m)$. Let $\widetilde{\boldsymbol{w}}$ be the frame obtained from $\boldsymbol{v}$ by the above procedure. We have the relation $v_j = \sum a_{i,j} \cdot \widetilde{v}_i$, where $a_{i,j}$ are holomorphic functions on $\mathcal{X}^{(2)}(\lambda_0, \epsilon_0)$, which induce the following relation:

$$w_j = \sum \widehat{a}_{i,j} \cdot \cdot \widetilde{w}_i, \qquad \widehat{a}_{i,j} := \psi^* a_{i,j} \cdot \prod_k z_k^{-n_k(v_j) + n_k(\widetilde{v}_i)}.$$

We put $T(i,j) := \underline{m} - T'(i,j)$, where $T'(i,j)$ is given as follows:

$$T'(i,j) = \bigl\{ p \in \underline{m} \,\big|\, {}^p\deg^{\mathbb{E}^{(\lambda_0)}}(\widetilde{v}_i) = {}^p\deg^{\mathbb{E}^{(\lambda_0)}}(v_j),\ {}^p\deg^{F^{(\lambda_0)}}(\widetilde{v}_i) \leq {}^p\deg^{F^{(\lambda_0)}}(v_j) \bigr\}.$$

We remark that $a_{i\,j\,|\,\mathcal{D}_p(\lambda_0, \epsilon_0)} = 0$ for any $p \in T(i,j)$. Hence there exists holomorphic functions $b_{i,j}$ on $\mathcal{X}^{(2)}(\lambda_0, \epsilon_0)$ such that the following holds:

$$a_{i,j} = b_{i,j} \cdot \prod_{p \in T(i,j)} \zeta_{i,j}.$$

Therefore we obtain the following:

$$\psi^* a_{i,j} \cdot \prod_k z_k^{-n_k(v_j) + n_k(\widetilde{v}_i)} = \psi^* b_{i,j} \cdot \prod_k z_k^{\sum_{p \in T(i,j)} c_{k\,p} - n_k(v_j) + n_k(\widetilde{v}_i)}.$$

Assume that ${}^k\deg(\widetilde{w}_i) = {}^k\deg(w_j)$ for any $k \in \underline{n}$. Then we obtain the following:

$$\sum_{p \in T(i,j)} c_{k\,p} + n_k(\widetilde{v}_i) - n_k(v_j) = \sum_{p \in \underline{m}} c_{k\,p} \cdot \Bigl(G_{T(i,j)}(p) + {}^p\deg^{F^{(\lambda_0)}}(v_j) - {}^p\deg^{F^{(\lambda_0)}}(\widetilde{v}_i) \Bigr).$$

Here $G_{T(i,j)}$ denotes the map $T(i,j) \longrightarrow \{0,1\}$ such that $G_{T(i,j)}(p) = 1$ $(p \in T(i,j))$ and $G_{T(i,j)}(p) = 0$ $(p \notin T(i,j))$. We always have $G_{T(i,j)}(p) + {}^{p}\deg^{F^{(\lambda_0)}}(v_j) - {}^{p}\deg^{F^{(\lambda_0)}}(\widetilde{v}_i) \geq 0$. It is easy to check $G_{T(i,j)}(p) + {}^{p}\deg^{F^{(\lambda_0)}}(v_j) - {}^{p}\deg^{F^{(\lambda_0)}}(\widetilde{v}_i) > 0$ in the case ${}^{p}\deg^{F^{(\lambda_0)}}(v_j) \neq {}^{p}\deg^{F^{(\lambda_0)}}(\widetilde{v}_i)$. Hence we obtain $\widehat{a}_{i,j\,|\,\mathcal{D}_{\underline{n}}(\lambda_0,\epsilon_0)} = 0$ in the case ${}^{\underline{m}}\deg^{F^{(\lambda_0)}}(v_j) \neq {}^{\underline{m}}\deg^{F^{(\lambda_0)}}(\widetilde{v}_i)$. It implies that the induced decomposition in the right hand side of (8.44) is canonical, and thus we are done. □

REMARK 8.126. The construction can be easier in the case $\lambda_0 \neq 0$. □

LEMMA 8.127. *Let $\boldsymbol{u}_1$ be an element of $\overline{\mathcal{KMS}}\big(\psi^{-1}\mathcal{E}^0, \underline{n}\big)$. We have the canonical isomorphism:*

$$\bigoplus_{\substack{\boldsymbol{u} \in \overline{\mathcal{KMS}}\left({}^{\diamond}\mathcal{E}^0, \underline{m}\right) \\ \boldsymbol{c}\cdot\boldsymbol{u} = \boldsymbol{u}_1}} {}^{\underline{m}}\mathcal{G}_{\boldsymbol{u}}(E) \simeq {}^{\underline{n}}\mathcal{G}_{\boldsymbol{u}_1}(\psi^{-1}E). \tag{8.45}$$

Proof The isomorphism (8.44) gives an isomorphism on a neighbourhood of λ_0. It can be glued, and we obtain the global isomorphism. □

COROLLARY 8.128. *Via the isomorphism (8.45), the nilpotent map induced by* $\mathrm{Res}_k(\psi^{-1}\mathbb{D})$ *on the right hand side is same as* $\sum_{p=1}^{m} c_{k\,p} \cdot \mathcal{N}_p$*, where* $\mathcal{N}_p$ *denote the nilpotent maps on the left hand side induced by* $\mathrm{Res}_p(\mathbb{D})$.

Proof It follows from (8.43). □

COROLLARY 8.129. *The correspondence $\boldsymbol{u} \longmapsto \boldsymbol{c}\cdot\boldsymbol{u}$ induces the surjective morphism $\overline{\mathcal{KMS}}(\mathcal{E}^0, \underline{l}) \longrightarrow \overline{\mathcal{KMS}}(\psi^*\mathcal{E}^0, \underline{n})$. We have $\sum_{\boldsymbol{c}\cdot\boldsymbol{u}=\boldsymbol{u}_1} \mathfrak{m}(0,\boldsymbol{u}) = \mathfrak{m}(0,\boldsymbol{u}_1)$.* □

8.10. Comparison of the norms for the family

8.10.1. Preliminary. Let X be an open subset of $\boldsymbol{C}$ with the usual Euclidean metric $dz \cdot d\bar{z}$. Let $(E, \overline{\partial}_E, \theta, h)$ be a harmonic bundle on X, and let f be a holomorphic section of $\mathcal{E}^{\lambda}$ on X. Let Δ'' denote the Laplacian of $\boldsymbol{C}$.

LEMMA 8.130 (Lemma 4.18 in [**65**]). *We have the inequality* $\Delta''|f|_h^2 \leq \big|\mathbb{D}^{\lambda} f\big|_h^2$. □

8.10.2. C^{∞}-family of tame harmonic bundles. Let Y be a C^{∞}-manifold. Let E be a C^{∞}-vector bundle on $\overline{\Delta}^* \times Y$ with $\overline{\Delta}^*$-holomorphic structure $\overline{\partial}_E$, i.e., $\overline{\partial}_E$ is a differential operator $C^{\infty}(E) \longrightarrow C^{\infty}(E \otimes \Omega^{0,1}_{\overline{\Delta}^*})$ such that $\overline{\partial}_E^2 = 0$. (See the subsection 3.1.1.) Let h be a C^{∞}-hermitian metric of E. Let θ be a relative Higgs field, i.e., θ is a $\overline{\Delta}^*$-holomorphic section of $\mathrm{End}(E) \otimes \Omega^{1,0}_{\overline{\Delta}^*}$ such that $\theta^2 = 0$.

DEFINITION 8.131. The tuple $(E, \overline{\partial}_E, \theta, h)$ is called a C^{∞}-family of tame harmonic bundle if the following holds:

- The restrictions $(E, \overline{\partial}_E, \theta, h)_{|\overline{\Delta}^* \times \{y\}}$ are tame harmonic bundles for any point $y \in Y$. □

Then we have the deformed $\overline{\Delta}^* \times \boldsymbol{C}_{\lambda}$-holomorphic bundle on $\overline{\Delta}^* \times \boldsymbol{C}_{\lambda} \times Y$:

$$\mathcal{E} = \big(p_{\lambda}^{-1}E,\ \overline{\partial}_{\lambda} + \overline{\partial}_E + \lambda \cdot \theta^{\dagger}\big).$$

We also have the family of λ-connections $\mathbb{D} = \overline{\partial}_E + \theta + \lambda(\partial_E + \theta^\dagger)$. The restrictions of $\mathcal{E}$ and $\mathbb{D}$ to $\overline{\Delta}^* \times \{(\lambda, y)\}$ are denoted by $\mathcal{E}_y^\lambda$ and $\mathbb{D}_y^\lambda$ respectively, for $x = (\lambda, y) \in \boldsymbol{C}_\lambda \times Y$. Similarly, E_y h_y and θ_y denote the restrictions to $\Delta^* \times \{y\}$. The restrictions of $\mathcal{E}$ and $\mathbb{D}$ to $\Delta^* \times \boldsymbol{C}_\lambda \times \{y\}$ are denoted by $\mathcal{E}_y$ and $\mathbb{D}_y$ respectively.

We impose the following conditions.

CONDITION 8.132. The family of the holomorphic bundles with the hermitian metrics $\{(\mathcal{E}_y^\lambda, h) \mid (\lambda, y) \in \boldsymbol{C} \times Y\}$ is uniformly acceptable, i.e., there exists a positive constant C, which is independent of $x = (\lambda, y) \in \boldsymbol{C} \times Y$ such that the curvatures $R(h, x) = R_0(h, x) \cdot dz \cdot d\bar{z}$ of $\mathcal{E}_y^\lambda$ is dominated as follows:

$$\bigl|R_0(h, x)\bigr|_h \leq C \cdot \frac{1}{|z|^2\bigl(-\log|z|\bigr)^2}.$$

□

CONDITION 8.133. The KMS-structures of $(E, \overline{\partial}_E, \theta, h)_y$ are independent of choices of $y \in Y$, i.e., $\mathcal{KMS}(\mathcal{E}_y^0) = \mathcal{KMS}(\mathcal{E}_{y'}^0)$ for any $y, y' \in Y$. Moreover the multiplicities are also same. □

For simplicity of the notation, we put $S(\lambda) := \mathcal{KMS}(\mathcal{E}_y^\lambda)$, which is independent of y. The multiplicity $\mathfrak{m}(\lambda, u)$ for $u \in S(\lambda)$ is defined similarly. We also impose the following condition.

CONDITION 8.134. Let λ_0 be any point of $\boldsymbol{C}_\lambda$, and let b be real number such that $b \notin S(\lambda_0)$. Then we can take a positive number ϵ_0, which is independent of $y \in Y$, such that the following holds:

- ${}_b\mathcal{E}_y$ is locally free on $\overline{\Delta} \times \Delta(\lambda_0, \epsilon_0) \times \{y\}$.
- On ${}_b\mathcal{E}_{y|\{O\} \times \Delta(\lambda_0, \epsilon_0) \times \{y\}}$, we have the filtration ${}^iF^{(\lambda_0)}$ and the decomposition ${}^i\mathbb{E}^{(\lambda_0)}$.

For such λ_0 and ϵ_0, we put $U(\lambda_0) := \Delta(\lambda_0, \epsilon_0)$. □

8.10.3. Uniform boundedness. Let V be an open subset of Y, and λ_0 be a point of $\boldsymbol{C}_\lambda$. Let f be a $\overline{\Delta}^*$-holomorphic section of $\mathcal{E}$ on $\overline{\Delta}^* \times U(\lambda_0) \times V$. Let $\partial\overline{\Delta}$ denote $\{z \mid |z| = 1\}$. For simplicity of the notation, $f_{|\overline{\Delta}^* \times \{x\}}$ is denoted by f_x for each $x \in U(\lambda_0) \times V$.

LEMMA 8.135. *Assume the following:*

- *For each $x \in U(\lambda_0) \times V$, we have the real number a_x such that $-\operatorname{ord}(f_x) \leq a_x$.*
- *$|f|_h$ is bounded on $\partial\overline{\Delta} \times U(\lambda_0) \times V$.*

Then there exist positive constants M and C, which is independent of $x \in U(\lambda_0) \times V$, such that $|f_x| \leq C \cdot |z|^{a_x} \cdot \bigl(-\log|z|\bigr)^M$. Here M depends only on the estimates of the curvatures $R(h, x)$ $(x \in U(\lambda_0) \times V)$, and C depends only on $\sup_{\partial\overline{\Delta} \times U(\lambda_0) \times V} |f|_h$.

Proof We put $h_M := h \cdot \bigl(-\log|z| + 1\bigr)^{-M}$ on $\Delta^* \times \boldsymbol{C} \times Y$ and $|f|_M := |f|_{h_M}$. Due to the uniform acceptability (Condition 8.132), there is a sufficiently positive number M such that $\Delta'' \log\bigl|f_x\bigr|_M \leq 0$ for each $x \in U(\lambda_0) \times V$ (see Lemma 2.49 and Corollary 2.50). Then the claim can be shown by the same argument as the proof of Corollary 2.53. □

Let f be a $\Delta^* \times U(\lambda_0)$-holomorphic section of $\mathcal{E}$ on $\overline{\Delta}^* \times U(\lambda_0) \times V$. For simplicity of the notation, let f_y denote the restriction of f to $\overline{\Delta}^* \times U(\lambda_0) \times \{y\}$ for each $y \in V$.

We impose some conditions to f, and we would like to derive the boundedness of f.

PROPOSITION 8.136. *Assume the following:*

- **(a):** $-\operatorname{ord}(f_{(\lambda,y)}) \leq 0$ *for each* $(\lambda, y) \in U(\lambda_0) \times V$, *and* $|f|_h$ *is bounded on* $\partial\overline{\Delta} \times U(\lambda_0) \times V$.
- **(b):** *Let* ϵ *be a small positive number such that* $]0,\epsilon[\cap S(\lambda_0) = \emptyset$. *We remark that* f_y $(y \in Y)$ *naturally gives the holomorphic section of* ${}_{\epsilon}\mathcal{E}_y$ *due to the first condition, and it induces the section* $[f_y]$ *of* $\operatorname{Gr}_0^{F^{(\lambda_0)}}({}_{\epsilon}\mathcal{E}_y)$. *Then we assume that* $\operatorname{Res}(\mathbb{D}_y)[f_y] = 0$ *in* $\operatorname{Gr}_0^{F^{(\lambda_0)}}({}_{\epsilon}\mathcal{E}_y)$.
- **(c):** $\mathbb{D}f$ *is bounded on* $\partial\overline{\Delta} \times U(\lambda_0) \times V$.

Then f *is bounded on* $\overline{\Delta}^* \times U(\lambda_0) \times V$.

Proof First we see the following lemma.

LEMMA 8.137. $f_{(\lambda,y)}$ *is bounded for each* $(\lambda, y) \in U(\lambda_0) \times V$.

Proof The condition (a) implies $f_{(\lambda,y)}$ is a section of ${}^{\diamond}\mathcal{E}_y^{\lambda}$, and it determines the element $[f_{(\lambda,y)}] \in \operatorname{Gr}_0^F({}^{\diamond}\mathcal{E}^{\lambda}{}_{y\,|\,O})$. The condition (b) implies implies $\operatorname{Res}(\mathbb{D}_y^{\lambda})([f_{(\lambda,y)}]) = 0$ in $\operatorname{Gr}_0^F({}^{\diamond}\mathcal{E}^{\lambda}{}_{y\,|\,O})$, which implies the degree of $[f_{(\lambda,y)}]$ with respect to the weight filtration is less than 0. Hence the claim follows from the norm estimate for a curve (Lemma 7.39). □

The $\overline{\Delta}^* \times U(\lambda_0)$-holomorphic section g is given by $\mathbb{D}f = g \cdot dz/z$.

LEMMA 8.138. *There exists a positive number* ϵ_1 *such that* $-\operatorname{ord}(g_{(\lambda,y)}) \leq -\epsilon_1$ *for any* $(\lambda, y) \in U(\lambda_0) \times V$.

Proof It follows from the condition (b). □

Let ϵ_2 be any positive number smaller than ϵ_1. By applying Lemma 8.135 to g, we obtain a positive constant C_1 such that $|g|_h \leq C_1 \cdot |z|^{\epsilon_2}$. Due to Lemma 8.130, we have the inequality $\Delta''|f_{(\lambda,y)}|_h^2 \leq \left|g_{(\lambda,y)}\right|^2 \cdot |z|^{-2} \leq C_1 \cdot |z|^{-2+2\epsilon'}$ for each x. Then the claim of Proposition 8.136 follows from Lemma 2.28. □

8.10.4. Comparison. Let $(E_i, \overline{\partial}_{E_i}, h_i, \theta_i)$ $(i = 1, 2)$ be C^{∞}-families of tame harmonic bundles on $\overline{\Delta}^* \times Y$, satisfying the conditions 8.132–8.134. We also assume that the KMS-structures of $(E_i, \overline{\partial}_{E_i}, h_i, \theta_i)$ are same.

Let V be an open subset of Y, and let $\Phi : \mathcal{E}_1 \longrightarrow \mathcal{E}_2$ be $\overline{\Delta}^* \times \boldsymbol{C}_{\lambda}$-holomorphic isomorphism defined on $\overline{\Delta}^* \times U(\lambda_0) \times V$. For simplicity of the notation, the restriction to $\overline{\Delta}^* \times U(\lambda_0) \times \{y\}$ is denoted by Φ_y for $y \in V$.

PROPOSITION 8.139. *Assume the following:*

- **(A):** *For any real number* $b \notin S(\lambda_0)$ *and for any* $y \in V$, *the morphism* Φ_y *can be prolonged to the holomorphic isomorphism* ${}_b\mathcal{E}_{1,y} \longrightarrow {}_b\mathcal{E}_{2,y}$, *which preserves the filtrations* $F^{(\lambda_0)}$ *and the decompositions* $\mathbb{E}^{(\lambda_0)}$. *Moreover* Φ *and* Φ^{-1} *are bounded on* $\partial\overline{\Delta} \times U(\lambda_0) \times V$.

(B): *On the associated graded vector bundle over* $\{O\} \times U(\lambda_0)$, *we have* $\mathrm{Gr}^{F^{(\lambda_0)}} \mathrm{Res}(\mathbb{D}_{2,y}) \circ \mathrm{Gr}^{F^{(\lambda_0)}} \Phi_{y\,|\,O} = \mathrm{Gr}^{F^{(\lambda_0)}} \Phi_{y\,|\,O} \circ \mathrm{Gr}^{F^{(\lambda_0)}} \mathrm{Res}(\mathbb{D}_{1,y})$. *Here "$|O$" means the restriction to* $\{O\} \times \Delta(\lambda_0, \epsilon_0)$.

(C): $\Phi \circ \mathbb{D}_1 - \mathbb{D}_2 \circ \Phi$ *is bounded on* $\partial\overline{\Delta} \times U(\lambda_0) \times V$, *and similar to* $\Phi^{-1} \circ \mathbb{D}_2 - \mathbb{D}_1 \circ \Phi^{-1}$.

Then Φ *and* Φ^{-1} *are bounded on* $\overline{\Delta}^* \times U(\lambda_0) \times V$.

Proof The bundles $Hom(E_1, E_2)$ and $Hom(E_2, E_1)$ with the induced structures naturally give the C^∞-family of tame harmonic bundles, and Φ and Φ^{-1} can be regarded as the sections of the deformed holomorphic bundles of them respectively.

Let f be the section of the deformed holomorphic bundle of $Hom(E_1, E_2)$ corresponding to Φ. The conditions (A), (B) and (C) for Φ imply the condition (a), (b) and (c) for f in Proposition 8.136, respectively. Hence we obtain the boundedness of f, which implies the boundedness of Φ. Similarly, we obtain the boundedness of Φ^{-1}. □

8.10.5. The norm estimate in one dimensional case. We put $X = \Delta$ and $D = \{O\}$. Let $(E, \overline{\partial}_E, \theta, h)$ be a tame harmonic bundle. Pick a point $\lambda_0 \in \boldsymbol{C}_\lambda$. Let b be a real number such that $b \notin \mathcal{KMS}(\mathcal{E}^{\lambda_0})$, and take a sufficiently small number ϵ_0 such that ${}_b\mathcal{E}$ is locally free on $\mathcal{X}(\lambda_0, \epsilon_0)$, and that $F^{(\lambda_0)}$ and $\mathbb{E}^{(\lambda_0)}$ are given.

The residue $\mathrm{Res}(\mathbb{D})$ induces the endomorphism of $\mathrm{Gr}^{F^{(\lambda_0)}}({}_b\mathcal{E})$. The nilpotent part is denoted by N, which is given on $\mathcal{D}(\lambda_0, \epsilon_0)$. Recall that the conjugacy classes are independent of $\lambda \in \mathcal{D}(\lambda_0, \epsilon_0)$ (Corollary 7.71). Therefore the weight filtration W is given in the category of vector bundles on $\mathcal{D}(\lambda_0, \epsilon_0)$.

Let $\boldsymbol{v}$ be a holomorphic frame of ${}_b\mathcal{E}$ on $\mathcal{X}(\lambda_0, \epsilon_0)$, which is compatible with the decomposition $\mathbb{E}^{(\lambda_0)}$ and the filtration $F^{(\lambda_0)}$. Moreover, we assume that the induced frame of $\mathrm{Gr}^F({}_b\mathcal{E})$ is compatible with the filtration W. For each v_i, we have the element $u(v_i) \in \mathcal{KMS}(\mathcal{E}^0)$ such that $\deg^{F^{(\lambda_0)}, \mathbb{E}^{(\lambda_0)}}(v_i) = \mathfrak{k}(\lambda_0, u(v_i))$. We remark that $\deg^F(v_{i\,|\,\mathcal{X}^\lambda}) = \mathfrak{p}(\lambda, u(v_i))$ (Lemma 8.109). We put $k(v_i) := \deg^W(v_i)$.

We put as follows:

$$v_i' := v_i \cdot |z|^{\mathfrak{p}(\lambda, u(v_i))} \cdot \big(-\log|z|\big)^{-k(v_i)/2}. \tag{8.46}$$

PROPOSITION 8.140. *The frame* $\boldsymbol{v}' = (v_i')$ *on* $(X - D) \times \Delta(\lambda_0, \epsilon_0)$ *is adapted.*

Proof For each $u \in \mathcal{KMS}({}^\diamond\mathcal{E}^0)$, we put $V_u = \mathrm{Gr}_u^{F,\mathbb{E}}({}^\diamond\mathcal{E}^0)$. The residue $\mathrm{Res}(\theta)$ induces the endomorphism whose nilpotent part is denoted by N_u. From (V_u, N_u), we obtain a model bundle $E(V_u, N_u)$ as in Lemma 6.9. Then we put as follows:

$$(E_0, \overline{\partial}_{E_0}, \theta_0, h_0) = \bigoplus E(V_u, N_u) \otimes L(u).$$

(See the section 6.1 for $L(u)$.) Then we obtain the deformed holomorphic bundle $\mathcal{E}_0$ with the λ-connection $\mathbb{D}_0$. Since the conjugacy classes of the nilpotent parts $\mathrm{Res}(\mathbb{D}^\lambda)$ is independent of λ in general, we can take an isomorphism $\Phi : {}_b\mathcal{E}_0 \longrightarrow {}_b\mathcal{E}$ on $\mathcal{X}(\lambda_0, \epsilon_0)$, such that the following holds:

- $\Phi_{|O}$ preserves the filtration $F^{(\lambda_0)}$ and the decomposition $\mathbb{E}^{(\lambda_0)}$.
- $\mathrm{Gr}^{F^{(\lambda_0)}} \mathrm{Res}(\mathbb{D}_2) \circ \mathrm{Gr}^{F^{(\lambda_0)}} \Phi_{|\mathcal{D}(\lambda_0, \epsilon_0)} = \mathrm{Gr}^{F^{(\lambda_0)}} \Phi_{\mathcal{D}(\lambda_0, \epsilon_0)} \circ \mathrm{Gr}^{F^{(\lambda_0)}} \mathrm{Res}(\mathbb{D}_1)$.

Due to Proposition 8.139, Φ and Φ^{-1} are bounded with respect to h and h_0.

Let $\boldsymbol{v}_0$ denote a canonical frame for $\mathcal{E}_0$. We may assume that $\Phi(\boldsymbol{v}_0) = \Phi(\boldsymbol{v})$. Let $\boldsymbol{v}_0'$ denote the C^∞-frame of $\mathcal{E}_0$ obtained from $\boldsymbol{v}_0$ as in (8.46). Then it is easy to check that $\boldsymbol{v}_0'$ is adapted with respect to h_0. Therefore we obtain $\boldsymbol{v}_0$ is adapted with respect to h_0. □

CHAPTER 9

The KMS-structure of the Space of the Multi-valued Flat Sections

9.1. The filtration ${}^i\mathcal{F}$

9.1.1. Preliminary. We put $X = \Delta^n$, $D_i = \{z_i = 0\}$ and $D = \bigcup_{i=1}^l D_i$. Let $(E, \overline{\partial}_E, \theta, h)$ be a tame harmonic bundle on $X - D$. We put $\mathcal{Y}^\lambda := \{\lambda\} \times Y$ for a complex variety Y.

Let $H(\mathcal{E}^\lambda)$ denote the space of the multi-valued flat sections of the flat bundle $\big(\mathcal{E}^\lambda, \mathbb{D}^{\lambda,f}\big)$. Let ${}^iM^\lambda$ denote the monodromy of $\mathcal{E}^\lambda$ with respect to the loop around the divisor $\mathcal{D}_i^\lambda$ with the anti-clockwise direction. We obtain the tuple of endomorphisms $\boldsymbol{M} = ({}^1M, \ldots, {}^lM)$. Then we obtain the generalized eigen decomposition (the subsection 2.7.2):

$$H(\mathcal{E}^\lambda) = \bigoplus_{\boldsymbol{\omega} \in \mathcal{S}p(\boldsymbol{M})} \mathbb{E}\big(H(\mathcal{E}^\lambda), \boldsymbol{\omega}\big).$$

We denote the restriction of $\boldsymbol{M}$ to $\mathbb{E}(H(\mathcal{E}^\lambda), \boldsymbol{\omega})$ by $\boldsymbol{M}_{\boldsymbol{\omega}}$. We obtain the tuple of endomorphisms $\boldsymbol{N}_{\boldsymbol{\omega}} = ({}^1N_{\boldsymbol{\omega}}, \ldots, {}^lN_{\boldsymbol{\omega}})$ given as follows:

$${}^iN_{\boldsymbol{\omega}} := \frac{-1}{2\pi\sqrt{-1}} \log {}^iM_{\boldsymbol{\omega}}^u.$$

Here ${}^iM_{\boldsymbol{\omega}}^u$ denotes the unipotent part of ${}^iM_{\boldsymbol{\omega}}$.

Let $\boldsymbol{b} = (b_1, \ldots, b_l)$ be an element of $\boldsymbol{R}^l$ and $\boldsymbol{\omega} = (\omega_1, \ldots, \omega_l)$ be an element of $\boldsymbol{C}^l$. We put as follows:

$$\boldsymbol{\alpha}(\boldsymbol{b}, \boldsymbol{\omega}) := \big(\alpha(b_1, \omega_1), \ldots, \alpha(b_l, \omega_l)\big).$$

Here $\alpha(b, \omega)$ for $(b, \omega) \in \boldsymbol{R} \times \boldsymbol{C}$ is given in the page 169.

Let s be an element of $H(\mathcal{E}^\lambda)$. If s is contained in $\mathbb{E}\big(H(\mathcal{E}^\lambda), \boldsymbol{\omega}\big)$, we put $\underline{l}\text{-}\deg^{\mathbb{E}}(s) = \boldsymbol{\omega}$. The i-th component is denoted by ${}^i\deg^{\mathbb{E}}(s)$.

9.1.2. The increasing order and the filtration ${}^i\mathcal{F}$. Let s be an element of $\mathbb{E}(H(\mathcal{E}^\lambda), \boldsymbol{\omega})$. Let P be an element of D_i°. Then we put ${}^i\operatorname{ord}(s) := \operatorname{ord}(s_{|\pi_i^{-1}(P)})$.

LEMMA 9.1. *The number ${}^i\operatorname{ord}(s)$ is independent of a choice of P.*

Proof Let P and P' be two points of D_i°. Let γ be a path in D_i° connecting P and P'. Since s is holomorphic with respect to $\overline{\partial}_{\mathcal{E}^\lambda}$, we have the equality $d(s,s)_h = (\partial_{\mathcal{E}^\lambda} s, s)_h + (s, \partial_{\mathcal{E}^\lambda} s) = 2\operatorname{Re}\big((\bar{\lambda} + \lambda^{-1}) \cdot (\theta s, s)\big)$. Hence we obtain the following equality:

$$d \log |s|_h^2 = -2\operatorname{Re}\Big((\bar{\lambda} + \lambda^{-1}) \cdot \frac{(\theta s, s)_h}{|s|_h^2}\Big).$$

Let q_i be the projection of $X = \Delta^n$ onto the i-th component. Let Q be any point of $\Delta - \{O\}$, and then $q_i^{-1}(Q)$ is a hyperplane of X. Due to the estimate of θ (Lemma 8.13, for example), there exists a positive constant C which is independent of Q, satisfying the following inequality on $q_i^{-1}(Q)$:

$$\left|d\log|s|^2_{h\,|\,q_i^{-1}(Q)}\right| \leq C\cdot\Big(\sum_{j\neq i}\frac{1}{|z_j|}\Big).$$

Thus we obtain the following inequality on the path $\gamma_Q := q_i^{-1}(Q)\cap\pi_i^{-1}(\gamma)$ for some constant C which is independent of Q:

$$\left|d\log|s_{|\gamma_Q}|^2_h\right| \leq C.$$

Hence we obtain the following estimate, which is independent of Q:

$$\left|\log\big|s_{|\pi_i^{-1}(P)}(Q)\big|_h - \log\big|s_{|\pi_i^{-1}(P')}(Q)\big|_h\right| < C'.$$

It implies our claim. □

We put as follows:

$${}^{i}\mathcal{F}_a\mathbb{E}(H(\mathcal{E}^\lambda),\boldsymbol{\omega}) = \big\{s\in\mathbb{E}(H(\mathcal{E}^\lambda),\boldsymbol{\omega})\,\big|\, -{}^{i}\mathrm{ord}(s)\leq a\big\}.$$

LEMMA 9.2. *The monodromies preserve the filtration ${}^{i}\mathcal{F}_a$.*

Proof It immediately follows from the definition of the filtration ${}^{i}\mathcal{F}$ and Lemma 9.1. □

Let ${}^{i}\mathrm{Gr}^{\mathcal{F}}$ denote the associated graded vector space to the filtration ${}^{i}\mathcal{F}$. We put as follows:

$$\mathcal{P}ar^{f}(\mathcal{E}^\lambda,i) := \big\{a\in\boldsymbol{R}\,\big|\,{}^{i}\mathrm{Gr}^{\mathcal{F}}_a\, H(\mathcal{E}^\lambda)\neq 0\big\}.$$

Let s be an element of $\mathbb{E}(H(\mathcal{E}^\lambda),\boldsymbol{\omega})$, and $\boldsymbol{b}$ be an element of $\boldsymbol{R}^l$. We put as follows:

$$F(s,\boldsymbol{b}) := \exp\Big(\sum_{i=1}^{l}\log z_i\cdot\big(\alpha(b_i,\omega_i)+{}^{i}N_{\boldsymbol{\omega}}\big)\Big)\cdot s. \tag{9.1}$$

Then $F(s,\boldsymbol{b})$ is a holomorphic section of $\mathcal{E}^\lambda$ over $X-D$. We put as follows:

$$c_i := -{}^{i}\,\mathrm{ord}(s) - \mathrm{Re}(\alpha(b_i,\omega_i)).$$

We put $\boldsymbol{c} := (c_i)$.

LEMMA 9.3. *$F(s,\boldsymbol{b})$ is a holomorphic section of ${}_{\boldsymbol{c}}\mathcal{E}^\lambda$.*

Proof Due to the result in the case of curves (Lemma 7.78), we obtain the following for any point $P\in D_i^\circ$:

$$-\,\mathrm{ord}\big(F(s,\boldsymbol{b})_{|\pi_i^{-1}(P)}\big)\leq c_i.$$

Then we obtain the result due to Corollary 2.53. □

Let s be an element of $H\big(\mathcal{E}^\lambda\big)$. The degree of s with respect to the filtration ${}^{i}\mathcal{F}$ is denoted by ${}^{i}\deg^{\mathcal{F}}(s)$. The tuple $\big({}^{1}\deg^{\mathcal{F}}(s),\ldots,{}^{l}\deg^{\mathcal{F}}(s)\big)$ is denoted by ${}^{\underline{l}}\deg^{\mathcal{F}}(s)$.

COROLLARY 9.4. *$F\big(s,{}^{\underline{l}}\deg^{\mathcal{F}}(s)\big)$ is a holomorphic section of ${}^{\diamond}\mathcal{E}^\lambda$.* □

9.1.3. Functoriality of ${}^i\mathcal{F}$. For a positive integer c, we have the morphism $\psi_{c\cdot\boldsymbol{\delta}_i} : X \longrightarrow X$ given by $(z_1,\ldots,z_n) \longmapsto (z_1,\ldots,z_{i-1},z_i^c,z_{i+1},\ldots,z_n)$. We have the natural isomorphism $\psi^*_{c\cdot\boldsymbol{\delta}_i} : H(\mathcal{E}^\lambda) \simeq H(\psi^{-1}_{c\cdot\boldsymbol{\delta}_i}\mathcal{E}^\lambda)$ via the pull back.

LEMMA 9.5. *Under the isomorphism, we have the following:*

$$\psi^*_{c\cdot\boldsymbol{\delta}}\Big({}^i\mathcal{F}_a H(\mathcal{E}^\lambda)\Big) = {}^i\mathcal{F}_{c\cdot a} H\big(\psi^{-1}_{c\cdot\boldsymbol{\delta}}\mathcal{E}^\lambda\big).$$

Proof It can be reduced to the case of curves (Lemma 7.106). □

We have the natural isomorphism $H(\mathcal{E}^{(1)\,\lambda}\otimes\mathcal{E}^{(2)\,\lambda}) \simeq H(\mathcal{E}^{(1)\,\lambda})\otimes H(\mathcal{E}^{(2)\,\lambda})$. We have the following isomorphism:

$$\mathbb{E}\big(H(\mathcal{E}^{(1)\,\lambda}\otimes\mathcal{E}^{(2)\,\lambda}),\boldsymbol{\omega}\big) \simeq \bigoplus_{\boldsymbol{\omega}_1\cdot\boldsymbol{\omega}_2=\boldsymbol{\omega}} \mathbb{E}\big(H(\mathcal{E}^{(1)\lambda}),\boldsymbol{\omega}_1\big)\otimes\mathbb{E}\big(H(\mathcal{E}^{(2)\,\lambda}),\boldsymbol{\omega}_2\big).$$

We have the two filtrations on $\mathbb{E}\big(H(\mathcal{E}^{(1)\,\lambda}\otimes\mathcal{E}^{(2)\,\lambda}),\boldsymbol{\omega}\big)$. One is ${}^i\mathcal{F}$ for $\mathbb{E}\big(H(\mathcal{E}^{(1)\,\lambda}\otimes\mathcal{E}^{(2)\,\lambda}),\boldsymbol{\omega}\big)$. The other is induced by ${}^i\mathcal{F}$ for $\mathbb{E}\big(H(\mathcal{E}^{(b)\lambda}),\boldsymbol{\omega}\big)$ for $b=1,2$.

LEMMA 9.6. *They are same. Namely the following holds:*

$$(9.2)\quad {}^i\mathcal{F}_a\Big(\mathbb{E}\big(H(\mathcal{E}^{(1)\,\lambda}\otimes\mathcal{E}^{(2)\,\lambda},\boldsymbol{\omega}\big)\Big) \simeq \bigoplus_{\boldsymbol{\omega}_1\cdot\boldsymbol{\omega}_2=\boldsymbol{\omega}}\sum_{a_1+a_2\leq a} {}^i\mathcal{F}_{a_1}\Big(\mathbb{E}\big(H(\mathcal{E}^{(b)\lambda}),\boldsymbol{\omega}_1\big)\Big)\otimes{}^i\mathcal{F}_{a_2}\Big(\mathbb{E}\big(H(\mathcal{E}^{(b)\lambda}),\boldsymbol{\omega}_2\big)\Big).$$

Proof Due to the result in the case of curves, we obtain the following:

$$(9.3)\quad {}^i\mathcal{F}_a\Big(\bigoplus_{q_i(\boldsymbol{\omega})=\omega}\mathbb{E}\big(H(\mathcal{E}^{(1)\,\lambda}\otimes\mathcal{E}^{(2)\,\lambda}),\boldsymbol{\omega}\big)\Big) \simeq \bigoplus_{\omega_1\cdot\omega_2=\omega}\sum_{a_1+a_2\leq a}{}^i\mathcal{F}_{a_1}\Big(\bigoplus_{q_i(\boldsymbol{\omega}_1)=\omega_1}\mathbb{E}\big(H(\mathcal{E}^{(1)\,\lambda}),\boldsymbol{\omega}_1\big)\Big)\otimes{}^i\mathcal{F}_{a_2}\Big(\bigoplus_{q_i(\boldsymbol{\omega}_2)=\omega_2}\mathbb{E}\big(H(\mathcal{E}^{(2)\,\lambda}),\boldsymbol{\omega}_2\big)\Big).$$

Since the monodromy endomorphisms M_j^λ $(j\neq i)$ preserve the filtration ${}^i\mathcal{F}$, the filtration is compatible with the generalized eigen decomposition. Therefore we are done. □

We have the following isomorphism:

$$(9.4)\qquad H\Big(\bigotimes^R\mathcal{E}^\lambda\Big)\simeq\bigotimes^R H(\mathcal{E}^\lambda).$$

We have the following isomorphism:

$$(9.5)\qquad \mathbb{E}\Big(H\Big(\bigotimes^R\mathcal{E}^\lambda\Big),\boldsymbol{\omega}\Big)\simeq\bigoplus_{f\in\mathcal{S}(\boldsymbol{\omega},R)}\bigotimes_{\boldsymbol{\omega}'\in\mathcal{S}p(\boldsymbol{M}^\lambda)}\bigotimes^{f(\boldsymbol{\omega}')}\mathbb{E}\big(H(\mathcal{E}^\lambda),\boldsymbol{\omega}'\big).$$

Here we put as follows:

$$\mathcal{S}(\boldsymbol{\omega},R):=\Big\{f:\mathcal{S}p(\boldsymbol{M}^\lambda)\longrightarrow\mathbb{Z}_{\geq 0}\,\Big|\,\sum f(\boldsymbol{\omega}')=R,\quad\prod\boldsymbol{\omega}'^{\,f(\boldsymbol{\omega}')}=\boldsymbol{\omega}\Big\}.$$

We naturally have the filtrations on the both sides of (9.4) and (9.5).

COROLLARY 9.7. *The filtrations of the both sides* (9.4) *and* (9.5) *are preserved by the isomorphisms.* □

We have the isomorphism $H(\mathcal{E}^{\vee\lambda}) \simeq H(\mathcal{E}^{\lambda})^{\vee}$ preserving the $\mathbb{E}$-decomposition.

$$\mathbb{E}\big(H(\mathcal{E}^{\vee\lambda}),\boldsymbol{\omega}\big) \simeq \mathbb{E}\big(H(\mathcal{E}^{\lambda})^{\vee},\boldsymbol{\omega}\big). \tag{9.6}$$

On the left hand side of (9.6), we have the filtration ${}^{i}\mathcal{F}$. On the right hand side of (9.6), we have the induced filtration ${}^{i}\mathcal{F}^{\vee}$ by ${}^{i}\mathcal{F}$, given as follows:

$${}^{i}\mathcal{F}^{\vee}_{a}H(\mathcal{E}^{\lambda})^{\vee} := \big\{f \in H(\mathcal{E}^{\lambda})^{\vee} \,\big|\, f({}^{i}\mathcal{F}_{b}H(\mathcal{E}^{\lambda})) \subset {}^{i}\mathcal{F}_{a+b}H(\mathcal{E}^{\lambda}),\ \forall b \in \boldsymbol{R}\big\}.$$

LEMMA 9.8. *The isomorphism preserves the filtrations.*

Proof It can be reduced to the case of curves (Lemma 7.105) by an argument similar to the proof of Lemma 9.6. □

9.1.4. The case λ is generic. Assume that λ is generic. Recall that $\mathfrak{e}^{f}(\lambda) : \overline{\mathcal{KMS}}(\mathcal{E}^{0},i) \longrightarrow \mathcal{S}p^{f}(\mathcal{E}^{\lambda},i)$ is isomorphic for any i, by definition.

LEMMA 9.9. *The filtration ${}^{i}\mathcal{F}$ on $\mathbb{E}\big(H(\mathcal{E}^{\lambda}),\boldsymbol{\omega}\big)$ is trivial in the following sense:*

> *For each i, we have the unique element $u_i \in \overline{\mathcal{KMS}}({}^{\diamond}\mathcal{E}^{0},i)$ such that $\omega_i = \mathfrak{e}^{f}(\lambda,u_i)$. Then ${}^{i}\operatorname{Gr}^{\mathcal{F}}_{b}\mathbb{E}(H(\mathcal{E}^{\lambda}),\boldsymbol{\omega}) \neq 0$ if and only if $b = \mathfrak{p}^{f}(\lambda,u_i)$.*

Proof It follows from the result in the case of curves (the subsection 7.5.2). □

COROLLARY 9.10. *If λ is generic, the tuple $({}^{i}\mathcal{F}\,|\,i = 1,\ldots,l)$ is compatible.*

Proof It immediately follows Lemma 9.9 . □

Let $\lambda \in \boldsymbol{C}^{*}_{\lambda}$ be generic, and let $\boldsymbol{\omega} = (\omega_1,\ldots,\omega_l)$ be an element of $\mathcal{S}p(\boldsymbol{M}^{\lambda})$. We have the unique element $u_i \in \mathcal{K}(\mathcal{E},\lambda,0,i) = \big\{u \in \mathcal{KMS}(\mathcal{E}^{0},i)\,\big|\, -1 < \mathfrak{p}(\lambda,u) \leq 0\big\}$ such that $\mathfrak{e}^{f}(\lambda,u_i) = \omega_i$.

LEMMA 9.11. *Assume that λ is generic. Let $\boldsymbol{\beta}$ be an element of $\boldsymbol{C}^{l}$ whose i-th component is $\beta_i = \mathfrak{e}(\lambda,u_i)$. Recall that we have the space $\mathbb{E}({}^{\diamond}\mathcal{E}^{\lambda}{}_{|O},\boldsymbol{\beta})$, which is a generalized eigenspace of the tuple of the residues $\operatorname{Res}_i(\mathbb{D}^{\lambda})$ $(i = 1,\ldots,l)$. Then we have the isomorphism:*

$$\mathbb{E}({}^{\diamond}\mathcal{E}^{\lambda}{}_{|O},\boldsymbol{\beta}) \simeq \mathbb{E}(H(\mathcal{E}^{\lambda}),\boldsymbol{\omega}).$$

Proof Let $\boldsymbol{s}$ be base of $H(\mathcal{E}^{\lambda})$, which is compatible with $\mathbb{E}$. We put $\boldsymbol{\omega}(s_j) := {}^{\underline{l}}\deg^{\mathbb{E}}(s_j)$ and $\boldsymbol{b}(s_j) := {}^{\underline{l}}\deg^{\mathcal{F}}(s_j)$.

We put $v_j := F(s_j,\boldsymbol{b}(s_j))$. Then we obtain the tuple $\boldsymbol{v} = (v_i)$ of sections of ${}^{\diamond}\mathcal{E}^{\lambda}$. Due to the result in the case of curves, $\boldsymbol{v}_{|\pi_i^{-1}(P)}$ is a frame of ${}^{\diamond}\mathcal{E}^{\lambda}_{|\pi_i^{-1}(P)}$ for any $i \in \underline{l}$ and $P \in D_i^{\circ}$. Thus $\boldsymbol{v}$ is a frame of ${}^{\diamond}\mathcal{E}^{\lambda}$. The frames $\boldsymbol{v}$ and $\boldsymbol{s}$ induce the isomorphism desired. □

COROLLARY 9.12. *Assume that λ is generic. For any $\boldsymbol{\omega} \in \mathcal{S}p(\boldsymbol{M}^{\lambda})$, we have the unique element $\boldsymbol{u} \in \overline{\mathcal{KMS}}(\mathcal{E}^{0},\underline{l})$ such that $\mathfrak{e}^{f}(\lambda,\boldsymbol{u}) = \boldsymbol{\omega}$.* □

COROLLARY 9.13. *Assume that λ is generic. Assume $\boldsymbol{\omega} = \mathfrak{e}^{f}(\lambda,\boldsymbol{u})$ and $\boldsymbol{b} = \mathfrak{p}^{f}(\lambda,\boldsymbol{u})$. We have the following equalities:*

$$\dim {}^{\underline{l}}\mathcal{F}_{\boldsymbol{b}}\Big(\mathbb{E}(H(\mathcal{E}^{\lambda}),\boldsymbol{\omega})\Big) = \begin{cases} \mathfrak{m}(0,\boldsymbol{u}), & (\boldsymbol{b} \geq \mathfrak{p}^{f}(\lambda,\boldsymbol{u})),\\[2ex] 0, & (\text{otherwise}). \end{cases}$$

$$\dim {}^{\underline{l}}\mathrm{Gr}^{\mathcal{F}}_{\boldsymbol{b}}\Bigl(\mathbb{E}(H(\mathcal{E}^{\lambda}),\boldsymbol{\omega})\Bigr)=\begin{cases}\mathfrak{m}(0,\boldsymbol{u}), & (\boldsymbol{b}=\mathfrak{p}^f(\lambda,\boldsymbol{u})),\\ 0, & (\text{otherwise}).\end{cases}$$

□

9.1.5. The family of the space of the multi-valued flat sections. Let $\mathcal{H}=\mathcal{H}(E)$ be the holomorphic vector bundle over $\boldsymbol{C}^*_{\lambda}$, obtained by $\mathcal{H}_{|\lambda}=H(\mathcal{E}^{\lambda})$. We have the monodromy endomorphisms $\boldsymbol{M}=(M_i\,|\,i=1,\ldots,l)$, where M_i denotes the monodromy with respect to the loop around D_i.

Let λ_0 be an element of $\boldsymbol{C}^*_{\lambda}$. Let ϵ_0 and ϵ_1 be sufficiently small numbers. Then we obtain the following decompositions, (see (2.21) for the notation $\mathbb{E}_{\epsilon_1}$):

$$\mathcal{H}_{|\Delta(\lambda_0,\epsilon_0)}=\bigoplus_{\boldsymbol{\omega}\in\mathcal{S}p(\boldsymbol{M}^{\lambda_0})}\mathcal{H}^{(\lambda_0)}_{\boldsymbol{\omega}},\qquad \mathcal{H}^{(\lambda_0)}_{\boldsymbol{\omega}}:={}^{\underline{l}}\mathbb{E}_{\epsilon_1}\bigl(H(\mathcal{E}^{\lambda}),\boldsymbol{\omega}\bigr).$$

We put $\mathcal{S}(\boldsymbol{\omega}):=\bigl\{\boldsymbol{u}\in\overline{\mathcal{KMS}}(\mathcal{E}^0,\underline{l})\,\big|\,\mathfrak{e}^f(\lambda_0,\boldsymbol{u})=\boldsymbol{\omega}\bigr\}$. We may assume that any $\lambda\in\Delta^*(\lambda_0,\epsilon_0)$ is generic. Then we have the following decomposition on $\Delta^*(\lambda_0,\epsilon_0)$:

$$\mathcal{H}^{(\lambda_0)}_{\boldsymbol{\omega}\,|\,\Delta^*(\lambda_0,\epsilon_0)}=\bigoplus_{\boldsymbol{u}\in\mathcal{S}(\boldsymbol{\omega})}\mathcal{H}_{\mathfrak{e}^f(\lambda,\boldsymbol{u})},\qquad \mathcal{H}_{\mathfrak{e}^f(\lambda,\boldsymbol{u})\,|\,\lambda}=\mathbb{E}(H(\mathcal{E}^{\lambda}),\mathfrak{e}^f(\lambda,\boldsymbol{u})).$$

As in the case of the curves, we consider the filtration ${}^i\mathcal{F}^{(\lambda_0)}$. We put as follows:

$$(9.7)\qquad {}^i\mathcal{F}^{(\lambda_0)}_d\mathcal{H}^{(\lambda_0)}_{\boldsymbol{\omega}\,|\,\Delta^*(\lambda_0,\epsilon_0)}=\bigoplus_{\substack{\boldsymbol{u}\in\mathcal{S}(\boldsymbol{\omega})\\ \mathfrak{p}^f(\lambda,q_i(\boldsymbol{u}))\le d}}\mathcal{H}_{\mathfrak{e}^f(\lambda,\boldsymbol{u})}.$$

Since it is given as the sum of the generalized eigenspaces, the filtration ${}^i\mathcal{F}^{(\lambda_0)}$ can be prolonged to the filtration of $\mathcal{H}^{(\lambda_0)}_{\omega}$, which we denote by ${}^i\mathcal{F}^{(\lambda_0)}$.

Note the following isomorphism for any $P\in D_i^{\circ}$ and for any $\lambda\in\Delta(\lambda_0,\epsilon_0)$:

$$H(\mathcal{E}^{\lambda})\simeq H(\mathcal{E}^{\lambda}_{|\pi_i^{-1}(P)}).$$

From the right hand side, we have the decomposition $\bigl\{\mathcal{E}^{(\lambda_0)}_{\omega}\,\big|\,\omega\in\mathcal{S}p^f(\mathcal{E}^{\lambda_0}_{|\pi_i^{-1}(P)})\bigr\}$ and the filtration $\mathcal{F}^{(\lambda_0)}$ of each $\mathcal{H}^{(\lambda_0)}_{\omega}$.

LEMMA 9.14.

- *Under the isomorphism above, we have* $\mathcal{H}^{(\lambda_0)}_{\omega}=\bigoplus_{q_i(\boldsymbol{\omega})=\omega}\mathcal{H}^{(\lambda_0)}_{\boldsymbol{\omega}}$.
- $\mathcal{F}^{(\lambda_0)}_d\mathcal{H}_{\omega}=\bigoplus_{q_i(\boldsymbol{\omega})=\omega}{}^i\mathcal{F}^{(\lambda_0)}_d\mathcal{H}^{(\lambda_0)}_{\boldsymbol{\omega}}$.
- *In particular, we have* $\bigl({}^i\mathcal{F}^{(\lambda_0)}_d\mathcal{H}^{(\lambda_0)}_{\boldsymbol{\omega}}\bigr)_{|\lambda_0}={}^i\mathcal{F}_d\bigl(\mathbb{E}\bigl(H(\mathcal{E}^{\lambda_0}),\boldsymbol{\omega}\bigr)\bigr)$.

Proof The first claim is clear. Since the monodromy endomorphisms preserve the filtration ${}^i\mathcal{F}$ (Lemma 9.2), we obtain the second claim. Then the corollary immediately follows from Proposition 7.119. □

9.2. The compatibility of the filtrations ${}^i\mathcal{F}$

9.2.1. Statement. We continue to use the setting in the section 9.1. The following theorem will be proved in the subsections 9.2.2–9.2.8.

THEOREM 9.15. *The tuple of the filtrations and the decompositions* $\bigl({}^i\mathcal{F},{}^i\mathbb{E}\,\big|\,i\in\underline{l}\bigr)$ *is compatible in the sense of Definition* 4.2.

We have already known that the filtration ${}^{i}\mathcal{F}$ is compatible with the generalized eigen decompositions ${}^{j}\mathbb{E}$ ($j \in \underline{l}$), we have only to check that the compatibility of the tuple of the filtrations $\bigl({}^{i}\mathcal{F} \bigm| i \in \underline{l}\bigr)$. We may and will assume that $l = n$ in the proof.

Before the proof, we give an immediate corollary.

COROLLARY 9.16.

- *The tuple of the filtrations and the decompositions* $\bigl({}^{i}\mathcal{F}^{(\lambda_0)}, {}^{j}\mathbb{E}^{(\lambda_0)} \bigm| i, j \in \underline{l}\bigr)$ *is compatible.*
- *We have the following decomposition on* $\Delta^*(\lambda_0, \epsilon_0)$*:*

$$^{\underline{n}}\mathcal{F}_{\boldsymbol{b}}^{(\lambda_0)}\mathcal{H}^{(\lambda_0)}_{\boldsymbol{\omega}\,|\,\Delta^*(\lambda_0,\epsilon_0)} = \bigoplus_{\boldsymbol{u}\in S(\boldsymbol{\omega},\boldsymbol{b})} \mathcal{H}_{\mathfrak{e}^f(\lambda,\boldsymbol{u})}.$$

Here we put $S(\boldsymbol{\omega}, \boldsymbol{b}) := \bigl\{\boldsymbol{u} \in \overline{\mathcal{KMS}}(\mathcal{E}^0, \underline{n}) \bigm| \mathfrak{e}^f(\lambda_0, \boldsymbol{u}) = \boldsymbol{\omega},\ \mathfrak{p}^f(\lambda_0, \boldsymbol{u}) \leq \boldsymbol{b}\bigr\}$.

Proof The compatibility of the tuple $\bigl({}^{i}\mathcal{F}^{(\lambda_0)}_{|\Delta^*(\lambda_0,\epsilon_0)} \bigm| i \in \underline{l}\bigr)$ and the second claim is obvious from the definition of the filtrations ${}^{i}\mathcal{F}^{(\lambda_0)}$. Hence the corollary immediately follows from Theorem 9.15 and the third claim of Lemma 9.14. □

We will also give another corollary on the weak norm estimate in the subsection 9.2.9.

9.2.2. The dimension and the virtual dimension. For any λ, pick $\boldsymbol{\omega} \in \mathcal{S}p(\boldsymbol{M}^{\lambda})$ and $\boldsymbol{a} \in \boldsymbol{R}^n$. We put as follows:

$$d(\lambda, \boldsymbol{\omega}, \boldsymbol{a}) := \dim {}^{\underline{n}}\mathcal{F}_{\boldsymbol{a}}\mathbb{E}(H(\mathcal{E}^{\lambda}), \boldsymbol{\omega}), \qquad v.d(\lambda, \boldsymbol{\omega}, \boldsymbol{a}) := \sum_{\substack{\boldsymbol{u}\in\mathcal{S}(\boldsymbol{\omega})\\ \mathfrak{p}^f(\lambda,\boldsymbol{u})\leq\boldsymbol{a}}} \mathfrak{m}(0, \boldsymbol{u}).$$

LEMMA 9.17. *We have the inequality* $d(\lambda_0, \boldsymbol{\omega}, \boldsymbol{a}) \geq v.d(\lambda_0, \boldsymbol{\omega}, \boldsymbol{a})$ *for any* $\lambda_0 \in \boldsymbol{C}$.

Proof Let us pick a sufficiently small positive number ϵ. Then we have the following inequality, for any $\lambda \in \Delta(\lambda_0, \epsilon_0)$.

$$d(\lambda_0, \boldsymbol{\omega}, \boldsymbol{a}) = \dim\Bigl(\bigcap_{i=1}^{n} {}^{i}\mathcal{F}_{a_i}^{(\lambda_0)}\mathcal{H}^{(\lambda_0)}_{\boldsymbol{\omega}\,|\,\lambda_0}\Bigr) \geq \dim\Bigl(\bigcap_{i=1}^{n} {}^{i}\mathcal{F}_{a_i}^{(\lambda_0)}\mathcal{H}^{(\lambda_0)}_{\boldsymbol{\omega}\,|\,\lambda}\Bigr). \tag{9.8}$$

For any generic $\lambda \in \Delta^*(\lambda_0, \epsilon_0)$, we have the following equality:

$${}^{i}\mathcal{F}_{a_i}^{(\lambda_0)}\mathcal{H}^{(\lambda_0)}_{\boldsymbol{\omega}\,|\,\lambda} = \bigoplus_{\substack{\boldsymbol{u}\in\mathcal{S}(\boldsymbol{\omega})\\ \mathfrak{p}^f(\lambda_0,u_i)\leq a_i}} \mathcal{H}^{(\lambda_0)}_{\mathfrak{e}^f(\lambda,\boldsymbol{u})\,|\,\lambda}.$$

Thus we obtain the following:

$$\bigcap_{i=1}^{n} {}^{i}\mathcal{F}_{a_i}^{(\lambda_0)}\mathcal{H}^{(\lambda_0)}_{\boldsymbol{\omega}\,|\,\lambda} = \bigoplus_{\substack{\boldsymbol{u}\in\mathcal{S}(\boldsymbol{\omega})\\ \mathfrak{p}^f(\lambda_0,\boldsymbol{u})\leq\boldsymbol{a}}} \mathcal{H}^{(\lambda_0)}_{\mathfrak{e}^f(\lambda,\boldsymbol{u})\,|\,\lambda}.$$

Therefore we obtain the following equality, due to Corollary 9.13:
(9.9)
$$\dim \bigcap_{i=1}^{n} {}^{i}\mathcal{F}^{(\lambda_0)}_{a_i} \mathcal{H}^{(\lambda_0)}_{\omega\,|\,\lambda} = \sum_{\substack{\boldsymbol{u}\in\mathcal{S}(\boldsymbol{\omega}) \\ \mathfrak{p}^f(\lambda_0,\boldsymbol{u})\leq \boldsymbol{a}}} \dim \mathcal{H}_{\mathfrak{e}^f(\lambda,\boldsymbol{u})\,|\,\lambda} = \sum_{\substack{\boldsymbol{u}\in\mathcal{S}(\boldsymbol{\omega}) \\ \mathfrak{p}^f(\lambda_0,\boldsymbol{u})\leq \boldsymbol{a}}} \mathfrak{m}(0,\boldsymbol{u}) = v.d(\lambda_0,\boldsymbol{\omega},\boldsymbol{a}).$$

The claim follows from (9.8) and (9.9). □

Later we will prove $d(\lambda,\boldsymbol{\omega},\boldsymbol{a}) = v.d(\lambda,\boldsymbol{\omega},\boldsymbol{a})$ (Proposition 9.21).

9.2.3. The filtration and the diagonal embedding. We put $X^{(1)} := \{(z_1,\ldots,z_n)\,|\,z_{n-1} = z_n\} \subset X$. We put $D_i^{(1)} := D_i \cap X^{(1)}$ for $i \leq n-1$. We have the natural isomorphism $X^{(1)} \simeq \Delta^{n-1} = \{(z_1,\ldots,z_{n-1}) \in \Delta^{n-1}\}$. We have the natural isomorphism $H(\mathcal{E}^\lambda) \longrightarrow H(\mathcal{E}^\lambda_{|X^{(1)}})$.

We have the tuple of the monodromy endomorphisms $\boldsymbol{M} := (M_1,\ldots,M_n)$ and $\boldsymbol{M}^{(1)} := (M_1^{(1)},\ldots,M_{n-1}^{(1)})$. Here we have the following:

$$M_i^{(1)} = \begin{cases} M_i, & (i \leq n-2), \\ M_{n-1}\circ M_n, & (i = n-1). \end{cases}$$

The tuple of eigenvalues $\phi(\boldsymbol{\omega}) \in \mathcal{S}p(\boldsymbol{M}^{(1)})$ is given as follows, for any $\boldsymbol{\omega} \in \mathcal{S}p(\boldsymbol{M})$:

$$q_i(\phi(\boldsymbol{\omega})) = \begin{cases} \omega_i, & (i \leq n-2), \\ \omega_{n-1}\cdot\omega_n, & (i = n-1). \end{cases}$$

The map $\phi : \boldsymbol{R}^n \longrightarrow \boldsymbol{R}^{n-1}$ is defined as follows:

$$q_i(\phi(\boldsymbol{a})) = \begin{cases} a_i, & (i \leq n-2), \\ a_{n-1}+a_n, & (i = n-1). \end{cases}$$

We have the filtrations ${}^{i}\mathcal{F}$ $(i = 1,\ldots,n)$ on $\mathbb{E}(H(\mathcal{E}^\lambda),\boldsymbol{\omega})$. For any $\boldsymbol{a} \in \boldsymbol{R}^n$, we have the subspace ${}^{\underline{n}}\mathcal{F}_{\boldsymbol{a}} := \bigcap_{i=1}^{n} {}^{i}\mathcal{F}_{a_i}$ of $H(\mathcal{E}^\lambda)$. We put as follows:

$$(9.10) \qquad {}^{\underline{n}}\mathcal{F}'_{\boldsymbol{a}} = \sum_{\boldsymbol{b}\lneq\boldsymbol{a}} {}^{\underline{n}}\mathcal{F}_{\boldsymbol{b}}.$$

Here $\boldsymbol{b} \lneq \boldsymbol{a}$ means $\boldsymbol{b} \leq \boldsymbol{a}$ and $\boldsymbol{b} \neq \boldsymbol{a}$.

We also have the filtrations on $\mathbb{E}\big(H(\mathcal{E}^\lambda_{|X^{(1)}}),\boldsymbol{\omega}^{(1)}\big)$, which we denote by ${}^{i}\mathcal{F}^{(1)}$ $(i = 1,\ldots,n-1)$. Similarly, we have ${}^{\underline{n-1}}\mathcal{F}^{(1)}_{\boldsymbol{a}^{(1)}}$ and ${}^{\underline{n-1}}\mathcal{F}'^{(1)}_{\boldsymbol{a}^{(1)}}$ for any $\boldsymbol{a}^{(1)} \in \boldsymbol{R}^{n-1}$.

LEMMA 9.18. *We have the following implication:*

$${}^{\underline{n}}\mathcal{F}_{\boldsymbol{a}} \cap \mathbb{E}(H(\mathcal{E}^\lambda),\boldsymbol{\omega}) \subset {}^{\underline{n-1}}\mathcal{F}^{(1)}_{\phi(\boldsymbol{a})} \cap \mathbb{E}\big(H(\mathcal{E}^\lambda_{|X^{(1)}}),\phi(\boldsymbol{\omega})\big).$$

Proof Let s be an element of the left hand side. We have the following for any $i = 1,\ldots,n$ and for any point P of D_i°:

$$-\operatorname{ord}\big(F(s,\boldsymbol{b})_{|\pi_i^{-1}(P)}\big) \leq a_i - \operatorname{Re}\big(\alpha(b_i,\omega_i)\big).$$

Let C be the subset of $X^{(1)} \simeq \Delta^{n-1}$ such that $C = \pi_{n-1}^{-1}(P)$ for a point $P \in D_{n-1}^{(1)} - \bigcup_{i<n-1} D_{n-1}^{(1)} \cap D_i^{(1)}$. The inclusion $C \subset X$ is obtained by the diagonal embedding $\{P\}\times\Delta_{n-1} \longrightarrow \{P\}\times\Delta_{n-1}\times\Delta_n$. We have the following, due to Corollary 2.53:

$$-\operatorname{ord}(F(s,\boldsymbol{b})_{|C}) \leq a_{n-1}+a_n - \operatorname{Re}(\alpha(b_n+b_{n-1},\omega_n\cdot\omega_{n-1})).$$

It implies $-\operatorname{ord}(s_{|C}) \leq a_{n-1}+a_n$ due to Lemma 7.78. Hence we have the inequality:

$$^{n-1}\deg^{\mathcal{F}^{(1)}}(s) \leq a_{n-1} + a_n.$$

Thus we are done. □

9.2.4. Preliminary proposition. Let us consider the following condition:

Assumption (D): For any $a, b \in \mathcal{P}ar^f(\mathcal{E}^\lambda, n)$ such that $a \neq b$, the following holds:

$$|a-b| > \sum_{c \in S} 2|c|, \qquad S = \bigcup_{j \leq n-1} \mathcal{P}ar^f(\mathcal{E}^\lambda, j).$$

We remark the following obvious lemma.

LEMMA 9.19. *Let c be a sufficiently large integer, and let ψ_c denote the morphism $X \longrightarrow X$ given by $\psi_c(z_1, \ldots, z_n) = (z_1, \ldots, z_{n-1}, z_n^c)$. Then the assumption (D) is satisfied for pull back $\psi_c^{-1}\mathcal{E}^\lambda$.*

Proof It follows from Lemma 9.5. □

The following lemma is also clear.

LEMMA 9.20. *Under the assumption (D), the following holds:*

(1) *The morphism $\mathcal{P}ar^f(\mathcal{E}^\lambda, n) \times \mathcal{P}ar^f(\mathcal{E}^\lambda, n-1) \longrightarrow \boldsymbol{R}$ given by $(a,b) \longmapsto a+b$ is injective. In particular, it induces the total order $\leq_1$ on the set $\mathcal{P}ar^f(\mathcal{E}^\lambda, n) \times \mathcal{P}ar^f(\mathcal{E}^\lambda, n-1)$.*
(2) *We have the natural orders on $\mathcal{P}ar^f(\mathcal{E}^\lambda, i)$ $(i = n, n-1)$, and we obtain the lexicographic order $\leq_2$ on $\mathcal{P}ar^f(\mathcal{E}^\lambda, n) \times \mathcal{P}ar^f(\mathcal{E}^\lambda, n-1)$. We have the equality $\leq_1 = \leq_2$.* □

PROPOSITION 9.21. *We use the notation in the subsection 9.2.3. The following holds.*

(A): $d(\lambda, \boldsymbol{\omega}, \boldsymbol{a}) = v.d(\lambda, \boldsymbol{\omega}, \boldsymbol{a})$.

(B): *If the assumption (D) is satisfied, the following morphism is isomorphic:*

$$\sum_{\substack{\phi(\boldsymbol{a}) \leq \boldsymbol{a}^{(1)} \\ \phi(\boldsymbol{\omega}) = \boldsymbol{\omega}^{(1)}}} \Big({}^{\underline{n}}\mathcal{F}_{\boldsymbol{a}} \cap \mathbb{E}\big(H(\mathcal{E}^\lambda), \boldsymbol{\omega}\big)\Big) \longrightarrow {}^{\underline{n-1}}\mathcal{F}^{(1)}_{\boldsymbol{a}^{(1)}} \cap \mathbb{E}\big(H(\mathcal{E}^\lambda_{|X^{(1)}}), \boldsymbol{\omega}^{(1)}\big).$$

(C): *If the assumption (D) is satisfied, then the following morphism is injective.*

$$\frac{{}^{\underline{n}}\mathcal{F}_{\boldsymbol{a}} \cap \mathbb{E}\big(H(\mathcal{E}^\lambda), \boldsymbol{\omega}\big)}{{}^{\underline{n}}\mathcal{F}'_{\boldsymbol{a}} \cap \mathbb{E}\big(H(\mathcal{E}^\lambda), \boldsymbol{\omega}\big)} \longrightarrow \frac{{}^{\underline{n-1}}\mathcal{F}^{(1)}_{\phi(\boldsymbol{a})} \cap \mathbb{E}\big(H(\mathcal{E}^\lambda_{|X^{(1)}}), \phi(\boldsymbol{\omega})\big)}{{}^{\underline{n-1}}\mathcal{F}'^{(1)}_{\phi(\boldsymbol{a})} \cap \mathbb{E}\big(H(\mathcal{E}^\lambda_{|X^{(1)}}), \phi(\boldsymbol{\omega})\big)}.$$

See (9.10) *for* $\mathcal{F}'$.

The proposition will be proved in the subsections 9.2.5–9.2.6.

9.2.5. A proof of the claims (A) and (B) of Proposition 9.21. Since the claim (A) is preserved by the pull back as in Lemma 9.19. Hence we have only to consider the case where the assumption (D) is satisfied.

Let $\boldsymbol{a}^{(1)}$ be an element of $\boldsymbol{R}^{n-1}$, whose i-th components are $a_i^{(1)}$. Then the element $(a_n^\circ, a_{n-1}^\circ) \in \mathcal{P}ar^f(\mathcal{E}^\lambda, n) \times \mathcal{P}ar^f(\mathcal{E}^\lambda, n-1)$ is determined as follows:

$$(a_n^\circ, a_{n-1}^\circ) := \max\bigl\{(b_n, b_{n-1}) \in \mathcal{P}ar^f(\mathcal{E}^\lambda, n) \times \mathcal{P}ar^f(\mathcal{E}^\lambda, n-1) \,\big|\, b_n + b_{n-1} \leq a_{n-1}^{(1)}\bigr\}.$$

LEMMA 9.22. *We have the following equality:*

$$\sum_{\phi(\boldsymbol{a}) \leq \boldsymbol{a}^{(1)}} {}^{\underline{n}}\mathcal{F}_{\boldsymbol{a}} = \Bigl({}^{\underline{n-2}}\mathcal{F}_{\boldsymbol{a}'} \cap {}^{n-1}\mathcal{F}_{a_{n-1}^\circ} \cap {}^{n}\mathcal{F}_{a_n^\circ}\Bigr) + \Bigl({}^{\underline{n-2}}\mathcal{F}_{\boldsymbol{a}'} \cap {}^{n}\mathcal{F}_{<a_n^\circ}\Bigr). \tag{9.11}$$

Here we put $\boldsymbol{a}' := (a_1^{(1)}, \ldots, a_{n-2}^{(1)}) \in \boldsymbol{R}^{n-2}$.

Proof It is clear that ${}^{\underline{n-2}}\mathcal{F}_{\boldsymbol{a}'} \cap {}^{n-1}\mathcal{F}_{a_{n-1}^\circ} \cap {}^{n}\mathcal{F}_{a_n^\circ}$ is contained in the left hand side of (9.11). Under the assumption (D), we have ${}^{n}\mathcal{F}_{<a_n^\circ} = {}^{n-1}\mathcal{F}_b \cap {}^{n}\mathcal{F}_{a_n^\circ - \eta}$ for some real numbers b and $\eta > 0$ such that $b + a_n^\circ - \eta < a_{n-1}^{(1)}$. Thus ${}^{\underline{n-2}}\mathcal{F}_{\boldsymbol{a}'} \cap {}^{n}\mathcal{F}_{<a_n^\circ}$ is contained in the left hand side of (9.11). Thus we obtain the implication $\supset$.

Next, we would like to show the implication $\subset$. We have only to show that ${}^{\underline{n}}\mathcal{F}_{\boldsymbol{a}}\mathbb{E}\bigl(H(\mathcal{E}^\lambda), \boldsymbol{\omega}\bigr)$ is contained in the right hand side when we have $\phi(\boldsymbol{a}) \leq \boldsymbol{a}^{(1)}$. Under the assumption D, the condition $\phi(\boldsymbol{a}) \leq \boldsymbol{a}^{(1)}$ implies the following:

$$a_i \leq a_i^{(1)}, \ (i \leq n-2), \qquad \text{and} \qquad \begin{cases} a_n < a_n^\circ, \\ \text{or}, \\ a_n = a_n^\circ,\ a_{n-1} \leq a_{n-1}^\circ. \end{cases}$$

Then the implication $\subset$ immediately follows. □

We will show the claims (A) and (B) in Proposition 9.21 by an induction on n. We assume that the claims (A) for $n-1$, and we will show that the claims (A) and (B) hold for n.

Let us consider the following claims:

- $(A, a, \leq)$: (A) holds for any $\boldsymbol{a}$ such that $a_n + a_{n-1} \leq a$.
- $(A, a, <)$: (A) holds for any $\boldsymbol{a}$ such that $a_n + a_{n-1} < a$.
- $(B, a, \leq)$: (B) holds for any $\boldsymbol{a}^{(1)}$ such that $a_{n-1}^{(1)} \leq a$.
- $(B, a, <)$: (B) holds for any $\boldsymbol{a}^{(1)}$ such that $a_{n-1}^{(1)} < a$.

If a is sufficiently negative, then $(A, a, \leq)$ and $(B, a, \leq)$ are true trivially.

LEMMA 9.23. *$(A, a, <)$ implies $(B, a, \leq)$ and $(A, a, \leq)$.*

Proof We have the following implication:

$$\Bigl({}^{\underline{n-2}}\mathcal{F}_{\boldsymbol{a}'} \cap {}^{n-1}\mathcal{F}_{a_{n-1}^\circ} \cap {}^{n}\mathcal{F}_{a_n^\circ}\Bigr) + \Bigl({}^{\underline{n-2}}\mathcal{F}_{\boldsymbol{a}'} \cap {}^{n}\mathcal{F}_{<a_n^\circ}\Bigr) \subset {}^{\underline{n-1}}\mathcal{F}^{(1)}_{\boldsymbol{a}^{(1)}}. \tag{9.12}$$

We also have the following:

$$\Bigl({}^{\underline{n-2}}\mathcal{F}_{\boldsymbol{a}'} \cap {}^{n-1}\mathcal{F}_{a_{n-1}^\circ} \cap {}^{n}\mathcal{F}_{a_n^\circ}\Bigr) \cap \Bigl({}^{\underline{n-2}}\mathcal{F}_{\boldsymbol{a}'} \cap {}^{n}\mathcal{F}_{<a_n^\circ}\Bigr) = {}^{\underline{n-2}}\mathcal{F}_{\boldsymbol{a}'} \cap {}^{n-1}\mathcal{F}_{a_{n-1}^\circ} \cap {}^{n}\mathcal{F}_{<a_n^\circ}.$$

Thus we have the following equality:

$$(9.13)\quad \dim\Bigl(\underline{{}^{n-2}}\mathcal{F}_{\boldsymbol{a}'}\cap{}^{n-1}\mathcal{F}_{a^{\circ}_{n-1}}\cap{}^{n}\mathcal{F}_{a^{\circ}_{n}}\cap\mathbb{E}\bigl(H(\mathcal{E}^{\lambda}),\boldsymbol{\omega}\bigr)\Bigr)\leq$$
$$\dim\bigl(\underline{{}^{n-1}}\mathcal{F}^{(1)}_{\boldsymbol{a}^{(1)}}\cap\mathbb{E}\bigl(H(\mathcal{E}^{\lambda}),\boldsymbol{\omega}\bigr)\bigr)+\dim\bigl(\underline{{}^{n-2}}\mathcal{F}_{\boldsymbol{a}'}\cap{}^{n-1}\mathcal{F}_{a^{\circ}_{n-1}}\cap{}^{n}\mathcal{F}_{<a^{\circ}_{n}}\cap\mathbb{E}\bigl(H(\mathcal{E}^{\lambda}),\boldsymbol{\omega}\bigr)\bigr)$$
$$-\dim\Bigl(\underline{{}^{n-2}}\mathcal{F}_{\boldsymbol{a}'}\cap{}^{n}\mathcal{F}_{<a^{\circ}_{n}}\cap\mathbb{E}\bigl(H(\mathcal{E}^{\lambda}),\boldsymbol{\omega}\bigr)\Bigr).$$

By using the assumption of the induction on n, or by using $(A,a,<)$, we obtain the following equality:

$$(9.14)\quad \begin{aligned}&\dim\Bigl(\underline{{}^{n-2}}\mathcal{F}_{\boldsymbol{a}'}\cap{}^{n}\mathcal{F}_{<a^{\circ}_{n}}\cap\mathbb{E}\bigl(H(\mathcal{E}^{\lambda}),\boldsymbol{\omega}\bigr)\Bigr)=\sum_{\boldsymbol{u}\in S_1(\boldsymbol{\omega})}\mathfrak{m}(0,\boldsymbol{u}),\\ &S_1(\boldsymbol{\omega})=\bigl\{\boldsymbol{u}\in\mathcal{S}(\boldsymbol{\omega})\,\big|\,\mathfrak{p}^{f}(\lambda,u_i)\leq a_i,\ (i\leq n-2),\ \mathfrak{p}^{f}(\lambda,u_n)<a^{\circ}_{n}\bigr\}.\end{aligned}$$

By using $(A,a,<)$, we obtain the following:

$$(9.15)\quad \begin{aligned}&\dim\Bigl(\underline{{}^{n-2}}\mathcal{F}_{\boldsymbol{a}'}\cap{}^{n-1}\mathcal{F}_{a^{\circ}_{n-1}}\cap{}^{n}\mathcal{F}_{<a^{\circ}_{n}}\cap\mathbb{E}\bigl(H(\mathcal{E}^{\lambda}),\boldsymbol{\omega}\bigr)\Bigr)=\sum_{\boldsymbol{u}\in S_2(\boldsymbol{\omega})}\mathfrak{m}(0,\boldsymbol{u}),\\ &S_2(\boldsymbol{\omega})=\left\{\boldsymbol{u}\in\mathcal{S}(\boldsymbol{\omega})\,\middle|\,\begin{array}{l}\mathfrak{p}^{f}(\lambda,u_i)\leq a_i,\,(i\leq n-2),\\ \mathfrak{p}^{f}(\lambda,u_{n-1})\leq a^{\circ}_{n-1},\ \mathfrak{p}^{f}(\lambda,u_n)<a^{\circ}_{n}\end{array}\right\}.\end{aligned}$$

Due to the assumption of the induction on n, we have the following:

(9.16)
$$\dim\bigl(\underline{{}^{n-1}}\mathcal{F}^{(1)}_{\boldsymbol{a}^{(1)}}\cap\mathbb{E}\bigl(H(\mathcal{E}^{\lambda}),\boldsymbol{\omega}^{(1)}\bigr)\bigr)=\sum_{\substack{\boldsymbol{u}\in\mathcal{S}(\boldsymbol{\omega}^{(1)})\\ \mathfrak{p}^{f}(\lambda,\boldsymbol{u})\leq\boldsymbol{a}^{(1)}}}\mathfrak{m}(0,\boldsymbol{u})=\sum_{\phi(\boldsymbol{\omega})=\boldsymbol{\omega}^{(1)}}\ \sum_{\boldsymbol{u}\in S_3(\boldsymbol{\omega})}\mathfrak{m}(0,\boldsymbol{u}),$$

$$S_3(\boldsymbol{\omega}):=\left\{\boldsymbol{u}\in\mathcal{S}(\boldsymbol{\omega})\,\middle|\,\begin{array}{l}\mathfrak{p}^{f}(\lambda,u_i)\leq a_i,\ (i\leq n-2),\\ \mathfrak{p}^{f}(\lambda,u_{n-1})+\mathfrak{p}^{f}(\lambda,u_n)\leq a^{(1)}_{n-1}\end{array}\right\}.$$

We put $S_4(\boldsymbol{\omega}):=S_3(\boldsymbol{\omega})-\bigl(S_1(\boldsymbol{\omega})-S_2(\boldsymbol{\omega})\bigr)$. It is easy to check the following:

$$S_4(\boldsymbol{\omega})=\bigl\{\boldsymbol{u}\in\mathcal{S}(\boldsymbol{\omega})\,\big|\,\mathfrak{p}^{f}(\lambda,u_i)\leq a_i\ (i\leq n-2),\ \mathfrak{p}^{f}(\lambda,u_i)\leq a^{\circ}_{i}\ (i=n-1,n)\bigr\}.$$

Then we obtain the following inequality by a direct calculation from (9.13), (9.14), (9.15) and (9.16):

$$(9.17)\quad \sum_{\phi(\boldsymbol{\omega})=\boldsymbol{\omega}'}\dim\Bigl(\underline{{}^{n-2}}\mathcal{F}_{\boldsymbol{a}'}\cap{}^{n-1}\mathcal{F}_{a^{\circ}_{n-1}}\cap{}^{n}\mathcal{F}_{a^{\circ}_{n}}\cap\mathbb{E}\bigl(H(\mathcal{E}^{\lambda}),\boldsymbol{\omega}\bigr)\Bigr)$$
$$\leq\sum_{\phi(\boldsymbol{\omega})=\boldsymbol{\omega}^{(1)}}\ \sum_{\boldsymbol{u}\in S_4(\boldsymbol{\omega})}\mathfrak{m}(0,\boldsymbol{u})=\sum_{\phi(\boldsymbol{\omega})=\boldsymbol{\omega}^{(1)}}v.d(\boldsymbol{a},\boldsymbol{\omega}).$$

Here $\boldsymbol{a}$ is determined by $q_i(\boldsymbol{a})=a^{(1)}_{i}$ for $i\leq n-2$ and $q_i(\boldsymbol{a})=a^{\circ}_{i}$ for $i=n-1,n$. On the other hand, we have already known the inequality (Lemma 9.17):

$$(9.18)\quad \dim\Bigl(\underline{{}^{n-2}}\mathcal{F}_{\boldsymbol{a}'}\cap{}^{n-1}\mathcal{F}_{a^{\circ}_{n-1}}\cap{}^{n}\mathcal{F}_{a^{\circ}_{n}}\cap\mathbb{E}\bigl(H(\mathcal{E}^{\lambda}),\boldsymbol{\omega}\bigr)\Bigr)\geq v.d(\lambda,\boldsymbol{a},\boldsymbol{\omega}).$$

From (9.17) and (9.18), we can conclude that the equality in (9.18) holds, which implies $(A,a,\leq)$. We can also conclude that the equality in (9.13) holds, which implies that the equality in (9.12) holds. Thus we obtain $(B,a,\leq)$. Thus the proof of Lemma 9.23 is accomplished. $\square$

For any a and some $\epsilon > 0$, the following implications are clear:

$$(B, a, \leq) \Longrightarrow (B, a+\epsilon, <),$$
$$(A, a, \leq) \Longrightarrow (A, a+\epsilon, <).$$

Hence we obtain (A, a) and (B, a) for any a. Thus the induction on n can proceed. Namely the proof of the claims (A) and (B) of Proposition 9.21 is accomplished.

9.2.6. A proof of the claim (C) of Proposition 9.21. Now we shall prove the claim (C). The following inclusion is surjective, due to (B):

$$\sum_{\phi(\boldsymbol{b}) \lneq \phi(\boldsymbol{a})} {}^{\underline{n}}\mathcal{F}_{\boldsymbol{b}} \cap \mathbb{E}\big(H(\mathcal{E}^{\lambda}), \boldsymbol{\omega}\big) \longrightarrow {}^{\underline{n-1}}\mathcal{F}'^{(1)}_{\phi(\boldsymbol{a})} \cap \mathbb{E}\big(H(\mathcal{E}^{\lambda}), \boldsymbol{\omega}\big).$$

(See (9.10) for $\mathcal{F}'$.) Thus it is isomorphic.

LEMMA 9.24. *Let $\boldsymbol{b}$ be an element of $\boldsymbol{R}^n$. Assume $\phi(\boldsymbol{b}) \lneq \phi(\boldsymbol{a})$. We put $\boldsymbol{a}' = (a_1, \ldots, a_{n-2}) \in \boldsymbol{R}^{n-2}$. Then we have one of the following:*

- ${}^{\underline{n}}\mathcal{F}_{\boldsymbol{b}} \subset {}^{\underline{n-2}}\mathcal{F}_{\boldsymbol{a}'} \cap {}^{n}\mathcal{F}_{<a_n}$.
- ${}^{\underline{n}}\mathcal{F}_{\boldsymbol{b}} \subset {}^{\underline{n-2}}\mathcal{F}_{\boldsymbol{a}'} \cap {}^{n-1}\mathcal{F}_{<a_{n-1}} \cap {}^{n}\mathcal{F}_{a_n}$.
- ${}^{\underline{n}}\mathcal{F}_{\boldsymbol{b}} \subset {}^{\underline{n-2}}\mathcal{F}'_{\boldsymbol{a}'} \cap {}^{n-1}\mathcal{F}_{a_{n-1}} \cap {}^{n}\mathcal{F}_{a_n}$.

Proof The condition implies $b_i \leq a_i$ $(i \leq n-2)$ and $b_{n-1} + b_n \leq a_{n-1} + a_n$, and at least one of the inequalities is not equality. Then we have at least one of the following:

- $b_n < a_n$.
- $b_n = a_n$ and $b_{n-1} < a_{n-1}$.
- $b_n = a_n$, $b_{n-1} = a_{n-1}$, and $(b_1, \ldots, b_{n-2}) \lneq \boldsymbol{a}'$.

Then the claim follows. □

LEMMA 9.25. *We have the following:*

$$(9.19) \qquad {}^{\underline{n}}\mathcal{F}_{\boldsymbol{a}} \cap \Big(\sum_{\phi(\boldsymbol{b}) \lneq \phi(\boldsymbol{a})} {}^{\underline{n}}\mathcal{F}_{\boldsymbol{b}} \Big) = \sum_{\boldsymbol{b} \lneq \boldsymbol{a}} {}^{\underline{n}}\mathcal{F}_{\boldsymbol{b}} = {}^{\underline{n}}\mathcal{F}'_{\boldsymbol{a}}.$$

Proof Due to Lemma 9.24, we have the following:

$$(9.20) \qquad \sum_{\phi(\boldsymbol{b}) \lneq \phi(\boldsymbol{a})} {}^{\underline{n}}\mathcal{F}_{\boldsymbol{b}} = \Big({}^{\underline{n-2}}\mathcal{F}_{\boldsymbol{a}'} \cap {}^{n}\mathcal{F}_{<a_n} \Big) + \Big({}^{\underline{n-2}}\mathcal{F}_{\boldsymbol{a}'} \cap {}^{n-1}\mathcal{F}_{<a_{n-1}} \cap {}^{n}\mathcal{F}_{a_n} + {}^{\underline{n-2}}\mathcal{F}'_{\boldsymbol{a}'} \cap {}^{n-1}\mathcal{F}_{a_{n-1}} \cap {}^{n}\mathcal{F}_{a_n} \Big).$$

Note that the second term in the right hand side of (9.20) is contained in ${}^{\underline{n}}\mathcal{F}_{\boldsymbol{a}}$. Let us pick elements: $x \in {}^{\underline{n-2}}\mathcal{F}_{\boldsymbol{a}'} \cap {}^{n-1}\mathcal{F}_{<a_{n-1}} \cap {}^{n}\mathcal{F}_{a_n} + {}^{\underline{n-2}}\mathcal{F}'_{\boldsymbol{a}'} \cap {}^{n-1}\mathcal{F}_{a_{n-1}} \cap {}^{n}\mathcal{F}_{a_n}$ and $y \in {}^{\underline{n-2}}\mathcal{F}_{\boldsymbol{a}'} \cap ({}^{n}\mathcal{F}_{<a_n})$. Assume $x + y \in {}^{\underline{n}}\mathcal{F}_{\boldsymbol{a}}$. Then we obtain $y \in {}^{\underline{n}}\mathcal{F}_{\boldsymbol{a}} \cap {}^{\underline{n-2}}\mathcal{F}_{\boldsymbol{a}'} \cap {}^{n}\mathcal{F}_{<a_n}$. Hence we have $y \in {}^{\underline{n-2}}\mathcal{F}_{\boldsymbol{a}'} \cap {}^{n-1}\mathcal{F}_{a_{n-1}} \cap {}^{n}\mathcal{F}_{<a_n}$. Thus the left hand side of (9.19) is as follows:

$$(9.21) \qquad {}^{\underline{n-2}}\mathcal{F}_{\boldsymbol{a}'} \cap {}^{n-1}\mathcal{F}_{a_{n-1}} \cap {}^{n}\mathcal{F}_{<a_n} + \Big({}^{\underline{n-2}}\mathcal{F}_{\boldsymbol{a}'} \cap {}^{n-1}\mathcal{F}_{<a_{n-1}} \cap {}^{n}\mathcal{F}_{a_n} + {}^{\underline{n-2}}\mathcal{F}'_{\boldsymbol{a}'} \cap {}^{n-1}\mathcal{F}_{a_{n-1}} \cap {}^{n}\mathcal{F}_{a_n} \Big).$$

It is same as the right hand side of (9.19). Thus we are done. □

The claim (C) immediately follows Lemma 9.25. Thus the proof of Proposition 9.21 is accomplished. □

9.2.7. Restriction to the diagonal curve. For each $1 \leq i \leq n$, we put as follows:

$$X^{(i)} := \{(z_1, \ldots, z_n) \in X \mid z_{n-i} = \cdots = z_n\}.$$

We put $X^{(0)} = X$. The dimension of $X^{(i)}$ is $n-i$, and $X^{(i)}$ is contained in $X^{(i-1)}$. We remark that $X^{(1)}$ has appeared in the subsection 9.2.3. We also remark that the result in the subsection 9.2.3 can be applied to the embedding $X^{(i)} \subset X^{(i-1)}$.

We have the natural isomorphisms $H(\mathcal{E}^\lambda) = H(\mathcal{E}^\lambda_{|X^{(i)}})$. The filtrations for the right hand side is denoted by ${}^{j}\mathcal{F}^{(i)}$ $(j = 1, \ldots, n-i)$.

In particular, we put $C_0 := X^{(n-1)}$ which is the diagonal curve, and the filtration for $H(\mathcal{E}^\lambda_{C_0})$ is denoted by ${}^{C_0}\mathcal{F}$.

Let us consider the following condition:

Assumption (E): The assumption (D) for $\mathcal{E}^\lambda_{|X^{(i)}}$ holds for each i. Namely, for any $a, b \in \mathcal{P}ar^f(\mathcal{E}^\lambda_{|X^{(i)}}, i)$ such that $a \neq b$, the following holds:

$$|a - b| > \sum_{c \in S(i)} 2 \cdot |c|, \qquad S(i) := \bigcup_{j \leq i-1} \mathcal{P}ar^f(\mathcal{E}^\lambda_{X^{(i)}}, j).$$

COROLLARY 9.26. *If the assumption (E) holds, then the following morphism is injective:*

$$\frac{{}^{\underline{n}}\mathcal{F}_{\boldsymbol{a}} \cap \mathbb{E}(H(\mathcal{E}^\lambda), \boldsymbol{\omega})}{{}^{\underline{n}}\mathcal{F}'_{\boldsymbol{a}} \cap \mathbb{E}(H(\mathcal{E}^\lambda), \boldsymbol{\omega})} \longrightarrow \frac{{}^{C_0}\mathcal{F}_{|\boldsymbol{a}|} \cap \mathbb{E}(H(\mathcal{E}^\lambda), \boldsymbol{\omega})}{{}^{C_0}\mathcal{F}'_{|\boldsymbol{a}|} \cap \mathbb{E}(H(\mathcal{E}^\lambda), \boldsymbol{\omega})}.$$

Here $|\boldsymbol{a}|$ denotes $\sum_{i=1}^n a_i$ for $\boldsymbol{a} = (a_1, \ldots, a_n)$.

Proof We have only to use the claims (C) in Proposition 9.21 inductively. □

For the use of Corollary 9.26, we remark the following lemma.

LEMMA 9.27. *Let us consider the morphisms $\psi_c^{(i)} : X \longrightarrow X$ for $i = 1, \ldots, n-1$ given as follows:*

$$\psi_c^{(i)}(z_1, \ldots, z_n) = (z_1, \ldots, z_{n-i}, z_{n-i+1}^c, \ldots, z_n^c).$$

If we take $c_1, \ldots, c_{n-1}$ appropriately, then the assumption (E) holds for the pull back of $\mathcal{E}^\lambda$ via $\prod_{i=1}^n \psi_{c_i}^{(i)}$.

Proof If the assumption (D) holds for $\mathcal{E}^\lambda$, then it holds for $\psi_c^{(i)*}\mathcal{E}^\lambda$ for any $c \geq 1$ and for any $i \geq 1$. We also remark that $\psi_c^{(i)}$ preserves $X^{(j)}$ for any $j < i$. Then we obtain that if the assumption (D) holds for $\mathcal{E}^\lambda_{|X^{(j)}}$, then it holds for $\psi_c^{(i)*}\mathcal{E}^\lambda_{|X^{(j)}}$ for any $c \geq 1$ and for any $j < i$. Therefore we have only to use Lemma 9.19 inductively. □

9.2.8. The end of the proof of Theorem 9.15. Let us return to the proof of Theorem 9.15. Due to Lemma 9.27 and Lemma 9.5, we have only to consider the case where the assumption (E) is satisfied.

Let s be an element of ${}^{\underline{n}}\mathcal{F}_{\boldsymbol{a}} \cap \mathbb{E}(H(\mathcal{E}^\lambda), \boldsymbol{\omega})$. Then we obtain the section $F(s, \boldsymbol{a})$ of ${}^\diamond\mathcal{E}^\lambda$, as is explained in the subsection 9.1.2 (Corollary 9.4). On the other hand, we have the element $\boldsymbol{u} \in \prod_{i=1}^n \mathcal{K}(\mathcal{E}, \lambda, \boldsymbol{0}, i)$ satisfying $\mathfrak{k}^f(\lambda, \boldsymbol{u}) = (\boldsymbol{a}, \boldsymbol{\omega})$. (See (8.41) for $\mathcal{K}(\mathcal{E}, \lambda, \boldsymbol{0}, i)$.) Due to the result in the case of curves (Lemma 7.78), we have the following for any $P \in D_i^\circ = D_i \setminus \bigcup_{j \neq i} D_j \cap D_i$:

$$-\operatorname{ord}\big(F(s, \boldsymbol{a})_{|\pi_i^{-1}(P)}\big) = \mathfrak{p}(\lambda, u_i) = a_i - \operatorname{Re}(\alpha(a_i, \omega_i)). \tag{9.22}$$

We put $c_i := \mathfrak{p}(\lambda, u_i)$ and $\boldsymbol{c} := (c_1, \ldots, c_n)$. We remark that $-1 < c_i \leq 0$. Hence $F(s, \boldsymbol{a})_{|O}$ is contained in ${}^{\underline{n}}F_{\boldsymbol{c}}(\mathcal{E}^{\lambda}_{|O})$. Therefore we obtain the following map:

$$\Phi : {}^{\underline{n}}\mathrm{Gr}^{\mathcal{F}}_{\boldsymbol{a}}\,\mathbb{E}(H(\mathcal{E}^{\lambda}), \boldsymbol{\omega}) \longrightarrow {}^{\underline{n}}\mathrm{Gr}^{F}_{\boldsymbol{c}}({}^{\diamond}\mathcal{E}^{\lambda}{}_{|O}), \quad s \longmapsto F(s, \boldsymbol{a})(O).$$

LEMMA 9.28. *The map Φ is injective.*

Proof Let $\boldsymbol{v}$ be a frame of ${}^{\diamond}\mathcal{E}^{\lambda}$, which is compatible with $({}^{i}F \,|\, i = 1, \ldots, n)$. We describe as follows:

$$F(s, \boldsymbol{a}) := \sum f_j \cdot v_j.$$

Assume that $\Phi(s) = 0$. We have $f_j(O) = 0$ unless ${}^{\underline{n}}\deg(v_j) \lneq \boldsymbol{c}$. It implies the following:

$$-\operatorname{ord}\big(F(s, \boldsymbol{a})_{|C_0}\big) < \sum c_i = |\boldsymbol{a}| - \sum_{i=1}^{n} \mathrm{Re}\big(\alpha(a_i, \omega_i)\big).$$

On the other hand, we have the following due to our construction of $F(s, \boldsymbol{a})_{|C_0}$ and Lemma 7.78:

$$-\operatorname{ord}\big(F(s, \boldsymbol{a})_{|C_0}\big) = -\operatorname{ord}(s) - \sum_{i=1}^{n} \mathrm{Re}\big(\alpha(a_i, \omega_i)\big).$$

Hence we obtain $-\operatorname{ord}(s_{|C_0}) < |\boldsymbol{a}|$, which implies $s_{|C_0} \in {}^{C_0}\mathcal{F}'_{|\boldsymbol{a}|}H(\mathcal{E}^{\lambda}_{|C_0})$. Then we obtain $s = 0$ in ${}^{\underline{n}}\mathrm{Gr}_{\boldsymbol{a}}\,\mathbb{E}(H(\mathcal{E}^{\lambda}), \boldsymbol{\omega})$, due to Corollary 9.26. □

We put $\gamma_i := \mathfrak{e}(\lambda, u_i)$ and $\boldsymbol{\gamma} := (\gamma_1, \ldots, \gamma_n)$.

LEMMA 9.29. *We have* $\mathrm{Im}(\Phi) \subset {}^{\underline{n}}\mathbb{E}\big({}^{\underline{n}}\mathrm{Gr}^{F}_{\boldsymbol{c}}(\mathrm{Res}_{\underline{n}}\,\mathbb{D}^{\lambda}), \boldsymbol{\gamma}\big)$.

Proof It suffices to check that $F(s, \boldsymbol{a})_{|(\lambda, P)}$ is contained in ${}^{i}\mathbb{E}\big({}^{i}\,\mathrm{Gr}^{F}_{c_i}({}^{\diamond}\mathcal{E}^{\lambda}{}_{|D_i}), \gamma_i\big)$ for any $P \in D_i^{\circ}$. It follows from the result in the case of curves (Corollary 7.81). □

Due to Lemma 9.29, we obtain the injection:

$$\Phi_{(\boldsymbol{a},\boldsymbol{\omega})} : {}^{\underline{n}}\mathrm{Gr}^{\mathcal{F}}_{\boldsymbol{a}}\big({}^{\underline{n}}\mathbb{E}\big(H(\mathcal{E}^{\lambda}), \boldsymbol{\omega}\big)\big) \longrightarrow {}^{\underline{n}}\mathbb{E}\big({}^{\underline{n}}\mathrm{Gr}^{F}_{\boldsymbol{c}}({}^{\diamond}\mathcal{E}^{\lambda}{}_{|O}), \boldsymbol{\gamma}\big).$$

Therefore we obtain the following injection:

$$\bigoplus_{(\boldsymbol{a},\boldsymbol{\omega})} \Phi_{(\boldsymbol{a},\boldsymbol{\omega})} : \bigoplus_{(\boldsymbol{a},\boldsymbol{\omega})} {}^{\underline{n}}\mathrm{Gr}^{\mathcal{F}}_{\boldsymbol{a}}\,{}^{\underline{n}}\mathbb{E}\big(H(\mathcal{E}^{\lambda}), \boldsymbol{\omega}\big) \longrightarrow \bigoplus_{(\boldsymbol{c},\boldsymbol{\gamma})} {}^{\underline{n}}\mathbb{E}\big({}^{\underline{n}}\mathrm{Gr}^{F}_{\boldsymbol{c}}(\mathrm{Res}_{\underline{n}}\,\mathbb{D}^{\lambda}), \boldsymbol{\gamma}\big).$$

PROPOSITION 9.30. *The morphisms $\Phi_{(\boldsymbol{a},\boldsymbol{\omega})}$ are isomorphic.*

Proof Since $\Phi_{(\boldsymbol{a},\boldsymbol{\omega})}$ is injective, we obtain the following inequalities:

$$\operatorname{rank}\mathcal{E}^{\lambda} \leq \sum_{(\boldsymbol{a},\boldsymbol{\omega})} \dim {}^{\underline{n}}\mathrm{Gr}^{\mathcal{F}}_{\boldsymbol{a}}\,\mathbb{E}(H(\mathcal{E}^{\lambda}), \boldsymbol{\omega}) \leq \sum_{(\boldsymbol{c},\boldsymbol{\gamma})} \dim \mathbb{E}({}^{\underline{n}}\mathrm{Gr}^{F}_{\boldsymbol{c}}(\mathrm{Res}_{\underline{n}}\,\mathbb{D}^{\lambda}), \boldsymbol{\gamma}). \tag{9.23}$$

Since we have already known that the tuple of the filtrations ${}^{i}F$ $(i \in \underline{n})$ of ${}^{\diamond}\mathcal{E}^{\lambda}{}_{|O}$ is compatible (Theorem 8.59), the right hand side is same as $\operatorname{rank}\mathcal{E}$. Then the proposition immediately follows. □

Due to Proposition 9.30, we have the following equality:

$$\sum_{\boldsymbol{a}} \operatorname{rank} {}^{\underline{n}}\mathrm{Gr}^{\mathcal{F}}_{\boldsymbol{a}}\,H(\mathcal{E}^{\lambda}) = \operatorname{rank} H(\mathcal{E}^{\lambda}).$$

It implies the compatibility of the filtrations, due to Lemma 4.4. Therefore the proof of Theorem 9.15 is accomplished. □

9.2.9. Weak norm estimate. We put $X = \Delta^n$, $D_i = \{z_i = 0\}$ and $D = \bigcup_{i=1}^n D_i$. Let λ be any point of $\boldsymbol{C}^*$. Let $\mathbb{H}$ denote the upper half plane. Let $\psi : \mathbb{H}^n \longrightarrow X - D$ be a universal covering given by $\zeta_i \longmapsto \exp\bigl(\sqrt{-1}\zeta_i\bigr) = z_i$. We use the real coordinate $x_i + \sqrt{-1}y_i = \zeta_i$ of $\mathbb{H}^n$, and we put $Z(C) := \bigl\{(\zeta_1, \ldots, \zeta_n) \in \mathbb{H}^n \bigm| |x_i| < C\bigr\}$.

Let $(E, \overline{\partial}_E, \theta, h)$ be a tame harmonic bundle on $X - D$. An element $H(\mathcal{E}^\lambda)$ naturally gives a flat section on $\psi^{-1}\mathcal{E}^\lambda$ on $\mathbb{H}^n$. We will use the identification in the following without mention.

Let $\boldsymbol{s} = (s_i)$ be a frame of $H(\mathcal{E}^\lambda)$ which is compatible with the decomposition ${}^{\underline{n}}\mathbb{E}$ and the filtrations ${}^j\mathcal{F}$ $(j = 1, \ldots, n)$. We put $b_j(s_i) = {}^j\deg^{\mathcal{F}}(s_i)$. Then we put as follows:

$$s_i' := s_i \cdot \prod_{j=1}^n |z_j|^{b_j(s_i)}. \tag{9.24}$$

LEMMA 9.31. *$\boldsymbol{s}'$ gives a C^∞-frame of $\psi^{-1}\mathcal{E}^\lambda$, whose restriction to $Z(C)$ is adapted up to the polynomial order of y_i $(i = 1, \ldots, n)$.*

Proof We give only a sketch of a proof. For each s_i, we have the element $\boldsymbol{u}(s_i) \in \mathcal{K}(\mathcal{E}, \lambda, \boldsymbol{0}, \underline{n})$ such that $\deg^{\mathcal{F},\mathbb{E}}(s_i) = \mathfrak{k}(\lambda, \boldsymbol{u}(s_i))$. We put $v_i = F(s_i, \boldsymbol{b}(s_i))$, where $F(s_i, \boldsymbol{b}(s_i))$ is given by the formula (9.1). Then the tuple of sections $\boldsymbol{v} = (v_i)$ gives a holomorphic frame of ${}^\diamond\mathcal{E}^\lambda$, which is compatible with ${}^iF$ and ${}^i\mathbb{E}$ $(i \in \underline{n})$. We have ${}^{\underline{n}}\deg^{F^{(\lambda_0)}}(v_i) = \mathfrak{p}(\lambda, \boldsymbol{u}(s_i))$. We have already known the following estimate for some positive numbers M_1 and C_1:

$$|v_i|_h \leq C_1 \cdot \prod_{j=1}^n |z_j|^{-\mathfrak{p}(\lambda, u_j(s_i))} \cdot \Bigl(-\sum \log|z_j|\Bigr)^{M_1}.$$

It is easy to derive the following estimate on $Z(C)$ for some positive constants C_2 and M_2, from (9.1):

$$|s_i|_h \leq C_2 \cdot \prod_{j=1}^n |z_j|^{-\mathfrak{p}^f(\lambda, u_j(s_i))} \cdot \Bigl(-\sum \log|z_j|\Bigr)^{M_2}. \tag{9.25}$$

Then we obtain the following estimate of the hermitian matrix valued function $H(h, \boldsymbol{s}')$ on $Z(C)$, for some positive constants C_3 and M_3:

$$H(h, \boldsymbol{s}') \leq C_3 \cdot \Bigl(-\sum_{j=1}^n \log|z_j|\Bigr)^{M_3}.$$

Let $\boldsymbol{s}^\vee$ be the dual frame of $\boldsymbol{s}$. Due to the result in the case of curves, we have ${}^j\deg^{\mathcal{F}}(s_i^\vee) = -{}^j\deg^{\mathcal{F}}(s_i)$ for each j. Let $\boldsymbol{s}^{\vee\prime}$ be the modification as in (9.24). We obtain the inequality $H(h, \boldsymbol{s}^{\vee\prime}) \leq C_4 \cdot \bigl(-\sum_{j=1}^n \log|z_j|\bigr)^{M_4}$ for some positive constants C_4 and M_4. Since we have $H(h, \boldsymbol{s}') \cdot H(h, \boldsymbol{s}^{\vee\prime}) = 1$, we are done. □

9.3. The induced objects

9.3.1. The induced vector bundle ${}^{\underline{l}}\mathcal{G}_{\boldsymbol{u}}(\mathcal{H})$. We continue to use the setting in the section 9.1. Let $\boldsymbol{u}$ be an element of $\mathcal{KMS}(\mathcal{E}^0, \underline{l})$. We put as follows:

$${}^{\underline{l}}\mathcal{G}_{\boldsymbol{u}}^{(\lambda_0)}\mathcal{H} := {}^{\underline{l}}\mathrm{Gr}^{\mathcal{F}^{(\lambda_0)}}_{\mathfrak{p}^f(\lambda_0, \boldsymbol{u})}\bigl(\mathcal{H}^{(\lambda_0)}_{\mathfrak{e}^f(\lambda_0, \boldsymbol{u})}\bigr).$$

Let $\lambda \in \Delta^*(\lambda_0, \epsilon_0)$ be generic. Let us pick $\epsilon_0' > 0$ such that $\Delta(\lambda, \epsilon_0') \subset \Delta^*(\lambda_0, \epsilon_0)$.

LEMMA 9.32. *We have the following isomorphism:*

$$\,^{\underline{l}}\mathcal{G}^{(\lambda_0)}_{\boldsymbol{u}}\mathcal{H}_{|\Delta(\lambda,\epsilon_0')} \simeq \,^{\underline{l}}\mathcal{G}^{(\lambda)}_{\boldsymbol{u}}\mathcal{H}.$$

Proof For $\boldsymbol{\omega} = \mathfrak{e}^f(\lambda_0, \boldsymbol{u})$, we have the following decomposition on $\Delta^*(\lambda_0, \epsilon_0)$:

$$\mathcal{H}^{(\lambda_0)}_{\boldsymbol{\omega}\,|\,\Delta^*(\lambda_0,\epsilon_0)} = \bigoplus_{\mathfrak{e}^f(\lambda_0,\boldsymbol{u})=\boldsymbol{\omega}} \mathcal{H}_{\mathfrak{e}^f(\lambda,\boldsymbol{u})}.$$

Then we obtain the natural isomorphisms:

$$\,^{\underline{l}}\mathcal{G}^{(\lambda_0)}_{\boldsymbol{u}}\mathcal{H}_{|\Delta(\lambda,\epsilon_0')} \simeq \,^{\underline{l}}\mathrm{Gr}^{\mathcal{F}^{(\lambda_0)}}_{\boldsymbol{\omega}} \mathcal{H}_{\mathfrak{e}^f(\lambda_0,\boldsymbol{u})\,|\,\Delta(\lambda,\epsilon_0')} \simeq \mathcal{H}_{\mathfrak{e}^f(\lambda,\boldsymbol{u})\,|\,\Delta(\lambda,\epsilon_0')} \simeq \,^{\underline{l}}\mathcal{G}^{(\lambda)}_{\boldsymbol{u}}\mathcal{H}.$$

Thus we are done. □

Due to Lemma 9.32, we obtain the vector bundle $\,^{\underline{l}}\mathcal{G}_{\boldsymbol{u}}\mathcal{H}$ over $\boldsymbol{C}^*_\lambda$.

9.3.2. The induced pairing and nilpotent maps. We have the natural pairing: $\mathcal{H}(E) \otimes \mathcal{H}(E^\vee) \longrightarrow \mathcal{O}_{\boldsymbol{C}^*_\lambda}$. Pick a point $\lambda_0 \in \boldsymbol{C}^*_\lambda$ and a sufficiently small positive number $\epsilon_0 > 0$. On $\Delta(\lambda_0, \epsilon_0)$, we have the filtrations $^i\mathcal{F}^{(\lambda_0)}$ and the decompositions $^i\mathbb{E}^{(\lambda_0)}$, which was preserved by the pairing. Hence we obtain the following induced pairing:

$$S : \,^{\underline{l}}\mathcal{G}_{\boldsymbol{u}}\mathcal{H}(E) \otimes \,^{\underline{l}}\mathcal{G}_{-\boldsymbol{u}}\mathcal{H}(E^\vee) \longrightarrow \mathcal{O}_{\boldsymbol{C}^*_\lambda}.$$

The monodromies M_i induces the endomorphisms $M_{i\,\boldsymbol{u}}$ on $\,^{\underline{l}}\mathcal{G}_{\boldsymbol{u}}\mathcal{H}$. We denote the unipotent part by $M^u_{i\,\boldsymbol{u}}$. Then we obtain the following nilpotent maps:

$$\mathcal{N}_{\boldsymbol{u},i} := \frac{-1}{2\pi\sqrt{-1}} \log M^u_{i\,\boldsymbol{u}}.$$

LEMMA 9.33. *We have the relation* $S\bigl(\mathcal{N}_{\boldsymbol{u}\,i} \otimes id\bigr) + S\bigl(id \otimes \mathcal{N}_{-\boldsymbol{u}\,i}\bigr) = 0$.

Proof Since the monodromy endomorphisms preserve the pairings, the claim is obtained. □

REMARK 9.34. For any element $\boldsymbol{u} \in \mathcal{KMS}(\mathcal{E}^0, \underline{l})$, we have the induced element $[\boldsymbol{u}] \in \overline{\mathcal{KMS}}(\mathcal{E}^0, \underline{l})$. The induced bundle $\,^{\underline{l}}\mathcal{G}_{[\boldsymbol{u}]}\mathcal{H}$ is often denoted by $\,^{\underline{l}}\mathcal{G}_{\boldsymbol{u}}\mathcal{H}$ for simplicity of the notation. □

9.3.3. Functoriality for the dual. Due to the subsection 9.3.2, we obtain the naturally defined morphism $\,^{\underline{l}}\mathcal{G}_{-\boldsymbol{u}}(\mathcal{H}(E^\vee)) \longrightarrow \,^{\underline{l}}\mathcal{G}_{\boldsymbol{u}}(\mathcal{H})^\vee$.

LEMMA 9.35. *The morphism* $\,^{\underline{l}}\mathcal{G}_{-\boldsymbol{u}}\bigl(\mathcal{H}(E^\vee)\bigr) \longrightarrow \,^{\underline{l}}\mathcal{G}_{\boldsymbol{u}}\bigl(\mathcal{H}(E)\bigr)^\vee$ *is isomorphic.*

Proof We have only to check that the restriction to each fiver over $\lambda \in \boldsymbol{C}^*$ is isomorphic. It easily follows from Lemma 9.31. □

9.3.4. Functoriality for tensor products. The naturally defined morphism $\mathcal{H}(E_1) \otimes \mathcal{H}(E_2) \longrightarrow \mathcal{H}(E_1 \otimes E_2)$ preserves the $\mathcal{F}^{(\lambda_0)}$ and $\mathbb{E}^{(\lambda_0)}$. We have the naturally defined morphism:

$$\bigoplus_{\substack{\boldsymbol{u}_i \in \overline{\mathcal{KMS}}(\mathcal{E}^0,\underline{l}),\\ \boldsymbol{u}_1+\boldsymbol{u}_2=\boldsymbol{u}}} \,^{\underline{l}}\mathcal{G}_{\boldsymbol{u}_1}\mathcal{H}(E_1) \otimes \,^{\underline{l}}\mathcal{G}_{\boldsymbol{u}_2}\mathcal{H}(E_2) \longrightarrow \,^{\underline{l}}\mathcal{G}_{\boldsymbol{u}}\mathcal{H}\bigl(E_1 \otimes E_2\bigr). \tag{9.26}$$

The nilpotent maps $\mathcal{N}_i$ of $\,^{\underline{l}}\mathcal{G}_{\boldsymbol{u}_i}\mathcal{H}(E_i)$ induce the nilpotent map $\mathcal{N}_1 \otimes 1 + 1 \otimes \mathcal{N}_2$ on the left hand side. On the other hand, we have the nilpotent map on the right hand side.

LEMMA 9.36. *The morphism (9.26) is isomorphic. It is compatible with the nilpotent maps.*

Proof We have only to check that the restriction to each fiver over $\lambda \in \boldsymbol{C}^*$ is isomorphic. It easily follows from Lemma 9.31. □

9.3.5. Functoriality for pull backs. We use the setting in the subsection 8.9.7. We have the naturally defined isomorphism $\mathcal{H}(E) \simeq \mathcal{H}(\psi^*(E))$.

LEMMA 9.37. *We have the naturally defined isomorphism:*

$$\bigoplus_{\psi_{\boldsymbol{c}}^*(\boldsymbol{u})=\boldsymbol{u}_1} {}^l\mathcal{G}_{\boldsymbol{u}}\big(\mathcal{H}(E)\big) \simeq {}^l\mathcal{G}_{\boldsymbol{u}_1}\big(\mathcal{H}(\psi^*E)\big).$$

Here $\psi_{\boldsymbol{c}}^*(\boldsymbol{u})$ denotes the element whose i-th component is $\boldsymbol{c}_i \cdot \boldsymbol{u} = \sum c_{i,j} \cdot u_j$ for $\boldsymbol{u} = (u_1, \ldots, u_l)$. It is compatible with the nilpotent maps.

Proof We have only to check that the restriction to each fiver over $\lambda \in \boldsymbol{C}^*$ is isomorphic. It easily follows from Lemma 9.31. □

CHAPTER 10

The Induced Regular λ-connection on $\Delta^n \times C^*$

10.1. The filtrations and the decompositions of $\mathcal{E}^\lambda$

10.1.1. The purpose. We put $X = \Delta^n$, $D_i = \{z_i = 0\}$ and $D = \bigcup_{i=1}^{l} D_i$. Let $(E, \overline{\partial}_E, \theta, h)$ be a tame harmonic bundle over $X - D$. Let λ be any element of C^*. Let $\boldsymbol{c}$ be an element of $\boldsymbol{R}^l$. In this section, we introduce the decompositions and the filtrations of $\mathcal{E}^\lambda$ and ${}_{\boldsymbol{c}}\mathcal{E}^\lambda$ on X.

Recall that we discussed the filtrations and the decompositions of ${}_{\boldsymbol{c}}\mathcal{E}^\lambda{}_{|D_I}$ in the sections 8.6 and 8.8. We also discussed the filtration and the decompositions of $H(\mathcal{E}^\lambda)$. We will see the relationships.

10.1.2. The compatibility of the decompositions. Let γ_i denote the loop around D_i with the anti-clockwise direction. It induces the monodromy endomorphisms M_i^λ on $\mathcal{E}^\lambda_{|P}$ for each $P \in X - D$. Hence we obtain the commuting tuple of the endomorphisms $\boldsymbol{M}^\lambda = \big(M_i^\lambda \,\big|\, i \in \underline{l}\big)$. Let $\mathcal{S}p(\boldsymbol{M}^\lambda)$ denote the set of the tuple of the eigenvalues of $\boldsymbol{M}^\lambda$. Then we have the following generalized eigen decomposition (see the subsection 2.7.2):

$$(10.1) \qquad \mathcal{E}^\lambda = \bigoplus_{\boldsymbol{\omega} \in \mathcal{S}p(\boldsymbol{M}^\lambda)} \mathcal{E}^\lambda_{\boldsymbol{\omega}}, \qquad \mathcal{E}^\lambda_{\boldsymbol{\omega}\,|\,P} = \mathbb{E}\big(\mathcal{E}^\lambda_{|P},\, \boldsymbol{\omega}\big).$$

Obviously we have $\mathbb{E}\big(H(\mathcal{E}^\lambda), \boldsymbol{\omega}\big) = H\big(\mathcal{E}^\lambda_{\boldsymbol{\omega}}\big)$.

Let $\boldsymbol{s} = (s_i)$ be a base of the space $H(\mathcal{E}^\lambda)$ of the multi-valued flat sections, which is compatible with the tuple of the decompositions and the filtrations $\big({}^i\mathbb{E}, {}^i\mathcal{F} \,\big|\, i = 1, \ldots, l\big)$. We put $\boldsymbol{a}_i := {}^{\underline{l}}\deg^{\mathcal{F}}(s_i)$. We put $v_i := F(s_i, \boldsymbol{a}_i - \boldsymbol{c})$, and then the tuples $\boldsymbol{v} = (v_i)$ obviously gives a frame of $\mathcal{E}^\lambda$ on $X - D$, which is compatible with the decomposition (10.1). We also know that $\boldsymbol{v} = (v_i)$ is a tuple of holomorphic sections of ${}_{\boldsymbol{c}}\mathcal{E}^\lambda$.

LEMMA 10.1. *The tuple $\boldsymbol{v}$ gives a frame of ${}_{\boldsymbol{c}}\mathcal{E}^\lambda$. The restriction of $\boldsymbol{v}$ to $\mathcal{D}_i(\lambda_0, \epsilon_0)$ is compatible with the decomposition ${}^i\mathbb{E}$ and the filtration ${}^iF$ of the vector bundle ${}_{\boldsymbol{c}}\mathcal{E}^\lambda{}_{|\mathcal{D}_i^\lambda}$.*

Proof We have only to show that $\boldsymbol{v}_{|\pi_j^{-1}(P)}$ gives a frame of ${}_{c_j}\big(\mathcal{E}^\lambda_{|\pi_j^{-1}(P)}\big)$ for any point $P \in D_j^\circ$. Then it follows from the result in the case of curves (the subsection 7.4.7). □

LEMMA 10.2. *The decomposition* (10.1) *is prolonged to the decomposition of the vector bundle ${}_{\boldsymbol{c}}\mathcal{E}^\lambda$. Namely, we have the following decomposition of ${}_{\boldsymbol{c}}\mathcal{E}^\lambda$:*

$${}_{\boldsymbol{c}}\mathcal{E}^\lambda = \bigoplus_{\boldsymbol{\omega}} {}_{\boldsymbol{c}}\mathcal{E}^\lambda_{\boldsymbol{\omega}}.$$

Proof We have only to use the frame $\boldsymbol{v}$ above. □

Let us compare two decompositions of ${}_{\boldsymbol{c}}\mathcal{E}_{|D_i}$. We have the restriction of the decomposition (10.1):

$$ {}_{\boldsymbol{c}}\mathcal{E}^{\lambda}{}_{|D_i} = \bigoplus_{\omega \in \mathcal{S}p(\boldsymbol{M}^\lambda)} {}_{\boldsymbol{c}}\mathcal{E}^{\lambda}{}_{\boldsymbol{\omega}\,|\,D_i}. $$

On the other hand, we have the decomposition of ${}_{\boldsymbol{c}}\mathcal{E}^{\lambda}{}_{|D_i}$ induced by the action of the residue $\mathrm{Res}_i(\mathbb{D}^\lambda)$:

$$ {}_{\boldsymbol{c}}\mathcal{E}^{\lambda}{}_{|D_i} = \bigoplus_{\beta \in \mathcal{S}p({}_{\boldsymbol{c}}\mathcal{E}^\lambda, i)} \mathbb{E}\big({}_{\boldsymbol{c}}\mathcal{E}^{\lambda}{}_{|D_i}, \beta\big). $$

The relation of the two decompositions are given in the following lemma.

LEMMA 10.3. *We have the following equality:*

$$ \bigoplus_{q_i(\boldsymbol{\omega})=\omega} {}_{\boldsymbol{c}}\mathcal{E}^{\lambda}{}_{\boldsymbol{\omega}\,|\,D_i} = \bigoplus_{\beta \in L(\lambda,\omega)} \mathbb{E}({}_{\boldsymbol{c}}\mathcal{E}^{\lambda}{}_{|D_i}, \beta). \tag{10.2} $$

Here we put $L(\lambda,\omega) := \big\{\beta \in \mathcal{S}p({}_{\boldsymbol{c}}\mathcal{E}^\lambda, i) \,\big|\, \exp(-2\pi\sqrt{-1}\lambda^{-1}\cdot\beta) = \omega\big\}$.

Proof Since the both sides of (10.2) are vector subbundles of ${}_{\boldsymbol{c}}\mathcal{E}^{\lambda}{}_{|D_i}$, we have only to show the equality for the fibers over the points $P \in D_i^\circ$. Thus we have only to consider the case $\dim(X) = 1$. Then it follows from Corollary 7.97. □

More generally, we have the two decompositions of ${}_{\boldsymbol{c}}\mathcal{E}^{\lambda}{}_{|D_I}$. The comparison immediately follows from Lemma 10.3.

COROLLARY 10.4. *Let I be a subset of $\underline{l}$. Let $\boldsymbol{\omega}^\circ = (\omega_i^\circ \,|\, i \in I)$ be an element of $\mathcal{S}p^f(\mathcal{E}^\lambda, I)$. Then we have the following:*

$$ \bigoplus_{q_I(\boldsymbol{\omega})=\boldsymbol{\omega}^\circ} {}_{\boldsymbol{c}}\mathcal{E}^{\lambda}{}_{\boldsymbol{\omega}\,|\,D_I} = \bigoplus_{\boldsymbol{\beta} \in L(\lambda,\boldsymbol{\omega}^\circ)} \mathbb{E}\big({}_{\boldsymbol{c}}\mathcal{E}^{\lambda}{}_{|D_I}, \boldsymbol{\beta}\big). $$

Here we put $L(\lambda,\boldsymbol{\omega}^\circ) := \big\{\boldsymbol{\beta} \in \mathcal{S}p({}_{\boldsymbol{c}}\mathcal{E}^\lambda, I) \,\big|\, \exp\big(-2\pi\sqrt{-1}\lambda^{-1}\cdot\beta_i\big) = \omega_i^\circ \ (i \in I)\big\}$. □

10.1.3. The compatibility of the filtrations. Recall that we have the filtrations ${}^i\mathcal{F}$ of $H\big(\mathcal{E}^\lambda\big)$, which is preserved by the monodromy endomorphisms. Therefore we have the induced filtrations ${}^i\mathcal{F}\big(\mathcal{E}^\lambda_{\boldsymbol{\omega}}\big)$ of $\mathcal{E}^\lambda_{\boldsymbol{\omega}}$ on $X - D$. Clearly we have ${}^{\underline{l}}\mathcal{F}_{\boldsymbol{a}}H(\mathcal{E}^\lambda) = H\big({}^{\underline{l}}\mathcal{F}_{\boldsymbol{a}}\mathcal{E}^\lambda\big)$. Let $\boldsymbol{s}$ and $\boldsymbol{v}$ be as in the subsection 10.1.2. Obviously, the frame $\boldsymbol{v}$ is compatible with the filtrations ${}^i\mathcal{F}$.

PROPOSITION 10.5.

- *The subbundle ${}^i\mathcal{F}_b\big(\mathcal{E}^\lambda_{\boldsymbol{\omega}}\big)$ is prolonged to the subbundle ${}^i\mathcal{F}_b\big({}_{\boldsymbol{c}}\mathcal{E}^\lambda_{\boldsymbol{\omega}}\big)$ of ${}_{\boldsymbol{c}}\mathcal{E}^\lambda_{\boldsymbol{\omega}}$.*
- *The family ${}^i\mathcal{F}\big({}_{\boldsymbol{c}}\mathcal{E}^\lambda_{\boldsymbol{\omega}}\big) = \big\{{}^i\mathcal{F}_a\big({}_{\boldsymbol{c}}\mathcal{E}^\lambda_{\boldsymbol{\omega}}\big) \,\big|\, a \in \boldsymbol{R}\big\}$ gives the filtration of the vector bundle ${}_{\boldsymbol{c}}\mathcal{E}^\lambda_{\boldsymbol{\omega}}$ in the category of vector bundles.*
- *The tuple of the filtrations $\big({}^i\mathcal{F}\big({}_{\boldsymbol{c}}\mathcal{E}^\lambda_{\boldsymbol{\omega}}\big) \,\big|\, i = 1,\ldots,l\big)$ of the vector bundle ${}_{\boldsymbol{c}}\mathcal{E}^\lambda$ is compatible.*

Proof It is easy to check the claims by using the frame $\boldsymbol{v}$ given in the first part of the subsection 10.1.2. □

Then we have the filtrations ${}^i\mathcal{F}\big({}_{\boldsymbol{c}}\mathcal{E}^\lambda_{\boldsymbol{\omega}}\big)_{|D_i}$ of the vector bundle ${}_{\boldsymbol{c}}\mathcal{E}^\lambda{}_{|D_i}$. On the other hand, we have the parabolic filtration ${}^iF$ of ${}_{\boldsymbol{c}}\mathcal{E}^\lambda{}_{|D_i}$. The relation of two filtrations is given in Proposition 10.6 below.

We recall that the number $\alpha(a,\omega) \in \boldsymbol{C}$ for $(a,\omega) \in \boldsymbol{R} \times \boldsymbol{C}$ is determined by the following conditions (the subsection 7.4.3):

$$\exp\big(-2\pi\sqrt{-1}\cdot\alpha(a,\omega)\big) = \omega, \qquad a \leq \operatorname{Re}\big(\alpha(a,\omega)\big) < a+1.$$

We put $d(a,\omega) := a - \operatorname{Re}(\alpha(a,\omega)) \in \boldsymbol{R}$.

PROPOSITION 10.6. *We have the following equality:*

$$\bigoplus_{q_i(\boldsymbol{\omega})=\omega} {}^i\mathcal{F}_b\big({}_{\boldsymbol{c}}\mathcal{E}^\lambda_{\boldsymbol{\omega}}\big)_{|D_i} = {}^iF_{d(b-c_i,\omega)}\big(\mathbb{E}\big({}_{\boldsymbol{c}}\mathcal{E}^\lambda{}_{|D_i}, \lambda\cdot\alpha(b-c_i,\omega)\big)\big) \oplus \bigoplus_{\beta\in K(\lambda,\omega,b)} \mathbb{E}\big({}_{\boldsymbol{c}}\mathcal{E}^\lambda{}_{|D_i}, \beta\big).$$

Here we put $K(\lambda,\omega,b) := \big\{\beta \in \mathcal{S}p(\mathcal{E}^\lambda, i) \,\big|\, \exp\big(-2\pi\sqrt{-1}\lambda^{-1}\cdot\beta\big) = \omega,\ \operatorname{Re}\big(\lambda^{-1}\cdot\beta\big) < \operatorname{Re}\big(\alpha(b-c_i,\omega)\big)\big\}$.

Proof The claim can be easily reduced to the case $\dim(X) = 1$ (Lemma 7.99) as in the proof of Lemma 10.3. □

10.1.4. The induced vector bundle ${}^{\underline{l}}\mathcal{G}_{\boldsymbol{u}}(\mathcal{E}^\lambda)$ on X. Let $\boldsymbol{u}$ be an element of $\mathcal{KMS}(\mathcal{E}^0, \underline{l})$. We put as follows:

$$\boldsymbol{b} := \mathfrak{p}^f(\lambda, \boldsymbol{u}), \qquad \boldsymbol{\omega} := \mathfrak{e}^f(\lambda, \boldsymbol{u}).$$

Let us take an element $\boldsymbol{c} \in \boldsymbol{R}^l$ such that $\operatorname{Gr}^F_{\mathfrak{p}(\lambda,\boldsymbol{u})}({}_{\boldsymbol{c}}\mathcal{E}^\lambda) \neq 0$. For example, we put $\boldsymbol{c} := \mathfrak{p}(\lambda, \boldsymbol{u})$. We obtain the holomorphic vector bundle ${}^{\underline{l}}\mathcal{G}_{\boldsymbol{u}}(\mathcal{E}^\lambda)$ over X, given as follows:

$${}^{\underline{l}}\mathcal{G}_{\boldsymbol{u}}(\mathcal{E}^\lambda) := \frac{{}^{\underline{l}}\mathcal{F}_{\boldsymbol{b}}\big({}_{\boldsymbol{c}}\mathcal{E}^\lambda_{\boldsymbol{\omega}}\big)}{\sum_{\boldsymbol{b}'\lneq\boldsymbol{b}} {}^{\underline{l}}\mathcal{F}_{\boldsymbol{b}'}\big({}_{\boldsymbol{c}}\mathcal{E}^\lambda_{\boldsymbol{\omega}}\big)}.$$

On the other hand, we have the vector bundle ${}^{\underline{n}}\mathcal{G}_{\boldsymbol{u}}(E)$ on $D_{\underline{l}} \times \boldsymbol{C}_\lambda$ (the section 8.9).

LEMMA 10.7. *We have the natural isomorphism:*

$${}^{\underline{l}}\mathcal{G}_{\boldsymbol{u}}\big(\mathcal{E}^\lambda\big)_{|D_{\underline{l}}} \simeq {}^{\underline{l}}\mathcal{G}_{\boldsymbol{u}}(E)_{|\{\lambda\}\times D_{\underline{l}}}.$$

Proof It follows from Lemma 10.3, Proposition 10.6 and the definitions. □

We have the induced λ-connection $\mathbb{D}^\lambda$ of ${}^{\underline{l}}\mathcal{G}_{\boldsymbol{u}}(\mathcal{E}^\lambda)$, which is flat and regular. Then we obtain the residues $\operatorname{Res}_i(\mathbb{D}^\lambda)$.

COROLLARY 10.8. $\operatorname{Res}_i(\mathbb{D}^\lambda)$ *has the unique eigenvalue* $\mathfrak{e}\big(\lambda, q_i(\boldsymbol{u})\big)$. □

Let $H\big({}^{\underline{l}}\mathcal{G}_{\boldsymbol{u}}(\mathcal{E}^\lambda)\big)$ be the space of the multi-valued flat sections of ${}^{\underline{l}}\mathcal{G}_{\boldsymbol{u}}(\mathcal{E}^\lambda)$. Naturally we have the following isomorphism:

$$H\big({}^{\underline{l}}\mathcal{G}_{\boldsymbol{u}}(\mathcal{E}^\lambda)\big) \simeq {}^{\underline{l}}\operatorname{Gr}^{\mathcal{F}}_{\boldsymbol{b}}\,\mathbb{E}\big(H(\mathcal{E}^\lambda), \boldsymbol{\omega}\big) = {}^{\underline{l}}\mathcal{G}_{\boldsymbol{u}}(\mathcal{H})_{|\lambda}.$$

10.2. The decompositions $\mathbb{E}^{(\lambda_0)}$ and the filtrations $F^{(\lambda_0)}$ on $\mathcal{E}$ for $\lambda_0 \neq 0$

10.2.1. The purpose. We continue to use the setting in the section 10.1. We recall that we put $\mathcal{Y}(\lambda_0, \epsilon_0) := Y \times \Delta(\lambda_0, \epsilon_0)$ and $\mathcal{Y}^\lambda := Y \times \{\lambda\}$ for a complex variety Y. Let λ_0 be any element of $\boldsymbol{C}^*$. Let $\boldsymbol{b}$ be an element of $\boldsymbol{R}^l$ such that $b_i \notin \mathcal{KMS}(\mathcal{E}^{\lambda_0})$ for each i. Then we have the locally free sheaf ${}_{\boldsymbol{b}}\mathcal{E}^\lambda$ on $\mathcal{X}(\lambda_0, \epsilon_0) = X \times \Delta(\lambda_0, \epsilon_0)$ for some sufficiently small positive number ϵ_0.

In this section, we would like to discuss the filtrations and the decompositions of $\mathcal{E}$ and ${}_{\boldsymbol{b}}\mathcal{E}$ on $\mathcal{X}(\lambda_0, \epsilon_0)$. We would like to compare them with the filtrations and the decompositions ${}^iF^{(\lambda_0)}$ and ${}^i\mathbb{E}^{(\lambda_0)}$ of ${}_{\boldsymbol{b}}\mathcal{E}_{|\mathcal{D}_i(\lambda_0,\epsilon_0)}$ in the section 8.8. We would also like to compare them with the filtrations and the decompositions ${}^i\mathcal{F}^{(\lambda_0)}$ and ${}^i\mathbb{E}^{(\lambda_0)}$ of $\mathcal{H}_{|\Delta(\lambda_0,\epsilon_0)}$ in the subsection 9.1.5.

10.2.2. The decompositions and the filtrations of $\mathcal{E}$. Recall that we have the decomposition $\mathcal{H} = \bigoplus_{\boldsymbol{\omega} \in \mathcal{S}p(\boldsymbol{M}^{\lambda_0})} \mathcal{H}_{\boldsymbol{\omega}}^{(\lambda_0)}$ and the filtrations ${}^i\mathcal{F}^{(\lambda_0)}$ $(i = 1, \ldots, l)$ of $\mathcal{H}_{|\Delta(\lambda_0,\epsilon_0)}$ for any sufficiently small positive number ϵ_0. Since they are preserved by the monodromy endomorphisms. they induce the decomposition $\mathcal{E} = \bigoplus_{\boldsymbol{\omega} \in \mathcal{S}p(\boldsymbol{M}^{\lambda_0})} \mathcal{E}_{\boldsymbol{\omega}}^{(\lambda_0)}$ and the filtrations ${}^i\mathcal{F}^{(\lambda_0)}(\mathcal{E}_{\boldsymbol{\omega}})$ over $\mathcal{X}(\lambda_0, \epsilon_0) - \mathcal{D}(\lambda_0, \epsilon_0)$.

REMARK 10.9. The decomposition $\mathcal{E} = \bigoplus_{\boldsymbol{\omega} \in \mathcal{S}p(\boldsymbol{M}^{\lambda_0})} \mathcal{E}_{\boldsymbol{\omega}}^{(\lambda_0)}$ will be often denoted by ${}^{\underline{l}}\mathbb{E}$. □

10.2.3. The prolongation of the decomposition and the filtrations. Let $\boldsymbol{b}$ be an element of $\boldsymbol{R}^l$ such that $b_i \notin \mathcal{P}ar(\mathcal{E}^{\lambda_0}, i)$. If ϵ_0 is sufficiently small, we have the locally free sheaf ${}_{\boldsymbol{b}}\mathcal{E}$ over $\mathcal{X}(\lambda_0, \epsilon_0)$. We may also assume that any $\lambda \in \Delta^*(\lambda_0, \epsilon_0)$ are generic.

PROPOSITION 10.10.

(1) *The vector subbundle* ${}^i\mathcal{F}_c^{(\lambda_0)}\big(\mathcal{E}_{\boldsymbol{\omega}}^{(\lambda_0)}\big)$ *of* $\mathcal{E}$ *is prolonged to the subbundle* ${}^i\mathcal{F}_c^{(\lambda_0)}\big({}_{\boldsymbol{b}}\mathcal{E}_{\boldsymbol{\omega}}^{(\lambda_0)}\big)$ *of* ${}_{\boldsymbol{b}}\mathcal{E}$ *on* $\mathcal{X}(\lambda_0, \epsilon_0)$.

(2) *We have* ${}_{\boldsymbol{b}}\mathcal{E}^{(\lambda_0)}_{\boldsymbol{\omega}\,|\,\mathcal{X}^{\lambda_0}} = {}_{\boldsymbol{b}}\mathcal{E}^{\lambda_0}_{\boldsymbol{\omega}}$. *On the other hand, we have the following decomposition, for any* $\lambda \in \Delta^*(\lambda_0, \epsilon_0)$:
$$ {}_{\boldsymbol{b}}\mathcal{E}^{(\lambda_0)}_{\boldsymbol{\omega}\,|\,\mathcal{X}^\lambda} = \bigoplus_{\boldsymbol{u} \in \mathcal{S}(\boldsymbol{\omega})} {}_{\boldsymbol{b}}\mathcal{E}^\lambda_{\mathfrak{e}^f(\lambda,\boldsymbol{u})}. $$
Here we put $\mathcal{S}(\boldsymbol{\omega}) := \big\{\boldsymbol{u} \in \overline{\mathcal{KMS}}(\mathcal{E}^0, \underline{l}) \,\big|\, \mathfrak{e}^f(\lambda_0, \boldsymbol{u}) = \boldsymbol{\omega}\big\}$.

(3) *We have* ${}^i\mathcal{F}^{(\lambda_0)}\big({}_{\boldsymbol{b}}\mathcal{E}_{\boldsymbol{\omega}}^{(\lambda_0)}\big)_{|\mathcal{X}^{\lambda_0}} = {}^i\mathcal{F}\big({}_{\boldsymbol{b}}\mathcal{E}_{\boldsymbol{\omega}}^{\lambda_0}\big)$. *On the other hand, we have the decomposition, for any* $\lambda \in \Delta^*(\lambda_0, \epsilon_0)$:
$$ {}^i\mathcal{F}_c^{(\lambda_0)}\big({}_{\boldsymbol{b}}\mathcal{E}_{\boldsymbol{\omega}}^{(\lambda_0)}\big)_{|\mathcal{X}^\lambda} = \bigoplus_{\boldsymbol{u} \in \mathcal{S}(\boldsymbol{\omega},c,i)} {}^i\mathcal{F}_{\mathfrak{p}^f(\lambda, q_i(\boldsymbol{u}))}\big({}_{\boldsymbol{b}}\mathcal{E}^\lambda_{\mathfrak{e}^f(\lambda,\boldsymbol{u})}\big) = \bigoplus_{\boldsymbol{u} \in \mathcal{S}(\boldsymbol{\omega},c,i)} {}_{\boldsymbol{b}}\mathcal{E}^\lambda_{\mathfrak{e}^f(\lambda,\boldsymbol{u})}. $$
Here we put as follows:
$$ \mathcal{S}(\boldsymbol{\omega}, c, i) := \big\{\boldsymbol{u} \in \overline{\mathcal{KMS}}(\mathcal{E}^0, \underline{l}) \,\big|\, \mathfrak{e}^f(\lambda_0, \boldsymbol{u}) = \boldsymbol{\omega},\ \mathfrak{p}^f\big(\lambda_0, q_i(\boldsymbol{u})\big) \leq c\big\}. $$

(4) *The tuple of the filtrations* $\big({}^i\mathcal{F}^{(\lambda_0)} \,\big|\, i = 1, \ldots, l\big)$ *of* ${}_{\boldsymbol{b}}\mathcal{E}$ *is compatible.*

Proof Once we know the claim (1), then the rest follows from the result at the specializations (Lemma 10.3, Proposition 10.5 and Proposition 10.6). Hence we have only to check the claim (1). We may assume that $\boldsymbol{b} = 0$. We remark that $0 \notin \mathcal{KMS}(\mathcal{E}^{\lambda_0}, i)$ in the case. We put $\mathcal{K}(\mathcal{E}^0, \lambda_0, \boldsymbol{0}, \underline{l}) := \big\{u \in \mathcal{KMS}(\mathcal{E}^0, \underline{l}) \,\big|\, -\boldsymbol{\delta} <$

$\mathfrak{p}(\lambda_0, u) < \mathbf{0}\}$. We also remark the natural bijection $\mathcal{K}(\mathcal{E}^0, \lambda_0, \mathbf{0}, \underline{l}) \simeq \overline{\mathcal{KMS}}(\mathcal{E}^0, \underline{l})$. We will use the following lemma.

LEMMA 10.11. *Let $\eta > 0$ be positive numbers such that* $\operatorname{rank}(E) \cdot \eta < 1$. *Assume $\mathcal{P}ar(^\diamond\mathcal{E}^\lambda, i) \subset] - \eta, 0[$ for any $\lambda \in \Delta(\lambda_0, \epsilon_0)$ and for any i. Then the filtration ${}^i\mathcal{F}_a^{(\lambda_0)}\mathcal{E}_{\boldsymbol{\omega}}$ can be prolonged to the subbundle of $^\diamond\mathcal{E}$ over $\mathcal{X}(\lambda_0, \epsilon_0)$.*

Proof Let $\boldsymbol{s}$ be a frame of $\mathcal{H}$ over $\Delta(\lambda_0, \epsilon_0)$ compatible with $\mathbb{E}^{(\lambda_0)}$ and $\mathcal{F}^{(\lambda_0)}$. We put $R := \operatorname{rank}\big({}^i\mathcal{F}_a^{(\lambda_0)}(\mathcal{E}_{\boldsymbol{\omega}}^{(\lambda_0)})\big)$. Due to the assumption of Lemma 10.11, we have $\bigwedge^R(^\diamond\mathcal{E}) = \left(^\diamond\bigwedge^R\mathcal{E}\right)$ over $\mathcal{X}(\lambda_0, \epsilon_0)$.

For each $s_j \in {}^i\mathcal{F}_a^{(\lambda_0)}\mathcal{H}_{\boldsymbol{\omega}}^{(\lambda_0)}$, there exist the element $\boldsymbol{u}'(s_j) \in \mathcal{S}(\boldsymbol{\omega}, c, i)$ such that $\deg^{\mathcal{F}^{(\lambda_0)}, \mathbb{E}^{(\lambda_0)}}(s_j) = \mathfrak{k}^f(\lambda_0, \boldsymbol{u}'(s_j))$. Let $\boldsymbol{u}(s_j)$ be the element corresponding to $\boldsymbol{u}'(s_j)$ via the bijection $\mathcal{K}(\mathcal{E}, \lambda_0, \mathbf{0}, i) \simeq \overline{\mathcal{KMS}}(\mathcal{E}^0, \underline{l})$. We put as follows:

$$\boldsymbol{u}_0 = \sum_{s_j \in {}^i\mathcal{F}_c^{(\lambda_0)}\mathcal{H}_{\boldsymbol{\omega}}^{(\lambda_0)}} \boldsymbol{u}(s_j).$$

We remark that $-\boldsymbol{\delta} < \mathfrak{p}(\lambda_0, \boldsymbol{u}_0) < \mathbf{0}$.

On the other hand, we obtain the naturally induced frame $\tilde{\boldsymbol{s}} = (\tilde{s}_j)$ of $\bigwedge^R\mathcal{H}$ over $\Delta(\lambda_0, \epsilon_0)$ from the frame $\boldsymbol{s}$ of $\mathcal{H}$ over $\Delta(\lambda_0, \epsilon_0)$. The frame $\tilde{\boldsymbol{s}}$ is compatible with $\mathbb{E}^{(\lambda_0)}$ and $\mathcal{F}^{(\lambda_0)}$. There exists j_0 such that $\tilde{s}_{j_0}$ gives a frame of the line bundle $\bigwedge^R({}^i\mathcal{F}_a^{(\lambda_0)}\mathcal{H}_{\boldsymbol{\omega}}^{(\lambda_0)})$. We remark $\deg^{\mathcal{F}^{(\lambda_0)}, \mathbb{E}^{(\lambda_0)}}(\tilde{s}_{j_0}) = \mathfrak{k}^f(\lambda_0, \boldsymbol{u}_0)$. Then we put as follows:

$$v = \exp\Big(\log z \cdot \big(\lambda^{-1} \cdot \mathfrak{e}(\lambda, \boldsymbol{u}_0)\big)\Big) \cdot \tilde{s}_{j_0}.$$

Due to $\mathfrak{p}(\lambda_0, \boldsymbol{u}_0) < 0$, the section v is a holomorphic section of $^\diamond\left(\bigwedge^R\mathcal{E}\right)$ over $\mathcal{X}(\lambda_0, \epsilon_0)$. It is easy to see that Lemma 10.11 can be reduced to the following lemma.

LEMMA 10.12. *There exists $\eta > 0$ and a neighbourhood U of O in X, such that $v_{|(\lambda, P)} \neq 0$ for any $(\lambda, P) \in \Delta(\lambda_0, \eta) \times U$.*

Proof The section $v_{|\mathcal{X}^{\lambda_0}}$ gives an element of the frame of $^\diamond\bigwedge^R\mathcal{E}^{\lambda_0}$ induced by $\tilde{\boldsymbol{s}}_{|(X-D)\times\{\lambda_0\}}$, as in the subsection 10.1.2. Then we obtain Lemma 10.12 and Lemma 10.11. □

Let us return to the proof of Proposition 10.10. We can pick $\boldsymbol{c} \in \mathbb{Z}_{>0}^l$ and $\boldsymbol{b} \in \boldsymbol{R}^l$ such that $^\diamond\psi_{\boldsymbol{c}}^{-1}\mathcal{E} \otimes L(\boldsymbol{b})$ satisfies the condition of Lemma 10.11 on $\Delta(\lambda_0, \epsilon_0)$. Hence $\psi_{\boldsymbol{c}}^{-1}({}^i\mathcal{F}_a^{(\lambda_0)}\mathcal{E}_{\boldsymbol{\omega}})$ can be prolonged to the $\mu_{\boldsymbol{c}}$-equivariant subbundle of $^\diamond\psi_{\boldsymbol{c}}^{-1}\mathcal{E}$. Then we pick the equivariant frame, and take the descent of the frame. (See the argument in the subsection 8.6.3). Then it follows that ${}^i\mathcal{F}^{(\lambda_0)}\mathcal{E}_{\boldsymbol{\omega}}^{(\lambda_0)}$ can be prolonged to the subbundle of $^\diamond\mathcal{E}_{\boldsymbol{\omega}}^{(\lambda_0)}$. Thus we obtain the claim (1) of Proposition 10.10. □

10.3. The induced regular λ-connection

10.3.1. The purpose. We continue to use the setting in the section 10.1. By taking the associated graded vector bundles, we obtain the vector bundle ${}^{\underline{l}}\mathcal{G}_{\boldsymbol{u}}(\mathcal{E})$ with regular λ-connection on $\mathcal{X}^\sharp = X \times C^*$, for each $\boldsymbol{u} \in \mathcal{KMS}(\mathcal{E}^0, \underline{l})$. In some sense, they give the interpolation of $\mathcal{G}_{\boldsymbol{u}}(\mathcal{H})$ and $\mathcal{G}_{\boldsymbol{u}}$.

10.3.2. The induced bundle ${}^{\underline{l}}\mathcal{G}_{\boldsymbol{u}}(\mathcal{E})$. Let $\boldsymbol{u}$ be an element of $\mathcal{KMS}(\mathcal{E}^0,\underline{l})$. Let λ_0 be any element of $\boldsymbol{C}_\lambda^*$. We put $(\boldsymbol{b},\boldsymbol{\omega}) = \mathfrak{k}^f(\lambda_0,\boldsymbol{u})$. We take any $\boldsymbol{c}\in\boldsymbol{R}^l$ such that ${}^{\underline{l}}\mathrm{Gr}^F_{\mathfrak{p}(\lambda_0,\boldsymbol{u})}\bigl({}_{\boldsymbol{c}}\mathcal{E}^{\lambda_0}\bigr)\neq 0$. Pick any sufficiently small ϵ_0. Then we have the following vector bundle over $\mathcal{X}(\lambda_0,\epsilon_0)$:

$$
{}^{\underline{l}}\mathcal{G}_{\boldsymbol{u}}^{(\lambda_0)}(\mathcal{E}) := {}^{\underline{l}}\mathrm{Gr}_{\boldsymbol{b}}^{\mathcal{F}^{(\lambda_0)}}\bigl({}_{\boldsymbol{c}}\mathcal{E}_{\boldsymbol{\omega}}^{(\lambda_0)}\bigr).
$$

The following lemma is clear from our construction.

LEMMA 10.13. *It is independent of a choice of* $\boldsymbol{c}$ *such that* ${}^{\underline{l}}\mathrm{Gr}^F_{\mathfrak{p}(\lambda_0,\boldsymbol{u})}\bigl({}_{\boldsymbol{c}}\mathcal{E}^{\lambda_0}\bigr)\neq 0$ *on a neighbourhood of* λ_0. □

We may assume that any point $\lambda\in\Delta^*(\lambda_0,\epsilon_0)$ is generic. Pick a positive number ϵ_0' such that $\Delta(\lambda,\epsilon_0')\subset\Delta^*(\lambda_0,\epsilon_0)$. We may assume that we have the vector bundle ${}^{\underline{l}}\mathcal{G}_{\boldsymbol{u}}^{(\lambda)}(\mathcal{E})$ on $\mathcal{X}(\lambda,\epsilon_0')$.

LEMMA 10.14. *We have the following natural isomorphism:*

$$
{}^{\underline{l}}\mathcal{G}_{\boldsymbol{u}}^{(\lambda_0)}(\mathcal{E})_{|\mathcal{X}(\lambda,\epsilon_0')} \simeq {}^{\underline{l}}\mathcal{G}_{\boldsymbol{u}}^{(\lambda)}(\mathcal{E}).
$$

Proof Let us take any element $\boldsymbol{c}\in\boldsymbol{R}^l$ such that ${}^{\underline{l}}\mathrm{Gr}^F_{\mathfrak{p}(\lambda_0,\boldsymbol{u})}\bigl({}_{\boldsymbol{c}}\mathcal{E}^{\lambda_0}\bigr)\neq 0$. We have the following decomposition:

$$
{}_{\boldsymbol{c}}\mathcal{E}_{\boldsymbol{\omega}}^{(\lambda_0)}{}_{|\mathcal{X}(\lambda,\epsilon_0')} = \bigoplus_{\boldsymbol{u}'\in\mathcal{S}(\boldsymbol{\omega})} {}_{\boldsymbol{c}}\mathcal{E}^{(\lambda)}_{\mathfrak{e}^f(\lambda,\boldsymbol{u}')},\qquad \mathcal{S}(\boldsymbol{\omega}) := \bigl\{\boldsymbol{u}'\in\overline{\mathcal{KMS}}(\mathcal{E},\underline{l})\bigm|\mathfrak{e}^f(\lambda_0,\boldsymbol{u}')=\boldsymbol{\omega}\bigr\}.
$$

Then we obtain the following isomorphism:

$$
{}^{\underline{l}}\mathcal{G}_{\boldsymbol{u}}^{(\lambda_0)}(\mathcal{E})_{|\mathcal{X}(\lambda,\epsilon_0')} \simeq {}^{\underline{l}}\mathrm{Gr}_{\boldsymbol{b}}^{\mathcal{F}^{(\lambda_0)}}\bigl({}_{\boldsymbol{c}}\mathcal{E}_{\boldsymbol{\omega}}^{(\lambda_0)}\bigr)_{|\mathcal{X}(\lambda,\epsilon_0')} \simeq {}_{\boldsymbol{c}}\mathcal{E}^{(\lambda)}_{\mathfrak{e}^f(\lambda,\boldsymbol{u})} \simeq {}^{\underline{l}}\mathcal{G}_{\boldsymbol{u}}^{(\lambda)}(\mathcal{E}).
$$

Hence we are done. □

Hence we obtain the vector bundle ${}^{\underline{l}}\mathcal{G}_{\boldsymbol{u}}(\mathcal{E})$ over $\mathcal{X}^\sharp = \boldsymbol{C}_\lambda^*\times X$ with the induced regular λ-connection $\mathbb{D}$.

LEMMA 10.15. *Let* $\boldsymbol{u}$ *be an element of* $\mathcal{KMS}(\mathcal{E}^0,\underline{l})$.

- *We have the following isomorphism on* $\mathcal{D}_{\underline{l}}^\sharp = D_{\underline{l}}\times\boldsymbol{C}_\lambda^*$:

$$
\bigl({}^{\underline{l}}\mathcal{G}_{\boldsymbol{u}}(\mathcal{E})_{|\mathcal{D}_{\underline{l}}^\sharp},\ \mathrm{Res}_i(\mathbb{D})\bigr) \simeq \bigl({}^{\underline{l}}\mathcal{G}_{\boldsymbol{u}},\ \mathcal{N}_i+\mathfrak{e}(\lambda,u_i)\bigr)_{|\mathcal{D}_{\underline{l}}^\sharp}.
$$

- *Taking the multi-valued flat sections of* ${}^{\underline{l}}\mathcal{G}_{\boldsymbol{u}}(\mathcal{E})$, *we obtain the vector bundle* ${}^{\underline{l}}\mathcal{G}_{\boldsymbol{u}}(\mathcal{H})$ *over* $\boldsymbol{C}_\lambda^*$.

Proof It immediately follows from our construction. □

10.3.3. Pairing. From the natural pairing $\mathcal{E}\otimes\mathcal{E}^\vee\longrightarrow\mathcal{O}_{\mathcal{X}-\mathcal{D}}$, we obtain the following morphism on $\mathcal{X}(\lambda_0,\epsilon_0)$ for $\lambda_0\in\boldsymbol{C}_\lambda^*$ and for any sufficiently small $\epsilon_0>0$:

$$
{}_{\boldsymbol{b}}\mathcal{E}\otimes{}_{-\boldsymbol{b}+(1-\epsilon)\boldsymbol{\delta}}\bigl(\mathcal{E}^\vee\bigr)\longrightarrow\mathcal{O}_{\mathcal{X}}.
$$

Since it preserves the filtrations $\mathcal{F}^{(\lambda_0)}$ and the decompositions $\mathbb{E}^{(\lambda_0)}$, we obtain the following morphism of regular λ-connections:

$$
{}^{\underline{l}}\mathcal{G}_{\boldsymbol{u}}(\mathcal{E})\otimes{}^{\underline{l}}\mathcal{G}_{-\boldsymbol{u}}(\mathcal{E}^\vee)\longrightarrow\mathcal{O}_{\mathcal{X}}. \tag{10.3}
$$

10.3.4. Functoriality. First we see the functoriality for dual.

LEMMA 10.16. *Let $\boldsymbol{u}$ be an element of $\mathcal{KMS}(\mathcal{E}^0,\underline{l})$. We have the naturally defined isomorphism of regular λ-connections:*

$$ {}^{\underline{l}}\mathcal{G}_{-\boldsymbol{u}}(\mathcal{E}^{\vee}) \longrightarrow {}^{\underline{l}}\mathcal{G}_{\boldsymbol{u}}(\mathcal{E})^{\vee}. $$

The restriction of the morphisms to $D_{\underline{l}} \times \boldsymbol{C}^*$ are same as the morphisms in Lemma 8.120 *under the identification in Lemma* 10.15.

Proof The morphism is induced by the pairing (10.3). We can check that it is isomorphic by using a local holomorphic frame and the weak norm estimate. □

To state the functoriality for tensor product, we begin with an obvious remark. Let $\boldsymbol{u}_i$ and $\boldsymbol{u}'_i$ be elements of $\mathcal{KMS}(\mathcal{E}_i^0,\underline{l})$ for $i = 1,2$. Assume that $\boldsymbol{u}_1 + \boldsymbol{u}_2 = \boldsymbol{u}'_1 + \boldsymbol{u}'_2$, then we have the naturally defined isomorphism of regular λ-connections:

$$ {}^{\underline{l}}\mathcal{G}_{\boldsymbol{u}_1}(\mathcal{E}_1) \otimes {}^{\underline{l}}\mathcal{G}_{\boldsymbol{u}_2}(\mathcal{E}_2) \simeq {}^{\underline{l}}\mathcal{G}_{\boldsymbol{u}'_1}(\mathcal{E}_1) \otimes {}^{\underline{l}}\mathcal{G}_{\boldsymbol{u}'_2}(\mathcal{E}_2). $$

Namely, the tensor product ${}^{\underline{l}}\mathcal{G}_{\boldsymbol{u}_1}(\mathcal{E}_1) \otimes {}^{\underline{l}}\mathcal{G}_{\boldsymbol{u}_2}(\mathcal{E}_2)$ depends only on the element:

$$ (\boldsymbol{u}_1,\boldsymbol{u}_2) \in T := \mathcal{KMS}(\mathcal{E}_1^0,\underline{l}) \times \mathcal{KMS}(\mathcal{E}_2^0,\underline{l}) \big/ \mathbb{Z}^l. \tag{10.4} $$

Here the action ρ of $\mathbb{Z}^l$ is given as follows: We have the natural actions ρ_i of $\mathbb{Z}^l$ on $\mathcal{KMS}(\mathcal{E}_1^0,\underline{l})$ and $\mathcal{KMS}(\mathcal{E}_2^0,\underline{l})$. Then ρ is given by $\rho_1 \times (-\rho_2)$.

Therefore we put as follows for any $\boldsymbol{u} \in \mathcal{KMS}(\mathcal{E}_1^0 \otimes \mathcal{E}_2^0)$:

$$ T(\boldsymbol{u}) := \big\{(\boldsymbol{u}_1,\boldsymbol{u}_2) \in T \,\big|\, \boldsymbol{u}_1 + \boldsymbol{u}_2 = \boldsymbol{u}\big\}. \tag{10.5} $$

LEMMA 10.17. *For any element $\boldsymbol{u} \in \mathcal{KMS}(\mathcal{E}_1^0 \otimes \mathcal{E}_2^0)$, the following morphism is isomorphic:*

$$ \bigoplus_{(\boldsymbol{u}_1,\boldsymbol{u}_2)\in T(\boldsymbol{u})} {}^{\underline{l}}\mathcal{G}_{\boldsymbol{u}_1}(\mathcal{E}_1) \otimes {}^{\underline{l}}\mathcal{G}_{\boldsymbol{u}_2}(\mathcal{E}_2) \longrightarrow {}^{\underline{l}}\mathcal{G}_{\boldsymbol{u}}(\mathcal{E}_1 \otimes \mathcal{E}_2). $$

Here the set $T(\boldsymbol{u})$ is given as in (10.5). *The restriction of the morphisms to $D_{\underline{l}} \times \boldsymbol{C}^*$ are same as the morphisms in Lemma* 8.121 *under the identification in Lemma* 10.15.

Proof By using the weak norm estimate, we can easily check that it is isomorphic on $\mathcal{X}^{\sharp}$. The compatibility follows from our construction. □

We also have the functoriality for the pull backs. It is slightly complicated to state precisely. We use the setting in the subsection 8.9.7. Let $\boldsymbol{u}_1$ be an element of $\mathcal{KMS}(\psi^{-1}\mathcal{E}^0,\underline{n})$. Let $T(\boldsymbol{u}_1)$ be the subset of $\mathcal{KMS}(\mathcal{E}^0,\underline{m})$ satisfying the following conditions:

- The i-th component of $\boldsymbol{u}_1 - \boldsymbol{c}\cdot\boldsymbol{u}$ is contained in $\{a \in \mathbb{Z} \,|\, 0 \leq a \leq c_i - 1\} \times \{0\} \subset \boldsymbol{R} \times \boldsymbol{C}$ for each i.

For each $\boldsymbol{u} \in T(\boldsymbol{u}_1)$, we put $\boldsymbol{n}(\boldsymbol{u}) := \boldsymbol{u}_1 - \boldsymbol{c}\cdot\boldsymbol{u}$ whose i-th component is denoted by $n_i(\boldsymbol{u})$. We also put $z^{-\boldsymbol{n}(\boldsymbol{u})} := \prod_{i=1}^{n} z_i^{-n_i(\boldsymbol{u})}$. We have the naturally induced injection $\psi^*\big({}^{\underline{m}}\mathcal{G}_{\boldsymbol{u}}(\mathcal{E})\big) \longrightarrow {}^{\underline{n}}\mathcal{G}_{\boldsymbol{u}_1}(\psi^*\mathcal{E})$. Moreover it is easy to see that we obtain the subsheaf $z^{-\boldsymbol{n}(\boldsymbol{u})}\cdot\psi^{*\underline{n}}\mathcal{G}_{\boldsymbol{u}}(\mathcal{E})$ of ${}^{\underline{n}}\mathcal{G}_{\boldsymbol{u}_1}(\psi^*\mathcal{E})$ which is preserved by the λ-connection. Therefore we obtain the following naturally defined morphism of λ-connections:

$$ \bigoplus_{\boldsymbol{u}\in T(\boldsymbol{u}_1)} z^{-\boldsymbol{n}(\boldsymbol{u})}\cdot\psi^{*\underline{m}}\mathcal{G}_{\boldsymbol{u}}(\mathcal{E}) \simeq {}^{\underline{n}}\mathcal{G}_{\boldsymbol{u}_1}(\psi^*\mathcal{E}). \tag{10.6} $$

LEMMA 10.18. *The morphism* (10.6) *is isomorphic. The restriction to* $D_{\underline{n}} \times \boldsymbol{C}^*$ *is same as the isomorphism in Lemma* 8.127 *under the identification in Lemma* 10.15.

Proof Due to Lemma 9.37, the morphism is isomorphic on $(X^{(1)} - D^{(1)}) \times \boldsymbol{C}^*$. It is easy to check the compatibility with the morphism in Lemma 8.127 by our construction. In particular, the morphism is isomorphic on $X^{(1)} \times \boldsymbol{C}^*$. □

10.4. Some morphisms between the induced vector bundles

10.4.1. The induced morphism $\Phi^{\mathrm{can}}_{\boldsymbol{u}}$. From the regular flat λ-connection $\big({}^{l}\mathcal{G}_{\boldsymbol{u}}(\mathcal{E}), \mathbb{D}\big)$, we obtain the isomorphism $\Phi^{\mathrm{can}}_{\boldsymbol{u}} : {}^{l}\mathcal{G}_{\boldsymbol{u}}\mathcal{H} \longrightarrow {}^{l}\mathcal{G}_{\boldsymbol{u}\,|\,\{O\}\times \boldsymbol{C}^*_{\lambda}}$, which we will explain. For any holomorphic section s of ${}^{l}\mathcal{G}_{\boldsymbol{u}}(\mathcal{H})$ over an open subset $U \subset \boldsymbol{C}^*$, we put as follows:

$$v = \exp\Big(\sum_j \log z_j \cdot \big(\lambda^{-1} \cdot \mathfrak{e}(\lambda, u_j) + \mathcal{N}_{j\,\boldsymbol{u}}\big)\Big) \cdot s.$$

Then it gives the holomorphic section of the vector bundle ${}^{l}\mathcal{G}_{\boldsymbol{u}}(\mathcal{E})$ over $U \times X$. Then the restriction $v_{|U\times\{O\}}$ is a section of ${}^{l}\mathcal{G}_{\boldsymbol{u}}$ over U. We put $\Phi^{\mathrm{can}}_{\boldsymbol{u}}(s) := v_{|U\times\{O\}}$, and then we obtain the isomorphism $\Phi^{\mathrm{can}}_{\boldsymbol{u}}$ desired.

The morphism $\Phi^{\mathrm{can}}_{\boldsymbol{u}}$ can be also seen as follows: Let s be a section of ${}^{l}\mathcal{G}_{\boldsymbol{u}}\mathcal{H}$. We have the expression as follows:

$$s = \exp\Big(-\sum_j \log z_j \cdot \big(\lambda^{-1} \cdot \mathfrak{e}(\lambda, u_j) + \mathcal{N}_{j,\boldsymbol{u}}\big)\Big) \cdot v = \prod_{j=1}^{l} z_j^{-\lambda^{-1}\mathfrak{e}(\lambda,u_j)} \sum_J (\log z)^J \cdot v_J.$$

Here $J = (j_1, \ldots, j_l)$ denote multi-indices and $(\log z)^J = \prod_{h=1}^{l} (\log z_h)^{j_h}$.

LEMMA 10.19. *We have* $\Phi^{\mathrm{can}}_{\boldsymbol{u}}(s) = v_0(O)$.

Proof It immediately follows from our construction. □

LEMMA 10.20. *The isomorphism* $\Phi^{\mathrm{can}}_{\boldsymbol{u}}$ *preserves the morphisms* $\mathcal{N}_{\boldsymbol{u}\,i}$ *and the pairing. It is compatible with duals.*

Proof It is clear from our construction. □

10.4.2. The induced morphisms $\Phi_{\boldsymbol{u},P,O}$. For any point $P \in X$, we have the isomorphisms of ${}^{l}\mathcal{G}_{\boldsymbol{u}}$ and ${}^{l}\mathcal{G}_{\boldsymbol{u}}(\mathcal{E})_{|\boldsymbol{C}^*\times\{P\}}$. For simplicity of notation, we denote ${}^{l}\mathcal{G}_{\boldsymbol{u}}(\mathcal{E})_{|\boldsymbol{C}^*\times\{P\}}$ by ${}^{l}\mathcal{G}_{\boldsymbol{u}}(\mathcal{E})_{|P}$.

Take a normalizing frame $\boldsymbol{v}$ of ${}^{l}\mathcal{G}_{\boldsymbol{u}}(\mathcal{E})$, namely we take a holomorphic frame $\boldsymbol{v}$ of ${}^{l}\mathcal{G}_{\boldsymbol{u}}(\mathcal{E})$ such that $\mathbb{D}\boldsymbol{v} = \boldsymbol{v} \cdot \sum A_i \frac{dz_i}{z_i}$ holds for some constant matrices A_i. For any points $P, Q \in X$, the trivialization $\boldsymbol{v}$ gives the isomorphism $\Phi_{\boldsymbol{u},P,Q} : {}^{l}\mathcal{G}_{\boldsymbol{u}}(\mathcal{E})_{|P} \longrightarrow {}^{l}\mathcal{G}_{\boldsymbol{u}}(\mathcal{E})_{|Q}$, by the correspondence $v_{i\,|\,P} \longmapsto v_{i\,|\,Q}$. If we fix the coordinate $(z_1, \ldots, z_n)$ of $X = \Delta^n$, the $\Phi_{\boldsymbol{u},P,Q}$ does not depend on a choice of normalizing frame. Note that we have the isomorphism:

$${}^{l}\mathcal{G}_{\boldsymbol{u}}(\mathcal{E})_{|\{O\}\times\boldsymbol{C}^*_{\lambda}} \simeq {}^{l}\mathcal{G}_{\boldsymbol{u}\,|\,\boldsymbol{C}^*_{\lambda}}.$$

Thus we obtain the isomorphism $\Phi_{\boldsymbol{u},P,O}$ of ${}^{l}\mathcal{G}_{\boldsymbol{u}}(\mathcal{E})_{|\{P\}\times\boldsymbol{C}^*_{\lambda}}$ and ${}^{l}\mathcal{G}_{\boldsymbol{u}\,|\,\boldsymbol{C}^*_{\lambda}}$.

LEMMA 10.21. *The isomorphism* $\Phi_{\boldsymbol{u},P,O}$ *preserves the morphisms* $\mathcal{N}_{\boldsymbol{u}\,i}$ *and the pairing. It is compatible with dual.* □

REMARK 10.22. In our previous paper [**65**], we used only the morphism $\Phi_{\boldsymbol{u},P,O}$ and did not use the morphism $\Phi^{\mathrm{can}}_{\boldsymbol{u}}$. □

10.4.3. Compatibility with the shift and the functoriality. We use the notation in the subsection 8.9.2. We remark the following compatibility, which is clear from our construction.

LEMMA 10.23. *Let* $\boldsymbol{u}$ *be an element of* $\mathcal{KMS}(\mathcal{E}^0, \underline{l})$. *The following diagrams are commutative:*

$$\begin{array}{ccc}
{}^{\underline{l}}\mathcal{G}_{\boldsymbol{u}}(\mathcal{H}) & \xrightarrow{\Phi^{\mathrm{can}}_{\boldsymbol{u}}} & {}^{\underline{l}}\mathcal{G}_{\boldsymbol{u}\,|\,\{O\}\times \boldsymbol{C}^*_\lambda} \\
= \downarrow & & \mathrm{sh} \downarrow \\
{}^{\underline{l}}\mathcal{G}_{\boldsymbol{u}+\boldsymbol{\delta}_{0,i}}(\mathcal{H}) & \xrightarrow{\Phi^{\mathrm{can}}_{\boldsymbol{u}+\boldsymbol{\delta}_{0,i}}} & {}^{\underline{l}}\mathcal{G}_{\boldsymbol{u}+\boldsymbol{\delta}_{0,i}\,|\,\{O\}\times \boldsymbol{C}^*_\lambda},
\end{array}$$

$$\begin{array}{ccc}
{}^{\underline{l}}\mathcal{G}_{\boldsymbol{u}}(\mathcal{E})_{|\{P\}\times \boldsymbol{C}^*_\lambda} & \xrightarrow{\Phi_{\boldsymbol{u},P,O}} & {}^{\underline{l}}\mathcal{G}_{\boldsymbol{u}\,|\,\{O\}\times \boldsymbol{C}^*_\lambda} \\
= \downarrow & & z_i(P)\cdot\mathrm{sh} \downarrow \\
{}^{\underline{l}}\mathcal{G}_{\boldsymbol{u}+\boldsymbol{\delta}_{0,i}}(\mathcal{E})_{|\{P\}\times \boldsymbol{C}^*_\lambda} & \xrightarrow{\Phi_{\boldsymbol{u},P,O}} & {}^{\underline{l}}\mathcal{G}_{\boldsymbol{u}+\boldsymbol{\delta}_{0,i}|\{O\}\times \boldsymbol{C}^*_\lambda}.
\end{array}$$

Here morphism sh *is the isomorphism given in the subsection* 8.9.2. □

We state the functoriality of the morphisms $\Phi^{\mathrm{can}}_{\boldsymbol{u}}$ and $\Phi_{\boldsymbol{u},P,O}$ for dual, tensor product and pull backs. It immediately follows from the previous remark and the construction.

LEMMA 10.24. *The construction of* $\Phi^{\mathrm{can}}_{\boldsymbol{u}}$ *is compatible with tensor product and pull back as in the subsection* 8.9.7. *The construction of* $\Phi_{\boldsymbol{u},P,O}$ *is compatible with tensor product and pull back as in the subsection* 8.9.7, *up to multiplication of some scalar depending only on* P. □

Part 3

Limiting Mixed Twistor Theorem and Some Consequence

CHAPTER 11

The Induced Vector Bundle over $\mathbb{P}^1$

11.1. The variation of pure twistor structures

11.1.1. Conjugate. Let X be a complex manifold. We denote the conjugate of X by $X^\dagger$. We put $\mathcal{X}^\dagger = \boldsymbol{C}_\mu \times X^\dagger$. Let $(E, \overline{\partial}_E, \theta, h)$ be a tame harmonic bundle over X. Then we obtain the tame harmonic bundle $(E, \partial_E, h, \theta^\dagger)$ over $X^\dagger$, and thus the deformed holomorphic bundle $\mathcal{E}^\dagger$ over $\mathcal{X}^\dagger - \mathcal{D}^\dagger$ and the μ-connection $\mathbb{D}^\dagger$ on $\mathcal{X}^\dagger$. We also have the associated family of flat connections $\mathbb{D}^{\dagger f}$.

Let $\sigma : \boldsymbol{C}_\mu \longrightarrow \boldsymbol{C}_\lambda$ be the morphism given by $\mu \longmapsto -\overline{\mu}$. It induces the anti-holomorphic map $\mathcal{X}^\dagger \longrightarrow \mathcal{X}$.

Let U be an open subset of $\mathcal{X}^\dagger$. We have the isomorphism $\sigma : U \longrightarrow \sigma(U)$. Let $\boldsymbol{v}$ be a frame of ${}_{\boldsymbol{b}}\mathcal{E}$ over $\sigma(U)$. Then we put as follows:

$$\boldsymbol{v}^\dagger := \sigma^*\Big(\boldsymbol{v} \cdot \overline{H(h,\boldsymbol{v})}^{-1}\Big). \tag{11.1}$$

Then $\boldsymbol{v}^\dagger$ is a tuple of C^∞-sections of $\mathcal{E}^\dagger$ on U.

LEMMA 11.1.

(1) *The tuple $\boldsymbol{v}^\dagger$ is a holomorphic frame of $\mathcal{E}^\dagger$ on U.*
(2) *Let $\mathcal{A}$ be the λ-connection one form of $\mathbb{D}$ with respect to the frame $\boldsymbol{v}$. Then the μ-connection one form of $\mathbb{D}^\dagger$ with respect to the frame $\boldsymbol{v}^\dagger$ is given by $\sigma^*\big({}^t\overline{\mathcal{A}}\big)$.*

Proof It can be checked by direct calculations. (See the subsubsection 3.1.6 in our previous paper [**65**], for example). □

11.1.2. The comparison of the flat connections. We identify $\boldsymbol{C}_\mu^*$ and $\boldsymbol{C}_\lambda^*$ by the relation $\mu = \lambda^{-1}$. It induces the identification of the C^∞-manifolds $\mathcal{X}^{\dagger\sharp} = \mathcal{X}^\sharp$.

We have the holomorphic family of the flat connections $(\mathcal{E}^\sharp, \mathbb{D}^f)$ on $\mathcal{X}^\sharp - \mathcal{D}^\sharp$. We also have the holomorphic family of the flat connections $(\mathcal{E}^{\dagger\sharp}, \mathbb{D}^{\dagger f})$ on $\mathcal{X}^{\sharp\dagger} - \mathcal{D}^{\sharp\dagger}$.

LEMMA 11.2. *We have $\big(\mathcal{E}^{\dagger\sharp}, \mathbb{D}^{\dagger f}\big) = \big(\mathcal{E}^\sharp, \mathbb{D}^f\big)$ over $\mathcal{X}^\sharp$, under the identification $\mathcal{X}^{\dagger\sharp} = \mathcal{X}^\sharp$ given above.*

Proof By definition of $\mathbb{D}$ and $\mathbb{D}^\dagger$, we obtain the following:

$$\mathbb{D}^f = \overline{\partial}_E + \lambda\theta^\dagger + \lambda^{-1}\cdot\Big(\lambda\partial_E + \theta\Big) = \partial_E + \mu\theta + \mu^{-1}\Big(\mu\cdot\overline{\partial}_E + \theta^\dagger\Big) = \mathbb{D}^{\dagger f}.$$

Thus we are done. □

11.1.3. The variation of pure twistor structures and the conjugate. Due to Lemma 11.2, we obtain the patched object as in the subsection 3.5.2. Thus we obtain the variation of pure twistor structures. It can be simply described as follows: Let $p : X \times \mathbb{P}^1 \longrightarrow X$ denote the projection. We put $\mathcal{E}^{\triangle} := p^{-1}(E)$, which is naturally $\mathbb{P}^1$-holomorphic bundle. For each $P \in X$, the restriction $p^{-1}(E)_{|\{P\}\times\mathbb{P}^1}$ is a pure twistor of weight 0. The differential operator $\mathbb{D}^{\triangle} : C^{\infty}\bigl(X \times \mathbb{P}^1, \mathcal{E}^{\triangle}\bigr) \longrightarrow C^{\infty}\bigl(X \times \mathbb{P}^1, \mathcal{E}^{\triangle} \otimes \xi\Omega^1_X\bigr)$ is given as follows:

$$\mathbb{D}^{\triangle} := \bigl(\overline{\partial}_E + \theta\bigr) \otimes \sqrt{-1} \cdot f_0^{(1)} + \bigl(\partial_E + \theta^{\dagger}\bigr) \otimes f_{\infty}^{(1)}.$$

LEMMA 11.3. *$\bigl(\mathcal{E}^{\triangle}, \mathbb{D}^{\triangle}\bigr)$ is a variation of pure twistor structures.*

Proof It can be checked by a direct calculation (Lemma 3.56). □

We obtain the conjugate $\bigl(\sigma^*\mathcal{E}^{\triangle}, \mathbb{D}^{\triangle}_{\sigma^*\mathcal{E}^{\triangle}}\bigr)$ (see the subsection 3.5.4). In this case, we have $\sigma^*\mathcal{E}^{\triangle} = p^{-1}(E)$.

LEMMA 11.4. *The $\mathbb{P}^1$-holomorphic structure d''_{λ} of $\sigma^*\mathcal{E}^{\triangle}$ is given as follows:*

$$d''_{\lambda}\sigma^* g = \sigma^*\Bigl(\frac{\partial g}{\partial \bar{\lambda}}\Bigr) \cdot (-d\bar{\lambda}).$$

(See the subsection 3.1.3 *for the formalism.)*

Proof By definition, we have the following:

$$d''_{\lambda}\sigma^* g = \varphi_0 \sigma^*\bigl(\overline{\partial}_{\lambda} g\bigr) = \sigma^*\Bigl(\frac{\partial g}{\partial \bar{\lambda}}\Bigr) \cdot (-d\bar{\lambda}).$$

Thus we are done. □

LEMMA 11.5. *Let g be a C^{∞}-section of $\mathcal{E}^{\triangle}$. The C^{∞}-sections A_i, B_i, C_i and D_i are determined as follows:*

$$\partial_E g = \sum A_i \cdot dz_i, \quad \theta^{\dagger} \cdot g = \sum B_i \cdot d\bar{z}_i, \quad \overline{\partial}_E g = \sum C_i \cdot d\bar{z}_i, \quad \theta \cdot g = \sum D_i \cdot dz_i.$$

Then we have the following formula:

(11.2) $\mathbb{D}_{\sigma^*\mathcal{E}^{\triangle}}(\sigma^* g) =$

$$\sum_i \Bigl(\sigma^* A_i \cdot d\bar{z}_i \otimes \sqrt{-1} f_0^{(1)} - \sigma^* B_i \cdot dz_i \otimes \sqrt{-1} f_0^{(1)} + \sigma^* C_i \cdot dz_i \otimes f_{\infty}^{(1)} - \sigma^* D_i \cdot d\bar{z}_i \otimes f_{\infty}^{(1)}\Bigr).$$

(See the subsection 3.5.4 *for the formalism.)*

Proof We have $\mathbb{D}(\sigma^* g) = \varphi_0 \sigma^*(\mathbb{D} g)$ by definition. We can check the formula (11.2) by using Lemma 3.62. □

11.1.4. Polarization. For any sections f and $\sigma^*(g)$ of $\mathcal{E}^{\triangle}$ and $\sigma^*\mathcal{E}^{\triangle}$, we have the C^{∞}-function $S\bigl(f, \sigma^*(g)\bigr) := h\bigl(f(\lambda, x), g(-\bar{\lambda}, x)\bigr)$. Thus we obtain the pairing $S : \mathcal{E}^{\triangle} \otimes \sigma^*\mathcal{E}^{\triangle} \longrightarrow \mathbb{T}(0)$.

LEMMA 11.6. *The pairing S is a morphism of $\mathbb{P}^1$-holomorphic bundles.*

Proof We have the following equality:

$$\overline{\partial}_{\lambda} S(f, \sigma^* g) = h\bigl(\overline{\partial}_{\lambda} f(\lambda, x), g(-\bar{\lambda}, x)\bigr) + h\bigl(f(\lambda, x), \partial_{\lambda}\bigl(g(-\bar{\lambda}, x)\bigr)\bigr).$$

The first term in the right hand side can be rewritten as $S\big(\overline{\partial}_\lambda f, \sigma^* g\big)$. The second term in the right hand side can be rewritten as follows:

$$h\Big(f(\lambda,x), \frac{\partial g}{\partial \bar{\lambda}}(-\bar{\lambda},x)\Big)\cdot(-d\bar{\lambda}) = S\Big(f, \sigma^*\frac{\partial g}{\partial \bar{\lambda}}\Big)\cdot(-d\bar{\lambda}) = S\Big(f, d''\sigma^* g\Big).$$

Thus we are done. □

LEMMA 11.7. *The pairing S is a morphism of variation of pure twistors.*

Proof We use the notation in Lemma 11.5. We have the following equalities:

(11.3)
$$\begin{aligned}
&\overline{\partial}_X S\big(f,\sigma^* g\big)\otimes\sqrt{-1}\cdot f_0^{(1)}\\
&= \Big(h\big(\overline{\partial}_E f(\lambda,x), g(-\bar{\lambda},x)\big) + h\big(f(\lambda,x), \partial_E g(-\bar{\lambda},x)\big)\Big)\otimes\sqrt{-1}\cdot f_0^{(1)}\\
&= \Big(h\big((\overline{\partial}_E+\theta)f(\lambda,x), g(-\bar{\lambda},x)\big) + h\big(f(\lambda,x), (\partial_E-\theta^\dagger)g(-\bar{\lambda},x)\big)\Big)\otimes\sqrt{-1}\cdot f_0^{(1)}\\
&= \left(S\big((\overline{\partial}_E+\theta)f,\sigma^* g\big) + \sum_i S(f,\sigma^* A_i)\cdot d\bar{z}_i - \sum_i S(f,\sigma^* B_i)\cdot dz_i\right)\otimes\sqrt{-1}f_0^{(1)}.
\end{aligned}$$

On the other hand, we also have the following:

(11.4)
$$\begin{aligned}
&\partial_X S\big(f,\sigma^* g\big)\otimes f_\infty^{(1)} = \Big(h\big(\partial_E f, g(-\bar{\lambda},x)\big) + h\big(f, \overline{\partial}_E g(-\bar{\lambda},x)\big)\Big)\otimes f_\infty^{(1)}\\
&= \Big(h\big((\partial_E+\theta^\dagger)f, g(-\bar{\lambda},x)\big) + h\big(f, (\overline{\partial}_E-\theta)g(-\bar{\lambda},x)\big)\Big)\otimes f_\infty^{(1)}\\
&= S\big((\partial_E+\theta^\dagger)f,\sigma^* g\big)\otimes f_\infty^{(1)} + \left(\sum_i S\big(f,\sigma^* C_i\big)\cdot d\bar{z}_i - \sum_i S\big(f,\sigma^* D_i\big)\cdot dz_i\right)\otimes f_\infty^{(1)}.
\end{aligned}$$

Then Lemma 11.7 immediately follows from Lemma 11.5. □

COROLLARY 11.8 (Simpson). *The tuple $(\mathcal{E}^\triangle, \mathbb{D}^\triangle, h)$ is a variation of polarized pure twistor structures.* □

LEMMA 11.9. *We obtain the isomorphism $\clubsuit : \sigma^*\mathcal{E}^\triangle \simeq \mathcal{E}^{\vee\,\triangle}$ of the variation of pure twistors. In particular, we obtain the isomorphisms $\sigma^*\mathcal{E}^\dagger \simeq \mathcal{E}^\vee$ and $\sigma^*\mathcal{E} \simeq \mathcal{E}^{\vee\,\dagger}$. We also denote them by $\clubsuit$.*

Proof Since the pairing S is perfect, it induces the isomorphism $\clubsuit$. □

11.2. The induced objects of the conjugate and the pairing

11.2.1. Compatible frame and the KMS-structure of the conjugate. We put $X = \Delta^n$, $D_i := \{z_i = 0\}$ and $D = \bigcup_{i=1}^l D_i$. We put $\mathcal{Y}(\lambda_0,\epsilon_0) = Y\times\Delta(\lambda_0,\epsilon_0)$ for a complex variety Y. For simplicity of the notation, we also put $(\mathcal{X}-\mathcal{D})(\lambda_0,\epsilon_0) := \mathcal{X}(\lambda_0,\epsilon_0) - \mathcal{D}(\lambda_0,\epsilon_0)$.

Let $(E,\overline{\partial}_E,\theta,h)$ be a tame harmonic bundle on $X-D$. Pick a point $\lambda_0\in\boldsymbol{C}_\lambda$. Let $\boldsymbol{b}$ be an element of $\boldsymbol{R}^l$ such that each i-th component b_i is not contained in $\mathcal{KMS}(\mathcal{E}^{\lambda_0}, i)$. There exists a sufficiently small positive number ϵ_0 such that the sheaf ${}_{\boldsymbol{b}}\mathcal{E}$ on $\mathcal{X}(\lambda_0,\epsilon_0)$ is locally free (the section 8.7). Let $\boldsymbol{v} = (v_i)$ be a frame of ${}_{\boldsymbol{b}}\mathcal{E}$, which is compatible with $\mathbb{E}^{(\lambda_0)}$ and $F^{(\lambda_0)}$. For each v_i, we have the element $\boldsymbol{u}(v_i)\in\mathcal{KMS}(\mathcal{E}^0,\underline{l})$ such that the following holds:

$$\mathfrak{k}(\lambda_0,\boldsymbol{u}(v_i)) = \deg^{\mathbb{E}^{(\lambda_0)},F^{(\lambda_0)}}(v_i)\in\mathcal{KMS}(\mathcal{E}^{\lambda_0},\underline{l}).$$

Let $u_j(v_i)\in\mathcal{KMS}(\mathcal{E}^0,j)$ denote the j-th component of $\boldsymbol{u}(v_i)$.

We denote the restriction $\boldsymbol{v}_{|(\mathcal{X}-\mathcal{D})(\lambda_0,\epsilon_0)}$ by $\boldsymbol{v}$, for simplicity of the notation. Then we obtain the holomorphic frame $\boldsymbol{v}^\dagger$ of $\mathcal{E}^\dagger$ over $(\mathcal{X}^\dagger-\mathcal{D}^\dagger)(-\overline{\lambda}_0,\epsilon_0)$, which is given by (11.1). (Note $\sigma(\mathcal{X}^\dagger(-\overline{\lambda}_0,\epsilon_0))=\mathcal{X}(\lambda_0,\epsilon_0)$.) We put as follows:

$$(11.5)\qquad \begin{aligned} v_i' &:= v_i\cdot\prod_{j=1}^{l}|z_j|^{\mathfrak{p}(\lambda,u_j(v_i))}, & \boldsymbol{v}' &:= (v_i'),\\ v_i^{\dagger\,\prime} &:= v_i^\dagger\cdot\prod_{j=1}^{l}|z_j|^{-\mathfrak{p}(\lambda,u_j(v_i))}, & \boldsymbol{v}^{\dagger\,\prime} &= (v_i^{\dagger\,\prime}).\end{aligned}$$

Then $\boldsymbol{v}'$ is a C^∞-frame of $\mathcal{E}$ over $(\mathcal{X}-\mathcal{D})(\lambda_0,\epsilon_0)$, and $\boldsymbol{v}^{\dagger\,\prime}$ is a C^∞-frame of $\mathcal{E}^\dagger$ over $(\mathcal{X}^\dagger-\mathcal{D}^\dagger)(-\overline{\lambda}_0,\epsilon_0)$.

LEMMA 11.10. *The frames $\boldsymbol{v}'$ and $\boldsymbol{v}^{\dagger\,\prime}$ are adapted up to log order.*

Proof The adaptedness of $\boldsymbol{v}'$ up to log order has already been shown in Proposition 8.110 (the subsection 8.8.7). Let L be the diagonal matrix such that $L_{i\,i}:=\prod_j|z_j|^{\mathfrak{p}(\lambda,u_j(v_i))}$. Then we have the relations $\boldsymbol{v}^{\dagger\,\prime}=\boldsymbol{v}^\dagger\cdot L^{-1}$ and $\boldsymbol{v}'=\boldsymbol{v}\cdot L$. Then we obtain the following:

$$(11.6)\quad \begin{aligned}\boldsymbol{v}^{\dagger\,\prime}=\boldsymbol{v}^\dagger\cdot L^{-1}=\boldsymbol{v}\cdot\overline{H(h,\boldsymbol{v})}^{-1}\cdot L^{-1}&=\boldsymbol{v}'\cdot L^{-1}\cdot\overline{H(h,\boldsymbol{v})}^{-1}\cdot L^{-1}\\ &=\boldsymbol{v}'\cdot\overline{(L\cdot H(h,\boldsymbol{v})\cdot L)}^{-1}=\boldsymbol{v}'\cdot\overline{H(h,\boldsymbol{v}')}^{-1}.\end{aligned}$$

Since $\boldsymbol{v}'$ is adapted up to log order, and since $H(h,\boldsymbol{v}')$ and $H(h,\boldsymbol{v}')^{-1}$ is bounded up to log order, $\boldsymbol{v}^{\dagger\,\prime}$ is adapted up to log order. □

Recall that we put $\boldsymbol{u}^\dagger:=(\overline{\boldsymbol{\alpha}},-\boldsymbol{b})$, for any element $\boldsymbol{u}=(\boldsymbol{\alpha},\boldsymbol{b})\in\boldsymbol{C}^l\times\boldsymbol{R}^l$.

COROLLARY 11.11.

- *We put $\boldsymbol{\delta}=(1,\ldots,1)\in\boldsymbol{R}^l$. Then $\boldsymbol{v}^\dagger$ gives a frame of ${}_{-\boldsymbol{b}+\boldsymbol{\delta}}\mathcal{E}^\dagger$ over $\mathcal{X}^\dagger(-\overline{\lambda}_0,\epsilon_0)$.*
- *The frame $\boldsymbol{v}^\dagger$ is compatible with the parabolic filtration $F^{(-\overline{\lambda}_0)}$ and the decomposition $\mathbb{E}^{(-\overline{\lambda}_0)}$ for ${}_{-\boldsymbol{b}+\boldsymbol{\delta}}\mathcal{E}^\dagger$.*
- *We have the following:*

$$\deg^{\mathbb{E}^{(-\overline{\lambda}_0)},F^{(-\overline{\lambda}_0)}}(v_i^\dagger)=\mathfrak{k}\big(-\overline{\lambda_0},\boldsymbol{u}(v_i)^\dagger\big).$$

Proof By using Lemma 2.8, we obtain the first claim. By using Lemma 2.9, we obtain that the frame $\boldsymbol{v}^\dagger$ is compatible with the filtration $F^{(-\overline{\lambda}_0)}$, and we have the following:

$$\deg^{F^{(-\overline{\lambda}_0)}}(v_i^\dagger)=-\mathfrak{p}(\lambda_0,\boldsymbol{u}(v_i))=\mathfrak{p}\big(-\overline{\lambda}_0,\boldsymbol{u}(v_i)^\dagger\big).$$

Here we have used Lemma 2.4.

Due to the claim (2) in Lemma 11.1, we obtain that the frame $\boldsymbol{v}^\dagger$ is compatible with $\mathbb{E}^{(-\overline{\lambda}_0)}$. Moreover we have the following:

$$\deg^{\mathbb{E}^{(-\overline{\lambda}_0)}}(v_i^\dagger)=\overline{\mathfrak{e}\big(\lambda_0,\boldsymbol{u}(v_i)\big)}=\mathfrak{e}\big(-\overline{\lambda}_0,\boldsymbol{u}(v_i)^\dagger\big).$$

Thus we obtain the second and the third claims in Corollary 11.11. □

COROLLARY 11.12. *By the correspondence $\boldsymbol{u}\longmapsto\boldsymbol{u}^\dagger$, we have the isomorphism preserving the multiplicity:*

$$\mathcal{KMS}(\mathcal{E}^\lambda,\underline{l})\longrightarrow\mathcal{KMS}(\mathcal{E}^{\dagger\,-\overline{\lambda}},\underline{l}).$$

In particular, we have the isomorphism $\mathcal{KMS}(\mathcal{E}^0,\underline{l})\longrightarrow\mathcal{KMS}(\mathcal{E}^{\dagger\,0},\underline{l})$. □

LEMMA 11.13. *Via the isomorphism* $\clubsuit : \sigma^*\mathcal{E}^\dagger \simeq \mathcal{E}^\vee$, *we have* $\clubsuit(\boldsymbol{v}^\dagger) = \boldsymbol{v}^\vee$.

Proof It can be shown by an elementary linear algebraic argument. (See the subsubsection 3.1.6 in our previous paper [**65**], for example.) □

11.2.2. The conjugate and the dual. Let λ_0 be a point of $\boldsymbol{C}_\lambda$. Let $\boldsymbol{b} = (b_1, \ldots, b_l)$ be an element of $\boldsymbol{R}^l$ such that $b_i \notin \mathcal{KMS}(\mathcal{E}^{\vee\,\lambda_0}, i)$ for any i. We take a sufficiently small positive number ϵ_0, then we have the locally free sheaf ${}_{\boldsymbol{b}}\mathcal{E}^\vee$ on $\mathcal{X}(\lambda_0, \epsilon_0)$.

COROLLARY 11.14. *The sheaf* ${}_{\boldsymbol{b}}\mathcal{E}^\dagger$ *on* $\mathcal{X}^\dagger(-\overline{\lambda}_0, \epsilon_0)$ *is locally free. We have the isomorphism* $\clubsuit : \sigma^*\big({}_{\boldsymbol{b}}\mathcal{E}^\dagger\big) \simeq {}_{\boldsymbol{b}}\mathcal{E}^\vee$.

Proof It follows from Lemma 11.13. □

The morphism $\clubsuit$ induces the isomorphism $\clubsuit_{D_i}$:

$$\clubsuit_{|D_i} : \sigma^*\big({}_{\boldsymbol{b}}\mathcal{E}^\dagger\big)_{|\mathcal{D}_i(-\bar{\lambda}_0, \epsilon_0)} \longrightarrow {}_{\boldsymbol{b}}\mathcal{E}^\vee{}_{|\mathcal{D}_i(\lambda_0, \epsilon_0)}.$$

Recall we have the decomposition ${}^i\mathbb{E}^{(\lambda_0)}$ and the filtration ${}^iF^{(\lambda_0)}$ of ${}_{\boldsymbol{b}}\mathcal{E}^\vee{}_{|\mathcal{D}_i(-\bar{\lambda}_0, \epsilon_0)}$, given in the subsection 8.8.1 and the subsection 8.8.5. Similarly, we have the decomposition and the filtration of $\sigma^*{}_{\boldsymbol{b}}\mathcal{E}^\dagger{}_{|\mathcal{D}_i(\lambda_0, \epsilon_0)}$.

LEMMA 11.15. *The morphism* $\clubsuit_{|D_i}$ *preserves the filtrations and the decompositions. In particular, the morphism* $\clubsuit_{|D_i}$ *induces the bijection* $\mathcal{KMS}({}_{\boldsymbol{b}}\mathcal{E}^{\dagger, -\overline{\lambda}}, i) \longrightarrow \mathcal{KMS}({}_{\boldsymbol{b}}\mathcal{E}^{\vee, \lambda}, i)$ *given by the correspondence* $(a, \alpha) \longmapsto (a, -\overline{\alpha})$.

Proof It can be shown by using the comparison of the degrees $\deg^{\mathbb{E},F}(v_i^\dagger)$ and $\deg^{\mathbb{E},F}(v_i^\vee)$ of the induced frames. □

Let us pick a point $\lambda_0 \in \boldsymbol{C}_\lambda^*$. Then we have the filtration and the decomposition of ${}_{\boldsymbol{b}}\mathcal{E}^\vee$ on $\mathcal{X}(\lambda_0, \epsilon_0)$, given in the subsection 10.2.2. Similarly, we have the filtration and the decomposition of ${}_{\boldsymbol{b}}\mathcal{E}^\dagger$ on $\mathcal{X}^\dagger(-\overline{\lambda}_0, \epsilon_0)$.

LEMMA 11.16. *The filtrations and the decompositions are preserved by the morphism* $\clubsuit$.

Proof Recall that the decompositions are induced by the monodromy endomorphisms. Since $\clubsuit$ preserves the flat connection, the decompositions are preserved.

Recall that the restriction of the filtrations to $\mathcal{X}^*(\lambda_0, \epsilon_0)$ have the canonical splittings, given by the generalized eigenspaces of the monodromy actions. (Here we may assume that any point λ of $\Delta^*(\lambda_0, \epsilon_0)$ is generic.) Thus the restriction of the filtrations to $\mathcal{X}^*(\lambda_0, \epsilon_0)$ are preserved. Then it follows that the filtrations are preserved on whole $\mathcal{X}(\lambda_0, \epsilon_0)$. □

By considering the spaces of the multi-valued flat sections, we obtain the vector bundle $\mathcal{H}(E, \partial_E, \theta^\dagger, h)$ on $\boldsymbol{C}_\mu^*$. We denote it by $\mathcal{H}^\dagger(E)$ for simplicity. Namely, $\mathcal{H}^\dagger_{|\mu}$ denotes the space of the multi-valued flat sections of $\mathcal{E}^{\dagger\,\mu}$. The isomorphism $\clubsuit$ induces the isomorphism $\sigma^*\mathcal{H}^\dagger \simeq \mathcal{H}^\vee$, which we denote also by $\clubsuit$. Let us pick a point $\lambda_0 \in \boldsymbol{C}_\lambda$ and a sufficiently small positive number ϵ_0. Then we have the filtrations and the decompositions of $\mathcal{H}^\vee$ on $\Delta_\lambda(\lambda_0, \epsilon_0)$, given in the subsections 9.1.1–9.1.2. Similarly, we have the filtrations and the decompositions of $\mathcal{H}^\dagger$ on $\Delta_\mu(-\overline{\lambda}_0, \epsilon_0)$.

LEMMA 11.17. *The morphism ♣ preserves the filtrations and the decompositions.*

Proof It can be shown by an argument similar to Lemma 11.16. □

11.2.3. The induced objects and the pairing. Let $\boldsymbol{u}^\dagger$ be an element of $\mathcal{KMS}(\mathcal{E}^{\dagger\,0}, \underline{l})$.

- By applying the constructions in the subsection 8.9.1, we obtain the vector bundle ${}^{\underline{l}}\mathcal{G}^\dagger_{\boldsymbol{u}^\dagger}$ on $\mathcal{D}^\dagger_{\underline{l}}$. We also have the endomorphisms $\mathrm{Res}_i(\mathbb{D}^\dagger)$ and the nilpotent parts $\mathcal{N}^\dagger_i$ $(i \in \underline{l})$.
- By applying the construction in the subsection 9.3.1, we obtain the holomorphic vector bundle ${}^{\underline{l}}\mathcal{G}^\dagger_{\boldsymbol{u}^\dagger}\mathcal{H}^\dagger$ on $\boldsymbol{C}^*_\mu$. We also have the monodromy endomorphisms, and the nilpotent parts $\mathcal{N}^\dagger_i$.
- By applying the construction in the subsection 10.3.2, we obtain the holomorphic vector bundle ${}^{\underline{l}}\mathcal{G}^\dagger_{\boldsymbol{u}^\dagger}(\mathcal{E}^\dagger)$ on $\mathcal{X}^{\dagger\,\sharp}$. We have the holomorphic family of the flat connections $\mathbb{D}^{\dagger\,f}$.

COROLLARY 11.18. *The morphism ♣ induces the following isomorphisms:*

$$\sigma^*{}^{\underline{l}}\mathcal{G}^\dagger_{\boldsymbol{u}^\dagger}(E) \simeq {}^{\underline{l}}\mathcal{G}_{-\boldsymbol{u}}(E^\vee), \quad \sigma^*{}^{\underline{l}}\mathcal{G}^\dagger_{\boldsymbol{u}^\dagger}\mathcal{H}^\dagger(E) \simeq {}^{\underline{l}}\mathcal{G}_{-\boldsymbol{u}}\mathcal{H}(E^\vee), \quad \sigma^*{}^{\underline{l}}\mathcal{G}^\dagger_{\boldsymbol{u}^\dagger}(\mathcal{E}^\dagger) \simeq {}^{\underline{l}}\mathcal{G}_{-\boldsymbol{u}}(\mathcal{E}^\vee).$$

In the first and the second isomorphisms, the isomorphisms reverse the signature of the nilpotent maps. The third isomorphism preserves the family of the flat connections.

Proof It immediately follows from Lemma 11.15, Lemma 11.16 and Lemma 11.17. □

COROLLARY 11.19. *We have the naturally defined pairings:*

$$
\begin{aligned}
&{}^{\underline{l}}\mathcal{G}_{\boldsymbol{u}} \otimes \sigma^*{}^{\underline{l}}\mathcal{G}^\dagger_{\boldsymbol{u}^\dagger} \longrightarrow \mathcal{O}_{\mathcal{D}_{\underline{l}}},\\
&{}^{\underline{l}}\mathcal{G}_{\boldsymbol{u}}\mathcal{H} \otimes \sigma^*{}^{\underline{l}}\mathcal{G}^\dagger_{\boldsymbol{u}^\dagger}\mathcal{H}^\dagger \longrightarrow \mathcal{O}_{\boldsymbol{C}^*_\lambda},\\
&{}^{\underline{l}}\mathcal{G}_{\boldsymbol{u}}(\mathcal{E}) \otimes \sigma^*{}^{\underline{l}}\mathcal{G}^\dagger_{\boldsymbol{u}^\dagger}(\mathcal{E}^\dagger) \longrightarrow \mathcal{O}_{\mathcal{X}^\sharp}.
\end{aligned}
$$

Proof It immediately follows from Corollary 11.18. □

11.3. The induced vector bundles over $\mathbb{P}^1$

11.3.1. The identification of the flat bundles with filtrations and the decompositions. We continue to use the setting in the section 11.2. Let M_i denote the monodromy endomorphism of $\mathcal{E}$ with respect to the loop γ_i around the divisor D_i with the anti-clockwise direction:

$$\gamma_i : [0,1] \longrightarrow \big(z_1, \ldots, z_{i-1}, e^{2\pi\sqrt{-1}t} \cdot z_i, z_{i+1}, \ldots, z_n\big). \tag{11.7}$$

Let $M^\dagger_i$ denote the monodromy endomorphism of $\mathcal{E}^\dagger$ of γ_i^{-1}. The following lemma immediately follows from Lemma 11.2.

LEMMA 11.20. *We have $M_i^{-1} = M_i^\dagger$.* □

Let us pick a point $\lambda_0 \in \boldsymbol{C}^*_\lambda$ and a small neighbourhood $U \subset \boldsymbol{C}^*_\lambda$ of λ_0. Then we have the filtration and the decomposition of $\mathcal{E}$ on $U \times (X - D)$, given in the subsection 10.2.2. We have the point $\lambda_0^{-1} \in \boldsymbol{C}^*_\mu$ and the neighbourhood

U' in $\boldsymbol{C}_\mu^*$, which is same as U by the identification $\lambda = \mu^{-1}$. Then we have the filtration and the decomposition of $\mathcal{E}^\dagger$ on $U' \times (X^\dagger - D^\dagger)$ similarly. As is noted in the subsection 11.1.2, we have $(\mathcal{E}^\sharp, \mathbb{D}^f) = (\mathcal{E}^{\dagger\,\sharp}, \mathbb{D}^{\dagger\,f})$ as flat bundles on $U \times (X - D) = U' \times (X^\dagger - D^\dagger)$.

COROLLARY 11.21. *The identification $\mathcal{E}^{\dagger\,\sharp} = \mathcal{E}^\sharp$ on $U \times (X - D)$ preserves the filtrations and the decompositions.*

Proof The decomposition is obtained from the generalized eigen decomposition of the monodromy endomorphisms. Thus the decompositions are preserved, due to Lemma 11.20. We put $U^* = U - \{\lambda_0\}$. We recall that the restriction of the filtrations to $(X - D) \times U^*$ have the splittings given by the generalized eigen decompositions as in (9.7). We also recall the relation $\mathfrak{p}^f(\lambda, u) = \mathfrak{p}^f(\lambda^{-1}, u^\dagger)$ (Lemma 2.5). Then we obtain that the restriction of the filtrations to $(X - D) \times U$ are preserved due to Lemma 11.20. Then it follows that the filtrations are preserved. □

We have the monodromy endomorphisms $M_i^\dagger$ of $\mathcal{H}^\dagger$ with respect to the loop γ_i^{-1}. We also have the monodromy endomorphisms M_i of $\mathcal{H}$ with respect to the loop γ_i.

LEMMA 11.22. *Under the identification $\boldsymbol{C}_\mu^* = \boldsymbol{C}_\lambda^*$ above, we have $\mathcal{H}^\dagger(E) = \mathcal{H}(E)$. We have $M_i^{-1} = M_i^\dagger$.*

Proof It follows from the coincidence of the flat connections $\mathbb{D}^{\mu,f} = \mathbb{D}^{\lambda,f}$. □

Let us pick a point $\lambda_0 \in \boldsymbol{C}_\lambda$ and an appropriate neighbourhood U of λ_0. Then we have the filtrations and the decompositions of $\mathcal{H}$ on U, given in the subsections 9.1.1–9.1.2. Similarly, we have the filtrations and the decompositions of $\mathcal{H}^\dagger$ on U.

COROLLARY 11.23. *The identification $\mathcal{H} = \mathcal{H}^\dagger$ preserves the filtrations and the decompositions.*

Proof It can be shown by an argument similar to the proof of Corollary 11.21. □

11.3.2. The identification of the induced objects. We have the holomorphic bundles ${}^l\mathcal{G}_{\boldsymbol{u}}\mathcal{H}$ on $\boldsymbol{C}_\lambda^*$. We have the tuple of the monodromy endomorphisms $\boldsymbol{M} = (M_1, \ldots, M_l)$. Here M_i denote the monodromy along the loop γ_i given in (11.7). We also have the holomorphic bundles ${}^l\mathcal{G}_{\boldsymbol{u}^\dagger}^\dagger\mathcal{H}^\dagger$ on $\boldsymbol{C}_\mu^*$. We have the tuple of the monodromy endomorphisms $\boldsymbol{M}^\dagger = (M_1^\dagger, \ldots, M_l^\dagger)$. Here $M_i^\dagger$ denotes the monodromy along the loop γ_i^{-1}.

We identify $\boldsymbol{C}_\lambda^*$ and $\boldsymbol{C}_\mu^*$ by the relation $\lambda = \mu^{-1}$.

LEMMA 11.24. *We have the natural identification ${}^l\mathcal{G}_{\boldsymbol{u}}\mathcal{H} = {}^l\mathcal{G}_{\boldsymbol{u}^\dagger}^\dagger\mathcal{H}$. We also have $M^{-1} = M^\dagger$.*

Proof Due to Lemma 11.22, we have the natural identification $\mathcal{H} = \mathcal{H}^\dagger$ over $\boldsymbol{C}_\lambda^*$, on which we have $M^{-1} = M^\dagger$. Due to Corollary 11.23 and Lemma 2.5, we obtain the result. □

We have the holomorphic vector bundle ${}^l\mathcal{G}_{\boldsymbol{u}}(\mathcal{E})$ on $\mathcal{X}^\sharp$. We denote the restriction of ${}^l\mathcal{G}_{\boldsymbol{u}}(\mathcal{E})$ to $\mathcal{X}^\sharp - \mathcal{D}^\sharp$ by the same notation. Then we have the holomorphic family of the regular connections $\mathbb{D}^f$.

Similarly, we also have the C^∞-bundle ${}^{l}\mathcal{G}^{\dagger}_{\boldsymbol{u}^\dagger}(\mathcal{E}^\dagger)$ with the holomorphic family of the flat connections $\mathbb{D}^{\dagger f}$ on $\mathcal{X}^{\dagger\sharp} - \mathcal{D}^{\dagger\sharp}$. By the relation $\lambda = \mu^{-1}$, we have $\mathcal{X}^\sharp - \mathcal{D}^\sharp = \mathcal{X}^{\dagger\sharp} - \mathcal{D}^{\dagger\sharp} = (X - D) \times \boldsymbol{C}^*_\lambda$.

LEMMA 11.25. *We have the natural identification* ${}^{l}\mathcal{G}^{\dagger}_{\boldsymbol{u}^\dagger}(\mathcal{E}^\dagger) = {}^{l}\mathcal{G}_{\boldsymbol{u}}(\mathcal{E})$ *and* $\mathbb{D}^f = \mathbb{D}^{\dagger f}$ *on* $\mathcal{X}^\sharp - \mathcal{D}^\sharp$.

Proof Recall Lemma 11.2 and Corollary 11.21. Then Lemma 11.25 can be shown by an argument similar to the proof of Lemma 11.24. □

11.3.3. The vector bundle $S_{\boldsymbol{u}}(E, P)$. Let $\boldsymbol{u} \in \mathcal{KMS}(\mathcal{E}^0, \underline{l})$. Then we obtain the holomorphic vector bundle ${}^{l}\mathcal{G}_{\boldsymbol{u}}$ on $\mathcal{D}_{\underline{l}}$. For simplicity, we denote the restriction of ${}^{l}\mathcal{G}_{\boldsymbol{u}}$ to $\boldsymbol{C}_\lambda \times \{O\}$ by the same notation. Namely, we have the holomorphic bundle ${}^{l}\mathcal{G}_{\boldsymbol{u}}$ on $\boldsymbol{C}_\lambda \simeq \boldsymbol{C}_\lambda \times \{O\}$. Similarly, we have the holomorphic bundle ${}^{l}\mathcal{G}^{\dagger}_{\boldsymbol{u}^\dagger}$ on $\boldsymbol{C}_\mu$.

For any point P of $X - D$, we have the isomorphism $\Phi_{P,O} : {}^{l}\mathcal{G}_{\boldsymbol{u}}(\mathcal{E})_{|\{P\}\times \boldsymbol{C}^*_\lambda} \simeq {}^{l}\mathcal{G}_{\boldsymbol{u}|\boldsymbol{C}^*_\lambda}$ over $\boldsymbol{C}^*_\lambda$, given in the subsection 10.4.2. Similarly, we have the isomorphism $\Phi^\dagger_{P,O} : {}^{l}\mathcal{G}_{\boldsymbol{u}^\dagger}(\mathcal{E}^\dagger)_{|\{P\}\times \boldsymbol{C}^*_\mu} \simeq {}^{l}\mathcal{G}_{\boldsymbol{u}^\dagger|\boldsymbol{C}^*_\mu}$ over $\boldsymbol{C}^*_\mu$. From the morphisms $\Phi_{P,O}$ and $\Phi^\dagger_{P,O}$, we obtain the isomorphism:

$$\Phi^\dagger_{P,O} \circ \left(\Phi_{P,O}\right)^{-1} : {}^{l}\mathcal{G}_{\boldsymbol{u}\,|\,\boldsymbol{C}^*_\lambda} \longrightarrow {}^{l}\mathcal{G}^{\dagger}_{\boldsymbol{u}^\dagger\,|\,\boldsymbol{C}^*_\mu}.$$

Then we obtain the vector bundle, which we denote by $S_{\boldsymbol{u}}(E, P)$ or simply by $S(P)$.

11.3.4. The vector bundle $S^{\mathrm{can}}_{\boldsymbol{u}}(E)$. As in the subsection 11.3.3, we denote the restriction of ${}^{l}\mathcal{G}_{\boldsymbol{u}}$ to $\boldsymbol{C}_\lambda \times \{O\}$ by the same notation. Similarly we have the following isomorphisms, given in the subsection 10.4.1:

$$\Phi^{\mathrm{can}} : {}^{l}\mathcal{G}_{\boldsymbol{u}}\mathcal{H} \longrightarrow {}^{l}\mathcal{G}_{\boldsymbol{u}\,|\,\boldsymbol{C}^*_\lambda}, \qquad \Phi^{\dagger\,\mathrm{can}} : {}^{l}\mathcal{G}^{\dagger}_{\boldsymbol{u}^\dagger}(\mathcal{H}^\dagger) \longrightarrow {}^{l}\mathcal{G}^{\dagger}_{\boldsymbol{u}^\dagger\,|\,\boldsymbol{C}^*_\mu}.$$

Since we have the canonical identification ${}^{l}\mathcal{G}_{\boldsymbol{u}}\mathcal{H} = {}^{l}\mathcal{G}^{\dagger}_{\boldsymbol{u}^\dagger}\mathcal{H}^\dagger$, we obtain the isomorphism:

$$\Phi^{\dagger\,\mathrm{can}} \circ \left(\Phi^{\mathrm{can}}\right)^{-1} : {}^{l}\mathcal{G}_{\boldsymbol{u}\,|\,\boldsymbol{C}^*_\lambda} \longrightarrow {}^{l}\mathcal{G}^{\dagger}_{\boldsymbol{u}^\dagger\,|\,\boldsymbol{C}^*_\mu}.$$

Thus we obtain the vector bundle, which we denote by $S^{\mathrm{can}}_{\boldsymbol{u}}(E)$.

11.3.5. Pairing.

LEMMA 11.26. *We have the natural isomorphisms* $\sigma^* S^{\mathrm{can}}_{\boldsymbol{u}}(E) \simeq S^{\mathrm{can}}_{\boldsymbol{u}}(E)^\vee$ *and* $\sigma^* S_{\boldsymbol{u}}(E, P) \simeq S_{\boldsymbol{u}}(E, P)^\vee$.

Proof It follows from Corollary 11.18 and our construction. □

COROLLARY 11.27. *Let $\boldsymbol{u}$ be an element of $\mathcal{KMS}(\mathcal{E}^0, \underline{l})$. We have the naturally induced pairings:*

$$S^{\mathrm{can}}_{\boldsymbol{u}}(E) \otimes \sigma^* S^{\mathrm{can}}_{\boldsymbol{u}}(E) \longrightarrow \mathbb{T}(0),$$

$$S_{\boldsymbol{u}}(E, P) \otimes \sigma^* S_{\boldsymbol{u}}(E, P) \longrightarrow \mathbb{T}(0).$$

They are perfect. (See the subsection 3.3.1 *for the Tate object* $\mathbb{T}(0)$.*)* □

11.3.6. Nilpotent maps. We have the nilpotent part of the residues $\mathcal{N}_{\boldsymbol{u}\,i\,|\,\boldsymbol{C}^*_\lambda}$ on ${}^{l}\mathcal{G}_{\boldsymbol{u}\,|\,\boldsymbol{C}^*_\lambda}$. We also have $\mathcal{N}^\dagger_{\boldsymbol{u}^\dagger\,i\,|\,\boldsymbol{C}^*_\lambda}$ on ${}^{l}\mathcal{G}^\dagger_{\boldsymbol{u}^\dagger\,|\,\boldsymbol{C}^*_\mu}$.

LEMMA 11.28. *Due to the isomorphisms* $\Phi^{\dagger\,\mathrm{can}}\circ\big(\Phi^{\mathrm{can}}\big)^{-1}$ *or* $\Phi^\dagger_{P,O}\circ\big(\Phi_{P,O}\big)^{-1}$, *we have the following:*

$$\lambda^{-1}\cdot\mathcal{N}_{\boldsymbol{u}\,i\,|\,\boldsymbol{C}^*_\lambda}=-\mu^{-1}\cdot\mathcal{N}^\dagger_{\boldsymbol{u}^\dagger\,i\,|\,\boldsymbol{C}^*_\mu}.$$

Proof Due to Lemma 11.20, we have the relation:

$$\exp\Big(2\pi\sqrt{-1}\lambda^{-1}\cdot\mathcal{N}_{\boldsymbol{u}\,i\,|\,\boldsymbol{C}^*_\lambda}\Big)=\exp\Big(-2\pi\sqrt{-1}\mu^{-1}\cdot\mathcal{N}^\dagger_{\boldsymbol{u}^\dagger\,i\,|\,\boldsymbol{C}^*_\mu}\Big).$$

Thus we are done. □

Thus we obtain the following morphisms:

$$\mathcal{N}^\triangle_i:S_{\boldsymbol{u}}(E,P)\longrightarrow S_{\boldsymbol{u}}(E,P)\otimes\mathbb{T}(-1),\quad \mathcal{N}^\triangle_i:S^{\mathrm{can}}_{\boldsymbol{u}}(E)\longrightarrow S^{\mathrm{can}}_{\boldsymbol{u}}(E)\otimes\mathbb{T}(-1).$$

Here we put $\mathcal{N}^\triangle_{i\,|\,\boldsymbol{C}_\lambda}:=\mathcal{N}_{\boldsymbol{u}\,i}\otimes t_0^{(-1)}$ and $\mathcal{N}^\triangle_{i\,|\,\boldsymbol{C}_\mu}:=\mathcal{N}^\dagger_{\boldsymbol{u}^\dagger\,i}\otimes t_\infty^{(-1)}$. Note that we have the relation $t_0^{(-1)}=-\lambda^2\cdot t_\infty^{(-1)}$. Thus $\mathcal{N}^\triangle_i$ is well defined.

LEMMA 11.29. *The isomorphisms* $\sigma^*S^{\mathrm{can}}_{\boldsymbol{u}}(E)\simeq S^{\mathrm{can}}_{-\boldsymbol{u}}(E^\vee)$ *and* $\sigma^*S_{\boldsymbol{u}}(E,P)\simeq S_{-\boldsymbol{u}}(E^\vee,P)$ *preserves the nilpotent morphisms.*

Proof It follows from Corollary 11.18 and the relation $\varphi_0(\sigma^*(t_\infty^{(-1)}))=-t_0^{(-1)}$. □

COROLLARY 11.30. *We have the relation* $S\big(\mathcal{N}^\triangle_i\otimes\mathrm{id}\big)+S\big(\mathrm{id}\otimes\sigma^*\mathcal{N}^\triangle_i\big)=0$. □

11.3.7. Shift. Let $\boldsymbol{\delta}_0$ denote the element $(1,0)\in\boldsymbol{R}\times\boldsymbol{C}$, and let $\boldsymbol{\delta}_{0,i}$ denote the element of $(\boldsymbol{R}\times\boldsymbol{C})^l$ such that $q_i(\boldsymbol{\delta}_{0,i})=\boldsymbol{\delta}_0$ and $q_j(\boldsymbol{\delta}_{0,i})=(0,0)$ for $j\neq i$. Here q_j denote the projection onto the j-th component as in the subsection 8.9.2. It is easy to see that we have the naturally induced isomorphisms $S^{\mathrm{can}}_{\boldsymbol{u}}(E)\simeq S^{\mathrm{can}}_{\boldsymbol{u}+\boldsymbol{\delta}_{0,i}}(E)$, which is compatible with the nilpotent maps and the pairing. We also have the naturally induced isomorphisms $S_{\boldsymbol{u}}(E,P)\simeq S_{\boldsymbol{u}+\boldsymbol{\delta}_{0,i}}(E,P)$, which is compatible with the nilpotent maps and the pairing. Therefore the induced bundles with the nilpotent maps and the pairing are defined for the elements of $\overline{\mathcal{KMS}}(\mathcal{E}^0,\underline{l})$. The same notation will be used for any element $\boldsymbol{u}\in\overline{\mathcal{KMS}}(\mathcal{E}^0,\underline{l})$.

11.3.8. Functoriality. The functoriality of the induced vector bundle can be easily obtained from the functorialities of ${}^{l}\mathcal{G}_{\boldsymbol{u}}(E)$, ${}^{l}\mathcal{G}_{\boldsymbol{u}}(\mathcal{H}(E))$, ${}^{l}\mathcal{G}_{\boldsymbol{u}}(\mathcal{E})$, and the morphisms Φ^{can} and $\Phi_{P,O}$. The lemmas in this subsection follows from the results in the subsections 8.9.5–8.9.7, 9.3.3–9.3.5, 10.3.4 and 10.4.1–10.4.2.

We have the naturally defined morphisms:

$$S^{\mathrm{can}}_{-\boldsymbol{u}}(E^\vee)\longrightarrow S^{\mathrm{can}}_{\boldsymbol{u}}(E)^\vee,\quad S_{-\boldsymbol{u}}(E^\vee,P)\longrightarrow S_{\boldsymbol{u}}(E,P)^\vee. \tag{11.8}$$

We have the naturally defined nilpotent maps $\mathcal{N}^\vee_i$ on $S^{\mathrm{can}}_{\boldsymbol{u}}(E)^\vee$ and $S_{\boldsymbol{u}}(E,P)^\vee$.

LEMMA 11.31. *The morphisms* (11.8) *are isomorphic. They are compatible with the pairing. The signature of the nilpotent map is reversed.* □

We have the naturally defined morphism:

$$\bigoplus_{\substack{\boldsymbol{u}_i\in\overline{\mathcal{KMS}}(\mathcal{E}_i^0,\underline{l}),\\ \boldsymbol{u}_1+\boldsymbol{u}_2\equiv\boldsymbol{u}}} S_{\boldsymbol{u}_1}^{\mathrm{can}}(E_1)\otimes S_{\boldsymbol{u}_2}^{\mathrm{can}}(E_2)\longrightarrow S_{\boldsymbol{u}}^{\mathrm{can}}(E_1\otimes E_2),$$

$$\bigoplus_{\substack{\boldsymbol{u}_i\in\overline{\mathcal{KMS}}(\mathcal{E}_i^0,\underline{l}),\\ \boldsymbol{u}_1+\boldsymbol{u}_2\equiv\boldsymbol{u}}} S_{\boldsymbol{u}_1}(E_1,P)\otimes S_{\boldsymbol{u}_2}(E_2,P)\longrightarrow S_{\boldsymbol{u}}(E_1\otimes E_2,P). \tag{11.9}$$

LEMMA 11.32. *The morphisms* (11.9) *are isomorphic. They are compatible with the pairings and the nilpotent maps.* □

We also have the functoriality for the pull backs. We use the setting in the subsection 8.9.7.

LEMMA 11.33. *Let $\boldsymbol{u}$ be an element of $\overline{\mathcal{KMS}}(\psi^*\mathcal{E}^0,\underline{n})$. We have the naturally defined isomorphism:*

$$\bigoplus_{\substack{\boldsymbol{u}\in\overline{\mathcal{KMS}}(\mathcal{E}^0,\underline{m})\\ \boldsymbol{c}\cdot\boldsymbol{u}=\boldsymbol{u}_1}} S_{\boldsymbol{u}}^{\mathrm{can}}(E)\simeq S_{\boldsymbol{u}_1}^{\mathrm{can}}(\psi^*E),\qquad \bigoplus_{\substack{\boldsymbol{u}\in\overline{\mathcal{KMS}}(\mathcal{E}^0,\underline{m})\\ \boldsymbol{c}\cdot\boldsymbol{u}=\boldsymbol{u}_1}} S_{\boldsymbol{u}}(E,\psi(P))\simeq S_{\boldsymbol{u}_1}(\psi^*E,P).$$

They are compatible with the pairings. Let $\mathcal{N}_i^{\triangle}$ $(i=1,\ldots,m)$ be the nilpotent maps for $S_{\boldsymbol{u}}^{\mathrm{can}}(E)$. Under the isomorphism, the nilpotent maps in the right hand side is expressed as $\sum_p c_{k,p}\cdot\mathcal{N}_p^{\triangle}$. □

11.4. $\mathrm{Gr}_h^W S_{\boldsymbol{u}}^{\mathrm{can}}(E)$ and $\mathrm{Gr}_h^W S_{\boldsymbol{u}}(E,P)$

11.4.1. The construction. This section is a preparation for the proof of the limiting mixed twistor theorem in one dimensional case (The section 12.1). Let us consider the case $X=\Delta$ and $D=\{O\}$. In this case, we have one nilpotent map $\mathcal{N}^{\triangle}$ on $S_u^{\mathrm{can}}(E)$ and $S_u(E,P)$. Due to Simpson, we know that the conjugacy classes of $\mathcal{N}_{|\lambda}^{\triangle}$ are independent of a choice of $\lambda\in\mathbb{P}^1$ (Corollary 7.71). Hence the weight filtrations are the filtration in the category of vector bundles. Thus we obtain the associated graded bundle $\mathrm{Gr}_h^W S_u^{\mathrm{can}}(E)$ and $\mathrm{Gr}_h^W S_u(E,P)$.

We have another construction of $\mathrm{Gr}_h^W S_u^{\mathrm{can}}(E)$ and $\mathrm{Gr}_h^W S_u(E,P)$.

- We have the vector bundles $\mathrm{Gr}_h^W \mathcal{G}_u(E)$ on $\boldsymbol{C}_\lambda$ and $\mathrm{Gr}_h^W \mathcal{G}_{u^\dagger}^\dagger(E)$ on $\boldsymbol{C}_\mu$.
- We have the vector bundles $\mathrm{Gr}_h^W \mathcal{G}_u(\mathcal{H})$ on $\boldsymbol{C}_\lambda^*$ and $\mathrm{Gr}_h^W \mathcal{G}_{u^\dagger}^\dagger(\mathcal{H}^\dagger)$ on $\boldsymbol{C}_\mu^*$.
- We have the vector bundles $\mathrm{Gr}_h^W \mathcal{G}_u(\mathcal{E})$ on $\mathcal{X}^\sharp$ and $\mathrm{Gr}_h^W \mathcal{G}_{u^\dagger}^\dagger(\mathcal{E}^\dagger)$ on $\mathcal{X}^{\sharp\,\dagger}$. Here the filtration W is the weight filtration of the logarithm of the unipotent part of the monodromy. We have the canonical isomorphisms:

$$\mathrm{Gr}_h^W \mathcal{G}_u(\mathcal{E})_{|\boldsymbol{C}_\lambda^*\times\{O\}}\simeq \mathrm{Gr}_h^W \mathcal{G}_u(E)_{|\boldsymbol{C}_\lambda^*},\quad \mathrm{Gr}_h^W \mathcal{G}_{u^\dagger}^\dagger(\mathcal{E}^\dagger)_{|\boldsymbol{C}_\mu^*\times\{O\}}\simeq \mathrm{Gr}_h^W \mathcal{G}_{u^\dagger}^\dagger(E)_{|\boldsymbol{C}_\mu^*}.$$

We have the family of the induced flat connections $\mathbb{D}^f$ and $\mathbb{D}^{\dagger\,f}$ on $\mathrm{Gr}_h^W \mathcal{G}_u(\mathcal{E})$ and $\mathrm{Gr}_h^W \mathcal{G}_{u^\dagger}^\dagger(\mathcal{E}^\dagger)$ respectively.

LEMMA 11.34. *Let us consider the monodromy endomorphisms for the family of flat bundles $\mathrm{Gr}_h^W \mathcal{G}_u(\mathcal{E})$ and $\mathrm{Gr}_h^W \mathcal{G}_{u^\dagger}^\dagger(\mathcal{E}^\dagger)$ with respect to the flat connections. They are the multiplication of some holomorphic function $F(\lambda)$.*

Proof Let P be a point of $X - D$. The monodromy of $\mathrm{Gr}_h^W {}^l\mathcal{G}_u(\mathcal{E})_{|(\lambda,P)}$ has the unique eigenvalue. Since we consider the associated graded object with respect to the monodromy weight filtration, the unipotent part is trivial. Hence we are done. □

As in the cases of $S_u^{\mathrm{can}}(E)$, we obtain the isomorphisms:

$$\Phi^{\mathrm{can}} : \mathrm{Gr}_h^W \mathcal{G}_u(\mathcal{H}) \longrightarrow \mathrm{Gr}_h^W \mathcal{G}_u(E)_{|\boldsymbol{C}_\lambda^*},$$

$$\Phi^{\mathrm{can}\,\dagger} : \mathrm{Gr}_h^W \mathcal{G}_{u^\dagger}^\dagger(\mathcal{H}^\dagger) \longrightarrow \mathrm{Gr}_h^W \mathcal{G}_{u^\dagger}^\dagger(E)_{|\boldsymbol{C}_\mu^*}.$$

Then we obtain the gluing of $\mathrm{Gr}_h^W \mathcal{G}_u(E)$ and $\mathrm{Gr}_h^W \mathcal{G}_{u^\dagger}^\dagger(E)$ via $\Phi^{\mathrm{can}\,\dagger} \circ \Phi^{\mathrm{can}\,-1}$. Thus we obtain the vector bundle, which is naturally isomorphic to $\mathrm{Gr}^W S_u^{\mathrm{can}}(E)$.

Similarly we obtain the gluing $\mathrm{Gr}_h^W \mathcal{G}_u(E)$ and $\mathrm{Gr}_h^W \mathcal{G}_{u^\dagger}^\dagger(E)$ via $\Phi_{O,P}^\dagger \circ \Phi_{O,P}^{-1}$. The resulted vector bundle is naturally isomorphic to $\mathrm{Gr}_h^W S_u(E,P)$.

11.4.2. The gluing matrices. For simplicity we put as follows:

$$\mathrm{Gr}^W \mathcal{G}(E) := \bigoplus_{k\in\mathbb{Z}} \bigoplus_{u\in\mathcal{KMS}({}^\diamond\mathcal{E}^0)} \mathrm{Gr}_k^W \mathcal{G}_u(E),$$

$$\mathrm{Gr}^W \mathcal{G}^\dagger(E) := \bigoplus_{k\in\mathbb{Z}} \bigoplus_{u\in\mathcal{KMS}({}^\diamond\mathcal{E}^0)} \mathrm{Gr}_k^W \mathcal{G}_{u^\dagger}^\dagger(E).$$

Let $\boldsymbol{w}$ be a frame of $\mathrm{Gr}^W \mathcal{G}(E)$ compatible with the grading. We denote the degree of w_i by $u(w_i) \in \mathcal{KMS}({}^\diamond\mathcal{E}^0)$, i.e., $w_i \in \mathrm{Gr}^W \mathcal{G}_{u(w_i)}(E)$. Let $\boldsymbol{w}^\dagger$ be a frame of $\mathrm{Gr}^W \mathcal{G}^\dagger(E)$. We denote the degree of $w_i^\dagger$ by $u(w_i^\dagger) \in \mathcal{KMS}({}^\diamond\mathcal{E}^0)$, i.e., $w_i^\dagger \in \mathrm{Gr}^W \mathcal{G}_{u(w_i^\dagger)^\dagger}(E)$.

REMARK 11.35. Note $u(w_i^\dagger)$ denotes an element of $\mathcal{KMS}({}^\diamond\mathcal{E}^0)$ not $\mathcal{KMS}({}^\diamond\mathcal{E}^{\dagger\,0})$. □

REMARK 11.36. Let $\boldsymbol{w}^\vee$ be the frame of $\left(\mathrm{Gr}^W \mathcal{G}(E)\right)^\vee$, which is the dual frame of $\boldsymbol{w}$. If $\boldsymbol{w}^\dagger$ is obtained from $\boldsymbol{w}^\vee$ via the isomorphism $\sigma^* \mathrm{Gr}^W \mathcal{G}^\dagger(E) \simeq \left(\mathrm{Gr}^W \mathcal{G}(E)\right)^\vee$, we have $u(w_i) = u(w_i^\dagger)$. □

By gluing $(\Phi^{\mathrm{can}\,\dagger}) \circ \Phi^{\mathrm{can}\,-1}$, we obtain the relation $\boldsymbol{w} = \boldsymbol{w}^\dagger \cdot A^{\mathrm{can}}$ for some holomorphic function $A^{\mathrm{can}} : \boldsymbol{C}_\lambda^* \longrightarrow GL(r)$. By gluing $\Phi_{P,O}^\dagger \circ \Phi_{P,O}^{-1}$, we obtain the relation $\boldsymbol{w} = \boldsymbol{w}^\dagger \cdot A_{P,O}$. We would like to give a method to calculate A^{can} and $A_{P,O}$.

We can take the normalizing frame $\widetilde{\boldsymbol{v}}$ of $\mathrm{Gr}^W \mathcal{G}(\mathcal{E})$ on $\mathcal{X}^\sharp$ which is a lift of $\boldsymbol{w}$, i.e., $\widetilde{\boldsymbol{v}}_{|\boldsymbol{C}_\lambda\times\{O\}} = \boldsymbol{w}$. We can also take the normalizing frame $\widetilde{\boldsymbol{v}}^\dagger$ of $\mathrm{Gr}^W \mathcal{G}^\dagger(\mathcal{E}^\dagger)$, which is a lift of $\boldsymbol{w}^\dagger$. Since we have the identification $\mathrm{Gr}^W \mathcal{G}(\mathcal{E}) = \mathrm{Gr}^W \mathcal{G}^\dagger(\mathcal{E}^\dagger)$ over $\mathcal{X}^\sharp - \mathcal{D}^\sharp$ as C^∞-bundles, we have the relation:

$$\widetilde{v}_i^\dagger = \sum_j \widetilde{v}_j \cdot \widetilde{J}_{j\,i}, \quad \text{i.e.,} \quad \widetilde{\boldsymbol{v}}^\dagger = \widetilde{\boldsymbol{v}} \cdot \widetilde{J}.$$

LEMMA 11.37. *We have $\widetilde{J}_{j\,i} = 0$ unless $u(w_i^\dagger) = u(w_j)$.*

Proof It follows from the compatibility of $\boldsymbol{w}$ and $\boldsymbol{w}^\dagger$ with the grading. □

LEMMA 11.38. *In the case $u(w_i^\dagger)=u(w_j)=u$, there exist holomorphic functions $K_{j\,i}$ on $\boldsymbol{C}_\lambda^*$ such that the following holds:*

$$\widetilde{J}_{j\,i}=\exp\Bigl(-\lambda^{-1}\mathfrak{e}(\lambda,u)\cdot\log|z|^2\Bigr)\cdot K_{j\,i}.$$

Proof We have $\mathbb{D}^f w_j=\lambda^{-1}\cdot\mathfrak{e}(\lambda,u(w_j))\cdot w_j$. Hence we have the following:

$$\mathbb{D}^f\Bigl(\exp\bigl(-\lambda^{-1}\cdot\mathfrak{e}(\lambda,u(w_j))\cdot\log z\bigr)\cdot w_j\Bigr)=0.$$

Similarly we have the following:

$$\mathbb{D}^{\dagger\,f}\Bigl(\exp\bigl(-\lambda\cdot\mathfrak{e}(\lambda^{-1},u(w_i^\dagger)^\dagger)\cdot\log\bar{z}\bigr)\cdot w_i^\dagger\Bigr)=0.$$

We have $\lambda\cdot\mathfrak{e}\bigl(\lambda^{-1},u(w_i^\dagger)^\dagger\bigr)=-\lambda^{-1}\cdot\mathfrak{e}(\lambda,u(w_i^\dagger))$. Then the claim immediately follows. □

COROLLARY 11.39. *There exists the $GL(r)$-valued holomorphic function K on $\boldsymbol{C}_\lambda^*$ such that $\widetilde{J}(\lambda,z)=C(\lambda,z)\cdot K(\lambda)$. Here $C(\lambda,z)$ is given as follows:*

$$C(\lambda,z)=\bigoplus_{\substack{u\in\mathcal{KMS}({}^\diamond\mathcal{E}^0),\\ k\in\mathbb{Z}}}\exp\Bigl(-\lambda^{-1}\mathfrak{e}(\lambda,u)\cdot\log|z|^2\Bigr)\cdot\mathrm{id}_{\mathrm{Gr}_k^W\mathcal{G}_u}\,. \tag{11.10}$$

Here $C(\lambda,z)$ is regarded as the endomorphism of $\bigoplus\mathrm{Gr}_k^W\mathcal{G}_u$ via the frame $\boldsymbol{w}$.

Proof It immediately follows from Lemma 11.38. □

LEMMA 11.40.

- *Via the gluing $\Phi^{\mathrm{can}\,\dagger}\circ\Phi^{\mathrm{can}\,-1}$, we have the relation $\boldsymbol{w}^\dagger=\boldsymbol{w}\cdot K$.*
- *Via the gluing $\Phi_{O,P}^\dagger\circ\Phi_{O,P}^{-1}$, we have the relation $\boldsymbol{w}^\dagger=\boldsymbol{w}\cdot C(\lambda,P)\cdot K(\lambda)$. Here $C(\lambda,P)$ denote the matrix given as follows:*

$$C(\lambda,P)=\bigoplus_{\substack{u\in\mathcal{KMS}({}^\diamond\mathcal{E}^0),\\ k\in\mathbb{Z}}}\exp\Bigl(-\lambda^{-1}\mathfrak{e}(\lambda,u)\cdot\log|z(P)|^2\Bigr)\cdot\mathrm{id}_{\mathrm{Gr}_k^W\mathcal{G}_u(E)}\,. \tag{11.11}$$

Here $C(\lambda,P)$ is regarded as the endomorphism of $\bigoplus\mathrm{Gr}_k^W\mathcal{G}_u$ via the frame $\boldsymbol{w}$.

Proof It follows from the definitions and Lemma 11.38. □

COROLLARY 11.41. *The vector bundles $\mathrm{Gr}_h^W S_u^{\mathrm{can}}(E)$ and $\mathrm{Gr}_h^W S_u(E,P)$ are isomorphic for any $u\in\mathcal{KMS}(\mathcal{E}^0)$ and $h\in\mathbb{Z}$.*

Proof For $u=(a,\alpha)$, we have the decomposition:

$$\exp\bigl(-\lambda^{-1}\cdot\mathfrak{e}(\lambda,u)\cdot A\bigr)=\exp\bigl(-\lambda^{-1}\cdot\alpha\cdot A\bigr)\cdot\exp\bigl(a\cdot A\bigr)\cdot\exp\bigl(\lambda\cdot\overline{\alpha}\cdot A\bigr).$$

Then the claim is obvious. □

11.4.3. Local lifting and the gluing matrices. Let $\boldsymbol{w}$ be a frame of the bundle $\mathrm{Gr}^W\mathcal{G}(E)$ over $\boldsymbol{C}_\lambda$, and let $\widetilde{\boldsymbol{v}}$ be a normalizing frame of the bundle $\mathrm{Gr}^W\mathcal{G}(\mathcal{E})$ on $\mathcal{X}^\sharp$ as in the subsection 11.4.2. Let us take any point $\lambda_0\in\boldsymbol{C}_\lambda^*$, and let $U(\lambda_0)$ be an appropriate neighbourhood of λ_0 in $\boldsymbol{C}_\lambda^*$.

Let us take a non-negative number $\epsilon=\epsilon(\lambda_0)$ satisfying the following:

- In the case $0\notin\mathcal{P}ar(\mathcal{E}^{\lambda_0})$, we put $\epsilon(\lambda_0)=0$.
- In the case $0\in\mathcal{P}ar(\mathcal{E}^{\lambda_0})$, $\epsilon(\lambda_0)$ is taken any positive number such that $]0,\epsilon(\lambda_0)]\cap\mathcal{P}ar(\mathcal{E}^{\lambda_0})=\emptyset$.

For each w_i, the integer $\nu(w_i) = \nu(w_i, \lambda_0)$ is determined by the condition $-1+\epsilon(\lambda_0) < \mathfrak{p}(\lambda_0, u(w_i)) + \nu(w_i) < \epsilon(\lambda_0)$.

Let $\boldsymbol{v}^{(\lambda_0)}$ be a frame of ${}_{\epsilon}\mathcal{E}$ on $X \times U(\lambda_0)$ satisfying the following:

- The frame $\boldsymbol{v}^{(\lambda_0)}$ is compatible with the filtration $\mathcal{F}^{(\lambda_0)}$, the decomposition $\mathbb{E}^{(\lambda_0)}$ (see the section 10.2) and the weight filtration W.
- We put $\bar{v}_i^{(\lambda_0)} := v_i^{(\lambda_0)} \cdot z^{\nu(w_i)}$. Then the tuple $\bar{\boldsymbol{v}}^{(\lambda_0)} = \big(\bar{v}_i^{(\lambda_0)}\big)$ gives a frame of $\mathcal{E}$ on $(X - D) \times U$, which is clearly compatible with $\mathcal{F}^{(\lambda_0)}$, $\mathbb{E}^{(\lambda_0)}$ and W. We remark $\bar{\boldsymbol{v}}^{(\lambda_0)}$ induces the frame of $\mathrm{Gr}^W \mathcal{G}(\mathcal{E})$ on $X \times U(\lambda_0)$.
- The induced frame of $\bar{\boldsymbol{v}}^{(\lambda_0)}$ is same as $\widetilde{\boldsymbol{v}}_{|X \times U(\lambda_0)}$.

Such $\boldsymbol{v}^{(\lambda_0)}$ is called a local lift of $\boldsymbol{w}$ around λ_0.

REMARK 11.42. In the case $\lambda_0 = 0$, a local lift $\boldsymbol{v}^{(0)}$ of $\boldsymbol{w}$ around 0 is defined to be a frame of ${}_{\epsilon(0)}\mathcal{E}$ on $U(0) \times X$ satisfying the following:

- $\boldsymbol{v}^{(0)}$ is compatible with the decomposition $\mathbb{E}^{(0)}$, the parabolic filtration $F^{(0)}$, and the weight filtration W.
- Then $\boldsymbol{v}^{(0)}$ induces the frame of $\mathrm{Gr}^W \mathcal{G}(E)_{|U(0)} = \mathrm{Gr}^E \mathcal{G}(\mathcal{E})_{|\{O\} \times U(0)}$. The induced frame is same as $\boldsymbol{w}_{|U(0)}$. □

Let $\boldsymbol{w}^{\dagger}$ be a frame of $\mathrm{Gr}^W \mathcal{G}^{\dagger}$ on $\boldsymbol{C}_{\mu}$, and let $\widetilde{\boldsymbol{v}}^{\dagger}$ be a normalizing frame of the bundle $\mathrm{Gr}^W \mathcal{G}^{\dagger}(\mathcal{E}^{\dagger})$ over $\mathcal{X}^{\sharp\,\dagger} = X^{\dagger} \times \boldsymbol{C}^*_{\mu}$ as in the subsection 11.4.2. Let μ_0 be any point of $\boldsymbol{C}^*_{\mu}$, and let $U'(\mu_0)$ be an appropriate neighbourhood of μ_0 in $\boldsymbol{C}^*_{\mu}$. We have a non-negative number $\epsilon' = \epsilon'(\mu_0)$ satisfying the following:

- In the case $0 \notin \mathcal{P}ar(\mathcal{E}^{\dagger\,\mu_0})$, we put $\epsilon'(\mu_0) = 0$.
- In the case $0 \in \mathcal{P}ar(\mathcal{E}^{\dagger\,\mu_0})$, $\epsilon'(\mu_0)$ is taken any positive number such that $]0, \epsilon'(\mu_0)] \cap \mathcal{P}ar(\mathcal{E}^{\mu_0}) = \emptyset$.

For each $w_i^{\dagger}$, the integer $\nu(w_i^{\dagger}) = \nu(w_i^{\dagger}, \mu_0)$ is determined by the condition $-\epsilon' < \mathfrak{p}(\lambda_0^{-1}, u(w_i^{\dagger})^{\dagger}) + \nu(w_i^{\dagger}) \leq 1 - \epsilon'$. We can take a frame $\boldsymbol{v}^{\dagger\,(\mu_0)}$ of ${}_{1-\epsilon'}\mathcal{E}^{\dagger}$ satisfying the following:

- The frame $\boldsymbol{v}^{\dagger\,(\mu_0)}$ is compatible with $\mathcal{F}^{(\mu_0)}$, $\mathbb{E}^{(\mu_0)}$ and W.
- We put $\bar{v}_i^{\dagger\,(\mu_0)} = \bar{z}^{\nu(w_i^{\dagger})} \cdot v_i^{\dagger\,(\mu_0)}$. Then the frame $\bar{\boldsymbol{v}}^{\dagger\,(\mu_0)}$ is clearly compatible with $\mathcal{F}^{(\mu_0)}$, $\mathbb{E}^{(\mu_0)}$ and W. Hence $\bar{\boldsymbol{v}}^{\dagger\,(\mu_0)}$ induces the frame of $\mathrm{Gr}^W \mathcal{G}^{\dagger}(\mathcal{E})$ on $X^{\dagger} \times U'(\mu_0)$.
- The induced frame of $\bar{\boldsymbol{v}}^{\dagger\,(\mu_0)}$ is same as $\boldsymbol{v}^{\dagger\,(\mu_0)}$.

Such $\boldsymbol{v}^{\dagger\,(\mu_0)}$ is called a local lift of $\widetilde{\boldsymbol{v}}^{\dagger}$ around μ_0.

REMARK 11.43. As in Remark 11.42, a local lift at $\mu_0 = 0$ is also defined, similarly. □

Let us consider the case $\lambda_0^{-1} = \mu_0$, and assume that we are given local lifts $\boldsymbol{v}^{(\lambda_0)}$ and $\boldsymbol{v}^{\dagger\,(\mu_0)}$. We also assume that $U(\lambda_0) = U'(\mu_0)$ via $\lambda = \mu^{-1}$. On $(X - D) \times U(\lambda_0) = (X^{\dagger} - D^{\dagger}) \times U'(\lambda_0^{-1})$, we have $\mathcal{E} = \mathcal{E}^{\dagger}$ as $U(\lambda_0)$-holomorphic bundles. Hence we obtain the relation:

$$v_i^{\dagger\,(\lambda_0^{-1})} = \sum J_{j\,i}^{(\lambda_0)} \cdot v_j^{(\lambda_0)}.$$

Here $J_{j\,i}^{(\lambda_0)}$ are U-holomorphic.

LEMMA 11.44. *In the case $J_{j\,i} \neq 0$, we have the following:*

$$\mathfrak{e}^f\big(\lambda_0^{-1}, u(w_i^\dagger)^\dagger\big) = \mathfrak{e}^f\big(\lambda_0, u(w_j)\big)^{-1}, \quad \mathfrak{p}^f\big(\lambda_0^{-1}, u(w_i^\dagger)^\dagger\big) \geq \mathfrak{p}^f\big(\lambda_0, u(w_j)\big),$$

$$\deg^W(w_i^\dagger) \geq \deg^W(w_j).$$

Here $\boldsymbol{w}$ and $\boldsymbol{w}^\dagger$ as in the subsection 11.4.2.

Proof It follows from the compatibility of the frames $\boldsymbol{v}^{(\lambda_0)}$, $\boldsymbol{v}^{\dagger\,(\lambda_0^{-1})}$ and the filtrations and the decompositions. □

LEMMA 11.45. *In the case $u(w_i^\dagger) = u(w_j)$ and $\deg^W(w_i^\dagger) = \deg^W(w_j)$, we have the relation for $\mu_0 = \lambda_0^{-1}$:*

$$J_{j\,i}^{(\lambda_0)} = \widetilde{J}_{j\,i} \cdot z^{\nu(w_j,\lambda_0)} \cdot \bar{z}^{-\nu(w_i^\dagger,\mu_0)}.$$

Proof It directly follows from the relation of lifts $\boldsymbol{v}^{(\lambda_0)}$ and $\boldsymbol{v}^{\dagger\,(\mu_0)}$ with the normalizing frames $\widetilde{\boldsymbol{v}}$ and $\widetilde{\boldsymbol{v}}^\dagger$. □

CHAPTER 12

Limiting Mixed Twistor Theorem

12.1. Limiting mixed twistor theorem in the case of curves

12.1.1. Statement. We put $X = \Delta$ and $D = \{O\}$. Let $(E, \overline{\partial}_E, \theta, h)$ be a tame harmonic bundle over $X - D$. Let u be any element of $\mathcal{KMS}(\mathcal{E}^0)$. We have the vector bundle $S_u^{\mathrm{can}}(E)$ and $S_u(E, P)$ with the nilpotent map and the pairing (the section 11.3). We will prove the following theorem in this section. (See the next section for the higher dimensional case.)

THEOREM 12.1. *The tuples* $(S_u^{\mathrm{can}}(E), W, N^{\triangle}, S)$ *and* $(S_u(E,P), W, N^{\triangle}, S)$ *are polarized mixed twistor structures of weight* 0 *in one variable, where* P *is any point of* $X - D$. □

(See Definition 3.48 for the definition of a polarized mixed twistor structure.)

12.1.2. Preliminary. We put $X = \Delta$ and $D = \{O\}$. Let $(E, \overline{\partial}_E, \theta, h)$ be a tame harmonic bundle over $X - D$. We can take a model bundle $(E_0, \overline{\partial}_{E_0}, \theta_0, h_0)$ as in the subsection 7.3.1. We denote the deformed holomorphic bundle of E_0 by $\mathcal{E}_0$.

Let $\boldsymbol{e}$ and $\boldsymbol{e}_0$ be holomorphic frames of ${}^{\diamond}E$ and ${}^{\diamond}E_0$ respectively, which are compatible with F, $\mathbb{E}$ and W. We take the isomorphism $\Phi : {}^{\diamond}E_0 \longrightarrow {}^{\diamond}E$ via the condition $\Phi(\boldsymbol{e}_0) = \boldsymbol{e}$. It satisfies the following:

- The morphism Φ is compatible with $\mathbb{E}$, F and W at the origin O.
- We have the induced isomorphism $\mathrm{Gr}^F \Phi : \mathrm{Gr}^F({}^{\diamond}E_0) \longrightarrow \mathrm{Gr}^F({}^{\diamond}E)$ and the endomorphisms $\mathrm{Gr}^F\big(\mathrm{Res}(\theta)\big)$ and $\mathrm{Gr}^F\big(\mathrm{Res}(\theta_0)\big)$. Under the isomorphism Gr^F, we have $\mathrm{Gr}^F\big(\mathrm{Res}(\theta)\big) = \mathrm{Gr}^F\big(\mathrm{Res}(\theta_0)\big)$.

We remark that we often regard Φ as a C^∞-isomorphism $p_\lambda^{-1}E_0 \longrightarrow p_\lambda^{-1}E$ on $(X - D) \times \boldsymbol{C}_\lambda$, where p_λ denote the natural projection $(X - D) \times \boldsymbol{C}_\lambda \longrightarrow X - D$ We also remark that Φ is $\boldsymbol{C}_\lambda$-holomorphic, in that case.

Recall that the morphisms Φ and Φ^{-1} give the isomorphism of E and E_0 which are bounded with respect to the metrics h and h_0, due to Simpson. (See the section 4.3.3 in our previous paper [**65**], for example.)

We use the setting of the subsection 11.4.3. Let $\boldsymbol{w}$ be a frame of $\mathrm{Gr}^W \mathcal{G}(E)$ over $\boldsymbol{C}_\lambda$ and $\widetilde{\boldsymbol{v}}$ be the normalizing frame of $\mathrm{Gr}^W \mathcal{G}(\mathcal{E})$ over $\mathcal{X}^\sharp$ as in the subsection 11.4.3. Let λ_0 be a point of $\boldsymbol{C}_\lambda$. Let $\epsilon = \epsilon(\lambda_0)$ be the non-negative number and $U(\lambda_0)$ be an appropriate neighbourhood of λ_0 as in the subsection 11.4.3. Let $\boldsymbol{v}^{(\lambda_0)}$ be a local lift of $\boldsymbol{w}$ on $X \times U(\lambda_0)$. In the case $\lambda_0 = 0$, we may assume $\boldsymbol{v}^{(0)}_{|X\times\{0\}} = \boldsymbol{e}$.

On the other hand, let $\boldsymbol{w}_0$ be a frame of $\mathrm{Gr}^W \mathcal{G}(E_0)$ over $\boldsymbol{C}_\lambda$. We may assume that we can take a canonical frame $\boldsymbol{v}_0^{(\lambda_0)}$ of ${}_{\epsilon(\lambda_0)}\mathcal{E}_0$ on $X \times U(\lambda_0)$, which is a local lift of $\boldsymbol{w}_0$ (See the subsection 6.2.7 for a canonical frame). It is compatible with

$\mathbb{E}^{(\lambda_0)}$, $\mathcal{F}^{(\lambda_0)}$ and W of ${}_{\epsilon(\lambda_0)}\mathcal{E}_0$, in the case $\lambda_0 \neq 0$ (see the section 10.2). In the case $\lambda_0 = 0$, it is compatible with $\mathbb{E}^{(0)}$, $F^{(0)}$ and W at D. We remark the following.

LEMMA 12.2. *We have $\nu(w_{0\,i}, \lambda_0) = \nu(w_j, \lambda_0)$ in the case $\deg^{F^{(\lambda_0)},\mathbb{E}^{(\lambda_0)}}(v_{0\,i}) = \deg^{F^{(\lambda_0)},\mathbb{E}^{(\lambda_0)}}(v_j)$ (See the subsection 11.4.3 for $\nu(w_j, \lambda_0)$).* □

We put as follows:

$$\boldsymbol{v}^{(\lambda_0)\,\prime} = \big(v_i^{(\lambda_0)\,\prime}\big), \quad v_i^{(\lambda_0)\prime} := v_i^{(\lambda_0)} \cdot |z|^{b(v_i^{(\lambda_0)})} \cdot \big(-\log|z|\big)^{-\frac{1}{2}k(v_i^{(\lambda_0)})},$$
$$(12.1)\qquad \boldsymbol{v}_0^{(\lambda_0)\,\prime} = \big(v_{0\,i}^{(\lambda_0)\,\prime}\big), \quad v_{0\,i}^{(\lambda_0)\prime} := v_{0\,i}^{(\lambda_0)} \cdot |z|^{b(v_{0\,i}^{(\lambda_0)})} \cdot \big(-\log|z|\big)^{-\frac{1}{2}k(v_{0\,i}^{(\lambda_0)})}.$$

Here $b(v_i^{(\lambda_0)})$ denote functions of λ given by $b(v_i^{(\lambda_0)})(\lambda) := \deg^F(v_{i\,|\,\mathcal{X}^\lambda}^{(\lambda_0)})$, and $k(v_i^{(\lambda_0)})$ denote $\deg^W(v_i^{(\lambda_0)})$. The meaning of $b(v_{0\,i}^{(\lambda_0)})$ and $k(v_{0\,i}^{(\lambda_0)})$ are similar. Recall that $\boldsymbol{v}^{(\lambda_0)\,\prime}$ and $\boldsymbol{v}_0^{(\lambda_0)\,\prime}$ are adapted (Proposition 8.140). We also put $\boldsymbol{e}' := \boldsymbol{v}^{(0)\,\prime}_{|(X-D)\times\{0\}}$ and $\boldsymbol{e}_0' := \boldsymbol{v}^{(0)\,\prime}_{0\,|\,(X-D)\times\{0\}}$.

We put $r := \operatorname{rank}(E)$. We have the $GL(r)$-valued C^∞-functions $B^{(\lambda_0)}$ and $B_0^{(\lambda_0)}$ defined in $U(\lambda_0) \times (X - D)$, determined by the following conditions:

$$(12.2)\qquad \boldsymbol{v}^{(\lambda_0)\,\prime} = \boldsymbol{e}' \cdot B^{(\lambda_0)}, \qquad \boldsymbol{v}_0^{(\lambda_0)\,\prime} = \boldsymbol{e}_0' \cdot B_0^{(\lambda_0)}.$$

Here we regard $\boldsymbol{e}'$ and $\boldsymbol{e}_0'$ as C^∞-frames of $\mathcal{E}_{|(X-D)\times U(\lambda_0)}$ and $\mathcal{E}_{0|(X-D)\times U(\lambda_0)}$ naturally.

The functions $I_{j\,i}^{(\lambda_0)\prime}$ are determined as follows:

$$\Phi(v_{0\,i}^{(\lambda_0)\prime}) = \sum I_{j\,i}^{(\lambda_0)\prime} \cdot v_j^{(\lambda_0)\,\prime}.$$

Then we obtain the $M(r)$-valued function $I^{(\lambda_0)\,\prime} := \big(I_{j\,i}^{(\lambda_0)\,\prime}\big) : U(\lambda_0) \times (X - D) \longrightarrow M(r)$. The following lemma is easy to see.

LEMMA 12.3. *The functions $I_{j\,i}^{(\lambda_0)\,\prime}$ are holomorphic with respect to the variable λ, in the case $\deg^{\mathbb{E}^{(\lambda_0)},F^{(\lambda_0)}}(v_{0\,i}^{(\lambda_0)}) = \deg^{\mathbb{E}^{(\lambda_0)},F^{(\lambda_0)}}(v_j^{(\lambda_0)})$ and $k(v_{0\,i}^{(\lambda_0)}) = k(v_j^{(\lambda_0)})$.*

Proof If we denote $\Phi(v_{0\,i}^{(\lambda_0)}) = \sum I_{j\,i}^{(\lambda_0)} \cdot v_j^{(\lambda_0)}$, then $I_{j\,i}^{(\lambda_0)}$ are holomorphic with respect to λ. Then the lemma immediately follows. □

LEMMA 12.4.

- *$I^{(\lambda_0)\,\prime}$ and $I^{(\lambda_0)\,\prime\,-1}$ are bounded.*
- *In the case $\deg^{\mathbb{E}^{(\lambda_0)},F^{(\lambda_0)}}(v_{0\,i}^{(\lambda_0)}) \neq \deg^{\mathbb{E}^{(\lambda_0)},F^{(\lambda_0)}}(v_j^{(\lambda_0)})$, we have the estimate $|I_{j\,i}^{(\lambda_0)\,\prime}| \leq C \cdot (-\log|z|)^{-1}$.*
- *In the case $\deg^W(v_{0\,i}^{(\lambda_0)}) \neq \deg^W(v_j^{(\lambda_0)})$, we have the following finiteness:*

$$\int_{U(\lambda_0)} ||I_{j\,i}^{(\lambda_0)\,\prime}||_W < \infty.$$

Proof It can be shown by an argument similar to the proof of Proposition 7.67. □

12.1.3. The construction of the isomorphism Ψ. Let $\mathrm{Gr}^W \mathcal{G}(E)$ and $\mathrm{Gr}^W \mathcal{G}(E_0)$ be as in the subsection 11.4.2. In this subsection, we would like to construct the holomorphic isomorphism $\Psi : \mathrm{Gr}^W \mathcal{G}(E) \longrightarrow \mathrm{Gr}^W \mathcal{G}(E_0)$.

For any point $\lambda_0 \in \boldsymbol{C}_\lambda$, we take a neighbourhood $U(\lambda_0)$ as in the subsection 11.4.3. Then we obtain the covering $\{U(\lambda_0) \,|\, \lambda_0 \in \boldsymbol{C}_\lambda\}$ of the complex plane $\boldsymbol{C}_\lambda$. Then we can take a discrete subset S of $\boldsymbol{C}_\lambda$ such that $\{U(\lambda_0) \,|\, \lambda_0 \in S\}$ is a covering of $\boldsymbol{C}_\lambda$. We may assume that $0 \in S$.

On each $X \times U(\lambda_0)$ ($\lambda_0 \in S$), we have the frames $\boldsymbol{v}^{(\lambda_0)}$ and $\boldsymbol{v}_0^{(\lambda_0)}$ of ${}_\epsilon\mathcal{E}$ and ${}_\epsilon\mathcal{E}_0$ respectively, as in the subsection 12.1.2. We recall that we have assumed that $\boldsymbol{v}^{(0)}_{|X\times\{0\}} = \boldsymbol{e}$, $\boldsymbol{v}^{(0)}_{0|X\times\{0\}} = \boldsymbol{e}_0$ and $\Phi(\boldsymbol{e}_0) = \boldsymbol{e}$. In the following, we identify E and E_0 via the isomorphism Φ as C^∞-vector bundles on $X - D$.

LEMMA 12.5. *We can take a sequence of subsets* $U_N \subset \{z \in \boldsymbol{C} \,|\, 0 < |z| < e^{-2}\}$ ($N = 1, 2, \ldots$) *satisfying the following conditions:*

(1) *The volume of U_N with respect to the following measure is infinite.*
$$\frac{|dz \cdot d\bar{z}|}{|z|^2 \cdot (-\log|z|) \cdot \log(-\log|z|)}.$$
(2) $U_N \supset U_{N+1}$.
(3) *Let P be any point of U_N and λ_0 be any point of $\{\lambda \in S \,\big|\, |\lambda| < N\}$. Then the inequality $\big|I^{(\lambda_0)\,\prime}_{j\,i}(\lambda, P)\big| \leq N^{-1}$ holds for any i and j such that* $\deg^{F^{(\lambda_0)},\mathbb{E}^{(\lambda_0)},W}\big(v^{(\lambda_0)}_{0\,i}\big) \neq \deg^{F^{(\lambda_0)},\mathbb{E}^{(\lambda_0)},W}\big(v^{(\lambda_0)}_{j}\big)$.
(4) *Let $\boldsymbol{e}$ be a frame of ${}^\diamond E$ as above. For any point $P \in U_N$, the inequality $|h(e'_i, e'_j)(P)| \leq N^{-1}$ holds, in the case* $\deg^{\mathbb{E},F,W}(e_i) \neq \deg^{\mathbb{E},F,W}(e_j)$.
(5) *For any point $P \in U_N$, the following inequalities hold:*
$$|z| \cdot (-\log|z|) \cdot \big|\theta_0 - \theta\big|_h(P) \leq N^{-1}, \qquad |z| \cdot (-\log|z|) \cdot \big|\theta_0^\dagger - \theta^\dagger\big|_h(P) \leq N^{-1}.$$

Proof It follows from Lemma 12.4, the asymptotic orthogonality (the subsections 7.7.1, 7.7.3 and 7.7.4), Lemma 7.66, and our choice $\mathrm{Gr}^F\big(\mathrm{Res}(\theta - \theta_0)\big) = 0$. □

Let us pick points $P_N \in U_N$ such that $\lim_{N\to\infty} P_N = O$.

LEMMA 12.6. *Note that we have the following for any $\lambda_0 \in S$:*

(1) *For any point $\lambda \in U(\lambda_0)$, we have $\lim_{N\to\infty} I^{(\lambda_0)\,\prime}_{j\,i}(\lambda, P_N) = 0$ in the case $\deg^{\mathbb{E},F,W}(v^{(\lambda_0)}_{0\,i}) \neq \deg^{\mathbb{E},F,W}(v^{(\lambda_0)}_{j})$. It follows from the claim* (3) *in Lemma* 12.5.
(2) *The $M(r)$-valued functions $I^{(\lambda_0)\,\prime}_{|U(\lambda_0)\times\{P_N\}}$ and $I^{(\lambda_0)\,\prime\,-1}_{|U(\lambda_0)\times\{P_N\}}$ are bounded. The functions $I^{(\lambda_0)\,\prime}_{i\,j\,|\,U(\lambda_0)\times\{P_N\}}$ are holomorphic with respect to the variable λ, in the case* $\deg^{\mathbb{E}^{(\lambda_0)},F^{(\lambda_0)}\,W}(v_{0\,i}) = \deg^{\mathbb{E}^{(\lambda_0)},F^{(\lambda_0)}\,W}(v_j)$.
(3) *The sequences of hermitian matrices $\{H(h, \boldsymbol{e}')_{|P_N}\}$ and $\{H(h, \boldsymbol{e}')^{-1}_{|P_N}\}$ are bounded.* □

LEMMA 12.7. *By taking a subsequence $\{N_i\}$ of $\{N\}$, we may assume that* $\{I^{(\lambda_0)\,\prime}_{|U(\lambda_0)\times\{P_{N_i}\}}\}$, $\{I^{(\lambda_0)\,\prime\,-1}_{|U(\lambda_0)\times\{P_{N_i}\}}\}$, $\{H(h, \boldsymbol{e}')_{|P_{N_i}}\}$ *and* $\{H(h, \boldsymbol{e}')^{-1}_{|P_{N_i}}\}$ *are convergent for any $\lambda_0 \in S$.*

Proof Since $\{H(h, \boldsymbol{e}')_{|P_N}\}$ and $\{H(h, \boldsymbol{e}')^{-1}_{|P_N}\}$ are bounded sequences of hermitian matrices, we can take a convergent subsequence, when we fix λ_0. Due to the claims (1) and (2) in Lemma 12.6, we can take a convergent subsequences of $\{I^{(\lambda_0)\,\prime\,-1}_{|U(\lambda_0)\times\{P_N\}}\}$ when we fix λ_0. Then we obtain the lemma by using the standard diagonal argument. □

We denote the limit by $I^{(\lambda_0)\prime}_{\infty}$ and $H^{(\lambda_0)}_{\infty}$. Then $I^{(\lambda_0)\prime}_{\infty}$ is holomorphic $M(r)$-valued function on $U(\lambda_0)$ (note the claim (2) in Lemma 12.6). Due to our construction and the claim (1) in Lemma 12.6, the $M(r)$-valued function $I^{(\lambda_0)\,\prime}_{\infty}$ can be regarded as the direct sum:

$$\bigoplus_{(u,k)\in\mathcal{KMS}(^{\circ}\mathcal{E}^0)\times\mathbb{Z}} I^{(\lambda_0)}_{u,k},$$

$$I^{(\lambda_0)}_{u,k} := \left(I^{(\lambda_0)\,\prime}_{\infty\;i,j} \;\middle|\; \begin{array}{l} \deg^{F^{(\lambda_0)}}(v^{(\lambda_0)}_{0\,i}) = \deg^{F^{(\lambda_0)}}(v^{(\lambda_0)}_{j}) = \mathfrak{p}(\lambda_0,u)+\nu(w_j,\lambda_0), \\ \deg^{\mathbb{E}^{(\lambda_0)}}(v^{(\lambda_0)}_{0\,i}) = \deg^{\mathbb{E}^{(\lambda_0)}}(v^{(\lambda_0)}_{j}) = \mathfrak{e}(\lambda_0,u)-\lambda_0\cdot\nu(w_j,\lambda_0), \\ \deg^{W}(v^{(\lambda_0)}_{0,i}) = \deg^{W}(v^{(\lambda_0)}_{j}) = k \end{array} \right).$$

See the subsection 11.4.3 for $\nu(w_j,\lambda_0)$. We have used Lemma 12.2.

Then the frames $\boldsymbol{w}$ and $\boldsymbol{w}_0$ and the function $I^{(\lambda_0)\,\prime}_{\infty}$ induce the isomorphism $\Psi^{(\lambda_0)}$ defined on $U(\lambda_0)$, which preserves the grading.

$$\Psi^{(\lambda_0)} : \mathrm{Gr}^W \mathcal{G}(E_0)_{|U(\lambda_0)} \longrightarrow \mathrm{Gr}^W \mathcal{G}(E)_{|U(\lambda_0)}.$$

It is easy to check the following lemma.

LEMMA 12.8.

- *Fix a subsequence $\{P_l\}$. Then the morphism $\Psi^{(\lambda_0)}$ is independent of choices of $\boldsymbol{v}^{(\lambda_0)}$, $\boldsymbol{v}^{(\lambda_0)}_0$, $\boldsymbol{w}$ and $\boldsymbol{w}_0$.*
- *In the case $A := U(\lambda_0)\cap U(\lambda_1)\neq\emptyset$, we have $\Psi^{(\lambda_0)}_{|A} = \Psi^{(\lambda_1)}_{|A}$.*
- *In particular, we obtain the global isomorphism on $\boldsymbol{C}_\lambda$:*

$$\Psi : \mathrm{Gr}^W \mathcal{G}(E_0) \longrightarrow \mathrm{Gr}^W \mathcal{G}(E).$$

Once we fix a subsequence $\{P_l\}$, then the morphism Ψ is independent of choices of $\boldsymbol{w}$, $\boldsymbol{w}_0$, $\boldsymbol{v}^{(\lambda_0)}$, $\boldsymbol{v}^{(\lambda_0)}_0$ and S. □

REMARK 12.9. In the following, we may and will assume that $\Psi(\boldsymbol{w}_0)=\boldsymbol{w}$. In the first part of the subsection 12.1.3, S was taken as a discrete set in $\boldsymbol{C}_\lambda$ for the convergence argument. Since we have already derived a convergence, we can take $S=\boldsymbol{C}_\lambda$, which is convenient for a formal argument. □

On the other hand, $H^{(\lambda_0)}_{\infty}$ is a positive definite hermitian matrix, and it can be regarded as a direct sum of the hermitian matrices $H^{(\lambda_0)}_u$ ($u\in\mathcal{KMS}({}_{\epsilon}\mathcal{E}^0)$), due to the claim (3) in Lemma 12.6. We have the induced hermitian metric $h_{u,k}$ on the vector space $\mathrm{Gr}^W_k \mathcal{G}_u(E)_{|0}$ induced by $H_{u,k}$.

12.1.4. Modification of the model bundle. We put $\theta = f_0\cdot dz/z$ and $\theta^\dagger = f_0^\dagger\cdot d\bar{z}/\bar{z}$. We pick ρ as in (7.1) in the page 146. Then the sequence of the endomorphisms $\{(-\log|z|)\cdot(f_0(P_N)-\rho(P_N))\}$ and $\{(-\log|z|)\cdot(f_0^\dagger(P_N)-\rho^\dagger(P_N))\}$ converges to the following morphisms, due to the condition (5) in Lemma 12.5 and Proposition 7.2:

$$f_\infty : \mathrm{Gr}^W_k \mathcal{G}_u(E)_{|0} \longrightarrow \mathrm{Gr}^W_{k-2}\mathcal{G}_u(E)_{|0},$$

$$f^\dagger_\infty : \mathrm{Gr}^W_k \mathcal{G}_u(E)_{|0} \longrightarrow \mathrm{Gr}^W_{k+2}\mathcal{G}_u(E)_{|0}.$$

By our construction f_∞ and $f_\infty^\dagger$ are mutually adjoint with respect to the metric $h_\infty = \bigoplus h_{u,k}$ of $\mathrm{Gr}^W \mathcal{G}(E)_{|0}$.

On the other hand, we have the metric $h_{0\,u\,k}$ on $\mathrm{Gr}_k^W(\mathcal{G}_u(E_0))$, which gives the metric $h_{0,\infty}$ of $\mathrm{Gr}^W \mathcal{G}(E_0)_{|0}$. We also have the morphisms $f_{0\,\infty}$ and $f_{0\,\infty}^\dagger$, which coincides f_∞ and $f_\infty^\dagger$ under the isomorphism Ψ, due to the conditions (4) and (5) in Lemma 12.5. In the following, we identify $\mathrm{Gr}_k^W(\mathcal{G}_u(E_0))$ and $\mathrm{Gr}_k^W(\mathcal{G}_u(E))$ via the isomorphism Ψ. We also identify $(f_\infty, f_\infty^\dagger)$ and $(f_{0\,\infty}, f_{0\,\infty}^\dagger)$.

We have the primitive decomposition $\mathrm{Gr}^W = \bigoplus P_h \mathrm{Gr}_k^W$ with respect to the morphisms $f_\infty = f_{0\,\infty}$. From the construction of the model bundle, the primitive decomposition is orthogonal with respect to the hermitian metric $h_{0\,\infty}$. In particular, the morphism $f_\infty^\dagger = f_{0\,\infty}^\dagger$ preserves the primitive decomposition.

LEMMA 12.10. *The primitive decomposition is orthogonal with respect to the hermitian metric h_∞.*

Proof Since we have $f_\infty^\dagger = f_{0\,\infty}^\dagger$, the morphism $f_\infty^\dagger$ preserves the primitive decomposition. It implies the orthogonality of the primitive decomposition with respect to the metric h_∞. □

It is easy to see that the model bundle $(E_0, \overline{\partial}_{E_0}, h_0, \theta_0)$ is isomorphic to the following:

$$\bigoplus_{k,u} (P_k \mathrm{Gr}_k^W, h_{0,u,k}) \otimes Mod(k) \otimes L(u).$$

We have the other model bundle $(E_1, \overline{\partial}_{E_1}, h_1, \theta_1)$ given as follows:

$$\bigoplus_{k,u} (P_k \mathrm{Gr}_k^W, h_{u,k}) \otimes Mod(k) \otimes L(u).$$

Note that we have the natural isomorphism of the deformed holomorphic bundles $\mathcal{E}_0$ and $\mathcal{E}_1$ compatible with λ-connections, which is mutually bounded, and thus we have $\mathcal{G}_u(E_1) \simeq \mathcal{G}_u(E_0)$, and then $H_{1\,u\,h} = H_{u\,h}$. Thus we may and will assume $H_{0\,u\,h} = H_{u\,h}$ from the beginning.

12.1.5. The morphism $\Psi^\dagger$. Since we have the isomorphism $\sigma^* \mathrm{Gr}^W \mathcal{G}(E) \simeq \left(\mathrm{Gr}^W \mathcal{G}(E)\right)^\vee$, the isomorphism Ψ induces $\Psi^\dagger : \mathrm{Gr}^W \mathcal{G}^\dagger(E_0) \simeq \mathrm{Gr}^W \mathcal{G}^\dagger(E)$. Clearly we have $\Psi^\dagger(\boldsymbol{w}_0^\dagger) = \boldsymbol{w}^\dagger$. We also have the direct construction of $\widehat{\Psi}^\dagger : \mathrm{Gr}^W \mathcal{G}^\dagger(E_0) \simeq \mathrm{Gr}^W \mathcal{G}^\dagger(E)$ like Ψ. We would like to show that $\Psi^\dagger = \widehat{\Psi}^\dagger$.

Let $\boldsymbol{w}^\dagger$ be the frame of $\mathrm{Gr}^W \mathcal{G}^\dagger(E)$ on $\boldsymbol{C}_\mu$, which is induced by the dual frame $\boldsymbol{w}^\vee$ of $\left(\mathrm{Gr}^W \mathcal{G}(E)\right)^\vee$ via the isomorphism $\sigma^* \mathrm{Gr}^W \mathcal{G} \simeq \left(\mathrm{Gr}^W \mathcal{G}\right)^\vee$, as in the subsection 11.4.2. Similarly we have the frame $\boldsymbol{w}_0^\dagger$ of $\mathrm{Gr}^W \mathcal{G}^\dagger(E_0)$.

On the other hand, we have the frame $\boldsymbol{v}^{(\lambda_0)\,\dagger}$ of ${}_{1-\epsilon(\lambda_0)}\mathcal{E}^\dagger$ over $\sigma(U(\lambda_0)) \times X$ obtained from a local lift $\boldsymbol{v}^{(\lambda_0)}$ of $\boldsymbol{w}$ around λ_0, given as in (11.1):

$$\boldsymbol{v}^{(\lambda_0)\,\dagger} = \sigma^*\Big(\boldsymbol{v}^{(\lambda_0)} \cdot \overline{H(h, \boldsymbol{v}^{(\lambda_0)})}^{-1}\Big). \tag{12.3}$$

It is easy to see that $\boldsymbol{v}^{(\lambda_0)\,\dagger}$ is a local lift of $\boldsymbol{w}^\dagger$ around $\mu_0 = -\overline{\lambda}_0$. Similarly we obtain the local lift $\boldsymbol{v}_0^{(\lambda_0)\,\dagger}$ of $\boldsymbol{w}_0^\dagger$ around $-\overline{\lambda}_0$ from $\boldsymbol{v}_0$.

The C^{∞}-frame $\boldsymbol{v}^{(\lambda_0)\,\dagger\,\prime}$ and $\boldsymbol{v}_0^{(\lambda_0)\,\dagger\,\prime}$ over $\sigma\bigl(U(\lambda_0)\bigr)\times(X-D)$ are given as in (12.1). From (12.2) and (12.3), we obtain the following:

$$(12.4)\quad \boldsymbol{v}^{(\lambda_0)\,\dagger\,\prime}=\sigma^*\Bigl(\boldsymbol{v}^{(\lambda_0)\,\prime}\cdot\overline{H(h,\boldsymbol{v}^{(\lambda_0)\,\prime})}^{-1}\Bigr)=\sigma^*\Bigl(\boldsymbol{e}'\cdot B^{(\lambda_0)}\cdot\overline{H(h,\boldsymbol{v}^{(\lambda_0)\,\prime})}^{-1}\Bigr)$$
$$=\boldsymbol{e}'\cdot\sigma^*\Bigl(\overline{H(h,\boldsymbol{e}')}^{-1}\cdot{}^t\overline{B}^{(\lambda_0)\,-1}\Bigr).$$

Similarly we have the following:

$$(12.5)\qquad \boldsymbol{v}_0^{(\lambda_0)\,\dagger\,\prime}=\boldsymbol{e}_0'\cdot\sigma^*\Bigl(\overline{H(h,\boldsymbol{e}_0')}^{-1}\cdot{}^t\overline{B}_0^{(\lambda_0)\,-1}\Bigr).$$

We determine the $GL(r)$-valued function $C^{(\lambda_0)}$ on $\sigma(U(\lambda_0))\times(X-D)$ by the following condition:

$$\Phi\bigl(\boldsymbol{v}_0^{(\lambda_0)\,\dagger\,\prime}\bigr)=\boldsymbol{v}^{(\lambda_0)\,\dagger\,\prime}\cdot C^{(\lambda_0)}.$$

LEMMA 12.11. *Let $\{P_l\}$ be a sequence as in Lemma 12.8. Then the sequence $\bigl\{C^{(\lambda_0)}(\lambda,P_l)\bigr\}$ converges to the identity matrix. Namely $\widehat{\Psi}^{\dagger}(\boldsymbol{w}_0^{\dagger})=\boldsymbol{w}^{\dagger}$, which implies $\Psi^{\dagger}=\widehat{\Psi}^{\dagger}$.*

Proof Recall that we have the following, due to $\Psi(\boldsymbol{w}_0)=\boldsymbol{w}$:

$$\lim_{N\to\infty}\bigl(B^{(\lambda_0)\,-1}\cdot B_0^{(\lambda_0)}\bigr)(\lambda,P_N)=\lim_{N\to\infty}I^{(\lambda_0)\,\prime}(\lambda,P_N)=\text{the identity matrix}$$

We also have the following, due to the modification in the subsection 12.1.4:

$$\lim_{N\to\infty}H(h,\boldsymbol{e}_0')_{|P_N}^{-1}\cdot H(h,\boldsymbol{e}')_{|P_N}=\text{the identity matrix.}$$

We also have the boundedness of the sequences $\bigl\{H(h,\boldsymbol{e}')_{|P_N}\bigr\}$, $\bigl\{H(h_0,\boldsymbol{e}_0')_{|P_N}\bigr\}$, $\bigl\{B^{(\lambda_0)\,-1}(\lambda,P_N)\bigr\}$ and $\bigl\{B_0^{(\lambda_0)}(\lambda,P_N)\bigr\}$. Then the lemma immediately follows from (12.4) and (12.5). □

12.1.6. The isomorphism of induced vector bundles. In the subsection 12.1.3, we constructed the isomorphism $\Psi:\mathrm{Gr}^W\mathcal{G}(E_0)\longrightarrow\mathrm{Gr}^W\mathcal{G}(E)$. In the subsection 12.1.5, we obtained the isomorphism $\Psi^{\dagger}:\mathrm{Gr}^W\mathcal{G}^{\dagger}(E_0)\longrightarrow\mathrm{Gr}^W\mathcal{G}^{\dagger}(E)$. They induce the isomorphisms for any $k\in\mathbb{Z}$ and $u\in\mathcal{KMS}({}^{\diamond}\mathcal{E}^0)$:

$$\Psi:\mathrm{Gr}_k^W\mathcal{G}_u(E_0)\longrightarrow\mathrm{Gr}_k^W\mathcal{G}_u(E),\quad \Psi^{\dagger}:\mathrm{Gr}_k^W\mathcal{G}^{\dagger}_{u^{\dagger}}(E_0)\longrightarrow\mathrm{Gr}_k^W\mathcal{G}^{\dagger}_{u^{\dagger}}(E).$$

PROPOSITION 12.12. *Let k be an integer and u be an element of $\mathcal{KMS}({}^{\diamond}\mathcal{E}^0)$. The isomorphisms Ψ and $\Psi^{\dagger}$ induce the isomorphisms $\Psi^{\triangle}:\mathrm{Gr}_k^W S_u^{\mathrm{can}}(E_0)\longrightarrow\mathrm{Gr}_k^W S_u^{\mathrm{can}}(E)$ and $\Psi^{\triangle}:\mathrm{Gr}_k^W S_u(E_0,P)\longrightarrow\mathrm{Gr}_k^W S_u(E,P)$.*

Proof We have only to show that the morphisms Ψ and $\Psi^{\dagger}$ are compatible with the gluings. For simplicity, we put as follows:

$$\mathrm{Gr}^W S^{\mathrm{can}}(E):=\bigoplus_{k\in\mathbb{Z}}\ \bigoplus_{u\in\mathcal{KMS}({}^{\diamond}\mathcal{E}^0)}\mathrm{Gr}_k^W S_u^{\mathrm{can}}(E).$$

Similarly, we have $\mathrm{Gr}^W S^{\mathrm{can}}(E_0)$. We have only to show that Ψ and $\Psi^{\dagger}$ induce the isomorphism of $\mathrm{Gr}^W S^{\mathrm{can}}(E_0)\longrightarrow\mathrm{Gr}^W S^{\mathrm{can}}(E)$.

Let $\boldsymbol{w}$ and $\boldsymbol{w}_0$ be frames of $\mathrm{Gr}^W\mathcal{G}(E)$ and $\mathrm{Gr}^W\mathcal{G}(E_0)$ respectively such that $\Psi(\boldsymbol{w}_0)=\boldsymbol{w}$. We have the induced frames $\boldsymbol{w}^{\dagger}$ and $\boldsymbol{w}_0^{\dagger}$ of $\mathrm{Gr}^W\mathcal{G}^{\dagger}(E)$ and $\mathrm{Gr}^W\mathcal{G}^{\dagger}(E_0)$ respectively, as in the subsection 11.4.2. We have $\Psi^{\dagger}(\boldsymbol{w}_0^{\dagger})=\boldsymbol{w}^{\dagger}$. We use $u(w_i)=u(w_{0\,i})=u(w_i^{\dagger})=u(w_{0\,i}^{\dagger})$ without mention. We also use $\deg^W(w_i)=\deg^W(w_{0\,i})$ and $\deg^W(w_i^{\dagger})=\deg^W(w_{0\,i}^{\dagger})$.

REMARK 12.13. We also have $\deg^W(w_i) = -\deg^W(w_i^{\dagger})$. □

We have the normalizing frames $\widetilde{\boldsymbol{v}}$ (resp. $\widetilde{\boldsymbol{v}}_0$) of $\operatorname{Gr}^W \mathcal{G}(\mathcal{E})$ (resp. $\operatorname{Gr}^W \mathcal{G}(\mathcal{E}_0)$) over $\mathcal{X}^{\sharp}$, which is a lift of $\boldsymbol{w}$ (resp. $\boldsymbol{w}_0$), i.e., $\widetilde{\boldsymbol{v}}_{|\{O\}\times \boldsymbol{C}_{\lambda}} = \boldsymbol{w}$ (resp. $\widetilde{\boldsymbol{v}}_{0\,|\{O\}\times \boldsymbol{C}_{\lambda}} = \boldsymbol{w}_0$). We also have the normalizing frame $\widetilde{\boldsymbol{v}}^{\dagger}$ (resp. $\widetilde{\boldsymbol{v}}_0^{\dagger}$) of $\operatorname{Gr}^W \mathcal{G}^{\dagger}(\mathcal{E}^{\dagger})$ (resp. $\operatorname{Gr}^W \mathcal{G}^{\dagger}(\mathcal{E}_0^{\dagger})$), which is a lift of $\boldsymbol{w}^{\dagger}$ (resp. $\boldsymbol{w}_0^{\dagger}$).

We have the relations $\widetilde{\boldsymbol{v}} = \widetilde{\boldsymbol{v}}^{\dagger}\cdot \widetilde{J}$ and $\widetilde{\boldsymbol{v}}_0 = \widetilde{\boldsymbol{v}}_0^{\dagger}\cdot \widetilde{J}_0$. It is clear that Proposition 12.12 can be reduced to the following lemma, due to Lemma 11.40.

LEMMA 12.14. *We have* $\widetilde{J} = \widetilde{J}_0$.

Proof Let λ_0 be any point of $\boldsymbol{C}_{\lambda}$. We can take local lifts $\boldsymbol{v}^{(\lambda_0)}$ and $\boldsymbol{v}_0^{(\lambda_0)}$ of $\boldsymbol{w}$ and $\boldsymbol{w}_0$ on $U(\lambda_0)$, which induce the normalizing frames $\widetilde{\boldsymbol{v}}$ and $\widetilde{\boldsymbol{v}}_0$ on $X\times U(\lambda_0)$. We take C^{∞}-frames $\boldsymbol{v}^{(\lambda_0)\,\prime}$ and $\boldsymbol{v}_0^{(\lambda_0)\,\prime}$ as in (12.1). We have the functions $B^{(\lambda_0)}$ and $B_0^{(\lambda_0)}$ from $(X-D)\times U(\lambda_0)$ to $GL(r)$ determined by the conditions $\boldsymbol{v}^{(\lambda_0)\,\prime} = \boldsymbol{e}'\cdot B^{(\lambda_0)}$ and $\boldsymbol{v}_0^{(\lambda_0)\,\prime} = \boldsymbol{e}_0'\cdot B_0^{(\lambda_0)}$, as in (12.2).

On the other hand, we put $U'(\lambda_0^{-1}) = \sigma\big(U(-\overline{\lambda}_0^{-1})\big)$ and $\boldsymbol{v}^{\dagger\,(\lambda_0^{-1})} := \big(\boldsymbol{v}^{(-\overline{\lambda}_0^{-1})}\big)^{\dagger}$ on $U'(\lambda_0^{-1})\times X^{\dagger}$ as in (12.3). Similarly we obtain $\boldsymbol{v}_0^{\dagger\,(\lambda_0^{-1})}$.

LEMMA 12.15. *The frames* $\boldsymbol{v}^{\dagger\,(\lambda_0^{-1})}$ *and* $\boldsymbol{v}_0^{\dagger\,(\lambda_0^{-1})}$ *are local lifts of* $\boldsymbol{w}^{\dagger}$ *and* $\boldsymbol{w}_0^{\dagger}$ *around* λ_0^{-1}*, respectively. They induce the normalizing frames* $\widetilde{\boldsymbol{v}}^{\dagger\,(\lambda_0^{-1})}$ *and* $\widetilde{\boldsymbol{v}}_0^{\dagger\,(\lambda_0^{-1})}$ *on* $U(\lambda_0^{-1})\times X^{\dagger}$ *respectively.*

Proof It is easy to check the claim from the definitions. □

We obtain $\boldsymbol{v}^{\dagger\,(\lambda_0^{-1})\,\prime}$ and $\boldsymbol{v}_0^{\dagger\,(\lambda_0^{-1})\,\prime}$ as in (12.1). Then the matrix valued functions $J^{(\lambda_0)\,\prime}$ and $J_0^{(\lambda_0)\,\prime}$ are determined by the following relations:

$$\boldsymbol{v}^{\dagger\,(\lambda_0^{-1})\,\prime} = \boldsymbol{v}^{(\lambda_0)\,\prime}\cdot J^{(\lambda_0)\prime},\qquad \boldsymbol{v}_0^{\dagger\,(\lambda_0^{-1})\,\prime} = \boldsymbol{v}_0^{(\lambda_0)\,\prime}\cdot J_0^{(\lambda_0)\prime}.$$

On the other hand, we have the following equalities:

(12.6)
$$\boldsymbol{v}^{\dagger\,(\lambda_0^{-1})\,\prime} = \sigma^*\big(\boldsymbol{v}^{(-\overline{\lambda}_0^{-1})\,\prime}\cdot \overline{H(h,\boldsymbol{v}^{(-\overline{\lambda}_0^{-1})\,\prime})^{-1}}\big) = \boldsymbol{e}'\cdot \overline{H(h,\boldsymbol{e}')}^{-1}\cdot \sigma^*\big(\overline{{}^tB^{(-\overline{\lambda}_0^{-1})}}\big)^{-1}$$
$$= \boldsymbol{v}^{(\lambda_0)\,\prime}\cdot B^{(\lambda_0)\,-1}\cdot \overline{H(h,\boldsymbol{e}')}^{-1}\cdot \sigma^*\big(\overline{{}^tB^{(-\overline{\lambda}_0^{-1})}}\big)^{-1}.$$

Hence we obtain the relation:

$$J^{(\lambda_0)\,\prime} = B^{(\lambda_0)\,-1}\cdot \overline{H(h,\boldsymbol{e}')}^{-1}\cdot \sigma^*\big(\overline{{}^tB^{(-\overline{\lambda}_0^{-1})}}\big)^{-1}.$$

Similarly we have the relation:

$$J_0^{(\lambda_0)\,\prime} = B_0^{(\lambda_0)\,-1}\cdot \overline{H(h_0,\boldsymbol{e}_0')}^{-1}\cdot \sigma^*\big(\overline{{}^tB_0^{(-\overline{\lambda}_0^{-1})}}\big)^{-1}.$$

LEMMA 12.16. *Let* $\{P_N\}$ *be as in the subsection* 12.1.3. *We have the following:*

$$\lim_{N\to\infty}\Big(\big(B_0^{(\lambda_0)}\big)^{-1}\cdot B^{(\lambda_0)}\Big)_{|U(\lambda_0)\times\{P_N\}} = \text{the identity matrix},$$

$$\lim_{N\to\infty}\Big(\big(B_0^{(-\overline{\lambda}_0^{-1})}\big)^{-1}\cdot B^{(-\overline{\lambda}_0^{-1})}\Big)_{|\sigma(U(\lambda))\times\{P_N\}} = \text{the identity matrix}.$$

Proof We have the relation: $\boldsymbol{v}^{(\lambda_0)\prime} = \boldsymbol{v}_0^{(\lambda_0)\prime} \cdot \big(B_0^{(\lambda_0)}\big)^{-1} \cdot B^{(\lambda_0)}$. Thus the first claim follows from our construction of the morphism Ψ and $\Psi(\boldsymbol{w}_0) = \boldsymbol{w}$. The second claim can be shown similarly. □

LEMMA 12.17. *Let* $\{P_N\}$ *be as in the subsection* 12.1.3. *We have the following:*

$$\lim_{N\to\infty} H(h,\boldsymbol{e}')_{|P_N} \cdot H(h_0,\boldsymbol{e}_0')^{-1}_{|P_N} = \text{the identity matrix.}$$

Proof It follows from our construction. (See the subsection 12.1.4). □

LEMMA 12.18. *We have the following:*

$$\lim_{N\to\infty}\Big(J^{(\lambda_0)\prime}\cdot\big(J_0^{(\lambda_0)\prime}\big)^{-1}\Big)_{|U(\lambda_0)\times\{P_N\}} = \text{the identity matrix.}$$

Proof It follows from Lemma 12.16, Lemma 12.17 and the boundedness of the sequences $\big\{H(h,\boldsymbol{e}')_{|P_N}\big\}$ etc. □

Let us return to the proof of Lemma 12.14. In the case $u(w_i) = u(w_j) = u$ and $\deg^W(w_i) = \deg^W(w_j) = h$, we can expand the (i,j)-component of $J^{(\lambda_0)\prime}\cdot \big(J^{(\lambda_0)\prime}\big)^{-1}$ as follows, by using Lemma 11.44:

$$\Big(J^{(\lambda_0)\prime}\cdot\big(J_0^{(\lambda_0)\prime}\big)^{-1}\Big)_{i\,j} = \sum_{\substack{u(w_k^\dagger)=u,\\ \deg^W(w_k^\dagger)=h}} J_{i,k}^{(\lambda_0)\prime}\cdot\big(J_0^{(\lambda_0)\prime}\big)^{-1}_{k,j}.$$

Due to Lemma 11.45 and our construction of $\boldsymbol{v}^{(\lambda_0)\prime}$ and $\boldsymbol{v}^{\dagger(\lambda_0^{-1})\prime}$, we have the following equality in the case $u(w_i) = u(w_j^\dagger)$ and $\deg^W(w_i) = \deg^W(w_j^\dagger)$:

$$J_{i\,j}^{(\lambda_0)\prime} = \tilde{J}_{i\,j}\cdot|z|^{-b(v_i^{(\lambda_0)})+b(v_j^{\dagger(\lambda_0^{-1})})}\cdot z^{\nu(w_i,\lambda_0)}\cdot \bar{z}^{-\nu(w_j^\dagger,\lambda_0^{-1})}.$$

We also have the following equality in the case $u(w_i^\dagger) = u(w_j)$, and $\deg^W(w_i^\dagger) = \deg^W(w_j)$:

$$\big(J_0^{(\lambda_0)\prime}\big)^{-1}_{i\,j} = \big(\tilde{J}_0^{-1}\big)_{i\,j}\cdot|z|^{-b(v_i^{\dagger(\lambda_0^{-1})})+b(v_j^{(\lambda_0)})}\cdot z^{-\nu(w_j,\lambda_0)}\cdot\bar{z}^{\nu(w_i^\dagger,\lambda_0^{-1})}. \tag{12.7}$$

We also remark $b(v_i^{(\lambda_0)}) = b(v_j^{(\lambda_0)})$ and $\nu(w_i,\lambda_0) = \nu(w_j,\lambda_0)$ in the case $u(w_i) = u(w_j)$. Then we obtain the following equality in the case $u(w_i) = u(w_j)$ and $\deg^W(w_i) = \deg^W(w_j)$:

$$\Big(J^{(\lambda_0)\prime}\cdot\big(J_0^{(\lambda_0)\prime}\big)^{-1}\Big)_{i\,j} = \Big(\tilde{J}\cdot\tilde{J}_0^{-1}\Big)_{i\,j}. \tag{12.8}$$

Due to (12.8) and Lemma 12.18, we obtain the following:

$$\lim_{N\to\infty}\Big(\tilde{J}\cdot\tilde{J}_0^{-1}\Big)_{|U(\lambda_0)\times\{P_N\}} = \text{the identity matrix.}$$

Due to Corollary 11.39, we have the $GL(r)$-valued holomorphic functions K and K_0 on $\boldsymbol{C}^*$ such that $\tilde{J}(\lambda,z) = K(\lambda)\cdot C(\lambda,z)$ and $\tilde{J}_0(\lambda,z) = K_0(\lambda)\cdot C(\lambda,z)$, where $C(\lambda,z)$ is given by the formula (11.10). Thus we obtain the following:

$$\begin{aligned}\text{the identity matrix} &= \lim_{N\to\infty}\big(K_0(\lambda)\cdot C(\lambda,P_N)\big)^{-1}\cdot K(\lambda)\cdot C(\lambda,P_N)\\ &= \lim_{N\to\infty} K_0(\lambda)^{-1}\cdot K(\lambda) = K_0(\lambda)^{-1}\cdot K(\lambda).\end{aligned} \tag{12.9}$$

Note $K(\lambda)$, $K_0(\lambda)$ and $C(\lambda, z)$ are commutative. Thus we obtain $K(\lambda) = K_0(\lambda)$. Then the equality $\widetilde{J}(\lambda, z) = \widetilde{J}_0(\lambda, z)$ holds. Therefore we obtain Lemma 12.14, and thus Proposition 12.12. □

12.1.7. The end of the proof of Theorem 12.1. We use the setting in the subsection 12.1.2.

PROPOSITION 12.19.
The tuples $\big(\mathrm{Gr}^W S_u^{\mathrm{can}}(E), W, N^{\triangle}, S\big)$ *and* $\big(\mathrm{Gr}^W S_u^{\mathrm{can}}(E_0), W, N^{\triangle}, S\big)$ *are isomorphic.*
The tuples $(\mathrm{Gr}^W S_u(E_0, P), W, N_0^{\triangle}, S_0)$ *and* $(\mathrm{Gr}^W S_u(E, P), W, N^{\triangle}, S)$ *are isomorphic.*

Proof We have already constructed the isomorphism $\Psi^{\triangle} : \mathrm{Gr}^W S_u^{\mathrm{can}}(E_0) \longrightarrow \mathrm{Gr}^W S_u^{\mathrm{can}}(E)$. Under the isomorphism, the restrictions of the following morphisms to the fibers over the point $0 \in \mathbb{P}^1$ are same:

$$N_0^{\triangle} : \mathrm{Gr}_k^W S_u^{\mathrm{can}}(E_0) \longrightarrow \mathrm{Gr}_{k-2}^W S_u^{\mathrm{can}}(E_0) \otimes \mathbb{T}(-1),$$

$$N^{\triangle} : \mathrm{Gr}_k^W S_u^{\mathrm{can}}(E) \longrightarrow \mathrm{Gr}_{k-2}^W S_u^{\mathrm{can}}(E) \otimes \mathbb{T}(-1).$$

Since $\mathrm{Gr}_k^W S_u^{\mathrm{can}}(E_0)$ and $\mathrm{Gr}_{k-2}^W S_u^{\mathrm{can}}(E_0) \otimes \mathbb{T}(-1)$ are pure twistors of weight k, we obtain the coincidence of the nilpotent morphisms $N_0^{\triangle} = N^{\triangle}$ over $\mathbb{P}^1$. Similarly we obtain the coincidence of the pairings $S_0 = S$ over $\mathbb{P}^1$. □

LEMMA 12.20. *The tuples obtained from a model bundle is a polarized mixed twistor structure.*

Proof It follows from Lemma 6.3, Corollary 6.7, the functoriality for the tensor products (Lemma 11.32) and the direct sums. □

Due to Proposition 12.19 and Lemma 12.20, the tuple $\big(\mathrm{Gr}^W S_u^{\mathrm{can}}(E), W, N^{\triangle}, S\big)$ is a polarized mixed twistor structure. Hence the tuple $\big(S_u^{\mathrm{can}}(E), W, N^{\triangle}, S\big)$ is also a polarized mixed twistor structure. Similarly, $\big(S_u(E, P), W, N^{\triangle}, S\big)$ is a polarized mixed twistor structure. Therefore the proof of Theorem 12.1 is accomplished. □

12.2. Limiting mixed twistor theorem in the higher dimensional case

12.2.1. Statements. We put $X = \Delta^n$, $D_i = \{z_i = 0\}$ and $D = \bigcup_{i=1}^{l} D_i$. Let $(E, \overline{\partial}_E, \theta, h)$ be a tame harmonic bundle over $X - D$. Let $\boldsymbol{u}$ be any element of $\mathcal{KMS}(\mathcal{E}^0, \underline{n})$. Recall that we have the vector bundles $S_{\boldsymbol{u}}^{\mathrm{can}}(E)$ and $S_{\boldsymbol{u}}(E, P)$, on which we have the commuting tuple of the nilpotent maps $\boldsymbol{N}^{\triangle} = (\mathcal{N}_1^{\triangle}, \ldots, \mathcal{N}_l^{\triangle})$. We also have the pairing $S : S_{\boldsymbol{u}}^{\mathrm{can}}(E) \otimes \sigma S_{\boldsymbol{u}}^{\mathrm{can}}(E) \longrightarrow \mathbb{T}(0)$ and $S : S_{\boldsymbol{u}}(E, P) \otimes \sigma S_{\boldsymbol{u}}(E, P) \longrightarrow \mathbb{T}(0)$. (See the subsections 11.3.3–11.3.6.)

For any element $\boldsymbol{a} \in \boldsymbol{R}_{>0}^{l}$, we put $\mathcal{N}^{\triangle}(\boldsymbol{a}) := \sum_{i=1}^{l} a_i \cdot \mathcal{N}_i^{\triangle}$. The weight filtration of $\mathcal{N}^{\triangle}(\boldsymbol{a})$ is denoted by $W(\boldsymbol{a})$. The following lemma will be proved in the rest of this section.

LEMMA 12.21. *The weight filtrations* $W(\boldsymbol{a})$ *are independent of a choice of* $\boldsymbol{a} \in \boldsymbol{R}_{>0}^{l}$.

The filtration is denoted by W. The following theorem is one of the main purpose in this paper, which will be proved in this section.

THEOREM 12.22. *The tuples* $(S_{\boldsymbol{u}}^{\mathrm{can}}(E), W, \boldsymbol{N}^{\triangle}, S)$ *and* $(S_{\boldsymbol{u}}(E,P), W, \boldsymbol{N}^{\triangle}, S)$ $(P \in X - D)$ *are polarized mixed twistor structures of weight* 0 *in* l*-variables.*

REMARK 12.23. Let P be a point of D_I°. Let $\boldsymbol{u}$ be an element of $\mathcal{KMS}(\mathcal{E}^0, I)$. Let $q_I : X \longrightarrow D_I$ be the projection. By considering the restriction of $(E, \overline{\partial}_E, \theta, h)$ to $q_I^{-1}(P)$, we obtain the polarized mixed twistor structure $(S_{\boldsymbol{u}}^{\mathrm{can}}, W, \boldsymbol{N}^{\triangle}, S)$ of weight 0 in $|I|$-variables due to Theorem 12.22. We also remark that we may and will assume that $l = n$ in the proof of Theorem 12.22. □

12.2.2. Weak result. We put $X = \Delta^n$ and $D = \bigcup_{i=1}^n D_i$. Let $(E, \overline{\partial}_E, \theta, h)$ be a tame harmonic bundle over $X - D$. Let $\boldsymbol{c}$ be any element of $\mathbb{Z}_{>0}^n$, and let us consider the morphism $\phi_{\boldsymbol{c}} : \Delta \longrightarrow X$ given by $z \longmapsto \left(z^{c_1}, \ldots, z^{c_n}\right)$. Via the pull back, we obtain the harmonic bundle $\phi_{\boldsymbol{c}}^{-1}(E, \overline{\partial}_E, \theta, h)$ over the punctured disc Δ^*. Due to the functoriality of the construction for the pull back (Lemma 11.33), we have the following lemma.

LEMMA 12.24. *Let* $\boldsymbol{u}_1$ *be any element of* $\mathcal{KMS}\left(\phi_{\boldsymbol{c}}^{-1}\mathcal{E}^0\right)$. *We have the isomorphisms:*

$$S_{\boldsymbol{u}_1}^{\mathrm{can}}\left(\phi_{\boldsymbol{c}}^{-1}E\right) \simeq \bigoplus_{\substack{\boldsymbol{u} \in \mathcal{KMS}\left({}^{\circ}\mathcal{E}^0\right) \\ \boldsymbol{c}\cdot\boldsymbol{u} \equiv \boldsymbol{u}_1}} S_{\boldsymbol{u}}^{\mathrm{can}}(E), \quad S_{\boldsymbol{u}_1}\left(\phi_{\boldsymbol{c}}^{-1}E, P\right) \simeq \bigoplus_{\substack{\boldsymbol{u} \in \mathcal{KMS}\left({}^{\circ}\mathcal{E}^0\right) \\ \boldsymbol{c}\cdot\boldsymbol{u} = \boldsymbol{u}_1}} S_{\boldsymbol{u}}\left(E, \phi_{\boldsymbol{c}}(P)\right).$$

The isomorphisms are compatible with the pairings. Under the isomorphisms, the nilpotent maps induced by the residue on $S_{\boldsymbol{u}_1}^{\mathrm{can}}\left(\phi_{\boldsymbol{c}}^{-1}E\right)$ *and* $S_{\boldsymbol{u}_1}\left(\phi_{\boldsymbol{c}}^{-1}E, P\right)$ *are given by* $\mathcal{N}^{\triangle}(\boldsymbol{c})$. □

COROLLARY 12.25. *Let* $\boldsymbol{u}$ *be any element of* $\mathcal{KMS}\left(\mathcal{E}^0, \underline{n}\right)$. *Let* $\boldsymbol{c}$ *be any element of* $\mathbb{Z}_{>0}^n$. *Then the following tuples are polarized mixed twistor structures:*

$$\left(S_{\boldsymbol{u}}^{\mathrm{can}}(E), W(\boldsymbol{c}), \mathcal{N}^{\triangle}(\boldsymbol{c}), S\right), \qquad \left(S_{\boldsymbol{u}}(E,P), W(\boldsymbol{c}), \mathcal{N}^{\triangle}(\boldsymbol{c}), S\right).$$

Proof Due to Lemma 12.24 and the limiting mixed twistor theorem in one variable case (Theorem 12.1), the direct sum $\bigoplus_{\boldsymbol{c}\cdot\boldsymbol{u}=\boldsymbol{u}_1}\left(S_{\boldsymbol{u}}^{\mathrm{can}}(E), W(\boldsymbol{c}), \mathcal{N}^{\triangle}(\boldsymbol{c}), S\right)$ is a polarized mixed twistor structure. Since it is easy to see that a direct summand of a polarized mixed twistor structure is also a polarized mixed twistor structure, we are done. □

Let λ be any point of $\boldsymbol{C}$. For any element $\boldsymbol{d} \in \boldsymbol{C}^n$, we put $\mathcal{N}^{\lambda}(\boldsymbol{d}) := \sum d_i \cdot \mathcal{N}_i^{\lambda}$. Then $\mathcal{N}^{\lambda}(\boldsymbol{d})$ is the endomorphism of the vector space ${}^{\underline{n}}\mathcal{G}_{\boldsymbol{u}\,|\,(\lambda,O)}$. We also put $\mathcal{N}^{\lambda}(\boldsymbol{t}) := \sum t_i \cdot \mathcal{N}_i^{\lambda}$ for variables t_i. Then $\mathcal{N}^{\lambda}(\boldsymbol{t})$ is the endomorphism of the vector space ${}^{\underline{n}}\mathcal{G}_{\boldsymbol{u}\,|\,(\lambda,O)} \otimes_{\boldsymbol{C}} \boldsymbol{C}(t_1, \ldots, t_n)$. We say that $\boldsymbol{d} \in \boldsymbol{C}^n$ is called generic with respect to the tuple $\left(\mathcal{N}_1^{\lambda_0}, \ldots, \mathcal{N}_n^{\lambda_0}\right)$, if the conjugacy classes of $\mathcal{N}^{\lambda}(\boldsymbol{d})$ and $\mathcal{N}^{\lambda}(\boldsymbol{t})$ are same. Let $T(\lambda)$ denote the set of $\boldsymbol{c} \in \mathbb{Z}_{>0}^n$ such that $\mathcal{N}^{\lambda}(\boldsymbol{c})$ is generic. Then $T(\lambda)$ is Zariski dense in $\boldsymbol{C}^n$.

LEMMA 12.26. *The set* $T(\lambda)$ *is independent of a choice of* $\lambda \in \boldsymbol{C}$.

Proof Let us fix any $\lambda_0 \in \boldsymbol{C}$, and we have only to show $T(\lambda_0) \subset T(\lambda)$ for any $\lambda \in \boldsymbol{C}$. Let $\boldsymbol{c}$ be any element of $T(\lambda_0)$. Due to Corollary 12.25, $\mathcal{N}^{\triangle}(\boldsymbol{c})$ is a morphism of mixed twistor structure, and hence the conjugacy classes of $\mathcal{N}^{\lambda}(\boldsymbol{c})$ and $\mathcal{N}^{\lambda_0}(\boldsymbol{c})$ are same. Since the conjugacy classes of $\mathcal{N}^{\lambda_0}(\boldsymbol{c})$ $(\boldsymbol{c} \in T(\lambda_0))$ are same, the conjugacy classes of $\mathcal{N}^{\lambda}(\boldsymbol{c})$ $(\boldsymbol{c} \in T(\lambda_0))$ are same. Since $T(\lambda_0)$ is Zariski dense in $\boldsymbol{C}^n$, we can conclude that $\boldsymbol{c} \in T(\lambda_0)$ are generic with respect to the tuple $\left(\mathcal{N}_1^{\lambda}, \ldots, \mathcal{N}_n^{\lambda}\right)$, i.e., $\boldsymbol{c} \in T(\lambda)$. Thus we are done. □

LEMMA 12.27. *For any i, we have $\mathcal{N}_i^{\triangle} \cdot W_h(\boldsymbol{c}) \subset W_{h-1}(\boldsymbol{c}) \otimes \mathbb{T}(-1)$ on both of $S^{\mathrm{can}}_{\boldsymbol{u}}(E)$ and $S_{\boldsymbol{u}}(E,P)$.*

Proof We have only to show $\mathcal{N}_i^{\lambda} W_h(\boldsymbol{c}) \subset W_{h-1}(\boldsymbol{c})$ for any λ. Since $\mathcal{N}^{\lambda}(\boldsymbol{c})$ is generic due to Lemma 12.26, we may apply a lemma of Cattani-Kaplan (Proposition 1.9 in [**10**].) □

LEMMA 12.28. *For any i, we have $\mathcal{N}_i^{\triangle} \cdot W_h(\boldsymbol{c}) \subset W_{h-2}(\boldsymbol{c}) \otimes \mathbb{T}(-1)$ on both of $S^{\mathrm{can}}_{\boldsymbol{u}}(E)$ and $S_{\boldsymbol{u}}(E,P)$.*

Proof We put $\theta = \sum f_i \cdot dz_i \cdot z_i^{-1}$, and we put ρ_i as in the page 193. We have the following inequality for some positive constant C, due to Simpson's Main estimate (see the subsection 8.2.2):

$$\left|(f_i - \rho_i)_{|\phi_c(\Delta)}\right|_h \leq C \cdot |z|^{-1} \cdot \big(-\log|z|\big)^{-1}. \tag{12.10}$$

Let us consider the restriction of $\mathcal{N}_i^{\triangle}$ to the fibers over $0 \in \mathbb{P}^1$. We denote the restriction by $\mathcal{N}_{i\,|\,0}^{\triangle}$. Since $\mathcal{N}_{i\,|\,0}^{\triangle}$ is given by $(f_i - \rho_i)_{|(0,O)}$, we obtain $\mathcal{N}_{i\,|\,0}^{\triangle} \cdot W_h(\boldsymbol{c})_{|0} \subset W_{h-2}(\boldsymbol{c})_{|0}$, due to (12.10) and the norm estimate in one dimensional case (see the subsection 7.2.1). Similarly, we obtain $\mathcal{N}_{i\,|\,\infty}^{\triangle} \cdot W_h(\boldsymbol{c})_{|\infty} \subset W_{h-2}(\boldsymbol{c})_{|\infty}$.

Due to Lemma 12.27, we have $\mathcal{N}^{\triangle} \cdot W_h(\boldsymbol{c}) \subset W_{h-1}(\boldsymbol{c}) \otimes \mathbb{T}(-1)$. Let us consider the induced morphism $\mathcal{N}^{\triangle} : \mathrm{Gr}_h^{W(\boldsymbol{c})} \longrightarrow \mathrm{Gr}_{h-1}^{W(\boldsymbol{c})} \otimes \mathbb{T}(-1)$. We have already shown the vanishing of the induced morphism at $(x = 0, \infty)$. Note that $\mathrm{Gr}_h^{W(\boldsymbol{c})}$ and $\mathrm{Gr}_{h-1}^{W(\boldsymbol{c})} \otimes \mathbb{T}(-1)$ are pure twistors of weight h and $h+1$ respectively. Thus the vanishing at $x = 0, \infty$ implies vanishing over $\mathbb{P}^1$. Thus we obtain the implication $\mathcal{N}_i^{\triangle} \cdot W_h(\boldsymbol{c}) \subset W_{h-2}(\boldsymbol{c}) \otimes \mathbb{T}(-1)$. □

Recall that $(S^{\mathrm{can}}_{\boldsymbol{u}}(E), W(\boldsymbol{c}))$ is mixed twistor structure for any $\boldsymbol{c} \in T$. Lemma 12.28 means that $\mathcal{N}_i^{\triangle}$ is a morphism of mixed twistor structure. We also obtain that $\mathcal{N}^{\triangle}(\boldsymbol{a})$ is a morphism of a mixed twistor structure for any $\boldsymbol{a} \in \boldsymbol{C}^n$. Hence the conjugacy classes of $\mathcal{N}^{\lambda}(\boldsymbol{a})$ is independent of a choice of $\lambda \in \mathbb{P}^1$ if we fix $\boldsymbol{a} \in \boldsymbol{C}^n$.

COROLLARY 12.29. *For any element $\boldsymbol{d} \in \boldsymbol{C}^n$ which is generic with respect to the tuple $(\mathcal{N}_1^{\triangle}, \ldots, \mathcal{N}_n^{\triangle})$, we have $W(\boldsymbol{d}) = W(\boldsymbol{c})$, where $\boldsymbol{c}$ is any element of $T(\lambda_0)$.*

Proof We put $W = W(\boldsymbol{c})$. Let us consider the morphism $\mathcal{N}(\boldsymbol{d})^h : \mathrm{Gr}_h^W S^{\mathrm{can}}_{\boldsymbol{u}} \longrightarrow \mathrm{Gr}_{-h}^W S^{\mathrm{can}}_{\boldsymbol{u}}(E) \otimes \mathbb{T}(-h)$. It is isomorphic in the case $\boldsymbol{d} = \boldsymbol{c}$. Hence there is a Zariski open subset U of $\boldsymbol{C}^n$ such that $\mathcal{N}(\boldsymbol{d})^h$ above is isomorphic for any $\boldsymbol{d} \in U$. Since we have already known that $\mathcal{N}^{\triangle}(\boldsymbol{d})$ decreases the degree with respect to W by two due to Lemma 12.28, we can conclude that W is the weight filtration of $\mathcal{N}^{\triangle}(\boldsymbol{d})$, due to the characterization of the weight filtration. □

In summary, we obtain the following.

COROLLARY 12.30. *On $S^{\mathrm{can}}_{\boldsymbol{u}}(E)$ and $S_{\boldsymbol{u}}(E,P)$, we have the filtrations W and the pairings S. The tuples $\big(S^{\mathrm{can}}_{\boldsymbol{u}}(E), W, \mathcal{N}^{\triangle}(\boldsymbol{c}), S\big)$ and $\big(S_{\boldsymbol{u}}(E,P), W, \mathcal{N}^{\triangle}(\boldsymbol{c}), S\big)$ are polarized mixed twistor structures. The maps $\mathcal{N}_i^{\triangle}$ $(i = 1, \ldots, n)$ are morphisms of mixed twistor structures.*

Proof It follows from Corollary 12.25 Lemma 12.28 and Corollary 12.29. □

12.2.3. Preliminary norm estimate. Let $(E, \overline{\partial}_E, \theta, h)$ be a tame harmonic bundle over $X - D$. Let us fix a point λ, which is generic, and we take a normalizing frame $\boldsymbol{v}$ of ${}^\diamond\mathcal{E}^\lambda$, which is compatible with $\mathbb{E}$ and F.

We have the matrices $A^k \in M_r(\boldsymbol{C})$ $(k = 1, \ldots, n)$ determined as follows:

$$\mathbb{D}^\lambda \boldsymbol{v} = \boldsymbol{v} \cdot \sum_{k=1}^{n} A^k \cdot \frac{dz_k}{z_k}.$$

Let f_{A^k} denote the endomorphism induced by A^k for the frame $\boldsymbol{v}$. We denote the nilpotent part of A^k by N^k. Similarly we obtain the endomorphisms f_{N^k}. We impose the following assumption in this subsection.

ASSUMPTION 12.31. The conjugacy classes of N^k are independent of $k = 1, \ldots, n$. □

Under the assumption 12.31, we obtain the weight filtration $W^{(0)}$ of the vector bundle ${}^\diamond\mathcal{E}^\lambda$ induced by f_{N^k}, which is independent of a choice of k. We may assume that $\boldsymbol{v}$ is compatible with the filtration $W^{(0)}$. We put $b_i(v_j) := {}^i\deg^F(v_j)$ and $\boldsymbol{b}(v_j) = \big(b_i(v_j) \,\big|\, i = 1, \ldots, n\big)$. We put $\beta_i(v_j) = {}^i\deg^{\mathbb{E}}(v_j)$ and $\boldsymbol{\beta}(v_j) = \big(\beta_i(v_j) \,\big|\, i = 1, \ldots, n\big)$. We put $k(v_j) := \deg^{W^{(0)}}(v_j)$.

Then we obtain the C^∞-frame $\boldsymbol{v}' = (v_i')$ of $\mathcal{E}^\lambda$, given as follows:

$$v_i' := v_i \cdot \prod_{h=1}^{n} |z_h|^{b_h(v_i)} \cdot \Big(-\sum_{h=1}^{n} \log|z_h|^2\Big)^{-k(v_i)/2}.$$

LEMMA 12.32. *The frame $\boldsymbol{v}'$ is adapted over $X - D$.*

Proof We put as follows:

$$Y_m := \Big\{(z_1, \ldots, z_n) \in X - D \,\Big|\, |z_h| = 1,\ (h \leq m)\Big\}.$$

We consider the following claim:

(P_m): The restriction of the frame $\boldsymbol{v}'_{|Y_m}$ is adapted over Y_m.

We show the claim (P_m) by a descending induction on m. Since Y_n is compact, the claim (P_n) holds. We assume that (P_{m+1}) holds, and we will derive (P_m).

Let us pick the elements $\boldsymbol{u}_1, \ldots, \boldsymbol{u}_a \in \mathcal{KMS}(\mathcal{E}^0, \underline{n})$ such that $\big\{\mathfrak{k}(\lambda, \boldsymbol{u}_i) \,\big|\, i = 1, \ldots, a\big\} = \mathcal{KMS}({}^\diamond\mathcal{E}^\lambda, \underline{n})$, in other words, $\mathcal{K}(\mathcal{E}, \lambda, 0, \underline{n}) = \{\boldsymbol{u}_1, \ldots, \boldsymbol{u}_a\}$. Then we have the generalized eigen decomposition $\mathcal{E}^\lambda = \bigoplus_i \mathcal{E}^\lambda_{\boldsymbol{u}_i}$ of f_{A^k} $(k = 1, \ldots, n)$. Here $\mathcal{E}^\lambda_{\boldsymbol{u}_i}$ denote the subbundles corresponding to the tuples of the eigenvalues $\mathfrak{e}(\lambda, \boldsymbol{u}_i) = \big(\mathfrak{e}(\lambda, q_1(\boldsymbol{u}_i)), \ldots, \mathfrak{e}(\lambda, q_n(\boldsymbol{u}_i))\big)$. Correspondingly we have the decomposition of the endomorphisms f_{A^m}:

$$f_{A^m} = \bigoplus_{\boldsymbol{u}_i} \Big(\mathfrak{e}(\lambda, q_m(\boldsymbol{u}_i)) + N^m_{\boldsymbol{u}_i}\Big).$$

We take the model bundle $Mod(N^m_{\boldsymbol{u}_i})$ on Δ^* corresponding to the nilpotent map $N^m_{\boldsymbol{u}_i}$ (see the section 6.2). We obtain the deformed holomorphic bundle $\mathcal{E}^\lambda_{\boldsymbol{u}_i}$, the λ-connection $\mathbb{D}^\lambda_{\boldsymbol{u}_i}$, the metric $h_{1\,\boldsymbol{u}_i}$, and the canonical frame $\boldsymbol{v}_{1\,\boldsymbol{u}_i}$. Let ϕ be the holomorphic map $X \longrightarrow \Delta$ given by $(z_1, \ldots, z_n) \longmapsto \prod_{i=1}^n z_i$. Then we obtain the harmonic bundle $\phi^{-1}Mod(N^m_{\boldsymbol{u}_i})$, the deformed holomorphic bundle $\phi^{-1}\mathcal{E}^\lambda_{\boldsymbol{u}_i}$, the λ-connection $\mathbb{D}^\lambda_{1\,\boldsymbol{u}_i}$, the metric $h_{1\,\boldsymbol{u}_i}$, and the canonical frame $\phi^{-1}\boldsymbol{u}_i$, over $X - D$.

On the other hand, we have the model bundle $L(\boldsymbol{u}_i)$ over $X - D$ of rank 1 (see the section 6.1). We obtain the deformed holomorphic bundle $\mathcal{L}^\lambda(\boldsymbol{u}_i)$, the λ-connection $\mathbb{D}^\lambda_{2\,\boldsymbol{u}_i}$, the metric $h_{2\,\boldsymbol{u}_i}$, and the canonical frame $e_{\boldsymbol{u}_i}$.

Then we obtain the following over $X - D$:

$$\mathcal{E}^\lambda_{0\,\boldsymbol{u}_i} := \mathcal{L}^\lambda(\boldsymbol{u}_i) \otimes \phi^{-1}\mathcal{E}^\lambda_{\boldsymbol{u}_i}, \quad \mathbb{D}^\lambda_{0\,\boldsymbol{u}_i} := \mathbb{D}^\lambda_2 \otimes 1 + 1 \otimes \phi^{-1}\mathbb{D}^\lambda_1,$$

$$h_{0\,\boldsymbol{u}_i} := h_{2\,\boldsymbol{u}_i} \otimes \phi^{-1} h_{1\,\boldsymbol{u}_i}, \quad \boldsymbol{v}_{0\,\boldsymbol{u}_i} := e_{\boldsymbol{u}_i} \otimes \phi^{-1}\boldsymbol{v}_{1\,\boldsymbol{u}_i}.$$

By taking a direct sum, we obtain $\mathcal{E}^\lambda_0$, $\mathbb{D}^\lambda_0$, h_0 and $\boldsymbol{v}_0$.

Moreover, by taking the dz_m-component, we obtain the λ-connections $\mathfrak{q}_m(\mathbb{D}^\lambda_0)$ and $\mathfrak{q}_m(\mathbb{D}^\lambda)$ along the z_m-direction. Note the following relations due to our construction:

$$\mathfrak{q}_m(\mathbb{D}^\lambda)\boldsymbol{v} = \boldsymbol{v} \cdot A^m \cdot \frac{dz_m}{z_m}, \qquad \mathfrak{q}_m(\mathbb{D}^\lambda_0)\boldsymbol{v}_0 = \boldsymbol{v}_0 \cdot A^m \cdot \frac{dz_m}{z_m}. \tag{12.11}$$

Let us consider the morphism $\Phi : \mathcal{E}^\lambda_0 \longrightarrow \mathcal{E}^\lambda$ given by the frames $\boldsymbol{v}_0$ and $\boldsymbol{v}$. The equalities (12.11) implies that Φ is flat with respect to the λ-connections along the z_m-direction. Moreover, the restriction $\Phi_{|Y_{m+1}}$ and the inverse $\Phi^{-1}_{|Y_{m+1}}$ are bounded, due to our assumption of the induction. Thus we obtain the boundedness of $\Phi_{|Y_m}$ and the inverse, and hence the adaptedness of $\boldsymbol{v}'$ on Y_m. (See the argument in the section 8.10, where we discussed the case where λ vary. It is easier to discuss the case where λ is fixed. See also the subsection 6.1 in [**65**].) Thus the induction can proceed. □

12.2.4. Preliminary constantness of the filtrations. We use the setting in the subsection 12.2.3, in particular, the assumption 12.31 is imposed. We also use the notation in the subsection 12.2.2. Recall that we put $\mathcal{N}^\lambda(\boldsymbol{a}) := \sum a_i \cdot \mathcal{N}^\lambda_i$, which is the endomorphism of ${}^{\underline{n}}\mathcal{G}_{\boldsymbol{u}\,|\,(\lambda,O)}$. We would like to show that the conjugacy classes of $\mathcal{N}^\lambda(\boldsymbol{a})$ are independent of $\boldsymbol{a} \in \boldsymbol{R}^n_{>0}$ (Lemma 12.37). But we need some preparation.

Let us pick the elements $\boldsymbol{u}_1, \ldots, \boldsymbol{u}_a \in \mathcal{KMS}(\mathcal{E}^0, \underline{n})$ such that $\{\mathfrak{k}(\lambda, \boldsymbol{u}_i) \,|\, i = 1, \ldots, a\} = \mathcal{KMS}({}^\diamond\mathcal{E}^\lambda, \underline{n})$. Then we have the generalized eigen decomposition ${}^\diamond\mathcal{E}^\lambda = \bigoplus_i {}^\diamond\mathcal{E}^\lambda_{\boldsymbol{u}_i}$ of f_{A^k} $(k = 1, \ldots, n)$. Here ${}^\diamond\mathcal{E}^\lambda_{\boldsymbol{u}_i}$ denote the subbundles corresponding to the eigenvalues $\mathfrak{e}(\lambda, \boldsymbol{u}_i) = \Bigl(\mathfrak{e}(\lambda, q_1(\boldsymbol{u}_i)), \ldots, \mathfrak{e}(\lambda, q_n(\boldsymbol{u}_i))\Bigr)$. Correspondingly we have the decomposition of A^k:

$$A^k = \bigoplus_{\boldsymbol{u}_i} \Bigl(\mathfrak{e}(\lambda, q_k(\boldsymbol{u}_i)) + N^k{\boldsymbol{u}_i}\Bigr).$$

The frame $\boldsymbol{v}$ induces the frame $\boldsymbol{v}_{\boldsymbol{u}_i}$ of ${}^\diamond\mathcal{E}^\lambda_{\boldsymbol{u}_i}$.

Let $\boldsymbol{a}$ be any element of $\boldsymbol{R}^n_{>0}$. Let $\mathbb{H}$ denote the upper half plane $\{\zeta \,|\, \mathrm{Im}(\zeta) > 0\}$. We put $y := \mathrm{Im}(\zeta)$. Let us consider the following morphism $\psi_{\boldsymbol{a}} : \mathbb{H} \longrightarrow X - D$, given by $z_i = \exp\bigl(\sqrt{-1} \cdot a_i \cdot \zeta\bigr)$ for $i = 1, \ldots, n$. We obtain the harmonic bundle $\psi_{\boldsymbol{a}}^{-1}(E, \overline{\partial}_E, \theta, h)$. We have the deformed holomorphic bundle $\psi_{\boldsymbol{a}}^{-1}\mathcal{E}^\lambda = \bigoplus_{\boldsymbol{u}_i} \psi_{\boldsymbol{a}}^{-1}\mathcal{E}^\lambda_{\boldsymbol{u}_i}$. We have the holomorphic frame $\boldsymbol{w}_{\boldsymbol{u}} := \psi_{\boldsymbol{a}}^{-1}(\boldsymbol{v}_{\boldsymbol{u}})$. We have the following relation:

$$\mathbb{D}\boldsymbol{w}_{\boldsymbol{u}} = \boldsymbol{w}_{\boldsymbol{u}} \cdot \Bigl(\sum_k \sqrt{-1} a_k \cdot A^k_{\boldsymbol{u}}\Bigr) \cdot d\zeta = \boldsymbol{w}_{\boldsymbol{u}} \cdot \sum_k a_k \cdot \Bigl(\mathfrak{e}(\lambda, q_k(\boldsymbol{u})) + N^k_{\boldsymbol{u}}\Bigr) \cdot \sqrt{-1} \cdot d\zeta. \tag{12.12}$$

We put $N_{\boldsymbol{u}}(\boldsymbol{a}) = \sum_k a_k N_{\boldsymbol{u}}^k$, and we take a model bundle $Mod(N_{\boldsymbol{u}}(\boldsymbol{a}))$ on Δ_z^* for $N_{\boldsymbol{u}}(\boldsymbol{a})$. We put $\boldsymbol{u} \cdot \boldsymbol{a} := \sum_k a_k \cdot q_k(\boldsymbol{u}) \in \boldsymbol{R} \times \boldsymbol{C}$, and we take a model bundle $L(\boldsymbol{u}\cdot\boldsymbol{a})$ on Δ_z^*. We put $E_{1,\boldsymbol{u}} := L(\boldsymbol{u}\cdot\boldsymbol{a}) \otimes Mod(N_{\boldsymbol{u}}(\boldsymbol{a}))$, and we denote the deformed holomorphic bundle by $\mathcal{E}_{1,\boldsymbol{u}}^{\lambda}$. We have the metric $h_{1\,\boldsymbol{u}}$ and the λ-connection $\mathbb{D}_{1\,\boldsymbol{u}}^{\lambda}$ on $\mathcal{E}_{1,\boldsymbol{u}}^{\lambda}$. We have the normalizing frame $\boldsymbol{v}_{1\,\boldsymbol{u}}$, such that the following relation holds on Δ_z^*:

$$\mathbb{D}_{\boldsymbol{u}}^{\lambda} \boldsymbol{v}_{1\,\boldsymbol{u}} = \boldsymbol{v}_{1\,\boldsymbol{u}} \cdot \big(\mathfrak{e}(\lambda, \boldsymbol{u}\cdot\boldsymbol{a}) + N_{\boldsymbol{u}}(\boldsymbol{a})\big) \cdot \frac{dz}{z}.$$

Let ϕ denote the holomorphic map $\mathbb{H} \longrightarrow \Delta^*$ given by $z = \exp(\sqrt{-1}\zeta)$. We put $\mathcal{E}_{0\,\boldsymbol{u}}^{\lambda} := \phi^* \mathcal{E}_{1\,\boldsymbol{u}}$ and the pull backs of $h_{1\,\boldsymbol{u}}$, $\mathbb{D}_{1\,\boldsymbol{u}}^{\lambda}$, and $\boldsymbol{v}_{1\,\boldsymbol{u}}$ are denoted by $h_{0,\boldsymbol{u}}$, $\mathbb{D}_{0\,\boldsymbol{u}}^{\lambda}$ and $\boldsymbol{v}_{0\,\boldsymbol{u}}$ respectively. On the upper half plane $\mathbb{H}$, we have the following:

$$\mathbb{D}_{0\,\boldsymbol{u}}^{\lambda} \boldsymbol{v}_{0\,\boldsymbol{u}} = \boldsymbol{v}_{0\,\boldsymbol{u}} \cdot \big(\mathfrak{e}(\lambda, \boldsymbol{u}\cdot\boldsymbol{a}) + N_{\boldsymbol{u}}(\boldsymbol{a})\big) \cdot \sqrt{-1}d\zeta. \tag{12.13}$$

Then we have the isomorphism $\Phi_{\boldsymbol{u}} : \mathcal{E}_{0\,\boldsymbol{u}}^{\lambda} \longrightarrow \psi_{\boldsymbol{a}}^{-1} \mathcal{E}_{\boldsymbol{u}}^{\lambda}$ given by the frames $\boldsymbol{v}_{0\,\boldsymbol{u}}$ and $\boldsymbol{w}_{\boldsymbol{u}}$. Then $\Phi_{\boldsymbol{u}}$ and $\Phi_{\boldsymbol{u}}^{-1}$ are compatible with λ-connections, due to (12.12) and (12.13).

We put $\mathcal{E}_0^{\lambda} := \bigoplus_{\boldsymbol{u}_i} \mathcal{E}_{0\,\boldsymbol{u}_i}^{\lambda}$. We have the induced metric h_0. We obtain the isomorphism $\Phi := \bigoplus_{\boldsymbol{u}} \Phi_{\boldsymbol{u}}$ from $\mathcal{E}_0^{\lambda}$ to $\psi_{\boldsymbol{a}}^{-1}\mathcal{E}^{\lambda}$. Then Φ and Φ^{-1} are compatible with the λ-connections. We regard them as the flat sections of $Hom\big(\mathcal{E}_0^{\lambda}, \psi_{\boldsymbol{a}}^{-1}\mathcal{E}^{\lambda}\big)$ and $Hom\big(\psi_{\boldsymbol{a}}^{-1}\mathcal{E}^{\lambda}, \mathcal{E}_0^{\lambda}\big)$. The metrics h_0 and $\psi_{\boldsymbol{a}}^{-1}h$ induce the metrics h_2 and h_3 of $Hom\big(\mathcal{E}_0^{\lambda}, \psi_{\boldsymbol{a}}^{-1}\mathcal{E}^{\lambda}\big)$ and $Hom\big(\psi_{\boldsymbol{a}}^{-1}\mathcal{E}^{\lambda}, \mathcal{E}_0^{\lambda}\big)$.

LEMMA 12.33. *We have the following estimate for some positive constants C_1 and C_2:*

$$\max\Big\{\log|\Phi|_{h_2},\ \log|\Phi^{-1}|_{h_3}\Big\} \leq C_1 + C_2 \cdot \log y. \tag{12.14}$$

Proof For each element $v_i \in \boldsymbol{v}_{\boldsymbol{u}}$, we put as follows:

$$v_i' := v_i \cdot \prod_{j=1}^{l} |z_j|^{\mathfrak{p}(\lambda, q_j(\boldsymbol{u}))}.$$

Then $\boldsymbol{v}' = (v_i')$ is C^∞-frame of $\mathcal{E}^\lambda$ over $X - D$, which is adapted up to log order. In particular, we have the following inequality on $\mathbb{H}$, for some positive constants C_i $(i = 1, 2)$ and M:

$$C_1 \cdot y^{-M} \leq H(h, \psi_{\boldsymbol{a}}^{-1}\boldsymbol{v}') \leq C_2 \cdot y^M.$$

We also have the following equality for $v_i \in \boldsymbol{v}_{\boldsymbol{u}}$:

$$\psi_{\boldsymbol{a}}^{-1} v_i' = w_i \cdot \exp\Big(-y \cdot \mathfrak{p}(\lambda_0, \boldsymbol{u}\cdot\boldsymbol{a})\Big).$$

On the other hand, we put as follows, for $v_{1\,j} \in \boldsymbol{v}_{1\,\boldsymbol{u}}$:

$$v_{1\,j}' := v_{1\,j} \cdot |z|^{\mathfrak{p}(\lambda_0, \boldsymbol{u}\cdot\boldsymbol{a})}.$$

Then the C^∞-frame $\boldsymbol{v}_1' = (v_{1\,j}')$ of $\bigoplus_{\boldsymbol{u}} \mathcal{E}_{1\,\boldsymbol{u}}^{\lambda}$ is adapted up to log order. We put $v_{0\,j}' := \phi^{-1} v_{1\,j}'$ and $\boldsymbol{v}_0' := (v_{0\,j}')$. Then we obtain the following inequalities for some positive constants C_i' $(i = 1, 2)$ and M':

$$C_1' \cdot y^{-M'} \leq H(h_0, \boldsymbol{v}_0') \leq C_2' \cdot y^{M'}.$$

We also have the following equality, for $v_{0\,i} \in \boldsymbol{v}_{0\,\boldsymbol{u}}$:

$$v_{0\,i}' = v_{0\,i} \cdot \exp\Big(-y \cdot \mathfrak{p}(\lambda_0, \boldsymbol{u}\cdot\boldsymbol{a})\Big).$$

Since Φ and Φ^{-1} are given by the frames $\psi_{\boldsymbol{a}}^{-1}\boldsymbol{v}_{\boldsymbol{u}}$ and $\boldsymbol{v}_{0\,\boldsymbol{u}}$, we have $\Phi(\boldsymbol{v}_0')=\psi_{\boldsymbol{a}}^{-1}\boldsymbol{v}'$, and hence we obtain the estimate $\max\bigl\{|\Phi|_{h_2},\,|\Phi^{-1}|_{h_3}\bigr\}\leq C''\cdot y^{M''}$ for some positive constants C'' and M''. Thus we are done. □

LEMMA 12.34. *The functions* $\log|\Phi|_{h_2}$ *and* $\log|\Phi^{-1}|_{h_3}$ *are subharmonic, i.e., we have the following inequalities:*

$$\Delta\log|\Phi|_{h_2}^2\leq 0,\qquad \Delta\log|\Phi^{-1}|_{h_3}^2\leq 0. \tag{12.15}$$

Proof It can be shown by an argument similar to the proof of Lemma 4.1 in [**81**]. □

The functions $\log|\Phi|_{h_2}^2$ and $\log|\Phi^{-1}|_{h_3}^2$ can be regarded as follows: We have the holomorphic bundle $\mathcal{E}_1^{\lambda}:=\bigoplus_{\boldsymbol{u}_i}\mathcal{E}_{1\,\boldsymbol{u}_i}^{\lambda}$ on Δ^*. We also have the frame $\boldsymbol{v}_1$ induced by the frames $\boldsymbol{v}_{1\,\boldsymbol{u}}$. Let q_1 and q_2 denote the projection of $\Delta^*\times(X-D)$ onto Δ^* and $X-D$ respectively. We have the holomorphic bundles $q_1^{-1}\mathcal{E}_1^{\lambda}$ and $q_2^{-1}\mathcal{E}^{\lambda}$. The frames $\boldsymbol{v}_1$ and $\boldsymbol{v}$ give the isomorphisms Φ' and Φ'^{-1} of $q_1^{-1}\mathcal{E}_1^{\lambda}$ and $q_2^{-1}\mathcal{E}^{\lambda}$. The metrics h_1 and h induce the metrics h_2' and h_3' of the bundles $Hom(q_1^*\mathcal{E}_1^{\lambda},q_2^*\mathcal{E}^{\lambda})$ and $Hom(q_2^*\mathcal{E}^{\lambda},q_1^*\mathcal{E}_1^{\lambda})$. The morphisms ϕ and $\psi_{\boldsymbol{a}}$ induce the morphism $F:\mathbb{H}\longrightarrow\Delta^*\times(X-D)$. Then we have the following equalities:

$$\log|\Phi|_{h_2}^2=F^{-1}\Bigl(\log|\Phi'|_{h_2'}^2\Bigr),\qquad \log|\Phi^{-1}|_{h_3}^2=F^{-1}\Bigl(\log|\Phi'^{-1}|_{h_3'}^2\Bigr). \tag{12.16}$$

Let us consider the functions $G_1:=\Xi\bigl(\log|\Phi|_{h_2}\bigr)$ and $G_2:=\Xi\bigl(\log|\Phi^{-1}|_{h_3}\bigr)$ on $\boldsymbol{R}_{>0}$ (the subsection 2.4.1).

LEMMA 12.35. *There exists a positive constants* C *such that the following inequality holds on the upper half plane* $\mathbb{H}$*:*

$$\max\bigl\{G_1,G_2\bigr\}\leq C.$$

Proof It follows from (12.15), (12.16) and Lemma 2.22 that the functions G_i $(i=1,2)$ are convex below. Then Lemma 12.35 follows from (12.14) and Lemma 2.23. □

We take the metric $h_{4\,\boldsymbol{u}}$ of $\psi_{\boldsymbol{a}}^{-1}\mathcal{E}_{\boldsymbol{u}}^{\lambda}$. For any $w_i,w_j\in\boldsymbol{w}_{\boldsymbol{u}}$, we put as follows:

$$h_{4\,\boldsymbol{u}}(w_i,w_j)=\delta_{i\,j}\cdot\exp\Bigl(2\cdot y\cdot\mathfrak{p}\bigl(\lambda,\boldsymbol{u}\cdot\boldsymbol{a}\bigr)\Bigr)\cdot y^{k(v_i)}.$$

Here $\delta_{i\,j}$ denotes 1 in the case $i=j$ and 0 in the case $i\neq j$. Recall that we put $k(v_i):=\deg^W(v_i)$. Taking the direct sum of $h_{4\,\boldsymbol{u}}$, we have the induced metric h_4 of $\psi_{\boldsymbol{a}}^{-1}\mathcal{E}^{\lambda}$.

LEMMA 12.36. *The metrics* $\psi_{\boldsymbol{a}}^{-1}h$ *and* h_4 *are mutually bounded*

Proof It immediately follows from Lemma 12.32. □

We have the weight filtration $W^{(1)}$ on $\mathcal{E}_{1\,\boldsymbol{u}}^{\lambda}$ induced by the logarithms of the unipotent part of the monodromy. We take the normalizing frame $\tilde{\boldsymbol{v}}_{1\,\boldsymbol{u}}$ of $\mathcal{E}_{1\,\boldsymbol{u}}^{\lambda}$, which is compatible with $W^{(1)}$. Let $\tilde{\boldsymbol{v}}_{0\,\boldsymbol{u}}$ denote the pull back of $\tilde{\boldsymbol{v}}_{1\,\boldsymbol{u}}$ via the morphism ϕ. We take the metric $h_{5\,\boldsymbol{u}}$ of $\mathcal{E}_{0\,\boldsymbol{u}}^{\lambda}$. For any $\tilde{v}_{0\,i},\tilde{v}_{0\,j}\in\tilde{\boldsymbol{v}}_{0\,\boldsymbol{u}}$, we put as follows:

$$h_{5\,\boldsymbol{u}}\bigl(\tilde{v}_{0\,i},\tilde{v}_{0\,j}\bigr)=\delta_{i\,j}\cdot\exp\Bigl(2\cdot y\cdot\mathfrak{p}\bigl(\lambda,\boldsymbol{u}\cdot\boldsymbol{a}\bigr)\Bigr)\cdot y^{k(\tilde{v}_{0\,i})}.$$

Then we have the induced metric h_5 of $\mathcal{E}_0^{\lambda}$, and the metrics h_5 and h_0 are mutually bounded.

The metrics h_4 and h_5 induce the metrics h_6 and h_7 of $Hom(\mathcal{E}_0^\lambda, \psi_{\boldsymbol{a}}^{-1}\mathcal{E}^\lambda)$ and $Hom(\psi_{\boldsymbol{a}}^{-1}\mathcal{E}^\lambda, \mathcal{E}_0^\lambda)$. Then h_6 and h_2 are mutually bounded, and h_7 and h_3 are mutually bounded. Hence we obtain the following inequality on $\mathbb{H}$, for some positive constants C_4:

$$\max\Big\{\Xi(\log|\Phi|^2_{h_6}), \Xi(\log|\Phi^{-1}|^2_{h_7})\Big\} \leq C_4.$$

Since the functions $\log|\Phi|_{h_6}$ and $\log|\Phi^{-1}|_{h_7}$ are independent of the real part of ζ, we have the equalities $\Xi(\log|\Phi|_{h_6}) = \log|\Phi|_{h_6}$ and $\Xi(\log|\Phi^{-1}|_{h_7}) = \log|\Phi^{-1}|_{h_7}$. Thus we obtain the following inequalities on $\mathbb{H}$, for some positive constant C_5:

$$\max\Big\{|\Phi|_{h_6},\ |\Phi^{-1}|_{h_7}\Big\} \leq C_5. \tag{12.17}$$

After the preparation above, we show the following lemma, which is the purpose in this subsection.

LEMMA 12.37. *Under Assumption* 12.31, *the weight filtration of* $\mathcal{N}^\lambda(\boldsymbol{a}) = \sum a_i \cdot \mathcal{N}_i^\lambda$ *are independent of a choice of* $\boldsymbol{a} \in \boldsymbol{R}^n_{>0}$ *for any* $\lambda \in \mathbb{P}^1$.

Proof We use the notation in the subsection 12.2.3. Recall that we have the filtration $W^{(0)}$ of the vector bundle $\mathcal{E}^\lambda$ under Assumption 12.31, which is the weight filtration of the nilpotent maps f_{N^k} for any k. We have only to show that the weight filtration of $\sum a_k \cdot f_{N^k}$ is same as $W^{(0)}$.

On the other hand, the weight filtration $W^{(1)}$ of $\mathcal{E}^\lambda_{1,\boldsymbol{u}}$ induces the filtration of $\mathcal{E}^\lambda_{0,\boldsymbol{u}}$ on $\mathbb{H}$, which is also denoted by $W^{(1)}$. Let g be a flat section of $\mathcal{E}^\lambda_{0\,\boldsymbol{u}}$ on $\mathbb{H}$. It gives the section of $W^{(1)}_k(\mathcal{E}^\lambda_{0\,\boldsymbol{u}})$ if and only if we have the following estimate, due to our choice of $h_{5\,\boldsymbol{u}}$:

$$\big|g\big|_{h_{5\,\boldsymbol{u}}} = O\Big(\exp(-y\cdot\mathfrak{p}(\lambda, \boldsymbol{u}\cdot\boldsymbol{a}))\cdot y^{k/2}\Big). \tag{12.18}$$

Due to (12.17), the estimate (12.18) of g with respect to h_5 is equivalent to the following estimate of $\Phi(g)$ with respect to h_4:

$$\big|\Phi(g)\big|_{h_{4\,\boldsymbol{u}}} = O\Big(\exp(-y\cdot\mathfrak{p}(\lambda, \boldsymbol{u}\cdot\boldsymbol{a}))\cdot y^{k/2}\Big). \tag{12.19}$$

From our construction of h_4, (12.19) means that $\Phi(g)$ gives the flat section of the subbundle $\psi_{\boldsymbol{a}}^{-1}W^{(0)}_k(\mathcal{E}^\lambda_{\boldsymbol{u}})$. Therefore, we obtain the coincidence of the filtrations $\psi_{\boldsymbol{a}}^{-1}W^{(0)}$ and $W^{(1)}$. Since $W^{(1)}$ is the weight filtration of $\sum a_i \cdot f_{N^i}$, Lemma 12.37 is proved. □

12.2.5. Constantness of the filtration at generic λ. We put $X = \Delta^n$, $D_i = \{z_i = 0\}$ and $D = \bigcup_{i=1}^n D_i$. Let $(E, \overline{\partial}_E, \theta, h)$ be a tame harmonic bundle over $X - D$, and let λ be generic with respect to $(E, \overline{\partial}_E, \theta, h)$. We would like to show that the conjugacy classes of $\mathcal{N}^\lambda(\boldsymbol{a})$ are independent of $\boldsymbol{a} \in \boldsymbol{R}^N$ without the assumption 12.31 (Lemma 12.38).

Let $\boldsymbol{v}$ be a normalizing frame of ${}^\diamond\mathcal{E}^\lambda$, compatible with the filtrations ${}^iF$ and the decompositions ${}^i\mathbb{E}$ $(i = 1, \ldots, n)$. We put $b_i(v_j) = {}^i\deg^F(v_j)$ and $\boldsymbol{b}(v_j) = (b_i(v_j)\,|\,i = 1, \ldots, n)$. We put $\beta_i(v_j) = {}^i\deg^{\mathbb{E}}(v_j)$ and $\boldsymbol{\beta}(v_j) = (\beta_i(v_j)\,|\,i = 1, \ldots, n)$. Then we obtain the matrices $A^k \in M_r(\boldsymbol{C})$ determined as follows:

$$\mathbb{D}^\lambda\boldsymbol{v} = \boldsymbol{v}\cdot\sum_{k=1}^n A^k\frac{d\zeta_k}{\zeta_k}.$$

We denote the nilpotent part of A^k by N^k.

We would like to reduce the problem to the case where the assumption 12.31 is satisfied. For that purpose, let us pick $\boldsymbol{c}_1,\ldots,\boldsymbol{c}_n\in\mathbb{Z}^n_{>0}$. We denote the tuple $(\boldsymbol{c}_1,\ldots,\boldsymbol{c}_n)$ by $\boldsymbol{C}$. Let $\phi_{\boldsymbol{C}}:X\longrightarrow X$ be the morphism as follows:

$$\phi_{\boldsymbol{C}}^{-1}(\zeta_k)=\prod_{h=1}^{n}z_h^{c_{h\,k}}.$$

Then we have the following:

$$\phi_{\boldsymbol{C}}^{-1}\Bigl(\sum A^k\frac{d\zeta_k}{\zeta_k}\Bigr)=\sum_{h=1}^{n}\Bigl(\sum_{k=1}^{n}c_{h\,k}\cdot A^k\Bigr)\frac{dz_h}{z_h}.$$

We put $\boldsymbol{c}_h\cdot\boldsymbol{b}(v_j):=\sum c_{h\,i}\cdot b_i(v_j)$. We decompose as follows:

$$\boldsymbol{c}_h\cdot\boldsymbol{b}(v_j)=n_{h\,j}+\kappa_{h\,j},\qquad n_{h\,j}\in\mathbb{Z},\ -1<\kappa_{h\,j}\leq 0.$$

We put $\tilde{v}_j:=v_j\cdot\prod_{h=1}^{n}z^{n_{h\,j}}$, and $\tilde{\boldsymbol{v}}=(\tilde{v}_j)$. Then $\tilde{\boldsymbol{v}}$ is a normalizing frame of ${}^\diamond\phi_{\boldsymbol{C}}^{-1}\mathcal{E}^\lambda$. We have $\mathbb{D}\tilde{\boldsymbol{v}}=\tilde{\boldsymbol{v}}\cdot\sum_h\tilde{A}^h\cdot dz_h/z_h$ for $\tilde{A}^h\in M_r(\boldsymbol{C})$. The components $\tilde{A}^h_{i\,j}$ is given as follows:

$$\tilde{A}^h_{i\,j}=\sum_k c_{h\,k}\cdot A^k_{i\,j}+\delta_{i\,j}\cdot n_{h\,j}\cdot\lambda.$$

Hence the nilpotent part $\tilde{N}_h$ of $\tilde{A}^h$ is given by $N(\boldsymbol{c}_h):=\sum c_{h\,k}\cdot N_k$. If $\boldsymbol{c}_h$ $(h=1,\ldots,n)$ are generic with respect to the tuple $(\mathcal{N}_1,\ldots,\mathcal{N}_n)$ for $(E,\overline{\partial}_E,\theta,h)$, then the conjugacy classes of $\tilde{N}_h$ are independent of $h=1,\ldots,n$. Namely, Assumption 12.31 is satisfied for ${}^\diamond\phi_{\boldsymbol{C}}^{-1}\mathcal{E}^\lambda$ and the frame $\tilde{\boldsymbol{v}}$.

LEMMA 12.38. *The conjugacy classes of $N(\boldsymbol{a})$ $(\boldsymbol{a}\in\boldsymbol{R}^n_{>0})$ are independent of a choice of $\boldsymbol{a}$. Namely, the conjugacy classes of the endomorphisms $\mathcal{N}^\lambda(\boldsymbol{a})$ of ${}^{\underline{n}}\mathcal{G}_{\boldsymbol{u}\,|\,(\lambda,O)}$ is independent of a choice of $\boldsymbol{a}\in\boldsymbol{R}^n_{>0}$ if λ is generic.*

Proof We can pick generic elements $\boldsymbol{c}_1,\ldots,\boldsymbol{c}_n\in\mathbb{Z}^n_{>0}$, such that $\boldsymbol{a}=\sum a'_h\boldsymbol{c}_h$ for some $\boldsymbol{a}'=(a'_h)\in\boldsymbol{R}^n_{>0}$. Due to the argument above, we have the following relation:

$$N(\boldsymbol{a})=\sum_k a_kN_k=\sum_{k,h}c_{h\,k}\cdot a'_h\cdot N_k=\sum_h a'_h\cdot N(\boldsymbol{c}_h).$$

Thus Lemma 12.38 can be reduced to Lemma 12.37. □

12.2.6. The end of the proof of Lemma 12.21. We will finish the proof of Lemma 12.21. We may assume that $l=n$. Due to our construction of W (the subsection 12.2.2), we have only to check that $W(\boldsymbol{a})$ is independent of a choice of $\boldsymbol{a}$, i.e., that the conjugacy classes of $\mathcal{N}^\triangle(\boldsymbol{a})$ is independent of a choice of $\boldsymbol{a}$. Since the conjugacy classes of $\mathcal{N}^\triangle(\boldsymbol{a})_{|\lambda}$ $(\lambda\in\mathbb{P}^1)$ is independent of a choice of λ, we have only to check the independence of $\mathcal{N}^\lambda(\boldsymbol{a})=\sum_{i=1}^n a_i\cdot\mathcal{N}_i^\lambda$ from $\boldsymbol{a}\in\boldsymbol{R}_{>0}$ for generic λ. Thus Lemma 12.21 is reduced to Lemma 12.38. □

12.2.7. The end of the proof of Theorem 12.22. We will finish the proof of Theorem 12.22. We may assume that $l=n$. Let us show that the tuple $(S^{\mathrm{can}}_{\boldsymbol{u}}(E),W,\boldsymbol{N}^\triangle,S)$ is a polarized mixed twistor structure. For that purpose, we have only to check that the tuples $\bigl(S^{\mathrm{can}}_{\boldsymbol{u}}(E),W,\mathcal{N}^\triangle(\boldsymbol{a}),S\bigr)$ are polarized mixed twistor structures for any $\boldsymbol{a}\in\boldsymbol{R}^n_{>0}$ of weight 0.

Since we have already known that W is the weight filtration of $\mathcal{N}^\triangle(\boldsymbol{a})$ (Lemma 12.21), we have only to check the following lemma.

LEMMA 12.39. *For any element* $\boldsymbol{a} \in \boldsymbol{R}^n_{>0}$, $S\big(\mathcal{N}^{\triangle}(\boldsymbol{a})^h \otimes \mathrm{id}\big)$ *gives a polarization on the primitive part* $P\,\mathrm{Gr}^W_h\, S^{\mathrm{can}}_{\boldsymbol{u}}(E)$.

Proof First we remark that the pairing $S\big(\mathcal{N}^{\triangle}(\boldsymbol{a})^h \otimes \mathrm{id}\big)$ gives the polarization for generic $\boldsymbol{a} \in \boldsymbol{Q}^n$, due to Corollary 12.30. It is one of the immediate consequences that the induced pairing $S : P_h\,\mathrm{Gr}^W_{-h}\, S^{\mathrm{can}}_{\boldsymbol{u}}(E) \otimes \sigma P_h\,\mathrm{Gr}^W_h\, S^{\mathrm{can}}_{\boldsymbol{u}}(E) \longrightarrow \mathbb{T}(0)$ is perfect.

Before going to the next step, we give some general notion. Let V be a pure twistor of weight n. Let $S' : V \otimes \sigma(V) \longrightarrow \mathbb{T}(-n)$ denote the $(-1)^n$-symmetric pairing. We say that S' is a semi-polarization, if the induced hermitian pairing on $H^0\big(\mathbb{P}^1, V \otimes \mathcal{O}(-n)\big)$ is positive semi-definite.

Let q_i denote the projection of $\mathbb{P}^1 \times \boldsymbol{R}^n_{>0}$ onto the i-th component. We have the C^∞-vector bundle $q_1^*\,\mathrm{Gr}^W_h$ on $\mathbb{P}^1 \times \boldsymbol{R}^n_{>0}$. Then $S\big(\mathcal{N}^{\triangle}(\boldsymbol{a})^h \otimes \mathrm{id}\big)$ gives a C^∞-family of pairings. Since $S\big(\mathcal{N}^{\triangle}(\boldsymbol{a})^h \otimes \mathrm{id}\big)$ is a polarization for generic $\boldsymbol{a} \in \boldsymbol{Q}^n_{>0}$ as is remarked, the pairing $S\big(\mathcal{N}^{\triangle}(\boldsymbol{a})^h \otimes \mathrm{id}\big)$ is a semi-polarization for any elements $\boldsymbol{a} \in \boldsymbol{R}^n_{>0}$.

Since W is a weight filtration for the nilpotent maps $\mathcal{N}^{\triangle}(\boldsymbol{a})$, the morphism $\mathcal{N}^{\triangle}(\boldsymbol{a})^h : P_h\,\mathrm{Gr}^W_h\, S^{\mathrm{can}}_{\boldsymbol{u}}(E) \longrightarrow P_h\,\mathrm{Gr}^W_{-h}\, S^{\mathrm{can}}_{\boldsymbol{u}}(E)$ is isomorphic for any $\boldsymbol{a} \in \boldsymbol{R}_{>0}$. Since the induced pairing $P_h\,\mathrm{Gr}^W_{-h}\, S^{\mathrm{can}}(E) \otimes \sigma P_h\,\mathrm{Gr}^W_h\, S^{\mathrm{can}}(E) \longrightarrow \mathbb{T}(0)$ is perfect as is remarked, the pairing $S\big(\mathcal{N}^{\triangle}(\boldsymbol{a})^h \otimes \mathrm{id}\big)$ is also perfect.

The perfectness and the semi-polarization-property implies the polarization-property. Hence we obtain Lemma 12.39. □

Therefore we can conclude that the tuple $\big(S^{\mathrm{can}}_{\boldsymbol{u}}(E), W, \boldsymbol{N}^{\triangle}, S\big)$ is a polarized mixed twistor structure. The claim for $\big(S_{\boldsymbol{u}}(E,P), W, \boldsymbol{N}^{\triangle}, S\big)$ can be shown similarly. Thus the proof of Theorem 12.22 is accomplished. □

12.3. Some straightforward consequences

12.3.1. Strong sequential compatibility on $S^{\mathrm{can}}_{\boldsymbol{u}}(E)$ and $S_{\boldsymbol{u}}(E,P)$. We continue to use the setting in the subsection 12.2.1. As is proved there, we obtain the polarized limiting mixed twistor structures from a harmonic bundle $(E, \overline{\partial}_E, \theta, h)$ over $X - D$:

$$(S^{\mathrm{can}}_{\boldsymbol{u}}(E), W, \boldsymbol{N}^{\triangle}, S), \quad (S_{\boldsymbol{u}}(E,P), W, \boldsymbol{N}^{\triangle}, S).$$

COROLLARY 12.40. *The tuple of nilpotent maps* $\boldsymbol{N}^{\triangle}$ *on* $S^{\mathrm{can}}_{\boldsymbol{u}}(E)$ *is strongly sequentially compatible.*

Proof It follows from Theorem 12.22 and Lemma 3.116. □

REMARK 12.41. We obtain the variation of $\mathbb{P}^1$-holomorphic bundle with the pairing from the associated graded tuple $\big(\mathrm{Gr}^W\, S^{\mathrm{can}}_{\boldsymbol{u}}(E), W, \boldsymbol{N}^{\triangle}, S\big)$ (see the subsections 3.5.3 and 3.6.1). In fact, it can be shown to be a harmonic bundle, i.e., a variation of polarized pure twistor structures, because it is a split polarized mixed twistor structure. (See the section 3.7.) Moreover, it is same as the limiting CVHS in our previous paper [**65**], in the case where $(E, \overline{\partial}_E, \theta, h)$ is nilpotent and with trivial parabolic structure. □

REMARK 12.42. Similarly, we obtain the variation of $\mathbb{P}^1$-holomorphic bundle $\mathcal{V}$ with the pairing, from the limiting mixed twistor structure $(S^{\mathrm{can}}_{\boldsymbol{u}}(E), \mathcal{N}_i, S)$ (the subsections 3.5.3 and 3.6.1). It is interesting to ask whether it give a variation of polarized pure twistor structures.

More generally, it can be asked whether any polarized mixed twistor structure is a twistor nilpotent orbit or not. In the one variable case, it is true (Lemma 3.99). In the two variable case, it is proved in [**68**]. In the Hodge case, it is proved in [**52**], in general. □

We recall that we obtain the vector bundle ${}^{\underline{l}}\mathcal{G}_{\boldsymbol{u}}(E)$ and the nilpotent maps $\mathcal{N}_i$ on $\mathcal{D}_{\underline{l}}$ (the section 8.9), from $(E, \overline{\partial}_E, \theta, h)$. We remark that we put $S^{\mathrm{can}}_{\boldsymbol{u}}(E)_{|\boldsymbol{C}_\lambda} = {}^{\underline{l}}\mathcal{G}_{\boldsymbol{u}}(E)_{|\{O\}\times \boldsymbol{C}_\lambda}$ in the construction of $S^{\mathrm{can}}_{\boldsymbol{u}}(E)$ (the subsection 11.3.4), where $O \in D_{\underline{l}}$ denotes the origin. However, the theory is clearly valid if we choose any point $P \in D_{\underline{l}}$ instead of O. We use it without mention.

COROLLARY 12.43. *Let $\boldsymbol{a}$ be an element of $\mathbb{Z}^l_{\geq 0}$ On ${}^{\underline{l}}\mathcal{G}_{\boldsymbol{u}\,|\,(\lambda,P)}$, the conjugacy classes of $\prod_i \mathcal{N}^{a_i}_{i\,|\,(\lambda,P)}$ is independent of a choice of $(\lambda, P) \in \boldsymbol{C}_\lambda \times D_{\underline{l}}$.*

Proof When we fix a point $P \in D_{\underline{l}}$, the independence follows from the fact that $\prod_i \mathcal{N}_i^{a_i}$ induces the morphisms of mixed twistor structures. When we fix a generic $\lambda \in \boldsymbol{C}^*$, we can obtain the independence of the conjugacy classes from P by using the normalizing frame. As a result, we obtain the independence of the conjugacy classes from (λ, P). □

Similarly we can show the following:

LEMMA 12.44. $\operatorname{Im}\prod_{i=1}^l \mathcal{N}_i^{a_i} \cap \operatorname{Ker}\prod_{i=1}^l \mathcal{N}_i^{b_i}$ *gives the vector subbundle of ${}^{\underline{l}}\mathcal{G}_{\boldsymbol{u}}$ on $\mathcal{D}_{\underline{l}}$ for any elements $\boldsymbol{a}$ and $\boldsymbol{b}$ of $\mathbb{Z}^l_{\geq 0}$.* □

COROLLARY 12.45. *On ${}^{\underline{l}}\mathcal{G}_{\boldsymbol{u}}$, the tuple of nilpotent maps $\mathcal{N}_1, \ldots, \mathcal{N}_l$ are strongly sequentially compatible.*

Proof This is a direct corollary of Corollary 12.40. □

12.3.2. Sequential compatibility of the tuple $\big(\mathcal{N}_i, {}^iF^{(\lambda_0)}, {}^i\mathbb{E}^{(\lambda_0)} \,\big|\, i \in \underline{l}\big)$. We continue to use the setting in the subsection 12.2.1. Let λ_0 be any point of $\boldsymbol{C}_\lambda$. Let $\boldsymbol{b}$ be an element of $\boldsymbol{R}^l$ such that $b_i \notin \mathcal{KMS}(\mathcal{E}^{\lambda_0}, i)$ for each i. Let ϵ_0 be a sufficiently small positive number. Then we may assume that ${}_{\boldsymbol{b}}\mathcal{E}$ over $\mathcal{X}(\lambda_0, \epsilon_0) = X \times \Delta(\lambda_0, \epsilon_0)$ is locally free, and that the filtrations ${}^iF^{(\lambda_0)}$ and the decompositions ${}^i\mathbb{E}^{(\lambda_0)}$ on ${}_{\boldsymbol{b}}\mathcal{E}_{|\mathcal{D}_i(\lambda_0,\epsilon_0)}$ are given ($i \in \underline{l}$). As is shown in Lemma 8.103, we have already known that the tuple $\big({}^iF^{(\lambda_0)}, {}^i\mathbb{E}^{(\lambda_0)} \,\big|\, i = 1, \ldots, l\big)$ is compatible in the sense of Definition 4.37.

We recall that the residues $\mathrm{Res}_i(\mathbb{D})$ induce the endomorphisms of the associated graded bundle ${}^i\operatorname{Gr}^{F^{(\lambda_0)}}\,{}^i\mathbb{E}^{(\lambda_0)}\big({}_{\boldsymbol{b}}\mathcal{E}, \beta\big)$, whose nilpotent part is well given on $\mathcal{D}_i(\lambda_0, \epsilon_0)$. The nilpotent part is denoted by $\mathcal{N}_i$. It induces the endomorphisms on the associated graded vector bundles with respect to the filtrations ${}^jF^{(\lambda_0)}$. We also use the notation $\mathcal{N}_i$ to denote the induced endomorphisms, for simplicity of the notation.

LEMMA 12.46. *Let P be any point of $D^\circ_{\underline{m}} = D_{\underline{m}} - \bigcup_{i>m}(D_i \cap D_{\underline{m}})$.*

- *For any $m_1 \leq m$, the tuple $(\mathcal{N}_1, \ldots, \mathcal{N}_{m_1}, {}^{m_1+1}F^{(\lambda_0)}, \ldots, {}^mF^{(\lambda_0)})$ is sequentially compatible on the bundle ${}^{\underline{m_1}}\operatorname{Gr}^{F^{(\lambda_0)}}\,{}^{\underline{m}}\mathbb{E}^{(\lambda_0)}\big({}_{\boldsymbol{b}}\mathcal{E}_{|P\times\Delta(\lambda_0,\epsilon_0)}, \boldsymbol{\beta}\big)$, in the sense of Definition 4.20.*
- *Let $\boldsymbol{a}$ be any element of $\mathbb{Z}^{m_1}_{\geq 0}$. Then the conjugacy classes of the nilpotent maps $\prod_i \mathcal{N}^{a_i}_{i\,|\,(\lambda,P)}$ on ${}^{\underline{m_1}}\operatorname{Gr}^{F^{(\lambda_0)}}\,{}^{\underline{m}}\mathbb{E}^{(\lambda_0)}\big({}_{\boldsymbol{b}}\mathcal{E}_{|(\lambda,P)}, \boldsymbol{\beta}\big)$ are independent of a choice of $\lambda \in \Delta(\lambda_0, \epsilon_0)$.*

Proof We know that ${}^{\underline{m}}\mathcal{N}_1, \ldots, {}^{\underline{m}}\mathcal{N}_m$ are sequentially compatible on the bundle ${}^{\underline{m}}\mathrm{Gr}^{F^{(\lambda_0)}}\,{}^{\underline{m}}\mathbb{E}^{(\lambda_0)}({}_{\boldsymbol{b}}\mathcal{E}_{|P\times\Delta(\lambda_0,\epsilon_0)}, \boldsymbol{\beta})$ and their conjugacy classes are independent of a choice of λ (Corollary 12.43 and Corollary 12.45).

On $\Delta^*(\lambda_0, \epsilon_0)$, the vector bundle ${}^{\underline{m_1}}\mathrm{Gr}^{F^{(\lambda_0)}}\,{}^{\underline{m}}\mathbb{E}^{(\lambda_0)}({}_{\boldsymbol{b}}\mathcal{E}_{|P\times\Delta(\lambda_0,\epsilon_0)}, \boldsymbol{\beta})$ is decomposed into the generalized eigen bundles of the endomorphisms $\mathrm{Res}_i(\mathbb{D})$ $(i = m_1 + 1, \ldots, m)$ (Lemma 8.102 and Lemma 8.109). It satisfies the following:

- The decomposition gives the splitting of the filtrations ${}^iF^{(\lambda_0)}$ $(i = m_1 + 1, \ldots, m)$.
- The decomposition is compatible with $\mathcal{N}_i$ $(i = 1, \ldots, m_1)$.

For any $\lambda \in \Delta(\lambda_0, \epsilon_0)$, let $\mathcal{R}$ denote the local ring of $\mathcal{O}_{\Delta(\lambda_0,\epsilon_0)}$ at λ. We put as follows:

$$V = {}^{\underline{m_1}}\mathrm{Gr}^{F^{(\lambda_0)}}\,{}^{\underline{m}}\mathbb{E}^{(\lambda_0)}\big({}_{\boldsymbol{b}}\mathcal{E}_{|P\times\Delta(\lambda_0,\epsilon_0)}, \boldsymbol{\beta}\big) \otimes_{\mathcal{O}_{\Delta(\lambda_0,\epsilon_0)}} \mathcal{R}.$$

Then we obtain the naturally induced filtrations ${}^iF^{(\lambda_0)}$ $(i = m_1 + 1, \ldots, m)$ and the nilpotent maps $\mathcal{N}_i$ $(i = 1, \ldots, m_1)$. Then we have only to apply Lemma 5.2 and Proposition 5.8. □

We have the bundle ${}^{\underline{m}}\mathrm{Gr}_{\boldsymbol{a}}^{F^{(\lambda_0)}}\,{}^{\underline{m}}\mathbb{E}^{(\lambda_0)}\big({}_{\boldsymbol{b}}\mathcal{E}_{|\mathcal{D}_{\underline{m}}}, \boldsymbol{\beta}\big)$ and the commuting tuple of nilpotent maps $\mathcal{N}_1, \ldots, \mathcal{N}_m$ on $\mathcal{D}_{\underline{m}}(\lambda_0, \epsilon_0)$.

LEMMA 12.47.

- *Let $\boldsymbol{a}$ be an element of $\mathbb{Z}^m_{\geq 0}$. The conjugacy classes of $\prod_i \mathcal{N}^{a_i}_{i\,|\,(\lambda,P)}$ are independent of a choice of $(\lambda, P) \in \mathcal{D}_{\underline{m}}(\lambda_0, \epsilon_0)$.*
- *The tuple of the nilpotent maps $\mathcal{N}_1, \ldots, \mathcal{N}_m$ is sequentially compatible.*

Proof First we remark that the claims have been already shown when we fix any point $P \in D_{\underline{m}}$. Namely, let us take $m' \geq m$ such that $P \in D^{\circ}_{\underline{m}'}$. By applying Lemma 12.46, we know that the conjugacy classes of $\prod_i \mathcal{N}^{a_i}_{i|(\lambda,P)}$ are independent of a choice of $\lambda \in \Delta(\lambda_0, \epsilon_0)$, and that the nilpotent maps $\mathcal{N}_1, \ldots, \mathcal{N}_m$ are sequentially compatible on $\{P\} \times \Delta(\lambda_0, \epsilon_0)$.

Let us take a generic $\lambda \in \Delta(\lambda_0, \epsilon_0)$ and a normalizing frame of ${}_{\boldsymbol{b}}\mathcal{E}^{\lambda}$. Then it is easy to see that the conjugacy classes of $\prod_i \mathcal{N}^{a_i}_{i|(\lambda,P)}$ are independent of a choice of $P \in D_{\underline{m}}$. Hence the first claim is proved.

To see the second claim, we put $I_{\boldsymbol{h}\,|\,(\lambda,P)} := \bigcap_{j=1}^m W(\underline{j})_{h_j\,|\,(\lambda,P)}$. We have only to show that $\big\{I_{\boldsymbol{h}\,|\,(\lambda,P)} \,\big|\, (\lambda,P) \in \mathcal{D}_{\underline{m}}\big\}$ forms a vector bundle for each $\boldsymbol{h} \in \mathbb{Z}^n$. For that purpose, we have only to show that the ranks of $I_{\boldsymbol{h}\,|\,(\lambda,P)}$ are independent of (λ, P). If we fix P, then they are independent of a choice of λ due to the previous lemma. Pick a generic λ, and then we have a normalizing frame for ${}_{\boldsymbol{b}}\mathcal{E}^{\lambda}$. Due to the normalizing frame, we obtain the isomorphism for any $P_1, P_2 \in D_{\underline{m}}$:

$$\big({}_{\boldsymbol{b}}\mathcal{E}^{\lambda}{}_{|P_1}, \mathrm{Res}_i(\mathbb{D}^{\lambda}) \,\big|\, i \in \underline{m}\big) \simeq \big({}_{\boldsymbol{b}}\mathcal{E}^{\lambda}{}_{|P_2}, \mathrm{Res}_i(\mathbb{D}^{\lambda}) \,\big|\, i \in \underline{m}\big).$$

Thus we are done. □

We obtain the tuple $\big(\mathcal{N}_i, {}^iF^{(\lambda_0)}, {}^i\mathbb{E}^{(\lambda_0)} \,\big|\, i \in \underline{l}\big)$ from $(E, \overline{\partial}_E, \theta, h)$, which is a data as in the subsection 4.5.3. In summary of this subsection, we obtain the following theorem.

THEOREM 12.48. *The tuple $\big(\mathcal{N}_i, {}^iF^{(\lambda_0)}, {}^i\mathbb{E}^{(\lambda_0)} \,\big|\, i \in \underline{l}\big)$ is sequentially compatible in the sense of Definition* 4.43.

Proof It follows from Lemma 12.47. □

12.3.3. Decomposition. We continue to use the setting in the subsection 12.3.2. Let us consider the nilpotent maps $\mathcal{N}_i$ ($i \in I$) on ${}^I\mathrm{Gr}^{F^{(\lambda_0)}}({}_{\boldsymbol{b}}\mathcal{E})$ over $\mathcal{D}_I(\lambda_0, \epsilon_0)$. Let J and K be subsets of I such that $J \cap K = \emptyset$. We put $\mathcal{N}_K := \prod_{k \in K} \mathcal{N}_k$. The conjugacy classes of $\mathcal{N}_{K\,|\,(\lambda,P)}$ are independent of $(\lambda, P) \in \mathcal{D}_I(\lambda_0, \epsilon_0)$ (Lemma 12.47). Thus we obtain the vector bundle $\mathrm{Im}(\mathcal{N}_K)$ on $\mathcal{D}_I(\lambda_0, \epsilon_0)$. We put as follows:

$$\mathcal{V}_J\big(\mathrm{Im}\,\mathcal{N}_K\big) := \mathrm{Im}\,\mathcal{N}_K[N] \Big/ \prod_{j \in J}(N - \mathcal{N}_j) \simeq \bigoplus_{i=0}^{|J|-1} \mathrm{Im}\,\mathcal{N}_K \cdot N^i.$$

Here N denotes a formal variable. (See the subsection 3.9.2.) $\mathcal{V}_J\big(\mathrm{Im}\,\mathcal{N}_K\big)$ is a vector bundle on $\mathcal{D}_I(\lambda_0, \epsilon_0)$. If we emphasize the dependence of I, we denote it by $\mathcal{V}_J^I\big(\mathrm{Im}\,\mathcal{N}_K^I\big)$. The multiplication of N induces the nilpotent map on $\mathcal{V}_J^I\big(\mathrm{Im}\,\mathcal{N}_K^I\big)$.

LEMMA 12.49. *The conjugacy classes of N is independent of a choice of $(\lambda, P) \in \mathcal{D}_I(\lambda_0, \epsilon_0)$.*

Proof Let I' be a subset of $\underline{l}$ such that $P \in D_{I'}^{\circ}$. Let $q_{I'} : X \longrightarrow D_{I'}$ denote the projection. Let $(E_1, \overline{\partial}_{E_1}, \theta_1, h_1)$ denote the restriction $(E, \overline{\partial}_E, \theta, h)_{|q_{I'}^{-1}(P)}$. Then we obtain the limiting mixed twistor structure $S^{\mathrm{can}}(E_1)$ with the nilpotent maps $N_i^{\triangle}$ ($i \in I'$) from $(E_1, \overline{\partial}_{E_1}, \theta_1, h_1)$. Recall that we obtain the polarized mixed twistor structure by applying Kashiwara's construction in the subsection 3.9.2 to $S^{\mathrm{can}}(E_1)$ and $N_i^{\triangle}$ ($i \in I'$). In particular, the multiplication of N induces the morphism of mixed twistor structure.

Then the claim of the lemma can be shown by an argument similar to the proof of Corollary 12.43. □

As in the subsection 3.9.4, we have the morphisms can_i and var_i. For any element $i \in J$, we put $J' := J - \{i\}$. Then the morphism $\mathrm{var}_i : \mathcal{V}_J(\mathrm{Im}\,\mathcal{N}_K) \longrightarrow \mathcal{V}_{J'}(\mathrm{Im}\,\mathcal{N}_K)$ is induced from the identity of $\mathrm{Im}\,\mathcal{N}_K$, and the morphism $\mathrm{can}_i : \mathcal{V}_{J'}(\mathrm{Im}\,\mathcal{N}_K) \longrightarrow \mathcal{V}_J(\mathrm{Im}\,\mathcal{N}_K)$ is induced from the multiplication of $N - \mathcal{N}_i$ on $\mathrm{Im}\,\mathcal{N}_K$. For any element $i \in K$, we put $K' := K - \{i\}$. Then the morphism $\mathrm{var}_i : \mathcal{V}_J(\mathrm{Im}\,\mathcal{N}_K) \longrightarrow \mathcal{V}_J(\mathrm{Im}\,\mathcal{N}_{K'})$ is induced by the natural inclusion $\mathrm{Im}(\mathcal{N}_K) \subset \mathrm{Im}\,\mathcal{N}_{K'}$ and the morphism $\mathrm{can}_i : \mathcal{V}_J(\mathrm{Im}\,\mathcal{N}_{K'}) \longrightarrow \mathcal{V}_J(\mathrm{Im}\,\mathcal{N}_K)$ is induced by the multiplication of $\mathcal{N}_i$. When we emphasize the dependence on I, we use the notation can_i^I and var_i^I.

LEMMA 12.50. *Let P be a point of $D_I^{\circ} = D_I \setminus \bigcup_{j \in \underline{l} - I} D_j \cap D_I$, and U denote the open set $\Delta(\lambda_0, \epsilon_0)$ for some $\epsilon_0 > 0$. Then we have the following decomposition of the primitive parts:*

$$P_k\,\mathrm{Gr}_k^{W(N)}(\mathcal{V}_J^I(\mathrm{Im}(\mathcal{N}_K^I)))_{|U \times \{P\}} = \mathrm{Im}\,\mathrm{can}^I_{i\,|\,U \times \{P\}} \oplus \mathrm{Ker}\,\mathrm{var}^I_{i\,|\,U \times \{P\}}\,.$$

Proof We have only to apply the limiting mixed twistor theorem (Theorem 12.22. See also Remark 12.23.) and Proposition 3.135. □

LEMMA 12.51. *Let P be any point of D_I, and let U denote the open set $\Delta(\lambda_0, \epsilon_0)$ for some $\epsilon_0 > 0$. Then we have the decomposition of the vector bundles:*

$$P_k\,\mathrm{Gr}_k^{W(N)}(\mathcal{V}_J^I(\mathrm{Im}\,\mathcal{N}_K^I))_{|U \times \{P\}} = \mathrm{Im}\,\mathrm{can}^I_{i\,|\,U \times \{P\}} \oplus \mathrm{Ker}\,\mathrm{var}^I_{i\,|\,U \times \{P\}}\,.$$

Proof We have the set $I_0 \subset \underline{l}$ such that $P \in D_{I_0}^{\circ}$ and $I \subset I_0$. By applying Lemma 12.50 to $\mathcal{V}_J^{I_0}\big(\mathrm{Im}\,\mathcal{N}_K^{I_0}\big)$, we know that the ranks of $\mathrm{can}_i^{I_0}$ and $\mathrm{var}_i^{I_0}$ are independent of λ. We also have the decomposition of $P_k \operatorname{Gr}_k^{W(N)}(\mathcal{V}_J^{I_0}(\operatorname{Im}\mathcal{N}_K^{I_0}))_{|(P,\lambda)}$ into the image of $\operatorname{can}_i^{I_0}$ and the kernel of $\operatorname{var}_i^{I_0}$.

We use the argument in the proof of Lemma 12.46. The filtrations ${}^iF^{(\lambda_0)}$ $(i \in I_0 - I)$ naturally induce the filtrations of $\operatorname{Gr}_k^{W(N)}(\mathcal{V}_J^I(\operatorname{Im}\mathcal{N}_K^I))_{|(P,\lambda)}$. The filtrations are compatible with the primitive decomposition due to Lemma 4.18, i.e., we obtain the filtrations of $P_k \operatorname{Gr}_k^{W(N)}(\mathcal{V}_J^I(\operatorname{Im}\mathcal{N}_K^I))_{|(P,\lambda)}$. The associated graded bundle is naturally isomorphic to $P_k \operatorname{Gr}_k^{W(N)}(\mathcal{V}_J^{I_0}(\operatorname{Im}\mathcal{N}_K^{I_0}))_{|(P,\lambda)}$, due to Lemma 5.2. The filtrations split on $\{P\} \times \Delta^*(\lambda_0, \epsilon_0)$, which are given by the generalized eigen decompositions. Then the claim of Lemma 12.51 follows from Lemma 12.50 and Lemma 5.15. □

PROPOSITION 12.52. *We have the following decomposition of the vector bundle on $\mathcal{D}_I(\lambda_0, \epsilon_0)$:*

$$P_k \operatorname{Gr}_k^{W(N)}\big(\mathcal{V}_J^I(\operatorname{Im}\mathcal{N}_K^I)\big) = \operatorname{Im}(\operatorname{can}_i^I) \oplus \operatorname{Ker}(\operatorname{var}_i^I).$$

Proof We have only to show that $\operatorname{Im}(\operatorname{can}_i^I)$ and $\operatorname{Ker}(\operatorname{var}_i^I)$ are vector bundles. Namely we have only to show that the ranks of the morphisms $\operatorname{can}_{i\,|\,(\lambda,P)}^I$ and $\operatorname{var}_{i\,|\,(\lambda,P)}^I$ are independent of $(\lambda, P) \in \mathcal{D}_i$. When we fix a point P, it follows from Lemma 12.51. When we pick a generic λ, we can show that the numbers are independent of P by using the normalizing frame. (See the last part of the proof of Lemma 12.47.) □

12.3.4. The induced polarized pure twistor structure. We give a remark which will be used in Part 4. Let us consider the case $l = n$. We use the notation in the subsection 3.9.2. Let $I \sqcup J = \underline{n}$ be a decomposition. We put $N_J^{\triangle} := \prod_{j \in J} N_j^{\triangle}$. Let $\boldsymbol{u}$ be any element of $\mathcal{KMS}(\mathcal{E}^0, \underline{n})$. Then we obtain the subbundle $\operatorname{Im} N_J^{\triangle} \subset S_{\boldsymbol{u}}^{\mathrm{can}}(E) \otimes \mathbb{T}(-|J|)$. Due to the limiting mixed twistor theorem, the vanishing cycle theorem (Proposition 3.126) and Kashiwara's lemma (Corollary 3.132), we obtain the polarized mixed twistor structure $\big(\mathcal{V}_I(\operatorname{Im} N_J^{\triangle}), W, \boldsymbol{N}, \mathcal{V}_I(S)\big)$ with weight $|J| - |I| + 1$ in $(n+1)$-variables. By applying the procedure explained in the first half of the subsection 3.10.4, we obtain the vector bundle $\mathcal{C}_h(I, J)$ of $P_h \operatorname{Gr}_h^{W(N)} S_{\boldsymbol{u}}^{\mathrm{can}}(E)$ for non-negative integer h, which is the pure twistor structure of weight h with the polarization.

REMARK 12.53. We have used the nilpotent part of the residues. In Part 4, we will consider the nilpotent part of the left action of $-\eth \cdot t$, or equivalently the right action of $t \cdot \eth_t$. Therefore the signature of the nilpotent maps will be opposite. □

CHAPTER 13

Norm Estimate

13.1. Preliminary for functoriality via pull backs

13.1.1. The general formula. We put $X = \Delta_\zeta^n$, and $D = \bigcup_{i=1}^l D_i$. Let $(E, \overline{\partial}_E, \theta, h)$ be a tame harmonic bundle over $X - D$. Let λ_0 be any point. We assume that $0 \notin \mathcal{P}ar(\mathcal{E}^{\lambda_0}, i)$ for each i. Then we can take a sufficiently small positive number ϵ_0, such that $(^\diamond\mathcal{E}, \mathbb{D})$ is locally free over $\mathcal{X}(\lambda_0, \epsilon_0) = X \times \Delta(\lambda_0, \epsilon_0)$. We also have the filtrations ${}^jF^{(\lambda_0)}$ and the decompositions ${}^j\mathbb{E}^{(\lambda_0)}$ for $j = 1, \ldots, l$.

Let $\boldsymbol{v} = (v_i)$ be a holomorphic frame of $^\diamond\mathcal{E}$ compatible with the filtrations $F^{(\lambda_0)}$ and the decompositions $\mathbb{E}^{(\lambda_0)}$. For each v_i, we have the element $\boldsymbol{u}(v_i) \in \mathcal{KMS}(\mathcal{E}^0, \underline{l})$ such that ${}^{\underline{l}}\deg^{\mathbb{E}^{(\lambda_0)}, F^{(\lambda_0)}}(v_i) = \mathfrak{k}(\lambda_0, \boldsymbol{u}(v_i))$. We put as follows:

$$v_i' := v_i \cdot \prod_{j=1}^{l} |\zeta_j|^{\mathfrak{p}(\lambda, \boldsymbol{u}(v_i))}, \qquad \boldsymbol{v}' = (v_i').$$

Then $\boldsymbol{v}'$ is adapted up to log order over $(X - D) \times \Delta(\lambda_0, \epsilon_0)$ (Proposition 8.110).

We pick any element $\boldsymbol{c} := (c_{j\,i} \,|\, j \in \underline{m}, i \in \underline{n}) \in \mathbb{Z}_{\geq 0}^{m \cdot n}$. We put $\tilde{X} = \Delta^m$. Then we have the morphism $\phi_{\boldsymbol{c}} : \tilde{X} \longrightarrow X$ determined as follows:

$$\phi_{\boldsymbol{c}}^* \zeta_i = \prod_{j=1}^{m} z_j^{c_{j\,i}}.$$

Then we obtain the C^∞-frame $\phi_{\boldsymbol{c}}^* \boldsymbol{v}'$ of $\phi_{\boldsymbol{c}}^* \mathcal{E}$ over $(\tilde{\mathcal{X}} - \tilde{\mathcal{D}})(\lambda_0, \epsilon_0) = (\tilde{X} - \tilde{D}) \times \Delta(\lambda_0, \epsilon_0)$, which is adapted up to log order. We have the following equalities:

(13.1)

$$\phi_{\boldsymbol{c}}^*(v_i') = \phi_{\boldsymbol{c}}^*(v_i) \cdot \prod_{j=1}^{l} \Bigl(\prod_{h=1}^{m} |z_h|^{c_{h\,j}}\Bigr)^{\mathfrak{p}(\lambda, u_j(v_i))} = \phi_{\boldsymbol{c}}^*(v_i) \cdot \prod_{h=1}^{m} |z_h|^{\mathfrak{p}\left(\lambda, \sum c_{h\,j} \cdot u_j(v_i)\right)}$$
$$= \phi_{\boldsymbol{c}}^*(v_i) \times \prod_{h=1}^{m} |z_h|^{\mathfrak{p}\left(\lambda, \boldsymbol{c}_h \cdot \boldsymbol{u}(v_i)\right)}.$$

Here we put $\boldsymbol{c}_h \cdot \boldsymbol{u}(v_i) = \sum c_{h\,j} \cdot u_j(v_i)$.

We use the notation ν and κ given in the subsection 2.1.5. For some small positive number ϵ_0' such that $\epsilon_0' \leq \epsilon_0$, the integer $\nu\bigl(\mathfrak{p}(\lambda, \boldsymbol{c}_h \cdot \boldsymbol{u}(v_i))\bigr)$ does not independent of $\lambda \in \Delta(\lambda_0, \epsilon_0')$ for any v_i and for any h. We replace ϵ_0 with ϵ_0', and we may and will assume that $\nu\bigl(\mathfrak{p}(\lambda, \boldsymbol{c}_h \cdot \boldsymbol{u}(v_i))\bigr)$ does not independent of $\lambda \in \Delta(\lambda_0, \epsilon_0)$. We put $\nu(v_i, h) := \nu\bigl(\mathfrak{p}(\lambda, \boldsymbol{c}_h \cdot \boldsymbol{u}(v_i))\bigr) \in \mathbb{Z}$. We put $\kappa(v_i, h) := \kappa\bigl(\mathfrak{p}(\lambda, \boldsymbol{c}_h \cdot \boldsymbol{u}(v_i))\bigr)$.

We remark that $\kappa(v_i,h)$ is a function of λ. Then we put as follows:

$$w_i := \phi_{\boldsymbol{c}}^* v_i \cdot \prod_{h=1}^{m} z_h^{\nu(v_i,h)}.$$

Then w_i is a holomorphic section of $^\diamond\phi_{\boldsymbol{c}}^*\mathcal{E}$. We put as follows:

$$w_i' := w_i \cdot \prod_{h=1}^{m} |z_h|^{\kappa(v_i,h)}, \qquad \boldsymbol{w}' := (w_i').$$

Then $\boldsymbol{w}'$ is adapted up to log order over $(\widetilde{\mathcal{X}}-\widetilde{\mathcal{D}})(\lambda_0,\epsilon_0')$, by our construction. Hence $\boldsymbol{w}=(w_i)$ is a holomorphic frame of $^\diamond\phi_{\boldsymbol{c}}^*\mathcal{E}$, which is compatible with the filtrations $F^{(\lambda_0)}$, due to Lemma 2.8 and Lemma 2.9. We also have the following:

$$^h\deg^{F^{(\lambda_0)}}(w_i) = \kappa\big(v_i,h\big)(\lambda_0). \tag{13.2}$$

Let $A=\sum_k A^k\cdot\zeta_k^{-1}\cdot d\zeta_k$ denote the λ-connection form of $\mathbb{D}$ with respect to the frame $\boldsymbol{v}$, i.e., $\mathbb{D}\boldsymbol{v}=\boldsymbol{v}\cdot A$ holds. We put $\tilde{\mathbb{D}}:=\phi_{\boldsymbol{c}}^*\mathbb{D}$. Then we have the following equalities:

$$\tilde{\mathbb{D}}\phi_{\boldsymbol{c}}^*v_i=\sum_j\phi_{\boldsymbol{c}}^*v_j\cdot\Big(\sum_k\phi_{\boldsymbol{c}}^*A^k_{j\,i}\cdot\phi_{\boldsymbol{c}}^*\frac{d\zeta_k}{\zeta_k}\Big)=\sum_j\phi_{\boldsymbol{c}}^*v_j\cdot\Big(\sum_h\frac{dz_h}{z_h}\cdot\sum_k c_{h\,k}\cdot\phi_{\boldsymbol{c}}^*A^k_{j\,i}\Big).$$

Thus we obtain the following:

$$\begin{aligned}\tilde{\mathbb{D}}w_i&=\tilde{\mathbb{D}}\Big(\phi_{\boldsymbol{c}}^*v_i\cdot\prod_{h=1}^m z_h^{\nu(v_i,h)}\Big)\\&=\tilde{\mathbb{D}}(\phi_{\boldsymbol{c}}^*v_i)\cdot\prod_{h=1}^m z_h^{\nu(v_i,h)}+\phi_{\boldsymbol{c}}^*v_i\cdot\prod_{h=1}^m z_h^{\nu(v_i,h)}\cdot\Big(\sum_{h=1}^m\nu(v_i,h)\cdot\lambda\cdot\frac{dz_h}{z_h}\Big)\\&=\sum_j w_j\cdot\Big(\sum_{h=1}^m\frac{dz_h}{z_h}\sum_{k=1}^n c_{h\,k}\cdot\phi^*A^k_{j\,i}\Big)\cdot\prod_{h=1}^m z^{\nu(v_i,h)-\nu(v_j,h)}+w_i\cdot\Big(\sum_{h=1}^m\nu(v_i,h)\cdot\lambda\cdot\frac{dz_h}{z_h}\Big).\end{aligned}\tag{13.3}$$

13.1.2. The special case (I). We continue to use the setting in the subsection 13.1.1. We often consider the restriction to $\mathcal{D}_i(\lambda_0,\epsilon_0)=D_i\times\Delta(\lambda_0,\epsilon_0)$. For simplicity of the notation, we often use the notation $\mathcal{D}_i$ instead of $\mathcal{D}_i(\lambda_0,\epsilon_0)$.

Let us consider the case that $\boldsymbol{c}$ is a diagonal matrix whose i-th diagonal component is c_i, that is, $\phi_{\boldsymbol{c}}^*\zeta_i=z_i^{c_i}$. Then we obtain the following formula from (13.3):

$$\begin{aligned}\tilde{\mathbb{D}}w_i&=\sum_j w_j\cdot\Big(\sum_{h=1}^m\frac{dz_h}{z_h}\cdot c_h\cdot\phi_{\boldsymbol{c}}^*A^h_{j\,i}\Big)\cdot\prod_{p=1}^m z_p^{\nu(c_p\cdot b_p(v_i))-\nu(c_p\cdot b_p(v_j))}\\&\quad+\lambda\cdot w_i\cdot\sum_{h=1}^m\nu(c_h\cdot b_h(v_i))\cdot\frac{dz_h}{z_h}=:\sum_{j,k}w_j\cdot\tilde{A}^h_{j\,i}\frac{dz_h}{z_h}.\end{aligned}\tag{13.4}$$

Here we put $b_p(v_i):=\mathfrak{p}(\lambda_0,u_p(v_i))$.

LEMMA 13.1. *We have the following vanishings:*

1. *In the case* $^k\deg^{F^{(\lambda_0)}}(v_i)<{}^k\deg^{F^{(\lambda_0)}}(v_j)$, *we have* $\tilde{A}^h_{j\,i\,|\mathcal{D}_k}=0$ *for any* h.
2. *Assume that* c_k *is sufficiently large and that the inequality* $^k\deg^{F^{(\lambda_0)}}(v_i)>{}^k\deg^{F^{(\lambda_0)}}(v_j)$ *holds, and then we have* $\tilde{A}^h_{j\,i\,|\mathcal{D}_k}=0$.

Proof In the case ${}^k\deg^{F^{(\lambda_0)}}(v_i) < {}^k\deg^{F^{(\lambda_0)}}(v_j)$, we have $A^h_{j\,i\,|\,\mathcal{D}_k} = 0$. Hence $A^h_{j\,i} \cdot \zeta_k^{-1}$ is holomorphic, and thus $\phi^*_{\boldsymbol{c}} A^h_{j\,i} \cdot z_k^{-c_k}$ is holomorphic. Since we have the following inequality: $-c_k < \nu\bigl(c_k \cdot b_k(v_i)\bigr) - \nu\bigl(c_k \cdot b_k(v_j)\bigr)$, we obtain the first claim.

Let us show the second claim. If c_k is sufficiently large and the inequality ${}^k\deg^{F^{(\lambda_0)}}(v_i) > {}^k\deg^{F^{(\lambda_0)}}(v_j)$ holds, then we obtain the inequality $\nu(c_k \cdot b_k(v_i)) > \nu(c_k \cdot b_k(v_j))$. Since $A^k_{j\,i}$ is holomorphic, we obtain the result. □

Assume that c_k $(k = 1, \dots, l)$ are sufficiently large. Then we have the following formula:

(13.5)

$$\begin{cases} \mathrm{Res}_k(\tilde{\mathbb{D}}) w_i = \sum w_j \cdot \tilde{A}^k_{j\,i\,|\,\mathcal{D}_k(\lambda_0,\epsilon_0)}, \\ \\ \bigl(\text{The case } {}^k\deg^{F^{(\lambda_0)},\mathbb{E}^{(\lambda_0)}}(v_i) = {}^k\deg^{\mathbb{F}^{(\lambda_0)},\mathbb{E}^{(\lambda_0)}}(v_j)\bigr) \\ \tilde{A}^k_{j\,i\,|\,\mathcal{D}_k} = \delta_{i,j} \cdot \lambda \cdot \nu(c_k \cdot b_k(v_i)) + c_k \cdot \phi^*_{\boldsymbol{c},k}(A^k_{j\,i\,|\,\mathcal{D}_k}) \prod_{a \neq k} z_a^{\nu(c_a \cdot b_a(v_i)) - \nu(c_a \cdot b_a(v_j))}, \\ \\ \bigl(\text{Otherwise}\bigr) \\ \tilde{A}^k_{j\,i\,|\,\mathcal{D}_k} = 0. \end{cases}$$

Here $\phi_{\boldsymbol{c},k} : \tilde{\mathcal{D}}_k \longrightarrow \mathcal{D}_k$ denotes the restriction of ϕ to $\tilde{\mathcal{D}}_k$, and $\delta_{i,j}$ denotes 1 $(i = j)$ or 0 $(i \neq j)$.

We put $\mathcal{K}(\mathcal{E}, \lambda_0, \boldsymbol{0}, k) := \bigl\{u \in \mathcal{KMS}(\mathcal{E}^0, i) \,\big|\, -1 < \mathfrak{p}(\lambda_0, u) < 0\bigr\}$. For any element $u \in \mathcal{K}(\mathcal{E}, \lambda_0, \boldsymbol{0}, k)$, let ${}^k\mathcal{K}'_u$ denote the subbundle of ${}^\diamond\phi^*_{\boldsymbol{c}}\mathcal{E}_{|\mathcal{D}_k}$ generated by the tuple of sections $\boldsymbol{w}_u := \bigl\{w_{i\,|\,\mathcal{D}_k} \,\big|\, \deg^{F^{(\lambda_0)},\mathbb{E}^{(\lambda_0)}}(v_i) = \mathfrak{k}(\lambda_0, u)\bigr\}$ of ${}^\diamond\phi^*_{\boldsymbol{c}}\mathcal{E}_{|\mathcal{D}_k}$.

Lemma 13.2.

- *The vector subbundle* ${}^k\mathcal{K}'_u$ *is preserved by the residue* $\mathrm{Res}_k(\tilde{\mathbb{D}})$.
- *The eigenvalue of* $\mathrm{Res}_k(\tilde{\mathbb{D}})_{|{}^k\mathcal{K}'_u}$ *is* $\lambda \cdot \nu(c_k \cdot \mathfrak{p}(\lambda_0, u)) + \mathfrak{e}(\lambda, u)$.
- *We remark that* ${}^k\deg^{F^{(\lambda_0)}}(w_i) = c_k \cdot \mathfrak{p}(\lambda_0, u) - \nu(c_k \cdot \mathfrak{p}(\lambda_0, u))$ *in the case* ${}^k\deg^{F^{(\lambda_0)},\mathbb{E}^{(\lambda_0)}}(v_i) = \mathfrak{k}(\lambda_0, u)$.

Proof The first two claims immediately follow from (13.5). The last claim immediately follows from the formula for w_i. □

We put $\mathcal{K}(\phi^*_{\boldsymbol{c}}\mathcal{E}, \lambda_0, \boldsymbol{0}, k) := \bigl\{u \in \mathcal{KMS}(\phi^*_{\boldsymbol{c}}\mathcal{E}^0, k) \,\big|\, -1 < \mathfrak{p}(\lambda_0, u) < 0\bigr\}$. We consider the map $\phi^*_{\boldsymbol{c},k} : \boldsymbol{R} \times \boldsymbol{C} \longrightarrow \boldsymbol{R} \times \boldsymbol{C}$ given as follows:

$$\phi^*_{\boldsymbol{c},k}(u) = c_k \cdot u + \nu(c_k \cdot \mathfrak{p}(\lambda_0, u)) \cdot \bigl(-1, 0\bigr) \in \boldsymbol{R} \times \boldsymbol{C}.$$

Then $\phi^*_{\boldsymbol{c},k}$ induces the map $\mathcal{K}(\mathcal{E}, \lambda_0, \boldsymbol{0}, k) \longrightarrow \mathcal{K}(\phi^*_{\boldsymbol{c}}\mathcal{E}, \lambda_0, \boldsymbol{0}, k)$, due to Lemma 13.2. We put as follows, for any $u \in \mathcal{K}(\phi^*_{\boldsymbol{c}}\mathcal{E}, \lambda_0, \boldsymbol{0}, k)$:

$${}^k\mathcal{K}_u := \bigoplus_{\substack{u_1 \in \mathcal{KMS}({}^\diamond\mathcal{E}^0, k) \\ \phi^*_{\boldsymbol{c},k}(u_1) = u}} {}^k\mathcal{K}'_{u_1}.$$

Lemma 13.3.

- *The tuple of the decompositions* ${}^k\mathcal{K}$ $(k = 1, \dots, l)$ *is compatible in the sense of Definition* 4.37.
- *The decomposition* ${}^k\mathcal{K}$ *is preserved by* $\mathrm{Res}_k(\mathbb{D})$ *for each* k. *The eigenvalue of* $\mathrm{Res}_k(\mathbb{D})$ *on* ${}^k\mathcal{K}_u$ $\bigl(u \in \mathcal{KMS}({}^\diamond\phi^*_{\boldsymbol{c}}\mathcal{E}, k)\bigr)$ *is* $\mathfrak{e}(\lambda, u)$.

- *Namely, when c_k $(k=1,\ldots,l)$ are sufficiently large, $\phi_c^*\mathcal{E}$ is (CA) in the sense of Definition* 13.4 *below.*

Proof It is clear from our construction. □

13.1.3. Convenient KMS-structure. Here we introduce the notion of convenient KMS-structure, which is related with the result in the subsection 13.1.2. The KMS-structure is briefly a combination of the parabolic filtration and the generalized eigen decomposition. If the parabolic filtration is split, it is easy to handle. A convenient KMS-structure is essentially such a KMS-structure.

We put $X=\Delta^n$, $D_i=\{z_i=0\}$ and $D=\bigcup_{i=1}^l D_i$. Let $(E,\overline{\partial}_E,\theta,h)$ be a tame harmonic bundle on $X-D$. Let λ_0 be a point of $\boldsymbol{C}_\lambda$. We assume $0\notin\mathcal{KMS}(\mathcal{E}^{\lambda_0},i)$ for each i. Let ϵ_0 be a sufficiently small positive number such that ${}^\diamond\mathcal{E}$ is locally free on $\mathcal{X}(\lambda_0,\epsilon_0)$ and that the filtrations ${}^iF^{(\lambda_0)}$ and ${}^i\mathbb{E}^{(\lambda_0)}$ are well given. We put $\mathcal{K}(\mathcal{E},\lambda_0,\boldsymbol{0},i):=\bigl\{u\in\mathcal{KMS}(\mathcal{E}^0,i)\bigm|-1<\mathfrak{p}(\lambda_0,u)<0\bigr\}$. Recall that we always have the following decomposition on $\mathcal{D}_i^*(\lambda_0,\epsilon_0)$:

$$ {}^\diamond\mathcal{E}_{|\mathcal{D}_i^*(\lambda_0,\epsilon_0)}=\bigoplus_{u\in\mathcal{K}(\mathcal{E},\lambda_0,\boldsymbol{0},i)}\mathbb{E}\bigl({}^\diamond\mathcal{E}_{|\mathcal{D}_i^*(\lambda_0,\epsilon_0)},\mathfrak{e}(\lambda,u)\bigr). $$

DEFINITION 13.4. $\mathcal{E}$ is called convenient at λ_0, if the following holds:

(A): The decomposition above is prolonged to the decomposition:

$$ {}^\diamond\mathcal{E}_{|\mathcal{D}_k(\lambda_0,\epsilon_0)}=\bigoplus_{u\in\mathcal{K}(\mathcal{E},\lambda_0,\boldsymbol{0},i)}{}^k\mathcal{K}_u. $$

Moreover the tuple of the decompositions $\bigl({}^k\mathcal{K}\bigm|k=1,\ldots,l\bigr)$ is compatible.

(B): There exists a sequence of positive numbers $\eta_1>\eta_2>\cdots>\eta_l>0$:

- $\sum\eta_i<1/2$.
- $\mathcal{P}ar({}^\diamond\mathcal{E}^{\lambda_0},i)\subset]-\eta_i,0[$.
- $\min\bigl\{|a-b|\bigm|a\neq b\in\mathcal{P}ar({}^\diamond\mathcal{E}^{\lambda_0},i)\cup\{0\}\bigr\}>2\cdot\sum_{j>i}\eta_j$.

$\mathcal{E}$ is called (CA) (resp. (CB)) at λ_0 if the condition (A) (resp. (B)) holds. □

We remark that the decompositions ${}^k\mathcal{K}$ are uniquely determined if $\mathcal{E}$ is convenient at λ_0.

LEMMA 13.5. *Assume that $\mathcal{E}$ is (CA) at λ_0. Then the induced connection ${}^i\mathbb{D}$ and the residue $\mathrm{Res}_i(\mathbb{D})$ preserve the decomposition ${}^i\mathcal{K}$. (See the subsection* 8.8.3 *for the definition of ${}^i\mathbb{D}$.)*

Proof Since the restrictions of ${}^i\mathbb{D}$ and $\mathrm{Res}_i(\mathbb{D})$ to $\mathcal{D}_i^*(\lambda_0,\epsilon_0)$ preserve ${}^i\mathcal{K}_{u\,|\,\mathcal{D}_i^*(\lambda_0,\epsilon_0)}$, they preserve ${}^i\mathcal{K}_u$ on $\mathcal{D}(\lambda_0,\epsilon_0)$. □

LEMMA 13.6. *Assume that $\mathcal{E}$ is (CA) at λ_0. We have the following decomposition:*

$$ {}^iF_b^{(\lambda_0)}\,{}^i\mathbb{E}^{(\lambda_0)}\bigl({}^\diamond\mathcal{E}_{|\mathcal{D}_i(\lambda_0,\epsilon_0)},\beta\bigr)=\bigoplus_{\substack{\mathfrak{e}(\lambda_0,u)=\beta\\-1<\mathfrak{p}(\lambda_0,u)\leq b}}{}^i\mathcal{K}_u. \tag{13.6} $$

Proof Due to Lemma 8.103, the restrictions of the both sides of (13.6) to $\mathcal{D}_i^*(\lambda_0,\epsilon_0)$ are same. Then Lemma 13.6 immediately follows. □

Hence, if $\mathcal{E}$ is (CA) at λ_0, we obtain the naturally defined isomorphism:

$$ {}^i\mathcal{K}_u\simeq{}^i\mathrm{Gr}^{F^{(\lambda_0)}}_{\mathfrak{p}(\lambda_0,u)}\,{}^i\mathbb{E}^{(\lambda_0)}\bigl({}^\diamond\mathcal{E}_{|\mathcal{D}_i(\lambda_0,\epsilon_0)},\mathfrak{e}(\lambda_0,u)\bigr). $$

Thus the nilpotent parts $\mathcal{N}_i$ of the residues $\mathrm{Res}_i(\mathbb{D})$ are well defined as elements of $\mathrm{End}\big({}^\diamond\mathcal{E}_{|\mathcal{D}_i(\lambda_0,\epsilon_0)}\big)$ for any $i=1,\ldots,l$.

LEMMA 13.7. *Assume that $\mathcal{E}$ is (CA) at λ_0. Then the tuple $\big({}^k\mathcal{K}_u,\mathcal{N}_i \,\big|\, k\in \underline{l}, i\in \underline{l}\big)$ is strongly sequentially compatible in the sense of Definition* 4.49.

Proof It immediately follows from the limiting mixed twistor theorem (Theorem 12.22 and Remark 12.23) and Lemma 3.116. □

Let $\boldsymbol{v}$ be a frame of ${}^\diamond\mathcal{E}$, which is compatible with the decompositions $({}^i\mathcal{K}\,|\,i=1,\ldots,l)$. Let $\sum A^h\cdot z_h^{-1}dz_h$ denote the λ-connection form of $\mathbb{D}$ with respect to the frame $\boldsymbol{v}$:

$$\mathbb{D}\boldsymbol{v}=\boldsymbol{v}\cdot\Big(\sum_h A^h\cdot\frac{dz_h}{z_h}\Big).$$

LEMMA 13.8. *Assume that $\mathcal{E}$ is (CA) at λ_0. We have $A^h_{i\,j\,|\,\mathcal{D}_k}=0$ in the case ${}^k\deg^{\mathcal{K}}(v_i)\neq {}^k\deg^{\mathcal{K}}(v_j)$.*

Proof It follows from Lemma 13.5. □

Assume that $\mathcal{E}$ is (CA) at λ_0, and ${}^k\mathcal{K}$ $(k=1,\ldots,l)$ be the decompositions as in Definition 13.4. We put $\mathcal{K}(\mathcal{E},\lambda_0,\boldsymbol{0},I):=\big\{\boldsymbol{u}\in\mathcal{KMS}(\mathcal{E}^0,I)\,\big|\,-\boldsymbol{\delta}<\mathfrak{p}(\lambda_0,\boldsymbol{u})<\boldsymbol{0}\big\}$ for any $I\subset\underline{l}$. Then we put as follows:

$${}^I\mathcal{K}_{\boldsymbol{u}}:=\bigcap_{i\in I}{}^i\mathcal{K}_{q_i(\boldsymbol{u})\,|\,\mathcal{D}_I(\lambda_0,\epsilon_0)}.$$

Then we obtain the decomposition of ${}^\diamond\mathcal{E}_{|\mathcal{D}_I(\lambda_0,\epsilon_0)}=\bigoplus {}^I\mathcal{K}_{\boldsymbol{u}}$.

13.1.4. The special case (II). Let us consider some functoriality for pull back, again. Let η be a positive number such that $(1+\eta)^{l-1}\cdot 2/3<1$. We put $\tilde{X}:=\Delta(1+\eta)^{l-1}\times\Delta(2/3)\times\Delta^{n-l}$. Let us consider the morphism $\phi:\tilde{X}\longrightarrow X$ given as follows:

$$\phi^*(\zeta_i)=\begin{cases}\prod_{j=i}^{l} z_j, & (i\leq l),\\ z_i, & (i>l).\end{cases}$$

We put $\tilde{D}:=\phi^{-1}(D)$.

LEMMA 13.9. *Assume that $\mathcal{E}$ is convenient at λ_0. Then we have the natural isomorphism ${}^\diamond(\phi^*\mathcal{E})\simeq\phi^*({}^\diamond\mathcal{E})$ on $\tilde{\mathcal{X}}(\lambda_0,\epsilon_0)$ for some $\epsilon_0>0$.*

Let $\boldsymbol{v}$ be a frame of ${}^\diamond\mathcal{E}$ which is compatible with the decompositions $\big({}^i\mathcal{K}\,\big|\,i\in\underline{l}\big)$. In this case, $w_i=\phi^*v_i$ holds (see the subsection 13.1.1*), and the degrees ${}^h\deg^{F^{(\lambda_0)}}(w_i)$ are given as follows:*

$${}^h\deg^{F^{(\lambda_0)}}(w_i)=\sum_{k\leq h}{}^k\deg^{F^{(\lambda_0)}}(v_i).$$

Proof It follows from the formula (13.2) for the degree of w_i. □

Let $\sum_k A^k\cdot d\zeta_k/\zeta_k$ be the λ-connection form of $\mathbb{D}$ with respect to the frame $\boldsymbol{v}$, i.e., $\mathbb{D}\boldsymbol{v}=\boldsymbol{v}\cdot(\sum_k A^k\cdot d\zeta_k/\zeta_k)$ holds. Then we have the following:

$$\tilde{\mathbb{D}}\boldsymbol{w}=\boldsymbol{w}\cdot\sum_h\Big(\sum_{k\leq h}\phi^*A^k\Big)\cdot\frac{dz_h}{z_h}=:\boldsymbol{w}\cdot\sum_h\tilde{A}^h\cdot\frac{dz_h}{z_h}.$$

Here we have the following:

$$\tilde{A}^h = \begin{cases} \sum_{k=1}^h \phi^* A^k, & (h \le l), \\ \phi^* A^h, & (h > l). \end{cases}$$

In particular, we obtain the following:

$$\tilde{A}^h_{|\tilde{\mathcal{D}}_i(\lambda_0,\epsilon_0)} = \sum_{k \le h} \phi_i^* A^k_{|\mathcal{D}_{\underline{i}}(\lambda_0,\epsilon_0)}.$$

Here ϕ_i denotes the morphism $\tilde{\mathcal{D}}_i \longrightarrow \mathcal{D}_{\underline{i}}$ induced by ϕ. In particular, we have the following formula:

$$\mathrm{Res}_i(\tilde{\mathbb{D}}) = \sum_{k \le i} \phi_i^* \,\mathrm{Res}_k(\mathbb{D})_{|\mathcal{D}_{\underline{i}}}.$$

We obtain the decomposition of the vector bundle ${}^\diamond\phi^*\mathcal{E}_{|\tilde{\mathcal{D}}_i(\lambda_0,\epsilon_0)} = \bigoplus \phi^{*\,\underline{i}}\mathcal{K}_{\boldsymbol{u}}$. We put as follows:

$${}^i\tilde{\mathcal{K}}_u = \bigoplus_{\phi_i^*(\boldsymbol{u})=u} \phi^{*\,\underline{i}}\mathcal{K}_{\boldsymbol{u}}.$$

Here we put $\phi_i^*(\boldsymbol{u}) := \sum_{j \le i} q_j(\boldsymbol{u})$. Then we obtain the decomposition:

$${}^\diamond\phi^*\mathcal{E}_{|\tilde{\mathcal{D}}_i(\lambda_0,\epsilon_0)} = \bigoplus {}^i\tilde{\mathcal{K}}_u.$$

Clearly the tuple of the decompositions $\left({}^i\tilde{\mathcal{K}} \,\middle|\, i = 1, \ldots, l\right)$ is compatible.

LEMMA 13.10. *The eigenvalue of* $\mathrm{Res}_i(\tilde{\mathbb{D}})$ *on* ${}^i\tilde{\mathcal{K}}_u$ *is* $\mathfrak{e}(\lambda, u)$. *In particular,* $\phi^*\mathcal{E}$ *is* (CA).

Proof It immediately follows from our construction. □

The following lemma is also seen easily from our construction.

LEMMA 13.11. $\phi_i^*\mathcal{N}(\underline{i}) = \sum_{j \le i} \phi_i^* \mathcal{N}_j$ *gives the nilpotent part* $\tilde{\mathcal{N}}_i$ *of the residue* $\mathrm{Res}_i(\tilde{\mathbb{D}})$. □

13.1.5. The special case (III). We put $\tilde{X} = \Delta_z^{n-1}$. Let us consider the morphism $\phi : \tilde{X} \longrightarrow X$, given as follows:

$$\phi^*(\zeta_i) = \begin{cases} z_1, & (i = 1, 2), \\ z_{i-1}, & (i \ge 3). \end{cases}$$

LEMMA 13.12. *Assume* $\mathcal{E}$ *is convenient at* λ_0. *Then we have the natural isomorphism* ${}^\diamond\phi^*\mathcal{E} \simeq \phi^{*\diamond}\mathcal{E}$ *on* $\tilde{\mathcal{X}}(\lambda_0, \epsilon_0)$ *for some* $\epsilon_0 > 0$.

Let $\boldsymbol{v}$ *be a frame of frame of* ${}^\diamond\mathcal{E}$ *compatible with the decompositions* ${}^k\mathcal{K}$ $(k = 1, \ldots, l)$. *In this case, we have* $w_i = \phi^* v_i$ *(see the subsection* 13.1.1*), and the degrees* ${}^h\deg^{F^{(\lambda_0)}}(w_i)$ *are given as follows:*

$${}^h\deg^{F^{(\lambda_0)}}(w_i) = \begin{cases} {}^1\deg^{F^{(\lambda_0)}}(v_i) + {}^2\deg^{F^{(\lambda_0)}}(v_i), & (h = 1), \\ {}^{h+1}\deg^{F^{(\lambda_0)}}(v_i), & (h \ge 2). \end{cases}$$

Proof If follows from the formula (13.2) for the degree of w_i. $\square$

Assume that $\mathcal{E}$ is convenient at λ_0. Let $\sum_k \mathcal{A}^k d\zeta_k/\zeta_k$ denote the λ-connection of $\mathbb{D}$ with respect to the frame $\boldsymbol{v}$, i.e., $\mathbb{D}\boldsymbol{v} = \boldsymbol{v}\cdot\big(\sum_k \mathcal{A}^k \cdot d\zeta_k/\zeta_k\big)$ holds. Then we have the following formula:

$$\tilde{\mathbb{D}}\boldsymbol{w} = \boldsymbol{w}\cdot\Big(\big(\phi^*\mathcal{A}^1 + \phi^*\mathcal{A}^2\big)\cdot\frac{dz_1}{z_1} + \sum_{h=2}^{n-1}\phi^*\mathcal{A}^{h+1}\cdot\frac{dz_h}{z_h}\Big).$$

Hence we have the following formula:

$$\mathrm{Res}_h(\tilde{\mathbb{D}}) = \begin{cases} \phi^*\,\mathrm{Res}_1(\mathbb{D})_{|\mathcal{D}_{\underline{2}}(\lambda_0,\epsilon_0)} + \phi^*\,\mathrm{Res}_2(\mathbb{D})_{|\mathcal{D}_{\underline{2}}(\lambda_0,\epsilon_0)}, & (h=1),\\[2ex] \phi^*\,\mathrm{Res}_{h+1}(\mathbb{D})_{\mathcal{D}_{h+1}(\lambda_0,\epsilon_0)}, & (h\geq 2).\end{cases}$$

Thus the nilpotent parts are as follows:

$$\tilde{\mathcal{N}_i} = \begin{cases} \phi^*(\mathcal{N}_1+\mathcal{N}_2), & (i=1),\\[2ex] \phi^*\mathcal{N}_{i+1}, & (i\geq 2).\end{cases}$$

Hence $\tilde{\mathcal{N}}(\underline{i}) = \phi^*\tilde{\mathcal{N}}(\underline{i+1})$. Let $\tilde{W}(\underline{i})$ denote the weight filtration of $\tilde{\mathcal{N}}(\underline{i})$, and then we have $\tilde{W}(\underline{i}) = \phi^* W(\underline{i+1})$.

On the divisor $\tilde{\mathcal{D}}_1(\lambda_0,\epsilon_0)$, we put as follows:

$$^1\tilde{\mathcal{K}}_u = \bigoplus_{u_1+u_2=u}\phi^{*2}\mathcal{K}_{(u_1,u_2)}.$$

On the divisors $\tilde{\mathcal{D}}_i(\lambda_0,\epsilon_0)$ $(i\geq 2)$, we put ${}^i\tilde{\mathcal{K}}_u = \phi^*({}^{i+1}\mathcal{K}_u)$. Then the decompositions ${}^i\tilde{\mathcal{K}}$ satisfies the condition (A) in Definition 13.4. It is also easy to check the condition (B) in Definition 13.4. Hence we have the following lemma.

LEMMA 13.13. *$\phi^*\mathcal{E}$ is convenient at λ_0.* $\square$

13.2. Preliminary norm estimate

13.2.1. Statement. We put $X = \Delta_\zeta^n = \big\{(\zeta_1,\ldots,\zeta_n)\,\big|\,|\zeta_i|<1\big\}$, $D_i = \{\zeta_i = 0\}$ and $D = \bigcup_{i=1}^l D_i$. Let $(E,\overline{\partial}_E,\theta,h)$ be a tame harmonic bundle over $X-D$. Let λ_0 be a point of $\boldsymbol{C}_\lambda$. We assume that $0\notin\mathcal{KMS}(\mathcal{E}^{\lambda_0},i)$ for each i. We take a sufficiently small positive number ϵ_0 such that $^\diamond\mathcal{E}$ is locally free over $\mathcal{X}(\lambda_0,\epsilon_0) = X\times\Delta(\lambda_0,\epsilon_0)$. In this section, we impose the following assumption.

ASSUMPTION 13.14. $\mathcal{E}$ is convenient at λ_0 in the sense of Definition 13.4. $\square$

Let $\boldsymbol{v}$ be a frame of $^\diamond\mathcal{E}$ compatible with the tuple $\big({}^i\mathcal{K}, W(\underline{m})\,\big|\, i\in\underline{l},\ m\in\underline{l}\big)$. Then $\boldsymbol{v}$ is compatible with the filtrations ${}^jF^{(\lambda_0)}$ and ${}^j\mathbb{E}^{(\lambda_0)}$ $(j=1,\ldots,l)$, and hence we have the element $\boldsymbol{u}(v_i) = \big(u_1(v_i),\ldots,u_l(v_i)\big)\in\mathcal{KMS}\big(\mathcal{E}^0,\underline{l}\big)$ for each v_i such that $\deg^{F^{(\lambda_0)},\mathbb{E}^{(\lambda_0)}}(v_i) = \mathfrak{k}\big(\lambda_0,\boldsymbol{u}(v_i)\big)$. We put as follows:

$$b_m(v_i) := \mathfrak{p}\big(\lambda, q_m(\boldsymbol{u}(v_i))\big),\quad h_m(v_i) := \frac{1}{2}\cdot\big(\deg^{W(\underline{m})}(v_i) - \deg^{W(\underline{m-1})}(v_i)\big).$$

We remark that $b_m(v_i)$ is a C^∞-function of λ. We put as follows:

$$v_i' := v_i\cdot\prod_{m=1}^{l}|\zeta_m|^{b_m(v_i)}\cdot\big(-\log|\zeta_m|\big)^{-h_m(v_i)}.$$

Then we obtain the C^∞-frame $\boldsymbol{v}' = (v'_i)$ of $\mathcal{E}$ over $(\mathcal{X} - \mathcal{D})(\lambda_0, \epsilon_0) = (X - D) \times \Delta(\lambda_0, \epsilon_0)$.

We consider the following subset Z:

$$Z := \big\{(\zeta_1, \dots, \zeta_n) \in X - D \,\big|\, |\zeta_{j-1}| \le |\zeta_j| \le 2^{-1},\ j \in \underline{l}\big\}. \tag{13.7}$$

For any $\epsilon > 0$, we put $\mathcal{Z}(\lambda_0, \epsilon) := Z \times \Delta(\lambda_0, \epsilon)$. The purpose of this section is to show the following proposition.

PROPOSITION 13.15. *The C^∞-frame $\boldsymbol{v}'$ is adapted on the region $\mathcal{Z}(\lambda_0, \epsilon'_0)$ for any $0 < \epsilon'_0 < \epsilon_0$.*

Let $\phi : \widetilde{X} \longrightarrow X$ be a morphism given in the subsection 13.1.4. We obtain the holomorphic frame $\boldsymbol{w} = \phi^*\boldsymbol{v}$ of ${}^\diamond\phi^*\mathcal{E}$. We also obtain the C^∞-frame $\boldsymbol{w}' = (w'_i) := \phi^*\boldsymbol{v}'$ over $(\widetilde{\mathcal{X}} - \widetilde{\mathcal{D}})(\lambda_0, \epsilon_0) = (\widetilde{X} - \widetilde{D}) \times \Delta(\lambda_0, \epsilon_0)$. It is easy to see the following:

$$\begin{aligned} w'_i &= w_i \cdot \prod_{m=1}^{l} |z_m|^{\sum_{j \le m} b_j(v_i)} \cdot \Big(-\sum_{t \ge m} \log|z_t|\Big)^{-h_m(v_i)} \\ &= w_i \cdot \prod_{m=1}^{l} |z_m|^{b_m(w_i)} \cdot \Big(-\sum_{t \ge m} \log|z_t|\Big)^{-h_m(w_i)}. \end{aligned} \tag{13.8}$$

Here we have used $b_m(w_i) = \sum_{j \le m} b_j(v_i)$ and $h_m(w_i) = h_m(v_i)$.

Let us consider the following subset $\widetilde{Z}$:

$$\widetilde{Z} := \big\{(z_1, \dots, z_n) \in \widetilde{X} - \widetilde{D} \,\big|\, |z_i| \le 1, (i \le l-1),\ |z_l| \le 2^{-1}\big\}.$$

For any $\epsilon > 0$, $\widetilde{\mathcal{Z}}(\lambda_0, \epsilon)$ are given similarly to $\mathcal{Z}(\lambda_0, \epsilon)$. We remark the $\phi(\widetilde{Z}) = Z$. Then it is easy to see that Proposition 13.15 is equivalent to the next Proposition 13.16.

PROPOSITION 13.16. *The C^∞-frame $\boldsymbol{w}'$ is adapted on $\widetilde{\mathcal{Z}}(\lambda_0, \epsilon'_0)$ for any $0 < \epsilon'_0 < \epsilon_0$.*

We will show Proposition 13.16, or equivalently Proposition 13.15 in the following subsections. We use an induction on the dimension of X. We assume that the propositions hold in the case $\dim X \le n - 1$, and we will prove the propositions hold in the case $\dim X = n$. The hypothesis of the induction will be used in Lemma 13.23.

13.2.2. Step 1. Independence of a choice of compatible frames.

LEMMA 13.17. *Let $\boldsymbol{v}$ be a frame of ${}^\diamond\mathcal{E}$ over $\mathcal{X}(\lambda_0, \epsilon_0)$, which is compatible with the tuple $\big({}^k\mathcal{K}, W(\underline{m}) \,\big|\, k \in \underline{l},\ m \in \underline{l}\big)$.*

Assume that the claim in Proposition 13.15 holds for $\boldsymbol{v}$. Then the same claim holds for any other frame of ${}^\diamond\mathcal{E}$ over $\mathcal{X}(\lambda_0, \epsilon_0)$, which is compatible with the tuple $\big({}^k\mathcal{K}, W(\underline{m}) \,\big|\, k \in \underline{l},\ m \in \underline{l}\big)$.

Proof Let $\boldsymbol{v}^{(1)}$ be other frame of ${}^\diamond\mathcal{E}$ over $\mathcal{X}(\lambda_0, \epsilon_0)$, which is compatible with the tuple $\big({}^k\mathcal{K}, W(\underline{m}) \,\big|\, k \in \underline{l},\ m \in \underline{l}\big)$. We have the relation of the form:

$$v_i^{(1)} = \sum B_{j\,i} \cdot v_j.$$

Here $B_{j\,i}$ are holomorphic on $\mathcal{X}(\lambda_0,\epsilon_0)$ and $B_{j\,i\,|\mathcal{D}_{\underline{k}}(\lambda_0,\epsilon_0)}=0$ unless the following holds:

$$(13.9)\quad {}^{\underline{k}}\deg^{\mathcal{K}}(v_i^{(1)})={}^{\underline{k}}\deg^{\mathcal{K}}(v_j),\qquad \deg^{W(\underline{m})}(v_i^{(1)})\geq\deg^{W(\underline{m})}(v_j)\ \ (\forall m\leq k).$$

We have the induced relation $w_i^{(1)}=\sum\phi^*B_{j\,i}\cdot w_j=\sum\widetilde{B}_{j\,i}\cdot w_j$. Then we have $\widetilde{B}_{j\,i\,|\,\widetilde{\mathcal{D}}_k(\lambda_0,\epsilon_0)}=0$ unless (13.9) holds for i and j. We also have the induced relation $w_i^{(1)\,\prime}=\sum\widetilde{B}'_{j\,i}\cdot w'_j$, and then we have the following:

$$\widetilde{B}'_{j\,i}=\widetilde{B}_{j\,i}\cdot\prod_p|z_p|^{b_p(w_i^{(1)})-b_p(w_j)}\cdot\prod_p\Bigl(-\sum_{t\geq p}\log|z_t|\Bigr)^{-h_p(w_i^{(1)})+h_p(w_j)}.$$

Once we obtain the boundedness of $\widetilde{B}'$, then we obtain the boundedness of $\widetilde{B}'^{-1}$ by symmetry. It implies the equivalence of the adaptedness of $\boldsymbol{w}^{(1)\,\prime}$ and $\boldsymbol{w}'$. So we have only to prove the boundedness of $\widetilde{B}'$. We give some remarks.

(i) Note that we have $\widetilde{B}_{j\,i\,|\widetilde{\mathcal{D}}_p(\lambda_0,\epsilon_0)}=0$ in the case $b_p(w_i^{(1)})-b_p(w_j)<0$. We also have $-1<b_p(w_i^{(1)})-b_p(w_j)$, due to convenience of $\mathcal{E}$ at λ_0. Thus $\widetilde{B}'_{j\,i\,|\widetilde{\mathcal{D}}_p(\lambda_0,\epsilon_0)}=0$.

(ii) We have the following equality:

$$\prod_p\Bigl(-\sum_{t\geq p}\log|z_t|\Bigr)^{-h_p(w_i^{(1)})+h_p(w_j)}=\prod_p\Bigl(\frac{-\sum_{t\geq p}\log|z_t|}{-\sum_{t\geq p+1}\log|z_t|}\Bigr)^{a_p}.$$

Here we put as follows:

$$a_p=-\frac{1}{2}\Bigl(\deg^{W(\underline{p})}(w_i^{(1)})-\deg^{W(\underline{p})}(w_j)\Bigr).$$

In the case $a_p\leq 0$, it is easy to see that $(-\sum_{t\geq p}\log|z_t|)^{a_p}\cdot(-\sum_{t\geq p+1}\log|z_t|)^{-a_p}$ is bounded. In the case $a_p>0$, we have $\widetilde{B}_{j\,i\,|\,\widetilde{\mathcal{D}}_p(\lambda_0,\epsilon_0)}=0$, and we also have the following inequality on $\widetilde{\mathcal{Z}}(\lambda_0,\epsilon_0)$, for some positive constant C:

$$\left(\frac{-\sum_{t\geq p+1}\log|z_t|-\log|z_p|}{-\sum_{t\geq p+1}\log|z_t|}\right)^{a_p}\leq\left(1+\frac{-\log|z_p|}{C}\right)^{a_p}.$$

Here we have used $-\sum_{t\geq p+1}\log|z_t|\geq-\log|z_l|\geq C$ for some positive constant C.

From (i) and (ii), the boundedness of $\widetilde{B}'_{i\,j}$ follows immediately. Thus the proof of Lemma 13.17 is accomplished. □

13.2.3. Step 2. Strongly compatible frame $\boldsymbol{v}$.

Thus we pick a frame $\boldsymbol{v}$ which is strongly compatible with $\bigl({}^k\mathcal{K},W(\underline{m})\,\big|\,k\in\underline{l},\ m\in\underline{l}\bigr)$ in the sense of Corollary 4.53. Namely we take a frame $\boldsymbol{v}$ satisfying the following conditions:

CONDITION 13.18.

- A compatible frame $\boldsymbol{v}$ consists of sections $v_{h,\boldsymbol{u},\boldsymbol{h},i}$. The indices h, $\boldsymbol{u}$ and $\boldsymbol{h}$ run through $\mathbb{Z}_{\geq 0}$, $\mathcal{K}(\mathcal{E},\lambda_0,\boldsymbol{0},\underline{l})$, and $\mathbb{Z}^l$. The index i runs through the set $S(h,\boldsymbol{u},\boldsymbol{h})$ depending on $h,\boldsymbol{u},\boldsymbol{h}$.
- We have $S(h,\boldsymbol{u},\boldsymbol{h})=S(h,\boldsymbol{u},\boldsymbol{h}-2\boldsymbol{\delta}_1)$ if $h-2q_1(\boldsymbol{h})$ is $2h'$ $(h'=0,1,\ldots,h)$, where q_1 denotes the projection onto the first component and $\boldsymbol{\delta}_1$ denotes the element $(1,0,\ldots,0)\in\mathbb{Z}^l$.

- The following holds on $\mathcal{D}_1(\lambda_0,\epsilon_0)$:

$$N(\underline{1})v_{h,\boldsymbol{u},\boldsymbol{h},i}=\begin{cases} v_{h,\boldsymbol{u},\boldsymbol{h}-2\boldsymbol{\delta}_1,i}, & (-h+2\leq q_1(\boldsymbol{h})\leq h,\ h-q_1(\boldsymbol{h})\text{ is even}),\\ 0, & (\text{otherwise}).\end{cases}$$

- We have ${}^j\deg^{\mathcal{K}}(v_{h,\boldsymbol{u},\boldsymbol{h},i})=q_j(\boldsymbol{u})$ and $2\cdot h_m(v_{h,\boldsymbol{u},\boldsymbol{h},i})=q_m(\boldsymbol{h})$. □

Then we obtain the frame $\boldsymbol{w}=\big(w_{h,\boldsymbol{u},\boldsymbol{h},i}\big)$ of ${}^\diamond\phi^*\mathcal{E}$ by pull back via ϕ.

Let q_1 denote the projection of $\widetilde{X}$ onto the first component $\Delta(1+\eta)$. We have the naturally defined projection $\Omega^{1,0}_{\widetilde{X}}\longrightarrow q_1^*\Omega^{1,0}_{\Delta(1+\eta)}$. Let $\mathfrak{q}_1(\widetilde{\mathbb{D}})$ denote the composite of the following:

$$\phi^*\mathcal{E}\xrightarrow{\ \widetilde{\mathbb{D}}\ }\phi^*\mathcal{E}\otimes\Omega^{1,0}_{\widetilde{X}}\longrightarrow\mathcal{E}\otimes q_1^*\Omega^{1,0}_{\Delta(1+\eta)}.$$

We give a general easy remark.

LEMMA 13.19. *Let $\boldsymbol{v}^{(1)}$ be a frame of ${}^\diamond\mathcal{E}$. For the frame $\boldsymbol{w}^{(1)}=\phi^*\boldsymbol{v}^{(1)}$ of ${}^\diamond\phi^*\mathcal{E}$, we have the following implication:*

$$\mathbb{D}v_i^{(1)}=\sum w_j^{(1)}\cdot\mathcal{A}^k_{j\,i}\frac{d\zeta_k}{\zeta_k}\Longrightarrow\mathfrak{q}_1(\widetilde{\mathbb{D}})w_i^{(1)}=\sum w_j^{(1)}\cdot\widetilde{\mathcal{A}}^1_{j\,i}\cdot\frac{dz_1}{z_1}=\sum w_j^{(1)}\cdot\phi^*\mathcal{A}^1_{j\,i}\cdot\frac{dz_1}{z_1}.$$

Proof It can be shown by a direct calculation. □

In particular, if $\boldsymbol{v}$ is as in Condition 13.18, we have the following equalities on $\mathcal{U}_k:=\widetilde{\mathcal{D}}_k(\lambda_0,\epsilon_0)$ $(2\leq k\leq l)$:

$$(13.10)\quad \mathfrak{q}_1(\widetilde{\mathbb{D}})w_{h,\boldsymbol{u},\boldsymbol{h},i\,|\,\mathcal{U}_k}=$$
$$\begin{cases}\Big(\mathfrak{e}(\lambda,u_1)\cdot w_{h,\boldsymbol{u},\boldsymbol{h},i}+w_{h,\boldsymbol{u},\boldsymbol{h}-2\boldsymbol{\delta}_1,i}\Big)\cdot\Big(\dfrac{dz_1}{z_1}\Big)_{|\mathcal{U}_k}, & \left(\begin{matrix}-h+2\leq \mathfrak{q}_1(\boldsymbol{h})\leq h,\\ h-q_1(\boldsymbol{h})\text{ even}\end{matrix}\right),\\ \mathfrak{e}(\lambda,u_1)\cdot w_{h,\boldsymbol{u},\boldsymbol{h},i}\Big(\dfrac{dz_1}{z_1}\Big)_{|\mathcal{U}_k}, & (q_1(\boldsymbol{h})=-h),\\ 0, & (\text{otherwise}).\end{cases}$$

On the divisor $\mathcal{U}_1:=\widetilde{\mathcal{D}}_1(\lambda_0,\epsilon_0)$, we have the following formula:

$$(13.11)\quad \operatorname{Res}(\mathfrak{q}_1\widetilde{\mathbb{D}})w_{h,\boldsymbol{u},\boldsymbol{h},i\,|\,\mathcal{U}_1}=$$
$$\begin{cases}\Big(\mathfrak{e}(\lambda,u_1)\cdot w_{h,\boldsymbol{u},\boldsymbol{h},i}+w_{h,\boldsymbol{u},\boldsymbol{h}-2\boldsymbol{\delta}_1,i}\Big)_{|\mathcal{U}_1}, & \left(\begin{matrix}-h+2\leq \mathfrak{q}_1(\boldsymbol{h})\leq h,\\ h-q_1(\boldsymbol{h})\text{ even}\end{matrix}\right),\\ \mathfrak{e}(\lambda,u_1)\cdot w_{h,\boldsymbol{u},\boldsymbol{h},i\,|\,\mathcal{U}_1}, & (q_1(\boldsymbol{h})=-h),\\ 0, & (\text{otherwise}).\end{cases}$$

13.2.4. Model bundle and the construction of the comparing morphism. Let h and $\boldsymbol{h}$ be elements of $\mathbb{Z}_{\geq 0}$ and $\mathbb{Z}^l$ respectively such that the first component of $\boldsymbol{h}$ is h. Let $\boldsymbol{u}$ be an element of $\mathcal{K}(\mathcal{E},\lambda_0,\boldsymbol{0},\underline{l})$. Recall that we have some index set $S(h,\boldsymbol{u},\boldsymbol{h})$ as in Condition 13.18. Let i be an element of $S(h,\boldsymbol{u},\boldsymbol{h})$.

Then we take the vector subspace of ${}^{\diamond}\mathcal{E}_{|(\lambda,O)}$ as follows:

$$V_{h,\boldsymbol{u},\boldsymbol{h},i} := \bigoplus_{a=0}^{h} \boldsymbol{C}\cdot v_{h,\boldsymbol{u},\boldsymbol{h}-2a\boldsymbol{\delta}_1,i} \subset {}^{\diamond}\mathcal{E}_{|(\lambda_0,O)}.$$

Then we have the decomposition: ${}^{\diamond}\mathcal{E}_{|(\lambda_0,O)} = \bigoplus V_{h,\boldsymbol{u},\boldsymbol{h},i}$. The decomposition is preserved by the endomorphism $\mathrm{Res}_1(\mathbb{D})$ and the nilpotent part $\mathcal{N}_1$. Let $\mathcal{N}_{h,\boldsymbol{u},\boldsymbol{h},i}$ denote the restriction of $\mathcal{N}_1$ to $V_{h,\boldsymbol{u},\boldsymbol{h},i}$.

Then we take the model bundle $E(V_{h,\boldsymbol{u},\boldsymbol{h},i},\mathcal{N}_{h,\boldsymbol{u},\boldsymbol{h},i})\otimes L(u_1)$ over Δ^* (see the subsection 6.2.7). We obtain the deformed holomorphic bundle $\mathcal{E}_{h,\boldsymbol{u},\boldsymbol{h},i}$ over $\Delta^*\times\boldsymbol{C}_\lambda$. The direct sum of $\mathcal{E}_{h,\boldsymbol{u},\boldsymbol{h},i}$ is denoted by $\mathcal{E}_0$. We have the natural metric h_0 on $\mathcal{E}_0$, which is a direct sum of the metrics $h_{h,\boldsymbol{u},\boldsymbol{h},i}$.

On $\Delta(\lambda_0,\epsilon_0)\times\Delta$, we have the prolongation ${}^{\diamond}\mathcal{E}_0$ and the canonical frame $\boldsymbol{w}^0 = (w^0_{h,\boldsymbol{u},\boldsymbol{h},i})$ such that the following holds:

$$(13.12)\quad \mathbb{D}_0 w^0_{h,\boldsymbol{u},\boldsymbol{h},i} := \begin{cases} \big(\mathfrak{e}(\lambda,u_1)\cdot w^0_{h,\boldsymbol{u},\boldsymbol{h},i} + w^0_{h,\boldsymbol{u},\boldsymbol{h}-2\boldsymbol{\delta}_1,i}\big)\cdot\dfrac{dz_1}{z_1}, & \left(\begin{array}{c} -h+2\leq q_1(\boldsymbol{h})\leq h,\\ h-q_1(\boldsymbol{h}) \text{ even}\end{array}\right),\\[2ex] \big(\mathfrak{e}(\lambda,u_1)\cdot w^0_{h,\boldsymbol{u},\boldsymbol{h},i}\big)\cdot\dfrac{dz_1}{z_1}, & (q_1(\boldsymbol{h})=-h),\\[2ex] 0, & (\text{otherwise}).\end{cases}$$

We also have the following formula:

$$(13.13)\quad \mathrm{Res}(\mathbb{D}_0) w^0_{h,\boldsymbol{u},\boldsymbol{h},i} := \begin{cases} \mathfrak{e}(\lambda,u_1)\cdot w^0_{h,\boldsymbol{u},\boldsymbol{h},i} + w^0_{h,\boldsymbol{u},\boldsymbol{h}-2\boldsymbol{\delta}_1,i}, & (-h+2\leq q_1(\boldsymbol{h})\leq h,\ h-q_1(\boldsymbol{h}) \text{ even }),\\ \mathfrak{e}(\lambda,u_1)\cdot w^0_{h,\boldsymbol{u},\boldsymbol{h},i}, & (q_1(\boldsymbol{h})=-h),\\ 0, & (\text{otherwise}).\end{cases}$$

We have the holomorphic vector bundle $\phi^*q_1^{*\diamond}\mathcal{E}_0$ over $\Delta(\lambda_0,\epsilon_0)\times\widetilde{X}$. We have the λ-connection $\widetilde{\mathbb{D}}_0 := \phi^*q_1^*\mathbb{D}_0$. The frames $\widetilde{\boldsymbol{w}}^0 := \phi^*q_1^*\boldsymbol{w}^0$ and $\boldsymbol{w}$ give the isomorphism $\Phi : \phi^*q_1^{*\diamond}\mathcal{E}_0 \longrightarrow {}^{\diamond}\mathcal{E}$ over $\widetilde{\mathcal{X}}(\lambda_0,\epsilon_0)$.

LEMMA 13.20. *We have the following equality on the divisor $\bigcup_{k=2}^{l}\widetilde{\mathcal{D}}_k(\lambda_0,\epsilon_0)$:*

$$\Phi\circ\mathfrak{q}_1\big(\widetilde{\mathbb{D}}_0\big) - \mathfrak{q}_1\big(\widetilde{\mathbb{D}}\big)\circ\Phi = 0.$$

We have the following equality on the divisor $\widetilde{\mathcal{D}_1}(\lambda_0,\epsilon_0)$:

$$\Phi\circ\mathrm{Res}\big(\mathfrak{q}_1(\widetilde{\mathbb{D}}_0)\big) - \mathrm{Res}\big(\mathfrak{q}_1(\widetilde{\mathbb{D}})\big)\circ\Phi = 0.$$

Proof It immediately follows from (13.10), (13.11), (13.12) and (13.13). □

13.2.5. Step 3. The metric on $\phi^*q_1^*\mathcal{E}_0$. We have the metric $\phi^*q_1^*h_{h,\boldsymbol{u},\boldsymbol{h},i}$ of $\phi^*q_1^*\mathcal{E}_{h,\boldsymbol{u},\boldsymbol{h},i}$ on $(\widetilde{\mathcal{X}}-\widetilde{\mathcal{D}})(\lambda_0,\epsilon_0)$. We modify it as follows:

$$\widetilde{h}_{h,\boldsymbol{u},\boldsymbol{h},i} := \phi^*q_1^*\big(h_{h,\boldsymbol{u},\boldsymbol{h},i}\big)\cdot\prod_{k=2}^{l}|z_k|^{-\sum_{2\leq t\leq k}\mathfrak{p}(\lambda,u_t)}\times\prod_{k=2}^{l}\Big(-\sum_{m\geq k}\log|z_m|^2\Big)^{q_k(\boldsymbol{h})}.$$

The metrics $\widetilde{h}_{h,\boldsymbol{u},\boldsymbol{h},i}$ induce the metric $\widetilde{h}_0$ of $\phi^* q_1^* \mathcal{E}_0$ over $(\widetilde{\mathcal{X}} - \widetilde{\mathcal{D}})(\lambda_0, \epsilon_0)$. The C^∞-frame $\widetilde{\boldsymbol{w}}^{0\,\prime}$ of $\phi^* q_1^* \mathcal{E}_0$ over $(\widetilde{\mathcal{X}} - \widetilde{\mathcal{D}})(\lambda_0, \epsilon_0)$ is obtained from $\widetilde{\boldsymbol{w}}^0 = \phi^* q_1^* \boldsymbol{w}^0$, as follows:

$$\widetilde{w}^{0\,\prime}_{h,\boldsymbol{u},\boldsymbol{h},i} := \widetilde{w}^0_{h,\boldsymbol{u},\boldsymbol{h},i} \cdot \prod_{k=1}^{l} |z_k|^{\sum_{1\leq t\leq k} \mathfrak{p}(\lambda, u_t)} \times \prod_{k=1}^{l} \Big(-\sum_{m\geq k} \log |z_t|^2\Big)^{-q_k(\boldsymbol{h})}.$$

LEMMA 13.21. *$\widetilde{\boldsymbol{w}}^{0\,\prime}$ is adapted $(\widetilde{\mathcal{X}} - \widetilde{\mathcal{D}})(\lambda_0, \epsilon_0)$ with respect to the metric $\widetilde{h}_0$.*

Proof The C^∞-frame $\boldsymbol{w}^{0\,\prime}$ of $\mathcal{E}_0$ on $\Delta^* \times \Delta(\lambda_0, \epsilon_0)$ is obtained from $\boldsymbol{w}^0$, as follows:

$$w^{0\,\prime}_{h,\boldsymbol{u},\boldsymbol{h},i} := w^0_{h,\boldsymbol{u},\boldsymbol{h},i} \cdot |z|^{\mathfrak{p}(\lambda, u_1)} \cdot (-\log |z|)^{-q_1(\boldsymbol{h})}.$$

Then the C^∞-metric $\boldsymbol{w}^{0\,\prime} = (w^{0\,\prime}_{h,\boldsymbol{u},\boldsymbol{h},i})$ over Δ^* is adapted with respect to h_0. Since $\phi^* q_1^* \boldsymbol{w}^{0\,\prime}$ is adapted with respect to the metric $\phi^* q_1^* h_0$, it is easy to check the adaptedness of $\widetilde{\boldsymbol{w}}^{0\,\prime}$ with respect to the metric $\widetilde{h}_0$. □

We put as follows:

(13.14)

$$^{\heartsuit}\widetilde{w}^{0\,\prime}_{h,\boldsymbol{u},\boldsymbol{h},i} := \widetilde{w}^0_{h,\boldsymbol{u},\boldsymbol{h},i} \times \prod_{k=2}^{l} |z_k|^{\sum_{1\leq t\leq k} \mathfrak{p}(\lambda, u_t)} \times \prod_{k=3}^{l} \Big(-\sum_{m\geq k} \log |z_m|^2\Big)^{-q_k(\boldsymbol{h})} \times \Big(-\sum_{m\geq 2} \log |z_m|^2\Big)^{-q_1(\boldsymbol{h})-q_2(\boldsymbol{h})}.$$

Then we obtain the frame C^∞-frame $^{\heartsuit}\widetilde{\boldsymbol{w}}^{0\,\prime}$ of $\phi^* q_1^* \mathcal{E}_0$ over $(\widetilde{\mathcal{X}} - \widetilde{\mathcal{D}})(\lambda_0, \epsilon_0)$. We put as follows:

$$Y := \big\{(z_1, \dots, z_n) \in \widetilde{Z} \,\big|\, |z_1| = 1\big\}. \tag{13.15}$$

COROLLARY 13.22. *The restriction $^{\heartsuit}\widetilde{\boldsymbol{w}}^{0\,\prime}_{|Y\times\Delta(\lambda_0,\epsilon_0)}$ is adapted with respect to the metric $\widetilde{h_{0|Y\times\Delta(\lambda_0,\epsilon_0)}}$.*

Proof It immediately follows from the comparison of the restrictions of $\widetilde{\boldsymbol{w}}^{0\,\prime}$ and $^{\heartsuit}\widetilde{\boldsymbol{w}}^{0\,\prime}$ to $Y \times \Delta(\lambda_0, \epsilon_0)$. □

13.2.6. Step 4. The end of the proof. From the frame $\boldsymbol{w} = \phi^* \boldsymbol{v}$ of $\phi^* \mathcal{E}$ in the subsection 13.2.3, we obtain the C^∞-frame $^{\heartsuit}\boldsymbol{w}'$ of $\phi^* \mathcal{E}$ over $(\widetilde{\mathcal{X}} - \widetilde{\mathcal{D}})(\lambda_0, \epsilon_0)$, given as follows:

$$^{\heartsuit}w'_{h,\boldsymbol{u},\boldsymbol{h},i} := w_{h,\boldsymbol{u},\boldsymbol{h},i} \cdot \prod_{k=2}^{l} |z_k|^{\sum_{1\leq t\leq k} \mathfrak{p}(\lambda, u_t)} \times \prod_{k=3}^{l} \Big(-\sum_{m\geq k} \log |z_m|^2\Big)^{-q_k(\boldsymbol{h})} \times \Big(-\sum_{m\geq 2} \log |z_m|^2\Big)^{-q_1(\boldsymbol{h})-q_2(\boldsymbol{h})}. \tag{13.16}$$

LEMMA 13.23. *Let Y be the set given in (13.15). The restriction $^{\heartsuit}\boldsymbol{w}'_{|Y\times\Delta(\lambda_0,\epsilon_0')}$ is adapted with respect to the metric $\phi^* h_{|Y\times\Delta(\lambda_0,\epsilon_0')}$ for any $\epsilon_0' < \epsilon_0$.*

Proof We put $\widetilde{X}_a := \{(a, z_2, \dots, z_n) \in \widetilde{X}\}$ and $X_a := \{(\zeta_1, \dots, \zeta_n) \in X \,|\, \zeta_1 = a\zeta_2\}$. Due to the result in the subsection 13.1.5, the restriction $\mathcal{E}_{|\mathcal{X}_a(\lambda_0,\epsilon_0)}$ is convenient at λ_0, and the frame $\boldsymbol{v}_{|\mathcal{X}_a(\lambda_0,\epsilon_0)}$ is compatible. Then the claim follows from the assumption of the induction. (See the last part of the subsection 13.2.1.) □

COROLLARY 13.24. *The restriction $\Phi_{|Y}$ is bounded over the set Y with respect to ϕ^*h and $\widetilde{h}_0$.*

Proof Due to our construction of ${}^{\heartsuit}\boldsymbol{w}'$, ${}^{\heartsuit}\widetilde{\boldsymbol{w}}^{0\prime}$, we have $\Phi({}^{\heartsuit}\widetilde{\boldsymbol{w}}^{0\prime}) = {}^{\heartsuit}\boldsymbol{w}'$. Then the claim immediately follows from Corollary 13.22 and Lemma 13.23. □

Then, by using the method explained in the section 6.1 of our previous paper [**65**] (the method is also explained in the section 8.10 of this paper), we obtain the boundedness of Φ on the region $\widetilde{\mathcal{Z}}(\lambda_0, \epsilon_0)$. Since we have $\Phi(\widetilde{\boldsymbol{w}^{0\prime}}) = \boldsymbol{w}'$, we obtain the adaptedness of $\boldsymbol{w}'$ due to Lemma 13.21. Thus the induction can proceed, and therefore we obtain Proposition 13.15 and Proposition 13.16. □

13.3. Norm estimate for holomorphic sections

13.3.1. Statement. We put $X := \Delta^n$, $D_i := \{z_i = 0\}$ and $D = \bigcup_{i=1}^{l} D_i$. Let $(E, \overline{\partial}_E, \theta, h)$ be a tame harmonic bundle over $X - D$. Pick any point $\lambda_0 \in \boldsymbol{C}$. Let $\boldsymbol{b}$ be an element of $\boldsymbol{R}^l$ such that $b_i \notin \mathcal{KMS}(\mathcal{E}^{\lambda_0}, i)$ for each $i = 1, \ldots, l$. Pick a sufficiently small positive number ϵ_0 such that ${}_{\boldsymbol{b}}\mathcal{E}$ is locally free over $\mathcal{X}(\lambda_0, \epsilon_0)$. We may also assume that the filtrations ${}^{i}F^{(\lambda_0)}$ and the decompositions ${}^{i}\mathbb{E}^{(\lambda_0)}$ are given for $i = 1, \ldots, l$. (See the subsections 8.7–8.8.)

Let $\boldsymbol{v} = (v_i)$ be a frame of ${}_{\boldsymbol{b}}\mathcal{E}$ on $\mathcal{X}(\lambda_0, \epsilon_0)$, which is compatible with the tuple $\bigl({}^{i}\mathbb{E}^{(\lambda_0)}, {}^{i}F^{(\lambda_0)}, W(\underline{i}) \bigm| i = 1, \ldots, l\bigr)$. We remark that we can take such a frame due to Corollary 4.47. For each v_i, we have the element $\boldsymbol{u}(v_i) \in \mathcal{KMS}(\mathcal{E}^0, \underline{l})$ such that $\deg^{\mathbb{E}^{(\lambda_0)}, F^{(\lambda_0)}}(v_i) = \mathfrak{k}\bigl(\lambda_0, \boldsymbol{u}(v_i)\bigr)$. We put as follows, for each v_i:

$$b_j(v_i) = \mathfrak{p}\bigl(\lambda, q_j(\boldsymbol{u}(v_i))\bigr), \quad h_j(v_i) = \frac{1}{2}\bigl(\deg^{W(\underline{j})}(v_i) - \deg^{W(\underline{j-1})}(v_i)\bigr).$$

We remark that $b_j(v_i)$ is a function of λ. Then we put as follows:

$$v_i' := v_i \cdot \prod_{j=1}^{l}\Bigl(|z_j|^{b_j(v_i)} \cdot (-\log|z_j|)^{-h_j(v_i)}\Bigr).$$

Then we obtain the C^∞-frame $\boldsymbol{v}' = (v_i')$ on $(\mathcal{X} - \mathcal{D})(\lambda_0, \epsilon_0) = (X - D) \times \Delta(\lambda_0, \epsilon_0)$. For any positive number C, we put as follows:

$$Z(C) := \bigl\{(z_1, \ldots, z_n) \in X - D \bigm| |z_{i-1}|^C \leq |z_i|, \ (i \in \underline{l})\bigr\}. \tag{13.17}$$

The following theorem will be proved in the rest of this section.

THEOREM 13.25. *Let C be any positive number. Then the C^∞-frame $\boldsymbol{v}'$ is adapted over the region $Z(C) \times \Delta(\lambda_0, \epsilon_0')$ for any $0 < \epsilon_0' < \epsilon_0$.*

13.3.2. Some reductions. First we give some easy reductions.

LEMMA 13.26. *For the proof of Theorem 13.25, we may assume $\boldsymbol{b} = \boldsymbol{0} = (0, \ldots, 0)$.*

Proof We take the model bundle $L(-\boldsymbol{b})$ over $X - D$, and prolongment ${}_{-\boldsymbol{b}}\mathcal{L}(-\boldsymbol{b})$ of the deformed holomorphic bundle over $\mathcal{X}$. We have the canonical frame e of ${}_{-\boldsymbol{b}}\mathcal{L}(-\boldsymbol{b})$ such that $|e|_h = \prod_{j=1}^{l}|z_j|^{q_j(\boldsymbol{b})}$.

We have the naturally defined isomorphism ${}_{\boldsymbol{b}}\mathcal{E} \otimes {}_{-\boldsymbol{b}}\mathcal{L}(-\boldsymbol{b}) \simeq {}^{\diamond}\bigl(\mathcal{E} \otimes \mathcal{L}(-\boldsymbol{b})\bigr)$. Once we show the claim of Theorem 13.25 for ${}^{\diamond}\bigl(\mathcal{E} \otimes \mathcal{L}(-\boldsymbol{b})\bigr)$, then we obtain the claim for ${}_{-\boldsymbol{b}}\mathcal{E}$, too. □

Therefore we will assume that $\boldsymbol{b} = \boldsymbol{0}$.

LEMMA 13.27. *Assume that we have already shown the following claim:*
(P): The C^∞-frame $\boldsymbol{v}'$ is adapted over $Z(C) \times \Delta(\lambda_0, \epsilon_0'')$ for some positive number ϵ_0''.
Then Theorem 13.25 is obtained.

Proof Let λ_1 be any point of $\Delta(\lambda_0, \epsilon_0)$. Due to the assumption of Lemma 13.27, we may assume that we have some positive number ϵ_1' such that $\boldsymbol{v}'$ is adapted over $Z(C) \times \Delta(\lambda_1, \epsilon_1)$. We may assume that we can take a finite subset $S \subset \Delta(\lambda_0, \epsilon_0)$ such that $\Delta(\lambda_0, \epsilon_0') \subset \bigcup_{\lambda_1 \in S} \Delta(\lambda_1, \epsilon_1)$. Then the adaptedness of $\boldsymbol{v}'$ over $\Delta(\lambda_1, \epsilon_1)$ for $\lambda_1 \in S$ implies the adaptedness of $\boldsymbol{v}'$ over $\Delta(\lambda_0, \epsilon_0')$. □

13.3.3. Proof of Theorem 13.25. Let us return to the proof of Theorem 13.25. Note we may freely replace a positive number ϵ_0 with a smaller one, due to Lemma 13.27. Let η_1 be a positive number such that $\eta_1 \cdot \operatorname{rank} \mathcal{E} < 1/3$.

LEMMA 13.28. *We can take numbers $a_i \in]-1, 0[$, $c_i \in \mathbb{Z}_{>0}$, and $\eta_i \in]0, 1/3[$ for $i = 1, \ldots, l$ as follows, inductively:*

- *We take $(a_1, c_1) \in]-1, 0[\times \mathbb{Z}_{>0}$ satisfying the following:*
  - *We put $S_1 := \{a_1 + \kappa(c_1 \cdot b) \,|\, b \in \mathcal{P}ar({}^\diamond\mathcal{E}^{\lambda_0}, 1)\}$, and then S_1 is contained in $]-\eta_1, 0[$.*

  To go to the next step, we put as follows:
$$\eta_2 := \frac{1}{3} \min\{|a - b| \,|\, a, b \in S_1 \cup \{0\},\ a \neq b\}.$$
- *Suppose that we have already taken $(a_j, c_j) \in]-1, 0[\times \mathbb{Z}_{>0}$ $(j < i)$ and $\eta_j \in]0, 1/3[$ $(j \leq i)$. Then we take $(a_i, c_i) \in]-1, 0[\times \mathbb{Z}_{>0}$ as follows:*
  - *The inequality $c_i > C \cdot c_{i-1}$ holds.*
  - *We put $S_i := \{a_i + \kappa(c_i \cdot b) \,|\, b \in \mathcal{P}ar({}^\diamond\mathcal{E}^{\lambda_0}, i)\}$, and then the set S_i is contained in $]-\eta_i, 0[$.*

  To proceed the inductive construction, we put as follows:
$$\eta_{i+1} := \frac{1}{3} \min\{|a - b| \,|\, a, b \in S_i \cup \{0\},\ a \neq b\}.$$

Moreover, we may assume that c_i is sufficiently large with respect to $\mathcal{KMS}({}^\diamond\mathcal{E}^{\lambda_0}, i)$ for each i, in the sense of Definition 2.1. □

Let $\boldsymbol{a} = (a_1, \ldots, a_l)$ and $\boldsymbol{c} = (c_1, \ldots, c_l)$ be as in Lemma 13.28. We use the notation in the subsections 13.1.1–13.1.2. By replacing ϵ_0 with a smaller positive number, we may assume that ${}^\diamond(\phi_{\boldsymbol{c}}^* \mathcal{E} \otimes L(\boldsymbol{a}))$ is locally free on $\mathcal{X}(\lambda_0, \epsilon_0)$. Due to our choice of $\boldsymbol{c}$, the KMS-structure of ${}^\diamond(\phi_{\boldsymbol{c}}^* \mathcal{E} \otimes L(\boldsymbol{a}))$ is convenient. (See Lemma 13.3 and Definition 13.4).

Let e be the canonical base of the deformed holomorphic bundle $\mathcal{L}(\boldsymbol{a})$ of the model bundle $L(\boldsymbol{a})$. We have $|e| = \prod_{j=1}^{l} |z_j|^{-a_j}$. We put $e' := e \cdot \prod_{j=1}^{l} |z_j|^{a_j}$, and then we have $|e'| = 1$. On the other hand, we have the frame $\boldsymbol{w} = (w_i)$ of ${}^\diamond\mathcal{E}$ over $\mathcal{X}(\lambda_0, \epsilon_0')$ as in the subsection 13.1.2:
$$w_i := \phi_{\boldsymbol{c}}^* v_i \cdot \prod_j z_j^{\nu(c_j \cdot b_j(v_i))}.$$

Then the tensor product $\boldsymbol{w} \otimes e = (w_i \otimes e)$ is a frame of the bundle ${}^\diamond\phi_{\boldsymbol{c}}^* \mathcal{E} \otimes L(\boldsymbol{a})$, which is compatible with $({}^k\mathcal{K}, W(\underline{m}) \,|\, k \in \underline{l},\ m \in \underline{l})$.

We take the C^∞-frame $\boldsymbol{w}' = (w'_i)$ of $\phi^*_{\boldsymbol{c}}\mathcal{E}$ over $(\mathcal{X}-\mathcal{D})(\lambda_0, \epsilon'_0)$:

$$w'_i := w_i \cdot \prod_{j=1}^{l} |z_j|^{\kappa(c_j \cdot b_j(v_i))}.$$

Due to Proposition 13.15, we obtain the adaptedness of the C^∞-frame $\boldsymbol{w}' \otimes e' = (w'_i \otimes e')$ on the region $Z \times \Delta(\lambda_0, \epsilon'_0)$ for any $0 < \epsilon'_0 < \epsilon_0$, where Z is the set given in (13.7). It is easy to see that we have the relation $w'_i = v'_i \cdot \omega_i$ for some C^∞-function ω_i on Z such that $|\omega_i| = 1$. Thus the frame $\phi^*_{\boldsymbol{c}}\boldsymbol{v}'$ is adapted on the region $Z \times \Delta(\lambda_0, \epsilon'_0)$. It implies the adaptedness of the frame $\boldsymbol{v}'$ on the region $Z(c_1, \ldots, c_l) \times \Delta(\lambda_0, \epsilon'_0)$, where $Z(c_1, \ldots, c_l)$ is given as follows:

$$Z(c_1, \ldots, c_l) := \big\{(z_1, \ldots, z_n) \in X - D \,\big|\, |z_{i-1}|^{c_{i-1}} < |z_i|^{c_i},\ i \in \underline{l}\big\}.$$

Since we have $C \cdot c_i < c_{i+1}$ due to our choice of $\boldsymbol{c} = (c_1, \ldots, c_l)$, we have the implication $Z(C) \subset Z(c_1, \ldots, c_l)$. Therefore the proof of Theorem 13.25 is accomplished. □

13.3.4. The case λ is fixed. We can show the norm estimate in the case where λ is fixed. It can be proved by an argument similar to the proof of Theorem 13.25. Otherwise, we can show it by extending the holomorphic sections of $\mathcal{E}^\lambda$ to the sections of $\mathcal{E}$ on $\mathcal{X}(\lambda, \epsilon)$ for some positive number $\epsilon > 0$. Therefore we state only the result.

We put $X := \Delta^n$, $D_i := \{z_i = 0\}$ and $D = \bigcup_{i=1}^{l} D_i$. Let $(E, \overline{\partial}_E, \theta, h)$ be a tame harmonic bundle over $X - D$. Let λ be any point of $\boldsymbol{C}_\lambda$, let $\boldsymbol{b}$ be any element of $\boldsymbol{R}^l$. Let $\boldsymbol{v} = (v_i)$ be a frame of ${}_{\boldsymbol{b}}\mathcal{E}^\lambda$ on X, which is compatible with the tuple $\big({}^jF, {}^j\mathbb{E}, W(\underline{j}) \,\big|\, j \in \underline{l}\big)$. For each v_i, we put as follows:

$$b_j(v_i) = {}^j\deg^F(v_i), \qquad h_j(v_i) = \frac{1}{2}\big(\deg^{W(\underline{j})}(v_i) - \deg^{W(\underline{j-1})}(v_i)\big).$$

Then we put as follows:

$$v'_i := v_i \cdot \prod_{j=1}^{l} \Big(|z_j|^{b_j(v_i)} \cdot (-\log|z_j|)^{-h_j(v_i)}\Big).$$

Then we obtain the C^∞-frame $\boldsymbol{v}' = (v'_i)$ on $X - D$. For any positive number C, $Z(C)$ denote the set given in (13.17). The following theorem is the norm estimate of holomorphic sections when λ is fixed.

THEOREM 13.29. *Let C be any positive number. Then the C^∞-frame $\boldsymbol{v}'$ is adapted over the region $Z(C)$.* □

13.4. Norm estimate for flat sections for fixed λ

13.4.1. Preliminary for the statement. We put $X = \Delta^n$, $D_i := \{z_i = 0\}$ and $D = \bigcup_{i=1}^{n} D_i$. Let $\mathbb{H}$ denote the upper half plane. We have the universal covering $\pi : \mathbb{H}^n \longrightarrow X - D$, given by $\zeta_i \longmapsto \exp\big(\sqrt{-1}\zeta_i\big)$.

Let $(E, \overline{\partial}_E, \theta, h)$ be a tame harmonic bundle over $X - D$. Let λ be a point of $\boldsymbol{C}^*_\lambda$. Let $H(\mathcal{E}^\lambda)$ denote the space of the flat sections of $\pi^{-1}\mathcal{E}^\lambda$ with the flat connection $\mathbb{D}^{\lambda,f}$.

Let γ_j denote the loop around D_j. Recall that we have the generalized eigen decomposition ${}^j\mathbb{E}$ of the monodromy endomorphism with respect to γ_j, and the filtration ${}^j\mathcal{F}$ given in the section 9.1. It is proved that the tuple of the filtrations and the decompositions $\big({}^j\mathcal{F}, {}^j\mathbb{E} \,\big|\, j \in \underline{n}\big)$ is compatible (Theorem 9.15). We have the associated graded vector space:

$$ {}^{\underline{n}}\mathrm{Gr}^{\mathcal{F},\mathbb{E}} H(\mathcal{E}^\lambda) = \bigoplus_{\boldsymbol{u}\in\overline{\mathcal{KMS}}^f(\mathcal{E}^\lambda,\underline{n})} {}^{\underline{n}}\mathrm{Gr}^{\mathcal{F},\mathbb{E}}_{\boldsymbol{u}} H(\mathcal{E}^\lambda). $$

Let M_j^u denote the unipotent part of the monodromy endomorphism of $\mathbb{D}^{\lambda,f}$ with respect to γ_j, and we take the logarithm:

$$ N_j := \frac{-1}{2\pi\sqrt{-1}} \log M_j^u. $$

It is the endomorphism of $H(\mathcal{E}^\lambda)$, preserving the decompositions ${}^i\mathbb{E}$ and the filtrations ${}^i\mathcal{F}$ (Lemma 9.2). Therefore, it induces the endomorphism of $\mathrm{Gr}^{\mathcal{F},\mathbb{E}} H(\mathcal{E}^\lambda)$, which is also denoted by N_j.

We remark the following corollary of the limiting mixed twistor theorem (Theorem 12.22) and Lemma 3.116.

COROLLARY 13.30. *The tuple of the nilpotent maps* $\big(N_1, \dots, N_n\big)$ *on the associated graded space* $\mathrm{Gr}^{\mathcal{F},\mathbb{E}} H(\mathcal{E}^\lambda)$ *is strongly sequentially compatible.* □

We put $N(\underline{k}) := \sum_{j\leq k} N_j$, and let $W(\underline{k})$ denote the weight filtration of $N(\underline{k})$ on $\mathrm{Gr}^{\mathcal{F},\mathbb{E}} H(\mathcal{E}^\lambda)$. We remark the compatibility of the filtrations $(W(\underline{1}), \dots, W(\underline{n}))$ on $\mathrm{Gr}^{\mathcal{F},\mathbb{E}} H(\mathcal{E}^\lambda)$.

13.4.2. The statement of the norm estimate for flat sections. Let $\boldsymbol{s} = (s_i)$ denote the frame of $H(\mathcal{E}^\lambda)$, which is compatible with $\big({}^i\mathcal{F}, W(\underline{i}) \,\big|\, i \in \underline{n}\big)$ in the sense the following sense:

- $\boldsymbol{s}$ is compatible with $\big({}^i\mathcal{F} \,\big|\, i \in \underline{n}\big)$.
- The induced frame of $\mathrm{Gr}^{\mathcal{F}} H(\mathcal{E}^\lambda)$ is compatible with the tuple of the filtrations $\big(W(\underline{i}) \,\big|\, i \in \underline{n}\big)$.

We put as follows:

$$ b_k(s_j) := {}^k\deg^{\mathcal{F}}(s_j), \qquad h_k(s_j) := \frac{1}{2}\big(\deg^{W(\underline{k})}(s_j) - \deg^{W(\underline{k-1})}(s_j)\big). $$

Here we put $\deg^{W(\underline{0})}(s_j) = 0$ for convenience sake. From the flat section s_j, we obtain the following C^∞-section on $\mathbb{H}^n$:

$$ s_j' := s_j \cdot \prod_{k=1}^{n} |z_k|^{b_k(s_j)} \cdot \big(-\log|z_k|\big)^{-h_k(s_j)}. \tag{13.18} $$

Then we obtain the C^∞-frame $\boldsymbol{s}' = (s_j')$ of $\mathcal{E}^\lambda$ over $\mathbb{H}^n$.

Let C_i $(i = 1, 2, 3)$ be positive numbers. We put as follows:

(13.19)

$$ \tilde{Z}(C_1, C_2, C_3) := \Big\{(\zeta_1, \dots, \zeta_n) \in \mathbb{H}^n \,\Big|\, |x_i| \leq C_1,\ y_{i+1} \leq C_2 \cdot y_i,\ (i \in \underline{n}),\ y_n \geq C_3\Big\}. $$

We prove the following theorem in the rest of this section.

THEOREM 13.31. *The frame* $\boldsymbol{s}'$ *is adapted on the region* $\tilde{Z}(C_1, C_2, C_3)$.

13.4.3. Preliminary for the proof. We begin with an elementary remark.

LEMMA 13.32. *Let (a_k, n_k) $(k = 1, \dots, l)$ be elements of $\boldsymbol{R} \times \mathbb{Z}$. We have the following equality:*

$$\prod_{k=1}^{l} y_k^{-a_k+a_{k-1}+n_k} = \prod_{k=1}^{l-1} \left(\frac{y_k}{y_{k+1}}\right)^{-a_k+\sum_{i\leq k} n_i} \times y_l^{-a_l+\sum_{i\leq l} n_i}. \tag{13.20}$$

Proof We use an induction on l. We assume that the following equality holds:

$$\prod_{k=1}^{l-1} y_k^{-a_k+a_{k-1}+n_k} = \prod_{k=1}^{l-2} \left(\frac{y_k}{y_{k+1}}\right)^{-a_k+\sum_{i\leq k} n_i} \times y_{l-1}^{-a_{l-1}+\sum_{i\leq l-1} n_i}. \tag{13.21}$$

By a direct calculation, we have the following:

$$y_{l-1}^{-a_{l-1}+\sum_{i\leq l-1} n_i} \times y_l^{-a_l+a_{l-1}+n_l} = \left(\frac{y_{l-1}}{y_l}\right)^{-a_{l-1}+\sum_{i\leq l-1} n_i} \times y_l^{-a_l+\sum_{i\leq l} n_i}. \tag{13.22}$$

From the equalities (13.21) and (13.22), we obtain the equality (13.20). Thus the induction can proceed. □

Let (a_k, n_k) $(k = 1, \dots, n)$ be elements of $\boldsymbol{R} \times \mathbb{Z}$. Let us consider the following function:

$$F_n := \prod_{k=1}^{n} y_k^{-a_k+a_{k-1}} \cdot \left|x_k + \sqrt{-1}y_k\right|^{n_k}.$$

LEMMA 13.33. *Let C_b $(b = 1, 2, 3)$ be any positive numbers. Assume $a_k \geq \sum_{i\leq k} n_i$. The function F_l is bounded on the region $\tilde{Z}(C_1, C_2, C_3)$.*

Proof We have only to show the boundedness of $\prod_{k=1}^{l} y_k^{-a_k+a_{k-1}+n_k}$ over the region $\tilde{Z}(C_1, C_2, C_3)$. Then Lemma 13.33 follows from Lemma 13.32. □

13.4.4. The independence of a choice of a compatible frame. Let $\widetilde{\boldsymbol{s}}$ be other frame of $H(\mathcal{E}^\lambda)$ compatible with the tuple $\left({}^i\mathcal{F}, W(\underline{i}) \,\middle|\, i \in \underline{n}\right)$. As in (13.18), we obtain the frame $\widetilde{\boldsymbol{s}}'$ of $\mathcal{E}^\lambda$ over $\mathbb{H}^n$.

LEMMA 13.34. *If $\boldsymbol{s}'$ is adapted on $\widetilde{Z}(C_1, C_2, C_3)$, then $\widetilde{\boldsymbol{s}}'$ is also adapted on $\widetilde{Z}(C_1, C_2, C_3)$.*

Proof We have the relation $s_j = \sum a_{i,j} \cdot \widetilde{s}_i$, where $a_{i,j} \in \boldsymbol{C}$. Then the following relation is obtained:

$$s'_j = \sum a_{i,j} \cdot \prod_{k=1}^{n} |z_k|^{b_k(s_j)-b_k(\widetilde{s}_i)} \cdot \left(-\log|z_k|\right)^{-h_k(s_j)+h_k(\widetilde{s}_i)} \cdot \widetilde{s}'_i.$$

We have only to show that the coefficients are bounded over $\widetilde{Z}(C_1, C_2, C_3)$. We remark $a_{i,j} = 0$ unless the following holds:

- $b_k(s_j) \geq b_k(\widetilde{s}_i)$ for any k.
- If $b_k(s_j) = b_k(\widetilde{s}_i)$ for some k, we have $\deg^{W(\underline{k})}(s_j) \geq \deg^{W(\underline{k})}(\widetilde{s}_i)$.

Let k_0 be the integer such that $b_k(s_j) - b_k(\widetilde{s}_i) = 0$ for any $k < k_0$ and $b_{k_0}(s_j) - b_{k_0}(\widetilde{s}_i) > 0$.

On $\widetilde{Z}(C_1, C_2, C_3)$, there are some positive constants C and M such that the following holds:

$$\prod_{k \geq k_0} |z_k|^{b_k(s_j) - b_k(\widetilde{s}_i)} \cdot \bigl(-\log|z_k|\bigr)^{-h_k(s_j) + h_k(\widetilde{s}_i)} \leq C \cdot |z_{k_0}|^{b_k(s_j) - b_k(\widetilde{s}_i)} \cdot \bigl(-\log|z_{k_0}|\bigr)^M.$$

On the other hand, we have the following formula, due to Lemma 13.32:

$$\prod_{k=1}^{k_0 - 1} \bigl(-\log|z_k|\bigr)^{-h_k(s_i) + h_k(s_j)} = \prod_{k=1}^{k_0 - 2} \left(\frac{y_k}{y_{k-1}}\right)^{-a_k} \cdot y_{k_0 - 1}^{-a_{k_0 - 1}}. \tag{13.23}$$

Here we put $a_k := \deg^{W(\underline{k})}(\widetilde{s}_i) - \deg^{W(\underline{k})}(s_j) \geq 0$. It is easy to see that (13.23) is bounded on $\widetilde{Z}(C_1, C_2, C_3)$. Therefore we obtain the desired boundedness. □

Due to Lemma 13.34, we may and will assume that the frame $\boldsymbol{s}$ is compatible with the decompositions ${}^i\mathbb{E}$ $(i = 1, \ldots, n)$.

13.4.5. Taking the holomorphic frame. The tuple $(b_1(s_i), \ldots, b_n(s_i)) \in \boldsymbol{R}^n$ is denoted by $\boldsymbol{b}(s_i)$. We put $v_i := F\bigl(s_i, \mathfrak{p}^f(\lambda, \boldsymbol{b}(s_i))\bigr)$. (See the formula (9.1) and Corollary 9.4.) Then $\boldsymbol{v} = (v_i)$ is a normalizing frame of ${}^\diamond\mathcal{E}^\lambda$ over X, which is compatible with the tuple $\bigl({}^i\mathbb{E}, {}^jF, W(\underline{m}) \bigm| i \in \underline{n},\ j \in \underline{n},\ m \in \underline{n}\bigr)$. We have the pull back $\pi^{-1}\boldsymbol{v}$ via the projection map $\pi : \mathbb{H}^n \longrightarrow X - D$, which is also denoted by $\boldsymbol{v}$.

We have the elements $\boldsymbol{u}(v_i) \in \mathcal{KMS}(\mathcal{E}^0, \underline{n})$ such that $\deg^{\mathbb{E},F}(v_i) = \mathfrak{k}(\lambda, \boldsymbol{u}(v_i))$. The j-th component of $\boldsymbol{u}(v_i)$ is denoted by $u_j(v_i)$. We put as follows:

$$b_k(v_i) := {}^k\deg^F(v_i) = \mathfrak{p}(\lambda, u_k(v_i)), \qquad h_k(v_i) := \frac{1}{2}\bigl(\deg^{W(\underline{k})}(v_i) - \deg^{W(\underline{k-1})}(v_i)\bigr).$$

Then we obtain the C^∞-sections of $\mathcal{E}^\lambda$ over $X - D$:

$$v_j' := v_j \cdot \prod_{k=1}^{n} |z_k|^{\mathfrak{p}(\lambda, u_k(v_j))} \cdot \bigl(-\log|z_k|\bigr)^{-h_k(v_j)}.$$

We have already known that $\boldsymbol{v}' = (v_j')$ is adapted over $Z(C)$, where $Z(C)$ is given in (13.17). If we take positive constant C appropriately, $\pi\bigl(\widetilde{Z}(C_1, C_2, C_3)\bigr)$ is contained in $Z(C)$ via the projection map $\pi : \mathbb{H}^n \longrightarrow X - D$. Hence the pull back $\pi^{-1}\boldsymbol{v}'$ is adapted on $\widetilde{Z}(C_1, C_2, C_3)$. In the following, $\pi^{-1}\boldsymbol{v}'$ is also denoted by $\boldsymbol{v}'$.

13.4.6. The comparison of the frames. The matrix valued function $B = (B_{i,j})$ is determined by the relation $v_i' = \sum B_{j,i} \cdot s_j'$. Since we have already known that $\boldsymbol{v}'$ is adapted over $\widetilde{Z}(C_1, C_2, C_3)$, we have only to show that B and B^{-1} are bounded over the region. For that purpose, we need some preparation.

We put as follows:

$$\alpha_k(s_i) := \frac{\mathfrak{e}(\lambda, u_k(v_i))}{\lambda}.$$

LEMMA 13.35. *We have the following relation.*

$$b_k(s_i) = b_k(v_i) + \mathrm{Re}\bigl(\alpha_k(s_i)\bigr).$$

We also have $h_k(v_i) = h_k(s_i)$.

Proof The first claim follows from $b_k(s_i) = \mathfrak{p}^f(\lambda, u_k(v_i))$, $b_k(v_i) = \mathfrak{p}(\lambda, u_k(v_i))$ and a general formula (2.2). The second claim follows from $\deg^{W(\underline{k})}(s_i) = \deg^{W(\underline{k})}(v_i)$. □

COROLLARY 13.36. *We put* $\omega_{i,k} := z^{\alpha_k(s_i)} \cdot |z|^{b_k(v_i)-b_k(s_i)}$. *Then* $|\omega_{i,k}| = 1$. □

For any $\boldsymbol{n} \in \mathbb{Z}^l_{\geq 0}$, we put as follows:

$$N^{\boldsymbol{n}} := \prod_{k=1}^{l} N_k^{n_k}.$$

The matrix $b(\boldsymbol{n}) := \big(b(\boldsymbol{n})_{j\,i}\big)$ is determined by the relation $N^{\boldsymbol{n}}\boldsymbol{s} = \boldsymbol{s} \cdot b(\boldsymbol{n})$, i.e., $N^{\boldsymbol{n}} s_i = \sum b(\boldsymbol{n})_{j\,i} \cdot s_j$.

LEMMA 13.37. *Assume that* $b(\boldsymbol{n})_{j\,i} \neq 0$. *Then we have the following:*

- $\deg^{\mathbb{E}}(s_i) = \deg^{\mathbb{E}}(s_j)$.
- ${}^k\deg^{\mathcal{F}}(s_i) \geq {}^k\deg^{\mathcal{F}}(s_j)$ *for any* $k = 1, \ldots, n$.
- *Let* l *be any integer such that* $1 \leq l \leq n$. *In the case* ${}^k\deg^{\mathbb{E},\mathcal{F}}(s_i) = {}^k\deg^{\mathbb{E}\mathcal{F}}(s_j)$ *for* $k \leq l$, *we also have the following, for any* $k \leq l$:

$$\deg^{W(\underline{k})}(s_i) \geq \deg^{W(\underline{k})}(s_j) + 2\sum_{t \leq k} n_t.$$

Proof It immediately follows from our choice of $\boldsymbol{s}$. □

LEMMA 13.38. *Let* $B = (B_{j,i})$ *denote the matrix valued function determined by* $v'_i = \sum B_{j\,i} \cdot s'_j$. *Then we have the following formula:*

(13.24)

$$B_{j\,i} := \sum_{\boldsymbol{n}} \frac{b(\boldsymbol{n})_{j\,i}}{\boldsymbol{n}!} \prod_k \omega_{i,k} \cdot z_k^{b_k(s_i)-b_k(s_j)} \times \big(-\log|z_k|\big)^{-h_k(s_i)+h_k(s_j)} \times \big(\log z_k\big)^{n_k}.$$

Proof By definition of $F\big(s_i, {}^{\underline{n}}\deg^{\mathcal{F}}(s_i)\big)$, we have the following formula:

$$v_i = \prod_{k=1}^{n} z_k^{\alpha_k(s_i)} \cdot \sum_j \sum_{\boldsymbol{n}} \frac{b(\boldsymbol{n})_{j\,i}}{\boldsymbol{n}!} \cdot \prod_{k=1}^{n} \big(\log z_k\big)^{n_k} \cdot s_j. \tag{13.25}$$

Hence we obtain the formula. □

LEMMA 13.39.

- *We have* $|B_{i\,i}| = 1$.
- *Assume* $B_{j\,i} \neq 0$ *and* $i \neq j$, *then the following holds:*
  1. $\deg^{\mathbb{E}}(s_i) = \deg^{\mathbb{E}}(s_j)$.
  2. ${}^k\deg^{\mathcal{F}}(s_i) \geq {}^k\deg^{\mathcal{F}}(s_j)$ *for any* $k = 1, \ldots, n$.
  3. *Let* l *be any integer such that* $1 \leq l \leq n$. *In the case* ${}^k\deg^{\mathbb{E},\mathcal{F}}(s_i) = {}^k\deg^{\mathbb{E}\mathcal{F}}(s_j)$ *for* $k \leq l$, *we also have the following, for any* $k \leq l$:

$$\deg^{W(\underline{k})}(s_i) > \deg^{W(\underline{k})}(s_j).$$

Namely the matrix B *is triangular, and the absolute values of the diagonal components are* 1.

Proof It immediately follows from Lemma 13.38. □

As is remarked in the first part of this subsection, the proof of Theorem 13.31 is reduced to the following lemma.

LEMMA 13.40. *The matrix valued functions B and B^{-1} are bounded over $\tilde{Z}(C_1, C_2, C_3)$.*

Proof Assume $b(\boldsymbol{n})_{j\,i} \neq 0$. Then we have $b_k(s_i) \geq b_k(s_j)$ for any k. Let h be the number such that $b_k(s_i) = b_k(s_j)$ for any $k < h$ and that $b_h(s_i) > b_h(s_j)$.

LEMMA 13.41. *Let h be as above. On the region $\tilde{Z}(C_1, C_2, C_3)$, we have the boundedness of the following:*

$$\prod_{k \geq h} z_k^{b_k(s_i)-b_k(s_j)} \times \big(-\log|z_k|\big)^{-h_k(s_i)+h_k(s_j)} \times (\log z_k)^{n_k}.$$

Proof We have only to compare the order of the functions $|z_h|^{b_h(s_i)-b_h(s_j)}$ and $\prod_{k \geq h}(-\log|z_k|)^M$. Note there exists a positive constant we have $|z_h|^C \leq |z_k|$ for any $k \geq h$ over $\widetilde{Z}(C_1, C_2, C_3)$. □

LEMMA 13.42. *Let h be as above. We have the boundedness of the following function over $\tilde{Z}(C_1, C_2, C_3)$:*

$$\prod_{k=1}^{h-1}(-\log|z_k|)^{-h_k(s_i)+h_k(s_j)} \times (\log z_k)^{n_k}.$$

Proof We put $a_k := 2^{-1}\big(\deg^{W(\underline{k})}(s_i) - \deg^{W(\underline{k})}(s_j)\big)$. Then we have $-h_k(s_i) + h_k(s_j) = -a_k + a_{k-1}$ and $-a_k + \sum_{i \leq k} n_i \leq 0$ for any $k \leq h$. Hence we obtain the desired boundedness from Lemma 13.33. Therefore we obtain Lemma 13.40 and thus Theorem 13.31. □

13.5. Corollary

As an explanation of the results, we give the estimate of the norms of primitive sections. We put $X = \Delta^n$, $D_i = \{z_i = 0\}$ and $D = \bigcup_{i=1}^n D_i$. The intersection $\bigcap_{i \in I} D_i$ is denoted by D_I. Recall that we put $\underline{m} = \{1, \ldots, m\}$. Let $(E, \overline{\partial}_E, \theta, h)$ be a tame harmonic bundle on $X - D$. Let λ be any complex number.

We have the prolongment $^\diamond\mathcal{E}^\lambda$. On each $D_{\underline{m}}$, we have the parabolic filtrations ^iF $(i \in \underline{m})$. The residues $\mathrm{Res}_i(\mathbb{D}^\lambda)$ induce the endomorphisms of $^{\underline{m}}\mathrm{Gr}^F(^\diamond\mathcal{E}^\lambda)$. The nilpotent parts are denoted by $\mathcal{N}_i$. We put $\mathcal{N}(\underline{j}) := \sum_{i=1}^j \mathcal{N}_i$ for $j \leq m$. They induce the weight filtrations $W(\underline{j})$. A holomorphic section v of $^\diamond\mathcal{E}^\lambda$ is called primitive, if the following holds for any $m \leq n$:

- There are $\boldsymbol{a} \in \boldsymbol{R}^n$ and $\boldsymbol{k} \in \mathbb{Z}^n$.
- $v_{|D_m} \in {}^mF_{a_m}$ for each m.
- $v_{|D_{\underline{m}}}$ is contained in $^{\underline{m}}F_{\rho_m(\boldsymbol{a})}$, where $\rho_m : \boldsymbol{R}^n \longrightarrow \boldsymbol{R}^m$ denotes the natural projection.
- The induced section $[v_{|D_{\underline{m}}}]$ of $^{\underline{m}}\mathrm{Gr}^F_{\rho_m(\boldsymbol{a})}$ is contained in $\bigcap_{i=1}^m W(\underline{j})_{k_j}$.
- The induced section $[[v_{|D_{\underline{m}}}]]$ of $\mathrm{Gr}^{W(\underline{1})}_{k_1}\mathrm{Gr}^{(W(\underline{2}))}_{k_2}\cdots\mathrm{Gr}^{W(\underline{m})}_{k_m}\,{}^{\underline{m}}\mathrm{Gr}^F_{\rho_m(\boldsymbol{a})}$ is not 0.

COROLLARY 13.43. *Let v be a primitive section, and let $\boldsymbol{a}$ and $\boldsymbol{k}$ be as above. Then there exist positive numbers A_1 and A_2 such that the following holds on $Z(C)$:*

$$A_1 \leq |v|_h^2 \cdot \prod_{j=1}^n |z_j|^{2a_j} \cdot \big(-\log|z_j|\big)^{-(k_j - k_{j-1})} \leq A_2.$$

Here we put $k_0 = 0$ for convenience sake, and $Z(C)$ is given in (13.17).

Proof It follows from Theorem 13.29 □

A multi-valued flat section s of $(\mathcal{E}^{\lambda}, \mathbb{D}^{\lambda,f})$ is called primitive, if the following holds:

- s is contained in ${}^{\underline{n}}\mathcal{F}_{\boldsymbol{a}} H(\mathcal{E}^{\lambda})$ and the induced element $[s] \in {}^{\underline{n}}\mathrm{Gr}^{\mathcal{F}}_{\boldsymbol{a}} H(\mathcal{E}^{\lambda})$ is not 0 for some $\boldsymbol{a} \in \boldsymbol{R}^n$.
- $[s]$ is contained in $\bigcap_{i=1}^{n} W(\underline{i})_{k_i}$ for some $\boldsymbol{k} = (k_i) \in \mathbb{Z}^n$, and the induced element $[[s]]$ is not 0 in $\mathrm{Gr}^{W(\underline{1})}_{k_1} \cdots \mathrm{Gr}^{W(\underline{n})}_{k_n} {}^{\underline{n}}\mathrm{Gr}^{\mathcal{F}}_{\boldsymbol{a}} H(\mathcal{E}^{\lambda})$.

COROLLARY 13.44. *Let $s \in H(\mathcal{E}^{\lambda})$ be primitive, and let $\boldsymbol{a}$ and $\boldsymbol{k}$ be as above. Then there exist positive constants B_1 and B_2 such that the following holds on $\tilde{Z}(C_1, C_2, C_3)$:*

$$B_1 \leq |s|_h^2 \cdot \prod_{i=1}^{n} |z_i|^{2a_i} \cdot (-\log|z_i|)^{-(k_i - k_{i-1})} \leq B_2.$$

Here we put $k_0 = 0$ for convenience sake, and $\tilde{Z}(C_1, C_2, C_3)$ is given in (13.19).

Proof It follows from Theorem 13.31. □

Bibliography

[1] L. V. Ahlfors, *An extension of Schwarz's lemma*, Trans. Amer. Math. Soc. **43** (1938), 359–364.

[2] A. Andreotti and E. Vesentini, *Carlman estimates for the Laplace-Beltrami equation on complex manifolds*, Inst. Hautes Etudes Sci. Publ. Math. **25**, 313–362 (1965).

[3] T. Aubin, *Nonlinear analysis on manifolds. Monge-Ampére equations*, Springer-Verlag, New York, (1982).

[4] D. Barlet and H. M. Maire, *Asymptotic expansion of complex integrals via Mellin transform*, J. Funct. Anal. **83**, (1989), 233–257.

[5] A. Beilinson, J. Bernstein, P. Deligne, *Faisceaux pervers*, Analysis and topology on singular spaces, I (Luminy, 1981), 5–171, Astèisque, **100**, (1982).

[6] O. Biquard, *Fibrés de Higgs et connexions intégrables: le cas logarithmique (diviseur lisse)*, Ann. Sci. École Norm. Sup. **30** (1997), 41–96.

[7] J.-E. Björk, *Analytic D-modules and applications*, Kluwer Academic Publisher, (1993).

[8] G. Boeckle and C. Khare, *Mod ℓ representations of arithmetic fundamental groups II (A conjecture of A. J. de Jong)*, math.NT/0312490.

[9] J. -L. Brylinski, and S. Zucker, *An Overview of Recent Advances in Hodge theory*, in *Several Complex Variables VI,* Encyclopedia of Mathematical Science, **69**, Springer-Verlag, (1990), 39–142.

[10] E. Cattani, and A. Kaplan, *A Polarized mixed Hodge structures and the local monodromy of variation of Hodge structure*, Invent. Math. **67** (1982), 101–115.

[11] E. Cattani, A. Kaplan and W. Schmid, *Degeneration of Hodge structures*, Ann. of Math. **123** (1986), 457–535.

[12] E. Cattani, A. Kaplan and W. Schmid, *L^2 and intersection cohomologies for a polarized variation of Hodge structure,* Invent. Math. **87** (1987), 217–252.

[13] K. Corlette, *Flat G-bundles with canonical metrics*, J. Differential Geom. **28** (1988), 361–382.

[14] K. Corlette, *Nonabelian Hodge theory. Differential geometry*, in geometry in mathematical physics and related topics (Los Angeles, CA, 1990), 125–144, Proc. Sympos. Pure Math., **54**, Amer. Math. Soc., Providence, RI, (1993).

[15] K. Corlette, *Rigid representations of Kählerian fundamental groups,* J. Differential Geom. **33** (1991), 239–252.

[16] K. Corlette, *Archimedean superrigidity and hyperbolic geometry*, Ann. of Math. **135** (1992), 165–182.

[17] M. Cornalba and P. Griffiths, *Analytic cycles and vector bundles on noncompact algebraic varieties*, Invent. Math. **28** (1975), 1–106.

[18] M. A. de Cataldo and L. Migliorini, *The Hodge theory of algebraic maps*, math.AG/0306030.

[19] P. Deligne, *Equation differentielles a points singularier reguliers,* Lectures Notes in Maths., vol. **163**, Springer, 1970.

[20] P. Deligne, *La conjecture de Weil II*, Publ. Math. IHES, **52**, (1980), 137–252.

[21] P. Deligne, *Un théorèm de finitude pour la monodromie*, Discrete Groups in Geometry and Analysis, Birhäuser, (1987), 1–19.

[22] P. Deligne, *Décompositions dans la catégorie dérivée*, Motives (Seattle, WA, 1991), 115–128, Proc. Sympos. Pure Math.,**55**, Amer. Math. Soc., Providence, RI, (1994).

[23] S. K. Donaldson, *Infinite determinants, stable bundles and curvature*, Duke Math. J. **54**, (1987), 231–247.

[24] S. K. Donaldson, *Twisted harmonic maps and the self-duality equations*, Proc. London Math. Soc. **55** (1987), 127–131.

[25] V. Drinfeld, *On a conjecture of Kashiwara*, math.AG/0108050.

[26] P. Eberlein, *Structure of manifolds of nonpositive curvature,* in *Global differential geometry and global analysis 1984*, 86–153, Lecture Notes in Math., **1156**, Springer, Berlin, 1985.

[27] J. Eelles and J. Sampson, *Harmonic mappings of Riemannian manifolds,* Amer. J. Math. **86** (1964), 109–160.

[28] H. Fujita and S. Kuroda, *Functional analysis I, II* (in Japanese), Iwanami Shoten, Tokyo, (1983).

[29] K. Fukaya, *The gauge theory and topology,* (in Japanese), Springer, Tokyo, (1995).

[30] D. Gaitsgory, *On de Jong's conjecture* math.AG/0402184.

[31] D. Gilbarg and N. Trudinger, *Elliptic partial differential equations of second order*, Second edition. Grundlehren Springer-Verlag, Berlin, (1983).

[32] F. Guillén and V. Navarro Aznar, *Sur le théorème local des cycles invariants*, Duke Math. J, **61**, (1990), 133–155.

[33] R. Hamilton, *Harmonic maps of manifolds with boundary*, Lecture Notes in Math. vol. **471**, Springer-Verlag, Berlin and New York. (1975)

[34] R. Hartshorne, *Algebraic geometry*, Springer-Verlag, New York-Heidelberg, (1977).

[35] L. Hörmander, *An introduction to complex analysis in several variables*, North-Holland Publishing Co., Amsterdam, 1990.

[36] R. Hotta and T. Tanisaki, *D-module and Algebraic group* (in Japanese), Springer-Verlag, Tokyo, (1995).

[37] S. Ito, *Functional Analysis*, (in Japanese), Iwanami Shoten, Tokyo, 1983.

[38] J. Jost, *Riemannian geometry and geometric analysis,* Springer-Verlag, Berlin.

[39] J. Jost and S. Y. Yau, *Harmonic mappings and Kähler manifolds,* Math. Ann. **262** (1983), 145–166.

[40] J. Jost and S. Y. Yau, *The strong rigidity of locally symmetric complex manifolds of rank one and finite volume,* Math. Ann. **275**, (1986), 291–304.

[41] J. Jost and S. Y. Yau, *On the rigidity of certain discrete groups and algebraic varieties,* Math. Ann. **278** (1987), 481–496.

[42] J. Jost and S. Y. Yau, *Harmonic maps and group representations*, in *Differential geometry,* (B. Lawson and K. Tenenblat ed.) Pitman Monogr. Surveys Pure Appl. Math., **52**, Longman Sci. Tech., Harlow, (1991) 241–259.

[43] J. Jost and K. Zuo, *Harmonic maps and* $\mathrm{Sl}(r, C)$*-representations of fundamental groups of quasiprojective manifolds*, J. Algebraic Geom. **5** (1996), 77–106.

[44] J. Jost and K. Zuo, *Harmonic maps of infinite energy and rigidity results for representations of fundamental groups of quasiprojective varieties*, J. Differential Geom. **47** (1997), 469–503.

[45] F. I. Karpelevic, *The geometry of geodesics and the eigenfunctions of the Beltrami-Laplace operator on symmetric spaces,* Trans. Moscow Math. Soc. **14** (1967), 51–199.

[46] M. Kashiwara, *The Riemann-Hilbert problem for holonomic systems*, Publ. Res. Inst. Math. Sci., **10**, 563–579 (1975).

[47] M. Kashiwara, *The asymptotic behavior of a variation of polarized Hodge str.* Publ. Res. Inst. Math. Sci **21** (1985), 853–875.

[48] M. Kashiwara, *A study of variation of mixed Hodge structure.* Publ. Res. Inst. Math. Sci. **22** (1986), 991–1024.

[49] M. Kashiwara, *Poincaré lemma for a variation of polarized Hodge structure*, Lecture Notes in Math., **1246**, Springer, Berlin, (1987), 115–124.

[50] M. Kashiwara, *Semisimple holonomic D-modules*, in *Topological Field Theory, Primitive Forms and Related Topics*, Progress in Math, vol **160**, Birkhäuser, (1998), 267–271.

[51] M. Kashiwara, *D-modules and microlocal calculus,* Translations of Mathematical Monographs, **217**. American Mathematical Society, Providence, (2003).

[52] M. Kashiwara and T. Kawai, *The Poincaré lemma for variations of polarized Hodge structure.* Publ. Res. Inst. Math. Sci. **23** (1987), 345–407.

[53] M. Kashiwara and P. Schapira, *Sheaves on Manifolds*, Springer, (1990).

[54] S. Kobayashi, *Differential geometry of complex vector bundles*, Princeton University Press, Princeton, NJ; Iwanami Shoten, Tokyo, (1987).

[55] K. Kodaira, *A differential-geometric method in the theory of analytic stacks* Proc. Nat. Acad. Sci. U. S. A. **39**, (1953). 1268–1273.

[56] F. Labourie, *Existence d'applications harmoniques tordues à valeurs dans les variétés á courbure négative*, Proc. Amer. Math. Soc. **111** (1991), 877–882.

[57] J. Lohkamp, *An existence theorem for harmonic maps,* Manuscripta Math. **67**, (1990), 21–23.

[58] M. Lübke, *Stability of Einstein-Hermitian vector bundles*, Manuscripta Math. **42** (1983), 245–257.
[59] T-W Ma, *Banach-Hilbert spaces, vector measures and group representations*, World Scientific, Singapore, (2002).
[60] M. Maruyama and K. Yokogawa, *Moduli of parabolic stable sheaves*, Math. Ann. **293**, 77–99 (1992).
[61] Z. Mebkhout, *Le formalisme des six opérations de Grothendieck pour les D-modules cohérent*, Hermann, Paris, (1989).
[62] Z. Mebkhout and C. Sabbah, *D-modules et cycles évanscents,* in *Le formalisme des six opérations de Grothendieck pour les D-modules cohérents*, Hermann, Paris, (1989), 201–239.
[63] J. Milnor, *Morse theory*, Annals of Mathematics Studies, **51**, Princeton University Press, (1963).
[64] S. Mizohata, *The theory of partial differential equations* (in Japanese), Iwanami Shoten, Tokyo, (1965).
[65] T. Mochizuki, *Asymptotic behaviour of tame nilpotent harmonic bundles with trivial parabolic structure*, J. Diff. Geometry, **62**, (2002), 351–559.
[66] *Asymptotic Behaviour of tame harmonic bundles and an application to pure twistor D-modules*, math.DG/0312230.
[67] T. Mochizuki, *A characterization of semisimple local system by tame pure imaginary pluri-harmonic metric*, math.DG/0402122.
[68] T. Mochizuki, *Kobayashi-Hitchin correspondence for tame harmonic bundles*, math.DG/0411300.
[69] J. Noguchi and T. Ochiai, *Geometric function theory in several complex variables*, Translations of Mathematical Monographs, **80**, American Mathematical Society, (1990).
[70] R. Palais, *Foundations of global non-linear analysis*, Benjamin, (1968).
[71] G. de Rham, *Differentiable manifolds. Forms, currents, harmonic forms.* Springer-Verlag, Berlin, (1984).
[72] C. Sabbah, *Polarizable twistor D-modules*, math.AG/0503038.
[73] M. Saito, *Modules de Hodge polarisables*, Publ. RIMS., **24** (1988), 849–995.
[74] M. Saito, *Duality for vanishing cycle functors*, Publ. RIMS., **25**, (1989), 889–921.
[75] M. Saito, *Mixed Hodge modules*, Publ. RIMS., **26**, (1990), 221-333.
[76] T. Sakai, *Riemannian geometry*, Translations of Mathematical Monographs, **149**, American Mathematical Society, (1996).
[77] W. Schmid, *Variation of Hodge structure: the singularities of the period mapping*, Invent. Math. **22** (1973), 211–319.
[78] R. Schoen and S. T. Yau, *Harmonic maps and the topology of stable hypersurfaces and manifolds with non-negative Ricci curvature*, Comment. Math. Helv. **51** (1976), 333–341.
[79] B. Shiffman and A. J. Sommese, *Vanishing Theorems on Complex Manifolds*, Progress in Math. **56**, Birkhäuser, (1985)
[80] C. Simpson, *Constructing variations of Hodge structure using Yang-Mills theory and application to uniformization*, J. Amer. Math. Soc. **1** (1988), 867–918.
[81] C. Simpson, *Harmonic bundles on non-compact curves*, J. Amer. Math. Soc. **3** (1990), 713–770.
[82] C. Simpson, *Higgs bundles and local systmes*, Publ. IHES, **75** (1992), 5–95.
[83] C. Simpson, *Mixed twistor structures*, math.AG/9705006.
[84] C. Simpson, *The Hodge filtration on nonabelian cohomology. Algebraic geometry—Santa Cruz 1995*, 217–281, Proc. Sympos. Pure Math., 62, Part 2, Amer. Math. Soc., Providence, RI, 1997.
[85] Y. T. Siu, *The complex-analyticity of harmonic maps and the strong rigidity of compact Kähler manifolds,* Ann. of Math. **112**, (1980), 73–111.
[86] K. Uhlenbeck, *Connections with L^p bounds on curvature*, Comm. Math. Phys. **83**, (1982) 31–42.
[87] K. Uhlenbeck and S. T. Yau, *On the existence of Hermitian Yang-Mills connections in stable bundles*, Comm. Pure Appl. Math., 39-S (1986), 257–293.
[88] H. Urakawa, *Calculus of Variations and Harmonic maps*, Translations of Mathematical Monographs, **132**, (1990).
[89] K. Yokogawa, *Compactification of moduli of parabolic sheaves and moduli of parabolic Higgs sheaves*, J. Math. Kyoto Univ. **33** (1993) 451–504.

[90] S. Zucker, *Hodge theory with degenerating coefficients: L^2 cohomology in the Poincaré metric*, Ann of Math. **109** (1979), 415–476.

[91] K. Zuo, *Representations of Fundamental Groups of Algebraic Varieties*, Lecture Notes in Mathematics, **1708**, Springer (1999).

Index

$(\cdot,\cdot)_{\boldsymbol{a},N}$, $\langle\langle\cdot,\cdot\rangle\rangle_{\boldsymbol{a},N}$, 44
$-\operatorname{ord}(f) \leq \boldsymbol{b}$, 26
$-\operatorname{ord}(s) \leq b$, 168
$<$, 25
$A_c^{p,q}(E)$, 41
$A_h^{p,q}(E)$, 41
$F(s,b)$, 169
$F\big(s,\boldsymbol{b}(\boldsymbol{\omega})\big)$, 196
$F_a({}_b\mathcal{E}^\lambda)$, 157
$H(\mathcal{E}^\lambda)$, 168, 236
$H(h,\boldsymbol{v})$, 27
$M(r)$, 24
$S_{\boldsymbol{u}}^{\mathrm{can}}(E)$, 269
$S_{(p,q)}$, 71
$S_{\boldsymbol{u}}(E,P)$, 269
$X^\dagger$, 24
Y-holomorphic
  section, 62
Y-holomorphic structure, 62
$[a,b]$, $[a,b[$, $]a,b]$, $]a,b[$, 24
$\mathrm{Bifilt}(X)$, 66
$\mathbb{D}_X^{\triangle}$, 78
$\Delta(z_0,C)$, $\Delta(C)$, Δ, 24
$\Delta^*(z_0,C)$, $\Delta^*(C)$, Δ^*, 24
Δ_z, Δ_i, 24
$\overline{\Delta}(C)$, $\overline{\Delta}$, $\overline{\Delta}^*(C)$, $\overline{\Delta}^*$, 24
$\mathbb{E}(\boldsymbol{f},\boldsymbol{\omega})$, $\mathcal{S}p(\boldsymbol{f})$, 39
$\mathbb{E}(f,\alpha)$, $\mathbb{E}(\alpha)$, $\mathbb{E}(V,\alpha)$, 39
$\mathbb{E}_\eta(f,a)$, 40
$\mathbb{E}^{(\lambda_0)}$, 179
$\mathrm{Equi}(X\times\mathbb{P}^1)$, 67
$\mathrm{Gr}_u^{F,\mathbb{E}}(\mathcal{E}^\lambda)$, 158
$\mathrm{Gr}^{\mathcal{F},\mathbb{E}}(\mathcal{E}^\lambda)$, 169
$\mathcal{KMS}(\mathcal{E}^\lambda)$, 158
$\mathcal{KMS}(\mathcal{E}^\lambda,i)$, 189
$\mathcal{KMS}({}_b\mathcal{E}^\lambda)$, 159
$\mathcal{KMS}(\psi_{t,u}\mathfrak{E}[\eth_t],i)$, 419
$\mathcal{KMS}(i_{g\,\dagger}\mathfrak{E}^0)$, $\mathcal{KMS}(\mathfrak{E}[\eth_t])$, 400
$\overline{\mathcal{KMS}}(\mathcal{E}^\lambda)$, 159
$\overline{\mathcal{KMS}}(\mathcal{E}^\lambda,i)$, 189
$\overline{\mathcal{KMS}}(\mathcal{M},X_0)$, $\overline{\mathcal{KMS}}(\mathcal{M},t)$, 336
$\overline{\mathcal{KMS}}^f(\mathcal{E}^\lambda)$, $\mathcal{P}ar^f(\mathcal{E}^\lambda)$, $\mathcal{S}p^f(\mathcal{E}^\lambda)$, 169
$\Omega(\boldsymbol{v})$, 197
$\mathcal{PH}(V)$, 24
$\mathcal{P}ar(\mathcal{E}^\lambda)$, $\mathcal{S}p(\mathcal{E}^\lambda)$, 158
$\mathcal{P}ar(\mathcal{E}^\lambda,i)$, 189
$\mathcal{P}ar({}_b\mathcal{E}^\lambda)$, 159
$\Phi_{\boldsymbol{u},P,O}$, 259
$\Phi_{\boldsymbol{u}}^{\mathrm{can}}$, 259
Ψ-Pol-MTS, 77
$\Psi_b^{(\lambda_0)}$, 399
$\mathcal{S}p(\mathcal{E}^\lambda,i)$, 189
$\mathcal{S}p({}_b\mathcal{E}^\lambda)$, 159
$\mathcal{S}p(f)$, 39
$\mathbb{T}^S(k)$, 109
$U^{(\lambda_0)}(\mathfrak{E}[\eth_t])$, 399
$\alpha(b,\omega)$, 169
$\boldsymbol{C}$, 24
ϵ-orthogonal, 40
$\mathrm{Equi}(X\times\boldsymbol{C}_\lambda)$, 66
$\mathrm{Filt}(X)$, 65
$\mathfrak{C}$, 445
$\mathfrak{E}$, 377
$\iota_{(p,q)}$, 70
κ_c, ν_c, κ, ν, 25
${}^IT^{(\lambda_0)}(\boldsymbol{c},\boldsymbol{d})$, 378
${}^I\,\mathrm{Prim}(f)$, 394
${}^IV^{(\lambda_0)}(\mathfrak{E})$, 378
${}^IV^{(\lambda_0)}\big({}^\square\mathcal{E}\big)$, ${}^iV^{(\lambda_0)}\big({}^\square\mathcal{E}\big)$, 370
${}^IV_{\boldsymbol{b}}^{(\lambda_0)}(\Psi_b^{(\lambda_0)})$, 409
${}^IV_{\boldsymbol{c}}^{(\lambda_0)}\big({}^\square\mathcal{E}^{(\lambda_0)}\big)$, 370
${}^I\mathcal{V}_{S,\boldsymbol{d}}(\mathfrak{E})$, ${}^I\mathcal{V}'_{S,\boldsymbol{d}}(\mathfrak{E})$, 386
${}^I\psi_{\boldsymbol{u}}^{(\lambda_0)}\,{}^JV_{\boldsymbol{d}}^{(\lambda_0)}({}^\square\mathcal{E})$, 372
${}^I\tilde{T}^{(\lambda_0)}(\boldsymbol{b},\boldsymbol{d})$, 373
${}^I\tilde{T}^{(\lambda_0)}(\boldsymbol{u},\boldsymbol{d})$, 372
${}^JF_{\boldsymbol{\eta}}$, 114
${}^J\mathbb{E}$, 115
${}^{\underline{l}}\mathcal{G}_{\boldsymbol{u}}$, 226
${}^{\underline{l}}\mathcal{G}_{\boldsymbol{u}}(\mathcal{E})$, ${}^{\underline{l}}\mathcal{G}_{\boldsymbol{u}}^{(\lambda_0)}(\mathcal{E})$, 257
${}^{\underline{l}}\mathcal{G}_{\boldsymbol{u}}(\mathcal{E}^\lambda)$, 254
${}^{\underline{l}}\mathcal{G}_{\boldsymbol{u}}^{(\lambda_0)}$, 226
${}^{\underline{n}}\,\mathrm{Prim}(f)$, 408
${}^{\underline{n}}V_{\boldsymbol{c}}^{(\lambda_0)}U_b^{(\lambda_0)}\big(\mathfrak{E}[\eth_t]\big)$, 401
${}^{\underline{n}}\overline{\mathcal{G}}_{\boldsymbol{u}}(\mathcal{E})$, 445

${}^{i}F_{b}$, 209
${}^{i}\mathbb{E}^{(\lambda_0)}\big({}_{\boldsymbol{b}}\mathcal{E},\beta\big)$, 222
${}^{i}F^{(\lambda_0)}({}_{\boldsymbol{b}}\mathcal{E})$, 218
${}^{i}\operatorname{Gr}^{F}$, 114
${}^{i}\deg^{F}$, 28
${}^{i}\mathcal{F}$, 236
${}^{i}\mathcal{F}^{(\lambda_0)}$, 240
$\lneq$, 25
μ_c, $\mu_{\boldsymbol{c}}$, 28
$\mathfrak{m}(\lambda,\boldsymbol{u},I)$, 222
$\mathfrak{m}(\lambda,u)$, 158
$\mathfrak{m}^{f}(\lambda,u)$, 169
${}^{\square}_{\boldsymbol{c}}\mathcal{E}^{(\lambda_0)}$, 369
${}^{\square}\mathcal{E}$, 369
$\underline{n}$, 24
$\mathcal{F}^{(\lambda_0)}$, 180
$\mathcal{F}_b(H(\mathcal{E}^{\lambda}))$, 168
$\mathcal{H}_r$, 24
$\mathcal{K}(A,c,\lambda_0)$, 332
$\mathcal{K}(\mathcal{E},\lambda,\boldsymbol{b},i)$, $\mathcal{T}(\mathcal{E},\lambda,c,i)$, 222
$\mathcal{K}(\mathcal{E},\lambda,\boldsymbol{0},i)$, 196
$\mathcal{K}(\mathcal{E},\lambda,c)$, 172
$\mathcal{K}(\mathcal{E},\lambda_0,i,c)$, $\mathcal{K}(\mathcal{E},\lambda_0,I,\boldsymbol{c})$, 371
$\mathcal{K}(i_{g\dagger}\mathfrak{C},\lambda_0,b)$, $\mathcal{K}(\mathfrak{C}[\eth_t],\lambda_0,b)$, 400
$\mathcal{K}\big(\psi_{t,u}\mathfrak{C},i,\lambda_0,c\big)$, 419
$\mathcal{O}(p,q)$, 70
$\mathcal{S}_l$, 24
$\mathcal{V}_I(S)$, 102
$\mathcal{V}_I(V,\boldsymbol{N})$, 101
$\mathcal{X}(\lambda_0,\epsilon_0)$, $\mathcal{X}^{*}(\lambda_0,\epsilon_0)$, $\mathcal{X}$, $\mathcal{X}^{\sharp}$, 24
$\mathfrak{p}$, $\mathfrak{e}$, $\mathfrak{k}$, 25
$\mathfrak{p}^{f}$, $\mathfrak{e}^{f}$, $\mathfrak{k}^{f}$, 26
${}_{\boldsymbol{b}}E$, ${}^{\diamond}E$, 27
$\psi_{t,u}C$, $\widetilde{\psi}_{t,u}C$, 352
$\psi_{t,u}\mathcal{M}$, 336
$\psi^{(\lambda_0)}_{t,u_0}C$, 350
$\psi^{(\lambda_0)}_{t,u}\mathcal{M}$, 336
$\boldsymbol{R}$, 24
$\rho_{(p,q)}$, 70
$\boldsymbol{Q}$, 24
$\mathbb{Z}$, 24
σ (the involution), 63
$\widetilde{\psi}_{t,u}(\mathcal{M})$, 340
$\widetilde{\psi}^{(\lambda_0)}_{t,u}C(m,\bar{\mu})$, 352
$\widetilde{\psi}^{(\lambda_0)}_{t,u}(\mathcal{M})$, 339
φ, φ_0, 63
$\varpi(\boldsymbol{c})$, $\vartheta(\boldsymbol{c})$, 410
$\boldsymbol{\delta}_0$, $\boldsymbol{\delta}_{0,j}$, 26
$\boldsymbol{\delta}_j$, 26
$\boldsymbol{v}^{\dagger}$, 262
$\xi(H;F)$, 65
$\xi(H;F,G)$, 66
$f_0^{(n)}$, $f_\infty^{(n)}$, 70
$f_0^{(p,q)}$, $f_\infty^{(p,q)}$, $f_1^{(p,q)}$, 70
q_J, 24
$t_0^{(n)}$, $t_\infty^{(n)}$, $t_1^{(n)}$, 69

$u^{\dagger}$, 25
${}^{I}\psi^{(\lambda_0)}_{\boldsymbol{u}}\big(\widetilde{\psi}_{t,u}(\mathfrak{C}[\eth_t])\big)$, 419
${}^{\underline{l}}\psi_{\boldsymbol{u}}(\mathfrak{C})$, ${}^{\underline{l}}\widetilde{\psi}_{\boldsymbol{u}}(\mathfrak{C})$, 385
${}^{i}\operatorname{Gr}^{F^{(\lambda_0)},\mathbb{E}^{(\lambda_0)}}_{\mathfrak{k}(\lambda_0,u)}\big({}_{\boldsymbol{b}}\mathcal{E}\big)$, 223
${}^{\underline{l}}\mathcal{G}_{\boldsymbol{u}}(\mathcal{H})$, ${}^{\underline{l}}\mathcal{G}^{(\lambda_0)}_{\boldsymbol{u}}(\mathcal{H})$, 249

acceptable, 43
adapted, 27
adapted
  up to log order, 27

compatible, 114–117, 126

defined over $\boldsymbol{R}$, 69

equivariant
  frame, 29
  lift, 29
  section, 28

generic, 177, 196
generically defined tame harmonic bundle, 457
generically defined tame variation of polarized pure twistor structure, 457

Hermitian adjoint, 108

Pol-MTS, 77
polarization, 360
Polarization of twistor structure, 72
polarized mixed twistor structure, 76
primitive decomposition, 393
primitive development, 409
primitive section, 394, 408
pure imaginary pure twistor D-module, 361
pure twistor D-module, 360

quasi adapted, 27
quasi canonical prolongment, 178

sequentially compatible, 119, 120, 127
specializable, 336
Split polarized mixed twistor structure, 89
split polarized mixed twistor structure, 89
splitting, 117, 118, 121, 126–128
strictly S-decomposable, 338
strictly specializable, 337
strongly sequentially compatible, 121, 128
sufficiently large, 25

tame, 189, 190
tame pure imaginary harmonic bundle, 507
Tate object, $\mathbb{T}(n)$, 69
The torus action ρ_0, 65
twistor nilpotent orbit, 85

Editorial Information

To be published in the *Memoirs*, a paper must be correct, new, nontrivial, and significant. Further, it must be well written and of interest to a substantial number of mathematicians. Piecemeal results, such as an inconclusive step toward an unproved major theorem or a minor variation on a known result, are in general not acceptable for publication. Papers appearing in *Memoirs* are generally at least 80 and not more than 200 published pages in length. Papers less than 80 or more than 200 published pages require the approval of the Managing Editor of the Transactions/Memoirs Editorial Board.

As of September 30, 2006, the backlog for this journal was approximately 11 volumes. This estimate is the result of dividing the number of manuscripts for this journal in the Providence office that have not yet gone to the printer on the above date by the average number of monographs per volume over the previous twelve months, reduced by the number of volumes published in four months (the time necessary for preparing a volume for the printer). (There are 6 volumes per year, each containing at least 4 numbers.)

A Consent to Publish and Copyright Agreement is required before a paper will be published in the *Memoirs*. After a paper is accepted for publication, the Providence office will send a Consent to Publish and Copyright Agreement to all authors of the paper. By submitting a paper to the *Memoirs*, authors certify that the results have not been submitted to nor are they under consideration for publication by another journal, conference proceedings, or similar publication.

Information for Authors

Memoirs are printed from camera copy fully prepared by the author. This means that the finished book will look exactly like the copy submitted.

The paper must contain a *descriptive title* and an *abstract* that summarizes the article in language suitable for workers in the general field (algebra, analysis, etc.). The *descriptive title* should be short, but informative; useless or vague phrases such as "some remarks about" or "concerning" should be avoided. The *abstract* should be at least one complete sentence, and at most 300 words. Included with the footnotes to the paper should be the 2000 *Mathematics Subject Classification* representing the primary and secondary subjects of the article. The classifications are accessible from `www.ams.org/msc/`. The list of classifications is also available in print starting with the 1999 annual index of *Mathematical Reviews*. The Mathematics Subject Classification footnote may be followed by a list of *key words and phrases* describing the subject matter of the article and taken from it. Journal abbreviations used in bibliographies are listed in the latest *Mathematical Reviews* annual index. The series abbreviations are also accessible from `www.ams.org/publications/`. To help in preparing and verifying references, the AMS offers MR Lookup, a Reference Tool for Linking, at `www.ams.org/mrlookup/`. When the manuscript is submitted, authors should supply the editor with electronic addresses if available. These will be printed after the postal address at the end of the article.

Electronically prepared manuscripts. The AMS encourages electronically prepared manuscripts, with a strong preference for AMS-LaTeX. To this end, the Society has prepared AMS-LaTeX author packages for each AMS publication. Author packages include instructions for preparing electronic manuscripts, the *AMS Author Handbook*, samples, and a style file that generates the particular design specifications of that publication series. Though AMS-LaTeX is the highly preferred format of TeX, author packages are also available in AMS-TeX.

Authors may retrieve an author package from e-MATH starting from `www.ams.org/tex/` or via FTP to `ftp.ams.org` (login as `anonymous`, enter username as password, and type `cd pub/author-info`). The *AMS Author Handbook* and the *Instruction Manual* are available in PDF format following the author packages link from `www.ams.org/tex/`. The author package can also be obtained free of charge by sending email to tech-support@ams.org (Internet) or from the Publication Division, American Mathematical Society, 201 Charles St., Providence, RI 02904-2294, USA. When requesting an author package, please specify AMS-LaTeX or AMS-TeX and the publication in which your paper will appear. Please be sure to include your complete mailing address.

Sending electronic files. After acceptance, the source file(s) should be sent to the Providence office (this includes any TeX source file, any graphics files, and the DVI or PostScript file).

Before sending the source file, be sure you have proofread your paper carefully. The files you send must be the EXACT files used to generate the proof copy that was accepted for publication. For all publications, authors are required to send a printed copy of their paper, which exactly matches the copy approved for publication, along with any graphics that will appear in the paper.

TeX files may be submitted by email, FTP, or on diskette. The DVI file(s) and PostScript files should be submitted only by FTP or on diskette unless they are encoded properly to submit through email. (DVI files are binary and PostScript files tend to be very large.)

Electronically prepared manuscripts can be sent via email to pub-submit@ams.org (Internet). The subject line of the message should include the publication code to identify it as a Memoir. TeX source files, DVI files, and PostScript files can be transferred over the Internet by FTP to the Internet node e-math.ams.org (130.44.1.100).

Electronic graphics. Comprehensive instructions on preparing graphics are available at www.ams.org/jourhtml/graphics.html. A few of the major requirements are given here.

Submit files for graphics as EPS (Encapsulated PostScript) files. This includes graphics originated via a graphics application as well as scanned photographs or other computer-generated images. If this is not possible, TIFF files are acceptable as long as they can be opened in Adobe Photoshop or Illustrator. No matter what method was used to produce the graphic, it is necessary to provide a paper copy to the AMS.

Authors using graphics packages for the creation of electronic art should also avoid the use of any lines thinner than 0.5 points in width. Many graphics packages allow the user to specify a "hairline" for a very thin line. Hairlines often look acceptable when proofed on a typical laser printer. However, when produced on a high-resolution laser imagesetter, hairlines become nearly invisible and will be lost entirely in the final printing process.

Screens should be set to values between 15% and 85%. Screens which fall outside of this range are too light or too dark to print correctly. Variations of screens within a graphic should be no less than 10%.

Inquiries. Any inquiries concerning a paper that has been accepted for publication should be sent directly to the Electronic Prepress Department, American Mathematical Society, 201 Charles St., Providence, RI 02904, USA.

Editors

This journal is designed particularly for long research papers, normally at least 80 pages in length, and groups of cognate papers in pure and applied mathematics. Papers intended for publication in the *Memoirs* should be addressed to one of the following editors. In principle the Memoirs welcomes electronic submissions, and some of the editors, those whose names appear below with an asterisk (*), have indicated that they prefer them. However, editors reserve the right to request hard copies after papers have been submitted electronically. Authors are advised to make preliminary email inquiries to editors about whether they are likely to be able to handle submissions in a particular electronic form.

*__Algebra__ to ALEXANDER KLESHCHEV, Department of Mathematics, University of Oregon, Eugene, OR 97403-1222; email: `ams@noether.uoregon.edu`

Algebra and its application to MINA TEICHER, Emmy Noether Research Institute for Mathematics, Bar-Ilan University, Ramat-Gan 52900, Israel; email: `teicher@macs.biu.ac.il`

Algebraic geometry to DAN ABRAMOVICH, Department of Mathematics, Brown University, Box 1917, Providence, RI 02912; email: `amsedit@math.brown.edu`

*__Algebraic number theory__ to V. KUMAR MURTY, Department of Mathematics, University of Toronto, 100 St. George Street, Toronto, ON M5S 1A1, Canada; email: `murty@math.toronto.edu`

*__Algebraic topology__ to ALEJANDRO ADEM, Department of Mathematics, University of British Columbia, Room 121, 1984 Mathematics Road, Vancouver, British Columbia, Canada V6T 1Z2; email: `adem@math.ubc.ca`

*__Combinatorics__ to JOHN R. STEMBRIDGE, Department of Mathematics, University of Michigan, Ann Arbor, Michigan 48109-1109; email: `FRS@umich.edu`

Complex analysis and harmonic analysis to ALEXANDER NAGEL, Department of Mathematics, University of Wisconsin, 480 Lincoln Drive, Madison, WI 53706-1313; email: `nagel@math.wisc.edu`

*__Differential geometry and global analysis__ to LISA C. JEFFREY, Department of Mathematics, University of Toronto, 100 St. George St., Toronto, ON Canada M5S 3G3; email: `jeffrey@math.toronto.edu`

Dynamical systems and ergodic theory to AMIE WILKINSON, Department of Mathematics, Northwestern University, 2033 Sheridan Road, Evanston, IL 60208-2730; email: `transactions@math.northwestern.edu`

*__Functional analysis and operator algebras__ to MARIUS DADARLAT, Department of Mathematics, Purdue University, 150 N. University St., West Lafayette, IN 47907-2067; email: `mdd@math.purdue.edu`

*__Geometric analysis__ to TOBIAS COLDING, Courant Institute, New York University, 251 Mercer St., New York, NY 10012; email: `traneditor@cims.nyu.edu`

*__Geometric analysis__ to MLADEN BESTVINA, Department of Mathematics, University of Utah, 155 South 1400 East, JWB 233, Salt Lake City, Utah 84112-0090; email: `bestvina@math.utah.edu`

Harmonic analysis, representation theory, and Lie theory to ROBERT J. STANTON, Department of Mathematics, The Ohio State University, 231 West 18th Avenue, Columbus, OH 43210-1174; email: `stanton@math.ohio-state.edu`

*__Logic__ to STEFFEN LEMPP, Department of Mathematics, University of Wisconsin, 480 Lincoln Drive, Madison, Wisconsin 53706-1388; email: `lempp@math.wisc.edu`

*__Ordinary differential equations, and applied mathematics__ to PETER W. BATES, Department of Mathematics, Michigan State University, East Lansing, MI 48824-1027; email: `bates@math.msu.edu`

*__Partial differential equations__ to GUSTAVO PONCE, Department of Mathematics, South Hall, Room 6607, University of California, Santa Barbara, CA 93106; email: `ponce@math.ucsb.edu`

*__Probability and statistics__ to KRZYSZTOF BURDZY, Department of Mathematics, University of Washington, Box 354350, Seattle, Washington 98195-4350; email: `burdzy@math.washington.edu`

*__Real analysis and partial differential equations__ to DANIEL TATARU, Department of Mathematics, University of California, Berkeley, Berkeley, CA 94720; email: `tataru@math.berkeley.edu`

All other communications to the editors should be addressed to the Managing Editor, ROBERT GURALNICK, Department of Mathematics, University of Southern California, Los Angeles, CA 90089-1113; email: `guralnic@math.usc.edu`.

Titles in This Series

870 **Takuro Mochizuki,** Asymptotic behaviour of tame harmonic bundles and an application to pure twistor D-modules, Part 2, 2007

869 **Takuro Mochizuki,** Asymptotic behaviour of tame harmonic bundles and an application to pure twistor D-modules, Part 1, 2007

868 **Gelu Popescu,** Entropy and multivariable interpolation, 2006

867 **Vilmos Totik,** Metric properties of harmonic measures, 2006

866 **William Craig,** Semigroups underlying first-order logic, 2006

865 **Nathanial P. Brown,** Invariant means and finite representation theory of $C*$-algebras, 2006

864 **John M. Lee,** Fredholm operators and Einstein metrics on conformally compact manifolds, 2006

863 **M. Lübke and A. Teleman,** The Universal Kobayashi-Hitchin correspondence on Hermitian manifolds, 2006

862 **Alberto Canonaco,** The Beilinson complex and canonical rings of irregular surfaces, 2006

861 **Leon A. Takhtajan and Lee-Peng Teo,** Weil-Petersson metric on the universal Teichmüller space, 2006

860 **Thomas M. Fiore,** Pseudo limits, biadjoints and pseudo algebras: Categorical foundations of conformal field theory, 2006

859 **N. Arcozzi, R. Rochberg, and E. Sawyer,** Carleson measures and interpolating sequences for Besov spaces on complex balls, 2006

858 **Enrico Valdinoci, Berardino Sciunzi, and Vasile Ovidiu Savin,** Flat level set regularity of p-Laplace phase transitions, 2006

857 **Donatella Danielli, Nocola Garofalo, and Duy-Minh Nhieu,** Non-doubling Ahlfors measures, perimeter measures, and the characterization of the trace spaces of Sobolev functions in Carnot-Carathéodory spaces, 2006

856 **Vladimir Bolotnikov and Harry Dym,** On boundary interpolation for matrix valued Schur functions, 2006

855 **Yevgenia Kashina, Yorck Sommerhäuser, and Yongchang Zhu,** On higher Frobenius-Schur indicators, 2006

854 **Noam Greenberg,** The role of true finiteness in the admissible recursively enumerable degrees, 2006

853 **Joachim Krieger,** Stability of spherically symmetric wave maps, 2006

852 **Viorel Barbu, Irena Lasiecka, and Roberto Triggiani,** Tangential boundary stabilization of Navier-Stokes equations, 2006

851 **Jie Wu,** On maps from loop suspensions to loop spaces and the shuffle relations on the Cohen groups, 2006

850 **Siegfried Echterhoff, S. Kaliszewski, John Quigg, and Iain Raeburn,** A categorical approach to imprimitivity theorems for C^*-dynamical systems, 2006

849 **Katsuhiko Kuribayashi, Mamoru Mimura, and Tetsu Nishimoto,** Twisted tensor products related to the cohomology of the classifying spaces of loop groups, 2006

848 **Bob Oliver,** Equivalences of classifying spaces completed at the prime two, 2006

847 **Eric T. Sawyer and Richard L. Wheeden,** Hölder continuity of weak solutions to subelliptic equations with rough coefficients, 2006

846 **Victor Beresnevich, Detta Dickinson, and Sanju Velani,** Measure theoretic laws for lim–sup sets, 2006

845 **Ehud Friedgut, Vojtech Rödl, Andrzej Ruciński, and Prasad V. Tetali,** A Sharp threshold for random graphs with a monochromatic triangle in every edge coloring, 2006

844 **Amadeu Delshams, Rafael de la Llave, and Tere M. Seara,** A geometric mechanism for diffusion in Hamiltonian systems overcoming the large gap problem: Heuristics and rigorous verification on a model, 2006